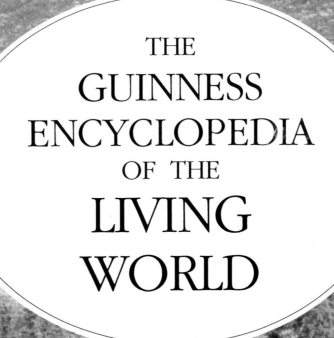

THE
GUINNESS
ENCYCLOPEDIA
OF THE
LIVING
WORLD

GUINNESS PUBLISHING

First published 1992

© Guinness Publishing Ltd. 1992

Published in Great Britain by Guinness Publishing Ltd., London Road, Enfield, Middlesex

Colour reproduction by Bright Arts (HK) Ltd., Hong Kong

Printed and bound in Italy by New Interlitho SpA, Milan

British Library Cataloguing in Publication Data:
A catalogue record for this book is available from the British Library.

ISBN 0–85112–963–3

Project Editor
Ben Dupré

Design
Amanda Sedge
Sarah Silvé

Systems Support
Alex Reid
Kathy Milligan

Picture Editing
P. Alexander Goldberg

Additional Picture Research
James Clift

Illustrators
Matthew Hillier FSCD
Peter Harper
Pat Gibbon
David McCarthy
Robert and Rhoda Burns
Ray Hutchins
Mike Long
Ad Vantage
Peter Bull
Suzanne Alexander

Art Director
David Roberts

Editorial Director
Ian Crofton

CONSULTANT EDITORS

Professor Brian Gardiner .....Professor of Vertebrate Palaeontology, King's College, University of London

Dr Peter Moore... Reader in Ecology, King's College, University of London

CONTRIBUTORS

BDT   Dr Bryan Turner, King's College, University of London
BG   Professor Brian Gardiner, King's College, University of London
BH   The late Dr Beverly Halstead, University of Reading
BSD   Ben Dupré
BTC   Dr Barry Clarke, The British Museum (Natural History), London
CH   Dr Colin Harrison
CW   Cyril Walker, The British Museum (Natural History), London
GS   Dr Gillian Sales, King's College, University of London
IDC   Ian Crofton
JAB   John A. Burton
JSC   Dr Sara Churchfield, King's College, University of London
JWC   Dr John Cowan, formerly of King's College, University of London
MB   Professor Michael Black, King's College, University of London

MEC   Dr Margaret Collinson, Royal Holloway and Bedford New College, University of London
MJB   Dr Michael Benton, University of Bristol
ML-E   Meredith Lloyd-Evans, BioBridge, Cambridge
OC   Oliver Crimmen, The British Museum (Natural History), London
PBG   Professor Peter Gahan, King's College, University of London
PDM   Dr Peter Moore, King's College, University of London
PF   Dr Peter Forey, The British Museum (Natural History), London
PM   Dr Paul Markham, King's College, University of London
PW   Dr Philip Whitfield, King's College, University of London
RHE   Dr Roland Emson, King's College, University of London
SM   Dr Stuart Milligan, King's College, University of London

The Guinness Encyclopedia of the Living World aims to give an accessible but authoritative overview of the world of nature, examining the fundamental processes that underlie it, and the plants and animals that inhabit it. The main part of the book follows the thematic approach of the *Guinness Encyclopedia*, and is arranged in a series of self-contained articles focusing on particular topics of interest within the natural world; the remainder of the book consists of an A-to-Z 'Factfinder' section, providing quick access to information on animal and plant groups, key concepts, technical terms, major historical figures, and many other topics. This dual format allows the reader to exploit the book in two different but complementary ways. The thematic section gives a broad account of the lives of plants and animals, explaining how they function, behave, and interact with their environment, and also examines the underlying processes and pressures that have shaped them. The Factfinder gives access to more detailed information about many more individual plants and animals, and is useful for finding a quick reference or checking a fact.

Using the Thematic Section

The thematic section of the Encyclopedia is divided up into six chapters. The first chapter, **The Living Planet**, deals with the fundamental geological, climatic and evolutionary processes that have shaped the natural world; the classificatory systems that help us to bring order to it; and the countless single-celled organisms, invisible to the eye but of incalculable significance in their effect. The second chapter, **The Plant Kingdom**, examines the fascinating world of plants, first explaining how they function and then proceeding, group by group, from the simple forms such as algae and lichens to the pinnacle of plant evolution, the flowering plants. **The Animal Kingdom** takes the reader on a similar journey through the world of animals, from the simplest invertebrates such as sponges and jellyfishes, to the enormously successful insects, and on to the vertebrates – the fishes, amphibians, reptiles, birds and, finally, mammals. The following chapter, **Animal Behaviour and Physiology**, takes a different perspective, explaining how and why animals behave as they do – why they migrate, how they communicate with one another, how they interact with one another to mate and to feed, and how they try to avoid falling prey to others. The second part of this chapter examines the physiological processes that underlie the functioning of animals – how their bodies work, the physical means by which they reproduce and grow, and the diverse ways in which they perceive the world around them. The various physical environments in which plants and animals live and interact are dealt with in **Ecosystems and Habitats**; first the central concepts of ecology are explained, and then each type of ecosystem, on land and in water, is reviewed in turn. The final section, **Threats to the Planet**, looks at some of the dangers facing the natural world, such as pollution and habitat destruction, and asks what measures we can take to save it.

Within the thematic section, the pages are colour-coded to indicate to which chapter they belong. The reader will quickly become familiar with these colours and be able to flick from chapter to chapter with ease. Most of the topics within the thematic section are in the form of double-page spreads, though some subjects are dealt with in longer or shorter articles. Within each article there is a 'See Also' box referring to related spreads within the same section or elsewhere. There are also cross-references within the text itself to guide the reader to pages where further relevant details will be found. If the user wishes to find out more about a particular animal or plant or some other subject to which only passing reference is made in the text, the appropriate entry can easily be found in the Factfinder section at the end of the book.

Using the Factfinder

The Factfinder consists of some 3000 alphabetically arranged entries providing at-a-glance information on a range of matters and issues relating to the natural world. Many of the entries relate to topics that receive some treatment in the main text; such entries tend to be brief, providing basic details such as a concise description of a particular plant or animal group (including distribution and numbers of species), together with a page reference to the relevant article in the thematic section. The Factfinder also contains numerous entries for items not covered in the main text – especially individual species, genera and families of notably diverse groups such as plants, insects, fishes, birds, bats, rodents, and so on. In addition there are entries providing further details on concepts or technical terms to which only brief allusion is made in the thematic section.

The Factfinder is linked with the main thematic section by means of a simple cross-referencing and indexing system. The references at the end of the entries refer to specific mentions of the topic in the main text: where there is a string of page references, the main one is indicated by means of bold type; italic type indicates a reference to an illustration or illustration caption. The words in small capital letters refer the user to important related entries within the Factfinder itself.

CONTENTS

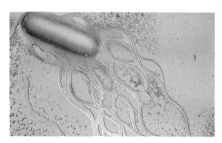

CONTENTS

4. ANIMAL BEHAVIOUR AND PHYSIOLOGY

5. ECOSYSTEMS AND HABITATS

6. THREATS TO THE PLANET

FACTFINDER

Title page illustration: Boreal forest and tundra in the Yukon, northwest Canada.

Illustration opposite: Sea anemone, Maldive Islands, Indian Ocean.

THE
GUINNESS
ENCYCLOPEDIA
OF THE
LIVING
WORLD

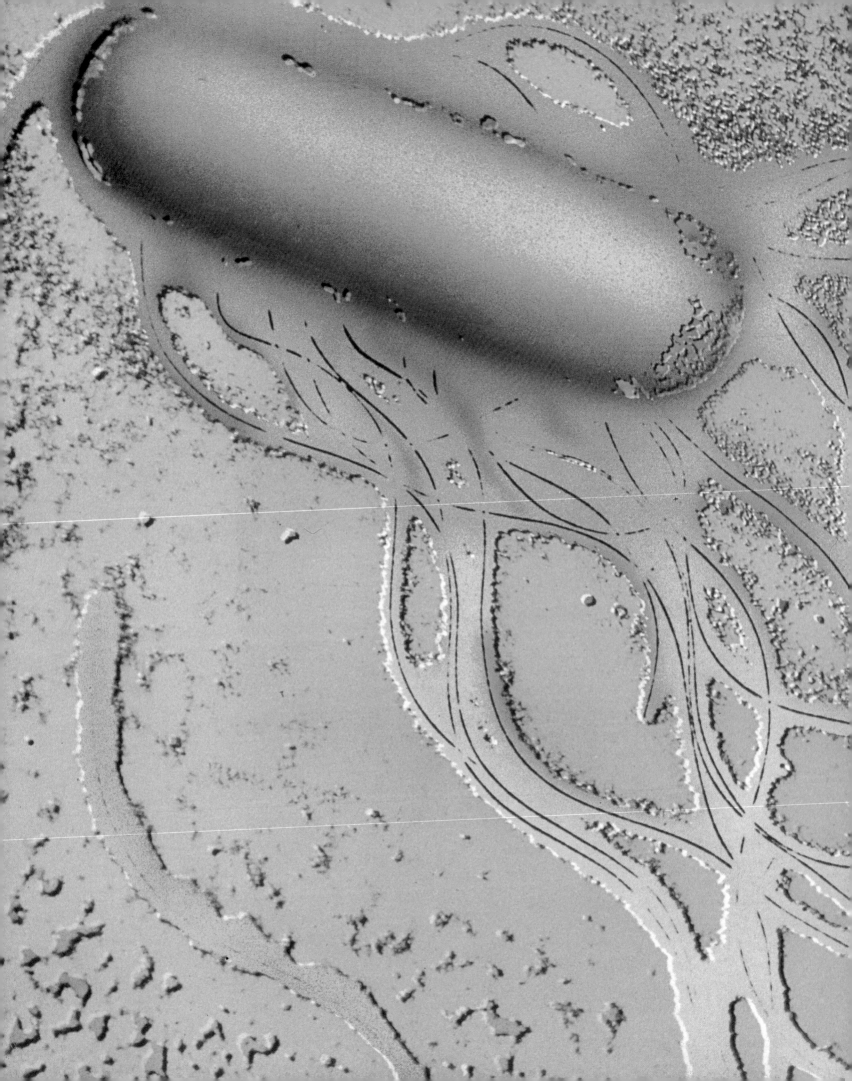

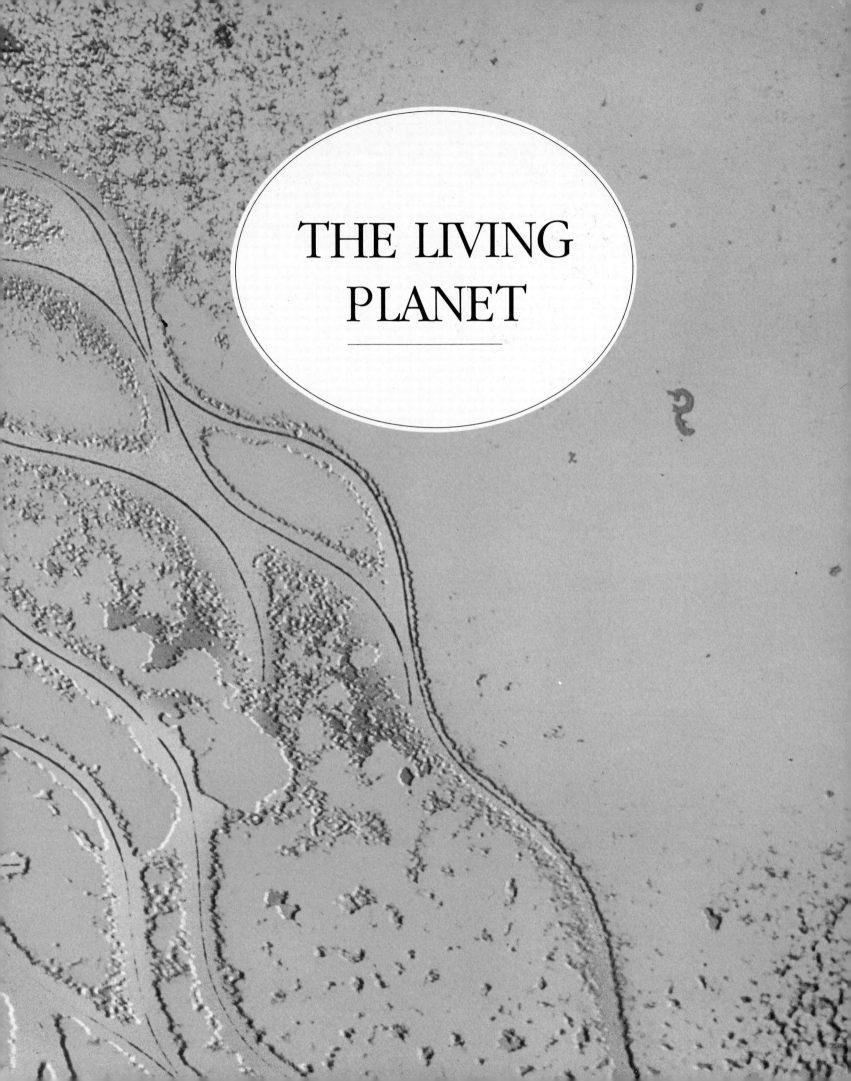

THE LIVING PLANET

The Ages of the Earth

The history of our planet is written in the rock around and beneath us. By studying rocks and the fossils preserved within them, scientists have been able to trace the momentous changes that the earth has undergone since its formation some 4·6 billion years ago.

Sedimentary rock is formed over vast expanses of time as layer upon layer of sediment is deposited on the sea bottom. Being formed in this way, such rock is usually arranged in a succession of horizontal bands or *strata*, with the oldest strata lying at the bottom. Moreover, each band will often contain the fossilized remains of the plants and animals that died at the time at which the sediment was originally laid down.

The strata of sedimentary rock are therefore like the pages of a book, each with a record of contemporary life etched upon it. Unfortunately, however, the record is far from complete. The process of sedimentation in any one place is invariably interrupted by periods in which new sediment is not laid down or existing sediment is eroded away. The succession of layers is further obscured as strata become twisted and folded, or even completely inverted, by enormous geological forces, such as those involved in mountain-building.

Dating methods

However patchy the record may be, any interpretation of it depends on first establishing a chronology. Relative dating is usually achieved by correlating strata on the basis of their fossil content, since a specific assemblage of fossils will be characteristic of a stratum of a particular age.

An absolute chronology, giving the ages of rock strata in thousands or millions of years before the present, is established by means of radiometric dating. Certain radioactive isotopes (known as 'parents'), such as potassium-40 and rubidium-87, occur naturally in rocks and decay at a constant rate to stable isotopes ('daughters') – in these cases, to argon-40 and strontium-87. Since the ratio of parent to daughter changes over time, it is possible – by measuring the proportion of each in a given rock – to calculate when the decay process began and hence the age of the rock.

A somewhat different radiometric dating method is based on the decay of radioactive carbon-14 to stable carbon isotopes. This method can be used to date significantly younger rocks, from the late Pleistocene onwards (approximately the last 70 000 years), and is also applicable to organic material such as wood, bone, hide, hair and shells.   BG

SEE ALSO

- WHAT IS LIFE? p. 6
- THE PLANET'S CHANGING FACE p. 8
- THE CLIMATIC FACTOR p. 10
- PLANTS p. 30
- ANIMALS p. 50

GEOLOGICAL TIMECHART

ERA	PERIOD	EPOCH	BEGAN MILLIONS OF YEARS AGO	GEOGRAPHY	
CENOZOIC	QUATERNARY	HOLOCENE	0.01	Retreat of glaciers leaves continents, seas and landscapes in more or less their present forms.	
		PLEISTOCENE	1.6	The thickest continental glaciers depressed the earth's crust to such an extent that large areas of northwestern Europe and North America are still rising at a considerable rate today (30 mm a year around Hudson Bay).	
	TERTIARY	PLIOCENE	5.3	Time of marked, often rapid change. Uplift of the Isthmus of Panama results in the connection of North and South America.	
		MIOCENE	23	Africa moves northwards into Eurasia. The Himalaya are raised as the Indian plate collides with Asia. The Red Sea opens and the Mediterranean has its origin.	
		OLIGOCENE	34	The main phase of Alpine mountain-building begins, followed soon afterwards by East African and Red Sea rifting. South America separates from Antarctica.	
		EOCENE	53	The Indian plate begins to collide with Eurasia (leading to the formation of the Himalaya in the Miocene). The Eurasian Basin opens as the final fragmentation of the Eurasian continent occurs. Pyrenean mountains form in the late Eocene. Australia separates from South America and Antarctica.	
		PALAEOCENE	65	Iberia converges on Europe. The Atlantic and Pacific Oceans are linked through the straits of Panama.	
MESOZOIC	CRETACEOUS		135	India separates from Antarctica, while Atlantic rifting brings about the separation of South America from Africa. Further opening of Atlantic and the separation of Greenland.	
	JURASSIC		205	At the start of the Jurassic, rifting occurs between Gondwanaland and Laurasia, initially separating southern Europe from Africa and eventually tearing Pangaea in two. Central Atlantic opens.	
	TRIASSIC		250	The start of the Mesozoic era sees all the major continents joined together. Consequently almost all of the earth's land surface is concentrated on one side of the globe, with the result that large areas lie far from the oceans and become very arid.	
PALAEOZOIC	PERMIAN		300	Pangaea is formed and in the late Permian Siberia collides with northern Pangaea to form the Ural mountains.	
	CARBONIFEROUS Divided in USA into Mississippian (early part) and Pennsylvanian (later part)		355	The major continents move closer and closer together, until, early in the next period, Laurasia collides with Gondwanaland to form the supercontinent Pangaea.	
	DEVONIAN		410	The gap between Laurasia and Gondwanaland has narrowed, but ocean levels remain high. *Insects*	
	SILURIAN		438	Laurasia forms as Laurentia and Baltica become welded together. Ocean levels are high, probably from melting of the ice cap.	
	ORDOVICIAN		510	Baltica moves from South Polar region towards the equator and closer to Laurentia. Meanwhile Gondwanaland moves towards the South Pole, and increased glaciation causes lowering of the sea level.	
	CAMBRIAN		570	Near the end of the Precambrian most landmasses fused into a giant supercontinent, but by the late Cambrian Gondwanaland, Siberia and Laurentia are separate continents and more or less sit astride the equator.	
	PRECAMBRIAN ERA		4600	North America, Greenland and Scotland united as Laurentia, equatorial in position. Gondwanaland in the southern hemisphere. *Stromatolites*	

CLIMATE	FLORA	FAUNA
Currently we are probably in an interglacial period.	Present-day disposition of floral regions established.	Present-day disposition of faunal regions established. Man increasingly comes to dominate the planet.
Cooler, drier and more seasonal, reflecting the rhythmic growth and decay of the northern-hemisphere continental glaciers. In the past million years there have been numerous 'ice ages', interspersed with warmer interglacial periods, the last ice age ending 10 000 years ago.	Savannahs and dry grassland expand and contract with each successive glaciation. *Homo erectus*	By the end of the epoch many large mammals had died out, including gigantic beavers, mammoths, and the giant ground sloth and armadillo of South America. The emergence of man, with the appearance of *Homo erectus* 1.6 million years ago and modern man about 30 000 years ago.
Sea levels rise early in the epoch. Global climate is initially warmer than today, but changes abruptly 3 million years ago with the development of polar glaciation in the northern hemisphere.	Flora largely similar to the flora of today. Demise of many components of the phytoplankton.	Origin of mammoths. Evolution of the first hominids (*Australopithecus*) 3-4 million years ago in Africa. *Australopithecus*
Polar cooling and local tectonic events cause climatic deterioration and many areas receive less rainfall. Generally cooler climates. Sea levels fall by up to 50 m (165 ft) at end of Miocene.	Spread of grasses and herbáceous plants; forests shrink. *Radiation of horses*	Radiation of apes and horses. Many groups, including frogs, snakes, rats and mice, have members largely indistinguishable from living forms.
Worldwide regression of the oceans early in the epoch. Major climatic cooling during the mid-Oligocene.	Diversity of planktonic flora is low. Spread of grasslands. *Earliest monkeys*	The radiation of mammals continues. Appearance of monkeys and extinction of the largest land mammals of all time – the rhinoceros-like titanotheres.
Generally warm, although temperatures fall by as much as 12°C (22°F) towards end of epoch, accompanied by increased glaciation and lower sea levels.	Forests are of distinctly modern appearance. Origin of continuous growth in grasses allows them to invade open country and withstand heavy grazing.	*Carnivores* Radiation of mammalian groups such as artiodactyls and perissodactyls, elephants, edentates, whales, rodents, carnivores and early higher primates (anthropoids). All bird orders have appeared by end of epoch.
The temperature at the end of the Cretaceous was probably 10°C (18°F) lower than at the start of that period, but a sharp rise in the early Palaeocene brings about warm climates.	Subtropical floras as far north as present-day southern England. Modern families of flowering plants evolve, thus flora is largely of modern aspect. Origin and spread of true grasses.	Explosive radiation of mammals, including the origin of primates. Many mammalian groups make their first appearance. *Primates*
Initially very warm. Sea levels rise, flooding much of Europe and western North America. Climate cools in late Cretaceous. *Flowering plants*	Origin of sea grasses. Angiosperms (the flowering plants) arise, and by the close of the period have already undergone an extensive radiation, so that forests in many areas are dominated by deciduous, broad-leaved trees. First grasses evolve late in the period.	*Mosasaurs* Teleosts undergo a major radiation, but the most spectacular vertebrates are the marine plesiosaurs, giant turtles, and the newly evolved mosasaurs. The end of the period sees widespread extinction, including ammonites, dinosaurs, ichthyosaurs and pterosaurs.
Tropical climates. Shallow seas flood large areas of Europe and North America.	Forests of cycads, conifers and ginkgoes. Diatoms (a kind of unicellular algae) evolve and radiate in the seas. *Birds*	Marine and freshwater turtles and crocodiles thrive. The dinosaurs further diversify, while towards the end of the Jurassic the first undisputed bird (*Archaeopteryx*) appears.
Relatively warm and dry even at high latitudes. Sea level generally rises. *Dinosaurs*	Coccolithophores (chalk-producing algae) make their first appearance. On land gymnosperm floras dominate (although ferns are still important), with all modern conifer families except the pines coming into existence. *Mammals*	The Triassic ends with the greatest mass extinction of all time, which eliminates many species of marine invertebrate as well as most families of synapsids. Nevertheless, this period sees the emergence of many groups: the teleosts (the dominant fish group today); enormous marine reptiles such as ichthyosaurs and plesiosaurs; thecodonts, which evolve into pterosaurs (flying reptiles) and dinosaurs; and the first true mammals. *Cycads*
The climate changes profoundly from region to region, and there are still steep climatic gradients. The glaciation of Gondwanaland slowly subsides with the warming of the higher latitudes. Hot, dry conditions in equatorial regions kill off coal-swamp floras.	Gymnosperms replace the spore-bearing plants, with conifers thriving under the drier conditions. The cycads – a highly diverse gymnosperm group – appear.	Trilobites disappear. The reptiles diversify, with the radiation of the mammal-like reptiles (synapsids).
In Gondwanaland continental glaciers reach northwards to within 30° of the ancient equator and are separated from coal swamps by steep temperature gradients. Near the equator warm moist conditions are prevalent. *Reptiles*	Plant fossils are more conspicuous than at any other time. Coal deposits develop in lowland swamps, and the first conifers appear. *Coal swamps*	Sharks and bony fishes thrive. Foraminifers (planktonic protozoans) undergo an enormous adaptive radiation. The earliest flying insects appear. Amphibians are much diversified but still mainly aquatic. The first reptiles (from which birds and mammals evolved) occur in the early Carboniferous. Extinction of graptolites.
Very warm and dry over Laurasia; cooler over southern Gondwanaland, but extensive warm-water reefs near the equator.	Progressive colonization of land by plants. By late Devonian there are large vascular plants such as lycopods (club mosses as big as trees) and calimites (horsetails with simple circlets of leaves). Ferns also appear, as do the first seed plants (gymnosperms). *Amphibians*	Within the warm lagoons of vast coral reefs, the bony fishes diversify. Elsewhere the shallow seas around the continents support new forms such as coiled ammonites. On land the first animals appear – scorpions and wingless insects – and somewhat later the earliest tetrapod (four-limbed animal) – a primitive amphibian. Great mass extinction at end of period eliminates much marine life.
Warm and in many areas relatively dry. Melting of ice cap raises sea levels. *Jawed fishes* *Sea scorpions*	On land the first undoubted vascular plants, such as club mosses, appear. Yellow-green algae (dinocysts) also make their first appearance. *Land plants*	The mass extinction at the end of the Ordovician is followed by a rediversification of marine life. Origin of the eurypterids (scorpion-like aquatic arthropods). The first true jawed fishes appear: the earliest shark-like fishes (chondrichthyans) and – late in the period – the first bony fishes (osteichthyans).
Ammonites Baltica becomes warmer, Gondwanaland much cooler. Glaciers spread over the south of Gondwanaland.	Spores resembling those of modern land plants found in latter part of the period. *Jawless fishes*	A tremendous adaptive radiation brings about highly diversified trilobites, lamp shells, and dendroid (tree-like) graptolites. New groups of gastropods, sea mats (bryozoans), the first sea urchins, and all five living orders of starfishes have their origin; also the first ammonites. The first vertebrate remains – jawless fishes (agnathans) – date from the transition from the Cambrian to the Ordovician.
Tropical or near tropical. Persistant sea-level rises bring about flooding of continents, followed by regional uplifts caused by mountain-building activity. *Graptolites*	Dense accumulations of single-celled planktonic algae. Multicellular phytoplankton such as acritarchs locally abundant. *Trilobites*	During the early Cambrian, evolution of external skeletons leads to the emergence of many larger marine animals, including hexactinellid sponges, gastropods, trilobites, lamp shells (brachiopods), and graptolites. Nautiloids appear towards the end of the period.
Volcanism widespread, but some glaciation at times over Gondwanaland. Photosynthetic algae saturating seas and lakes bring about the steady build-up of atmospheric oxygen. *Soft-bodied animals*	The earliest unmistakable record of life, dating back 3.4 billion years, is in the form of stromatolites – layers of calcium carbonate built up by blue-green algae. By 1.5 billion years the first algae and the earliest fungi have appeared.	Many microfossils dating back 3.25 billion years were perhaps produced by bacteria and by blue-green algae, while the first eukaryotes (single-celled organisms with nuclei) appear at about 1.5 billion. By 800 million we find predatory protists, and by 590 million the first multicellular animals, including coelenterates, annelid worms and arthropods.

What is Life?

Over 2500 years ago Aristotle defined life as the ability of a thing to feed itself, grow and decay. All of these attributes are indeed characteristics of living things – but not exclusively so. If, for example, a string is suspended in a strong solution of sugar and water, a crystal will grow on it, and will keep on growing ('feeding itself') until all the sugar is used up. The crystal can also be dissolved back into the water, and so can 'decay' and 'die'.

Until the last few decades one could say that only living things could react to environmental stimuli, but with the advent of computer-programmed robots, this is no longer valid. The ability to self-replicate – to reproduce 'copies' of oneself – and the ability to repair oneself would both seem to be exclusive attributes of living things, but even here there are problems. It is actually theoretically possible to construct and program a self-replicating machine, and even one that could repair itself, by drawing on outside materials made available to it.

Living cells

The crucial difference between living things and the non-living processes described above is to be found in the way that living things carry out such operations as self-replication and repair. No scientist has come anywhere near creating in the laboratory the basic unit of all life, the cell. The cell ($\triangleright$ p. 20) has the unique property of being able to take in external raw materials and to alter them chemically within itself into an enormous variety of more complex chemical compounds. In this way a single cell, a tiny fraction of a millimetre in diameter, can divide and redivide, and this is how organisms replicate, grow, and repair themselves when injured.

Essential chemical components

Living cells are built up of various groups of complex molecules ($\triangleright$ box), which in turn are largely composed of much simpler components. Perhaps the most important of these are water and carbon. Water (H_2O) is essential to all life, not only as a contributor of hydrogen and oxygen to more complex compounds, but also in its liquid form: on average, over two thirds of all cells – and hence of all organisms – are made up of water.

THE MOLECULES OF LIFE

In all cells there are four groups of complex molecules that are essential to the structure and function of the cell – nucleic acids, proteins, carbohydrates, and lipids.

NUCLEIC ACIDS

In most cells both of the nucleic acids (DNA and RNA) are present. DNA is the encoder of genetic information ($\triangleright$ p. 22), while different forms of RNA have different functions: m-RNA acts as the messenger of DNA, t-RNA transports the amino acids to the newly forming proteins ($\triangleright$ below), and r-RNA ribosomes act as sites for the construction of the proteins.

PROTEINS

A living organism contains more than 10 000 different kinds of protein, of which many are involved in the structure of the cell, while the others act as enzymes – catalysts that help to drive chemical reactions in the cell without themselves being changed. Proteins are constructed at the r-RNA construction sites, where the amino acids are linked together in chains; one or more such chains, when folded in a particular way, creates a particular protein molecule, with its own special form and function.

CARBOHYDRATES

The carbohydrates are a series of simple sugars linked into short or long chains. Carbohydrates are used by organisms either as energy sources or as building blocks; some also have a role in molecular recognition ($\triangleright$ illustration, p. 20). The simplest sugar, glucose, is formed by plants and some bacteria through the process of photosynthesis ($\triangleright$ p. 33), and therefore all animals are completely dependent on plants for their supplies of carbohydrates.

LIPIDS

Lipids, the final essential group, are crucial components of cell membranes ($\triangleright$ pp. 20–1). Like carbohydrates, lipids are simple molecules, being formed from glycerol, fatty acids and phosphate, to which may be added a sugar or an amino acid.

Carbon is unique among the elements in that it can combine with other elements to form an almost infinite variety of molecules. Because of its fundamental importance to life, molecules containing carbon are called organic compounds. Part of the versatility of carbon is due to its ability to form rings and chains, and many carbon compounds are soluble in water.

The beginnings of life

The earth – together with the sun and the rest of the solar system – formed some 4·6 billion years ago. The early atmosphere was formed by the gases expelled in volcanic eruptions, including water vapour (H_2O), hydrogen (H_2), nitrogen (N_2), carbon dioxide (CO_2) and carbon mon-

Stromatolites in Shark Bay, Western Australia. Stromatolites – some of which are up to 3·4 billion years old – are fossilized colonies of cyanobacteria, the first organisms capable of photosynthesis. As a result of their photosynthetic activity, oxygen entered the earth's atmosphere for the first time, and virtually all forms of life since then have been dependent on oxygen or respiration.

oxide (CO). As this early atmosphere cooled, the hydrogen reacted with the oxides of carbon to form methane (CH_4) and with nitrogen to form ammonia (NH_3).

Ammonia and methane are able to combine with water and carbon dioxide if they are subjected to ultraviolet light (as emitted by the sun) and if an electrical spark (such as lightning) is passed through a mixture of these gases. From such reactions are formed simple amino acids, which, if subsequently heated, link up into linear chains – the characteristic structure of proteins (▷ box). Similar reactions involving atmospheric hydrogen cyanide (HCN) and ammonia possibly formed purines and pyramidines – the building blocks of DNA (▷ p. box).

An alternative theory suggests that amino acids did not originate on earth at all. Small organic spheres containing mixtures of amino acids and fatty acids (a component of lipids; ▷ box) have been recovered from meteorites, and it has been proposed that these essential components of life in fact arrived here from outer space.

Whatever the origin of the amino acids, it is certain that of all the planets in the solar system only our earth could have supported the subsequent development of life as we know it, as only here can water exist in its liquid form. The oceans were formed by the condensation of water vapour in the atmosphere, and soon accumulated a wide range of dissolved minerals and gases, resulting in the primeval 'broth' necessary for the evolution of the organic molecules of life.

The first crucial step towards life-like forms, perhaps occurring some 4 billion years ago, is thought to have been the appearance in the oceans of molecules – such as DNA – capable of self-replication. With the ozone layer as yet unformed, the ultraviolet radiation from the sun would have been intense, causing frequent mutations (▷ p. 16) during molecular replication, and at this point it is probable that something like natural selection might have begun to operate. Frequent and accurate replicators would have multiplied successfully, as would those that were better at protecting themselves, for example by using other molecules as the basis of protective coverings round the central replicator molecule.

It is possible that this latter process may have involved an early form of predation or symbiosis, but an alternative scenario has been provided by experiments that show that in conditions of intense volcanic activity, followed by rapid cooling with cold water, amino acids can be turned into cell-like structures with double-layered membranes.

In one or other of these ways arose the first forms of what we would unequivocally recognize as life, in the form of single-celled prokaryotic organisms (▷ p. 20) such as bacteria. These first simple life forms reproduced by splitting into two new cells, and fed on the abundance of organic molecules available in the early oceans.

Early evolutionary stages

However, around 3·4 billion years ago, as the organic molecules began to be used up, new forms of bacteria began to emerge, the cyanobacteria or blue-green algae (▷ p. 25). Cyanobacteria are capable of photosynthesis, i.e. able to convert simple inorganic molecules into complex organic compounds using the energy of sunlight (▷ p. 33), rather than depending on ready-made sources. The by-product of photosynthesis is oxygen, which began to accumulate in the atmosphere in large quantities. Previously oxygen had been poisonous to all living things, but now bacteria began to evolve that depended on oxygen for respiration.

About 1·5 billion years ago new kinds of single-celled organisms began to appear that – although still microscopic – were many times larger than the bacteria. These were the first protists (▷ p. 26), some of which were animal-like consumers (protozoans), and others plant-like photosynthesizers. Protists were the first organisms to possess eukaryotic cells (▷ p. 20), which are much more complex than the prokaryotic cells of bacteria. For one thing, eukaryotic cells have organelles (▷ box, p. 20), miniature 'organs' with specialized functions. Certain of these organelles bear a striking resemblance to various bacteria (▷ illustration) – and indeed two of them actually contain their own DNA – so it has been suggested that the first eukaryotic cells may have resulted from the cooperative coming together of simpler prokaryotic cells.

The main defining feature of eukaryote cells is their possession of a central nucleus containing the chromosomes, the structures into which their DNA is organized. It seems likely that sex-cell division (meiosis; ▷ p. 22) must have quickly followed. The development of sex – the exchange of genetic material between two individual organisms – provided a level of variation that speeded up the process of natural selection, and thus of evolution.

The next important development was the emergence of multicellular organisms, which occurred around 600 million years ago. The first multicellular animals may have been like the sponges (▷ p. 52), within which there are several types of cell, each of which is capable of functioning independently, as well as in concert with other cells in the 'colony'. Some of the specialized cells are in fact virtually indistinguishable from certain protozoans, which may indicate how multicellular organisms first evolved. Within a few million years of the appearance of the first multicellular animals, most of the main groups of invertebrates had appeared. The first vertebrates – in the form of the jawless fishes – appeared around 500 million years ago, and it was from them that all the other fishes, together with the amphibians, reptiles, birds and mammals, eventually descended. IDC/BG

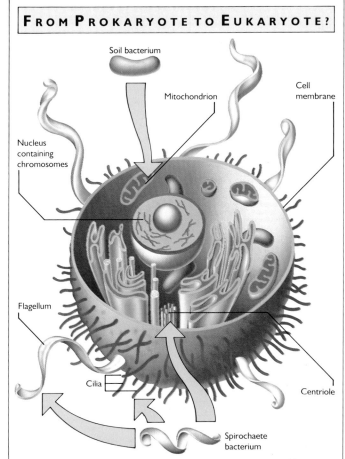

FROM PROKARYOTE TO EUKARYOTE?

Soil bacterium

Mitochondrion

Cell membrane

Nucleus containing chromosomes

Flagellum

Cilia

Centriole

Spirochaete bacterium

There is a striking resemblance between certain bacteria – which are simple prokaryotic organisms – and the organelles (miniature 'organs') found in the more complex eukaryotic cells of higher organisms. This has led some scientists to suggest that the first eukaryotic cells were the result of a cooperative coming together of various prokaryotic organisms.

Deep in the oceans there are areas where hot water rich in hydrogen sulphide is released from volcanic vents. Round these hydrothermal vents are found certain species of bacteria that are capable of using sulphur as a source of energy. The bacteria in turn provide nutrition for bizarre communities of animals (including these pogonophore worms), the only known communities of organisms that do not ultimately rely on the energy of the sun via photosynthesis. Some scientists believe that life may have first evolved in conditions similar to those found round these deep-sea vents.

The Planet's Changing Face

For centuries, it was assumed that the continents were fixed in their relative positions, and although various people had observed the 'jigsaw' fit of Africa and South America, serious evidence to support the idea that the continents are drifting apart was not forthcoming until 1915.

It was in that year that the German meteorologist Alfred Wegener pointed out that many of the rocks, mountain chains and even fossils along the coast of Africa are similar to those along the east coast of South America. He also noted that the long-extinct plant *Glossopteris* was even more widespread, occurring not only in South America and South Africa but also in Madagascar, India and Australia.

Continental drift

From this and other evidence Wegener deduced that there had once been an immense super-continent, which he called Pangaea. He concluded that this land mass had begun to split up in the Mesozoic era (250–65 million years ago), and that the various smaller continents had separated from it and moved apart at various times. They had then drifted to their present positions.

Wegener's theory was subsequently elaborated by the South African geologist

Alexander du Toit, who proposed that there had previously been two supercontinents, a southern one, which he called Gondwanaland (after a region in India), and a northern one, Laurasia, and that Pangaea had been formed by the collision of these continents in the Carboniferous period (355–300 million years ago). By the 1960s the geological evidence was overwhelming in support of the view that the continents are drifting apart at the rate of a few centimetres a year.

Plate tectonics

All that was now needed was a hypothesis to explain the phenomenon. This was provided in 1962 by the American geologist Harry Hess. He suggested that new oceanic crust was constantly being formed by the extrusion and solidification of molten material from the earth's mantle along the lengths of the mid-ocean ridges. At the same time oceanic crust is recycled back into the mantle by means of subduction along deep-sea trenches (⊳ diagram).

By the late 1960s, continental drift and ocean-floor spreading had come to be seen as two aspects of a wider phenomenon – *plate tectonics*. Today we recognize that the Earth's crust is divided into 15 major plates of various sizes, which float on the molten magma and have the freedom to move horizontally. Some of the plates (such as the Pacific plate) are almost entirely oceanic, but most consist of both oceanic and continental crust.

Continents, distribution and dispersal

The past and present juxtaposition of the continental plates has profoundly influenced the distribution and dispersal of most forms of life on this planet, and hence has played an important role in evolution.

At first, life on the planet was restricted to the oceans. But even in the oceans, early life forms found the warm shallow coastal seas of the continental shelves the most conducive environment – and hence the distribution of these primitive organisms and their dispersal was influenced by continental drift.

By the Ordovician period (510–438 million years ago) two major faunal provinces can be recognized in the Atlantic region – the American (or Laurentian) and the European (or Baltican), each with its own characteristic assemblage of trilobites, brachiopods, conodonts and graptolites. But during the Silurian period (438–410 million years ago) the gap between Baltica and Laurentia became progressively narrower, and as a result the larval stages of trilobites and brachiopods could now cross the Proto-Atlantic, and the invertebrate faunas became the same on both sides of the ocean.

The Silurian period also witnessed the first colonization of freshwater habitats by larger animals (jawless fishes and the scorpion-like eurypterids). At about the same time small, spore-bearing vascular plants colonized the land. By the end of

the Devonian period (410–355 million years ago), the marshy fringes of streams, rivers and lakes had been successfully colonized by new groups of spore-bearing plants, various arthropods such as primitive insects, and the first four-legged vertebrates (tetrapods). Life on land was firmly established and, once there, more prone than ever to the effects of continental drift.

The spore-bearing plants that created the great forests of the Carboniferous period (355–300 million years ago) were restricted by their dependence on water for fertilization (⊳ p. 36) to the coastal swamps, as were the amphibians, which were similarly dependent on water (⊳ p. 78). The first seed-bearing plants (primitive gymnosperms), much less dependent on water for reproduction (⊳ p. 38), were able to colonize the drier, higher interiors of the continents, as were the first amniotes, the reptiles (⊳ p. 82). The colonization of dry land was under way.

The formation at the end of the Carboniferous of the supercontinent Pangaea had a profound effect on the oceanic circulatory patterns, and hence on the global climate. In the Permian period (300–250 million years ago) these climatic changes brought about the formation of two distinct floral provinces (Gondwana and Euramerica). The hot, dry conditions in the equatorial regions killed off the coal-swamp floras, which were replaced by conifers and cycads.

At the start of the Mesozoic era (250 million years ago) all the major continents were joined together, with the result that much of the land lay far from the oceans and was quite arid. This did not, however, prove too much of a problem for the reptiles, adapted as they were to drier environments, and the Triassic period (250–205 million years ago) saw the emergence and diversification of the dinosaurs (⊳ p. 82) and the origin of the mammals (⊳ p. 98).

The Jurassic period (205–135 million years ago) heralded the start of the break-up of Pangaea, with North America, Greenland and Eurasia remaining connected to one another but separating from Gondwanaland. East Gondwanaland (India, Australia and Antarctica) then parted from West Gondwanaland (Africa and South America). Despite the eventual break-up of Pangaea in the Cretaceous period (135–65 million years ago), most major groups of dinosaurs had already appeared by the early Jurassic. The dinosaurs thus managed to obtain a worldwide distribution, dominating life on earth until their sudden extinction around 65 million years ago (⊳ pp. 82–4).

Most mammals were small until the early Eocene epoch (53–34 million years ago), which witnessed the appearance and radiation of many larger mammals, including ungulates (hooved animals), carnivores and early anthropoids. Initially these mammals were virtually identical in North America and Europe, but they were isolated from one another in the middle of the epoch by the separa-

---

CREATION AND DESTRUCTION

Tectonic plates are made up of crust and the upper, solid part of the earth's mantle, which together form the *lithosphere*. The plates float on the *asthenosphere*, the lower, molten part of the mantle.

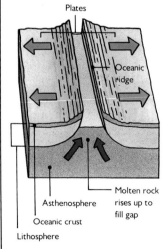

Crust is created at mid-ocean ridges by molten rock rising from the asthenosphere.

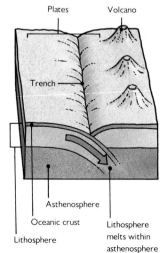

Crust is destroyed where an oceanic plate is formed under a continental plate.

THE AMERICAN FAUNAL INTERCHANGE

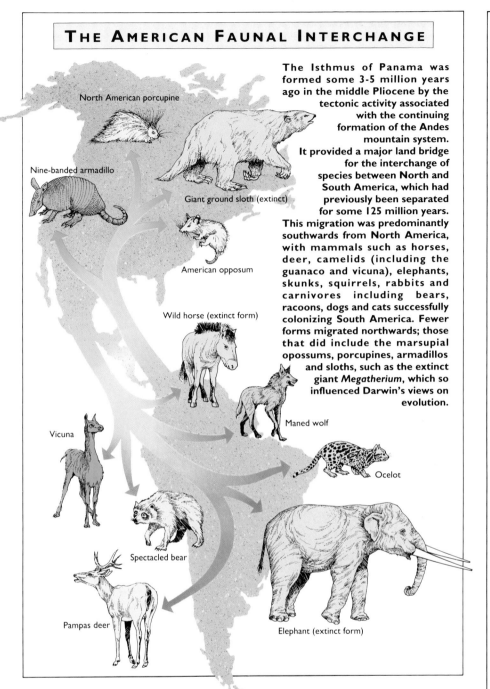

North American porcupine

Nine-banded armadillo

Giant ground sloth (extinct)

American opposum

Wild horse (extinct form)

Maned wolf

Vicuna

Ocelot

Spectacled bear

Pampas deer

Elephant (extinct form)

The Isthmus of Panama was formed some 3-5 million years ago in the middle Pliocene by the tectonic activity associated with the continuing formation of the Andes mountain system. It provided a major land bridge for the interchange of species between North and South America, which had previously been separated for some 125 million years. This migration was predominantly southwards from North America, with mammals such as horses, deer, camelids (including the guanaco and vicuna), elephants, skunks, squirrels, rabbits and carnivores including bears, racoons, dogs and cats successfully colonizing South America. Fewer forms migrated northwards; those that did include the marsupial opossums, porcupines, armadillos and sloths, such as the extinct giant *Megatherium*, which so influenced Darwin's views on evolution.

CONTINENTAL DRIFT

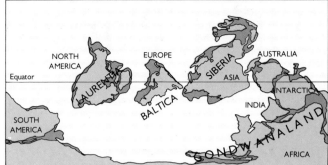

EARLY ORDOVICIAN PERIOD 510–438 million years ago

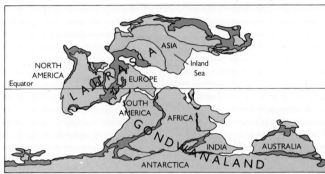

MID-CARBONIFEROUS PERIOD 355–300 million years ago

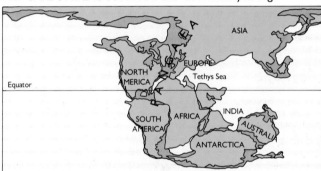

LATE TRIASSIC PERIOD 250–205 million years ago

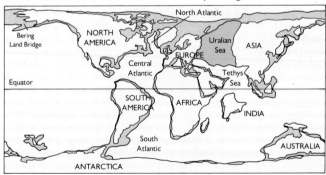

PALAEOCENE EPOCH 65–53 million years ago

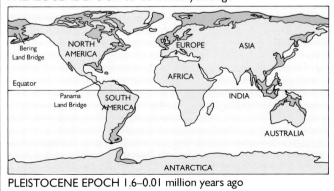

PLEISTOCENE EPOCH 1.6–0.01 million years ago

tion of Greenland and Scandinavia. The European fauna was also separated from the Asian one by the north–south Uralian Sea. Eurasia was similarly separated from Africa by the Tethys Sea. However, the Bering land bridge, a neck of continental crust, allowed faunal interchange between North America and Asia. The closure of the Uralian Sea in the late Oligocene epoch (34–23 million years ago) permitted trans-Eurasian migration of mammalian faunas, but towards the end of the Oligocene more frequent inundations of the sea prevented the exchange of species via the Bering land bridge. During the Miocene epoch (23–5.3 million years ago) Eurasia and Africa became united, allowing an interchange that included the immigration of apes, elephants, cattle and pigs into Eurasia. The Bering land bridge opened from time to time, allowing elephants to reach North

America, while horses entered Eurasia from North America.

Late in the Pliocene epoch (5.3–1.6 million years ago) the northern polar glaciation commenced, causing sea levels to fall and the Bering land bridge to re-open, which enabled mammoths to enter North America. The short phase of mountain building at the end of the Pliocene elevated the Andes and brought about the formation of the Isthmus of Panama, connecting North and South America for the first time since the Cretaceous (⇨ box). By the start of the Pleistocene epoch (1.6–0.01 million years ago), early man (*Homo erectus*) had evolved in Africa, from where he spread into Europe and Asia. During the Pleistocene glaciations, the Bering land bridge was intermittently open, enabling modern man (*Homo sapiens*) to cross from Asia into North America.   BG

The Climatic Factor

All animals and plants are affected by climate. Each species has a preferred range of temperatures and humidities within which it can live and reproduce most effectively. Towards the edge of its temperature and humidity range it becomes more sensitive to the effects of competition simply because it is living in suboptimal – less than ideal – conditions. A species may, in fact, be unable to fill its full potential range in the wild simply because of its inability to maintain a population in the face of competition from better adapted, more efficient species. So the geographical limits that we observe when studying the distribution of organisms may not reflect precisely their climatic limits, but the range in which they can compete efficiently.

For example, various palms can be grown in the mild climate of western Scotland, far to the north of their natural distribution, but they could not survive in the long term outside a garden because they would have to cope with other more efficient species and simply could not compete. Palms evolved in the tropics, and their current distribution as a family is almost completely tropical.

Other species operate most efficiently at the other end of the climatic spectrum, being most successful under cold polar conditions. The white-billed diver, for example, is a large loon-like species of bird that is rarely seen outside the Arctic Circle and breeds in tundra vegetation around the edge of the Arctic Ocean, spending its winters in the coastal regions.

The legacy of the ice ages

The pattern of species distributions across the face of the earth can often be related to their climatic requirements, but some patterns are clearly more complicated and demand a more sophisticated explanation. Sometimes we can understand distribution patterns only by considering past climatic changes and the disruption that these have caused to the animals and plants concerned. Over the

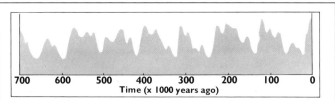

Climatic fluctuations over the last 700 000 years are indicated by fluctuations in the relative proportions of oxygen isotopes derived from ocean sediments, as plotted on this graph. The high points indicate warm (interglacial) periods, while the lower points indicate glaciations.

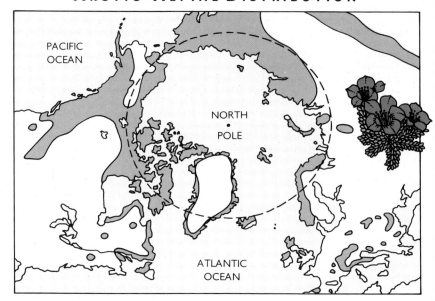

The purple saxifrage (*Saxifraga oppositifolia*) is just one example of the many plants and animals with an arctic–alpine distribution. Such species had widespread distributions in the northern hemisphere during the last ice age, but with the retreat of the glaciers some 10 000 years ago, distribution patterns became more fragmented, with some populations moving north with the tundra, while others were left stranded in high mountain areas further south.

past million years the climate has been extremely unstable, and for the majority of that time the earth has been much colder than at present, perhaps 5 °C (9 °F) colder on the whole. But the cold of the ice ages has been periodically disrupted by episodes of warmth (interglacial periods) and we are fortunately experiencing one of these at the moment.

The influence of these changes on the vegetation of the earth has been reconstructed from fossils, most particularly from the preserved pollen found in lake sediments and peat deposits. A sufficient number of sites has now been investigated for maps of changing vegetation to be drawn with some degree of accuracy, and it is evident that the cold periods in which ice has advanced have been associated with a retreat of boreal and deciduous forests (▷ pp. 207–9) from much of North America, Europe and northern Asia and their survival in more southerly refuges. The record of animal fossils from these cold and warm episodes are less abundant than is the case with plants, but they evidently followed a similar course and changed their range with the changing climate.

The end of the megafauna

Many species of plant and animal have become extinct during these climatic fluctuations, unable to adapt or to disperse or migrate rapidly enough to survive the instability. One puzzle that faces us is why so many species of large mammal – the mammoth, the cave bear, the giant Irish elk, the shasta ground sloth, to name but a few – became extinct in the early stages of the current warm period (the Holocene epoch, starting 10 000 years ago) and yet evidently survived previous episodes of warmth.

The coincident extinction of these mammals – sometimes referred to as the *megafauna* – has posed a difficult problem. Some zoogeographers propose that the one feature differentiating our present warm period from previous interglacials is the fact that human populations have expanded so strongly during this time. The extinction of the megafauna may well be a consequence of their failure to cope with mankind rather than an inability to adapt to the changing climate. The possible impact of man on plant and animal distributions in the current interglacial must always be taken into account.

It is also evident that the pattern of species distribution that we find in the modern world is simply a frozen moment in a time of constant tumoil. Distributions are never static, but always dynamic. They may never actually catch up and achieve equilibrium with the changing climate that they inevitably follow.

Temperature and sea levels

Some species have become fragmented in their distributions as a result of these changes. The tiny springtail (*Tetracanthella arctica*, a primitive wingless insect), for example, is found around the coast of Greenland and in Iceland and Spitzbergen, but also in the Tatra Mountains of Czechoslovakia. It seems likely that this arctic invertebrate was more widespread during the last cold period and has since become more fragmented in distribution as the world has become warmer. A further factor has added to its disruption, namely the rising level of the ocean during the present warm phase. This is due to the melting of the ice caps in the warmer conditions, raising ocean levels overall by about 100 metres (330 ft) during the last 10 000 years. So movement between landmasses was often a simpler matter during and shortly after the last ice advance.

Many plants show similar patterns of distribution to the springtail, and have what is termed an *arctic–alpine* distribution, being found around the northern tundra and also on high mountain peaks in more southerly latitudes. The purple saxifrage (*Saxifraga oppositifolia*), for example, is found right around the Arctic region just like the white-billed diver, but is also found in the European Alps and other southerly mountain chains.

The rising sea levels have provided a serious barrier to the dispersal of some species and this may mask their true climatic demands. In the mainland of Europe, for example, the stoat is found throughout the region apart from the south. The similar mammal, the weasel, is generally even more widespread with one notable exception – it is completely absent from Ireland. Clearly the Irish Sea must have been well established by the time this animal arrived in the western extremities of Europe and it never managed to make the crossing, whereas the stoat achieved this, either by early arrival or a chance crossing, and it now has Ireland to itself.

Habitat changes

But some species have distribution patterns that are more difficult to explain. For example, the azure-winged magpie has an extraordinary distribution, being found in Spain and also in eastern Asia. A simple climatic explanation is inadequate to account for this wide separation in populations. But a consideration of climatic history may begin to offer a way of understanding the problem. The bird is currently most at home in a warm temperate climate and in a habitat where small stands of trees form a parkland in open grassland or heath. Some 10 000 to 8000 years ago, immediately after the end of the last glaciation, this type of habitat was widespread throughout the Middle East and much of southern Asia, so the azure-winged magpie may well have been a widespread species at that time. But more recently much of this area has become drier and is now covered by desert, which presents a barrier to dispersal for the magpie. The two present populations are now effectively isolated and will proceed to evolve along their own paths independently of one another because no more cross-breeding is possible between them.

Climate and speciation

This process of separation is extremely important in permitting the splitting of a single species into separate ones (the process called *speciation*). Take the example of pied and collared flycatchers in Europe. These two species are very similar in appearance and their ranges overlap, though the pied flycatcher is generally more northern and is more pre-

pared to accept coniferous forest. The collared flycatcher is centred on southern and eastern Europe and prefers deciduous woodland. The cause of their separation seems to be associated with their migration pattern. Both species leave Europe during the winter and migrate to the southern side of the Sahara Desert in Africa; however, they generally adopt rather different migration routes, the pied species heading west through Spain and the western Mediterranean, while the collared species moves to the east, overland through Asia Minor.

In the course of geological history these two migration routes may reflect a total separation of the two populations, possibly during the last glaciation. They may well have been totally separated for tens of thousands of years, but have now come together again as a result of climatic change and the spread of forest back into Europe. But, even though their current distributions overlap, their behaviour patterns have changed and they still retain their separate identity as species, even to the point of sticking to their old migration routes.

The future

The study of how populations of animals and plants have behaved in response to the climatic changes of the past provides some basis for predicting how they may respond to the climatic changes of the future, some of which are being de-

termined by the activities of mankind. One thing that emerges from studies of the biogeography of the past is that each species behaves according to its own demands and limitations. We should not think of whole communities migrating in consort, because this has not happened in the past. The current assemblage of species that we like to divide into different types of community is merely a momentary juxtaposition of species in terms of geological time. In a world with a different climate we shall find new combinations of species – new communities.

As yet our predictions about the possible pattern of future climates (▷ p. 224) is still vague, so specific statements about possible plant and animal patterns are even more speculative. Many of the semi-arid parts of the world are likely to become hotter and drier, leading to the further spread of deserts (▷ p. 204). Some of the frost-sensitive species of plant, particularly weed species, may well spread northwards and become familiar pests of the temperate areas. The tundra (▷ p. 210) may well become invaded by trees, as spruce, birch and larch move northwards in America, Europe and Siberia. This could lead to the extinction of some tundra plant and animal species, but the rate of extinction resulting from climatic change is unlikely to be significant when compared to the extinctions being caused by human destruction of habitats (▷ p. 228) like the tropical rain forest.   PDM

CLIMATE AND SPECIATION

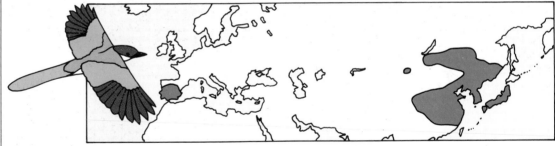

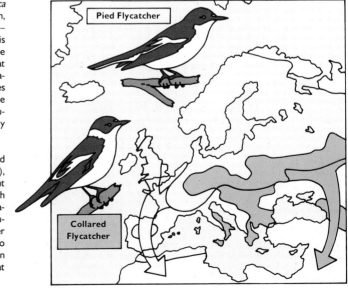

The azure-winged magpie (*Cyanopica cyanus*) (above) has a most unusual distribution, its two populations – in Spain and eastern Asia – being separated by some 9000 km (5600 mi). It is likely that 10 000 to 8000 years ago, after the end of the last ice age, the bird's favoured habitat extended between its two present-day locations. However, subsequent climatic changes have rendered the intervening area unfavourable to the species, and the two separated populations, unable to interbreed, will eventually evolve into different species (▷ text).

Two closely related species, the pied flycatcher and the collared flycatcher (right), evolved from a common ancestor. It is thought that two populations of the original species, each following different migratory routes, were separated by climatic changes, possibly tens of thousands of years ago during the last ice age. Further climatic changes have enabled the two species to extend their ranges back into central and eastern Europe, but their long separation means that they can no longer interbreed (▷ text).

Evolution
The Origin of Species

In the history of science there have been a handful of break-throughs that have proved to be of fundamental importance to our understanding of ourselves and of the universe. To Copernicus's theory of a heliocentric universe, Newton's laws of motion and gravity, Einstein's theories of relativity and Max Planck's origination of quantum theory must be added Darwin's theory of evolution by natural selection.

Outside the field of science, Darwin's theory ranks alongside that of Copernicus in its impact on changing the way we think about humanity and its place in the universe. By putting the sun and not the earth at the centre of the universe, Copernicus reduced the special status given to humanity in Christian cosmology. Similarly, by providing a mechanism, in the form of natural selection, by which evolution must have occurred, Darwin demolished the traditional Christian view that the living world – including humanity – was the result of a single, more or less instantaneous creation.

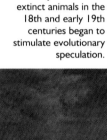

The discovery and study of fossils of extinct animals in the 18th and early 19th centuries began to stimulate evolutionary speculation.

Why evolution?

The clear-cut account of creation in the Bible meant that for centuries speculation on any alternative theory was firmly discouraged. However, during the 18th century scientists began to amass increasing amounts of fossils of animals that no longer existed on earth. Arguments raged as to whether these bones and shells embedded in solid rock were genuine organic remains that had been petrified in former times, perhaps when such strata had been on the sea bed, or whether – like minerals and crystals – they had been produced by the great creator of nature. Moreover, since fossils often appeared to represent creatures and plants of a kind no longer in existence, the question arose as to how a benevolent creator could have allowed them to become extinct.

By the early 1800s many naturalists believed that the fossil record showed a gradual hierarchical progression from the simpler to the more complex forms of life, and thus revealed the rational plan of creation through time. This explanation, however, presented a number of problems. Why, for example, is there such a diversity of forms alive today? Why have such lowly forms as the bacterium, sponge, slime mould and liverwort persisted alongside the more derived or highly evolved forms of life?

Early theories

One of the earliest proponents of the concept of organic change was Erasmus Darwin, Charles's grandfather. According to his theory of 'transmutation', briefly outlined in the final decade of the 18th century, organisms are designed to be self-improving and can develop new features and even organs in their effort to combat environmental changes. Thus Erasmus Darwin subscribed to the view that is now usually associated with the name of the Frenchman Jean Baptiste Lamarck – the 'inheritance of acquired characters'.

Lamarck's theory, put forward in the early years of the 19th century, stressed the creative power of nature to change in response to new conditions. He proposed a hierarchical arrangement of all living forms that was assumed to represent the historical pattern along which life had advanced from simpler ancestors to more complex animals. This evolutionary diversity, he believed, was the result of each organism's response to environmental conditions. Imagining that an animal's needs determine the organs that its body will develop, Lamarck particularly stressed those changes he assumed resulted from the use and disuse of body parts: thus the giraffe developed its long neck from its habit of browsing on tall trees. Lamarck believed that those characters acquired during the lifetime of the animal as a result of its efforts could be transmitted to its offspring, and so be carried over into future generations.

Catastrophism and uniformitarianism

Lamarck's theory attracted few followers at the time and he died in obscurity. The reason for his eclipse can be directly attributed to his arch rival and fellow countryman, Baron Georges Cuvier. Rejecting both transmutation and Lamarck's hierarchical view of the animal kingdom, Cuvier proposed just four types of animal organization – Vertebrata, Mollusca, Articulata and Radiata. He did not believe that extinct fossil animals could have evolved into or shared an ancestry with their modern counterparts. Instead he proposed that all extinct and living animals had at one time coexisted, and that a series of strictly localized catastrophes had subsequently wiped many of them out.

As Cuvier's view did not contradict the Bible's account of creation, it became extremely popular. Its popularity was reinforced by the geological theory of *catastrophism*, which held that all geological changes are the result of sudden, short-lived catastrophes such as floods. This idea not only reflected the Biblical flood, it also allowed for a relatively recent creation of the earth – put at 4004 BC by Bishop James Ussher in the 17th century, on the basis of the genealogies in the Bible.

Catastrophism was firmly refuted by the Scottish geologist Sir Charles Lyell, who in the 1830s popularized the theory of *uniformitarianism*, which states that all geological change is slow, gradual and continuous, involving processes such as heat and erosion; thus rocks – and the fossils in them – are many millions of years old, rather than just a few thousand. Lyell's theory eventually became universally accepted.

Charles Darwin

In 1831 Charles Darwin – who had initially studied medicine and then started to train for the ministry – became the naturalist on HMS *Beagle* on her five-year voyage of discovery (1831–6). The extensive observations he made on this voyage not only won him over to uniformitarianism, but also led him to the conception of evolution by natural selection. Eventually he committed his ideas to paper, but instead of publishing them immediately he continued to cogitate on his theory for a further two years until 1858, when he received a letter from a young naturalist, Alfred Russell Wallace. This provided Darwin with the stimulus to publish, since Wallace's letter contained the essence of his theory of natural selection.

A paper written jointly by Wallace and Darwin and entitled 'On the Tendency of Species to form Varieties; and on the Perpetuation of Varieties and Species by Natural Means of Selection' was read out at the Linnean Society in London on 1 July 1858, but it caused little or no immediate reaction within the scientific community. However, public interest was quickly aroused by the appearance on 24 November 1859 of Darwin's *On the Origin of the Species by Means of Natural Selection*, all 1250 copies being sold on the day of issue. The heavy guns of the established Church soon began to rumble against Darwinism as a heretical doctrine intended 'to limit God's glory in creation', and the barrage reached a climax in the famous Oxford debate of the following summer (▷ box).

Natural selection

In his *Origin of Species* Darwin showed how the process of adaptation of organisms to their environment, and hence of evolution, was due to the blind operation of everyday laws of nature, by means of a mechanism he termed 'natural selection'. If Darwin was right, it meant an end to the claim that the course of evolution either had any purpose or that it was subject to or preordained by a divine will.

Although a major part of Darwin's achievement was to amass an overwhelming amount of evidence in support of his theory, his genius lay rather in the conclusions he was able to draw from readily available evidence. There was nothing novel in noting that there are considerable variations between individuals of a particular species, nor that more offspring are produced in each generation than survive to reproduce.

Yet it is an inference from these observations that provides the very cornerstone of Darwin's theory – the notion that, in the 'struggle for existence' resulting from competition among individuals, variations in attributes (some of them due to heredity) will affect success in survival and reproduction. In other words, a fitter individual – one that is in some way better adapted to its surroundings – will be more likely to survive and leave descendants than a less fit one, and hence will tend to perpetuate within the population those inherited differences to which it owed its success.

THE GREAT EVOLUTION DEBATE

The first real occasion for the established Church of England to openly debate Darwin's views presented itself on 30 June 1860, when the meeting of the British Association was held in Oxford. By this time most people were primed and eager for the coming encounter.

An audience of nearly a thousand crammed into the long room that later became part of the Old Ratcliffe Library. After a long and rather dreary paper read by one Dr Draper, Samuel Wilberforce, bishop of Oxford, rose 'and spoke for full half an hour with inimitable spirit, emptiness and unfairness', finally asking Thomas Huxley, one of Darwin's leading supporters, whether he was 'related by his grandfather's or grandmother's side to an ape'. There are many versions of Huxley's reply, one of the most vivid and colourful being that of John Green in a letter to his friend William Boyd Dawkins. Green describes Huxley as 'young, cool, quiet, sarcastic, scientific in fact and in treatment', and records his words thus: 'I asserted, and I repeat, that a man has no reason to be ashamed of having an ape for his grandfather. If there were an ancestor whom I should feel shame in recalling, it would rather be a *man*, a man of restless and versatile intellect, who, not content with an equivocal success in his own sphere of activity, plunges into scientific questions with which he has no real acquaintance, only to obscure them by an aimless rhetoric, and distract the attention of his hearers from the real point at issue by eloquent digressions, and skilled appeals to religious prejudice.'

No one who was present doubted Huxley's meaning and the effect was tremendous. One lady fainted and had to be carried out. As for Huxley, he for ever after became known as 'Darwin's bulldog'.

Caricature of Charles Darwin, from the *London Sketch Book*, 1874.

In this way natural selection can act over time to change or diversify the characteristics of a population. In other words there is what Darwin called 'descent with modification', dubbed by Herbert Spencer 'the survival of the fittest'. Thus many populations continuously improve their adaptations to the environment to which they are subjected, while less well-adapted populations become extinct.

Gaps in Darwin's theory

Darwin realized that evolution was only a theory, and though he argued at length that the fossil record was compatible with it, he realized that it did not show the expected transformations of structure through time; instead, it seemed to show the origin of new types or classes fully formed. Darwin attributed this to the 'imperfection of the geological record', and in fact, despite its incompleteness, the fossil record fits the major predictions based on comparative studies of living organisms (▷ p. 14).

Another problem facing natural selection was the length of time necessary for it to have brought about the diversity and complexity of life today. There was plenty of evidence by Darwin's time of the speed with which domesticated species could be made to change using selective breeding, but it was obvious to Darwin that the non-directed, almost random trial-and-error nature of natural selection would take very much longer to bring about noticeable changes. Thus Darwin was somewhat put out in 1862 when the great physicist, Lord Kelvin, argued that the sun – and thus the solar system – could not be more than 24 million years old. However, Kelvin assumed the sun burned a fuel similar to coal, and we now know that in fact it is powered by nuclear fusion reactions. We also know that the solar system, including the earth, is around 4·6 billion years old – sufficient time for natural selection to have brought us to where we are now.

One of the key objections to Darwin's theory was that selection could only operate on variations that already existed and did not explain how new changes arise in the first instance. Moreover, Darwin, like most of his contemporaries, subscribed to the idea of 'blending' inheritance, imagining the characteristics of an offspring are a blend or average of those of the parents. When in 1867 it was demonstrated mathematically that natural selection could not work if it was based on the blending model, it was something of a set-back for Darwin.

However, we now know that heredity works according to the laws devised in 1865 by Gregor Mendel (but not rediscovered until 1909) by which a particular trait is either passed on or not – there is no question of blending (▷ p. 22). Furthermore, once the function of the DNA molecule in inheritance was established in 1953 (▷ p. 22), it was discovered that it was very common for an accidental change to occur when DNA replicates itself. Such changes – known as *mutations* – usually have little effect, but in some cases they can be harmful, while in others they can lead to the development of a new and desirable characteristic that enables the individual concerned to compete more effectively than others of its species. The modern genetic theory of natural selection (▷ pp. 16–17) has strongly reinforced Darwin's theory, and genetic methods have also provided scientists with new means of establishing the probable lines of descent of modern species and the degree to which different species are related (▷ p. 15).   BG

The Evidence for Evolution

The fossil remains of long-extinct plants and animals that prompted the revival of evolutionary theories prior to Darwin (▷ p. 12) have also provided much evidence of the way individual groups of organisms have evolved. Although the fossil record is far from complete, it has consistently supported Darwin's view of evolution as an ever-branching process in which each branch represents a separate group evolving in its own particular way.

SEE ALSO

- WHAT IS LIFE? p. 6
- EVOLUTION: THE ORIGIN OF SPECIES p. 12
- EVOLUTION IN ACTION p. 16
- THE CLASSIFICATION OF LIFE p. 18

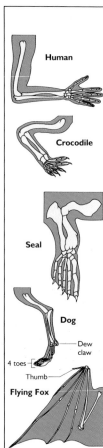

The forelimbs of five different animals, showing that the pentadactyl (five-digit) limb is common to them all. Such basically similar structures are described as *homologous*, and the existence of homologous structures in different organisms suggests they have all evolved from a common ancestor.

Some of the most convincing support for evolution comes from comparative anatomy, the study of similarities and differences between living organisms. The major predictions of comparative anatomy as to the likely descent of groups have been borne out by the fossil evidence. Further support has come more recently from molecular evidence, in which comparisons of proteins and nucleic acids (DNA and RNA) in different groups have broadly reinforced the anatomical and fossil evidence.

Finally – and this is what really convinced Darwin – there is the need to explain the correspondence between speciation and geographical isolation. This phenomenon, like so many other observations that have been made of the living world, can most simply be explained by Darwin's evolutionary theory. Any other explanations are forced into incredible complexities and contradictions, and the creation theory only holds good if, as one commentator has written, 'the creator deliberately set out to deceive us'.

Homologous structures

The aim of comparative anatomy is to distinguish between resemblances resulting from shared ancestry (*homologies* or *homologous structures*) from those that have arisen independently in unrelated groups (*analogies* or *analogous structures*). Thus anatomical similarity *may* be evidence of common descent, but it may also be a coincidence.

A knowledge of how anatomical structures develop is essential in distinguishing homologies and analogies. For example, the five-digit (pentadactyl) limb is a structure shared by all four-limbed vertebrates (tetrapods), and we can deduce that the very different-looking fore limbs of a human, crocodile, seal, dog and bat are all homologous structures because they are all built on a common skeletal plan (▷ illustration). Their differences are at a more superficial level, the fore limb of each being adapted to different functions. In many cases, however, superficially similar structures have evolved in unrelated groups to perform the same or similar functions. In effect,

the process of evolution has independently found the same solution to the same problem – a process known as *convergent evolution* (▷ pp. 16–17). For instance, wings have evolved separately in both birds and insects (unrelated groups), but they do not share a common underlying plan and are therefore analogous structures.

Care is necessary in making this kind of comparison, since structures may be homologous at one level but not at another. For example, the wing of a bird and that of a bat are homologous at the level of fore limbs (since they are based on a common skeletal plan), but at a higher level (that of the development of flight) they are merely analogous. No one would doubt that birds and mammals share a common ancestor if one goes back far enough, but bats are descendants of non-flying terrestrial mammals while birds evolved directly from reptiles, so each group must have acquired wings and the ability to fly independently.

The embryonic evidence

The relationships suggested by homologous structures are corroborated by the similarities observed between embryos of different species. The embryos of vertebrates, for example, show remarkable similarities at comparable stages of development, and the earlier the stage, the more similar is the resemblance (▷ illustration).

The observation of this process by the German biologist Ernst Haeckel (1834–1919) gave important support to Darwin. Haeckel himself summarized his findings in the memorable phrase 'ontogeny recapitulates phylogeny' – ontogeny being the sequence of events in the development of

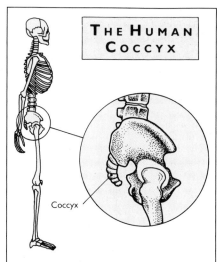

THE HUMAN COCCYX

One piece of evidence that humans and monkeys had a common ancestor is the existence in humans of the coccyx – the bone at the base of the spine. The coccyx is the remains of a tail, and is an example of a vestigial structure, i.e. one that no longer functions. In very rare cases humans are actually born with a stump of a tail.

an individual organism, and phylogeny the sequence of events in the evolution of a species. The phrase is in fact somewhat misleading, as it is only certain embryonic stages of a species (not the adult stage) that resemble the embryonic stages of another – at no point does a chicken embryo look like an adult fish. The closer the relationship of the two species, the longer their embryos are virtually identical.

Vestigial structures

Many animals possess structures for which there is no apparent function. If such structures are very much reduced in comparison with similar structures in related animals, they are known as *vestigial structures*. Examples include the isolated pelvic bones found in whales (▷ illustration, p. 142), the pair of tiny 'claws' about two-thirds of the way down the bodies of certain snakes (indicating that the ancestors of the snakes may have had hind limbs like other reptiles), and the hidden 'tail' of the great apes (the coccyx in *Homo sapiens*; ▷ illustration).

It would be very difficult indeed to explain the presence of these vestigial structures without the theory of evolution. With it, and with the aid of comparative anatomy, they can be shown to be homologous with similar structures in other animals. The evolutionary assumption is that the possessor of the vestigial structure shares a common ancestor with those animals where the homologous structure still has a function; the vestigial structure is presumed to have lost some or all of its function and to be degenerating because it serves no purpose. Thus whales, no longer needing to support some of their weight on their hind limbs – which have evolved into a paddle-like structure (the fluke) – have no use for a pelvis. Similarly, snakes employ a hunting strategy in which crawling has proved a more undetectable way of stalking certain prey

PARASITISM AND EVOLUTION

Darwin's analysis of parasitism in barnacles not only presented more evidence of evolution but also helped him to solve the problem of vestigial structures (▷ text). Darwin found that though the majority of barnacles are hermaphrodite (i.e. they possess both male and female reproductive organs), in a few species there are minute males that live parasitically upon the female. He also discovered other hermaphrodite species with dwarf males inside them. Darwin presumed that this whole group of barnacles had originally been hermaphrodite and that some species had later evolved separate males. Eventually in this second group the male parts of the hermaphrodite became superfluous and hence degenerated.

Many groups of organisms are like barnacles in having parasitic as well as free-living members. These range from parasitic bacteria and protists to crustaceans, vertebrates and plants. Parasitism, Darwin concluded, is itself evidence for evolution, since it is hard to envisage how parasites could have arisen other than by adaptation – in other words by evolution.

than walking, and their body form allows them to perform the final strike at lightning speed.

Geographical isolation

One of the phenomena of the natural world that would be very difficult to explain without evolution is *discontinuous distribution* – the existence of similar, but not identical, animals and plants in different parts of the world. Evolution, which seeks to explain how different species arise, postulates that these different species diverged from a common ancestor, populations of which somehow became isolated and therefore drifted slowly apart over time, unable to interbreed with other populations.

On a large scale this process of divergence by isolation can be seen in the case of the big cats, which take different forms on different continents: for example, the tiger in Asia, the lion in Africa, and the jaguar in South America. These animals are similar enough for them to be placed in the same genus (*Panthera*; ▷ p. 124), indicating that they share a common ancestor. But how do we explain their differences? The explanation provided by evolution is that the common ancestor of the big cats lived when the continents were joined together, and that populations of this ancestor were separated when the continents began to drift apart (▷ pp. 8–9). These isolated populations then evolved along slightly different paths. Isolation of populations can also occur as a result of climatic change, as in the case of the pied and collared flycatchers (▷ p. 11).

However, the most dramatic cases of divergence have occurred in isolated oceanic island groups such as the Galápagos Archipelago. Being volcanic in origin, these islands, which formed only 4–5 million years ago, were never connected to the mainland, yet each island

has its own unique constellation of species. There are, for example, 13 different species of Darwin's finch on the Galápagos Islands, with a further species on Cocos Island several hundred kilometres to the north. In addition there are several unique species of mocking bird and a variety of endemic plants and animals (including insects, lizards and tortoises) that have apparently evolved in isolation on these islands. This pattern is also characteristic of other island groups, such as Hawaii (▷ illustrations, pp. 16 and 214) and the Seychelles. Darwin's explanation for this group of distinct but closely related species was that an ancestral South American mocking bird and finch had somehow been transported to these islands and had then evolved in

different directions on each island (a process now known as *adaptive radiation*; ▷ pp. 16–17). The only alternative explanation available to Darwin was to assume that the creator had set about work in a different way in many different parts of the world to no apparent purpose.   BG

The fossil record consistently supports the theory of evolution by natural selection. These fossilized dinosaur bones are being carefully uncovered at Dinosaur National Monument in Utah.

THE MOLECULAR RECORD

Since Darwin's day the greatest advance in evolutionary biology came with the discovery of the structure of DNA and the realization of its role in heredity (▷ p. 22). Once it was understood that DNA acts as the blueprint for the manufacture of proteins, it was apparent that every organism carries a record of its history encoded in its DNA and in the sequences of amino acids that make up its proteins.

The molecular record, however, is not readily accessible. It was not until the 1970s that techniques for analysing the amino-acid sequences in proteins became available, showing, for example, that chimpanzees and human beings have identical amino-acid sequences in three proteins. For more comprehensive comparisons it became necessary to find proteins that are widely distributed in different groups of organisms. It was also necessary that there be parts of these proteins where the exact details of the structure are irrelevant to the functioning of the proteins. The reason for this is that

surviving mutations in the amino acids in these parts would be neutral – i.e. neither harmful nor beneficial (▷ p. 16).

Such neutral mutations being random, it is mathematically probable that they accumulate at a regular rate. One can therefore surmise that in comparing two groups of organisms, the number of differences in the amino-acid sequences in such proteins reflects the period of time since the two groups diverged. Correlating the number of differences with the fossil record, it is possible to create a 'molecular clock' of evolution, an average protein undergoing a change in one amino-acid site approximately once every million years.

Protein sequencing has now been superseded with the development in the 1980s of methods for the rapid sequencing of RNA and DNA. Vast amounts of data are being generated by these new techniques, but, by and large, molecules and morphology (anatomical characteristics) reveal similar patterns of development in the history of life on earth.

COMPARATIVE EMBRYO DEVELOPMENT

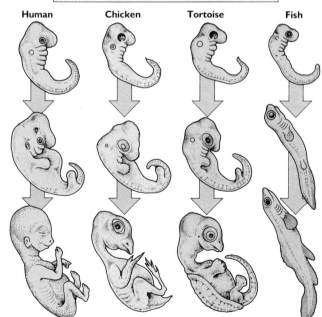

Human   Chicken   Tortoise   Fish

The similarities between the embryos of vertebrates at comparable stages of development have provided considerable support to the theory of evolution (▷ text). Thus fishes, amphibians (not shown), reptiles, birds and mammals all start with a similar number of gill arches (the folds below the head) and a similar vertebral column; even the early human embryo has a 'tail'. However, as the embryos develop, the similarities decrease and the individual species become more and more differentiated.

Evolution in Action

The relatively new science of genetics has not only under-scored Darwin's basic theory of evolution, it has also elaborated it in some rather unpredictable ways. In establishing how genetic mutations work it has provided the missing key as to how new characteristics arise on which natural selection can work. It has also provided us with the concept of natural selection acting on genes rather than individuals, the idea that genes use living organisms for their propagation rather than the other way round.

Genetics has also shown that not all change comes about by natural selection, and that some characteristics may be entirely arbitrary, providing neither benefit nor disadvantage. This reinforces the fundamental truth that evolution is not purposeful: although natural selection would seem to have made certain organisms 'perfectly' adapted to their environment, it can never achieve a permanent point of 'perfection'. Not only will arbitrary mutations continue to happen, but the environment to which natural selection adapts a species can never remain forever stable.

Mutation, the gene pool and genetic drift

Each of the physical characteristics of an organism is dictated by a gene (a section of DNA; ▷ p. 22) or a collection of genes. Many behavioural characteristics are also dictated in this way, although environmental influences can modify behaviour in varying degrees. During sexual reproduction DNA replicates itself, and during this replication process random errors can arise. Such errors are known as *mutations*. In the vast majority of mutations, no observable difference is evident in the offspring. However, in some cases the mutation will give rise to a harmful new characteristic, while in others a new characteristic may prove beneficial; natural selection will act to 'weed out' individuals with a harmful characteristic and, conversely, perpetuate the beneficial characteristic.

At the genetic level, one can regard an interbreeding population of plants or animals as a *gene pool*, an 'environment' in which different genes compete to survive. Genes that make bodies that are better at surviving and reproducing will perpetuate themselves and will have a high frequency (i.e. be common) in the gene pool, while genes that are not so good in these respects will be rarer. As the make-up of the gene pool changes over time, so will the characteristics of the individuals of the species.

However, most mutations are in fact 'neutral' – i.e. neither harmful nor beneficial – and the random shuffling of parental genes that occurs during sexual reproduction (▷ p. 23) may purely by chance perpetuate through the generations such neutral mutations. This process is called *genetic drift*, and it is possible over a large timescale that genetic drift may play as important a part in evolution as natural selection, but we have as yet no means of determining this.

Certainly genetic drift does help to explain characteristics that do not seem to have an adaptive purpose. Examples of this include the different numbers of horns on Indian and African rhinoceroses: having a horn is important for defence, but the number makes no discernible difference.

Speciation

Genetic drift provides one of the answers to why species diverge – in other words, why new species arise – when populations are isolated. However, in most cases the geographical isolation of populations – whether through continental drift (▷ p. 8), climatic change (▷ p. 10) or being stranded on an island by accident – is likely to present the separated populations with new environmental circumstances to which each population will

THE PEPPERED MOTH

The light-coloured form of the peppered moth, well camouflaged against an unpolluted tree trunk.

Darwin himself had no example of natural selection actually taking place in nature. But since his time a classic case history of evolution in action has been observed in the case of the British peppered moth.

The peppered moth has two forms, a light-coloured form and a mutant dark form. Before the Industrial Revolution light-coloured moths were the most prevalent (judging from private insect collections) and dark forms virtually non-existent. During the next 200 years the tree trunks on which the moths alight became darkened by increasing amounts of soot in the vicinity of the major cities. In these areas there was a marked increase in the dark forms, while in the unpolluted countryside the light-coloured forms predominated. Various scientists demonstrated by mark-and-recapture experiments that in the sooty areas the dark forms were less liable to be spotted – and eaten – by birds, whereas in the non-polluted areas the dark forms suffered heavy predation. With the Clean Air Act of 1956 (and the resulting reduction in smoke pollution), the frequency of the dark moths decreased significantly.

Since these studies many other cases of natural selection have been observed, particularly among organisms where the turnover of generations is rapid. Examples include the evolution of forms of mosquito that are resistant to DDT, and bacteria that are resistant to antibiotics (▷ pp. 227 and 231).

adapt in its own way. The new circumstances may be regarded as challenges (e.g. a harsher or more competitive environment) or as opportunities (such as a newly colonized habitat in which there are a number of vacant ecological niches) – although 'challenges' and 'opportunities' really amount to the same thing if a species successfully adapts.

The process of adaptation by natural selection acting in different ways on isolated populations, together with genetic drift, will result in the gradual evolutionary divergence of the populations. The end result will be that the gene

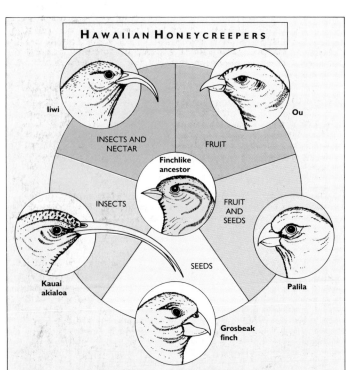

HAWAIIAN HONEYCREEPERS

Iiwi — INSECTS AND NECTAR

Ou — FRUIT

Finchlike ancestor

INSECTS

FRUIT AND SEEDS

SEEDS

Kauai akialoa

Palila

Grosbeak finch

Hawaiian honeycreepers (constituting the family Drepanididae) provide one of the most spectacular examples of adaptive radiation. It is thought that the ancestor of the honeycreepers may have been a finchlike bird that arrived on the newly formed volcanic islands several million years ago. With a wide range of vacant ecological niches available, the ancestral bird rapidly evolved into a large number of species, each with a distinctive shape of beak suited to its favoured food. However, when exotic species are introduced onto such islands, the native species may find themselves unfit to compete; such highly specialized species are also particularly prone to any human disturbance of the ecological balance of their habitats. The Hawaiian honeycreepers have found themselves vulnerable to such changes, and many are now extinct, including the grosbeak finch and possibly the ou.

frequency in the gene pools of the respective populations will eventually be so different that there comes a point when they can no longer successfully interbreed even if their isolation is ended: they have become two different species.

There can, however, be certain intermediary stages where the distinctions between species are somewhat blurred. Around the Arctic, for example, there is a 'ring' of species and subspecies of gull of the genus *Larus*, each of which can interbreed with its neighbour. However, when the ring is 'completed' in northwest Europe, the two extreme representatives, the herring gull and the black-backed gull, are clearly distinct species and cannot interbreed, even though they are now found in the same area.

Where geographical overlaps such as this occur it is actually very important for species to signal their differences. An individual that mated with an individual of another species would either produce no offspring or sterile offspring (the mule, the result of crossing a horse with a donkey, exemplifies the latter case). Not only would the individual have wasted much energy in courtship rituals – or even in giving birth and raising young – it would also be a complete dead-end for that individual's genes. Therefore species develop distinct characteristics – such as markings and coloration or certain patterns of behaviour – that may have no other function than to ensure that the members of the species can recognize each other for mating purposes. Such differences – known as *isolating mechanisms* – may in some cases have had their origins in neutral mutations, but when species are no longer isolated natural selection will act to heighten the difference.

The selfish gene

On the face of it one might assume that natural selection, the 'survival of the fittest', exclusively acts in favour of self-preservation of individual organisms. However, there are many characteristics of species, both physical and behavioural, that are difficult or impossible to explain on this basis. If, alternatively, one thinks of natural selection acting exclusively in favour of the self-preservation of individual genes, then many things begin to fall into place.

Competition between individuals is not just for resources (food, territory, etc.), it is also for mates. Darwin regarded this process, called *sexual selection*, to be an evolutionary force acting in parallel with natural selection (▷ Sexual Dimorphism, p. 159). However, it can be regarded as natural selection from the point of view of the individual gene. In many species, such as deer, males expend much energy in fighting each other to gain as large a 'harem' of females as possible. In other species where the females do the selecting, males often develop extravagant appendages used for display – the magnificent plumage of male peacocks being a prime example. The need to out-

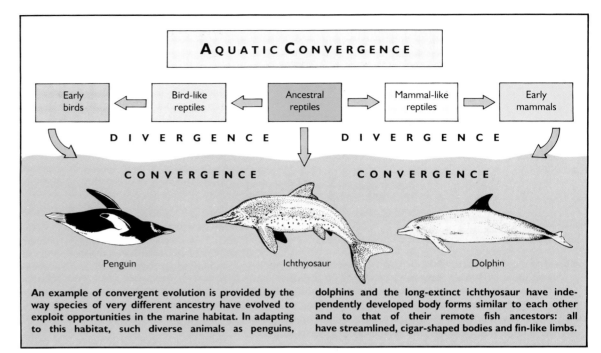

AQUATIC CONVERGENCE

Early birds ← Bird-like reptiles ← Ancestral reptiles → Mammal-like reptiles → Early mammals

DIVERGENCE   DIVERGENCE

CONVERGENCE   CONVERGENCE

Penguin   Ichthyosaur   Dolphin

An example of convergent evolution is provided by the way species of very different ancestry have evolved to exploit opportunities in the marine habitat. In adapting to this habitat, such diverse animals as penguins, dolphins and the long-extinct ichthyosaur have independently developed body forms similar to each other and to that of their remote fish ancestors: all have streamlined, cigar-shaped bodies and fin-like limbs.

display rivals means that males not only put a lot of energy resources into growing these structures, but also that they are a much more obvious target for predators. The point is that from the selfish gene's point of view, it is no good being in a non-aggressive deer or a tailless peacock. Even though such individuals would probably live much longer than other males, they would have no chance of ever mating – and so the gene for non-aggression or taillessness would die out. Only genes that favour their own propagation survive, whatever the cost to the individual organism that carries them.

Another problem that long puzzled zoologists was the existence of apparently altruistic phenomena in the animal world, such as the prominent tail flashes of rabbits fleeing from a predator, (▷ Altruism, p. 149). However, it has been established that in all such circumstances the altruistic individual is fairly closely related to – and thus has a relatively high probability of sharing its genes with – the other members of the group on whose behalf it is acting. This process is called *kin selection*. Thus one can say that a gene has a better chance of survival if it can make its bearer act altruistically on behalf of other individuals that carry the same gene.

Many observations have also been made of altruistic behaviour between non-related individuals, such as vampire bats sharing food with neighbours. This *reciprocal behaviour* is not in fact true altruism, because the animal helping the other individual can expect the favour to be returned at a later date. The interests of the gene and the individual here coincide. Reciprocal behaviour is not uncommon in social animals – particularly humans – and one might say in such cases that 'Do unto others as you would have them do unto you' is an evolutionary necessity as well as a moral injunction.
IDC/BG

EVOLUTIONARY VARIATIONS

Nature provides many variations on the theme of evolution by adaptation to a changing environment.

CONVERGENT EVOLUTION
Convergent evolution is where different organisms have evolved similar solutions to similar problems, even though they have evolved from very different ancestors. Such solutions are usually anatomical modifications, which are known as analogous structures – for example, the wings of insects, birds and bats, all of which arose independently (▷ p. 14). Other examples of convergent evolution include swallows and swifts (both of which look and behave very similarly, but which in fact belong to entirely different orders), and the adaptations that various reptiles, birds and mammals have made in returning to the seas (▷ illustration).

PARALLEL EVOLUTION
Similar in some respects to convergent evolution is parallel evolution, whereby unrelated plants and animals will adapt in similar ways to fill ecological niches in similar but geographically separated ecosystems. Thus the marsupial mammals of Australia – which was separated from the other continents 45 million years ago – have evolved very similar forms to the placental mammals elsewhere. Thus there is, for example, a marsupial 'mole', 'rat', 'wolf', etc.

ADAPTIVE RADIATION
Adaptive radiation is a speeding up of the normal process of evolutionary divergence of species, and may arise in two kinds of situation. The first is where a highly successful mutation or series of mutations occurs, bringing about new forms that outcompete existing forms, and which rapidly take over a range of ecological niches. A good example of this is the success of the flowering plants over the conifers and their allies (▷ pp. 39–41). The flowering plants rapidly came to dominate the world's flora, and show an incredible diversification of forms (▷ pp. 44–7). The second kind of situation is where a new habitat arises with an array of vacant ecological niches. Famous examples of this include the colonization of relatively recent volcanic island groups such as the Galápagos and Hawaii, where birds such as Darwin's finches (▷ p. 15) and the honeycreepers (▷ illustration), and plants such as the tarweeds (▷ photo, p. 214), have evolved a wide array of types from a single ancestral species.

COEVOLUTION
Coevolution is where two or more species evolve in continuous adaptation to each other. This is seen in a variety of kinds of relationship – perhaps most obviously between symbiotic partners and between parasite and host (▷ p. 168). It is also apparent in the 'arms race' between predators and their prey (▷ p. 164), and in the deceits practised by both predators and prey pretending to be something other than they are (▷ p. 166). Finally, the whole evolution of flowers and the animals – especially insects – that pollinate them (▷ p. 42) provides numerous examples of astonishingly specialized examples of coevolution.

The Classification of Life

The study of biological classification, or *taxonomy*, aims to provide a rational framework in which to organize our knowledge of the great diversity of living and extinct organisms. Today, comparative biology seeks to understand the living world by searching for the underlying order that exists in nature.

The first well-authenticated system of classification goes back to Aristotle (384–322 BC), but it was not until the 17th century that the Englishman John Ray (1628–1705) proposed the first *natural classification* – an arrangement based on presumed relationships, rather than an artificial scheme aiming merely to facilitate correct identification of species. In the 18th century the Swedish naturalist Carl Linnaeus (1707–78) produced a rational system of classification based on patterns of similarity between different organisms. Linnaeus developed one of the first comprehensive subordinated schemes of taxonomy – a scheme that places each organism in a group that is itself part of a larger group.

The Linnaean system

The essential feature of Linnaeus's scheme is that it is *binomial*. He gave every distinct type or *species* of organism (e.g. the lion) a two-part (binomial) name (e.g. *Felis leo*) in which the second element identified the individual species, while the first element placed the species in a particular *genus* – a group comprising all those species that showed obvious similarities with one another. For instance, he grouped together all cat-like animals in the genus *Felis*: *Felis leo* – the lion; *Felis tigris* – the tiger; *Felis pardus* – the leopard; and so on. Having thus subordinated each species to a particular genus, he went on to place groups of genera in a higher rank or category called a *family*, then families in *orders*, and so on through *classes*, *phyla* (or *divisions*, for plants) and finally *kingdoms*. Within this hierarchy, each successive rank is thus more embracing than the last, each containing a greater number of organisms with fewer characteristics in common.

Although Linnaeus's genera and families were for the most part natural, his higher ranks were of necessity artificial in order to deal with the vast numbers of new organisms being discovered. Linnaeus used his highest taxonomic rank (kingdom) to separate plants and animals, but it is now clear that this simple subdivision is untenable, because certain groups such as bacteria, protists and fungi fit into neither category – and are now each assigned their own kingdom. Nevertheless, the binomial system has remained unchanged to this day and every newly discovered organism is given a Latin or Latinized binomial name.

Another difficulty with the Linnaean system is that it is too simplistic to reflect the complexity of classification that is implied by Darwin's theory of evolution (▷ pp. 12–17). Since it is generally believed that every organism on this planet has arisen through a unique historical process of descent with modification, it follows that all of them should fall into uncontradicted patterns of groups within groups. Thus, although we shall go on using such terms as family, order and class, we will need to introduce additional ranks as the analysis of the pattern of life is further elucidated. For example, the classification of living organisms illustrated opposite has at least 24 major branching points or divisions, each of which would need to be assigned a separate hierarchical category – 12 of them higher, or more embracing, than that of class.

Modern classification systems

The change from essentially artificial systems of classification, such as that used by Linnaeus, came about as taxonomists realized that there was a natural order underlying all living things; this order was furthermore provided with a theoretical background by Darwin in 1859 (▷ p. 12). Today's taxonomist endeavours to provide the most natural classification possible by the use of one of the following systems:

Phenetic classification aims to incorporate as much information as possible about organisms, and then groups them together on the basis of their overall similarity. All characteristics are given equal weight and no account is taken of the evolutionary relationships of the groups. This approach is amenable to numerical analysis (*numerical taxonomy*) and may be useful for large groups of simple organisms such as bacteria.

Orthodox (or *phylogenetic*) *classification* provides the basis for most text-book classifications. The system is based on presumed evolutionary relationships, and – as with phenetics – as much information as possible is taken into account. Organisms are grouped by features they share in common and that seem to reflect their common ancestry.

Cladistics is generally believed to be the most precise and natural method of classification, and is used by most progressive taxonomists. The only groups formally recognized are *clades*, each clade being defined by features that are shared by all its members and that are found in members of no other group. These groups

THE CLASSIFICATION OF MAN

RANK	NAME	MEMBERS	DISTINGUISHING FEATURES
KINGDOM	ANIMALIA	All multicellular organisms except plants and fungi	Nervous system
PHYLUM	CHORDATA	Lancelet, sea squirts, all vertebrates	Notochord (central nerve cord)
SUBPHYLUM	VERTEBRATA	All vertebrates, i.e. fish, amphibians, reptiles, birds, mammals	Backbone (protecting notochord)
SUPER-DIVISION	GNATHO-STOMATA	All vertebrates except lampreys and hagfishes	Jaws
DIVISION	OSTEICHTHYES	All vertebrates except lampreys, hagfishes and cartilaginous fishes	Principal component of skeleton is bone (not cartilage)
SUPERCLASS	TETRAPODA	All vertebrates except fish	Four limbs, the front pair being pentadactyl (i.e. having five digits)
CLASS	MAMMALIA	All mammals	Milk and sweat glands, hair
SUBCLASS	EUTHERIA	All mammals except monotremes and marsupials	Placenta
ORDER	PRIMATES	Lemurs, lorises, pottos, galagos, tarsier, marmosets, tamarins, monkeys, apes	Fingers with sensitive pads and nails
SUPERFAMILY	HOMINOIDEA	Apes (gibbons, orang-utan, gorilla, chimpanzees, humans)	No tail, broad chest, shoulder blades at back rather than sides
FAMILY	HOMINIDAE	Gorilla, chimpanzees, humans	Upright posture, flat face, large brain; similar blood plasma protein
SUBFAMILY	HOMININAE	Chimpanzees, humans	Similar brain shape and anatomical and protein structures; tool and weapon users
GENUS	*Homo*	*Homo habilis*, *H. erectus* (both extinct), *H. sapiens*	Exclusively bipedal (walking on two feet), manual dexterity
SPECIES	*SAPIENS*	Neanderthal man (*H. sapiens neanderthalensis*; extinct), modern man	Double-curved spine
SUBSPECIES	*SAPIENS*	Modern man (*Homo sapiens sapiens*)	Well-developed chin

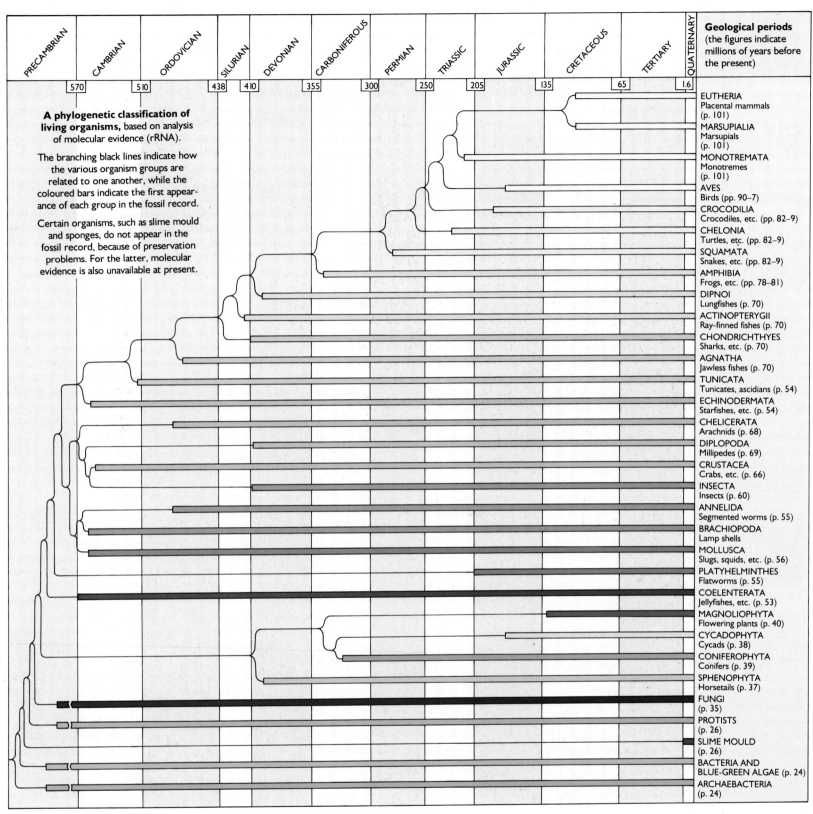

A phylogenetic classification of living organisms, based on analysis of molecular evidence (rRNA).

The branching black lines indicate how the various organism groups are related to one another, while the coloured bars indicate the first appearance of each group in the fossil record.

Certain organisms, such as slime mould and sponges, do not appear in the fossil record, because of preservation problems. For the latter, molecular evidence is also unavailable at present.

Geological periods (the figures indicate millions of years before the present)

PRECAMBRIAN | CAMBRIAN | ORDOVICIAN | SILURIAN | DEVONIAN | CARBONIFEROUS | PERMIAN | TRIASSIC | JURASSIC | CRETACEOUS | TERTIARY | QUATERNARY

570 510 438 410 355 300 250 205 135 65 1.6

EUTHERIA
Placental mammals (p. 101)

MARSUPIALIA
Marsupials (p. 101)

MONOTREMATA
Monotremes (p. 101)

AVES
Birds (pp. 90–7)

CROCODILIA
Crocodiles, etc. (pp. 82–9)

CHELONIA
Turtles, etc. (pp. 82–9)

SQUAMATA
Snakes, etc. (pp. 82–9)

AMPHIBIA
Frogs, etc. (pp. 78–81)

DIPNOI
Lungfishes (p. 70)

ACTINOPTERYGII
Ray-finned fishes (p. 70)

CHONDRICHTHYES
Sharks, etc. (p. 70)

AGNATHA
Jawless fishes (p. 70)

TUNICATA
Tunicates, ascidians (p. 54)

ECHINODERMATA
Starfishes, etc. (p. 54)

CHELICERATA
Arachnids (p. 68)

DIPLOPODA
Millipedes (p. 69)

CRUSTACEA
Crabs, etc. (p. 66)

INSECTA
Insects (p. 60)

ANNELIDA
Segmented worms (p. 55)

BRACHIOPODA
Lamp shells

MOLLUSCA
Slugs, squids, etc. (p. 56)

PLATYHELMINTHES
Flatworms (p. 55)

COELENTERATA
Jellyfishes, etc. (p. 53)

MAGNOLIOPHYTA
Flowering plants (p. 40)

CYCADOPHYTA
Cycads (p. 38)

CONIFEROPHYTA
Conifers (p. 39)

SPHENOPHYTA
Horsetails (p. 37)

FUNGI
(p. 35)

PROTISTS
(p. 26)

SLIME MOULD
(p. 26)

BACTERIA AND
BLUE-GREEN ALGAE (p. 24)

ARCHAEBACTERIA
(p. 24)

may be represented on a branching diagram or *cladogram* (such as the large diagram shown here), which shows how particular organisms are grouped together and how the various clades form a hierarchical distribution of groups within groups. In evolutionary terms, a clade is a *monophyletic group* – a group consisting of species descended from a single ancestor and including all the most recent common ancestors of all its members. Thus cladograms can literally be seen as

evolutionary trees. However, cladistics may or may not consider the evolutionary history of the organisms concerned.

The evidence used to classify organisms has for the most part been *morphological*, i.e. derived from the study of structural features such as bones and teeth. *Physiological* evidence is also employed but is of limited use. *Embryological* evidence, on the other hand, is of greater importance, since the developmental stages of an embryo may mirror the evolution-

ary progression by which a species has reached its present form (⇨ pp. 14–15). More recently, *molecular evidence* has proved to be of paramount importance in solving the more difficult problems of phylogenetic relationship. Using new techniques, scientists are able to compare DNA, RNA and proteins from different organisms, so helping to establish degrees of relationship, and providing a new and more comprehensive means of comparing the different forms of life on Earth.   BG

SEE ALSO

● BACTERIA p. 24
● PROTISTS p. 26
● PLANTS p. 30
● FUNGI p. 35
● FLOWERING PLANTS p. 45
● ARTHROPODS p. 58
● INSECTS p. 60
● FISHES p. 70
● AMPHIBIANS p. 78
● REPTILES p. 82
● BIRDS p. 90
● MAMMALS p. 98

Cells
The Basic Units of Life

Cells are the basic biological units of all living things, and a single cell is the smallest component of an organism that is able to function independently. Virtually every cell in an individual organism – there are an estimated 10^{13-14} in the adult human body – carries a complete inherited genetic blueprint for the formation and development of that organism. Cells range in size from 0·0003 mm in the case of certain bacteria, to the egg – the female sex cell or ovum – of the ostrich, which averages 15–20 cm (6–8 in) in length.

Many organisms – the bacteria (▷ p. 24) and protists (▷ p. 26) – are *unicellular*, i.e. the individual organism consists of only one cell. All higher plants and animals are *multicellular*, i.e. they consist of an assemblage of cells. In multicellular organisms, groups of cells have different functions: for example, in animals there are nerve cells, blood cells, muscle cells, and so on (▷ p. 186). The lifespan of cells varies from a few days to – in the case of muscle cells – the lifetime of the organism.

At the molecular level, all cells are constructed from four groups of organic

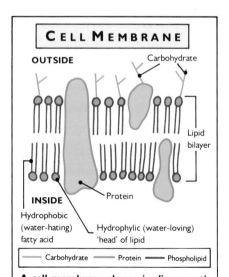

CELL MEMBRANE

OUTSIDE

Carbohydrate

Lipid bilayer

Protein

INSIDE

Hydrophobic (water-hating) fatty acid

Hydrophylic (water-loving) 'head' of lipid

— Carbohydrate — Protein — Phospholipid

A cell membrane shown in diagramatic cross-section. A membrane is a highly dynamic structure composed of two layers of phospholipids arranged with their hydrophobic (water-hating) 'tails' inwards and their hydrophilic (water-loving) 'heads' facing outwards. Protein molecules are interposed in lipid layers, some being in either the outer or inner lipid layer whilst others traverse the whole membrane. These molecules are mobile and can be moved within the membrane. Carbohydrate molecules are attached to the outer surface of the membrane and are involved with molecular recognition. This total structure is called a unit membrane and is essentially typical in its general form for all membranes in living organisms.

COMPARTMENT	MAJOR FUNCTIONS AND CHARACTERISTICS
EUKARYOTE ORGANELLES	
Cytosol	The solution between the organelles, containing ions, dissolved molecules and enzymes; site of protein synthesis and glycolysis.
Nucleus	Contains the genetic resources of the cell; makes m-RNA, t-RNA, r-RNA; regulates cell multiplication and differentiation.
Mitochondria	Possess own genes for partial replication; produce energy-containing molecules NADH⁺ and ATP in tricarboxylic acid cycle; break down fatty acids.
Plastids	Possess own genes for partial replication; plastids form (a) *chloroplasts* in plants, which contain the chlorophyll responsible for photosynthesis, and which are the site of the conversion of nitrates to ammonia as the first stage of protein synthesis (▷ pp. 31–3), (b) colourless *leucoplasts*, which store starch, protein and oil, and (c) *chromoplasts*, which contain pigments (yellow, red and orange carotenes).
Endoplasmic reticulum	Intracellular transport system; responsible for synthesis of lipids and sugars; attached *ribosomes* make protein for export; forms vacuoles and storage vacuoles (▷ below) in plants.
Golgi apparatus	Manipulates and packages material from endoplasmic reticulum for secretion; enzymes packaged for transport to lysosomes (▷ below).
Lysosomes	Contain acid hydrolase enzymes for controlled digestion of molecules, bacteria, red blood cells and worn-out organelles such as mitochondria.
Microbodies	Comprise *peroxisomes*, which combat oxygen poisoning; and *glyoxysomes*, which break down fatty acids during seed germination.
Vacuoles	Control osmosis and water pressure in some animal and all plant cells; act as lysosomes in plants.
Centrioles	Self-replicating organelles in animal cells, but rarely in plants; focal point of production of part of cytoskeleton.

molecules: the nucleic acids (DNA and RNA), proteins, carbohydrates (which include the sugars) and lipids (▷ p. 6). These molecules were first produced some 4·5 billion years ago, and the first simple cells evolved from them around 3·5 billion years ago (▷ p. 6). All cells are surrounded by a membrane; plant cells also have an outer cell wall, principally made of cellulose. Membranes also play various crucial structural and biochemical roles within the cell (▷ below).

Prokaryote and eukaryote cells

The manner in which membranes evolved and developed has resulted in the formation of two basic types of cell, the prokaryotes and eukaryotes. Prokaryote cells – which include all forms of bacteria (▷ p. 24) – are the most primitive form of cell, consisting essentially of an outer membrane with some rudimentary membrane systems within. The material of the cell within the membrane is called the *protoplasm*. The DNA of prokaryote cells resides in a special area, the *nucleoid*. Unlike the eukaryotic nucleus (▷ below), the nucleoid is not surrounded by a membrane, but it remains free of all structures except DNA. Again unlike eukaryotes, prokaryotes contain no organelles (▷ below), the nearest equivalents being the ribosomes, the mesosome, and, in some, the presence of photosynthetic systems (▷ pp. 24 and 31–3) in the cell membrane. A further distinction from eukaryotic cells is the way that prokaryotic cells divide and duplicate (▷ p. 24).

Eukaryotic cells evolved later than the prokaryotes, probably by the fusion of prokaryotes into symbiotic relationships (▷ p. 7). They are generally larger than prokaryotes (up to 0.01 mm on average, though some can be bigger than 1 mm), and exist as single cells (protists; ▷ p. 26), colonies of cells (e.g. sponges; ▷ p. 52), or as constructs in the form of multicellular organisms (plants and animals). Their increased size and ability to form complex colonies and organisms is largely due to three features.

Firstly, the protoplasm in a eukaryote cell is differentiated into the *nucleus* (surrounded by a membrane and containing the genetic material in the form of chromosomes) and the rest of the cell, the *cytoplasm*. Secondly, there are membrane-bound *organelles* – miniature organs with specialized functions, which permit a greater range of metabolic processes (▷ table and diagram). Some organelles – for example, mitochondria, chloroplasts and centrioles – are self-replicating, containing their own DNA and RNA. However, this DNA represents only part of the genes needed for organellar replication, the rest residing in the nucleus. Thirdly, eukaryote cells possess a *cytoskeleton*, mainly around the peripheries of the cell and of the nucleus. The cytoskeleton, which consists of many long microtubules and filaments, has many functions in the cell, including giving the cell its shape and providing a monorail-type system along which the organelles and transport vesicles are moved about the cytoplasm. This is especially important for the movement of chromosomes during cell division (mitosis and meiosis; ▷ p. 22). The cytoskeleton is also involved with animal cell movement and some of its components form the basis of the mechanism for both muscle contraction in animals and cilia and flagella motion in protists (▷ p. 27). Finally, eukaryotes have also developed the true sexual reproductive process with egg/sperm fertilization (⊳ pp. 22–3 and 184–7).

Membranes and compartments

The membrane in all living cells shares a similar configuration, based on a dynamic molecular structure. The interaction of lipids forms the basic membrane, into which protein molecules are inserted and carbohydrate molecules attached (⊳ diagram). In addition to creating a membrane that separates the inside of the cell from its environment (the *plasmamembrane*), this basic membrane structure also provides a variety of compartments in eukaryote cells – the organelles (⊳ table and diagram). Each organelle is surrounded by a membrane, with the nucleus having a second membrane surrounding it. This second membrane is linked to a complex network of membranes – the *endoplasmic reticulum* – which in turn is in communication with another organelle called the Golgi apparatus to form an interlinked membrane system from the nucleus to the plasmamembrane. Two other kinds of organelle – the mitochondria and the chloroplasts (⊳ table) – have an outer membrane that separates them from the *cytosol*, the solution within which all the organelles of the cell are suspended, while an inner membrane forms additional compartments within these organelles. Thus the membrane provides a large surface inside the cell on which chemical reactions can occur, and permits specialized transport systems into, through and out of the cell while separating the different activities of the cell so that they do not interfere with each other.

These transport systems are responsible for importing (*endocytosis*) into the cell materials such as nutrients for digestion or chemicals for processing such as cholesterol, and for exporting (*exocytosis*) from the cell materials such as polysaccharides (long-chain carbohydrate molecules) to make plant cell walls and mucopolysaccharides to protect the lining of the intestine. In order to trigger these transport processes the membrane needs to perceive signals from outside the cell, which it does by special receptor protein and carbohydrate molecules. Although plant cells are surrounded by a cell wall, molecules can pass through the wall to the plasmamembrane, which responds to signals as does an animal cell. On receipt of signals from materials such as hormones and antibodies, the membrane can be stimulated to endocytose or exocytose, or biochemical and physiological responses, such as differentiation (⊳ below) or mitosis (⊳ p. 22) may be triggered within the cell.

Energy production

Energy obtained from sunlight is transformed in plants and some bacteria by photosynthesis (⊳ pp. 31–3) into chemical energy utilizable by the cells. The basic energy-rich molecule is glucose (the simplest sugar), which is stored in plants as starch or transported as sucrose, or stored in animals as glycogen. One glucose molecule contains 686 kilocalories

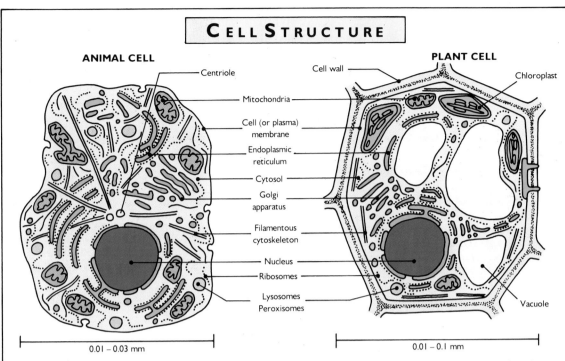

CELL STRUCTURE

ANIMAL CELL — Centriole, Mitochondria, Cell (or plasma) membrane, Endoplasmic reticulum, Cytosol, Golgi apparatus, Filamentous cytoskeleton, Nucleus, Ribosomes, Lysosomes Peroxisomes

PLANT CELL — Cell wall, Chloroplast, Vacuole

0.01 – 0.03 mm

0.01 – 0.1 mm

These diagrams of thin sections of generalized animal and plant cells show the many similarities of structure between the two basic types of eukaryote cell. Blue links the components of the *endomembrane system* (i.e. the principal linked membrane apart from the outer cell membrane), comprising the nuclear envelope, endoplasmic reticulum, Golgi apparatus, and transport and secretory vesicles.

(2·9 million joules) of energy and the cell has to transform this energy into a utilizable form – for example, as ATP (adenosine triphosphate) or NADH (reduced nicotinamide adenine dinucleotide) molecules – so as not to allow a total release of such energy as heat. A multistep system is used, involving two major metabolic pathways, known as glycolysis and the Krebs or tricarboxylic acid cycle (TCA cycle), both of which involve a series of highly complex chemical reactions, and which are also involved in the synthesis of carbohydrates, lipids, proteins and nucleic acids. Although glycolysis can occur under anaerobic conditions (i.e. without oxygen), the TCA cycle occurs only under aerobic conditions (i.e. in the presence of oxygen), so forming the basis of cell respiration.

Approximately half of the glucose energy appears in ATP, as opposed to some 15–20% efficiency of many petrol engines. Some plant cells convert the NADH energy directly to heat: for example, the flowering part of *Arum maculatum* (lords and ladies) can generate temperatures of 39 °C (102·2 °F) to create odour molecules to attract insects for pollination, while other plants can use heat to overcome low external root temperatures.

Cell differentiation

Fertilized eggs from multicellular organisms (both plants and animals) rapidly divide to form an embryo (⊳ p. 186). During the first series of divisions it is possible to separate the cells and each one can be induced to form a new individual organism, i.e. the cells are *totipotent*. After this initial development period, each cell will have become changed genetically so that it can only form one of the many types of specialized cells seen in

multicellular organisms. Once cells have become specialized or differentiated, the process is essentially irreversible, and in the case of mammals only a limited number of cell types are capable of responding to environmental stimuli to change. In addition, in animals there is a permanent *germ cell line*, i.e. cells that will form eggs and sperm are established very early on in the development of the embryo.

In contrast, the flowering plants form no continuous germ cell line. Instead, the vegetative shoot apex that normally produces leaves receives signals, such as light and cold, from the environment, and such signals cause it to produce flowers. It is in the flowers that the egg- and sperm-making cells will form (⊳ p. 40). Plant cells also appear to be more flexible than most animal cells since apart from the xylem and phloem cells of the vascular tissue (⊳ pp. 31–3), which are irreversibly committed to differentiate, many other types of plant cells can redifferentiate. Thus the mesophyll cells of the leaf can be transformed into xylem elements. More importantly, many plant cells can remain totipotent, each cell giving rise to whole plants. This is the basis of plant propagation *in vitro* and enables the genetic make-up of a plant to be altered by special methods of gene transfer to produce, for example, different flower colours, or disease- or herbicide-resistant plants. This means that it is possible to generate whole plants in large numbers – all of them being clones (i.e. genetically identical) – and at a rate some three times faster than by conventional plant-breeding methods. It is also possible to breed virus-free plants by such new biotechnological methods. **PBG**

SEE ALSO

● WHAT IS LIFE? p. 6
● EVOLUTION pp. 12–17
● GENETICS p. 22
● VIRUSES AND BACTERIA p. 24
● PROTISTS p. 26
● HOW PLANTS FUNCTION pp. 31–3
● ANIMAL PHYSIOLOGY pp. 178–91

The Blueprint of Life: Genetics

Although theories of inheritance or heredity were put forward at least as early as the 5th and 4th centuries BC, genetics – the scientific study of inheritance – only truly began in the 18th and 19th centuries. Observations were made of how specific characteristics of plants and animals were passed from one generation to the next, to provide a rational basis for the improvement of crop plants and livestock.

The most significant breakthrough in genetics was made by the Austrian monk Gregor Mendel (1822–84). He observed specific features of the pea plant and counted the number of individuals in which each characteristic appeared through several generations. By concentrating on just a few features and determining what proportion of each generation received them, he was able to demonstrate specific patterns of inheri-tance. The discrete nature and independent segregation of genetic character-istics that he observed became known as Mendel's laws of inheritance, and have been shown to apply to most genetic systems (▷ box).

DNA

By the start of the 20th century it was clear that organisms inherited charac-teristics by the reassortment and redis-tribution of many apparently independent factors, but the identity of the material that carried this information was un-known. To code for such a large amount of information, any type of molecule would have to be highly variable, and proteins were thought to be the most likely candi-date. In 1944, however, the American microbiologist Oswald T. Avery (1877–1955) demonstrated that the inheritable characteristics of a certain bacterium could be altered by *deoxyribonucleic acid* (DNA) taken up from outside the cell.

To understand how genetic information was encoded required the structure of DNA to be determined. In 1953 the Ameri-can James Watson (1928–) and the Englishman Francis Crick (1916–) reported that DNA is a large molecule in the shape of a double helix (▷ box). The genetic code is contained in the sequence of paired *nucleotide bases* that lie in the central region of the molecule, with each *triplet* (or *codon*) of bases specifying a particular amino acid. The comple-mentary pairing of bases on each of the two strands also explains how DNA rep-lication can take place (▷ below). Incor-rect pairing disrupts the structure of the DNA molecule and may be a source of mutation, but is usually recognized and changed by an enzyme correction system.

Genes, the genome and mutation

Mendel discovered that characteristics are passed from generation to generation in the form of discrete units. Once the structure of DNA was established, these units, called *genes*, could be understood at the molecular level. A gene is a linear section of a DNA molecule that includes all the information for the structure of a particular protein or *ribonucleic acid* (RNA) molecule. The sum of all an organ-ism's genetic information is called its *genome*.

A *mutation* occurs when there is a change in the sequence of nucleotide bases in a piece of DNA. Such changes may occur naturally, as bases are added, deleted or exchanged. The rate at which this process (known as *mutagenesis*) takes place may be accelerated by exposure to chemicals or radiation, and mutations may disrupt or prevent the production of proteins, thus disturbing the functioning of the organism as a whole. However, natural mutations passed on from generation to generation may also confer benefits, and are indeed essential to the process of evolution (▷ pp. 12–17).

Every sequence of nucleotide bases that makes up a single gene is called an *allele*. There are usually several different alleles available for each gene. Thus, if the eye colour of a given organism is dependent on a single gene, different eye colours will be dictated by various alleles of the gene responsible.

Reading of genes and protein synthesis

The first stage in the process by which cells use genetic information stored in DNA is to make an RNA copy of a gene. This is called *gene transcription*. Most of the RNA copies (known as messenger RNA or mRNA) travel from the nucleus of the cell (where the genes are located) to the *ribosomes* – particles in the cytoplasm (▷ pp. 20–1) – where proteins are manu-factured or *synthesized*. Ribosomes make proteins by joining together amino acids in the sequence dictated by the order of triplet groups in the mRNA. Proteins may become part of the cell structure or they may be enzymes, which catalyse bio-chemical reactions.

Chromosomes

A *chromosome* consists of several dif-ferent types of protein tightly associated with a single DNA molecule, and each chromosome carries a large number of genes. Most eukaryotic organisms (or-ganisms with cell nucleuses) have several chromosomes, but bacteria, which do not have a nucleus, have only one.

All the cells of a particular organism have the same number of different chromo-somes, but numbers vary widely between different species. There is no clear pattern to this, but plants tend to have fewer different chromosomes than animals. Humans have 23 different chromosomes per cell, but there are two copies of each (called *homologous* chromosomes), mak-ing 46 altogether.

DNA replication and cell division

As organisms grow, their body cells divide and multiply. By the time a cell divides, its DNA will have doubled by a process called *DNA replication*. The hydrogen bonds linking the two strands of a DNA molecule break apart, and each strand uses nucleotides present in the nucleus to synthesize a new strand complementary to itself; the result is that two daughter molecules are produced, each identical to the parent molecule. This process is com-pleted just before cells divide, so that each chromosome contains two DNA mole-cules instead of the usual one. A mechan-ism called *mitosis* operates during cell division to ensure that each of the daughter cells receives one of the DNA molecules from every chromosome.

Most organisms are *diploid* – like humans, they have two copies of each different chromosome in normal body cells. However, sex cells or *gametes* (sperm and egg cells in most organisms) are *haploid* – they contain only one of each chromosome. Sex cells are produced from body cells by a process called *meio-sis*. DNA replication takes place as before mitosis, but an extra stage of chromosome separation results in four sex cells being produced from a single body cell. As sex

---

THE STRUCTURE OF DNA

Deoxyribonucleic acid (DNA) is the basic genetic material of most living organisms. Although a large and apparently complex molecule, the structure of DNA is in fact astonishingly simple.

A single DNA molecule consists of two separate strands wound around each other to form a double-helical (spiral) structure. Each strand is made up of a com-bination of just four chemical com-ponents known as *nucleotides*, all of which have the same basic com-position: each nucleotide consists of a sugar molecule (deoxyri-bose) linked to a phosphate group to form the helical backbone; different nucleotides are distin-guished only by the identity of the nitrogen-based unit (called a *nucleotide base*) bonded to the sugar molecule. The four bases – adenine (A), cytosine (C), guanine (G) and thymine (T) – lie in the central region of the double helix, with each base linked by hydrogen bonds to a specific complementary base on the partner strand: A pairs only with T, and G only with C.

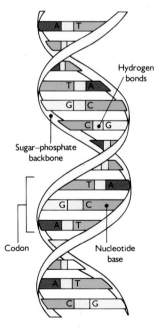

Hydrogen bonds

Sugar–phosphate backbone

Codon

Nucleotide base

This simple structure explains the two key properties of DNA – how it codes for the manufacture of amino acids (from which proteins are formed) and its capacity to replicate itself. Each combination of three bases (known as a *triplet* or a *codon*) within a DNA molecule codes for a particular amino acid, while the specific pairing of bases explains how two identical DNA molecules can be produced by the separation of the two strands of the parent molecule.

A SINGLE-FACTOR INHERITANCE

The simplicity of Mendel's laws is illustrated by a single-factor inheritance, in which an inheritable characteristic is determined by the action of a pair of dominant and recessive alleles of a single gene.

The coloration of rats is an example of such a characteristic, with the allele for black being dominant over the allele for white. If a homozygous black rat (i.e. with two identical alleles) mates with a homozygous white rat, all the offspring will be heterozygous black – each similar in appearance to the black parent, but with a recessive (and unexpressed) white allele. However, mating between these offspring will result (potentially) in a mixture of black and white rats. The ratio of black rats to white will be on average 3 to 1, any white rat having inherited one recessive white allele from each of its parents.

The same pattern is found in the inheritance of certain human characteristics, including recessive genetic diseases such as cystic fibrosis. The inheritance of characteristics dependent on a number of different genes is more complex, but the underlying principle is the same in all genetic interactions.

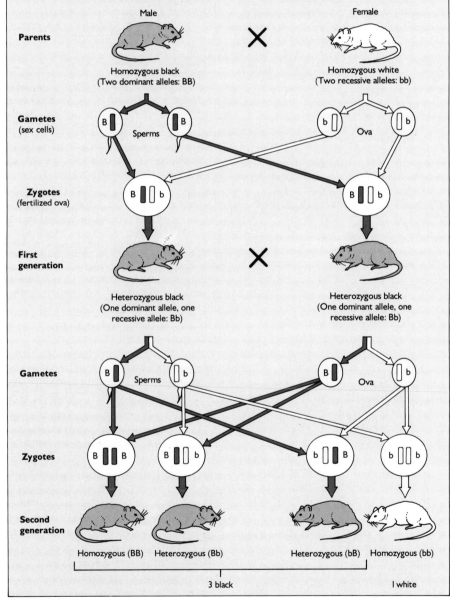

similar form of sexual determination is found in most sexually differentiated animals.

In organisms with several different chromosomes, the offspring receive some chromosomes from each parent and so a combination of genes different from either. Genetic diversity is further increased by a process known as *crossing-over*: during meiosis, homologous chromosomes can exchange bits of DNA between themselves, and so move genes into new combinations. However, the closer together genes are on the same chromosome, the more likely it is that they will be inherited together. Sexual interaction therefore greatly accelerates the rate at which genes are moved into new and potentially beneficial combinations; as such it is crucial to the evolutionary development of species.

Dominant and recessive genes

In diploid organisms there are two copies of every gene, one in each member of every chromosome pair (the only exception being the sex chromosomes of the male). When both copies of a particular gene have identical alleles, the individual is said to be *homozygous* for that gene. However, many genes will have different alleles in each copy and the individual is said to be *heterozygous* for such genes. One allele may be *dominant* in that the gene product it codes for is used by the cell in preference to that of the other allele, which is called *recessive* (▷ box). The outcome of such interactions between the alleles of all the different genes produces the characteristics of an individual, known collectively as the *phenotype*. The entire set of alleles in the genome of an individual is known as its *genotype*.

Molecular genetics and the future

The basis of *genetic engineering* is the use of certain bacterial enzymes that can cut DNA into small fragments. These fragments can be joined together in almost any combination, using a range of other enzymes. This can be done with DNA from any source – for example, human genes can be put into bacterial DNA. Bacteria are easy to grow in large quantities and can make useful products if they contain suitable genes (▷ p. 25).

Genetic-engineering techniques can also be used to determine the exact sequence of nucleotide bases in short stretches of DNA. Many genes from different organisms have been sequenced, and this has, among other things, aided the understanding of the degrees of relationship between different species (▷ p. 19).

A worldwide project to sequence the human genome has begun and will take about 30 years to complete. It will reveal the location of many genes and probably unexpected information about human genetic organization. This type of information is likely to allow *gene-replacement therapy* – the replacement of defective genes with normal copies – to be used in the 21st century to cure many inherited diseases. PM

cells fuse at fertilization, a single diploid cell (called a *zygote*) is created, with the full complement of chromosomes.

Sex chromosomes and recombination

In humans, one of the pairs of chromosomes – the sex chromosomes – is responsible for determining the sex of an individual, and can have two different forms, X and Y. Females have a pair made up of two X chromosomes, while a male has an X and a Y. Thus the sex cells of a female always carry an X chromosome, while a male's sex cells may contain either an X or a Y. The sex of a child is therefore determined by the type of chromosome passed on by the father. A

Simple Life Forms: Bacteria and Viruses

Of the five kingdoms of living organisms, the most primitive is the kingdom Monera, which includes the archaebacteria, bacteria and mycoplasms. An individual bacterium consists of a single primitive form of cell, called a *prokaryote* cell, which is distinguished from the *eukaryote* cells of higher organisms by the lack of a nucleus in which to hold its genetic material (▷ p. 20). The archaebacteria probably derived some 3·25 billion years ago from an ancestral archetypal prokaryote, which had itself developed from pre-cellular forms (▷ pp. 6 and 20). This ancestral prokaryote also gave rise to mycoplasms, eubacteria, cyanobacteria (blue-green algae) and chloroxybacteria (green prokaryote algae).

The viruses are even simpler forms of life than bacteria. They seem to have developed directly from the plasmids present in bacteria (▷ below), with which they remain the most simple forms of life: indeed, viruses are very much in a grey area between living and non-living things. The fact that they are completely dependent on living cells as vehicles in which they can reproduce indicates that they probably evolved later than the earliest cellular organisms.

Viruses

It was not until 1892, when the tobacco mosaic virus (TMV) was detected, that it was shown that there are infectious agents smaller than bacteria. The first virus to affect bacteria, the bacteriophage, was discovered 20 years later.

A large range of viruses are now known to exist. Many cause diseases – from AIDS to the common cold – and many such diseases remain incurable. Viruses have also been implicated in evolution as transmitters of blocks of genetic material from one organism to another. Viruses range in size from 0·000018 to 0·0006 of a millimetre, and it was not until the advent of the electron microscope in the 1930s that anyone actually 'saw' one, revealing variations in shape from almost spherical to rod-shaped to icosahedral (having 20 faces).

At about the same time, biochemical analysis showed that viruses consist simply of a single nucleic acid – either DNA (as in the bacteriophage) or RNA (as in the TMV) – surrounded by a protein layer, together forming the *nucleocapsid*. The whole is often surrounded by a further bilipid layer containing proteins. The layer surrounding the nucleic-acid core has three main functions: to protect the nucleic-acid from environmental factors; to help in viral transmission from host to host, often via a carrier; and to initiate nucleic-acid replication in the host cell. To assist in this last process, some viruses carry specially needed enzymes that are not found in the host cell.

Viral replication

On infecting a host cell, a virus can do one of three things. The first two involve reproduction using the mechanisms of the host cell in order to make viral nucleic acids and proteins from which new virus particles can be constructed; either the cell is then caused to destroy itself in order to free the viruses or the new particles can leave the cell without damaging it. Thirdly, the virus can enter a cell and insert itself into the host chromosome, where it can remain silent for a period of time after which it leaves the chromosome, replicates and leaves the host cell.

Retroviruses and viroids

Retroviruses are a group of RNA viruses, some of which are tumour inducing. They enter the host cell and make copies of DNA from their own RNA using their own special enzyme. This DNA then integrates into the host chromosome, which is ultimately used to make viral RNA and proteins to form new viruses, and these are released non-lethally from the host cell. However, such viruses can leave the host cell with a permanent genetic change, since the presence of the viral DNA in the chromosome results in the production of new proteins by the host cell. The virus itself may also become defective, having left some of its genetic material behind in the host chromosome.

Some infections such as potato spindle disease are caused by *viroids*, particles that are even smaller than viruses, and which are essentially small infectious single-stranded RNA molecules, probably coding for about 55 amino acids. The degree of virulence of a viroid can be changed by altering a single base in the RNA.

Mycoplasms

Mycoplasms are the smallest known free-living organisms (0·0001–0·001 mm), differing from true bacteria by the

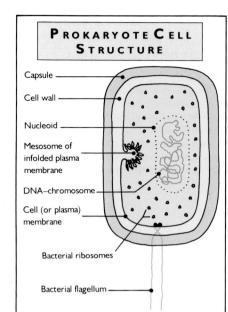

Capsule

Cell wall

Nucleoid

Mesosome of infolded plasma membrane

DNA–chromosome

Cell (or plasma) membrane

Bacterial ribosomes

Bacterial flagellum

Diagram of the main components of a prokaryote cell. It is surrounded by a *plasma* (or *cell*) *membrane* (▷ pp. 20–1), which is infolded at the region of the *mesosome* to create extra membrane surface for various biochemical processes. The plasma membrane abuts the *cell wall*, which (unlike plant cell walls) does not contain cellulose. Outside the cell wall there is a capsule. Inside the plasma membrane is the *protoplasm*, which contains ions, enzymes and other dissolved molecules, together with free *ribosomes* for the synthesis of proteins. These ribosomes compare in size to those found in the chloroplasts and mitochondria of eukaryotic cells, but are bigger than those found in the eukaryotic cytoplasm. Within the protoplasm, the *nucleoid* (the clear space with no limiting membrane) houses the normally circular pieces of DNA. Some prokaryote cells are mobile and swim with the aid of *flagella* made from a special protein, flagellin. Such flagella do not have the same construction as the cilia and flagella of eukaryote cells (▷ pp. 26–7).

absence of a cell wall and mesosomes. They were first found in cattle, where they caused a pneumonia type of fatal infection, and can cause a number of diseases both in cattle and man. Some saprophytic forms (i.e. forms feeding on dead matter) are found growing in sewage.

Bacteria

These form the majority of the prokaryote species of unicellular organisms. They vary in shape from almost spherical (*coccus*) to coma-shaped (*vibrio*) to rod-like (*bacillus*) to corkscrew-shaped (*spirochaete*) or spiral (*spirillium*). Some, such as *Salmonella*, are motile (i.e. capable of spontaneous and independent movement), while others are non-motile, being grouped in packets (*Sarcina*) or chains (*Streptococcus*). Bacteria vary considerably in size: the smallest, such as the chlamydia, are no more than 0·0003 mm, while the largest, including the spirochaetes, can be more than 0·02 mm. Bacteria cause a wide range of diseases, including cholera, food poisoning, leprosy, plague, syphilis, tetanus, tuberculosis and typhoid. Most of these diseases

Some viruses, such as the T4 bacteriophages seen in this electron micrograph, rely on the reproductive apparatus of bacteria in order to replicate. They do so by attaching to the cell wall of the bacterium using the spidery tail fibres visible here. The tail is used to inject the contents of the head, the genetic material (DNA), into the cell. The genetic apparatus of the host is commandeered, and within 20 minutes progeny fill the cell.

can be successfully treated by antibiotic drugs.

Environmental extremes

Although we think of bacteria as disease-causing parasites, they are very diverse in both their habitat and biology. Although many are parasites, there are also saprophytes (living and feeding on dead matter), free-living organisms, and bacteria that have developed symbiotic relationships with plants (▷ below) and animals, in many of which they aid digestion (▷ pp. 169 and 178). A carry-over from their evolutionary origins has led to the appearance of bacteria that have varying sensitivities to oxygen. Some groups have an absolute need for oxygen (*aerobic bacteria*), others can grow only in the absence of oxygen (*anaerobic bacteria*), yet others can shift readily between these two states, and a final group, those involved in fermentation, although tolerating either the presence or absence of oxygen, retain the fundamental biochemistry for living in the absence of oxygen.

Most bacteria grow well at temperatures of either 37–40 °C (98·6–104 °F), for example in the mammalian body, or at about 30 °C (86 °F) in the general environment. Nevertheless, some of the archaebacteria are found in extreme, hostile habitats: bacilli grow well at 50–55 °C (122–131 °F), for example in compost heaps, and anaerobic sulphur bacteria even tolerate temperatures as high as the 90 °C (194 °F) found in hot springs. At the other end of the temperature spectrum, some bacteria will even multiply at 0 °C (32 °F). Other archaebacteria are very tolerant to high salt concentrations, and as well as living in the oceans are also found in the waters of salt flats – and in pickling solutions.

Clearly, not all bacteria are able to tolerate extreme conditions, and a number have developed the habit of forming *spores*, reproductive bodies formed by cell division. Spores have an impervious protein coat, which protects them from desiccation, heat, freezing, toxic chemicals and radiation. Spores remain dormant until more favourable conditions return, and viable spores have been found to germinate after storing for more than 50 years. However, under ideal conditions, bacteria employ a rather different reproductive mechanism (▷ below).

Metabolic diversity

The diversity of habitats to which bacteria have adapted is linked to the diversity of their metabolic capabilities – and both of these features reflect the broad evolutionary background of the bacteria. Bacteria can live either on organic molecules produced by other living organisms (*heterotrophic metabolism*, as also seen in animals), or by making their own organic molecules from inorganic sources (*autotrophic metabolism*, as also seen in plants). Autotrophic bacteria can be divided into those performing photosynthesis (*phototrophic bacteria*) and those performing other chemical transformations (*chemotrophic bacteria*).

The photosynthetic bacteria comprise two types of purple bacteria, those using sulphur or those using other inorganic molecules such as hydrogen or hydrogen sulphide. The chemistry of bacterial photosynthesis is fundamentally different from photosynthesis in green plants, including the algae, (▷ pp. 31–3): the photosynthetic bacteria tend to be anaerobic and so do not use water or release oxygen in producing energy in the form of ATP molecules (▷ p. 21). Carbon from carbon dioxide is then fixed by a process similar to that seen in higher plants (▷ pp. 31–3).

One group of aerobic bacteria, the nitrogen-fixing bacteria, are the only living organisms that are capable of converting the rather volatile ammonia or nitrous acid derived from decomposing organic matter, especially in the soil, to a more stable form of nitrogen, nitrates. In turn, these can be exploited by higher plants to form organic amino compounds, the components of protein. These then become available to animals, which can perform neither of these processes. Thus bacteria play an absolutely vital role in the nitrogen cycle (▷ p. 196), and so in the maintenance of virtually all life on earth. In some cases, the bacteria live in nodules on the roots or shoots of certain plants – notably the legumes (▷ pp. 31–3).

Other aerobic bacteria use sulphur, iron or hydrogen as their sources of energy, in some cases producing sulphuric acid and surviving in this very acid environment. The hydrogen-tapping bacteria are exploited in sewage works, where they reduce the accumulation of free hydrogen, which would otherwise form a dangerously explosive mixture with oxygen from the atmosphere.

Nevertheless, many bacteria, especially parasites and saprophytes, exploit organic molecules such as glucose for their energy source and handle them via fermentation (glycolysis; ▷ p. 21) under anaerobic conditions to yield ATP and either an alcohol such as ethanol or an acid such as lactic or formic acid. Under aerobic conditions, there are different by-products. The fermentation process has been exploited commercially for the production of a number of compounds that may be used directly, such as ethanol (the alcohol in drinks) and vinegar, or as the starting point for other industrial syntheses.

Reproduction and the role of plasmids

Although under hostile conditions bacteria can reproduce by producing spores (▷ above), under favourable conditions bacteria have a very high reproductive rate, a single bacterium being able to divide once every 20 minutes. As with all prokaryotes, cell division occurs by simple fission (division) of the cells and not mitosis (the process used by higher organisms; ▷ p. 22). The DNA replicates and the bacteria divides into two, each daughter bacterium having a similar piece of DNA representing the bacterial chromosome.

Many bacteria contain one or more extra, small, circular pieces of DNA/RNA (*plasmids*) in addition to the main chromosome. The plasmids carry genes equivalent to a sex factor. Normally bacteria do not undergo eukaryotic-type sexual reproduction, but exchange genetic material between bacteria through a process controlled by the plasmid genes.

Genes for drug resistance and special metabolic processes can also be found on the plasmid. In plants, the plasmids carried by *Agrobacterium tumefaciens* and *A. rhyzogenes* are capable of inducing the formation of tumours in plants. Since such plasmids can integrate into the chromosomes of both prokaryotes and eukaryotes, they are exploited in biotechnology. Specific genes are inserted into the inactivated plasmid, which then carries the gene into the host chromosome in a way that permits expression of that gene in the host to make specific proteins; for example, human insulin is manufactured in this way using the bacterium *Escherichia coli*.

Cyanobacteria

Standing somewhat apart from the main group of bacteria are the cyanobacteria or blue-green algae. They occur either as small rounded cells or mobile filaments of cells, and can be free-living or present in symbiotic relationships either together with fungi to form lichens (▷ p. 34) or inside the cells of higher plants such as the roots of the ginkgo tree.

Cyanobacteria are able to fix nitrogen and to perform photosynthesis in the same way as green plants – i.e. producing oxygen as a by-product – and it was their proliferation some 2·6 billion years ago that first converted the earth's atmosphere into an oxygen-rich environment (▷ pp. 6–7) – so allowing the evolution of aerobic bacteria and all subsequent organisms that depend on oxygen. It is thought that by a symbiotic process certain cyanobacteria became the first chloroplasts, the organelles within the cells of plants and plant-like protists that contain the chlorophyll pigments responsible for photosynthesis (▷ pp. 31–3). **PBG**

SEE ALSO

● THE AGES OF THE EARTH p. 4
● WHAT IS LIFE? p. 6
● EVOLUTION pp. 12–17
● CELLS p. 20
● GENETICS p. 22
● PROTISTS p. 26

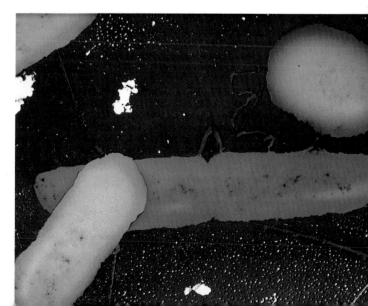

Bacterial conjugation between two *Escherichia coli* bacteria, shown here in a false-colour electron micrograph. Conjugation involves genes moving from one bacterium (here coloured magenta) to another (coloured green) via a pilus (red). The genes transferred are plasmids, small autonomously replicating pieces of DNA.

Simple Life Forms:
Protists

The evolution of prokaryote cells such as bacteria into eukaryote cells – cells with a nucleus – involved an increased complexity of cell structure and function (▷ p. 20), accompanied by the development of a great variety of shapes and sizes. This evolutionary pathway proceeded via single-celled eukaryotes, which eventually gave rise to multicellular plants, animals and fungi.

It is the single-celled eukaryotes – the protists – that form the kingdom Protista. This includes the single-celled algae and protozoa, slime moulds and nets, all of which have developed incredible degrees of complexity and sophistication. Although remaining as single cells some of them form the first approach to multicellular organisms through the construction of colonies.

Unlike the majority of multicellular organisms, which have well-defined shapes, the protozoal and algal protists have variable shapes, being round, oval or cup-shaped. Most of the protists are very small organisms, ranging from 0·00001 to not more than 1 mm in length, though the exceptions to this include the plasmodia, which can be several centimetres long (▷ box). Algal forms include the euglenoids, dinoflagellates, diatoms, and some green and brown algae.

Habitats

The protists are ubiquitous on land and in water (where they are a major component of plankton), as well as inside other organisms as parasites or symbionts (▷ p. 168). Free-living forms of algae and protozoa can be found in the soil – at a density of up to 200 000 per gram – and also occur in both fresh and sea water. Slime moulds favour fallen leaves and rotting logs. Symbiotic forms of algae occur in other protists, and in multicellular organisms such as hydra and termites. Parasitic forms are complex in their life-cycles, often having a secondary host (▷ p. 168) that transmits them from one primary host to another, while permitting life-cycle changes to occur. Some parasitic forms cause well-known diseases of man and animals, including malaria

A dinoflagellate, a member of the flagellate group of protozoa. Found in both freshwater and salt water, dinoflagellates can feed either like plants (by photosynthesis) or like animals (by taking in preformed food such as algae or diatoms).

(plasmodia), sleeping sickness and leishmaniasis (trypanosomes), and dysentery (*Entamoeba histolitica*). Plants are also affected by slime moulds, which cause such diseases as club root in cabbage and finger-and-toe disease in turnips.

Environmental adaptations

The many protists that live in an aqueous environment – be it marine or freshwater, or even inside another organism – have to be able to swim and to control the amount of water entering them. In addition, all protists need to be able to respond to environmental signals and pressures such as acidity, temperature, light, salinity, food sources, other chemicals in their immediate environment, and even attack from the defence mechanisms of the host they are parasitizing.

Some photosynthetic protists have specialized structures, eye-spots, with which to measure light intensity, enabling the organism to react accordingly. Similarly, protozoa are able to sense temperature, and if exposed to a gradient of temperature from high to low they will move to the more acceptable temperature point in the gradient. For most protists the optimal temperature range is 20–30 °C (68–86 °F), but some protozoa live in hot springs at temperatures of 44 °C (111 °F), while diatoms can be found in the cold waters of the Antarctic.

Nutrition

Only a few of the protists, the primitive algal forms, are *autotrophic*, i.e. able (like the plants and some bacteria) to convert simple inorganic molecules into complex organic molecules – principally carbohydrates and proteins – by photosynthesis (▷ pp. 24–5 and 31–3). However, the majority of protists are *heterotrophic* organisms, i.e. dependent on the consumption of other organic matter as a source of these complex molecules. Whichever the form of nutrition, food material is often stored within the cell. This food material can take the form of carbohydrates such as starch in the free-living forms or glycogen in the parasitic forms. There are also reserves of protein, and, although they are not known for their ability to digest fats, a number of species actually store oil droplets. All this stored food material is of value for those periods when the organism is unable to find food easily or when a large supply of energy is needed for processes such as rapid reproduction.

The mechanisms by which the heterotrophic organisms obtain their food depends to a certain extent upon the form of the exterior of the cell. Amoebae, for example, only have a membrane on the outside of their cells (▷ p. 20), and are thus able to engulf their food source by the process of *phagocytosis*. The free-moving amoeba is attracted by chemical emanations towards the food source, and these emanations when strong enough stimulate the extension of the amoeba's cell surface, which engulfs the food particle. The food particle is taken up inside a vacuole (cavity), where hydrolytic enzymes digest the contents, and the breakdown products are released through the vacuole mem-

brane into the surrounding cytosol (▷ diagram, pp. 20–1). Not all of the material contained in the food vacuole will be digested, the remains being excreted from the surface of the cell. While the digestion process occurs, the food vacuole circulates within the cell.

Some protists have more complex surfaces, and so the ability to phagocytose at any part of the surface is lost. Instead, a permanent point on the surface of the organism is created for the purpose of food uptake. This means that although a free-swimming cell such as *Paramecium* can be attracted in the same way as an amoeba towards its food source, the food must be trapped and guided to the point of uptake. This task is undertaken by hair-like structures on the cell surface, the cilia (▷ motility, below), some of which are specialized to trap and others to hold food. The food is then guided by other cilia to the gullet, down which it is wafted by more cilia lining the gullet, and so to the point where it can be ingested via a food vacuole. These food vacuoles also circulate within the cell, often along a circumscribed route. Excretion of the undigested matter can only occur at a specific site at the cell surface – the anus.

In more complex protists, the mechanisms for trapping food become increasingly involved. In some cases 'darts' are shot out to spear the prey, for protozoa are not averse to eating other living organisms including other protozoa. A primitive form of filter-feeding is also adopted by a number of protists; in such cases the food is wafted to the gullet either by a group of flagella (another form of hair-like structure; ▷ below) or by cilia banded together to form an oral membrane, which retains the very big particles too large to be ingested, but permits the passsage of smaller particles such as bacteria. Other forms of protist have a more structured feeding apparatus that allows them to swallow any passing foreign cell.

The protist outer surface

The outer surface of protists varies in its structure, depending on the environment in which the particular organism lives and the functions required of the surface. Clearly, the surface must protect the organism, but it is also necessary that substances and gases can move across the surface between the cell and its environment. Some protozoa such as amoebae just have the cell membrane between them and their aqueous environment, so permitting ready exchanges across the membrane. In contrast, some other forms have elaborated layers outside the membrane: *Paramecium*, a free-swimming ciliate, has a hard protective outer layer called a pellicle, while the dinoflagellates are plant-like in having plates of cellulose. Some protists are even surrounded by a shell: this can be made of different kinds of material, including silica, nitrogenous matter, calcarious material (as in most foraminfera), a chitin-like substance (similar to the principal component of arthropod exoskeletons), and cellulose (in the Mycetozoa).

Osmosis and respiration

Many aquatic protists, especially those living in fresh water, possess organelles (miniature organs) known as *contractile vacuoles*. These are primarily used to remove excess water. In a freshwater environment, water is able to pass across the outer membrane by osmosis, so resulting in the presence of too much water in the cell. This excess water is passed, sometimes via long tubes, to the contractile vacuole, which when full contracts and ejects the water from the cell. This water may pass directly to the outside or may be ejected into the gullet.

As with virtually all living organisms, protists need oxygen. This would appear to diffuse from the environment across the whole surface of the organism. Equally, unwanted gases such as carbon dioxide probably also diffuse out in a similar way, although it is possible that some dissolved carbon dioxide is extruded by the contractile vacuoles.

Motility

Many protists are mobile, and progress either by what is called amoeboid movement or by swimming with the aid of cilia or flagella. In *amoeboid movement*, progress is made by a complex series of changes to the amoeba's shape. On the leading edge of the underside of the amoeba a temporary projection called a *pseudopodium* ('false foot') attaches, with the aid of the cytoskeleton (▷ p. 20), to the surface, and pulls the rest of the cell after it. The pseudopodium then releases the surface, while a new leading edge forms a pseudopodium, which attaches to the surface and repeats the process. The overall effect is that the amoeba 'rolls' forward, with the cell contents flowing forward to the point of attachment and beyond.

Cilia and *flagella* (from the Latin meaning 'eyelash' and 'whip', respectively) are specialized structures used for locomotion and other functions (▷ above). They are thought to have evolved from the spirochaetes (a group of prokaryote bacteria), which when attached in groups to a host tend to beat in synchrony. Cilia are usually present in large numbers and tend to beat in a controlled fashion resembling the movement of wind across a field of wheat. This kind of movement is also

seen in a group of spirochaetes. Flagella may also be present in large numbers, but more often in pairs or singly. The protist flagellum is of the same basic structure as flagella in animal and plant cells. Although its outer appearance may vary, internally it is built of nine pairs of longitudinal tubules linked into a ring around a pair of central tubules; this internal structure is similar to that of cilia.

Cysts

Different forms of cysts occur amongst the protists, depending on the conditions inducing their formation. Usually some form of stress is the trigger: adverse environmental conditions prompt the cells to form 'resistance cysts', which are coated with jelly or stronger (usually organic) material. 'Resting cysts' are formed when the organism wants to spend an undisturbed period digesting and conserving energy. Special 'reproductive' cysts can also be formed (▷ below).

Reproduction

As in multicellular plants and animals, two forms of reproduction can occur in protists, namely asexual and sexual. Asexual reproduction is by cell fission (division). There are a number of types of fission, the basic event being the division of the cell nucleus (mitosis; ▷ p. 22). Amoeba and *Paramecium* are among the protists that undergo binary fission, which results in the production of two, usually similar, individuals. In some cases there is repeated fission, which results in the production of more than four nuclei; on completion of the process, the original cell breaks up, the number of individuals so created being equal to the number of nuclei produced. Repeated fission is usual in the formation of spores (which serve a similar function in protists as they do in bacteria; ▷ pp. 24–5). Fission can take place either inside or in the absence of a cyst (▷ above), the products being either naked or coated.

Sexual reproduction is also important and helps to mark the protists from prokaryotes. Although it is not possible to distinguish separate sexes, it is clear that in many cases fusion occurs between cells from different individuals. In this process a series of nuclear divisions occur, resulting in the formation of haploid gametes (sex cells, carrying only one copy of each chromosome), which are released into the watery environment and fuse in pairs to form diploid zygotes (cells with the full complement of chromosomes; ▷ p. 22).

Some of the members of the protozoa have two nuclei known as the micronucleus and macronucleus. It is the former that is concerned with the genetic and reproductive aspects of such cells. In some species of both algae and protozoa, reproduction occurs through *conjugation*, in which two individuals join together. If the macronucleus is present it is dispersed and the micronucleus divides to yield a number of nuclei. All but one of the nuclei disappear, and the remaining 'male' pronucleus moves from one cell into the neighbouring cell, where it fuses with the 'female'

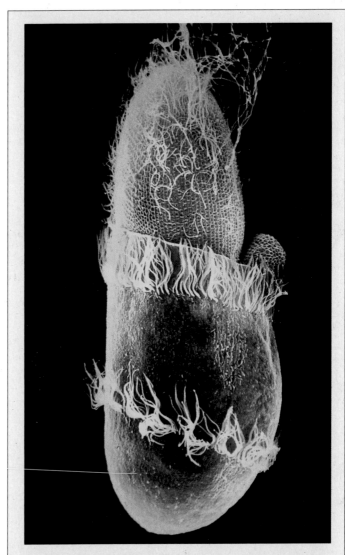

Protist eats protist. In this false-colour electron micrograph, the ciliate protozoan *Didinium* (bottom) is consuming fellow ciliate *Paramecium*. The barrel-shaped *Didinium* has two girdles of locomotory cilia, and a prominent snout that is used as a probing and seizing organ during feeding. Here the *Didinium's* snout is moving downwards within the interior of the cell, drawing the *Paramecium* into its body cavity. The *Paramecium* will completely fill the interior of the *Didinium*. Once digestion is complete, the snout returns to the front of the cell, ready to feed again.

pronucleus to form a zygote (akin to a fertilized egg in higher organisms). The zygotic nucleus then divides to form a micronucleus, and when relevant, a new macronucleus. Parthenogenesis – by which unfertilized eggs develop into individuals with a haploid (single) set of chromosomes (▷ p. 22) – is known to occur in a few species of protists.   PBG

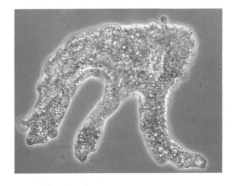

An unidentified species of amoeba showing the presence of three pseupodia ('false feet'), temporary projections used for locomotion and for feeding.

THE PLANT KINGDOM

Plants

The hundreds of thousands of species of plants comprise the kingdom Plantae, one of the five kingdoms of living things. Plants have colonized virtually every habitat on the planet, from the northern tundra to the equatorial forests, and from the driest desert to the oceans. In so doing, plants have evolved an astonishing variety of adaptations, resulting in one of the most diverse – and beautiful – groups of living organisms.

The ability of virtually all plants to photosynthesize (▷ p. 33) distinguishes them from all other life forms apart from certain bacteria (▷ p. 25). This ability to synthesize the complex organic molecules necessary for life from simple inorganic molecules – using the energy of the sun – means that nearly all other living organisms are directly or ultimately dependent on plants as a source of organic molecules (▷ p. 194).

Distinguishing features

Although superficially plants have much in common with fungi (kingdom Fungi; ▷ p. 35), the latter cannot photosynthesize and have cell walls made of chitin rather than cellulose. Plants are usually defined as multicellular organisms, so those unicellular algae capable of photosynthesis are often placed with other unicellular eukaryote organisms in the kingdom Protista (▷ p. 26).

SEE ALSO

● WHAT IS LIFE? p. 6
● CELLS p. 20
● SIMPLE LIFE FORMS pp. 24–7
● HOW PLANTS FUNCTION p. 31
● FUNGI p. 35
● MAIN PLANT GROUPS pp. 34–47
● ANIMALS p. 50
● THE BIOSPHERE pp. 194–7
● CLIMATE AND VEGETATION p. 198
● ECOSYSTEMS pp. 200–21

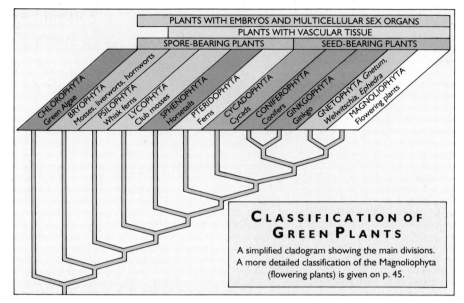

CLASSIFICATION OF GREEN PLANTS

A simplified cladogram showing the main divisions. A more detailed classification of the Magnoliophyta (flowering plants) is given on p. 45.

PLANTS WITH EMBRYOS AND MULTICELLULAR SEX ORGANS
PLANTS WITH VASCULAR TISSUE
SPORE-BEARING PLANTS   SEED-BEARING PLANTS

CHLOROPHYTA Green Algae · BRYOPHYTA Mosses, liverworts, hornworts · PSILOPHYTA Whisk ferns · LYCOPHYTA Club mosses · SPHENOPHYTA Horsetails · PTERIDOPHYTA Ferns · CYCADOPHYTA Cycads · CONIFEROPHYTA Conifers · GINKGOPHYTA Ginkgo · GNETOPHYTA Gnetum, Welwitschia, Ephedra · MAGNOLIOPHYTA Flowering plants

The largest flower belongs to *Rafflesia arnoldii*, the 'stinking-corpse lily'. The flowers of this tropical parasite may be up to 91 cm (3 ft) across.

Plants are distinguished from animals by various characteristics. Plant cells are different from those of animals in one fundamental respect, in that they possess cell walls (made of cellulose; ▷ illustration, p. 21). Many kinds of plant cell are also capable of giving rise to a whole new plant (the basis of vegetative reproduction; ▷ p. 33). Perhaps the most fundamental difference is the lack in plants of a nervous system – although plants do have various means of responding actively to their environment (growing towards the light, for example; ▷ p. 33). Another obvious difference is the inability of plants to move in a purposeful fashion, although plants can alter their growth direction in response to environmental stimuli, are capable of spreading over large areas by vegetative means, and can colonize new sites by passive dispersal of spores or seeds.

Extremes of size and age

The creosote bush of southwest California spreads very slowly from its original position to form an ever-widening circle of clones, and in the case of one particular plant the size of the circle indicates an age of 11 700 years. It has also been estimated that Antarctic lichens (symbiotic colonies of algae and fungi) more than 100 mm (4 in) in diameter are probably at least 10 000 years old. The oldest distinct individual plants are the bristlecone pines of North America; one of these growing in Nevada was aged (using growth rings) at 5100 years old when cut down.

The smallest flowering plant is an Australian duckweed (*Wolffia angusta*), which is only 0·6 mm (0·025 in) long and 0·33 mm (0·013 in) wide, and weighs 0·00015 g (100 000 to the oz). In contrast, the Sierra or giant redwoods of western North America are the most massive living things, one specimen, named 'General Sherman', weighing an estimated 2500 tonnes (tons). A related species, the coast redwood, holds the record for the tallest living tree, one specimen having a height of 113·7 m (373 ft). However, the tallest tree ever recorded was a giant gum (*Eucalyptus regnans*) growing in Victoria, Australia, which in 1872 was reported to have a height of 132·6 m (435 ft), and is thought to have originally measured over 150 m (500 ft).   IDC

FLORAL REGIONS

The world can be divided into six distinct floral (or phytogeographic) regions, principally on the basis of the similarity of the flowering plant inhabitants that each contains. The distribution of plants is based on a number of factors, including past and present climatic conditions (▷ pp. 10 and 198), the historical distribution of landmasses (▷ p. 8), and the evolutionary history of the plant groups concerned. The floral regions do not exactly coincide with the faunal regions (▷ p. 50), partly because of the greater ease with which plants are able to disperse themselves across physical barriers.

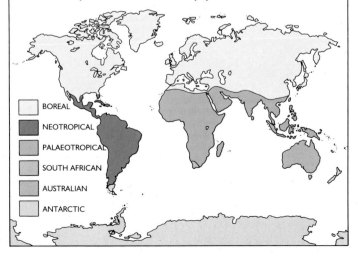

BOREAL
NEOTROPICAL
PALAEOTROPICAL
SOUTH AFRICAN
AUSTRALIAN
ANTARCTIC

How Plants Function

The same basic functions occur in plants as in all living organisms. They grow, and therefore they assimilate material from the environment, and this material provides the chemicals of which they are made. To maintain life they require energy; and they need water, which like the substances for growth and energy is taken up from their surroundings and transported around the plant body. Growth and the various other functions are spatially and temporarily regulated and coordinated. Plants, like animals, perceive their environment and react towards it. They also reproduce. But there are unique properties of green plants – principally their structure and the fact that they are fixed in one location – that determine the way they carry out these functions.

In general, the material for energy and growth is not obtained from organic material as it is in animals. Green plants together with some bacteria (▷ p. 24) are the only *autotrophic* organisms on earth, i.e. they are the only organisms that can make these materials from simple inorganic sources of carbon, nitrogen, hydrogen, oxygen and the other elements that are found in living things. To do this they need energy, and they derive this energy from sunlight. In doing so, they are converting simple inorganic chemicals into forms upon which the whole of animal, fungal and some bacterial life ultimately depends. Because they live in a fixed position the higher plants have no escape from their environment, and their physiological systems must enable them to withstand its rigours, to detect changes, and to bring about the appropriate responses.

Plant nutrition

Life is based on the chemical element carbon. Green plants obtain this from the air in the form of carbon dioxide, which they change (synthesize) into more elaborate chemicals – various carbohydrates such as sugars (glucose and sucrose) and starch (▷ p. 6). The energy required to drive this transformation is obtained from sunlight (although artificial lamps will work as well). This process is called *photosynthesis* (▷ box). Water is also essential as a source of the hydrogen atoms in the sugars, and, as a by-product, molecules of oxygen are given off. This oxygen is essential to virtually all life on earth.

The carbohydrates from photosynthesis provide the carbon from which nearly all the constituents of the plant body are made. Especially important among these is protein (▷ p. 6). Protein contains nitrogen (N), an element that is taken up by plants from the soil or water, generally in the form of nitrates (NO_3^- compounds; ▷ the nitrogen cycle, p. 196). In the majority of higher plants nitrates are converted to ammonium (NH_4^+) in the leaves; the ammonium is then combined with the carbon coming from the photosynthetic products to produce amino acids, the molecules from which proteins are constructed. This is another example of how plants are able to convert relatively simple substances from the environment into more complex organic forms. All animal life depends on this property, as all animals ultimately derive their protein from plants (▷ p. 196). Certain plants, prominently the legumes (e.g. peas, beans, soybean, groundnut), get most of their nitrogen from a beneficial association with bacteria that live in their roots; these bacteria convert gaseous nitrogen from the air into a form that is used by the host plant. This process, called fixation (▷ box), plays a very important part in agriculture through its ability to enrich the soil.

Other elements that plants need as components of chemical constituents (e.g. magnesium in chlorophyll, sulphur in proteins) or to promote chemical reactions (e.g. iron for chlorophyll synthesis) are taken up from the soil by the roots and transported to regions where they are utilized (▷ below).

Relatively massive production of carbohydrate (e.g. starch), proteins and vegetable oils occurs in plant reproductive structures – in seeds or other organs such as tubers (e.g. the potato). These materials are laid down as food reserves that are later digested to support the growth of the seedling or sprout until photosynthesis and other synthetic processes become well established. Animals and humans exploit these reserves as extremely valuable parts of the diet (cereal grains, pulses, etc.).

Uptake and transport

In higher land plants the roots take up water, mineral elements, nitrates and sulphates, sometimes using energy to do so. Water enters partly by osmosis, but more importantly by a process that is ultimately driven by *transpiration*, the evaporation of water through the stomata of the leaves (▷ box). Water exists in plants as an integral column in the conducting cells, the *xylem*. The xylem, one of the principal components of the vascular system, is continuous from the root, through the stem, the leaves and other parts of the plant. (In trees, the xylem, strengthened by a material called lignin, is the major component of the wood.) At the ends of the column, in the leaves, water is lost by transpiration and more water moves up to take its place. This effect is transmitted all the way down the stem, into the root, and eventually it is responsible for 'sucking' water in from the soil.

If the soil water becomes depleted (in a drought, for example) it may become

---

STOMATA

Stomata are microscopic pores on the surface of leaves. They are located mostly on the lower surface of the leaf at a frequency ranging from 20 to over 1000 per square millimetre depending on the species. The stomata allow carbon dioxide for photosynthesis into the leaf, and water vapour to escape during transpiration.

Each pore is bordered by two 'guard cells', which are typically about 0·045 mm long and about 0·012 mm wide. The guard cells can open and close the stomatal pore, so regulating the entry and exit of gases. They do this by taking up or losing water. When water enters the guard cell it swells up into a crescent shape, so that when both guard cells are swollen, the pore that they border becomes bigger. Conversely, when water is lost, the cells collapse and block off the pore.

What regulates these water movements? Water enters and leaves by osmosis, passing from a solution that is less concentrated to one that is stronger. Changes in potassium ion concentration occur in the guard cells: when the concentration increases, water is subsequently drawn in and the guard cell swells into its unique shape. As potassium ions enter, hydrogen ions (protons) leave: the hydrogen ions come from certain organic acids (malic acid), produced by starch breakdown.

What causes these events to take place? Stomata of most plants open in the light and close in darkness, so clearly light plays a role. In the light, photosynthesis leads to a fall in the carbon dioxide concentration in the leaf, and it is this that brings about the biochemical changes leading to water uptake by the guard cell. Superimposed on this are effects of the plant hormone, abscisic acid, whose concentration increases when leaves begin to wilt: abscisic acid causes the guard cells to close the pore, thus preventing further water loss.

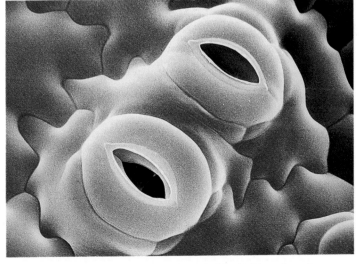

Open stomata on the surface of a tobacco leaf. In this false-colour electron micrograph the guard cells on either side of the stomatal opening are clearly visible.

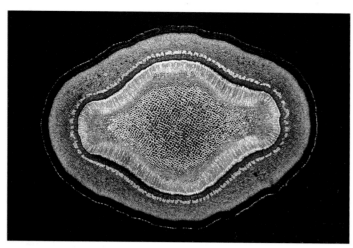

The concentric layers of different types of tissue of a young ash stem, seen in cross-section through a light microscope. Starting from the centre and working outwards the layers are: the pith (structural material, here coloured blue), xylem (pink with fibrous appearance), cambium (thin white layer), phloem (dark band), strengthening phloem fibres (broken pink band), cortex (structural material, blue), cork (protective and insulating material, dark band), and the epidermis (protective material, thin outer 'skin').

impossible for the plant to take up enough to offset transpiration, and then the total water content of the leaf drops, i.e. it wilts. There is, however, a physiological mechanism to offer protection against this, through the production of a wilt-induced chemical hormone, abscisic acid, which causes the stomata to close, and therefore reduces further loss of water.

The substances taken up from the soil are also transported in the xylem, so the liquid in the xylem – the sap – is actually a very dilute solution of mineral elements, nitrates, etc. But the substances made within the plant, the carbohydrates and amino acids, are carried in another vascular tissue, the *phloem* (part of the bark in woody plants). This process is called *translocation*. Carbohydrate is transported almost exclusively as sucrose (the

kind of sugar in sugar cane and sugar beet), which is taken to the flowers, developing seeds, roots and all tissues and organs of the plant. The driving force for this movement is set up by the relatively high concentration of sugars in the leaf cells. This force is also responsible for the movement of nitrogen-containing chemicals, principally amino acids. All these substances are carried around the plant to furnish material for growth (e.g. the production of new cell walls and protoplasm) and energy.

Energy and respiration

Although the energy that plants (and ultimately animals, too) need to maintain themselves is derived originally from sunlight, it is first transformed into the chemical energy residing in sucrose and other compounds, such as reserve starch, protein and oil. The energy is later released by the process of respiration – the use of oxygen to 'burn' (oxidize) various compounds. This oxidation occurs in all living cells of the plant, partly in the cytoplasm and partly in the mitochondria ($\triangleright$ p. 20). Respiration is used to support the synthesis of various chemical compounds (e.g. new proteins, nucleic acids, cellulose), growth, and the uptake and accumulation of various mineral elements. Before the starch in, say, a potato tuber can be used to provide the energy for the sprout growing from it, it must first be converted into a transportable form, namely sucrose; the same applies to the starch or oil in a germinated seed. This sucrose, like the sucrose formed in photosynthesis, moves to other parts of the plant, where it is broken down by respiration, via glucose, to release energy. We can therefore picture a flow of energy around the plant from the leaves or storage organs, principally in the form of transported sucrose. In broader terms this flow occurs as: sunlight (energy source) → photosynthesis (energy conversion) → intermediate compounds (energy storage) → respiration (energy conversion) → energy utilization.

Plant growth and development

Plants grow by increasing the number and size of their cells. New cells are produced by cell division (mitosis; $\triangleright$ p. 22) in specialized areas, the *meristems*. The *primary meristems* are found at the tips (apexes) of shoots and roots, and in newly formed leaves and other organs. When cells in already established tissues start dividing they form the *secondary meristems*: these form the *cambium*, a layer between the xylem and the phloem, and are responsible, for example, for the growth in girth of a stem or root.

The new cells formed at the root apex gradually push back those produced previously. Some time after their formation cells begin to enlarge, an event that occurs when they are just a few millimetres behind the very tip itself: this is therefore the zone where the greatest elongation is taking place. A similar process occurs in the shoot, but here it is complicated by the production of new organs – the leaves or the flower parts. These organs are themselves meristema-

tic at first, and in the case of the leaves, grow out of the stem at the *nodes*. As the leaves grow, the lengths of stem between the nodes – the *internodes* – elongate, and this internode extension contributes a far higher proportion of lengthwise growth than the growth at the tip.

Cells are enlarged mostly by an increase in their water content, which swells the volume of the cell cavities called vacuoles ($\triangleright$ p. 20). New protoplasm is also subsequently made. For the cell to increase in size the cell wall (comprising cellulose, other carbohydrates and protein) must be able to extend, so an increase in wall stretchability is an essential component of cell growth. This happens when the chemical properties of the wall are altered under the influence of a chemical hormone (or growth regulator) called auxin.

NITROGEN FIXATION

Almost all plant species in the legume family (Fabaceae or Leguminosae) that have been examined have entered into symbiotic relationships with certain bacteria living in their roots. These bacteria, which invade the roots from the soil, can fix gaseous or molecular nitrogen. The legume family includes many important crop plants (peas, beans, soybean, groundnut) that produce high-protein seeds, and it has been estimated that up to 50% of the nitrogen in the protein comes from the activity of the bacteria. Several species in other families also have nitrogen-fixing bacteria: these are generally plants, such as bog myrtle and alder, that typically colonize soils poor in nitrates.

The bacteria infecting the leguminous plants belong to the genus *Rhizobium*, and each species of legume plant has a particular species of *Rhizobium* that will infect no other legume. Infection occurs on a damaged root hair, from where the bacteria spread into the centre of the root. There, root cells are stimulated to divide to produce a *nodule*, inside whose cells the modified *Rhizobium* bacteria (now called *bacteroids*) live. The bacteroids require oxygen, and this is carried to the bacteria by a red pigment made by the plant, leghaemoglobin (like the haemoglobin of blood; $\triangleright$ p. 180), which confers a pinkish colour to the interior of the nodule. Using carbohydrate (in the form of sucrose) provided by the host plant, the bacteroids convert the nitrogen that has diffused into the nodule into ammonia, which is then changed by the nodule cells into amino acids and other nitrogenous compounds. These compounds pass into the xylem of the vascular tissue, to be carried up the shoot where they are used for protein synthesis.

Nitrogen fixation in legume crops plays an extremely important part in the nitrogen economy: these plants need far less nitrogen fertilizer than other crops, and do not deplete the soil of this essential element. In addition, when ploughed in, legumes enrich the soil, and are therefore commonly used in four-crop rotation.

UPTAKE AND TRANSPORT

——▶**Uptake**: Water and minerals 'sucked up' from soil via roots and transported in xylem to all parts of plant.

——▶**Translocation**: Amino acids (the components of proteins) and sugars transported in phloem from leaves to all parts of plant. In some cases (e.g. some trees in springtime) sugars from the roots are transported upwards.

Other hormones, gibberellins, also act on the cell wall; and these, together with cytokinins, also promote cell division.

An array of growth and development processes are affected by these three plant hormones and by two others, ethylene and abscisic acid. Such processes include cell differentiation ($\triangleright$ p. 21), seed development and germination, flowering, root initiation, fruit growth, and responses to drought-induced stress.

Environmental stimuli

Plant growth and development are also greatly influenced by environmental factors – light, temperature, and gravity. Plant stems respond to the direction of light, growing towards it – an ability called *phototropism*. Stem extension is regulated by the brightness and quality of light, which are perceived by light-absorbing molecules (pigments). One of these pigments, phytochrome, is sensitive to the different parts of the red region of the spectrum and can tell the plant if it is shaded by the green leaves of its neighbours; if it is, the plant increases its stem growth to carry it out of the shade. This same property of phytochrome enables seeds (which also contain the pigment) to detect how deeply they are buried in the soil, or if they are in vegetational shade, and they can modify their germination behaviour accordingly.

Sensitivity to light also enables plants to judge seasonal changes in daylight hours, and this regulates the time of year when plants flower – an ability called *photoperiodism*. Temperature is effective in many ways, including the regulation of reproduction. *Vernalization* is the process by which a period of cold induces many plant species – such as winter wheat and barley – to form flowers. The direction of plant growth is influenced by gravity (*gravitropism*), roots reacting positively and stems negatively. Although it is not known precisely how this happens it is clear that a change in orientation rapidly causes electrical disturbances in a root, and these stimulate an alteration in directional growth.

Reproduction

Reproduction operates in different ways in different groups of plants, and details of these methods are given on the following pages. However, a useful broad distinction can be made between the primitive and more advanced plants.

Many primitive plants – including mosses, liverworts, ferns and some algae – exhibit an obvious *alternation of generations*, a sexually reproducing (*gametophyte*) generation being followed by an asexual (*sporophyte*) generation, both generations (in most cases) being free-living. The gametophyte generation (in genetic terms, a haploid generation; $\triangleright$ p. 22) develops male and female organs, which in turn produce male and female sex cells (the *gametes*). Moisture is always needed to carry male sperm to female egg, and when these fuse they produce the sporophyte generation (in genetic terms, a diploid generation; $\triangleright$ p. 22), which in turn produces asexual

PLANTS AND THE ATMOSPHERE

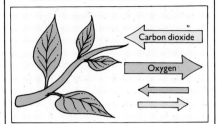

Day: plants absorb more carbon dioxide from the atmosphere by photosynthesis than they give out by respiration. They also give out more oxygen by photosynthesis than they absorb from the atmosphere for respiration.

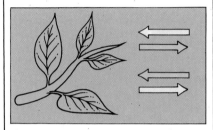

Dawn and dusk: with less light available for photosynthesis, plants give out similar amounts of oxygen and carbon dioxide as they take in.

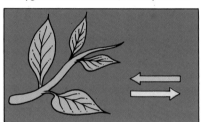

Night: with no light available, photosynthesis ceases. However, respiration continues, with oxygen being absorbed and carbon dioxide being given out.

spores that develop into the next gametophyte generation ($\triangleright$ p. 36).

In contrast, the seed-bearing plants – comprising the gymnosperms (conifers, etc.) and the angiosperms (the flowering plants) – have no obvious alternation of generations, but there are nevertheless gametophyte and sporophyte stages: the gametophytes (the female ovule and the male pollen grain) are much reduced, and are nutritionally dependent on the adult sporophyte plant. Crucially, water is not necessary for fertilization to take place ($\triangleright$ pp. 38 and 42) and the sporophyte embryo so formed is contained within a resistant seed that is dispersed from the plant, and that can (if conditions are unfavourable) remain dormant for many years before finally germinating and growing into an adult sporophyte plant ($\triangleright$ p. 38).

Many of the flowering plants are also able to reproduce asexually, by a process called *vegetative reproduction*. Special shoots extend from the parent plant, either underground or overground, and from these shoots daughter plants (genetically identical to the parent) bud and

PHOTOSYNTHESIS

The chemicals in plants that give them their green colour are the *chlorophylls*. The chlorophylls are particularly abundant in leaves, and are contained in *chloroplasts*, which are miniature organs – or organelles ($\triangleright$ p. 20) – within cells. Chlorophyll strongly absorbs the blue and red regions of the light spectrum, and this ability is used to drive the most important single chemical reaction on Earth. This reaction, which maintains both plant and animal life, is photosynthesis.

The chloroplasts in individual cells acquire carbon dioxide by diffusion through the intercellular spaces after it has entered the leaf through the microscopic pores on the leaf surface, the stomata. In photosynthesis, the carbon dioxide is then transformed into carbohydrates, initially the sugar glucose, and eventually sucrose and starch. This transformation is powered by the energy from sunlight absorbed by the chlorophylls. This is how the plant obtains the organic carbon needed for the synthesis of the materials of which it is composed. As vegetation is the primary food of all animal food chains, it is by photosynthesis that almost all of the carbon enters the living world: hence it sustains life on earth.

Overall, the chemical reaction of photosynthesis is:

$$\overset{\text{Sunlight}}{6CO_2 + 6H_2O \rightarrow C_6H_{12}O_6 + 6O_2}$$
$$\text{carbon dioxide} \quad \text{water} \quad \text{glucose} \quad \text{oxygen}$$

But this simple equation hides the complex chemical nature of photosynthesis, which comprises a set of reactions involved with the absorption of light (the *light reactions*) and a set that can take place in darkness (the *dark reactions*). The essential feature of the light reactions is that the light energy absorbed by chlorophyll is used to split water molecules into hydrogen and oxygen. The oxygen ultimately released by plants is the source of all oxygen in the atmosphere of this planet.

In the dark reactions carbon dioxide is converted into glucose by a complex cycle of chemical transformations, some of which use the hydrogen generated from water, and the chemical energy that has been produced from light energy.

As plants acquire their carbon dioxide, they also give off water vapour through the open stomata. As this water loss is potentially a problem in arid climates, many plants living in such environments have evolved physiological adaptations to overcome the danger. Many succulent plants, for example, keep their stomata closed during the day but open them at night when there is no drying effect of the sun. The carbon dioxide is assimilated (or 'fixed') at night not into sugars but into certain organic acids, which later, during daylight hours, release the carbon dioxide within the leaf when the stomata are closed; the carbon dioxide then participates in photosynthesis in the normal way. In other types of plant adapted to semi-arid climates (such as maize) the carbon dioxide is again fixed very efficiently into organic acids, but during the day, even by leaves with partially closed stomata; these organic acids then move across to the inner cells of the leaf where the carbon dioxide is liberated and used in photosynthesis.

Because in photosynthesis carbon dioxide is used up and oxygen given out as a waste product – the reverse of plant and animal respiration – the overall effect of plants and animals living together is to keep the atmospheric levels of these gases more or less constant. The felling and burning of large areas of tropical forest, combined with industrial processes that produce vast quantities of carbon dioxide, may contribute to an imbalance in the atmosphere, and hence to the 'greenhouse effect' ($\triangleright$ p. 224).

send down their own roots. There is a variety of different kinds of shoot that perform this function, including underground rhizomes (as in irises), tubers (as in potatoes), bulbs (as in daffodils), and overground runners (as in strawberries). Some of these vegetative shoots may grow faster when soil conditions are poor, in order to arrive at richer sites; they are said to exhibit 'foraging behaviour'. In some plants (such as begonias), daughter plants may grow from a broken leaf that becomes embedded in the soil.   MB

SEE ALSO

- WHAT IS LIFE? p. 6
- CELLS p. 20
- GENETICS p. 22
- PLANT GROUPS pp. 34–41
- POLLINATION AND SEED DISPERSAL p. 42
- THE BIOSPHERE pp. 194–7

Algae

The term 'algae' is a broad one, generally applied to lower plants that do not possess any special modifications for life outside water. Such plants first evolved in the oceans 1·5 billion years ago. Although the larger algae – the seaweeds – are always considered as plants, not all scientists agree as to how the single-celled algae should be classified. The latter are often grouped with other unicellular eukaryotic organisms in the kingdom Protista (▷ p. 26), even though many of them possess, like plants, the ability to photosynthesize. Less debatably, the so-called blue-green algae – the cyanobacteria – are usually grouped with the bacteria and other unicellular prokaryotic organisms in the kingdom Monera (▷ p. 25).

SEE ALSO

- WHAT IS LIFE? p. 6
- CELLS p. 20
- GENETICS p. 22
- PROTISTS p. 26
- HOW PLANTS FUNCTION p. 31
- PARASITISM AND SYMBIOSIS p. 168

The majority of algae – including all the seaweeds – live in sea water, but others live in lakes, rivers, soils, damp tree trunks, hot springs, and snow and ice. Some unicellular forms are symbiotic with animals such as corals, while lichens (▷ box) are a symbiotic association of an alga and a fungus. There are over 6000 species of seaweeds in three divisions, and over 12 000 species of unicellular algae in eight divisions.

Single-celled or protistan algae

Some single-celled algae are *autotrophic*, i.e. like the plants they can use photosynthesis (▷ pp. 31–3) to synthesize the complex organic molecules necessary for life from simple inorganic molecules. As such, they are vital primary producers (▷ p. 194) throughout the world's oceans, and, as the principal component of plankton, they are at the base of all marine food chains. Furthermore, by absorbing carbon dioxide and expelling oxygen in the photosynthetic process, they play a vital role in maintaining the balance of these gases in the atmosphere, helping to reduce the 'greenhouse effect' (▷ p. 224).

Autotrophic forms include the green algae (division Chlorophyta), which are the most like land plants of all the single-celled algae, possessing the same types of

Seaweeds at low tide in southwest England. Examples of red, green and brown algae are all visible.

chlorophyll, storing starch in chloroplasts (▷ p. 20), and having a cellulose cell wall. Another important autotrophic group are the diatoms (a class within the division Chrysophyta): it has been estimated that they are responsible for up to 25% of the world's primary productivity.

Other single-celled algae are *heterotrophic*, i.e. like the animals they are dependent on the consumption of other organic matter. The most animal-like of the algae are the euglenoids, which lack a cell wall, and have a distinctive movement. Yet other single-celled algae can exist with either an animal- or plant-like mode of nutrition. For other aspects of single-celled algae, including motility and reproduction, ▷ Protists, p. 26.

The larger algae – seaweeds

The seaweeds belong to three divisions of algae, each named after its most obvious pigmentation: red algae (Rhodophyta), brown algae (Phaeophyta), and green algae (Chlorophyta); the green algae also include many single-celled algae (▷ above). Seaweeds, like single-celled algae and the fungi, are described as *thalloid*, i.e. the plant body, the *thallus*, is not differentiated into root, stem and leaves. The large kelps have a tough stalk (*stipe*) and a frond-like blade, the smaller wracks have a tough thallus, while green forms are often soft.

Seaweeds are at their most abundant between latitudes 20° and 60° N and S, and most seaweeds prefer the area between high and low tide (the intertidal zone) on rocky shores. However, communities vary according to the nature of the shoreline, substrate (rock, gravel, sand, etc.), tidal range, slope of the shore, and wave action. Some species are restricted to subtidal zones (areas below low tide) – the large kelps form the canopy in subtidal 'forests', which may extend up to 10 km (6 mi) offshore – while other species only grow where they are occasionally splashed by the highest spring tides. One species, the Pacific giant kelp, grows up to 45 cm (18 in) per day – perhaps the fastest growth in the plant kingdom – and may reach lengths of 60 m (196 ft).

The seaweeds exhibit a complex variety of life cycles, with various types of alternation of generations (▷ p. 33) between diploid and haploid forms (▷ p. 22). In some cases the obvious seaweed plant may be either a diploid sporophyte or a haploid gametophyte (▷ p. 33), both forms appearing identical. More frequently (and also typical of all higher plants) the sporophytes and gametophytes look very different: the obvious seaweed plant is often the sporophyte, and the gametophyte is a scarcely visible filamentous or encrusting form.

In most cases the seaweed gametophyte produces gametes (sex cells), which may either be identical or distinct. In the latter case, the larger is called the female. In most brown seaweeds (including kelps and wracks) the female gamete (the ovum or egg) is non-motile and supplied with food materials, while the smaller male gamete (the spermatozoid or sperm) is

---

THE LICHENS

A fruticose lichen, *Usnea florida* (old man's beard; left), and a foliose lichen, *Platismatia glauca* (right), growing as epiphytes on the branch of a tree. Disc-like ascomycete fruiting bodies are visible on the *U. florida*.

Lichens are composite organisms formed by the symbiotic relationship of a fungus (in most cases an ascomycete, but in a few a basidiomycete) and an alga or a cyanobacterium (▷ p. 22). The fungus provides a protective environment for the growth of the green partner, which provides the fungus with the products of photosynthesis. The metabolism of lichens is suspended while they are dry or exposed to the heat of the sun, but soon active again when conditions are moist. This intermittent activity gives rise to a very slow growth rate, but enables lichens to colonize inhospitable habitats such as rock surfaces, even at extreme altitudes and latitudes – one species has been found at over 86° S in Antarctica.

In form lichens range from *crustose*, where the growth is crust-like and closely pressed to the surface, through *foliose*, where the growth is lobed and strap-like and loosely attached to the surface, to *fruticose*, where growth is bushy and hair-like, giving a very shaggy appearance.

Asexual reproduction is achieved by the production of *soredia*, clumps of algal cells surrounded by fungus. Thus fungus and alga are disseminated together. The fungus also produces ascomycete or basidiomycete fruiting bodies. In a few cases algal cells are attached to ascospores, but spores without these passengers would depend on germinating in the vicinity of the appropriate alga to reform the lichen.

Many lichens are highly sensitive to pollution, particularly to sulphur dioxide, and can therefore be used as pollution monitors.   JWC

---

motile and is attracted to the ovum by chemical signals. Fusion of two gametes (fertilization) results in a diploid product, the zygote. This is usually unprotected and merely drifts to a site where it germinates. If meiosis (the formation of new sex cells; ▷ p. 22) does not occur prior to germination, a diploid sporophyte will form. When mature this produces structures called *sporangia*, in which haploid asexual spores are formed by meiosis. These spores are often motile, and when released swim to a new site where they settle and grow into a gametophyte plant.   MEC

THE FUNGI

The fungi are not in fact plants, but constitute an entirely separate kingdom, the Myceteae, which first evolved around 1·5 billion years ago. Unlike the plants, the fungi lack chlorophyll and hence cannot photosynthesize. They therefore have to obtain the carbon and energy necessary for life from other sources.

In most fungal groups there are examples of parasites (▷ p. 168), which grow on living animals, plants or other fungi, and also of *saprotrophs*, which grow on their dead remains. The function of the latter as decomposers is a very important one, assisting in the recycling of materials needed for life (▷ pp. 194–7). Some parasitic fungi continue to live as saprotrophs on the dead remains of their once-live hosts. Other fungi have developed symbiotic relationships (▷ p. 169) with other organisms: for example, some are *micorrhizal*, i.e. living in association with the roots of higher plants (such as conifers; ▷ p. 39), while others associate with algae to form lichens (▷ box).

Although some fungi, such as yeasts, are unicellular, most fungi consist of a mass of filaments. Aggregates of these filaments may give rise to quite large fruiting bodies – the obvious visible parts of fungi such as toadstools. An individual filament is called a *hypha*, and a mass of hyphae is called a *mycelium*. Although in some fungi the cell walls – like those of plants – contain cellulose, in most fungi the cell walls contain chitin (also the principal material in the exoskeletons of insects and other arthropods). Reproduction is by spores, which may be produced sexually or asexually. The spores are usually dispersed by wind, but sometimes by water or insects.

Excluding the slime moulds (usually considered as protists; ▷ p. 26), the true fungi fall into two divisions: the Mastigomycota and the Amastigomycota. The Mastigomycota are either unicellular or have non-septate (unpartitioned) mycelia, and their spores are generally motile, dependent on water for their dispersal. The division includes various parasites, including *Phytophthora infestans*, the potato blight fungus, which, like many specialized parasitic fungi, develops specialized structures that absorb nutrients from the living cells of its host.

The Amastigomycota include both unicellular and filamentous species. Reproduction is either sexual or asexual. The latter may involve *budding* (fragmentation of the mycelium), the production of spores in specialized structures called *sporangia*, or the production of *conidia*, which are spores formed (often in chains) at the ends of special hyphae called *conidiophores* (▷ diagram). There are four subdivisions: Zygomycotina, Ascomycotina, Basidiomycotina and Deuteromycotina.

ZYGOMYCETES

The Zygomycetes are one of the two classes of the Zygomycotina. They produce asexual spores (usually in sporangia), while in sexual reproduction, two special hyphae fuse and a *zygospore* is produced. This is a resting stage, but the zygospore eventually germinates to produce a sporangium containing many spores. One familiar zygomycete is *Rhizopus stolonifer*, a common mould on damp bread.

ASCOMYCETES

The Ascomycotina comprise one class, the Ascomycetes. There are both unicellular and filamentous species. The spores that result from meiosis (▷ p. 22) are contained in a sac or *ascus*, the spores being referred to as *ascospores* (▷ diagram). In most species a mitotic division (▷ p. 22) follows meiosis so that each ascus contains eight spores, which may be explosively discharged several centimetres into the air.

Microscopic forms include the unicellular yeast, *Saccharomyces*, which is important in the making of bread and alcoholic beverages, and the filamentous *Ceratocystis ulmi*, the cause of Dutch elm disease. Larger forms include *Daldinia concentrica*, King Alfred's cakes (which has a more-or-less spherical, black fruit body), and *Aleuria aurantia*, orange-peel fungus, one of the cup fungi.

BASIDIOMYCETES

The Basidiomycotina comprise one class, the Basidiomycetes. This contains most of the fungi generally thought of as mushrooms and toadstools – though it also includes microscopic plant parasites, the rusts and smuts. For much of their life cycle, there are two haploid nuclei (▷ p. 22) per compartment in the mycelium. In the fruiting body these two nuclei fuse in special cells referred to as *basidia*. Meiosis (▷ p. 22) follows to produce four nuclei, which migrate into swellings on little stalks on the basidium to form the *basidiospores* (▷ diagram). In a mushroom the basidia are borne on the gills on the underside of the 'cap', and, although the basidiospores are discharged explosively, the distance travelled is tiny in comparison to that of ascospores. The basidiospore must clear the gill surface, yet not hit the adjacent gill, and then fall to be carried away by air currents below the mushroom cap.

A number of good edible fungi such as *Pleurotus ostreatus*, the oyster mushroom, are to be found in the Basidiomycetes, but there are also some, such as *Psilocybe semilanceata*, the liberty cap, that are hallucinogenic, and others, such as *Amanita phalloides*, the death cap, that are deadly poisonous. Many are important as decomposers, being wood-rotting or leaf-litter fungi. Others are parasitic, such as *Armillaria mellea*, the honey fungus; and yet others are mycorrhizal, gaining carbohydrates from the trees with which they are associated and conferring on the tree a more efficient uptake of nutrients from the soil.

FUNGI IMPERFECTI

The Deuteromycotina is a 'dustbin' group of fungi, whose true classification cannot be decided because they do not possess a sexual stage, which forms the basis of classification into Ascomycotina and Basidiomycotina. Included in the group are some important plant parasites such as *Botrytis* (grey mould) and *Verticillium* (vascular wilt).   JWC

A moth in Papua New Guinea infected by *Akanthomyces*, a genus of the Deuteromycotina.

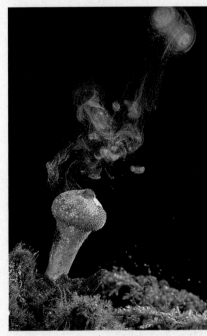

Common puffball (*Lycoperdon perlatum*), a basidiomycete in which the basidiospores mature inside the toadstool. This functions like a bellows so that anything, such as a rain drop, touching the thin container causes a puff of spores to be discharged through an opening in the top.

Psathyrella pennata, a basidiomycete fungus characteristic of old bonfire sites. The basidiospores are borne on the gills, which are clearly seen on the uprooted specimen.

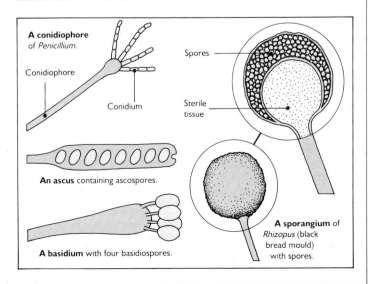

A conidiophore of *Penicillium*.

Conidiophore

Conidium

An ascus containing ascospores.

A basidium with four basidiospores.

Spores

Sterile tissue

A sporangium of *Rhizopus* (black bread mould) with spores.

Spore-bearing Plants

Although the algae had thrived in the oceans for hundreds of millions of years, the first signs of spores resembling those of modern land plants date from the later part of the Ordovician period (510–438 million years ago). By the end of the Silurian period (438–410 million years ago) the first undoubted vascular plants (▷ below), such as club mosses, had established themselves on the land.

Prior to the colonization of the land by plants, the soil would have been largely devoid of organic matter – and hence low in nutrients. With no plant roots to bind the soil or to provide shelter, land surfaces would have been very unstable, with high erosion and drastic fluctuations in humidity and temperature. Life in water protects plants from these problems, and enables nutrient and gaseous exchange all over the surface of the plant cells. To cope with conditions on land, plants had to evolve several crucial adaptations.

Adaptations to life on land

Out of water the spore, the dispersal unit of non-seed-bearing plants, needs extra protection. This is achieved by a depos-ition on the spore wall of *sporopollenin*, one of the most resistant compounds in the plant world. Once spores germinate, the young plant needs tolerance of, or protection from, the rigours of the land environment. Early land colonizers may well have adopted the strategy still used by some bryophytes (▷ below), which when dry become dormant. However, pro-tection is preferable to inactivity as a solution, and this is achieved by a protec-tive 'skin', the cuticle. Such a largely impermeable barrier would have pre-vented the plant from undertaking gas-eous exchange (crucial for respiration and photosynthesis; ▷ pp. 31–3), so thicker cuticles need pores for this pur-pose. Such pores, with controllable open-ings, are the *stomata* (▷ p. 31).

As the land surface became covered, plants gained height to outcompete neigh-bours for the sunlight required for photo-synthesis. With this gain in height, greater structural strengthening is re-quired, and water and nutrients need to be transported round the plants. Various related solutions evolved in plants in response to these problems. The walls of plant cells became thicker with the depos-ition of lignin on parts of the cellulose cell wall, and tubular cells thus strengthened – constituting the *xylem* – are able to conduct water and nutrients up from the roots without collapsing (▷ pp. 31–2). Other kinds of cell – constituting the *phloem* – transport chemicals synthesized within the plant to different locations (▷ p. 32). Xylem and phloem together comprise the *vascular tissue* that is char-acteristic of all higher land plants – which are thus known as the *vascular plants* or *tracheophytes*. Roots – also important for both anchoring and nutrient acquisition – also evolved.

Increased surface area for light intercep-tion – hence increased productivity – was achieved early on through the evolution of simple outgrowths from the stem. These early leaves only had a singular vascular strand (*microphyllous* leaves), so restrict-ing their size, but larger leaves (like those of ferns and flowering plants) with exten-sive networks of vascular tissue (veins) eventually emerged; these are called *macrophyllous* leaves. Increasingly, com-petition for light at lower levels led to the evolution of trees, which, by accumu-lating old growth as wood (▷ p. 31), are able to achieve great heights. The earliest trees emerged some 370 million years ago, and by around 320 million years ago, in the Carboniferous period, club mosses and horsetails grew as trees in the great coal swamps – although today these plants are only herbaceous.

Reproduction in spore-bearing plants

Spore-bearing plants exhibit an obvious alternation of generations, with both gametophyte and sporophyte generations being for the most part free-living (▷ p. 33). The gametophyte plant has egg-producing organs (*archegonia*) and sperm-producing organs (*antheridia*). The egg is retained on the gametophyte plant, and the flagellate sperms need free water to enable them to swim to the egg for fertilization to take place. The *zygote* (fertilized egg) is retained in the arche-gonium, where it develops into an *embryo* (a diminutive young sporophyte plant).

Unlike the embryo of a seed-bearing plant (▷ pp. 38 and 43), the embryo of a spore-bearing plant cannot be dispersed, and must grow wherever the gametophyte grows, however unsuitable the site. Furthermore, it has only the limited resources of a small gametophyte on which to rely for development. When the sporophyte plant matures it develops structures called *sporangia*, in which meiosis (▷ p. 22) gives rise to spores. The spores are dispersed in air currents, and if they land on a suitable site they germi-nate to form gametophyte plants. In most cases the spores are undifferentiated (the *homosporous* condition), but in a few species two spore types of different sexes are produced (the *heterosporous* condi-tion); it is from the ancestors of the latter that the seed-bearing plants probably evolved (▷ p. 38).

In the bryophytes (mosses, etc.) it is the gametophyte stage that is the obvious plant, the sporophyte plant consisting only of a foot, stalk and capsule. Spores are produced in the capsule. The sporo-phyte obtains nutrients from its gameto-phyte parent through its foot, and is therefore never entirely independent. In contrast, in the other spore-bearing plants (ferns, club mosses, horsetails, etc.), it is the sporophyte stage that is the obvious plant, with the sporangia (spore-producing organs) usually borne on the leaves. The gametophyte plants are in-dependent and mostly capable of photo-

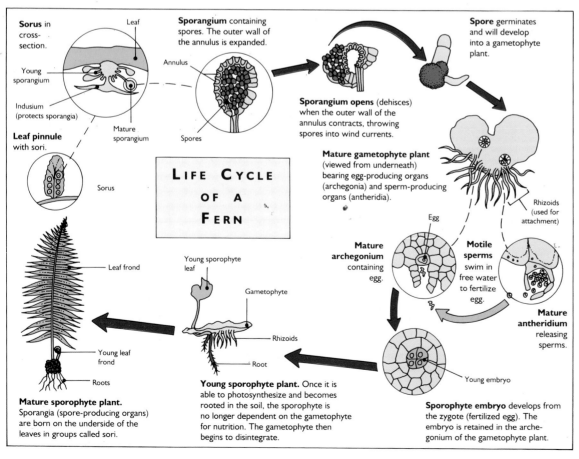

Sorus in cross-section.

Leaf

Young sporangium

Indusium (protects sporangia)

Mature sporangium

Leaf pinnule with sori.

Sorus

Sporangium containing spores. The outer wall of the annulus is expanded.

Annulus

Spores

Sporangium opens (dehisces) when the outer wall of the annulus contracts, throwing spores into wind currents.

Spore germinates and will develop into a gametophyte plant.

Mature gametophyte plant (viewed from underneath) bearing egg-producing organs (archegonia) and sperm-producing organs (antheridia).

Rhizoids (used for attachment)

Egg

Mature archegonium containing egg.

Motile sperms swim in free water to fertilize egg.

Mature antheridium releasing sperms.

LIFE CYCLE OF A FERN

Leaf frond

Young sporophyte leaf

Gametophyte

Rhizoids

Root

Young leaf frond

Roots

Mature sporophyte plant. Sporangia (spore-producing organs) are born on the underside of the leaves in groups called sori.

Young sporophyte plant. Once it is able to photosynthesize and becomes rooted in the soil, the sporophyte is no longer dependent on the gametophyte for nutrition. The gametophyte then begins to disintegrate.

Young embryo

Sporophyte embryo develops from the zygote (fertilized egg). The embryo is retained in the arche-gonium of the gametophyte plant.

Mosses, like all spore-bearing plants, require moist conditions to complete their life cycle: the presence of free water is essential for the sperms to swim to the egg.

synthesis, but they are very small and lack cuticle or conducting tissues. They are restricted to moist sites and can easily be swamped by leaf litter or other plants. In some cases (e.g. in whisk ferns) the gametophytes avoid this vulnerability by growing underground; however, they depend on a symbiotic association with a fungus and are therefore restricted to sites where the fungus is available.

Mosses, liverworts and hornworts – the bryophytes

The division Bryophyta comprises three classes: the Musci (mosses), Hepaticae (liverworts) and Anthocerotae (hornworts). Although rather diminutive plants, they show remarkable diversity in both form and habit. The majority are leafy, with leaves for the most part only one cell thick, so that they very easily dry out. Water transport is mostly external, relying on capillary action – for example in spaces between leaves and stem. A few of the larger mosses have conducting tissue similar to that of the vascular plants. Bryophytes have no roots, but attach themselves by means of fine hair-like structures called *rhizoids*.

All mosses are leafy and their growth habit may be either erect, often forming tufts or cushions, or creeping, forming a weft over the surface. Although most grow in moist habitats, some grow in exposed places liable to desiccation, and in the latter case there is a physiological adaptation that enables them to resume photosynthesis shortly after becoming moistened by mist or rain. The sporophytes of mosses take months to develop and are often photosynthetic when young. The capsules of most develop an opening surrounded by teeth; these teeth are sensitive to humidity changes, opening when dry and closing when wet. Thus not all the spores are shed at one time, and when they are shed the conditions are optimum for dispersal. The spores germinate to produce a branching green thread; young moss plants develop from buds on this thread, so facilitating colonization.

The liverworts that are most distinct from mosses are *thalloid* (like seaweeds; ▷ p. 34), that is they are flat and have no leaves. Most liverworts, though, are leafy, but distinct from mosses in having their leaves in two or three ranks rather than in a spiral. Most liverworts are more or less prostrate and the leaves tend to be flattened in one plane. In contrast to mosses, the sporophytes show rapid development and are not photosynthetic. Spores are released with the aid of humidity-sensitive cells interspersed among them.

The hornworts are so called because of their distinctive cylindrical sporophyte. This grows continuously from its base, which is embedded in a thalloid gametophyte.

Whisk ferns – the psilophytes

The division Psilophyta is a tropical to warm-temperate group comprising two genera of herbaceous plants. Many species are epiphytes, i.e. they grow on other plants without damaging them. The short aerial stems have vascular tissue and diminutive microphyllous leaves. In place of absorptive roots, they are anchored by rhizoids (like the bryophytes) and have creeping stems that obtain nutrients in symbiotic association with a fungus. Lobed sporangia are borne on some of the leaves.

Club mosses – the lycophytes

The division Lycophyta is widespread and diverse. The plants are generally small, herbaceous, vascular, and possess roots. The leaves are microphyllous, some bearing sporangia. In most cases the leaves bearing sporangia differ from the other leaves and are grouped into a cone-like structure at the tip of the branch.

The lycophytes can be subdivided into three major families. The Isoetaceae (quillworts) have very short upright stems from which arise crowns of elongate leaves. Superficially they resemble clumps of rushes or grasses, and grow mainly in wet areas. The Lycopodiaceae and Selaginellaceae are both commonly called club mosses. They are distinguished primarily by their reproductive biology, the latter being heterosporous (▷ above). *Lycopodium* does look like a moss as the short stems are obscured by tiny crowded leaves. Some species of *Selaginella* branch extensively, trailing to cover large areas, and giving the appearance of a large fern frond.

Horsetails – the sphenophytes

The division Sphenophyta comprises a single genus, *Equisetum* (horsetails), with about 15 species, which grow mainly in temperate zones. Horsetails are herbaceous plants distinguished by their hollow, segmented stems, round which diminutive leaves and aerial branches are borne in whorls. Most are small, but one tropical species reaches 8 m (26 ft) in height. Even tiny portions of the underground rhizome are able to regenerate, making the horsetail a pernicious weed that can penetrate through many metres of soil and even tarmac and concrete.

VARIETY IN FERNS

TREE FERNS can reach up to 25 m (82 ft) in height. They have a crown of leaves topping an unbranched trunk. The trunks have no wood but rely on fibres strengthened by lignin, persistent leaf bases and a mantle of roots growing from the trunk for their structural strength and support. Tree ferns may form a major part of the leafy canopy along rivers in wet tropical and sub-tropical forests.

EPIPHYTIC FERNS grow on other plants, often trees, without damaging them. The bird's-nest fern (*Asplenium nidus*) has large shiny leaves arranged in a rosette. Water and leaf debris falling from the canopy above accumulates inside the base of the rosette where it can be utilized by the plant. The stag's-horn fern (*Platycerium*) utilizes some of its own dead leaves to accumulate moisture.

CLIMBING FERNS such as *Polypodium* use creeping rhizomes to climb trunks, while *Lygodium* climbs by means of evergrowing leaves that twine around objects they touch. This ability is very rare outside the flowering plants.

WATER FERNS such as *Azolla* grow free-floating on the water surface. The reduced leaves have a special waxy coating to prevent water swamping the plant. *Azolla* has a symbiotic association with a nitrogen-fixing cyanobacterium, and is used as a green fertilizer.

FILMY FERNS have delicate leaves only one cell thick. For those growing as epiphytes in wet forest even short exposure to the sun's rays penetrating the canopy can be fatal.

MARATTIA has one of the largest leaves in the plant world, reaching up to 7 m (23 ft) in length. There are massive leaf stalks (the thickness of an adult's lower arm) to support this enormous structure.
                                                                          MEC

Sporangia are borne on specialized structures grouped in whorls into cones at the tip of the main stem.

Ferns – the pteridophytes

The ferns – constituting the division Pteridophyta – are distinguished from other spore-bearing plants by their large megaphyllous leaves, known as fronds. These leaves have an extensive network of conducting strands and are often intricately subdivided. Sporangia are borne on the leaves in groups called *sori*, which may be protected by a structure called an *indusium* (▷ diagram). In the more advanced ferns, modified cells (the *annulus*) in the sporangial wall enable spores to be thrown into air currents. Ferns include over 9000 species, making up about 90% of the diversity of the vascular spore-bearing plants.
                                            MEC/JWC

SEE ALSO

● PLANTS p. 30
● HOW PLANTS FUNCTION p. 31
● ALGAE p. 34
● CONIFERS AND THEIR ALLIES p. 38

Horsetails today are generally small, but in the Carboniferous period (355–300 million years ago) their extinct relatives reached heights of 30 m (100 ft). Club mosses also grew as trees, and together they dominated the great coal swamps.

Conifers and their Allies

Conifers and their allies, together with the flowering plants, are distinguished from the spore-bearing plants (▷ p. 36) by the production of seeds rather than spores as the units of dispersal. Seed-bearing has a number of advantages over spore-bearing in successfully spreading and increasing the species, and in ensuring survival if the parents are short-lived or suffer some catastrophe.

Seed plants (sometimes called spermatophytes) are usually subdivided into five divisions: Cycadophyta (cycads), Ginkgophyta (ginkgo or maidenhair tree), Coniferophyta (conifers), Gnetophyta (*Gnetum*, *Welwitschia* and *Ephedra*), and Magnoliophyta or Anthophyta (flowering plants; ▷ pp. 40–7). The first four of these divisions are often grouped together as the *gymnosperms*, the common characteristic being that they bear naked seeds, i.e. the seeds are not enclosed by an ovary. In contrast, the flowering plants – the *angiosperms* – have their seeds enclosed in an ovary, which has a special extension (the stigma) for receiving pollen (▷ p. 40). However, the different gymnosperm groups are highly varied, and cladistic analyses suggest that only the ginkgo and the conifers share a common ancestor, while the other two groups evolved independently.

The evolution and biology of the seed

Most spore-bearing plants are *homosporous* – carrying only one form of spore – but a few living forms are *heterosporous*, i.e. they produce two spore types of different sexes. The heterosporous condition was formerly widespread, and it is via this characteristic, early in the history of plant life on land, that ovules (which when fertilized mature into seeds) are thought to have evolved.

In heterosporous plants, the female egg is contained in a larger spore (a *megaspore*), while a small spore (a *microspore*) contains the male sperm. When the megaspore ruptures, the female sex organ (the *archegonium*) is exposed; when the microspore ruptures, male sperm are released. The male sperm are motile, and must swim in free water to the female archegonia; because megaspore and microspore are not necessarily in close proximity, fertilization occurs somewhat by chance. The fertilized egg develops inside the wall of the megaspore, which provides protection and includes some food reserves that assist the development of the young plant or embryo.

Seed-bearing plants have developed the modification found in heterosporous plants in several ways. Crucially, in seed-bearing plants, the megaspore is retained on the parent plant, while the microspores (pollen grains) are carried to it by wind or (predominantly in the case of flowering plants) by pollinating animals – principally insects (▷ p. 42). The vital difference here is that in seed-bearing plants, unlike spore-bearing plants, the egg and developing young plant are provided with food, water and protection by the parent plant. Equally important is the fact that free water is no longer required for fertilization to take place.

Structurally and functionally, the female organs of seed-bearing plants are different from those of spore-bearing plants. Within the ovule the single megaspore is enclosed by the *nucellus* (equivalent to a sporangium that does not open; ▷ pp. 36–7), which is enveloped in additional structures, the *integuments* (one in conifers and allies, two in flowering plants). At the junction of the integument lobes is another structure, the *micropyle*, through which pollen (or the pollen tube) reaches the nucellus (▷ below). The integuments develop into the seed coat, which is crucial in protecting the young plant contained within the seed when the seed is dispersed.

The microspores of seed-bearing plants, the pollen grains, are also modified, consisting of only a few vegetative cells with a protective wall, and two or more male sex cells. Thus pollen is reduced almost to a minimum (the absolute minimum is attained in the flowering plants), to enable easy, safe transport of male gametes to the megaspore and hence the egg.

In primitive living seed plants (cycads and the ginkgo) the nucellus tissue breaks down into a fluid in which flagellate sperm, released by rupture of the pollen wall, swim to the egg, thus recalling their spore-bearing ancestry. In all other seed plants pollination triggers the growth of a pollen tube, which extends through the nucellus to carry non-motile sperm to the egg. After pollination and fertilization the ovule develops into the seed.

The evolution of the ovule and seed did not transform the landscape rapidly like the later evolution of the flowering plants (▷ p. 40). Seed plants remained subordinate to spore-bearing plants for almost 100 million years; their rise to dominance occurred during the Permian period (300–250 million years ago), when the arid climate was hostile to the spore-bearing plants, dependent as they are on free water for the completion of their sexual cycle.

The cycads

The cycadophytes (cycads) include 11 modern genera. Most of these have very restricted distributions within the tropics and sub-tropics; only *Cycas*, which ranges from Africa across Asia to Japan and Australia, and *Zamia*, which is found in the Americas, are relatively widespread. The cycads emerged about 270 million years ago, and during the Mesozoic era (beginning 250 million years ago) they and their extinct relatives dominated land vegetation. However, they had declined drastically by 100 million years ago following the evolution of the flowering plants.

Cycads resemble small palms with a short trunk, although some species of *Macroza-*

SEE ALSO

- THE AGES OF THE EARTH p. 4
- GENETICS p. 22
- PLANTS p. 30
- HOW PLANTS FUNCTION p. 31
- SPORE-BEARING PLANTS p. 36
- THE EMERGENCE OF FLOWERING PLANTS p. 40
- BOREAL FOREST p. 208

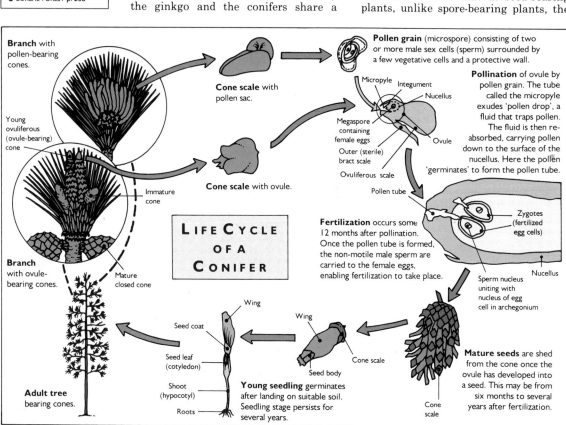

LIFE CYCLE OF A CONIFER

Branch with pollen-bearing cones.

Cone scale with pollen sac.

Pollen grain (microspore) consisting of two or more male sex cells (sperm) surrounded by a few vegetative cells and a protective wall.

Pollination of ovule by pollen grain. The tube called the micropyle exudes 'pollen drop', a fluid that traps pollen. The fluid is then re-absorbed, carrying pollen down to the surface of the nucellus. Here the pollen 'germinates' to form the pollen tube.

Micropyle Integument Nucellus

Megaspore containing female eggs

Outer (sterile) bract scale

Ovuliferous scale

Ovule

Pollen tube

Young ovuliferous (ovule-bearing) cone

Cone scale with ovule.

Immature cone

Fertilization occurs some 12 months after pollination. Once the pollen tube is formed, the non-motile male sperm are carried to the female eggs, enabling fertilization to take place.

Zygotes (fertilized egg cells)

Sperm nucleus uniting with nucleus of egg cell in archegonium

Nucellus

Branch with ovule-bearing cones.

Mature closed cone

Wing

Wing

Seed coat

Seed leaf (cotyledon)

Shoot (hypocotyl)

Roots

Seed body

Young seedling germinates after landing on suitable soil. Seedling stage persists for several years.

Cone scale

Adult tree bearing cones.

Cone scale

Mature seeds are shed from the cone once the ovule has developed into a seed. This may be from six months to several years after fertilization.

mia (a genus endemic to Australia) can reach up to 18 m (60 ft) in height, and others are recorded at 10 m (33 ft) or more. The trunk is covered with the bases of old leaves, and tipped with a crown of large pinnate leaves. Cycads usually occur scattered in the understorey or ground vegetation of forests.

Cycads are *dioecious*, i.e. male and female reproductive organs are located on separate plants. The pollen sacs and ovules are borne on scales, which are usually grouped together in cones. Some cycads are pollinated by insects, though wind undoubtedly plays an important part in pollination. Cycad seeds are large, having well-developed food reserves and fleshy outer seed coats that may be attractive to animal dispersers (▷ p. 43).

The ginkgo

The division Ginkgophyta – formerly much more diverse and widespread – is represented by a single modern species, *Ginkgo biloba* (the maidenhair tree), which in its wild state today is restricted to a small region in eastern China. The ginkgo, like the cycads, is dioecious (▷ above), and it is the male tree that is commonly grown in gardens; the female is less popular owing to an objectionable smell produced by the flesh of the seeds.

The ginkgo is a large, much branched tree, distinguished from the related conifers by its fan-shaped leaf. The branches bear long and short shoots, the reproductive structures being situated on the latter. Ovules are borne on long stalk-like structures, and pollen sacs are borne on scales in catkin-like structures. After pollination, the ovules remain on the parent plant, maturing to a large size; fertilization occurs after the ovule has been shed and lies on the ground.

The conifers

Conifers first emerged during the coal-forming period of the Carboniferous over 300 million years ago. They are the most widespread and diverse seed-plant group (apart from the flowering plants) in existence today. Modern conifers are grouped into 6–9 families with about 50 genera and some 500 species.

Most conifers are tall forest trees. Their leaves are typically needle-like, usually with a single vascular strand. The xylem (▷ pp. 31–3) in conifers – and in other trees and shrubs – is stiffened by lignin to form wood, and the wood is protected and insulated by bark. Resin is secreted by most conifers as an unwanted end product of metabolism.

The reproductive structures are grouped in cones. In the mature female cone there are pairs of papery or woody scales, comprising a sterile outer or bract scale, and an inner ovule-carrying or *ovuli-ferous* scale. In the male cone, pollen sacs are borne on the underside of cone scales. Most conifers have a very long life cycle, with several static periods. The period between pollination and fertilization often exceeds one year, and up to five years can intervene between cone initiation and shedding of mature seeds, so that on one tree cones of several different ages are often found at any given time. Seedlings are very slow to establish, the seedling stage persisting for a number of years. The slowness of the conifer life cycle partially explains why they have been out-competed by the flowering plants in many parts of the world.

In the northern hemisphere coniferous trees are the main constituents of the high-latitude boreal forests of North America and Eurasia (▷ p. 208). The boreal forests are dominated by members of the Pinaceae family, including *Larix* (larch), *Pinus* (pines), *Picea* (spruces), *Abies* (firs), *Pseudotsuga* (Douglas fir) and *Tsuga* (western hemlock). These high-latitude areas are characterized by a short growing season, a low angle of sunlight, high winds, extreme cold, frozen ground, snow falls, and very slow nutrient recycling. Similar conditions also exist at lower latitudes in high-altitude forests, which are often also dominated by conifers.

The conifers show several modifications that enable them to exist under these harsh conditions. The conical shape encourages shedding of snow rather than accumulation on the canopy, and also optimizes interception of low-angled light. To cope with the drying effects of frosts and high winds, the needle-like leaves are xeromorphic (drought-resistant) with a low ratio of surface area to volume, and a thick outer layer. Most leaves are evergreen, enabling rapid establishment of productivity as soon as conditions are suitable, but during the winter wax plugs seal the leaf pores and insulate the leaves, while the cell sap is chemically modified to lower its freezing point. Because in the northern part of their range the subsoil is always frozen, conifer roots are shallow. The inhospitable conditions reduce the available fauna, and hence wind is the main agent of both pollination and dispersal. Nutrient recycling is aided by symbiotic fungi growing on the roots or by periodic fires, to which some conifers are especially adapted, having resistant bark and cones that only open to release seeds following fires.

In the southern hemisphere there are no major landmasses at appropriate latitudes, and hence no extensive conifer forests. However, two conifer families are today found in the southern hemisphere: the Araucariaceae (kauris and monkey puzzles) and the Podocarpaceae (podocarps). These conifers often form large emergent trees in forests otherwise domi-nated by flowering plants, and some are the dominant trees in limited areas. Their leaves are broader than typical Pinaceae and show less xeromorphy. *Agathis* (kauri) trees – which reach heights of 75 m (250 ft) and girths of 23 m (75 ft) – were formerly dominant over much of northern North Island, New Zealand, but vast areas have been felled. Monkey puzzles are native to Chile and Argentina, and are widely grown as garden trees. The podocarps are found in Australasia, Africa and South America, and include over 125 species of trees and shrubs.

Not all conifers are high-latitude or high-altitude plants. Two members of the Taxodiaceae (the redwood family) grow in wetland or swamp settings. These are the swamp cypresses *Taxodium* of the Florida Everglades and *Glyptostrobus* from China. Ancient relatives of these conifers were important elements in the swamp vegetation of 65–10 million years ago, and formed much of the brown-coal reserves originating in this period.

The gnetophytes

The gnetophytes include only three living genera, all of which are strongly adapted to distinctive habitats and thus appear very different from one another. Their unifying features are rather obscure aspects of reproductive biology and anatomy. They share a number of features with the flowering plants, some of which – notably the flower-like organization of reproductive structures – are thought to indicate that the gnetophytes are the closest living relatives of the flowering plants. Gnetophytes appear to have emerged in the Triassic period (250–205 million years ago), but only three genera have survived competition from the flowering plants.

Of the 35 species of *Gnetum* most are tropical woody climbers (lianas), but a few are small trees. They have broad leaves and the seeds of some species form an important food resource in Southeast Asia. *Welwitschia* includes only one species, a bizarre desert plant endemic to southwestern Africa, especially the Namib Desert (▷ photo). *Ephedra* includes about 35 species of xeromorphic, much-branched shrubs. The stems are green and photosynthetic, while the whorled leaves are reduced to scale-like form. These plants occur in the Mediterranean area and other extratropical dry sites in America and China. *Ephedra* yields the alkaloid ephedrine, which is used as a drug in the treatment of asthma and to constrict blood vessels. MEC

Cycads (left) are normally scattered, but in eastern Australia cycads of the genus *Cycas* (shown here) grow densely in the understorey of eucalyptus forests. Among other unique features, cycads possess the largest male sperm known, the cells reaching 0·3 mm in diameter, and produce a chemical, cycasin, that is harmful to cattle if they eat the leaves.

Welwitschia mirabilis growing in the Namib Desert. Contrary to appearances, there are only two leaves, but, as these grow continuously, only the first 2 m (6·5 ft) are intact, the remaining 4 m (13 ft) or so becoming tattered. Individual plants can be up to 2000 years old.

The Emergence of Flowering Plants

Fossils over 200 million years old show that certain seed-bearing plants – the gnetophytes (▷ p. 39) and the extinct bennettites – evolved flower-like arrangements of reproductive organs long before the emergence of the flowering plants themselves. In contrast to the reliance on wind found in the conifers, or the dependence on free water of the spore-bearing plants (▷ p. 36), some gnetophytes, together with some cycads (▷ p. 38), also exploit animals (especially insects) for pollination – a characteristic otherwise only found in the flowering plants (▷ p. 42).

Although there are fundamental differences between the reproductive organs of these gymnosperm groups (▷ p. 38) and those of the flowering plants (the angiosperms; division Magnoliophyta), the similarity of organ arrangement is significant in terms of probable evolutionary pathways. The flowering-plant flower is the only flower form that is diverse and abundant today – flowering plants account for over 99% of all species of seed-bearing plants – and thus it may be said to be the most successful. The reasons for this lie with certain unique features of the organs themselves (▷ below) rather than in the mere organization into flowers or exploitation of animal pollinators. The rapidity of the flowering-plant life cycle compared to that of the conifers and other gymnosperms (▷ p. 39) is another important reason for their success.

Early flowering plants

The earliest evidence for the presence of flowering plants comes from fossil pollen grains dating from the early Cretaceous period, around 118 million years ago. Some of the pollen grains are similar to those of certain members of the subclass Magnoliidae (magnolias, etc.; ▷ classification box, p. 45). These early flowering-plant pollen grains represent only 1% of the pollen and spore assemblages in which they occur. As an individual flowering plant contains millions of pollen grains, it is probable that these first flowering plants were very rare. Within 20 million years, however, flowering-plant pollen represents up to 70% of pollen and spore assemblages. Furthermore, by this time a great variety of pollen forms are found, including those similar to the pollen of members of the Hamamelidae (e.g. witch hazels), and, slightly later, to that of members of the Rosidae (e.g. roses) and Dilleniidae (e.g. violets, primroses).

No fossils of flowers have been found with the oldest flowering-plant pollen – not surprisingly given the assumed rarity of the first flowering plants. Flowers, being delicate compound structures, are rarely fossilized, but in recent years many exciting finds have been made, providing considerable evidence of flower variety in the mid-Cretaceous period, some 95–85 million years ago (▷ illustrations). Much of this evidence comes from flowers that have been fossilized by conversion to charcoal during wild fires; charcoal, being almost pure carbon, is inert and resists subsequent decomposition. Although many modern flower-types are present at this stage, other forms did not arise until after the second major diversification of flowering plants at the Cretaceous/Tertiary boundary around 65 million years ago.

The fossil evidence – which is by no means unequivocal – suggests that the flowering plants first evolved in the lower latitudes, and subsequently spread into higher latitudes. These early flowering plants appear to have been small woody plants or herbs, which preferred disturbed habitats such as stream sides and open plains. They probably invaded such areas much like modern weeds invade cleared land; their ability to reproduce much more rapidly than the gymnosperms would have enabled them to become established in such disturbed habitats before gymnosperm plants and so outcompete them for light and nutrients. The first evidence for flowering plants attaining tree stature dates from around 70 million years ago, and it was not until after the Cretaceous/Tertiary boundary that flowering plants formed dense forests, and biomes (major ecological communities) similar to those we know today began to evolve.

The structure and function of flowers

In true flowering plants the complete or perfect flower (▷ diagram) consists of a series of four organs arranged in succession on the reproductive axis or *receptacle*. From base to apex these are sepals, petals, stamens and carpels. They may occur in a continuous spiral (e.g. in magnolias), sometimes with a series of gradual changes between organs (as in the sepals, petals and stamens of the water lily). Alternatively each organ occurs in a discrete, distinct whorl, often with fusion between the elements of the whorl (e.g. in heathers, dead-nettles and primroses).

FOSSIL FLOWERS

The range of forms shown by early fossil flowers about 95–85 million years ago. All are regular in shape, and many are very small. A notable exception is the large flower resembling a modern magnolia (2), which is also exceptional in having numerous free organs. All the others have a reduced number of free (3 and 4) or partly fused organs. Both 3 and 4 were unisexual (having only either male or female organs on the one flower), and probably exploited both winds and insects for pollination. All the others were bisexual.

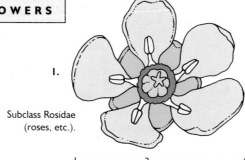

1.

Subclass Rosidae (roses, etc.).

— 3 cm —

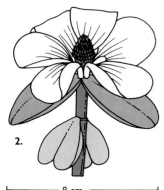

2.

— 8 cm —
Subclass Magnoliidae (magnolias, etc.).

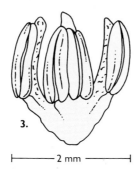

3.

— 2 mm —
Family Chloranthaceae, subclass Magnoliidae.

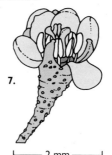

4.

— 5 mm —

Family Platanaceae (plane trees), subclass Hamamelidae. This is an inflorescence, i.e. a flower head with numerous individual flowers.

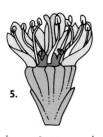

5.

— 2 mm —
Order Juglandales (walnut, hickory, etc.), subclass Hamamelidae.

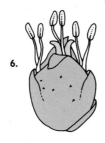

6.

— 1.5 mm —
Order Juglandales (walnut, hickory, etc.), subclass Hamamelidae.

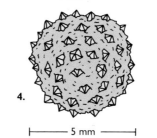

7.

— 2 mm —
Order Saxifragales (saxifrages, etc.), subclass Rosidae.

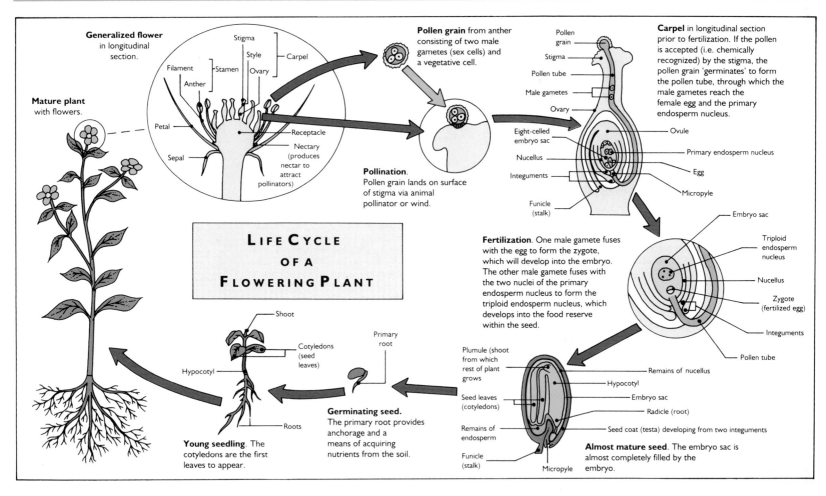

Generalized flower in longitudinal section.

Stigma
Style
Carpel
Ovary
Filament — Stamen
Anther
Mature plant with flowers.
Petal
Sepal
Receptacle
Nectary (produces nectar to attract pollinators)

Pollen grain from anther consisting of two male gametes (sex cells) and a vegetative cell.

Pollination. Pollen grain lands on surface of stigma via animal pollinator or wind.

Pollen grain
Stigma
Pollen tube
Male gametes
Ovary
Eight-celled embryo sac
Nucellus
Integuments
Funicle (stalk)

Carpel in longitudinal section prior to fertilization. If the pollen is accepted (i.e. chemically recognized) by the stigma, the pollen grain 'germinates' to form the pollen tube, through which the male gametes reach the female egg and the primary endosperm nucleus.

Ovule
Primary endosperm nucleus
Egg
Micropyle

LIFE CYCLE OF A FLOWERING PLANT

Fertilization. One male gamete fuses with the egg to form the zygote, which will develop into the embryo. The other male gamete fuses with the two nuclei of the primary endosperm nucleus to form the triploid endosperm nucleus, which develops into the food reserve within the seed.

Embryo sac
Triploid endosperm nucleus
Nucellus
Zygote (fertilized egg)
Integuments
Pollen tube

Shoot
Cotyledons (seed leaves)
Primary root
Hypocotyl
Roots

Young seedling. The cotyledons are the first leaves to appear.

Germinating seed. The primary root provides anchorage and a means of acquiring nutrients from the soil.

Plumule (shoot from which rest of plant grows)
Seed leaves (cotyledons)
Remains of endosperm
Funicle (stalk)
Micropyle
Remains of nucellus
Hypocotyl
Embryo sac
Radicle (root)
Seed coat (testa) developing from two integuments

Almost mature seed. The embryo sac is almost completely filled by the embryo.

Although frequently modified, the fundamental structure and function of each of the four floral organs is the same in all flowering plants.

Sepals are usually small, tough and green. Their function is one of protection, especially in the bud stage, when they enclose the unopened flower. Later, in the open flower, sepals help to prevent damage to mature ovules (⊳ below) by herbivores or nectar robbers attacking flowers from beneath.

Petals are usually large, delicate and colourful. Their function is one of advertisement and attraction of pollinators (⊳ p. 42).

Stamens consist of a stalk or filament that carries pollen-containing sacs called *anthers*. Their function is pollen production and release at the appropriate time and place. Pollen contains the male gametes (male sex cells).

Carpels consist of an ovary surmounted by a style and a stigma. The *stigma* is the specialized surface at which pollen is received and recognized (⊳ below). The *style* serves to position the stigma within the flower at an appropriate place to receive pollen from wind or animal pollinators. The *ovary* contains one or more *ovules*, within each of which is a female gamete (the ovum or egg). The ovary protects the ovules and subsequently the developing seeds, becoming modified to form the great variety of fruit that are so important for the dispersal of seeds (⊳ p. 43).

Features unique to flowering plants

Two obvious external features are unique to flowering-plant flowers. The first is the presence of the carpel enclosing the ovules; this is the key distinction between the gymnosperms (the conifers and their allies; ⊳ p. 38) and the angiosperms (the flowering plants). The second unique feature is the presence, on this carpel, of the receptive stigmatic surface. Taken together these two features permit control of pollen germination and hence of fertilization. Pollen must be accepted, i.e recognized chemically, by the stigma before the pollen tube 'germinates'. The pollen tube grows through carpel tissue to reach the ovule, where the male gametes are released to fertilize the egg. This control means that self-fertilization can be avoided and cross-fertilization or outbreeding enhanced or even enforced – with the resultant benefits in fitness and variation of progeny. In addition, the carpel tissue can develop into a variety of fruits, which enable seeds to be dispersed to a wide variety of sites. These features of flowers and fruits have led to the huge variety of animal/flowering plant interactions in pollination and dispersal (⊳ pp. 42–3).

In addition to these obvious external features, two aspects of internal organization are also unique to the flowering plants. As in the gymnosperms, the alternation of generations between sporophyte and gametophyte stages (⊳ p. 33)

is not at all obvious, but in the flowering plants both female and male gametophytes are even more reduced. The pollen grain (the male gametophyte) consists of only two gametes and one vegetative cell, while the embryo sac (the female gametophyte) typically contains only eight cells. This extreme reduction minimizes the energy required for gametophyte production and avoids excess waste if fertilization fails. The second unique internal feature – the double fertilization in the flowering plant ovule – adds to this efficiency. One male gamete fuses with the egg in the embryo sac to form the *zygote*, which develops into the new young sporophyte plant (the *embryo*). The other male gamete fuses with two nuclei from the embryo sac to form the *triploid endosperm nucleus*, which develops into the food reserve within the seed. Thus, this food reserve is only produced after fertilization when it is certain to be required, in contrast to other seed-bearing plants where the food reserve is produced prior to fertilization and hence can be wasted. In addition, the food package is in existence for a shorter time and is thus less vulnerable to predation. This vulnerability is further reduced by protection within the developing fruit. To deter predators, the developing fruit sometimes remains tough until it is ripe, or, if it is fleshy, often contains toxins. Ripeness – and therefore readiness for dispersal – is often signalled to the consumers that will disperse the seed by a colour change, such as green to red (⊳ p. 43). MEC

SEE ALSO

- THE AGES OF THE EARTH p. 4
- GENETICS p. 22
- PLANTS p. 30
- HOW PLANTS FUNCTION p. 31
- CONIFERS AND THEIR ALLIES p. 38
- POLLINATION AND SEED DISPERSAL p. 42
- FLOWERING-PLANT DIVERSITY p. 44
- ECOSYSTEMS pp. 200–21

Pollination and Seed Dispersal

The flowering plants are so diverse in form and have spread to such varied habitats that there can be little doubt that their evolutionary success has been undiminished by their immobility. Water and minerals are absorbed through the roots, while energy-giving sugars are produced by the process of photosynthesis in the leaves (▷ p. 31). Nevertheless, when it comes to reproduction, movement of pollen from one flower to another is necessary, and seeds must be dispersed if suitable habitats are to be exploited.

VARIETIES OF POLLINATION

Pollen produced in the loosely hanging male catkins of hazel (above left) is readily carried on the wind, and some will be caught on the sticky red stigmas of the female flowers.

The Shirley poppy (above right) has an open bowl-shaped flower, with its abundant pollen available to all comers, such as the flies feasting here.

In contrast, Rothschild's slipper orchid (right) relies exclusively on a single species of hover fly for pollination. The fly is attracted to the central part of the flower, which has a covering of short, white hairs, which are thought to resemble the eggs of aphids on which the fly's larvae normally feed. The fly lays its eggs here, and in doing so frequently falls off into the pouch of the orchid lying directly below. The only way out of the slippery pouch is by means of a ladder of hairs that leads the fly beneath the stigma and anthers; as it brushes past these, the fly collects fresh pollen and deposits any pollen picked up in the course of a previous escape.

The passion flower *Passiflora racemosa* (top) is red – a colour distinguishable by birds. Its anthers and style are so positioned that the former dust pollen onto hummingbirds seeking nectar, while the latter receives pollen from visiting birds.

For transport of both pollen and seeds, plants make use of wind, water and animals. In many cases, plant and animal have evolved together, so that each benefits from association with the other.

Pollination

Sexual reproduction in the great majority of flowering plants depends on the successful transfer of pollen from anther (male organ) to stigma (female organ) (▷ p. 40). Most flowers are hermaphrodite, having both of these organs, so there would seem to be little problem in achieving this. However, although some flowers are naturally self-pollinated, there is a genetic bonus to be gained from cross-pollination, which can maintain or even increase the vigour of the population.

Some plants have a genetic incompatibility system that prevents pollen functioning on stigmas of the same or closely related plants. A well-known example of this is the Cox's Orange Pippin, which requires another apple variety as a pollinator. Other plants have physical mechanisms that reduce the likelihood of self-pollination, such as a difference in the time at which anthers and stigmas mature. An absolute barrier to self-pollination is seen in those species that have separate male and female plants, such as willow and dog's mercury.

Animal pollination

A flower that depends on an animal to

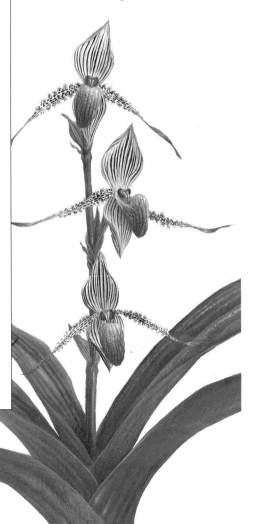

carry pollen to another plant needs first to advertise its wares. To this end many flowers are conspicuous and may also be scented. Some have large and colourful petals or sepals (or both), while in others there are *inflorescences* – masses of small flowers clustered together. Colourful foliage near the flowers may also assist or even take over this function. Flower colour is related to the colour vision of the pollinating animal. Bees do not distinguish red, but birds do; so red is a common colour for bird-pollinated flowers. Thus among the passion flowers *Passiflora racemosa* is red (▷ box), but the bee-pollinated *P. incarnata* is purple. Bees, on the other hand, are sensitive to ultraviolet light and will be attracted to a red poppy, not by the red, which they cannot see, but by the ultraviolet reflection. Flowers frequently have spots or lines on the petals, known as *nectar guides*, to direct the pollinator to the source of nectar (see below). Even flowers that are apparently of a single colour may actually have nectar guides, invisible to us but visible to a bee with its ultraviolet vision.

As well as acting as an advertisement, a flower must also provide – or seem to provide – a reward. The most common rewards are a share in the pollen itself, which is rich in protein, and *nectar*, a sugary liquid that serves only to attract the interest of a pollinator. A few plants, such as the yellow loosestrife, supply oil as a reward, while the fig provides a place to rear young (▷ box opposite). In a number of cases the pollinator is deceived, for the promise of a reward is not fulfilled. A flower that does not produce nectar may mimic one that does, as in the case of the orchid *Epidendrum*, the coloration of which is similar to that of another flower (*Lantana*). Species such as the cactus-like *Stapelia* and the massive tropical *Rafflesia* (▷ p. 201) have flowers that look and smell like carrion, so that flies are attracted to visit and lay eggs. Some plants, such as the hammer orchid, take advantage of the sexual urge of insects by mimicking the shape and colouring of the female of the species, and even produce the appropriate pheromone to attract males.

For successful pollination, the pollen brought by the pollinator must be from the same species of plant, and this in part explains the great diversity that is to be seen in floral structure. Some flowers are quite open, as in the case of poppies, where flies and beetles can easily feast on the abundant pollen (▷ box). Others have a nectar reward accessible only to particular animals, such as bees that can push their way into the flower. An individual honeybee will be constant to a particular flower species, so assisting pollination, but at the same time it gains efficiency in nectar collection by learning to harvest from a particular flower. Flowers with long corolla tubes or long spurs are geared to long-tongued insects such as moths and butterflies. Others are adapted for pollination by birds (notably hummingbirds) or bats, or even by non-flying mammals such as the honey possum.

Wind pollination

Flowers that rely on the wind for pollination naturally do not need to expend energy on attracting and rewarding animal pollinators. However, they must produce vast quantities of pollen to saturate the air and ensure that any sticky stigma exposed to air currents will pick up pollen of the same species (⊳ box). Wind pollination may seem a chancy business, but its success is evident enough, not least from the vast quantities of birch seeds resulting from successful pollination that may readily be found blown into lofts and window casements.

Grasses are also wind-pollinated. Their anthers, dangling on long filaments outside the flowers, shed pollen that is carried to feathery stigmas held in a similar position on other plants of the same species. Unfortunately humans also pick up pollen from grass and other wind-pollinated plants on the mucous membranes of the nose, giving rise to hay fever in some people.

Seed dispersal by wind and water

Just as with pollination, seed dispersal may be achieved by wind or water, or by animals. Seeds contain the embryo (the young sporophyte plant; ⊳ p. 33), and most also contain a food reserve to sustain the early growth of the plant. Thus even the smallest seed is heavy compared to a pollen grain.

Modifications in the shape and surface sculpture of wind-dispersed seeds help to slow down the rate of fall and so to increase the chances of being caught on an air current. Such features include thin, papery wings and fine hairs, which may be developed from various parts of the seed, fruit or flower. In old man's beard, the hairy appendages are derived from the style of the carpel, whereas in willowherb the hairs arise from the surface of the seed coat (⊳ photo). Seed-coat hairs are exploited by man as cotton from *Gossypium* seeds. The wings on maples and birch are derived from the carpel, whereas those on lime are modified floral bracts.

Some of the smallest seeds (dust seeds) are little bigger than dust particles, weighing only a few micrograms. These occur in the heathers and orchids. Although easily wind-dispersed, they carry little or no food reserves and rely on an association with a fungus to provide nutrients during germination. Numerous seeds must be produced, as many will land at sites where the appropriate fungus is lacking. This situation also creates problems in rearing orchids from seed. For this reason tissue-culture methods are particularly important for commercial and conservation purposes, since such methods allow many new plants to be obtained from one original plant without having to use seeds.

If the dust seed is at one extreme of seed size, the Seychelles or double coconut is at the other (⊳ photo). While this simply falls to the ground, edible coconuts are dispersed only by floating on sea water, the fibrous outer layers preventing penetration of salt water and trapping air to give buoyancy. These examples illustrate an important phenomenon in seeds. The smallest have the best dispersal potential but the least chance of successful establishment of the young plant, whereas the largest, with the embryo well developed and a large food reserve, have limited dispersal potential but the best chance of establishment. Broadly speaking, the smaller seeds are typical of colonizing species (so-called 'r-strategists'), such as those we call weeds in our gardens. These establish themselves in open sites where competition is limited. The larger-seeded climax species ('k-strategists'), such as forest trees, grow in closed settings, where the habitat is already fully colonized.

Dispersal by animals

Variation in seed size also influences animal dispersal. Animals disperse seeds by carrying them externally (on fur, feet, or beaks) or internally (in the gut, passing out in the faeces). For this reason, large seeds can only be dispersed by proportionately large animals. Seeds that become attached to animals have hooks or spines or exude sticky substances, but generally rely on a chance encounter with an animal.

Internally dispersed seeds, on the other hand, must offer a reward to an animal in return for dispersal. This reward comes in the form of sweet and nutritious fleshy tissues, which are rich in energy. The flesh is usually brightly coloured with a pleasant, attractive aroma when ripe, and it too may be developed from all parts of the flower. In a pineapple, it is derived from an inflorescence, or whole head of flowers, in rose hips from the floral cup, in strawberries and figs from the receptacle, and in tomatoes, grapes and plums from the carpel wall.

The size and often the number of seeds within a fruit may vary according to its particular dispersal strategy. Tomatoes and gooseberries are examples of *berries*, which usually contain many small seeds. When a large primate such as man eats a tomato, the seeds remain viable after passing through the gut (as evidenced by the growth of tomatoes on sewage farms), and thus may be dispersed in the faeces. In contrast, *drupes*, such as peaches, cherries and plums, contain one or a small number of relatively large seeds ('stones') protected by a hard outer layer. As swallowing a plum stone is an unpleasant experience, an animal will tend to gather the plum and remove it to a safe place, where it eats the fleshy part and discards the stone, thus ensuring its dispersal. However, a small mammal would

Rosebay willowherb (above right) has plumed seeds that are readily picked up and carried by the wind; these seeds often establish themselves in ground cleared by fire, hence the plant's other name – fireweed. The seeds of the **rose** (right) are contained in brightly coloured fleshy hips; these provide a meal for birds, which disperse the seeds in their faeces or simply scatter them as they eat. The enormous **Seychelles** or **double coconut** (top) is the largest of all seeds: it weighs up to 18 kg (40 lb) and is said to be 'gravity-dispersed' – it simply falls from the tree to the ground beneath.

be an inappropriate agent of dispersal in these cases, since it would probably crush the tomato seeds, thus killing them, and would not be able to carry the plum away from the parent plant.

Animal hoarding habits are also exploited by plants. Squirrels and jays both bury or cache nuts, such as hazel, in times of plenty for use in times of scarcity. Although a nut has no flesh, it has a large stored food reserve, which may be eaten by the animal, thus killing the seed. On the other hand, if seeds are not needed or if the hiding place is forgotten, they are ready planted in suitable new growth sites. This is a good example of the frequent compromise between dispersal and predation, in which one or other partner may benefit at different times.   JWC/MEC

SEE ALSO

● PLANT PHYSIOLOGY pp. 31–3
● THE EMERGENCE OF FLOWERING PLANTS p. 40
● FLOWERING PLANT DIVERSITY pp. 44–7

THE FIG AND THE FIG WASP

The young fig is a hollow, pear-shaped structure, with many tiny flowers covering the inner surface. The only access to these flowers is through a small opening at one end (the apex), by which a female fig wasp enters with difficulty, carrying pollen from another fig. Inside the plant there are separate male and female flowers; when the wasp arrives, only the latter are mature, their stigmas forming a carpet upon which she walks. The wasp deposits pollen on the stigmas and also lays eggs in the ovaries of the flowers, but not in all of them, for some have styles longer than her ovipositor, so that their ovaries are beyond her reach. Thus the long-styled flowers produce seed, while the short-styled ones form galls and produce a new generation of fig wasps.

The wingless male wasps emerge first and use their powerful mandibles to gnaw their way through to the young females, which they proceed to fertilize while still within their gall flowers. The males then act in concert to gnaw a tunnel through the wall of the fig to the outside world, before retreating within the fig to die. By the time the females emerge, the male flowers are mature; the female wasps gather pollen and then depart through the tunnel made for them by their mates, to seek young figs in which to lay their eggs.

Flowering Plant Diversity

The flowering plants are the most diverse, widespread and abundant group of multicellular plants living today. It is estimated that there are between 250 000 and 300 000 species – and probably many more that have not yet been collected from the remote areas of the tropical rainforests. Flowering plants grow in almost all areas of the world, under a wide variety of conditions. On land they are only absent from extremely cold habitats, such as permanent snowfields, or extremely hot habitats, such as geysers (where the algae alone among the plant world are able to survive); they are also excluded from very high latitudes and very high altitudes (where moss-lichen vegetation prevails). Almost everywhere else in the world, apart from the oceans and the great northern coniferous forests, the vegetation is dominated by the flowering plants.

To cope with this wide range of habitats, and to compete effectively with other plant species, the flowering plants have evolved an enormous number of adaptations. The flowering plants are also of crucial economic importance: not only are they a vital food source, but they also supply many other commercially useful products.

Types of flowering plant

Various terms are commonly used to describe a variety of different types of life cycle in the flowering plants. *Annuals* are plants such as marigolds and petunias that complete their life cycles from seed germination to death in less than one year. *Biennials* – such as foxgloves, cabbages and carrots – complete their life cycles in more than one year but in less than two; flowering and seed production usually occur in the second year. *Herbs* or *herbaceous perennials* are plants such as daisies and dandelions that lack woody cells and die back to the roots (or to other perennating organs; ⊳ below) at the onset of frost or drought; on the return of spring or rain new growth is produced above ground. *Woody perennials* – trees, shrubs and woody climbers – take longer than two years to complete their life cycles, and do not die back to their roots.

These terms are not used as the basis of classification, however, as it is common to find more than one type of life cycle within a particular taxonomic group; for example, the family Rosaceae includes apples (woody perennials) and strawberries (herbaceous perennials), and the Fabaceae includes peas (annuals) and brooms (woody perennials).

Biennials and herbaceous perennials rely on a variety of underground food-storage organs – collectively known as *perennat-*

The baobab is one of the most curious-looking of all trees, with a barrel-like trunk that may have a diameter up to 9 m (30 ft). Two related species occur in Africa and Australia.

ing organs – to survive through frost or drought. *Taproots* (e.g. dandelions, carrots) are large single roots growing vertically downwards, with smaller lateral roots growing outwards from them. *Bulbs* (e.g. onions, daffodils) are compact underground stems bearing fleshy leaves, enclosing an underdeveloped flower or buds. *Corms* (e.g. crocuses, gladioli) are swollen stem bases surrounded by scale leaves, with one or more buds. *Rhizomes* (e.g. irises) are underground, horizontal stems. *Tubers* (e.g. potatoes) are much-swollen underground stems that support the sprouting of buds. In many cases these perennating organs are also used for vegetative reproduction (⊳ p. 33).

Epiphytes and climbers

There are many plants that rely on other plants without damaging them. *Epiphytes* are plants that grow on other plants – for example, on the branches of trees – in order to obtain a position nearer to the light. They rely almost entirely on rain water as a source of minerals, taking nothing from their 'host' plants, and often have specialized absorptive cells on the surfaces of their leaves and roots. Quite a few spore-bearing plants (including many ferns) are epiphytes, and many flowering-plant families – notably the orchids and bromeliads – have epiphytic species, mostly in tropical regions.

Another strategy for reaching the light before competitors is to adopt a climbing habit. Although a few ferns are climbers, as are most species of the gymnosperm genus *Gnetum* (⊳ p. 39), most climbers are found among the flowering plants. Flowering-plant climbers include both herbaceous and woody species, the latter often called lianas. Flowering plants climb by means of touch-sensitive stems or leaves, many of which are specially modified in the form of tendrils. These organs twine around objects with which they come into contact, such as tree trunks and other climbers. Mature lianas often have a large woody stem, part of which hangs freely in the understorey of the forest, while the leafy branches intermingle with the tree canopies high above. Two of the best-known groups of climbers are the rattan palms and the grape-vine family (Vitaceae).

Trees

Most forest and woodland communities are dominated by flowering-plant trees – the only exception being the coniferous forests of high latitudes and altitudes (▷ p. 208). Most of these trees – such as beeches, oaks, figs, mahoganies and eucalyptuses – are dicotyledons (▷ box, p. 45), and are often referred to as hardwood or broadleaved trees. Some are evergreen, while others are deciduous (▷ pp. 200 and 207). All broadleaved trees possess an extensive growth of wood and bark, and many are highly prized for their timber (more durable than the 'softwoods' of conifers). This has led to widespread deforestation, especially in tropical regions (▷ p. 201). Broadleaved trees also provide other commercially valued products in addition to timber, most notably fruit (▷ below), spices (bay, cinnamon, etc.), cork (from the cork oak), and latex for making into rubber (from the rubber tree).

The major group of monocotyledonous trees are the palms (family Arecaceae, formerly Palmae), which grow mainly in the tropics as understorey plants (▷ illustration, p. 201). Palm trees do not possess wood or bark, but are supported by other tissues strengthened by lignin, and by various other structural characteristics. Commercially valued products include sago, dates, coconuts, edible and soap-making oils, fibres and waxes. The fibrous fruit wall of coconuts is now being exploited as a renewable alternative to peat (▷ p. 217) for horticultural purposes.

Grasses

The grasses – of which there are about 9000 species – constitute the monocotyledonous family Poaceae (formerly Graminae). This group includes the cereal or grain crops that provide staple foods for much of the world's population: wheat, rice, millet, maize (corn), barley, rye and oats. In addition, sugar cane is the source of half of the world's sugar. Sugar cane is also fermented to produce ethanol (an alcohol), which is used to power most motor vehicles in Brazil.

Grasslands dominate large areas of both the tropics and the temperate regions (▷ pp. 202 and 206). In grasses typical of grassland habitats the growing points of the stems are below ground and are protected by encircling leaves or leaf sheaths. The leaves themselves grow, not

THE CLASSIFICATION OF FLOWERING PLANTS

Plants are classified into the same hierarchical groupings as other forms of life, although the term 'division' is used instead of 'phylum' (▷ p. 18). Thus the genus Magnolia belongs to the family Magnoliaceae in the order Magnoliales, subclass Magnoliidae, class Magnoliopsida, division Magnoliophyta (the flowering plants). There are standard endings that indicate the status of the grouping: -phyta (division), -opsida (class), -idae (subclass), -ales (order), -aceae (family). The endings for genus and species are not standardized.

The detailed classification of the flowering plants is based on a variety of elements: growth form, vegetative biology, reproductive biology (including floral organization and the structures of pollen, ovule, seed and fruit), wood anatomy, and chemical characteristics. A complete cladistic classification of the flowering plants has yet to be achieved, although the major groupings are generally accepted by most authorities.

The division Magnoliophyta (the flowering plants or angiosperms) is divided into two classes, the Magnoliopsida (dicotyledons) and the Liliopsida (monocotyledons). All the subclasses are listed below, but only a selection of representative families are given for each, with representative plants in brackets.

Class LILIOPSIDA

The monocotyledons, comprising less than 30% of flowering-plant species. Distinguishing features include: a single seed leaf (cotyledon) in the embryo and young plant; parallel veins in leaves (which are usually long, narrow and pointed); floral organs in threes; and fibrous root systems. Most monocots are herbaceous; those that are trees (e.g. palms) lack wood and therefore cannot support a branching canopy. There are five subclasses:

Subclass ALISMATIDAE: Alismataceae (water plantains), Potamogetonaceae (pondweed).
Subclass ARECIDAE: Arecaceae (formerly Palmae: palms).
Subclass COMMELINIDAE: Commelinaceae (tradescantias), Bromeliaceae (bromeliads), Poaceae (formerly Graminae: grasses – including cereals – bamboo, reeds).
Subclass ZINGIBERIDAE: Zingiberaceae (gingers).
Subclass LILIIDAE: Liliaceae (lilies, tulips, bluebells), Alliaceae (onions, garlic), Amaryllidaceae (daffodils, snowdrops), Iridaceae (irises, crocuses, gladioli), Agavaceae (agaves), Asparagaceae (asparagus), Orchidaceae (orchids).

Class MAGNOLIOPSIDA

The dicotyledons, comprising over 70% of flowering-plant species. Distinguishing features include: two seed leaves (cotyledons); leaves with branching main veins connected by net-like venation; floral organs in fours or fives; and a persistent primary root system. Those that are trees are woody and can therefore support extensively branched canopies. There are six subclasses:

Subclass MAGNOLIIDAE: Magnoliaceae (magnolias), Lauraceae (laurels, bay, avocado), Nymphaeaceae (water lilies), Ranunculaceae (buttercups, clematis, delphiniums, anemones), Papaveraceae (poppies).
Subclass HAMAMELIDAE: Hamamelidaceae (witch hazels), Platanaceae (plane trees, known as sycamores in the USA), Ulmaceae (elms), Urticaceae (nettles), Fagaceae (oaks, beeches), Betulaceae (birches, alders, hazels), Juglandaceae (walnuts).
Subclass CARYOPHYLLIDAE: Caryophyllaceae (pinks), Chenopodiaceae (beets, spinach), Cactaceae (cacti), Polygonaceae (dock, knotgrass, buckwheat).
Subclass DILLENIIDAE: Paeoniaceae (peonies), Theaceae (camellias, tea plant), Violaceae (violets, pansies), Begoniaceae (begonias), Brassicaceae (formerly Cruciferae: cabbage, turnip, rape, mustard, wallflower), Salicaceae (willows, poplars, aspen), Ericaceae (heathers, rhododendrons, bilberry, cranberry), Primulaceae (primroses), Tiliaceae (lime trees), Euphorbiaceae (euphorbias, spurges).
Subclass ROSIDAE: Rosaceae (roses, strawberry, blackberry, raspberry, apple, cherry, plum, pear, hawthorn, rowan), Grossulariaceae (currants, gooseberries), Hydrangeaceae (hydrangeas), Saxifragaceae (saxifrages), Mimosaceae (acacias, mimosas), Fabaceae or Papilionaceae (formerly part of the Leguminosae, the legume family: peas, beans, soybean, peanut, clovers, vetches, gorse, brooms, lupins, lucerne), Myrtaceae (eucalyptuses), Rutaceae (citrus fruits: orange, grapefruit, lemon, lime), Hippocastanaceae (horse chestnut), Aceraceae (maples, sycamore), Geraniaceae (geraniums), Araliaceae (ivies, ginseng), Apiaceae (formerly Umbilliferae: carrot, parsley, coriander, caraway, hog-weeds, hemlock, celery, parsnip, fennel, lovage), Aquifoliaceae (hollies), Vitaceae (grape vine, virginia creeper), Oleaceae (olive, ash).
Subclass ASTERIDAE: Asteraceae (formerly Compositae: daisies, marigold, yarrow, chrysanthemum, thistles, dandelion, lettuce), Caprifoliaceae (honeysuckle, elder), Gentianaceae (gentians), Polemoniaceae (phloxes), Convolvulaceae (convolvulus, morning glory), Boraginaceae (borage, forget-me-not), Solanaceae (potato, tobacco, tomato, nightshades), Lamiaceae (formerly Labiatae: mints, dead-nettles, basil, marjoram, thyme), Campanulaceae (bellflowers: campanulas, harebell).

A typical monocotyledon, *Watsonia* (family Iridaceae), growing in the Drakensberg Mountains, South Africa.

A typical dicotyledon, the Californian poppy. The poppy family Papaveraceae has a widespread distribution in Europe, Asia and the Americas.

from their tips, but from zones within the leaf, and this means that the grazed upper portions of the leaf are simply replaced by growth from below – as anyone will know who has a lawn to mow. Grasses can spread rapidly by seeds or by sending out rhizomes (a form of vegetative reproduction; ⊳ p. 33), and their fibrous root systems are good soil binders. Grasses contain deterrents to grazers in the form of silica deposits in their surface cells, and the grinding effect of these helps to explain the presence of very high-crowned teeth in grazing mammals (⊳ box, p. 110). Humans have extensively exploited the special relationship between grasses and grazers such as cattle and sheep (⊳ p. 212).

Not all grasses form grassland communities. Many occur as scattered plants or as ground cover in woodlands, and some, like cord grass, form extensive swards in salt marshes. Others like bamboos can grow as trees up to 40 m (130 ft) tall and form forests.

Green vegetables and herbs

Many flowering plants are valued as sources of vegetables and flavourings (herbs). The most important vegetable plants belong to the family Brassicaceae (formerly Cruciferae); cultivated forms of the wild cabbage (*Brassica oleracea*) not only yield cabbage, but also broccoli, sprouts and cauliflowers. Spinach is a member of the Chenopodiaceae or goose-

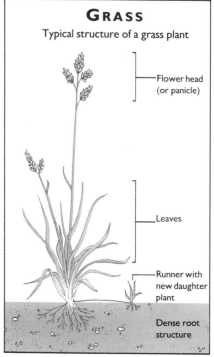

GRASS

Typical structure of a grass plant

— Flower head (or panicle)

— Leaves

— Runner with new daughter plant

Dense root structure

foot family, while lettuces, cress and water cress belong to the daisy family (Asteraceae, formerly Compositae). The lily family (Liliaceae), well known for beautiful ornamental plants, also includes the genus *Allium*, species of which produce onions, leeks, chives and garlic.

Many aromatic plants are found within

the dead-nettle family (Lamiaceae, formerly Labiatae). These include mints, sage, marjoram, oregano, thyme, rosemary and basil. The flavours and aromas are due to the presence of essential oils secreted by glands on the leaves, which may play a part in defending the plant from insect predation. Several aromatic seeds are found among species of the umbel family (Apiaceae, formerly Umbelliferae), including caraway, coriander and cumin. Other important flavourings are derived from tubers and roots, for example, ginger and turmeric are obtained from the rhizomes of members of the tropical family Zingiberaceae.

Fleshy fruits

Many of the fleshy fruits of flowering plants that encourage animal dispersers (⊳ p. 43) are, predictably, also attractive to human consumers. Such fruits include many members of the rose family (Rosaceae), such as plums, peaches, apricots and cherries (*Prunus* species), blackberries and raspberries (*Rubus* species), strawberries (*Fragaria*), apples (*Malus*), and pears (*Pyrus*). Another important fruit-producing family is the Rutaceae or citrus family, which includes oranges, tangerines, lemons, limes and grapefruit, all species or hybrids within the genus *Citrus*. Bananas are fleshy fruits from the tropical monocotyledon genus *Musa*, which has huge, leathery, strap-shaped leaves. The wild banana has small fruits with numerous hard seeds; the cultivated seedless form originated following the chance discovery of a sterile hybrid plant. Grapes are probably one of the most commercially important fruits, both for eating and for wine production. Grapes are produced by the woody vine *Vitis vinifera* and its many cultivars (family Vitaceae). Dried grapes – raisins, currants and sultanas – are also commercially important. Pineapples – which come from the tropical American plant *Ananas comosus* (a member of the Bromeliaceae) – are multiple organs derived from the entire inflorescence, all of which becomes fleshy in fruiting condition. Numerous tropical flowering plants produce fleshy fruit that are exploited locally, and many of these, such as mangoes, lychees and custard apples, are gradually becoming more widely available.

Other families are very important producers of vegetable fruit. The Cucurbitaceae is a large family of climbing and trailing plants growing mainly in warmer regions, and species include cucumbers, marrows, courgettes (zucchini), pumpkins and other gourds, as well as the melons. One other very important family is the Solanaceae, amongst whose products are the potatoes, a dietary staple (along with bread) in northern temperate regions. The potato is a tuber, but other members of the family produce edible fruit such as tomatoes, aubergines (egg plant), and many varieties of peppers.

Tea, coffee and cocoa

The tea plant (*Camellia sinensis*; family Theaceae) is a small tree that grows wild

---

ORCHIDS

The orchids (family Orchidaceae) are one of the largest of flowering-plant families, with about 750 genera and 18 000 species. They exhibit a wide variety of flower forms, and many are grown as ornamentals. Almost all have a specialized association with a particular animal pollinator: flowers of some species even mimic female insects with which male insects try to mate, thus transferring pollen, while others actually trap insects in the flower. Pollen grains are transferred in the complete anther clump, and each ovary contains thousands of ovules, with the result that one pollination by a single insect can produce thousands of new plants. This partly accounts for the large number of orchid species. Many orchids live as scattered epiphytes (⊳ text) in dense tropical rain forests, while others are lianas, for example *Vanilla*, which is used for flavouring. There are also many ground-dwelling species, such as the bee and fly orchids.

An epiphytic orchid of the genus *Cattleya*.

CARNIVOROUS AND PARASITIC PLANTS

Not all plants obtain their essential mineral requirements through the normal methods of uptake from the soil (▷ pp. 31–2). Some plants, including the sundews, Venus flytrap, pitcher plants, butterworts and bladderworts, have turned to a carnivorous way of life, trapping and digesting insects and other arthropods. Many of these plants grow in low-nutrient environments on acid soils, so their prey may provide their main source of nitrogen, phosphorus and certain trace elements.

Insects are attracted to the carnivorous plants by various visual stimuli, including colour patterns, and pitcher plants also encourage prey by providing nectar secretions as a potential reward. The mechanisms for trapping prey vary considerably. The simplest form of trap – as found in sundews and butterworts – is the adhesive trap, where glandular hairs on the leaf surface produce a secretion to which the insect sticks and cannot escape. The Venus flytrap has a 'snap trap', a two-lobed leaf that can snap closed when triggered by long, touch-sensitive hairs on the leaf surface. Pitcher plants live up to their name in having leaves where part or all of the leaf is modified to form a deep pitcher-shaped container into

which prey falls. The prey is prevented from escaping by downwardly directed hairs and surface waxes. The bladderworts also have a pitcher-like trap in the form of a 'bladder' with a door that opens inwards. The door opens suddenly when triggered and the trap walls, released from tension, expand, inflating the trap and sucking the prey inside; the door then snaps closed and prevents the prey swimming out. Whatever the mechanism, once trapped the prey is slowly digested by the release of enzymes from special secretory glands.

Just as a few plants have adopted carnivorous habits to obtain essential minerals, so a few others – such as mistletoe and the giant-flowered *Rafflesia* – have found they can obtain nutrients by parasitizing other plants. A native of the jungles of Southeast Asia, *Rafflesia* – the 'stinking-corpse lily' – parasitizes the roots of certain vines. Its flower, borne at ground level, can reach up to 91 cm (3 ft) in diameter, and attain a weight of 7 kg (15 lb). The flower smells of carrion, which attracts the flies that are thought to pollinate it (▷ photo, p. 30).

Many parasites have a climbing habit, such as this *Cassytha filiformis* in the Pacific island group of Kiribati.

A mossy pitcher plant at around 2700 m (9000 ft) on Mount Kinabalu, Borneo.

SEE ALSO

● PLANTS p. 30
● HOW PLANTS FUNCTION p. 31
● THE EMERGENCE OF FLOWERING PLANTS p. 40
● POLLINATION AND SEED DISPERSAL p. 42
● CLIMATE AND VEGETATION p. 198
● ECOSYSTEMS pp. 200–21

from China to India. China types have narrower leaves than Indian types. The trees are pruned to encourage leaf production, and the terminal leaves on shoots are picked then fermented, dried and broken to produce leaf teas. Green teas are produced by heating leaves at an early stage to prevent fermentation. Varieties of teas come from different growing areas, some of the best coming from high-altitude sites. The tea leaves contain tannins and the alkaloid caffeine, which provide the body and the stimulating effect of the drink.

Coffee and cocoa are both produced from

fruits rather than leaves. Coffee (genus *Coffea*, family Rubiaceae) grows naturally in tropical Africa, but much of the world's production comes from tropical American countries such as Brazil and Colombia. The coffee fruit is a berry, and this is pulped (or dried and the outer covering removed) to release the hard seed or 'bean'. The beans must be roasted to develop the characteristic aroma and flavour. Like tea, coffee contains caffeine.

Cocoa comes from the cacao tree (*Theobroma cacoa*, family Sterculiaceae), a small tree native to the American tropics but now cultivated elsewhere, especially

in West Africa. The large fruits are produced on the trunk, and contain an edible pulp within which are hard seeds. The pulp and seeds are fermented and the seeds dried to form the cocoa 'beans', which have a high content of fat called cocoa butter. Most of this fat is removed in the production of drinking cocoa, but the drink retains a high food value and is also a mild stimulant owing to the presence of caffeine and a related alkaloid, theobromine. For chocolate manufacture additional cocoa butter is added as well as other substances such as sugar and milk.   MEC

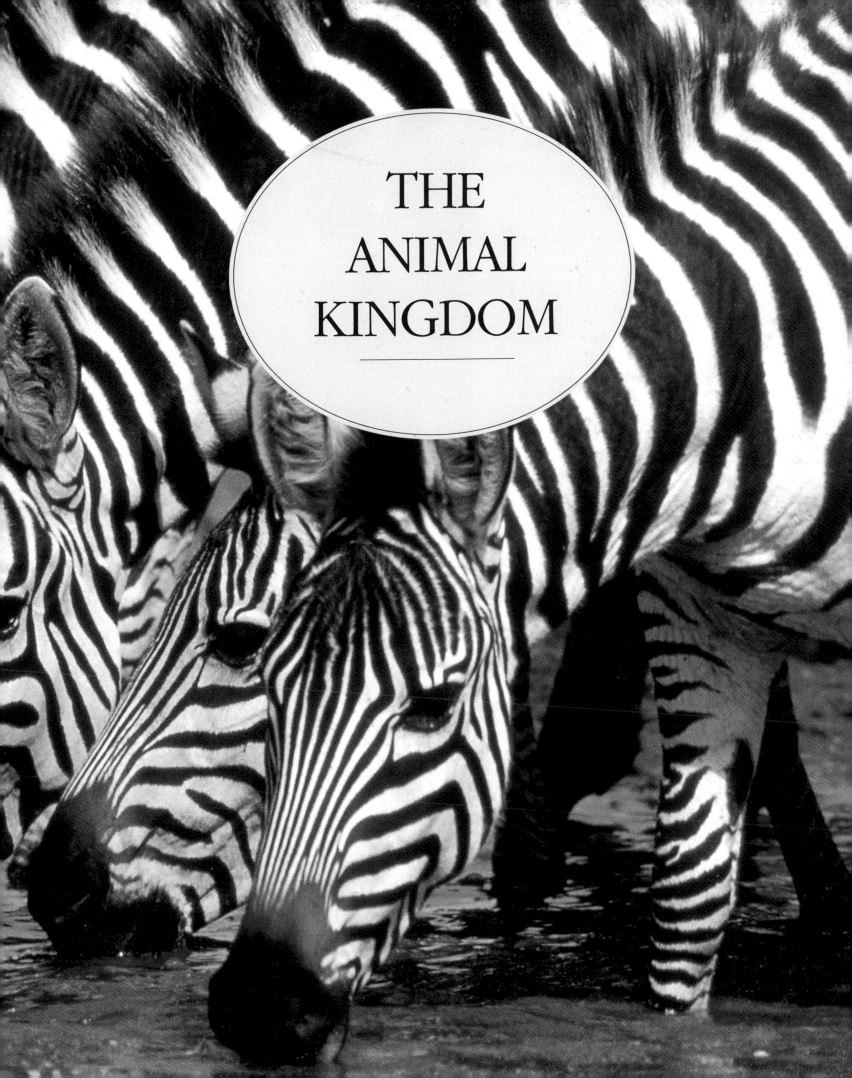

THE
ANIMAL
KINGDOM

Animals

In spite of the pre-eminent importance of the animal kingdom and the almost unbounded interest in it among professional scientists and amateurs alike, there is surprisingly little agreement about the precise features that characterize animals and consequently about the exact boundaries of the animal kingdom itself. The kingdom Animalia is conventionally restricted to multicellular forms, but there is a multitude of single-celled protozoans, generally classified with other unicellular life in the kingdom Protista (⊳ p. 26), that show all or most of the features normally regarded as defining animals and that might reasonably be classified with them. Even with this restriction, the animal kingdom is unsurpassed in size and diversity. It ranges from simple sponges and corals, through a host of more complex invertebrates such as insects and squid, to vertebrates, which include the most advanced animals of all, the birds and mammals.

We can only guess at the number of species contained in this enormous kingdom. Perhaps one and a half million species have been scientifically described, but this must represent no more than a fraction of the animals that actually exist. The majority of the species described – over a million – are insects, but some estimates suggest that the true figure for this group alone may be closer to 30 million. It is doubtless true that few if any large animals remain to be discovered, although the ocean depths might yet hold some surprises, and some large land mammals, such as the okapi and the giant forest hog, remained unknown to science until the early 20th century. But this takes no account of the millions of smaller, less conspicuous species that have so far eluded discovery.

Distinguishing features

Although animals can be distinguished from plants at a fundamental level by the absence in the former of cell walls (⊳

SEE ALSO

- EVOLUTION pp. 12–17
- ORDER IN THE LIVING WORLD p. 18
- ANIMAL GROUPS pp. 52–145
- ANIMAL BEHAVIOUR pp. 148–73
- ANIMAL PHYSIOLOGY pp. 174–91

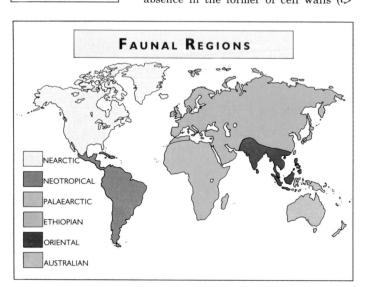

FAUNAL REGIONS

- NEARCTIC
- NEOTROPICAL
- PALAEARCTIC
- ETHIOPIAN
- ORIENTAL
- AUSTRALIAN

Quintessential animal: with sophisticated nervous and sensory systems, great mobility, and a predatory feeding habit, lions exemplify the principal features by which animals are distinguished.

illustration, p. 21), perhaps the most significant difference between them lies in the manner in which they obtain food. While virtually all plants are able to harness the energy of the sun to synthesize the complex organic compounds they require from simple inorganic molecules (⊳ photosynthesis, p. 33), animals rely on ready-made sources of such compounds – in other words, they are dependent on plants, either eating them directly or feeding on the body tissues of animals that have done so (⊳ p. 194).

Other differences between plants and animals can be related to this central distinction. Although some aquatic animals such as sea anemones and clams are more or less immobile and can rely on their food coming to them, the majority of animals must actively find it or seize it: parasitic worms search out the appropriate hosts, butterflies seek out nectar, cheetahs pursue antelope. Unlike plants, therefore, which can function normally while literally rooted to the spot, animals generally require at least some degree of mobility. Animals have devised a great diversity of locomotive systems (⊳ p. 176); in most cases these involve a system of musculature working against some form of rigid skeleton, whether external (as in insects and other arthropods), internal (as in vertebrates), or hydrostatic (as in earthworms).

With mobility comes a range of other requirements that are not shared by plants.

Movement in multicellular animals requires a great deal of coordination between different body parts; in all but the simplest animals, such as jellyfishes and other coelenterates, this entails a reasonably sophisticated system of nervous control (⊳ p. 182). Furthermore, an animal needs to move not only in a coordinated fashion but also in a particular direction; thus it requires some form of sensory apparatus tied in with its nervous system. Although plants can detect various environmental stimuli and react accordingly (⊳ p. 33), animals have evolved an astonishing range of systems by which they can perceive what is happening in the world around them (⊳ p. 188).

These various differences between plants and animals naturally have a profound effect on the overall structure or 'architecture' of the two groups. Most animals have some form of digestive tract or cavity in which the complex molecules on which they feed are broken down (⊳ p. 178), and this generally forms a central point or axis around which the rest of their body cells are arranged. Likewise, all but the simplest nervous systems necessitate some central coordinating point – i.e. some form of 'brain' – and this in turn leads to the development of a distinct

Higher and lower animals

Throughout the Middle Ages and beyond, man's understanding of his fellow animals was often founded on the idea of a *scala naturae* or 'ladder of nature'. According to this notion, which goes back at least to Aristotle, animals can be arranged in a neat linear series, with 'lower' animals such as sponges and amoebas on the bottom rungs, 'higher' animals such as mammals on the top ones, and all other animals at various levels in between. This idea of course runs counter to the theory of evolution, in which the pattern of relationships between different animals, living and extinct, forms a branching tree rather than a ladder. As such any living species is at the tip of a 'branch', however long the branch itself may be. Indeed, far from being 'inferior', as implied by the ladder concept, an animal at the tip of a long branch, such as the tuatara, the coelacanth or a horseshoe crab, is one that has been so successful that it has survived with little change over many millions of years.

In the evolutionary 'tree of life', greater complexity or sophistication often comes with progress along a branch, but this need not be so. As species evolve over time, they sometimes 'degenerate' or lose some of the adaptations seen in their ancestors; this is often the case when free-living animals adapt to a parasitic way of life. When we speak of a 'higher' animal, we should mean no more than that it shows a relatively complex physical organization, not that it is in any sense 'superior'. In the same way, a 'primitive' animal – one that is believed to have departed little from the form of its ancestors – is not inferior to an 'advanced' one; if it has fewer adaptations, it is presumably because it has not needed to 'improve' on those of its ancestors.

The distribution of animals

Animals are to be found in all bodies of water and on all the world's continents, where they have successfully adapted even to the most hostile environments. Nevertheless, the world can be divided into a number of discrete *faunal* or *zoogeographical regions*, each of which is characterized by a distinctive fauna (▷ map). The precise boundaries of these regions are to some extent artificial – there is inevitably a degree of blurring as one area merges into another – but the regions themselves are more than a matter of convention.

The characteristic mix of animals to be found in each faunal region is the product of a number of factors, the most important of which are past and present climatic conditions (▷ pp. 10 and 198) and the successive conjunction and separation of the earth's landmasses over geological time (▷ p. 8). Broadly speaking, the most distinctive faunas are to be found in those regions, such as the Australian and the Neotropical, that have enjoyed the longest periods of separation from other landmasses.   BSD

head. The physical constraints imposed by the need to move under the force of gravity, combined with the relative lack of rigidity in animal cells (due to the absence of cell walls), also set a more or less determinate limit on the size to which animals can grow. These pressures are less intense in the buoyant medium of water – all the largest animals are aquatic – but some plants attain very much greater proportions than even the largest animal (▷ p. 30).

The basic feeding method of plants – uptake from the soil combined with photosynthesis – is common to nearly all species; much of the diversity that exists among them can therefore be attributed to the pressures of adapting to a wide range of different habitats. Animals, on the other hand, not only need to adapt to as diverse a range of habitats, but generally must also specialize in one of a potentially enormous range of feeding options. Thus in a diverse ecosystem such as a tropical forest, although there may be numerous plant species, there are actually relatively few *types* of plant – mainly trees, climbers, epiphytes and parasites (▷ pp. 44–7). In contrast, the animal species present may show an almost limitless diversity in their adaptations to the vast range of available niches. It is revealing that the most diverse group of plants – the flowering plants – owe much of their diversity to the enormous range of animals that may be attracted to pollinate them (▷ p. 42).

THE INVERTEBRATE–VERTEBRATE DIVIDE

The familiar division of animals into vertebrates and invertebrates dates back to the early years of the 19th century and is now enshrined in popular usage, but it is important to see the limitations of the distinction. Apart from the obvious fact that the animals grouped together as invertebrates lack a backbone – the column of segments or vertebrae supporting the central nerve cord in vertebrates – invertebrates actually have very little else in common. While vertebrates (fishes, amphibians, reptiles, birds and mammals) do indeed form a natural group – the subphylum Vertebrata of the phylum Chordata (▷ table, p. 18) – the corresponding 'group', Invertebrata, no longer has any currency in modern classifications. The term 'invertebrate' is in effect a negative definition, applicable to any animal that is not a vertebrate. This may be understandable, given the intrinsic interest of the vertebrates (and given the fact that we ourselves are among them), but it must be admitted that the distinction itself is hopelessly unbalanced. Well over 90% of all animal species are invertebrates, and with the exception of one part of one phylum they account for all of the many recognized phyla of animals.

However illogical the distinction may be, we may still ask how the important group of animals known as vertebrates evolved from invertebrate ancestors. A clue to answering this question seems to lie in the apparent relationship between the starfishes, sea urchins and other echinoderms (▷ p. 54) and the chordates. During the early stages of their embryological development, echinoderms – like chordates but unlike all other invertebrates – develop by means of radial (rather than spiral) cleavage of their cells, and the first opening to appear in the blastula (the ball of dividing cells) is the anus, rather than the mouth (▷ p. 186); echinoderms and chordates are consequently referred to as deuterostomes ('mouth second'), while all other animals are protostomes ('mouth first'). On this basis it is suggested that the echinoderms are our closest relatives among the invertebrates; even so, they are distant relatives at best – the first chordates must have branched off from an echinoderm-like ancestor hundreds of millions of years ago.

Further clues to vertebrate origins are offered by the protochordates, a group of rather obscure marine animals that together with the vertebrates make up the phylum Chordata: the hemichordates (acorn worms), the cephalochordates (including the lancelet or amphioxus), and the tunicates (including sea squirts). Further corroboration of the echinoderm–chordate link is provided by the extraordinary similarity between hemichordate and echinoderm larvae. Amphioxus has a tough notochord (the rod of cartilage that is incorporated into the backbone of adult vertebrates; ▷ p. 187), gill slits, and can swim by fish-like movements of its body, and thus has been suggested as the type of chordate that may have given rise to the vertebrates; other experts regard the larvae of the tunicates as a more likely candidate (▷ photo).

Adult sea squirts or tunicates are immobile filter-feeders – not, it would appear, a promising model of how our remote ancestors may have looked. Their larvae, however, swim like tadpoles and have a well-developed notochord, and for this reason some experts believe that the first vertebrates – jawless fishes – may have evolved from tunicate-like chordates.

Sponges

Sponges, which are widespread and abundant in aquatic habitats, are the simplest and most primitive of multicellular animals. Although not particularly animal-like in appearance, they actually have all the properties associated with animals. They feed on other organic matter (i.e. as heterotrophs; ⇨ p. 194), and they reproduce sexually by means of an ovum (egg) and a spermatozoan. Movement in mature sponges is limited to slight contraction of muscle-like cells, but the larval forms are free-swimming. There is also some degree of coordination of activity, even though there is no real nervous system.

Tube sponges (right), belonging to the class Demospongia, the largest and most successful group of sponges, are prominent features of tropical reefs. Long-lived and often brightly coloured, they reach a height of up to 1 m (39 in). Periodically they undergo a phase of spectacular reproduction, releasing clouds of sperm into the surrounding water.

Sponges are found in marine habitats ranging from the intertidal zone to the greatest ocean depths, and also in freshwater lakes and rivers. Some form flat, brightly coloured incrustations, others a variety of single or multiple structures in the shape of vases, chimneys or purses. Some sponges are only 1 cm (⅖ in) long, but various species in the tropics and the Antarctic can reach more than 1 m (39 in) in height and breadth. In the Antarctic extensive areas of the sea bed are dominated by large sponges.

General characteristics

Adult sponges are *sessile* – they spend their lives attached to rocks or other hard surfaces – and they feed by extracting small particles from the surrounding water. A typical sponge has a honeycomb-like structure of canals separated by cells and a hard skeleton. Water is drawn in through pores in the external surface by the beating of the flagella (whip-like projections) of special cells lining chambers opening off the canals. These flagellated cells are also responsible for removing food particles from the water. Filtered water, along with any waste products, is passed through to a single opening at the top of the sponge and expelled. The loose structure of sponges makes them an ideal shelter for other species, and a single sponge may have thousands of other animals living within it.

In some sponges the skeleton is formed from pin-like or complex star-shaped rods (or *spicules*) of silica or calcite, minerals based on silicon and calcium carbonate respectively. Others have a combination of these plus a meshwork of a fibrous protein called *spongin*, while various species, including the bath sponges, have

SEE ALSO

- WHAT IS LIFE? p. 6
- FILTER-FEEDING (BOX) p. 57
- ANIMAL BUILDERS p. 160
- CORAL REEFS p. 215
- MARINE ECOSYSTEMS p. 220

spongin alone. A few have no skeleton at all. In some forms the skeleton may be highly organized, resulting in a complex and highly specific lattice structure. The deep-sea species known as Venus's flower baskets, for instance, have an intricate skeleton of interwoven six-rayed spicules. The skeleton makes sponges rather unattractive as food, and only a few animals feed on them. Nevertheless, many sponges, like many other sessile marine animals, have developed distasteful defensive chemicals to deter predators and other animals that might otherwise settle on their outer surfaces. One group has developed the ability to bore into the shells of clams and even into limestone rocks; this is achieved by a combination of an acid secretion and physical etching by specialized cell extensions. This group is a major cause of the erosion of coral reefs and a serious pest of oysters.

Sponges disperse themselves and colonize new habitats through their larvae. Reproduction is sexual: eggs contained within the body of a female sponge are fertilized by sperm drawn in with the feeding currents. Free-swimming larvae develop and are then released into the sea. These disperse, settle on suitable surfaces, and metamorphose into young sponges. If two young sponges of identical genetic composition settle close to each other and come into contact as they grow, they will fuse and become a single organism. Equally, if a sponge becomes separated into several different pieces, these will reorganize themselves into new sponges. This ability to reorganize is so great that a sponge forced through a fine net and thus broken up into its constituent cells will form a new sponge from the collection of individual cells. Freshwater sponges may suffer seasonal food shortages and die back in winter. They form asexual repro-

ductive bodies called *gemmules* – balls of cells with a resistant outer wall capable of withstanding periods of stress.

The major groups of sponges

Within the phylum Porifera to which all sponges belong, four classes are recognized, each of which is differentiated principally on the nature of the internal skeleton. There are some 5000 described species in total, of which only about 150 live in fresh water.

The class Calcarea contains small, exclusively marine species with a skeleton of calcite that often protrudes through the surface of the body. The simplest of the sponges, they are usually purse- or vase-shaped, white or pale pastel in colour, and are often found in clumps. Some form branching structures from which upright parts arise.

The glass sponges (class Hexactinellida) are deep-water species that have complex silica skeletons, often formed into intricate and beautiful structures. The Venus's flower baskets belong to this group.

The class Demospongia is the largest and most widespread and successful group. It is also the only group to have colonized fresh water. These sponges have a variety of different kinds of spicule, sometimes augmented or replaced by spongin, and show almost limitless variation in body shape, size and colour. The group includes the bath sponges and the giant Antarctic species.

The coralline sponges (class Sclerospongia) were thought to be extinct until the 1960s, when they were found to exist in deep water on coral reefs. They have a skeleton that combines calcite, silica spicules and spongin.   RHE

THE VALUE OF SPONGES

The traditional use of sponges for washing and bathing has continued for thousands of years, but in recent times they have become the subject of much scientific study, because of certain other, previously unforeseen properties. Various tropical sponges have been found to produce chemicals that inhibit cell division, and manufactured analogues of these chemicals have been used in the treatment of cancers. These discoveries have prompted searches for sponges and other organisms that may have defensive chemicals with potentially beneficial properties.

Coelenterates

Whereas sponges are little more than colonies of individual cells, with very limited coordination or dependence between cells, all higher animals have two or three distinct cell layers, which provide the basis for any amount of structural sophistication (▷ p. 186). The simplest of the animals with differentiated cell layers (and the only group with just two such layers) are the coelenterates. There are three classes within the phylum Coelenterata, with a total of some 9000 species: the free-swimming jellyfishes (Scyphozoa), the sea anemones and the soft and hard corals (Anthozoa), and the hydrozoans (Hydrozoa), which include the familiar freshwater hydra as well as many simple or branched forms found in the sea.

SEE ALSO

● ANIMAL BUILDERS p. 160
● DIGESTION p. 178
● NERVOUS SYSTEMS p. 182
● CORAL REEFS p. 215
● MARINE ECOSYSTEMS p. 220

Coelenterates are often colonial, forming flower-like structures or complex free-floating colonies. Some species build up colonies of vast proportions; amongst these are the largest living structures on earth – the coral barrier reefs of Australia and Belize (▷ p. 215).

General characteristics

All coelenterates have a simple contractible body wall of two cell layers separated by a jelly-like layer (the *mesoglea*). Two basic coelenterate body plans exist: the polyp and the medusa. A *polyp* is a simple sac enclosing the gut cavity, with a single opening surrounded by a number of short tentacles. This opening serves both for the intake of food and for the expulsion of wastes; the end opposite the 'mouth' is attached to an external surface. A *medusa* is essentially an inverted free-swimming polyp, with a central downward-pointing mouth. Medusae are bell- or umbrella-shaped, with long tentacles fringing the bell. Some coelenterates undergo an *alternation of generations*, in which one form succeeds the other, with the medusa acting as the dispersal phase; others have only one form or the other. This simplicity of structure means that asexual reproduction by budding or fission (▷ p. 184) is easy, and this kind of reproduction is very characteristic of the group. Most conspicuously, it is the means by which coral colonies increase in size.

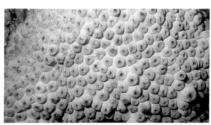

Playing safe: during the day, when the risk of predation is high, corals keep their polyps tightly closed (above). They wait for the relative safety of night before extending their tentacles into the water to intercept drifting zooplankton, which also emerge at night to feed (below).

The luminescent jellyfish *Pelagia noctiluca.* The luminescence seen in some jellyfishes is thought to act as a deterrent, warning would-be predators that they have a dangerous sting and are best left alone. Alternatively, the glow may attract inquisitive fishes, which then become ensnared in the jellyfish's stinging tentacles.

Coelenterates are mostly predatory, using their tentacles to capture food. While the tentacles can be contracted and may be used to grip prey, it is the batteries of unique gripping and stinging cells (*nematocysts*) carried on the tentacles that are the principal weapon. These are triggered as they make contact with prey, shooting out their contents of thread-like filaments. Some nematocysts produce sticky threads that grip prey, but the most important type are hollow and armed with sharp spines; these penetrate body tissues and discharge poison into the prey. As the tentacles wrap around the prey, more and more nematocysts are discharged to subdue it further. The stinging cells of some jellyfishes are capable of penetrating human skin, causing pain and occasionally even death.

Sea anemones and many hydrozoans are sedentary for most of their lives, but jellyfishes and highly advanced hydrozoans called siphonophores (including the Portuguese man-of-war) swim by means of pulsations of the bell.

Anthozoans

Sea anemones are never colonial and spend their entire adult lives as polyps. Although common in temperate latitudes, they are more diverse in warm waters. They can grow as large as 20 cm (8 in) across the disc and may live for 50 years or more. Anemones are normally sessile (anchored to a surface) and dependent upon food coming to them, but most are capable of very slow crawling and a few of floating or swimming. Some species enter symbiotic relationships with crabs in which they protect the crab in return for scraps of discarded food (▷ p. 67). Others harbour algal cells within their tissues in a cooperative association similar to that between corals and algae (▷ below). Although asexual reproduction is common, sea anemones – like all anthozoans

– also reproduce sexually, producing a free-swimming larva (the *planula*), which settles and metamorphoses into a polyp.

Stony corals are anemone-like polyps that secrete a limestone (calcium carbonate) skeleton that gives both support and protection. Although a few species are solitary, most form colonies (▷ p. 215). Feeding in corals has two components. Most are capable of trapping plankton with their tentacles, but a part of their nutrition is supplied by photosynthetic algal cells in their tissues. These algal cells are also vital in the process of calcification by which reefs are built up, so reef-forming corals are found only in shallow well-lit waters.

The octocorals, including sea fans, sea whips, sea pens and pipe corals, are polyps bearing eight pinnate (feather-like) tentacles, but they vary very considerably in shape. Sea fans and sea whips are found principally on coral reefs; they have a core skeleton of a flexible horny material that allows them to move with the currents and feed on small plankton. Sea pens are mud-dwellers with a bulb-like supporting skeleton at the base and a feather-like structure protruding at the top. They too feed on plankton.

In crowded situations corals and sea anemones vigorously defend their living space against rivals. They fight using a variety of weapons: toxic chemicals, nematocysts on specially elongated tentacles, and special gut extensions that can digest the tissues of a competitor.

Jellyfishes

All jellyfishes are free-swimming, with the medusa phase forming the principal part of their life history. Most inhabit coastal waters, and are particularly abundant in summer, when huge shoals may form. Some species may reach 60 cm (24 in) or more in diameter. Most swim by leisurely pulsations of the bell and feed by trailing long tentacles bearing nematocyst batteries through the water, but some pump water through themselves and extract small particles. The polyp phase appears after sexual reproduction: fertilized eggs give rise to a planula, which settles and metamorphoses into a polyp on the bottom; this grows and feeds, and then buds off a series of young medusae.

Hydrozoans

The hydrozoans are a very diverse group, with some 2700 species displaying all kinds of life history. The siphonophores, including the jellyfish-like Portuguese man-of-war, are complex colonial hydrozoans. They display extreme *polymorphism*, with different individuals performing different functions within the colony: some medusae pulsate to provide the propulsion for swimming, others have a purely reproductive function; of the several kinds of polyp present, some are adapted for feeding, others for defence. The anemone-like hydra is a solitary freshwater polyp (▷ diagrams, pp. 178 and 183), while various closely related marine colonial forms have both a medusa and a polyp phase.   RHE

Echinoderms

Amongst the echinoderms are some of the more familiar marine animals, such as starfishes, sea urchins and brittle-stars. The animals of this group are extremely distinctive, since the most striking feature that they have in common – five-part radial symmetry – is virtually unparalleled in the animal kingdom. Although exclusively marine, echinoderms form an extremely successful group, often dominating the communities in which they occur, and they are found in every habitat, from the intertidal zone to the greatest ocean depths.

Five-part variation (clockwise from top left): a sea urchin (*Echinometra viridis*) on coral; the ochre sea star (*Pisaster ochraceus*), a common intertidal species from California to Alaska; a brittlestar (*Astrophyton muricatum*), a nocturnal filter-feeder; an Indo-Pacific sea cucumber (*Cucumaria tricolor*), another filter-feeder.

There are some 6000 species in the phylum Echinodermata, which is subdivided into six classes. As well as five-part symmetry, the common features of the group are an internal skeleton based on calcite (a mineral composed of calcium carbonate) and small, hollow walking and feeding organs known as *tube feet*. The latter are often equipped with suckers and are connected to a unique system of internal body cavities. Frequently echinoderms are very spiny (their scientific name means 'spiny-skinned'). In spite of these shared characteristics, the members of each class are very distinctive in appearance. Although essentially simple animals lacking many of the organ systems that characterize more complex groups, the echinoderms in fact have certain features linking them with the chordates, the group that includes ourselves and other vertebrates (➢ box, p. 51).

Echinoderms reproduce sexually and disperse through free-swimming larvae, but they are also able to multiply their numbers by asexual means. Remarkable powers of regeneration are characteristic of the group; individuals that split in half or are severely damaged can replace missing parts without difficulty.

Sea lilies

The echinoderms are an ancient group, with origins going back at least to the Cambrian period, 570–510 million years ago. The earliest forms are thought to have resembled the sea lilies or feather stars (class Crinoidea), the most primitive living echinoderms. Two groups of sea lilies exist today: stalked, sedentary forms found exclusively in deep water, and mobile, stalkless species occurring principally in shallow waters. Both feed by sieving small particles from passing currents by means of many long tube feet. As a large surface area is required for effective sieving, sea lilies have evolved numerous branched arms and may have 10, 20 or 40 arms to create a huge crown or *calyx*, which may measure up to 40 cm (16 in) across. In areas where there are reliable currents, these animals may carpet the sea bottom.

Sea urchins

Sea urchins (class Echinoidea) come in a variety of sizes and shapes – globular, oval or disc-shaped. They have a complete skeleton (or *test*) of strong linked plates, and are densely covered with mobile spines used in locomotion and defence, and with small pincer-like organs (*pedicellaria*), also used in defence. Both spines and pedicellaria may be poisonous. Their tube feet are arranged in five rows around the test and are variously used in movement, attachment, food collection and sensory perception. Some live on hard rocky surfaces and graze on small plants, which they consume by means of a complex jaw apparatus. The forms known as sand dollars and sea potatoes live in or on sand and have short spines, which they use for burrowing; they are deposit-feeders, swallowing sand and digesting any organic particles.

Starfishes

The familiar starfishes or sea stars (class Asteroidea) usually have five hollow arms linked to an ill-defined central disc. Some, however, can have six or more arms, while some sunstars have as many as forty. A few reach 1 m (39 in) in diameter, but most are very much smaller. They are common from the intertidal zone to the ocean depths, and most abundant in the shallow waters of the continental shelves. Small spines and pedicellaria cover the upper surface, while the arm edges are often set with large defensive plates. On the underside there are five rows of tube feet stretching from the central mouth to the arm tips, which often bear eye spots. The tube feet are used for walking, digging and grasping prey, and those near the arm tips are sensory. Starfishes are mostly scavengers or predators, finding their food by smell, and frequently feed on clams, which they prise open with their tube feet. The gut may be turned inside out over prey and digestion then takes place outside the body. The infamous coral-reef pest, the crown-of-thorns starfish, eats large quantities of coral in this way. A few species that feed by collecting small particles have very short, almost indistinguishable arms. Damaged starfishes readily regenerate lost parts; in a few cases a whole new starfish can grow from a short piece of arm.

Brittlestars

Brittlestars or serpent stars (class Ophiuroidea) have long, thin, spiny arms that are flexible and linked to a compact central disc. They are very mobile and found in huge numbers in all marine habitats. Their common name reflects their readiness to cast off arms and regrow them. If a brittlestar is attacked, the tendons linking its arm segments become fluid and irreversibly weakened at a particular point, causing the arm to drop off. Brittlestars are principally suspension-feeders or scavengers, although some feed on organic particles deposited in or on the sea bottom. They move by flexing their arms and rowing themselves across the sea bed, reserving their tube feet exclusively for feeding and sensory purposes.

Sea cucumbers

Worm- or sausage-like in appearance, the sea cucumbers (class Holothuroidea) have little internal skeleton and lack spines. Although they can become rigid by altering the state of their internal tissues, they are normally soft and extremely flexible; this allows a range of movement and some can swim actively. Deep-sea forms resemble jellyfishes in being highly gelatinous, which makes them very light and able to move from place to place with little effort. They live in or on soft bottoms from shallow water to the greatest ocean depths, feeding by means of tentacles (enlarged, branched tube feet) that surround the mouth. Many species feed by spreading out their sticky tentacles to pick up particles as they move slowly over the sea bottom. Others are sedentary, extending their tentacles into water currents to intercept food particles.   RHE

SEE ALSO

● THE INVERTEBRATE–VERTEBRATE DIVIDE (BOX) p. 51
● FILTER-FEEDING (BOX) p. 57
● MARINE ECOSYSTEMS p. 220

Worms

The many invertebrate animals known as 'worms' are in fact of very diverse origins; what they share is a typically slender, elongated body form, not a common ancestry. Amongst these animals are a number of more or less obscure groups, such as horsehair worms, spiny-headed worms, peanut worms, and arrow worms, but three groups or phyla are of major importance: the flatworms, the roundworms or nematodes, and the annelids.

SEE ALSO

● PARASITISM AND SYMBIOSIS p. 168
● HOW ANIMALS MOVE p. 176
● DIGESTION p. 178
● NERVOUS SYSTEMS p. 182

Worms are successful animals, and they are extremely numerous in a wide range of habitats. Many are parasitic, but there are also a large number of free-living forms occurring in the soil and in aquatic habitats. Although essentially simple animals, the internal organization of worms is generally more complex than that of coelenterates such as corals and jelly-fishes. They (and all 'higher' animals) have bodies composed of three cell layers (▷ p. 186), which allows greater structural sophistication. In contrast to the blind-ending gut of coelenterates, the digestive system of most worms is a continuous tube with a mouth and anus (▷ p. 178). The nervous system is also better developed (▷ pp. 182–3), and the coordination of different body parts is therefore improved, allowing more elaborate and efficient modes of locomotion.

Flatworms

Flatworms (phylum Platyhelminthes) are simple, bilaterally symmetrical animals with an upper and a lower surface. They have a distinct head with a simple brain (▷ illustration, p. 183) and simple organ systems, but (as in coelenterates) the gut is blind-ending – food is taken in and wastes expelled through a single opening. The simple structure of these worms means that they can regenerate lost parts with ease. Thus cutting a flatworm in half does not kill it but results in a relatively short time in two smaller but perfectly formed worms.

The free-living *turbellarians* (class Turbellaria) are small, flattened, leaf-like worms found in aquatic and damp terrestrial habitats, where they live as predators or scavengers. They move by beating rows of cilia (tiny hair-like projections) in a fine film of mucus. In the largest group, the *planarians*, the mouth is central on the underside and part of the gut may be forced outside the body to make contact with food. Turbellarians are hermaphrodite – each individual has male and female organs – and reproduction is usually sexual.

The *flukes* or *trematodes* (class Trematoda) are similar in shape to the turbellarians, but they live as external and internal parasites and are equipped with suckers to grip their host. They have complex life cycles, usually involving an aquatic snail and at least one other species. Often hermaphrodite, they produce a series of larvae, each of which may multiply asexually to increase its infective potential. Many are parasitic on domesticated animals or humans, causing severe health problems (▷ p. 168).

The *tapeworms* (class Cestoda) are ribbon-like worms living a parasitic life in the guts of vertebrates. Their bodies are made up of many repeated, egg-producing segments, and some may reach several metres in length. They are unusual in having no gut, all food being absorbed through the body wall. Like trematodes, they have a complex life history.

Roundworms

The roundworms or nematodes (phylum Nematoda), which include eelworms, threadworms and hookworms, are a little-known but economically important group found in large numbers in all environments. These worms have long, thin bodies that are sharply pointed at one or both ends and a thick, resistant cuticle ('skin'), which is often transparent. They range in size from the microscopic to 1 m (39 in) in length. Unlike flatworms, nematodes have a second opening to the gut (an anus) and an internal cavity separating the body wall and the gut, allowing both to function independently and thus more efficiently. The fluid in this cavity is under pressure, which makes the body stiff, and they can only perform writhing movements. Only 15 000 species have been scientifically described, but probably half a million exist.

Most nematodes are harmless, but almost all animals and plants harbour parasitic forms. Many soil-dwelling forms are important crop pests, causing damage by consumption of plant tissue and by facilitating the entry of viruses and other diseases into plants. Many of those parasitic in animals cause debilitation or death. Hookworms are thought to be the commonest cause of human illness, affecting a quarter of the world's population.

Annelid worms

The annelids or segmented worms (phylum Annelida) have long, thin bodies with distinct head and tail ends. The body is made up of a series of separate segments, each of which can act semi-independently and usually bears limb-like structures (*parapodia*) or hair-like protrusions (*chaetae*). Internally an extensive body cavity separates the gut from the muscular body wall, and the nervous, circulatory and muscle systems are well developed. This combination of features allows great structural diversity and complex movement patterns, with the result that annelids have been able to become highly specialized and adopt many different ways of life.

Of the three classes of annelids, the oldest and most diverse are the marine *polychaetes* or *bristleworms* (Polychaeta), which include paddleworms, lugworms, ragworms and fan worms. They are characterized by a pair of crawling limbs (parapodia) on each segment. Most live in or on the sea bed, where they feed on all kinds of food, but a very few swim in the upper waters. Highly mobile predatory and scavenging forms, such as ragworms, have small heads, sensory tentacles and large retractable jaws. Others are sedentary and have complicated head structures – either masses of sticky tentacles with which they pick up food from the bottom, or a stiff crown of tentacles, which are held out in water currents to catch floating particles (▷ photo).

The *earthworms* and their freshwater relatives (class Oligochaeta) burrow their way through earth or detritus, feeding on decaying vegetable matter. They are long, thin worms without external structures except retractable chaetae used in burrowing (▷ p. 176), and a reproductive structure (the *clitellum*). Earthworms are of considerable ecological and economic importance because of their role in recycling nutrients in the soil.

Leeches (class Hirudinea) are the most advanced annelids. Predatory or parasitic, they are common in aquatic habitats and in moist situations on land. Suckers at each end of the body are used to grasp prey and in the characteristic movement pattern of these animals (▷ p. 176). All species have a well-developed sense of smell and terrestrial leeches parasitic on mammals have specialized heat-sensitive organs. Parasitic leeches have saliva that contains an anaesthetic to prevent detection and an anticoagulant so that blood remains fluid in the gut and can be easily digested. RHE

Two polychaetes, belonging to the most diverse group of annelid (segmented) worms. Fan worms (above left) are sedentary particle-feeders, extending their crowns of tentacles into water currents to intercept food; if alarmed, they withdraw into their protective tubes with astonishing speed. Fireworms (below left) are mobile predators or detritus-feeders; they are so named because of their white hair-like projections (chaetae), which inject poison on contact.

Molluscs

**Slugs and snails, oysters and clams, octopuses and squid –
all are members of the phylum Mollusca, a highly diverse
group that encompasses seven classes and contains over
50 000 species. The primitive ancestral mollusc is thought to
have looked somewhat like a modern limpet (a gastropod) –
that is, it had a protective shell under which there was a
body divisible into two distinct sections: the muscular head-
foot and the body mass, containing all the other organs.**

**The African giant
snail,** the largest land
gastropod. Although not
much different from
common garden snails in
appearance, these outsize
versions have shells mea-
suring around 20 cm
(8 in) in length and con-
siderably larger specimens
have been known. They
have been introduced to
many areas outside their
natural range in Africa,
principally for their
potential food value, but
their voracious appetites
for all kinds of plant
matter and their ability to
reproduce rapidly have
meant that they have
become serious pests,
particularly in
Southeast Asia.

Although each
class of molluscs has
become specialized for particular ways of
life and departed greatly from the ancestral 'model', there are still a number of
common characteristics, some or all of
which are found in all molluscs. These
include a broad locomotory foot, a protective shell, a tongue-like feeding organ
(the *radula*), and special respiratory gills
(the *ctenidia*).

Gastropods

The largest group of molluscs, the class
Gastropoda, includes such diverse creatures as slugs, snails, periwinkles, top-
shells and limpets (⊳ photo). Members of
this group have not only exploited marine
and freshwater habitats but also terrestrial ones. Their diversity extends to both
feeding habit and structure. Among the
35 000 or so species there are herbivores, carnivores, parasites and particle-
feeders; they range in size from tiny forms
that live between sand grains to giant
conchs up to 70 cm (28 in) long.

All gastropods have a muscular foot for
walking or swimming (a feature reflected
in their scientific name, which means
'stomach-foot'), and almost all have a
radula. In carnivorous species the radula
has sharp, pointed teeth for tearing, while
in the specialized cone shells it is modified
into a single harpoon-like weapon armed
with a poison gland. Most gastropods
have external shells, which are strong and
versatile. Usually there is a hard, outer
protein layer, a mixed protein-chalk layer,
and an inner chalky layer. Some shells are
simple cones, but most are complex spirals; this shape has proved the best 'design'
for growth and compactness. Most gastropods reproduce sexually, producing a
free-swimming larva called a *veliger*, but
some lay large eggs protected by tough
capsules, which develop directly into
miniature adults.

Most marine gastropods live on the sea
bottom and crawl by means of the broad,
flat foot, but some have become adapted to
life in the upper waters in order to exploit
the rich planktonic food resources available there. In these species the shell is
reduced and the foot is modified into two
wing-like swimming organs. Sea slugs are
predatory gastropods that lack an external shell and are frequently brightly coloured. This is warning coloration (⊳
p. 167): the skins of such animals may
contain distasteful acids, while some have
special organs containing stinging cells
stolen from the coelenterates on which
they feed (⊳ p. 53); these are arranged as
batteries of offensive weapons for the
mollusc's own defence.

The adaptations of land gastropods for life
in dry air include the development of a
closeable lung in place of the gills and
a form of reproduction that involves
copulation. Each of the small number of
embryos is provided with its own food
supply and is deposited in a moist habitat
to develop. In order to avoid the desiccat-
ing effect of the daytime heat, land gastropods are typically active by night.

Bivalve molluscs

Bivalves (class Bivalvia), as their name
implies, are molluscs whose shell is made
of two parts (or 'valves') connected by an
elastic hinge. The group contains some
14 000 species, including clams, mussels,

The common limpet (above), a familiar inhabitant of rocky shorelines. Limpets feed by rasping
tiny green algae from rocks with their radula (the
characteristic molluscan feeding organ); in these
gastropods this takes the form of a broad tongue
with many rows of teeth, which are iron-
hardened and made of chitin (the protein also
found in arthropod exoskeletons; ⊳ p. 58). Like
miniature lawnmowers, they move systematically
back and forth across the surface of a rock,
leaving distinct trails in their wake (below).

oysters and cockles. The whole body of
these animals can be withdrawn into the
shell for protection.

Bivalves are essentially living filter
pumps, with no radula and no distinct
head. Large volumes of water are drawn
into the shell through a siphon by the
rhythmic beating of thousands of tiny
projections (cilia) on the massive gills,
and small floating particles are sieved off
as the water passes over the gills. These
particles are sorted and the organic ones
retained as food; the filtered water,
together with unwanted particles, are
then forced out through a second siphon.
Giant clams employ a different feeding
method (⊳ photo): like corals (⊳ p. 53),
they have a symbiotic relationship with
photosynthetic algae and obtain nutrition
from the algae's surplus production.

Although a few are capable of active
swimming (⊳ photo), most bivalves are
more or less immobile. Many live permanently attached to rocks or other hard
surfaces, while others have a large foot for
burrowing into sand or mud. Because of
their limited mobility, many bivalves are
prime targets for predators such as starfishes, and can use their foot to perform
rapid, violent escape responses. Some,
such as scallops, can clap the valves of
their shell together and so spring away
from an attacker. Bivalves disperse themselves and colonize new locations through
their larvae. They shed tens of millions of
eggs into the water, where they are fertilized and develop into planktonic larvae.
These are then carried by currents to new
sites for settlement.

As bivalves grow quickly and are nutritious and easy to collect, they are widely
cultivated for food. Some species encase
the tiny grit particles they take in with
their food in shell material to form pearls.
This can of course be deliberately induced
to form artificial pearls of various shapes.
Although marine and freshwater mussels
form pearls, only oyster pearls are generally of commercial value.

Cephalopods

In total contrast to the snails and clams
are the cephalopods (class Cephalopoda),
which include squid, cuttlefishes and octopuses. These animals have evolved as
highly specialized predators – they are
fast-moving, alert and more intelligent
than any other invertebrates. They grow
quickly and – strangely for such complex
creatures – reproduce only once before
dying.

Squid have long, streamlined bodies and
no shell. At one end is the head, which
bears two large eyes and ten tentacles –
eight suckered arms and two long feeding
tentacles – developed from the modified
foot. At the other end, near the pointed
extremity, is a pair of lateral fins, which
provide stability. Squid have developed a
novel and highly effective means of
moving through the water – they can hang
motionless in the water or dart rapidly in
any direction by means of a form of jet
propulsion. Water is forced from a cavity
within the body through a funnel just
behind the head by the contraction of

Bivalve diversity. The massive giant clams of tropical waters (left) are unusual among bivalves in obtaining their nutrition by a symbiotic association with photosynthetic algae (▷ text). They are the largest living bivalves, growing to around 1 m (39 in) in length; in contrast, the smallest species are less than 1 mm (¹⁄₂₅ in) when adult. The fileshell (above) is one of relatively few bivalves with the ability to swim – most species are more or less immobile.

muscles in the body wall. As the funnel can be pointed in any direction, a squid can control its movements with great precision.

In accord with their predatory life style, squid have very well-developed sensory systems associated with vision and balance. Their eyes are very like our own in structure, with a similar arrangement of the lens, cornea, iris and retina (▷ p. 189). Information from the eyes and the other sense organs is fed to a large, highly organized brain, which directs and co-ordinates the animal's behaviour.

Squid live in the open sea, often in large shoals, feeding on crustaceans and fish. If prey is within reach, the two long tentacles are shot out to seize it and drag it into the grip of the eight arms. Captured prey is subdued by a bite from the strong beak and by injection of poison. It is then chewed and swallowed. Within the mouth cavity is a classic molluscan radula, which is used to fragment food. In the South Atlantic and elsewhere squid are harvested by large commercial fisheries.

Cuttlefishes are similar to squid in most respects, but they are inshore species that mostly swim and feed near the bottom. One major difference is the presence of an internal shell, which acts as a variable flotation device, allowing these animals to maintain neutral buoyancy. All cephalopods have the ability to change colour (▷ box, p. 67), but it is the cuttlefishes that display the most dazzling array of colour patterns. This is achieved by thousands of small contractible pigment sacs in the skin, each of which is one of three colours. These can be expanded or shrunk at will to produce shimmering colour changes or long-term colour patterns. Specific patterns are used to deter predators, in courtship displays, and to distract prey animals. Strangely, cuttlefishes do not themselves have colour vision.

Octopuses differ from both squid and cuttlefishes in having only eight tentacles. Most are bottom-dwelling and relatively small, although some may reach a length of 1·8 m (6 ft). Studies of octopuses have revealed the extraordinary sophistication of their nervous systems. They can learn to associate particular visual and chemical cues with either a reward or a punishment and are able to retain this memory for weeks (▷ p. 150). Their ability to discriminate between objects using visual or tactile information is considerable.

Despite their sophistication, present-day cephalopods are only the remaining vestige of a previously superabundant group. Before the evolution of bony fishes, sharks and rays (▷ p. 70), cephalopods were the major marine predators of the upper waters and dominated the seas for millions of years. The vast array of ammonites and belemnites found as fossils in rocks bears witness to how successful they were. Today only a few highly specialized forms are able to compete successfully with fishes.   RHE

SEE ALSO

- COLOUR CHANGE (BOX) p. 67
- LEARNING AND INTELLIGENCE p. 150
- PREDATOR AND PREY p. 164
- DEFENCE, DISGUISE AND DECEIT p. 166
- HOW ANIMALS MOVE p. 176
- MARINE ECOSYSTEMS p. 220

The giant squids are the largest invertebrates known: individuals over 18 m (59 ft) in length having been washed up on beaches. They are deep-water species, however, and are rarely seen.

Arthropods

Judged by the number and variety of species, the arthropods are the most successful of all living organisms on earth. Crustaceans such as crabs and shrimps are the dominant arthropods in the sea, while on land insects can be found in most available habitats. Albeit to a lesser degree than insects, the other important arthropod groups – arachnids (spiders, mites and scorpions) and myriapods (millipedes and centipedes) – are highly successful land animals.

Whatever their evolutionary origins (▷ below), the arthropods alive today represent a highly diverse collection of animals. As a group they are characterized by having external segmented skeletons and pairs of jointed limbs.

Success and limitations

The success of the arthropods lies in the external skeleton (*exoskeleton*), which both supports and protects the animal. The exoskeleton is formed by a protective layer known as the *cuticle*, which covers the animal's soft underlying parts. The cuticle is based on a complex molecule called *chitin*. When mixed with differing proportions of proteins and inorganic salts, such as calcium carbonate, the arthropod skeleton shows remarkable versatility, varying from the soft elastic bag that surrounds a caterpillar to the immensely hard claws of a lobster.

Chitin is based on long-chain molecules of the sugar acetyl glucosamine; these become cross-linked to give tough fibres, which are embedded in a protein matrix. The exoskeleton is further strengthened by a process known as *tanning*. The darkening of the exoskeleton that com-

SEE ALSO

● INSECTS pp. 59–65
● CRUSTACEANS p. 66
● ARACHNIDS p. 68

THE MISSING LINK?

Of all the arthropods the velvet worms, or onychophorans, are the most bizarre and uncharacteristic. They have features found both in annelid worms (▷ p. 55) and in other arthropods, and it has been suggested that these animals represent a missing evolutionary link between the worms and the arthropods. To what extent this is true is still uncertain, and the precise relationship of velvet worms to other arthropods will not be clear until there is greater agreement on the origins of arthropods as a whole.

Velvet worms are generally small, although the biggest species reach up to 15 cm (6 in) in length. They have rudimentary eyes, which are probably only light-sensitive, and a pair of fleshy antennae.

Onychophoran cuticle, although containing chitin, is thin and flexible and has no segmentation, while the limbs are unjointed and fleshy. Breathing takes place through tracheae (air tubes) like those of insects, and the layout of the internal organs is similar to the normal arthropod pattern. Desiccation is a major danger for velvet worms, so they are restricted to moist tropical and subtropical habitats around the world. They are carnivorous, feeding at night on insects and other small animals; they also feed on carrion. If disturbed, velvet worms protect themselves by ejecting a liquid from a pair of short appendages by the side of the mouth. On contact with the air, this fluid becomes viscous and covers the intruder with sticky threads.

---

monly accompanies this process is due to an excess of the organic compound quinone, which turns into the pigment melanin after hardening of the proteins is complete. Movement is possible only where the exoskeleton is not hardened, hence the need for joints. These are composed of flexible cuticle, and allow as varied a set of movements as that found in animals with internal skeletons.

Despite its obvious success, the exoskeleton has its drawbacks. The most important of these is that the hard, external skeleton limits the size of the animal. Growth can take place only if the old skeleton is discarded and a new, larger one grown in its place. This process, known as *moulting* (▷ p. 62), has to occur a number of times in all arthropods as they grow to full adult size. At each moult the animal is vulnerable to attack by predators, parasites or disease, and – in the case of terrestrial species – there is also a danger of dehydration. Much of the old exoskeleton is absorbed and re-used in

the new one, but even so, at each moult there is the loss of the cast 'skin'. The need to moult may also influence the absolute size that arthropods can reach, since in the course of moulting an arthropod must be able to hold its shape until the new cuticle has hardened.

Arthropod evolution

The evolutionary history of the arthropods has long been an area of active debate. The fact that all arthropods have segmented exoskeletons and jointed limbs was once thought sufficient to show that they had evolved from a single ancestral stock in which these features occurred. However, recent comparative analyses of the chemical composition of arthropod exoskeletons and of the detailed structure of joints, mouthparts and legs suggest that this may not be so – that several distinct evolutionary lines may have branched off from a segmented worm-like ancestor and that the basic arthropod features evolved several times over.

According to this theory, one evolutionary line (Uniramia) gave rise to the insects, the myriapods, and the velvet worms (▷ box). These all have simple (uniramous) legs that operate in essentially the same way. A second line (Biramia) gave rise to the crustaceans, in which the legs are two-branched (biramous), and in addition to the normal leg there is an outer part, often involved in feeding or swimming. A third line gave rise to the chelicerates (spiders, scorpions and mites) and the now extinct trilobites. Marine species of this group have biramous limbs, but of a different design to those of crustaceans, while land species have uniramous legs.

This view is not universally accepted, however. Current research into the molecular basis of arthropods (based on ribosomal RNA) suggests that the idea of a three-line evolution may be wrong, and that arthropods do constitute a single (monophyletic) group after all. With persuasive evidence on either side, it will be some time yet before a clear picture of the evolutionary history of the arthropods emerges.　　　　　　　　　　　BDT

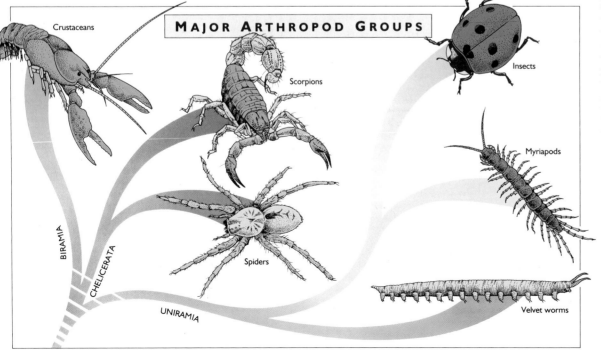

MAJOR ARTHROPOD GROUPS

Crustaceans

Scorpions

Insects

Myriapods

Spiders

Velvet worms

BIRAMIA

CHELICERATA

UNIRAMIA

Insects
Form and Structure

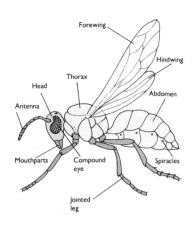

The body form of the insect has proved to be the most successful of any living organism. According to some estimates, over 80% of all plant and animal species are insects, and they have colonized almost every habitat except the sea, where crustaceans are the dominant arthropods. The success of insects stems from the light, strong, wax-covered cuticle (▷ opposite), which is highly protective, particularly against dehydration. For this reason, they are the only invertebrate group to have made a full conquest of dry land.

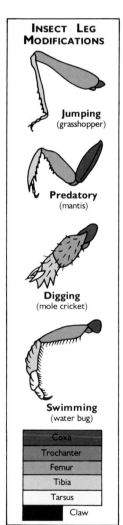

INSECT LEG MODIFICATIONS

Jumping
(grasshopper)

Predatory
(mantis)

Digging
(mole cricket)

Swimming
(water bug)

| Coxa |
| Trochanter |
| Femur |
| Tibia |
| Tarsus |
| Claw |

The small size of insects has meant that the amount of living space available to them is almost infinite. To an elephant a tree may provide part of a meal as it browses along a forest edge, but to insects only a few millimetres long it represents a multitude of habitats, enough for hundreds of species.

Economies of size

This host of suitable habitats can only be exploited by insects if they can reach them. One problem with small size is that distances become relatively vast and that walking is a very inefficient method of covering the ground. To overcome this, insects – uniquely among the invertebrates – have developed powers of flight.

Small size brings other advantages. It is a general rule of nature that the lifespan of smaller animals is shorter than that of larger ones. This means that in a given time a small animal such as an insect will pass through more generations than a larger one. A consequence of this rapid succession of generations is that the evolutionary processes of natural selection are accelerated, allowing a proliferation of species specially adapted to fill a wide range of habitats.

Head and thorax

In general, insects are easily recognizable by their three body sections – head, thorax and abdomen. The *head* usually has a pair of compound eyes for vision, and often two or three small single eyes (ocelli), which respond to varying light levels (▷ pp. 188–91). Two antennae and two pairs of palps (short sensory appendages) at the mouth provide information on both touch and taste. The mouthparts are typically adapted for biting, but in many species they are modified for other functions (▷ box). Within the head is a bilobed brain, which closely connects with the optic lobes of the compound eyes; two nerve cords pass from the brain to the first of a series of segmentally arranged ganglia that extend along the ventral (front) side of the insect (▷ p. 182).

The middle section of the insect body – the *thorax* – is the locomotory centre. It consists of a rigid box containing the muscles that operate the two pairs of wings. The wings are usually thin yet strong, with the wing membrane stiffened by a network of veins. Some insects move their wings by contraction of muscles attached directly to them, while others use an indirect 'click' mechanism (▷ p. 177). Wings may have other functions besides flight. In some insects, such as beetles and grasshoppers, the forewings are hardened, covering and protecting the large hindwings beneath. Crickets use their wings to produce sound, while in many insects the wings are coloured or patterned and used in courtship display, or as camouflage.

Three pairs of legs are attached to the lower part of the thorax. Some of the leg muscles are in the limbs themselves, but there are others within the thorax; indeed, in some insects muscles are shared by the wings and the legs. Insect legs are highly varied in form, ranging from the strong digging forelegs of mole crickets to the long hindlegs of grasshoppers, which are adapted for jumping. In the praying mantises and mantispid flies, the forelegs are enlarged and scissor-like, and used for grasping prey. Aquatic insects have legs modified for swimming. Diving beetles and water boatmen, for instance, have flattened hindlegs fringed with long hairs, which are used like oars (▷ box).

The abdomen

The digestive and reproductive systems occupy the rear body section – the *abdomen*. In most insects the abdomen is clearly segmented, and there are sets of small paired appendages at the end, which are used in reproduction. Males have claspers with which to hold the female during sperm transfer, while females often have a structure called the *ovipositor*, which is used to place fertilized eggs in chosen sites, from crevices in bark to the tissues of another insect. In many bees, ants and wasps the ovipositor forms a sting, which is used in defence or to kill or paralyse prey.

The insect gut consists of three sections. The foregut starts at the mouth, passes through the thorax as a narrow oesophagus, and then swells in the abdomen to form a crop for food storage. The midgut, which often has pockets lined with secretory cells, is the place where digestion and absorption of food occur. An insect's excretory organs, equivalent to our kidneys, are the *Malpighian tubules*, which are connected to the alimentary canal at the junction of the mid- and hindguts. In the hindgut water is extracted from waste food and excretory products. The resulting mixture of dry faecal and excretory material is passed out of the anus at the end of the abdomen.

The blood system of insects is relatively simple. The organs float in a blood-filled cavity called the *haemocoel*. The blood (or *haemolymph*) is stirred by a tubular heart, which lies along the dorsal (back) side of the abdomen. A single blood vessel extends forwards from the heart to the head. The blood is pumped to the head and then percolates back through the thorax to the abdomen. The blood carries nutrients, waste products and carbon dioxide, and also contains corpuscles that counter invading bacteria, but (in contrast to vertebrates) it does not carry oxygen.

Tissues are supplied with oxygen through a series of branched tubes (tracheae), which extend from holes (spiracles) in the exoskeleton (▷ p. 180). Whilst this system is clearly successful in insects, it would be unsuitable for larger animals, since the spiracles and tracheae would have to be unmanageably large. This method of respiration is probably another reason (in addition to moulting; ▷ opposite) why insects are limited in size. BDT

Forewing
Hindwing
Head
Thorax
Abdomen
Antenna
Mouthparts
Compound eye
Spiracles
Jointed leg

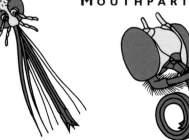

MOUTHPART SPECIALIZATIONS

Piercing-sucking
Mosquitoes have long, slender mouthparts (here shown separated out) that fit together to form a hollow needle that is inserted into a host and used to suck up bodily fluids.

Siphoning
Butterflies and moths have a long, coiled proboscis, which is unfurled to suck up nectar and other liquid food.

Sponging
Houseflies have a 'tongue' (labium) with which they lap up liquid food that has been pre-digested by salivary enzymes exuded onto it.

The Classification of Insects

The number of living insect species currently known to science amounts to perhaps two million, but there are many millions of others – perhaps ten or twelve times this figure – that await discovery. Despite this enormous proliferation of species, all living insects can be divided into just 29 groups or orders. These groups are of varying size and importance, and on the basis of various common features can be further subdivided in a number of ways.

At a high level of generality, insects can be divided into two groups – those that are wingless (the *apterygotes*) and those with wings (the *pterygotes*). Even at this basic level, however, a degree of caution is necessary, since some insects have become highly specialized and lost their wings in the course of evolution (⇨ box).

Apterygote insects

The four orders of apterygote insects (bristletails, diplurans, springtails or collembolans, and proturans) are quite different from each other, and the last two are rather unlike insects in general. This has led to the suggestion that all the six-legged animals that were formerly thought of as insects should be grouped as hexapods, with the springtails and proturans forming two separate classes and the insects a third.

The apterygotes have a number of features that suggest links with the other

---

PRIMITIVE OR ADVANCED?

At the most basic level, insects are divided into two groups, according to whether or not they possess wings. Thus a springtail is classed as an apterygote, together with various other insects, all of which lack wings and are considered to be 'primitive' – relatively unspecialized and presumably little (or less) changed from the ancestral forms from which insects evolved.

According to this criterion, literally interpreted, fleas would be classed as apterygotes, since they are wingless. This is very far from the truth, however. Fleas are in fact highly 'advanced' (specialized) insects that have lost their wings during the course of evolution, as an adaptation to their parasitic life style. They are therefore said to be 'secondarily' wingless.

This is a good example of a difficulty frequently encountered in taxonomy. The absence of a structure or feature in a group may indicate that it is primitive and that the group originated before the structure had evolved, but it is also possible that the group is highly advanced and that the feature, originally present in the evolutionary history, has been lost. Only by careful consideration of many strands of evidence can such questions be answered.

---

uniramous arthropods (⇨ p. 58), especially the myriapods (centipedes and millipedes). Like the myriapods, they tend to grow continuously during their life through a succession of moults, even after they have reached sexual maturity. Myriapods have a pair of limbs on every segment, while the apterygotes have two vestigial limb-like structures on each body segment. In contrast, the pterygotes grow through a series of juvenile stages, with no further growth once they become adult, and have no appendages along most of the abdomen.

Pterygote insects

The 25 orders of pterygote insects can be subdivided into two groups – the exopterygotes and the endopterygotes – depending on the way they grow and develop to adults (⇨ p. 63). In the 16 orders of *exopterygotes*, the wings develop externally (as is indicated by their scientific name). The juvenile stages, sometimes called *nymphs*, are essentially smaller, wingless versions of the adults. As they grow by moulting from one stage to the next, their wings become increasingly evident as small buds on the thorax. When individuals are fully grown, they moult for the last time to become adults, and the wing buds are replaced by wings. The reproductive system also becomes fully developed at this time. Thus the change from juvenile to adult is gradual and is not marked by dramatic changes in body form. For this reason, these insect groups are often said to undergo *incomplete metamorphosis*. Examples of insects that develop in this way are grasshoppers and crickets (Orthoptera), dragonflies (Odonata), earwigs (Dermaptera), and aphids and other bugs (Hemiptera).

In contrast, the nine orders of *endopterygotes* (meaning 'internal wings') have juvenile stages that are quite unlike the adults. They differ not only in body form but often in diet and habitat too. Since the juveniles, called *larvae*, are so unlike the adults, they have to pass through an intervening stage in their development (the *pupal stage*). It is at this time that the juvenile is transformed into the adult, acquiring the mature body form, including wings and reproductive organs. Thus in these cases the wings develop inside the pupa (hence the scientific name of the group). Because of the major change in body form, these insects are said to undergo *complete metamorphosis*. Some well-known examples of endopterygotes are the beetles (Coleoptera), butterflies and moths (Lepidoptera), flies (Diptera), and the bees, ants and wasps (Hymenoptera).

Although there are nearly twice as many orders of exopterygotes, they represent only 12% of all known insect species. The majority of insects belong to a few endopterygote orders, the most important of which is the beetles. This one order contains over half of all known insects and a quarter of all known animal and plant species. The endopterygote life style, with its larval feeding stage separated from the winged adult by a resting pupal stage, is clearly a very successful recipe for colon-

izing the earth. It has several obvious advantages over the exopterygote life history. Adults and larvae frequently exploit different habitats, and may have completely different mouthparts and so use different foods. This prevents competition between adults and larvae, and thus increases the resources available to the species as a whole. Furthermore, the timing of the resting pupal stage is often such that it coincides with periods when food is absent, during drought or winter conditions, so that seasonal habitats can also be exploited.

Insect evolution

The oldest fossil hexapod comes from the Devonian period (410–355 million years ago), and is a collembolan called *Rhyniella praecursor*. A number of other fossils, which appear to be somewhat like present-day apterygotes, have been found in deposits from Carboniferous and Permian times (355–250 million years ago).

The fossil history of the pterygotes shows that by the end of the Carboniferous period, some 300 million years ago, there was already a diversity of winged species. The power of flight thus developed very early in the evolution of insects. Some of the oldest fossil pterygotes have flattened plates running along the length of the body. It has been suggested that these lateral plates may have originally helped the insects to glide, perhaps assisting escape from predators. Gliding would be enhanced if the plates could be tilted by muscles, and from there it is a short step to flapping flight. However flight developed, it clearly gave huge advantages to insects and led to the enormous radiation of the group on land.

The earliest winged insects could not fold their wings up over the abdomen. This restriction is still to be seen today in the mayflies and dragonflies, which also have a complex net-like venation in the wings, another feature seen in early fossils. These insects are thus thought to be the surviving remnants of a large and diverse fauna that flourished in the Palaeozoic era (ending 250 million years ago) and are distinguished from other insects as the *palaeopteran* ('ancient-winged') orders. In all other pterygotes (the *neopteran* orders, meaning 'new-winged'), the wings can be folded. Nevertheless, among the neopterans there are certain distinctive features that suggest that, in addition to the palaeopterans, there were at least three other major branches of insect evolution. The roots of the *orthopteroid* orders also go back to the Carboniferous period, to cockroach-like ancestors. In these orders the forewings are often hardened and smaller than the folding hindwings which they protect. The *hemipteroid* orders go back to the Permian period. Amongst the specializations of this group are piercing mouthparts for fluid-feeding.

The palaeopteran, orthopteroid and hemipteroid orders are all exopterygotes. The endopterygote orders are not readily divisible and probably represent evolution from a single source.　　　　　BDT

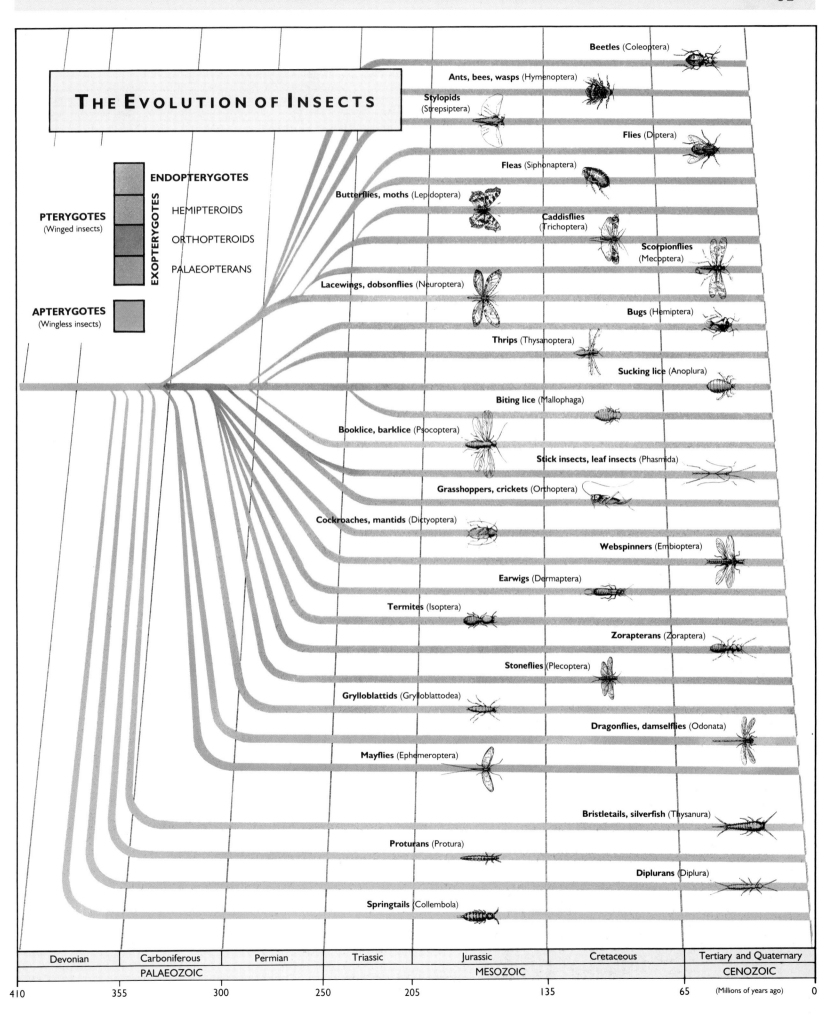

THE EVOLUTION OF INSECTS

ENDOPTERYGOTES

PTERYGOTES
(Winged insects)

HEMIPTEROIDS

ORTHOPTEROIDS

PALAEOPTERANS

EXOPTERYGOTES

APTERYGOTES
(Wingless insects)

Beetles (Coleoptera)

Ants, bees, wasps (Hymenoptera)

Stylopids (Strepsiptera)

Flies (Diptera)

Fleas (Siphonaptera)

Butterflies, moths (Lepidoptera)

Caddisflies (Trichoptera)

Scorpionflies (Mecoptera)

Lacewings, dobsonflies (Neuroptera)

Bugs (Hemiptera)

Thrips (Thysanoptera)

Sucking lice (Anoplura)

Biting lice (Mallophaga)

Booklice, barklice (Psocoptera)

Stick insects, leaf insects (Phasmida)

Grasshoppers, crickets (Orthoptera)

Cockroaches, mantids (Dictyoptera)

Webspinners (Embioptera)

Earwigs (Dermaptera)

Termites (Isoptera)

Zorapterans (Zoraptera)

Stoneflies (Plecoptera)

Grylloblattids (Grylloblattodea)

Dragonflies, damselflies (Odonata)

Mayflies (Ephemeroptera)

Bristletails, silverfish (Thysanura)

Proturans (Protura)

Diplurans (Diplura)

Springtails (Collembola)

Devonian	Carboniferous	Permian	Triassic	Jurassic	Cretaceous	Tertiary and Quaternary
PALAEOZOIC				MESOZOIC		CENOZOIC

410 355 300 250 205 135 65 (Millions of years ago) 0

Insects
Life Cycles

A large part of the phenomenal success of insects as a group can be ascribed to their astonishing diversity in size, form and habit. This diversity has enabled them to adapt to an astounding array of terrestrial and freshwater habitats all over the globe, from tundra and deserts to swamps and tropical forests, and to exploit a vast range of food sources.

The capacity merely to subsist in a given habitat is not sufficient, however. Permanent colonization depends also on the ability to breed successfully. In this respect, too, insects have proved preeminent: a testimony to their remarkable versatility is seen in the enormous variety of life-cycle patterns that they have evolved.

Finding a mate

Down at the scale of an insect, the world it inhabits must seem highly complex. How does a male find a female? Vision is used in some cases. Male speckled wood butter-

Insect growth is achieved by a process known as moulting. The female dragonfly shown here with its recently cast 'skin' is waiting for the wing cuticle to harden before taking to the air for the first time.

flies defend small sunny spots as territories and wait for passing females to be attracted to them. At night in the tropics male fireflies send out flashes by means of luminous organs on the underside of the abdomen, which contain a compound called luciferin that emits light when chemically triggered; females respond to these flashes with flashes of their own. In Southeast Asia synchronous flashes of light produced by countless fireflies create a stunning firework display among the trees.

More commonly chemicals are used to attract mates. Either males or females may emit a volatile substance that is attractive to the opposite sex. These chemicals, called pheromones (▷ pp. 152 –5), do not merely smell pleasant – they actually affect the physiology of the target insect and change its behaviour. Such signals can be effective over considerable distances as they are readily carried on wind currents. Female moths produce attractive pheromones that are wafted on the wind and detected by males by means of their fine, feathery antennae (▷ pp. 188–9).

Other insects use sound. The ticking sound of the wood-boring deathwatch beetle is made as the male bangs his head on the tunnel walls to attract the attention of a female. Grasshoppers produce a complex repertoire of songs by rubbing their legs or wings against a set of small pegs on the inside of the femur, while crickets achieve similar effects by rubbing their wings together (a process known as *stridulation*). Cicadas produce very loud sounds by muscular vibration of a membranous 'drumskin' called a tymbal.

Courtship and mating

When males come close to females, they may begin a courtship display. Such displays are often highly stereotyped and follow an exact sequence of actions. Any interruption usually means that the male has to start at the beginning again. Courtship often takes the form of a circling dance by the male, which may use its wings to embellish the display. The male grayling butterfly grasps the female's antennae between its wings and anoints them with an aphrodisiac pheromone from special glandular scent scales. Grasshoppers have a specific courtship song to entice females to mate.

Internal fertilization is common to all successful groups of land animals, and the insects are no exception. Insects adopt several different copulatory positions. Very commonly they attach themselves end to end, locked together by a complex genital mechanism. In most insects sperm from the male's testes is packaged in a container called a *spermatophore*, which is

passed across to the female during copulation. Once in the female's reproductive tract, the spermatophore dissolves to release the sperm, which is kept in a special structure called the *spermatheca* until required. As eggs pass down the oviducts from the ovaries, sperm is expressed to fertilize them. A bizarre variant is seen in the hemipteran family Cimicidae, which includes the blood-feeding bedbug. The male punctures the female's abdomen and injects the sperm directly into the blood cavity (haemocoel). From there the sperm migrates to the ovaries, where the eggs are fertilized.

Growth and development

Insects usually develop from eggs laid by the female, which in time hatch to give juveniles. There are exceptions to this, however. In *viviparous* species, the female lays larvae. The aphids often bypass the egg stage in order to increase the number of generations they can fit into a season. The larvae of the tsetse fly are nurtured within the female, in a womb-like expansion of the oviduct. Produced one at a time, the larvae are so far advanced in their development when they are laid that they immediately burrow into the soil and pupate.

Insect growth takes place during the juvenile or larval phase. During this time insects pass through a series of stages called *instars*, the number of such stages depending on the species. Typically there are five or six during the juvenile period, but there may be 15 or more in species with a long larval life, such as mayflies.

Insects grow from one instar to the next by *moulting* (▷ photo). During each moult a complex sequence of events takes place under hormonal control in the insect's outer layer of cells (the *epidermis*). One hormone, called ecdysone, initiates the moulting process, while another, called juvenile hormone, ensures that the epidermal cells continue to fashion juvenile rather than adult or pupal cuticle. The cells first divide, increasing their number so that they can cover the new enlarged instar. This causes the old exoskeleton to become detached from the epidermis. The epidermal cells then begin to lay down the new exoskeleton and secrete enzymes that digest much of the old one.

When the new exoskeleton has formed, the insect expands in size by taking in air. This causes the old cuticle to split along special lines of weakness, and the insect struggles free. The insect retains its new size by internal pressure until the new cuticle has hardened and can fully support the internal organs. During each moult everything that is composed of cuticle is shed. This includes the linings of the fore- and hindguts and the tracheae (air tubes), which can often be seen as wispy strands inside a cast 'skin'.

When juvenile growth is complete, exopterygote insects (▷ p. 60) go through one final moult to reach adulthood, but the endopterygotes (⇨ p. 60) moult to give the pupal stage. Before pupating, many larvae search out a suitable site: some burrow down into the soil, while others find a sheltered crevice. During the final moult, low levels of juvenile hormone cause a change in the type of cuticle made by the epidermal cells. In the course of the pupal stage much of the internal organization of the insect is broken down to be re-formed as adult tissues. The adult structures grow from small pads of embryonic tissue called *imaginal discs*, which were present but inactive throughout the juvenile phase. When fully developed, the adult cuticle is separated from the pupal skin. The adult (often called an *imago*) expands by taking in air, thus splitting the pupal cuticle and allowing it to escape. In winged species blood is pumped along the veins of the wings to expand them before the cuticle hardens.

Reproductive variations

In most insect species, both males and females are found. Reproduction involves the fusion of sperm and egg cells, each contributing half of the genetic material to the fertilized egg, which may develop into either a male or a female. In *parthenogenetic* species, however, males do not occur and females are able to produce fertile eggs unassisted. In such cases, all the genetic material comes from the mother and only female offspring are produced.

In bees, ants and wasps (Hymenoptera), there is frequently a reproductive system that allows the sex of the offspring to be determined. Mated females have a supply of sperm to fertilize their eggs. If the eggs are fertilized, they give rise to female offspring; if not, they give rise to males. If males are scarce, then some females will be unmated and so produce only males, thus increasing the proportion of males in the next generation. Queen honeybees have a valve on the spermatheca and so can decide whether to produce males by choosing whether to fertilize the eggs.

In some parasitic wasps there is an unusual condition known as *polyembryony*. The wasp lays a single egg that contains up to 3000 embryos, each of which develops into an individual. Although this would appear to be a very efficient way of reproducing, especially if laying eggs is hazardous, it is very rare.

Some insects are effectively castrated to prevent them from reproducing. In the social insects only the queens produce eggs. Worker ants, bees or termites are female, but they are prevented from developing functional reproductive systems during their growth. In termites a pheromone exuded by the queen is thought to keep all other females sterile. If the queen dies, then the pheromone disappears and new females with functional reproductive systems develop, one of which will replace the dead queen. In honeybees the reproductive system of a female only develops if the larva is fed on royal jelly (⇨ p. 65).

In one order of insects, the stylops (Strepsiptera), females are strange, maggot-like creatures that live as parasites in the abdomens of host insects. The fertilized eggs develop within the female's body and then emerge as larvae to colonize other insects. The host becomes quite modified by the presence of the stylops. Its coloration changes, structures such as antennae and genitalia become abnormal, and the reproductive system withers away. Such insects are said to be 'stylopized'.

A number of insects have life cycles that show several reproductive patterns. Many aphids switch from one plant host to another throughout the year. The cherry-oat aphid spends the winter as an egg on the bark of cherry trees. In the spring it grows and reproduces parthenogenetically several times. In early summer a winged generation is produced, which flies to grasses, including cereal crops, where several more generations are produced parthenogenetically. As autumn approaches, winged males and females are produced. They fly to the cherry trees, mate, and lay the fertilized eggs that will begin the cycle again. The change from parthenogenetic to sexual reproduction appears to be induced by a combination of changes in daylength and temperature.   BDT

INSECT DIVERSITY

Two stag beetles, with the male on the left. Sexual dimorphism occurs in many insects and adds another dimension to their extraordinary diversity.

Insects display an extraordinary range of diversity. In terms of size alone, they range from minute beetles less than 0.25 mm (¹/₁₀₀ in) long to stick insects that are around 1200 times as big, growing to over 30 cm (12 in) in length. This diversity of form is not limited by the number of species, as there is often variation even within a single species. For example, there are over 30 recognized colour morphs (types) of the common two-spot ladybird.

Sexual dimorphism, where males and females are quite different from one another (⇨ p. 159), is common in insects and adds yet more diversity. A familiar example is the stag beetle, in which the males have enormously enlarged mandibles, which they use to wrestle with other males, while the females have normal-sized ones. Another striking example is seen in coccids – small insects that feed on plants, frequently in large numbers. Female coccids are flattened, scale-like animals, which are anchored to plants by their long, sap-sucking mouthparts and are often covered in a fluffy cushion of wax filaments. They are wingless and virtually immobile. In contrast, the males have wings and are small and active, flitting from female to female and mating with each in turn.

COMPLETE METAMORPHOSIS

Complete metamorphosis is found in endopterygote insects such as bees, beetles and butterflies (⇨ pp. 60–1).

As it hatches from the egg, the *larva* (called a caterpillar in the case of moths and butterflies) bears no resemblance to the adult form.

The larva commonly lives in a different environment from the adult, and may depend on a different food source. The larva feeds and grows, often for long periods, finally becoming quiescent and entering the protected pupal stage.

Within the *pupa* (called a chrysalis in moths and butterflies), the insect undergoes complete metamorphosis – wings appear externally for the first time and the insect takes on the appearance of an adult.

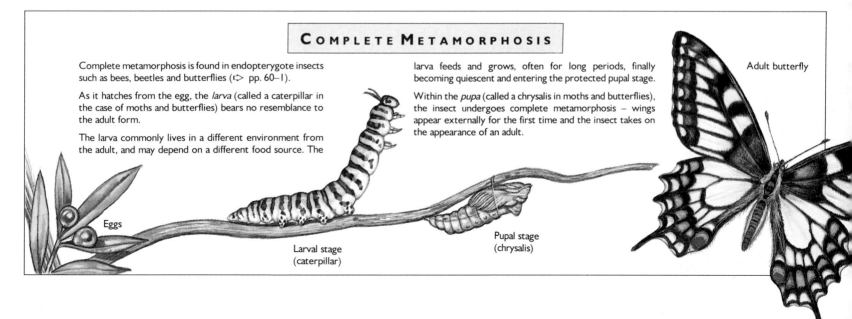

Adult butterfly

Eggs

Larval stage
(caterpillar)

Pupal stage
(chrysalis)

Insect Interactions

With such an abundance of species it is hardly surprising that insects have evolved many complex interactions between themselves and with plants and animals, including man. Such interactions take virtually every conceivable form, from highly damaging parasitic relationships to symbiotic associations that may be vital to the well-being of the species concerned.

For the most part insects are their own worst enemies. Without the very large numbers of predatory and parasitic insects, we would have a far greater problem with insect pests than we do. It is often forgotten that in a balanced natural system pests as such do not exist (⊳ p. 230). The artificial agricultural ecosystems that dominate much of the landscape are either too short-lived or too closely manipulated, for example by the use of pesticides, to allow the development of a balanced community. Predatory insects can only survive where there are suitable prey. The many insects that feed on crops become serious pests because predatory insects need time to find them and then to increase their numbers to a level that can effectively reduce the pests' numbers.

Impact on man: negative...

The repercussions of insect activities on man are not limited to crop damage. Certain insects have evolved to colonize the woody tissue of trees, and as such are important in the natural breakdown of dead wood in forests. Such insects do not discriminate between dead trees in a woodland and timber used in construction. So in temperate regions man wages an endless war against several beetle species, including deathwatch beetle and woodworm, and in the tropics against termites. Man's food stores form convenient concentrations of flour and grain, and are attacked by weevils and other seed-eating insects. Clothing and furniture are consumed by insects such as fur beetles, museum beetles and clothes moths that normally help to break down the remains of dead animals.

Some insects use man as a food source. The human botfly of tropical America lays its eggs on the skin and the hatching larvae burrow into the skin and feed on body fluids. It is related to the warble flies and botflies that attack cattle. Mosquitoes, sandflies, black flies, tsetse flies, bed bugs, fleas, body and head lice – all feed on human blood, which they extract by means of their fine piercing mouthparts. The feeding process alone, especially if the insects are abundant (as for example mosquitoes in the arctic summer), can be extremely irritating and even life-threatening. A far greater danger, however, particularly in tropical areas, arises from the fact that most of these blood feeders can transmit diseases from human to human or animal to human (⊳ photo). Malaria, sleeping sickness, yellow fever, river blindness, leishmaniasis, viral encephalitis, and epidemic typhus are just some of the diseases spread by blood-feeding insects.

...and positive

Besides their invaluable role as predators and parasites on other insects, a small number of insects are used directly by man and actively cultivated. Before the advent of synthetic varnishes, articles were lacquered to impart a transparent shiny finish. The lacquer was made from the resinous protective scale produced by the sedentary female Indian lac insect, a species of scale insect that feeds on plant sap. Another scale insect, a native of Mexico that feeds on cacti, is the source of the red food colorant cochineal.

Despite the enormous advances in synthetic fibres for clothing, silk is still highly valued for its luxurious feel, warmth and lightness. Silk comes from silk moths, or more precisely from the fully developed larva – the silkworm – which spins a silken cocoon in which to pupate (⊳ photo and p. 160). The larvae are farmed on large trays of the mulberry leaves on which they feed. Moths have been selected for their high yields of silk, with the result that the domesticated silkworm moth now produces a cocoon that is too dense for it to escape from naturally.

Honeybees provide us with a number of useful products. Bees have been kept for honey and honeycomb for thousands of years. Before the production of sugar from sugar cane and sugar beet, honey was the only sweetening agent available. Bees also provide beeswax, which is used as furniture polish and in the manufacture of high-quality candles.

Bees are able to produce both honey and wax from nectar, which they collect from flowers as a reward for carrying pollen from one flower to another (⊳ p. 42). During the flowering period of crops such as fruit trees, hives of bees are moved from orchard to orchard to ensure successful pollination and to increase the likelihood of a good harvest.

In recent years there has been a growth in the popularity of royal jelly as a health product. This is the substance that is fed to larvae so that they develop into queens rather than workers. Royal jelly is produced by glands around the mouth of worker bees and is deposited in cells containing the developing queen larvae. Despite considerable research into royal jelly, it is still not understood how this substance affects the development of larvae. Still less is there any evidence that royal jelly has any effect on humans.

Social insects

The usefulness of honeybees is in large part attributable to the way that the individuals that make up a colony exist and interact together. Along with other social insects – termites, ants, bumblebees and wasps – honeybees have developed a complex society in which individuals do different jobs for the good of the colony as a whole. Workers go through a succession of jobs in a honeybee colony. Just after they emerge as adults, they help in nursing the young larvae and feeding the queen, whose sole function is to lay eggs. After about a week, by which time their wax glands have begun to function, the workers become involved in constructing and maintaining the combs (⊳ p. 161). At the end of around three weeks, the workers take on the role of guards and start to forage for nectar and pollen. In termites (and to a lesser extent ants)

Blood-feeding insects such as mosquitoes are of enormous economic significance as the intermediary hosts of many of the most serious diseases afflicting humans and domestic livestock. Intensive research has been carried out into the biology and behaviour of these insects in attempts to reduce the severity of their impact.

themselves both as larvae and later as adults. Such species are usually brightly coloured in black, white and yellow or orange – colour combinations that have become recognized throughout the animal kingdom as warning colours (▷ p. 166). For example, cinnabar moth caterpillars, which store the cardiac glycosides from their host plant, have orange and black stripes and the adults red and black ones.

Although many plants lose out to insect herbivores, some derive considerable benefits from having insects on them. The most important of these is that of pollination symbiosis (▷ p. 42), but it is far from being the only one. Tailor ants sew leaves together with silk obtained from their larvae (▷ p. 160). This reduces the effectiveness of some of the plant's leaves, but these ants are highly protective of their nests and therefore of the plant too. Some plants, including various species of acacia, go further than this and actually provide suitable structures for ants to live in. These so-called ant plants provide hollow structures for the ants to nest in and feed them with nectar-secreting glands on leaves and with specialized food bodies. The ants concerned are usually fairly large and armed either with a sting or with chemicals. They are predatory and vigorously protect their nest plant against herbivores.

A similar system of protection is seen between ants and aphids. The aphids feed on plant sap and have an excess of sugar and water in their diet. This waste, known as honeydew, is 'milked' by the ants by stroking the aphids' abdomens. In return, the ants offer protection against predators and parasites. Colonies of aphids protected in this way grow more quickly and can therefore spread more rapidly to other plants. BDT

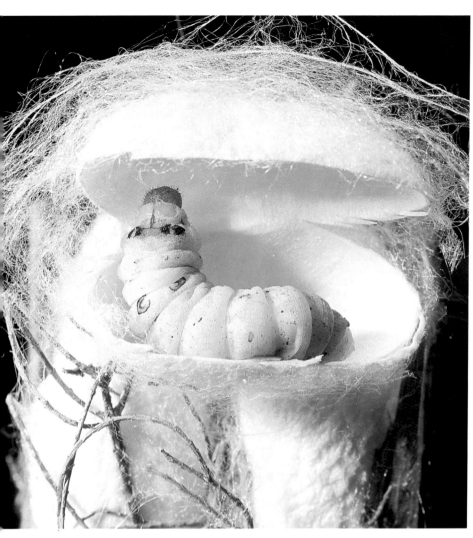

The cocoon of a silk moth, cut open to show the larva or 'silkworm' within. After hatching, the larvae are transferred to trays where they are fed on a diet of mulberry leaves. To prevent damage to the cocoons, the larvae are killed in high-temperature ovens before they pupate. Around 200 kg (440 lb) of leaves are needed to produce just one kilogram of silk.

there are several castes that fulfil these roles without change (▷ p. 156).

In the leaf-cutting ants a caste of small individuals known as minors look after special areas called fungus gardens. The ants cut off pieces of leaf and bring them back to the nest, where they are shredded, chewed with saliva, and then added to the garden. A particular kind of fungus grows on this material and produces small knobbed structures that are eaten by the ants. A relationship similar to this also takes place between a fungus and certain species of termite (▷ p. 160).

Associations with fungi are not restricted to social insects. Wood-boring beetles use fungi to assist them in obtaining nutrients from wood. These fungi have cellulase enzymes, not present in the insects, which break down cellulose to sugars that the insect can digest. When a female beetle lays an egg, she anoints its specially roughened end with fungus. When the young larva hatches from the egg, it eats through the roughened end and in doing so picks up the fungus.

Insects and plants

Plants and insects have evolved together for some 400 million years, ever since the latter began to colonize dry land in the Devonian period. Since well over half of all insects are herbivores depending on vegetation for food, many remarkable interactions have developed. One likely cause of the very considerable variety of insect (and plant) species is the 'arms race' that has carried on between the two groups. Insects eat plants, but in the course of evolutionary time the latter have developed chemicals, called 'secondary plant compounds', which have an important role in making them unpalatable to insects and so reducing the amount of insect damage. As plants evolved new chemical protection, so insects evolved biochemical mechanisms to detoxify them. Thus those insects with the necessary biochemical apparatus to cope with specific secondary plant products had those plants to themselves. Over time this has led to a high degree of specialization in both plants and insects.

The toxic substances produced by plants are sometimes used by insects for their own protection. A number of caterpillars store the secondary plant compounds they consume from their host plant to protect

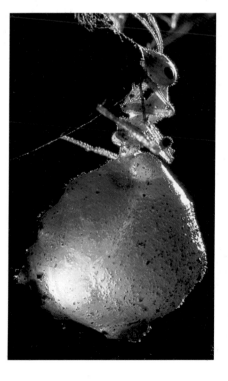

All for the good of the community... In colonies of honey ants, special workers known as 'repletes' are fed by their sisters on nectar and honeydew until their abdomens become grotesquely bloated. Hardly able to move, they usually remain hanging from the ceiling of a cavity in the nest. At times of scarcity, these living honeypots regurgitate their contents for the benefit of their needy fellows.

Crustaceans

Crustaceans have been called 'the insects of the sea', for their success as marine invertebrates is comparable to that of insects on land. They are the major component of the marine zooplankton – the vast mass of small animals that drift through the upper waters of the sea – and they are extremely abundant elsewhere. The freshwater species, though numerous, suffer competition from insects and are less abundant. A very few – land crabs and woodlice – have successfully adapted to life on land.

The crustaceans (subphylum Crustacea) form a very varied group of arthropods (▷ p. 58), with some 42 000 species divided into 10 classes. Of these the most important are fairy shrimps (Branchiopoda), copepods (Copepoda), barnacles (Cirripedia), crabs, shrimps and prawns (Decapoda), sand hoppers (Amphipoda), and woodlice (Isopoda).

Most shrimps blend in with their environment by means of camouflage, so reducing the impact of predation. A few species, however, are brightly coloured: this is either warning coloration, advertising their unpalatability to potential predators (▷ p. 167), or it may indicate that they are cleaner shrimps, providing a valeting service for fishes (▷ p. 169).

General characteristics

Crustaceans are typically elongated animals with many-segmented bodies clearly divided into a head/thorax region (the *cephalothorax*) and a tail or abdomen. The cephalothorax is often covered by a protective shell plate (the *carapace*), which is made of chitin strengthened with calcium carbonate to form a hard shell (▷ p. 58). Each segment bears a pair of limbs. The head segments bear sensory antennae, compound eyes and mouthparts, and behind these in primitive forms is a series of similar jointed limbs. In more advanced crustaceans the limbs have different functions and are structurally modified. The limbs on the thorax are generally used for walking, digging or swimming, while the abdominal limbs are usually adapted for respiratory and reproductive purposes.

Primitive crustaceans, such as fairy shrimps, feed by means of limbs fringed with hairs. As the animal moves, these automatically sieve the water and collect small food particles. Most planktonic forms and all larvae also use this sieving technique, but use only a few specialized limbs. More advanced crustaceans have adopted every conceivable feeding habit. Their strong mouthparts mean that they are well suited to chewing and biting, and many are predatory. The front limbs of crustaceans often take the form of claws specialized for food capture and defence.

As in other arthropods, growth in crustaceans can only occur at moulting. In this hormonally controlled process a new, soft cuticle is laid down beneath the old one, which by this time has been weakened by reabsorption of most of its valuable material. As the animal breaks out of the old cuticle, its body swells by intake of water or air to the new size before the new cuticle begins to harden. For many large crustaceans such as crabs this is a time to seek shelter, as soft crabs cannot defend themselves. Moulting of this kind occurs periodically throughout the lives of most crustaceans.

Most crustaceans are sexually differentiated, and as their sperm is non-motile, mating is necessary to achieve fertilization. For many this is only possible when the female has recently moulted,

Sea slaters, related to the familiar terrestrial woodlice, are important scavengers on rocky shores throughout the world. They emerge at night to feed on fragments of seaweed left behind by the receding tide.

and she may produce chemicals (pheromones) at this time to attract a mate. Female lobsters seek out a suitable male in his lair and enter. Secretion of pheromones subdues his aggressive tendencies and induces courtship. The female moults and mating occurs, after which the male protects the female until her carapace hardens and she can leave to spawn. Marine forms produce free-swimming larvae that disperse to feed and grow in the plankton. All crustaceans have the same primary larva, the *nauplius*, but later stages may be different. Eventually a special moult transforms the larva to a juvenile.

Barnacles

Barnacles are unusual but highly successful crustaceans that have become adapted for a sedentary life in which they feed on tiny planktonic plants. As larvae they resemble other larval crustaceans, swimming actively in the plankton. Later, however, they settle on a carefully chosen surface, usually a rock, although oceanic species will attach themselves to virtually anything, including whales and turtles. Once settled – preferably close to other members of the same species so as to allow mating – they glue themselves down by means of head glands. All organs except the thoracic limbs become reduced in size, and the animal secretes a protective shell, complete with closeable lid. From this the thoracic limbs, which are fringed with hairs, are protruded to sieve plankton from the water. The success of this way of life can be judged by their abundance on rocky seashores. Most barnacles are hermaphrodites, but in a few species there is a separate, minute male that lives parasitically on the female (▷ box, p. 14).

Copepods and krill

Copepods are insignificant in size, the biggest being less than 1 cm (⅖ in) in length, but they are undoubtedly the most numerous crustaceans and probably the most important. Their importance lies in the ecology of the sea, since they and their larvae are the principal grazers of the tiny plants that make up the marine phytoplankton; they are the food both of commercially important fishes such as herring and anchovy and of huge numbers of

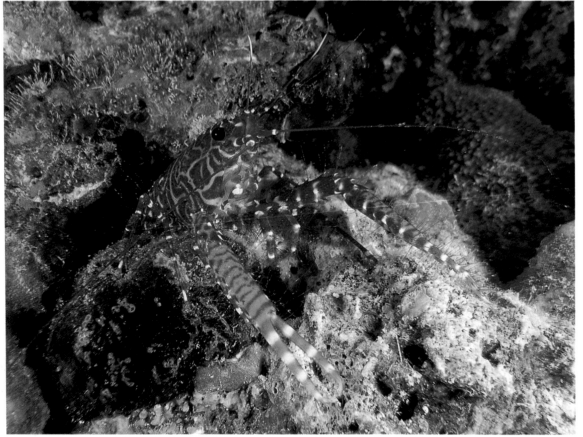

other animals, including those that form the food of many other fishes and whales. They are also important in the food chains of lakes and ponds.

Krill are larger shrimp-like crustaceans, growing to around 5 m (2 in) in length, and they feed in a similar way to copepods. They form the main food source of the baleen whales of the southern oceans (▷ p. 145), but the decimation of whale populations has led to an enormous build-up of these crustaceans around Antarctica and resulted in an increase in other vertebrates. Today krill are harvested directly by man as food for domestic animals and as fertilizer.

Fairy shrimps, water fleas and amphipods

The fairy shrimps or brine shrimps live in temporary pools where they escape predation. They are classic primitive crustaceans with rows of similar limbs used for swimming and feeding. They sieve particles from water with fringes of hair on the limbs.

Water fleas are related, more advanced forms common in fresh water. Although these tiny creatures have all the limbs except the large swimming antennae enclosed in a shell, they still feed in the same way by sieving particles.

Amphipods are small shrimp-like crustaceans found in enormous numbers in marine and freshwater habitats. Most familiar are the sand hoppers or water shrimps. These feed on rotting seaweed, and on moist nights may travel some distance inland to feed (▷ p. 170).

Woodlice

Although isopods are common in fresh and salt water, the group also includes the most highly adapted land crustaceans – the woodlice or pillbugs. Woodlice are characteristic scavengers of temperate woodland litter. They need never go near water, but they lose water rapidly in dry air and thus are mostly nocturnal. Any water that they require is quickly taken up from dew or rainfall. The most highly adapted of these leaf-scavenging forms have developed respiratory tubes similar to the tracheae of insects (▷ p. 180).

Lobsters, crabs and prawns

The decapods are not only the largest group of crustaceans, with some 10 000 species, but also the most diverse, exhibiting many different life styles and body patterns. The group includes the largest crustaceans and most of the well-known commercially important species, such as crabs, spider crabs, lobsters, crayfish, shrimps and prawns. The scientific name Decapoda ('ten feet') refers to the five pairs of walking limbs seen on the thorax of these animals. Most species are scavengers or predators, but the fiddler crabs feed on detritus, and there are some small particle-feeders.

The most characteristic feature of the true lobsters is the massive, powerful claws. These animals are typically large, and the American or North Atlantic lobster is the heaviest of all crustaceans, one enormous specimen weighing just over 20 kg (44 lb). Lobsters are secretive,

scavenging creatures of shallow waters, which spend most of their time concealed under rocks. They are easily caught because they are attracted to dead fish, and their numbers have dwindled so much in some areas that artificial culture may be profitable. Crayfish are very similar in life style but are found in fresh water. Spiny lobsters or crawfish also resemble lobsters, except that they have huge antennae and simple legs instead of claws. The antennae are strong and spiny, and are used to ward off predatory fishes. They are nocturnal in habit and omnivorous. Spiny lobsters undertake lengthy breeding migrations, often in long chains.

The prawns and shrimps of marine and estuarine waters are similar in form to lobsters but are generally smaller, and unlike lobsters, which walk on the sea bed, they swim using their abdominal limbs. They display a wide range of life styles – some are bottom-dwelling, others live exclusively in the upper waters. The cleaner shrimps, which feed by removing parasites from fishes, are brightly coloured, but most species are camouflaged for protection (▷ photo). Many feed on plant or animal remains and are thus found in huge numbers in highly productive estuarine areas, where they are extensively fished. They are generally highly tolerant of varying conditions and their tastes are catholic, so many species have proved suitable for commercial cultivation.

In crabs the abdomen is reduced and concealed under the short, broad cephalothorax (combined head and thorax). The large, characteristic pincers are used for feeding, the particular size and structure reflecting a crab's feeding habit: heavy, blunt claws are generally suitable for crushing prey such as clams; light claws with serrated edges for grasping slippery prey and cutting; and slim, delicate claws for picking up small food items. Claws also function as sex symbols. Males often have one claw much enlarged and strongly marked, which can be brandished as a threat to a rival or to impress a mate. Crabs may also use their claws defensively, and most will stand their ground rather than retreat when threatened. Although the majority of crabs walk on the sea bed, some have their back legs modified as paddles and can swim.

The largest known crabs are the Japanese spider crabs, which – with outstretched

Tropical land crabs are the largest terrestrial crustaceans. Some species need never enter water, but they are confined to moist forest conditions, where they are effective scavengers.

claws – can span around 2·5 m (8¼ ft). These and Alaskan snow crabs are the commercially exploited representatives of a large group of curious long-legged but slow-moving crabs, many of which are specialist predators. Many disguise themselves by attaching seaweed or hydroids to the back of their shell. Dead or inappropriate items are periodically removed and replaced.

The hermit crabs are unusual in having a soft abdomen, which they must protect with a mollusc shell. In order to fit in their borrowed lodgings, their abdomens are spirally coiled and asymmetrically developed. As they grow, they need to find ever larger shells, and they are frequently to be seen trying out shells for size with their claws. If suitable, the hermit releases its tight grip on the old shell and rapidly slips its abdomen into the new one. These crabs are often associated in a symbiotic relationship with sea anemones, deriving vicarious protection from the anemone's stinging cells, and they will painstakingly transport their resident anemone from one shell to the next. The so-called sponge crabs hold carefully selected sponges over their backs for protection and may even use an unfortunate sea urchin for the task. RHE

SEE ALSO

- PARASITISM AND EVOLUTION (BOX) p. 14
- FILTER-FEEDING (BOX) p. 57
- ARTHROPODS p. 58
- THE RHYTHMS OF LIFE p. 174
- SENSES AND PERCEPTION p. 188
- FRESHWATER ECOSYSTEMS p. 218
- MARINE ECOSYSTEMS p. 220

COLOUR CHANGE

Many animals match the colour of the background against which they live, and camouflage is a common way by which otherwise-vulnerable species evade predation (▷ p. 166). The ability to change colour against different backgrounds and still remain hidden from predators is clearly advantageous. Two types of colour change are seen in crustaceans. Some species can change colour rapidly as a result of the presence of a variety of hormonally controlled pigment sacs or *chromatophores* in the skin. These respond to changes in the background as perceived by the eyes and concentrate or disperse the pigment accordingly, thereby providing a rapid method of blending with the environment. Dusk- and dawn-active forms have black and white chromatophores, while day-active forms have a range of colours including red, yellow, blue and black. Colour change also plays an important part in the lives of cuttlefishes and other cephalopods (▷ p. 57). Many thousands of variously coloured chromatophores, each separately controlled by the nervous system, are found in the skin, and by expanding and contracting them a cephalopod can change colour instantly for concealment or display.

Colour changes are also seen in a number of vertebrates. The best known of these are the chameleons (▷ p. 88–9), but the ability is also well developed in some other groups, especially the flatfishes (▷ p. 76). Some birds and mammals, such as ptarmigans, arctic foxes and snowshoe hares, undergo seasonal colour changes in their plumage or fur (▷ p. 174).

Spiders and other Arachnids

With over 60 000 species, the arachnids (class Arachnida) are the largest group of arthropods (▷ p. 58) after the insects. Scorpions, spiders, and ticks and mites are the most important members of this group, which also includes harvestmen and pseudoscorpions.

Although far less numerous than insects, many arachnids are common and familiar terrestrial animals, and – unlike the insects – several members of the group have successfully colonized marine habitats. Arachnids also have considerable economic importance: spiders play a significant role in controlling insect numbers, while ticks and mites are involved in the transmission of a number of diseases in humans and in domestic animals.

Scorpions

Scorpions (order Xiphosura) were among the earliest terrestrial animals, first appearing on land in the Devonian period, 410–355 million years ago. The modern forms are secretive and nocturnal, usually hiding under stones or logs by day. Although often thought of as desert-dwellers, this is not their only habitat and they are found in most warm countries. They are highly distinctive in appearance, with large pincers on a squat eight-legged body and a long abdomen culminating in a prominent pointed sting. Mostly 'sit-and-wait' predators (▷ p. 164), scorpions sense their prey by means

of highly sensitive vibration and odour receptors on the underside. They then wait for it to move into range, grip it with their massive claws, and inject a paralysing venom with the sting. The venom of some species may cause death in humans, but most are not dangerous to man. Food is chewed and digestive enzymes secreted onto it outside the body; it is then swallowed in fluid form.

Scorpions undertake elaborate courtship rituals in which a male and female lock pincers and move to and fro for hours on end. Eventually the male deposits a packet of sperm (spermatophore) on the ground and drags the female to it, whereupon fertilization occurs. The female then broods the young, usually carrying them about on her back until they grow large enough to take care of themselves.

MYRIAPODS

A giant centipede (*Scolopendra heros*), a desert-dwelling species growing to a length of around 25 cm (10 in).

There are four groups of myriapods, or 'many-legged' arthropods, with a total of over 10 000 species. These animals are immediately recognizable by the highly distinctive form of their body, which is composed of a head and a long trunk with many leg-bearing segments. The various groups are not thought to be closely related, however, and only two are of significance.

The centipedes (class Chilopoda) are secretive predatory arthropods, and are found throughout the world in damp terrestrial habitats – beneath bark, logs, stones, and so on. There are some 2800 species ranging in size from 1 to 30 cm (0·4–12 in). They are restricted to such

habitats by their vulnerability to water loss in dry air, but their flattened body shape makes it easy for them to move about in confined spaces. The head has large jaws concealed beneath the modified first pair of legs, which form poison fangs. Behind the head are 15 or so similar segments with strong walking legs. Nocturnal in habit and virtually sightless, centipedes sense their prey by means of chemoreceptors on the head (▷ p. 188) and kill it with the poison fangs. They can move very quickly, using the long legs in a highly coordinated fashion (▷ p. 177).

The millipedes (class Diplopoda) number some 8000 species and are worldwide in distribution. All are herbivorous scavengers found in similar habitats to centipedes. They may be extremely long with up to 750 legs, but 100 is more common and some have as few as 12. The segments are very short, which gives greater strength, and are fused in pairs, so that it appears that there are two pairs of legs per segment. Millipedes can only move slowly over the ground, but their legs and movement patterns are also suitable for burrowing. Rapid escape from predators is impossible, so these animals have chemical defences in the form of phenols and hydrogen cyanide. They are also heavily armoured, and some short forms can roll into a ball.

Ticks and mites

Ticks and mites (order Acari) are the most widespread of arachnids and economically important as parasites, vectors (carriers) of several serious diseases, and food pests. Some 30 000 species have been scientifically described, but this is probably less than a tenth of those in existence, and they are found in enormous numbers in all environments. They are able to live in dry environments, but their skins are less waterproof than those of insects. They compensate for water loss by periodic retreats to damp habitats where they can absorb water. Unlike insects, mites have invaded the sea successfully and are found at all depths.

Most ticks and mites are tiny – adult mites are often less than 1 mm (1/25 in) in length, although ticks may be as large as 6 mm (1/4 in). They have a distinctive oval body with four pairs of legs, and mouthparts adapted for biting, sucking or sawing. All are liquid-feeders, but they have adopted many different life styles. Plant parasites suck cell sap, while animal parasites either suck blood or secrete digestive enzymes into body tissue and suck it up in digested form. They are unusual among arthropods in having a cuticle capable of great expansion during a meal. This is due to the cuticle being highly folded, and terrestrial ticks at least can fast for long periods between meals. Many feed and then drop from the host to digest the food in safety on the ground.

Ticks are important vectors of a variety of diseases in humans and other animals, especially cattle. They may transmit ence-

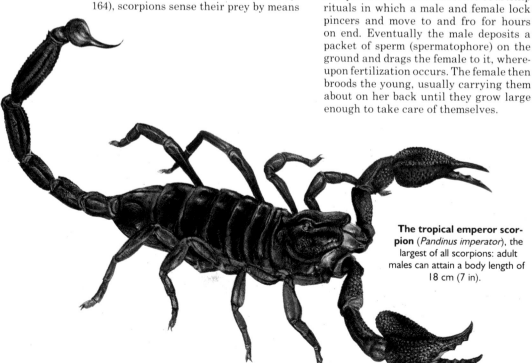

The tropical emperor scorpion (*Pandinus imperator*), the largest of all scorpions: adult males can attain a body length of 18 cm (7 in).

SEE ALSO

● ARTHROPODS p. 58
● ANIMAL BUILDERS p. 160
● PREDATOR AND PREY p. 164

phalitis viruses, rickettsias such as Rocky Mountain spotted fever, bacteria, spirochaetes and protozoans. Red water fever of cattle is a tick-borne disease. Mites are responsible for skin diseases such as scabies in humans and similar diseases in other animals. Others are important crop pests and may frequently cause problems in greenhouses.

Spiders

After the insects, spiders (order Araneae) are the most successful group of land-living arthropods – at least if judged by numbers of species. They are compact, highly specialized predators, with a body clearly divided into a combined head and thorax (the *cephalothorax*) and a large abdomen. The cephalothorax bears four pairs of walking legs and two other pairs of appendages: one is a pair of hollow fangs used to inject venom into prey; the other is leg-like in females, and a complicated reproductive structure in males. As many as eight large eyes may be seen at the front end.

At the tip of the abdomen there are a number of glands (developed only in females) that are used to produce silks. Proteinaceous fluids secreted by these glands dry as threads as they are forced out through three pairs of external nozzles (*spinnerets*). These secretions are used in different combinations to manufacture different types of silk. Two glands produce trapping threads, which are coated by secretions from four glue glands on expulsion. Dry threads, used by a spider to anchor its webs or to travel from place to place, are produced by four 'bottle' glands (the name describing their characteristic shape). There are six special glands for egg-cocoon thread, two for parcel-wrapping prey, and two for fastening dry threads to surfaces.

Spiders feeding on insects have evolved an amazing variety of techniques for capturing their prey (▷ p. 164). Day-active forms include jumping spiders, which detect and identify prey visually and then leap upon it from a distance, and camouflaged crab spiders, which sit in flowers and ambush visiting insects. Trapdoor spiders leave radiating silken trap lines and rush out from their burrow when an insect stumbles over one. A wide range of webs are constructed to catch crawling, hopping or flying insects. Sheet and hammock webs trap insects that crawl or hop onto them, while the familiar orb webs slung across gaps catch flying insects. In fact relatively few spiders can weave orb webs, but this design is very efficient, catching as many as a hundred insects a day. Orb-web spiders are wary of larger prey and may attempt to immobilize it first by wrapping it in silk. Although spiders feed principally on insects, the largest species, which are the size of a human hand, can tackle small birds and mammals (▷ illustration).

Female spiders often do not distinguish between mates and prey, and as the female is usually larger, elaborate rituals are

HORSESHOE CRABS

The horseshoe crabs or king crabs (Merostomata) are not crabs at all but the remnants of an ancient group closely related to the arachnids. These bizarre creatures have existed virtually unchanged since the Carboniferous period (355–300 million years ago). The four surviving species are found in shallow water on sandy and muddy shores, where they feed on small

Horseshoe crabs gathering together to lay their eggs.

burrowing animals. Concealed beneath a horseshoe crab's protective shell, the body has a combined head and thorax with five pairs of legs and an abdomen with a long, stiff tail. The abdomen bears a series of flat plate-like gills.

gone through to allow successful mating. Semaphore-like signalling from a distance combined with a distinctively striped body pattern may be necessary for a female to recognize a male of her own species and in order to suppress her predatory tendencies. Offering a silk-wrapped food item may give a male enough time to mate and depart before the female has finished eating, while some shrewd males present a parcel that is nothing more than silk wrapping. Female spiders make better mothers than wives, however, and will fiercely protect the silken sac in which their young develop.   RHE

A bird-eating spider (*Theraphosa blondi*), one of several South American species that can attain a leg span of around 25 cm (10 in). Although these spiders do indeed prey on small birds, they will attack any animal of an appropriate size, including insects and small mammals.

Fishes
Evolution and Classification

Almost three quarters of the globe is covered with water, and the depth of the deepest oceans exceeds the greatest mountain heights. In this vast space and in the enormous range of habitats that it provides, fishes have become extraordinarily diverse. Indeed, the number of known fish species is roughly equivalent to the combined total for all the other vertebrate groups put together.

The earliest fossil fishes do not offer any clues to the invertebrate ancestors from which they may have evolved, and there are no fossil forms intermediate between these primitive fishes and any invertebrate group. Comparison with living forms suggests that the ancestor of the vertebrates was a chordate with many of the features seen in tunicate larvae and in the amphioxus or lancelet, while various clues point to the echinoderms, such as starfishes and sea urchins, as the invertebrate group most closely related to the vertebrates (▷ box, p. 51).

The first fishes

The earliest fossil fishes date from the late Cambrian/early Ordovician period, some 510 million years ago. The fragmentary remains consist of bony plates that fitted together to form an armour covering the head and front part of the body of jawless fishes (*agnathans*) known as *heterostracans*. The microstructure of these fragments indicates that these fishes had already developed complex dermal (i.e. non-skeletal) bone tissue similar to that found in later vertebrates. Heterostracan fossils are relatively rare in Ordovician rocks but they become more numerous and quite diverse in form around the middle of the Silurian, around 420 million years ago.

At about this time the remains of other jawless fishes called *thelodonts* and *osteostracans* appear. The former were entirely covered in small *denticles* (tooth-like scales) like those found in the skins of modern sharks and rays, while the latter had their heads encased in a single-piece head shield. The freshwater *anaspids*, which appear later in the Silurian, about 410 million years ago, had no conspicuous armour but were covered with scale-like bony plates. All these groups of agnathans became extinct at various times in the Devonian period (410–355 million years ago). The only jawless fishes alive today are the hagfishes and lampreys, whose remains are first found in the Carboniferous period (355–300 million years ago).

An important development in the evolution of fishes was the development of jaws, which are first seen in the *acanthodians* of the mid-Silurian, before the major radiation of the agnathans but nearly 100 million years after the first fishes had appeared. The acanthodians also had several other features, including paired pectoral and pelvic fins, that gave them an appearance similar to the majority of the fishes we know today. By about 200 million years ago, however, this group had died out and been replaced by more advanced fishes. Later in the Silurian there also emerged a curious group of heavily armoured jawed fishes, the *placoderms*, which disappeared in the early Carboniferous.

At the beginning of the Devonian, the *chondrichthyans* – the successful group of cartilaginous fishes represented today by the sharks, rays and chimaeras – make their first appearance in the fossil record. Even the earliest of these fishes had many features in common with their modern representatives. They practised internal fertilization (like the living forms, they had structures known as *claspers* to maintain contact during copulation), and they had a covering of denticles. Although the distinctive chimaeras were numerous and widespread during the Carboniferous period, today they are represented by relatively few species, perhaps 30 in total.

The bony fishes

The name 'bony fish' is somewhat confusing, since true bone is found in some of the armoured fossil groups described above. However, the term is intended to indicate the group that has an internal skeleton of endochondrial bone, i.e. bone that gradually replaces cartilage during development – a feature found only in the bony fishes (*osteichthyans*) and all higher vertebrates.

There are fragmentary fossil remains of bony-fish skeletons from the late Silurian period, although more complete fossils are not found until the Devonian, by which time the group appears to have already become quite diverse and successful. The appearance of bony fishes in the fossil record thus pre-dates that of the chondrichthyans, although it is probable that the latter in fact evolved earlier.

By the end of the Devonian the bony fishes had become the dominant fish group, diverging in the process to form two distinct evolutionary lines. The first of these groups, the ray-finned fishes or *actinopterygians*, subsequently underwent a major diversification, and this group contains the great majority of the fishes alive today. Its living representatives include a number of small, isolated subgroups – the sturgeons and paddlefishes (25 species), garfishes (7 species) and the bowfin (1 species) – which are the survivors of larger groups that were quite successful in the past but most of whose members had died out by the end of the Cretaceous period, 65 million years ago. Their demise coincided with the spectacular radiation of the *teleosts*, by far the largest subgroup of bony fishes and the largest of all vertebrate groups. This one group contains over 20 000 species – over 95% of all living fish species.

The other group of bony fishes has only a handful of living species – the coelacanth and the lungfishes – but is of enormous interest, since it is the group from which the tetrapods or terrestrial vertebrates (including ourselves) are believed to have arisen. For this reason it has been the subject of the most intense scientific study and debate (▷ box).   OC

THE COELACANTH

In 1938 a living species of coelacanth was discovered and soon rocketed to a level of fame that has been matched by few animals in the history of biology. Fossil coelacanths or lobe-finned fishes (Crossopterygii) were discovered as early as 1839, and some 80 species were described in the following hundred years. They had clearly been a successful group, widely distributed in marine and freshwater habitats. However, there were no fossil coelacanths found after the Cretaceous period and it was believed that the group had died out some 80 million years ago.

The first specimen, retrieved from a trawler in December 1938, was found near the mouth of the Chalumna River off the coast of South Africa. The curator of a local museum, Marjorie Courtney-Latimer, realized that the fish was unusual, and after some difficulties, partly due to the Christmas holidays, managed to bring it to the attention of the South African ichthyologist Professor J.L.B. Smith. He recognized its great significance and named the fish *Latimeria chalumnae* in honour of Miss Courtney-Latimer and of the locality where it was captured. Despite intensive efforts by Smith and others, including a widely publicized offer of a reward, it was 14 years before another specimen was captured. This was partly due to the fact that the first specimen turns out to have strayed widely from the region that coelacanths chiefly inhabit, and partly due to their relatively inaccessible deep-water habitat. Since then 130 captures have been recorded and live specimens have recently been filmed in the wild.

Why is the coelacanth so important? Much of its significance lies in its supposed relationship to land vertebrates or tetrapods. Although many experts now regard the lungfishes as the most likely ancestral stock from which the first amphibians arose (▷ p. 78), the coelacanth still has its champions and it certainly stands in a close relationship to tetrapods. As the sole survivor of what was thought to be a long-extinct group, this fish has much intrinsic interest. Its discovery was virtually equivalent to finding a living dinosaur, and it prompts us to wonder what other surprises the oceans may hold in store.

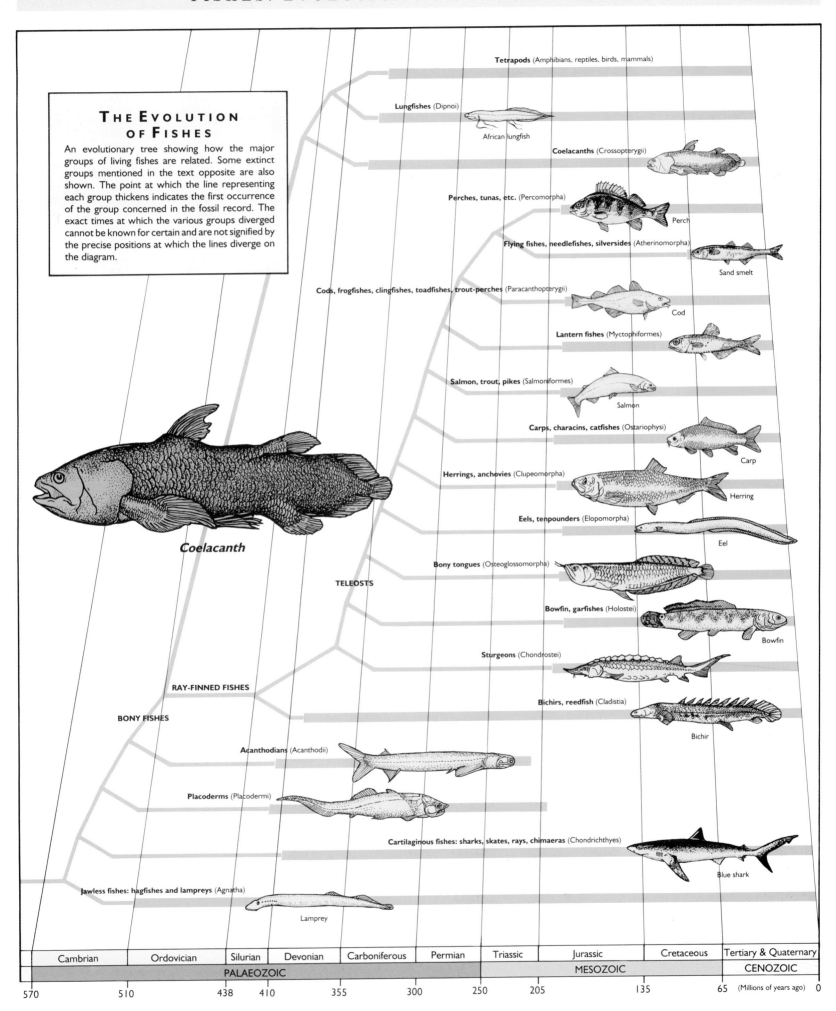

THE EVOLUTION OF FISHES

An evolutionary tree showing how the major groups of living fishes are related. Some extinct groups mentioned in the text opposite are also shown. The point at which the line representing each group thickens indicates the first occurrence of the group concerned in the fossil record. The exact times at which the various groups diverged cannot be known for certain and are not signified by the precise positions at which the lines diverge on the diagram.

Tetrapods (Amphibians, reptiles, birds, mammals)

Lungfishes (Dipnoi)

African lungfish

Coelacanths (Crossopterygii)

Perches, tunas, etc. (Percomorpha)

Perch

Flying fishes, needlefishes, silversides (Atherinomorpha)

Sand smelt

Cods, frogfishes, clingfishes, toadfishes, trout-perches (Paracanthopterygii)

Cod

Lantern fishes (Myctophiformes)

Salmon, trout, pikes (Salmoniformes)

Salmon

Carps, characins, catfishes (Ostariophysi)

Carp

Herrings, anchovies (Clupeomorpha)

Herring

Eels, tenpounders (Elopomorpha)

Eel

Bony tongues (Osteoglossomorpha)

TELEOSTS

Bowfin, garfishes (Holostei)

Bowfin

Sturgeons (Chondrostei)

RAY-FINNED FISHES

Bichirs, reedfish (Cladistia)

Bichir

BONY FISHES

Acanthodians (Acanthodii)

Placoderms (Placodermi)

Cartilaginous fishes: sharks, skates, rays, chimaeras (Chondrichthyes)

Blue shark

Jawless fishes: hagfishes and lampreys (Agnatha)

Lamprey

Coelacanth

Cambrian	Ordovician	Silurian	Devonian	Carboniferous	Permian	Triassic	Jurassic	Cretaceous	Tertiary & Quaternary
		PALAEOZOIC					MESOZOIC		CENOZOIC

570   510   438   410   355   300   250   205   135   65   (Millions of years ago)   0

Fishes
Form and Structure

The characteristic adaptations of fishes are related to the need to propel themselves through water – a much denser medium than air – and to extract oxygen from water. An elongated, streamlined body shape, a well-muscled hind end of the body and a powerful tail fin are all adaptations to propulsion through water, while the gills absorb oxygen with great efficiency.

Fishes typically have a covering of protective scales, the most familiar being the disc-like overlapping plates seen in the majority of bony fishes. The forward edge of each scale is embedded in the dermis (the inner layer of skin) and all the scales are covered by a continuous sheet of epidermis. Although this is the most common type of scale, highly modified forms are found, for example, in sturgeons, bichirs and swordfishes. Some groups, such as mackerels, tunnies and eels, have very small scales, while others, including lampreys, some catfishes and deep-sea fishes, have none at all.

In addition to the vertical tail fin, most fishes have vertically positioned dorsal fins on the back and an anal fin behind the anus. These can usually be raised or lowered, as can the paired horizontal pectoral and pelvic fins, which correspond to the fore and hind limbs of land vertebrates.

Movement

While terrestrial animals must lift their weight clear of the ground as well as provide directional thrust, fishes have only to overcome a relatively small gravitational force. Some can achieve near-neutral buoyancy by means of the *swim bladder*, a gas-filled sac in the upper part of the body cavity, which is thought to have developed from the lungs found in the most primitive bony fishes (⊳ below). This is either filled with air taken in at the surface of the water via the mouth or charged with gases from the blood by means of a special gas gland. All locomotory effort can thus be used to produce and control directional movement.

In most fishes forward propulsion is produced by a series of spinal contractions that cause undulations in the lateral curvature of the body. As these waves pass backwards down the body, the effect is to present a series of moving, inclined planes pushing outwards and backwards against the water. These waves are of very small amplitude at the head end, which oscillates only slightly from side to side as a fish swims, but they become more pronounced towards the tail. The body is often laterally compressed towards the tail, so presenting a larger surface to the water. The tail fin may contribute up to 90% of the total thrust, but some fishes, such as eels, have very small tails and it is the undulations of the body alone that provide the propulsive force.

Many fishes depart quite widely from the usual swimming method. Some, such as rays and flatfishes, which live on the sea bed, have greatly flattened bodies. Flatfishes (⊳ p. 74 and photo, p. 76) undulate the whole body and/or the vertical fins (now in the horizontal plane) to produce the waves that generate thrust. The rays have greatly enlarged pectoral fins, the edges of which move in waves – the tail is relatively small and thin, and plays little part in forward locomotion (⊳ photo, p. 75). These bottom-living fishes cannot swim fast and tend to rely on concealment to escape their enemies, many of them being well camouflaged or able to change colour to blend with that of the sea floor.

Other fishes, such as boxfishes and seahorses, have sacrificed some ability to swim fast but have become relatively inedible. Their bodies are encased in an armour of bony plates, which makes undulation of the body impossible. They have well-developed swim bladders for near-neutral buoyancy, and propulsion is provided by movement of the fins. This makes for a very stately but rather slow progression through the water.

Except in these armoured fishes, the fins are generally used for steering and balance but they can contribute to propulsion. In fishes that can maintain near-neutral buoyancy, including most teleosts (⊳ p. 70), the paired pectoral and pelvic fins can also be turned so as to act as brakes. This accounts for the precise swimming movements of aquarium fishes, which can come to an abrupt stop in their forward motion and even swim backwards away from an obstacle. By contrast, sharks have no swim bladder, and although their large oily livers contribute towards buoyancy, they tend to sink if not swimming forward. Their pectoral and pelvic fins are relatively rigid, so they cannot be turned to face the water. Instead of braking when it encounters an obstacle, a shark has to avoid it by turning the head and steering to one side.

Respiration

Although most of their internal organs are similar to those of terrestrial vertebrates, fishes generally obtain oxygen by means of gills rather than lungs (⊳ p. 180). Since water contains around 30 times less oxygen than air, fishes' gills have to be very efficient. Although a number of shark species have six or seven pairs of gills and some hagfishes have 16, most fishes have four gills on each side of the head, which look like a row of V-shaped bars. On the hind edge of each bar there is a closely packed band of soft red filaments, which extract oxygen from the water flowing over them. Their absorptive power is increased by their large surface area, which may be about three times greater in an ocean-swimming fish such as a mackerel than in less active fishes.

Flying fishes accomplish their extraordinary gliding flights by swimming rapidly at the surface and oscillating the extended lower limb of the tail fin to provide lift and thrust. When the body is out of the water, the paired fins are spread out like wings and the fish lifts clear of the surface. Depending on wind and updraughts, the fish may remain airborne for 30 seconds, reaching a height of 10 m (33 ft) and covering a distance of 400 m (1300 ft). For this reason flying fishes sometimes crash onto the decks of ships.

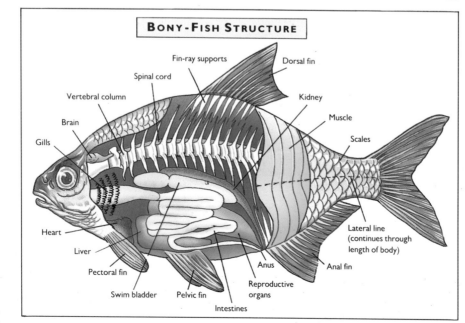

BONY-FISH STRUCTURE

Fin-ray supports
Spinal cord
Vertebral column
Brain
Gills
Heart
Liver
Pectoral fin
Swim bladder
Pelvic fin
Intestines
Reproductive organs
Anus
Dorsal fin
Kidney
Muscle
Scales
Lateral line (continues through length of body)
Anal fin

Blood passes through the gill filaments in the opposite direction to the flow of water, thus meeting successively 'fresher' water (containing more and more oxygen) as it does so. Thanks to this arrangement (the so-called 'counter-current principle'), the gills are able to absorb up to 80% of the oxygen passing over them. By contrast, humans can only absorb about 25% of the oxygen taken into the lungs.

Water for respiration is generally taken in through the mouth and expelled once it has passed over the gills. In most teleosts a sucking action by the gill cavity combined with compression of the water by the mouth ensures a pressurized flow of water to the gills, so increasing the oxygen supply. In many fast swimmers, however, the water simply flows into the mouth and freely over the gills. These include open-ocean swimmers such as sharks, tunnies and mackerel.

Some fishes have accessory air-breathing organs. In shallow or stagnant waters, poor in oxygen, a few fishes can gulp air at the surface. Some do this merely to pass the oxygen-rich surface film of water over their gills, but others have pouches with gas-absorbent linings associated with the gill chamber and retain bubbles of air from which they extract oxygen. In one species, the air-sac catfish, the air chambers extend the length of the body, while in some loaches oxygen is absorbed through the intestine, and in *Domitator latifrons*, a sleeper fish from Central America, oxygen is absorbed through the top of the head, which is richly supplied with blood vessels. The African and South American lungfishes have paired lungs joined by a canal to the floor of the mouth (the Australian lungfish has a single lung), and they are able to survive in oxygen-poor environments. At least one of the African species, for example, is able to survive periods in which its habitat dries up completely by burrowing into the mud, making a mucus-lined chamber underground, and breathing shallowly through a tubular passage running up to the surface.

Salt and water balance

The salt concentration of a fish's blood and body fluids is lower than that of sea water but higher than that of fresh water. As a result, sea fishes face the danger of water seeping out of their body (by osmosis), largely through the gills, the mouth and the gut lining, and they must drink copious amounts of sea water to compensate for this. In so doing, they also take in a great deal of extra salt, and this

has to be excreted before it builds up in the body. In contrast, in freshwater fishes there is a tendency for water to seep into the body, and this too must be excreted. Sharks, rays and chimaeras have an ingenious solution to this problem: they top up their blood with urea until it reaches a concentration closer to that of sea water, so neutralizing the tendency of water to seep in or out of their bodies. Some fishes, such as salmon and eels, can move between rivers and the sea, but they are only capable of making this adjustment gradually. A sudden transfer from one type of water to the other would kill them.

Body temperature

Fishes are 'cold-blooded' or poikilothermic (⊳ p. 87): they do not alter their body temperature to compensate for the temperature fluctuations in their environment as birds and mammals do. Although the range of temperatures encountered in bodies of water is smaller than the extremes experienced on land, some Antarctic fishes – the so-called 'ice-fishes' – live in water that is below freezing point. Crystallization of their body fluids would be fatal, but this is prevented by means of special 'anti-freeze' compounds in the blood. These fishes lack the usual oxygen-carrier, haemoglobin, which gives blood its red colour, so their blood is almost transparent. At the other extreme, some freshwater fishes can live in the waters of hot springs at nearly 40 °C (104 °F). Perhaps the most rugged individuals, however, are those that live in shore pools in temperate seas: being exposed in sunlit pools at low tide, they have to survive a very wide range of temperatures.

Some active ocean-swimmers, such as tunnies, can maintain their body temperatures above that of the water by means of a heat-exchange mechanism between closely associated veins and arteries. OC

═══════ FISH SENSES ═══════

The eyes of fishes are basically similar to those of other vertebrates (⊳ p. 189), but the transparent lens is more nearly spherical than those of land animals, which gives a wider angle of vision. In addition, fishes' eyes are usually positioned on the sides of the head so that they can see backwards down the flanks of the body and forwards down each side of the snout. When looking straight ahead fishes have binocular vision. The lens is relatively fixed in shape and must be moved forwards or backwards in order to focus on near or distant objects. This contrasts with most land animals, which focus by altering the curvature of the lens. The four-eyed fish has eyes divided into upper and lower compartments with separate retinas. The fish swims at the surface of coastal and estuarine waters with the upper compartments protruding above the surface and the lower ones submerged.

Fishes perceive sound via a pair of internal ear cavities at the back of the skull, which are divided into three chambers (⊳ p. 191). A considerable number of species are equipped with some sort of connecting apparatus that transmits vibrations picked up by the swim bladder to the ear, thus amplifying their reception. Fishes can hear sounds over a greater distance than they can see, and they have a special head-flicking mechanism that allows them to ascertain the direction from which a sound is coming even though the waves reach the closely spaced ear mechanisms at almost the same time. The lateral-line system (⊳ p. 190) helps them to detect movements in the water, such as those made by other fishes swimming.

As in land vertebrates, fishes perceive odours via their nostrils and tastes by cells lining their mouths. However, some pufferfishes are notable for their complete lack of an olfactory apparatus. In some species taste buds may be found on the fin rays or barbels, or in patches elsewhere on the skin where they may be useful in detecting food in the water (⊳ illustration, p. 188).

The ocean-swimming **sailfish** is generally thought to be the fastest species of fish over short distances. A speed of 109 km/h (68 mph) was recorded over a short three-second burst – faster than a cheetah can move on land. The sailfish is a popular sport fish, partly on account of its ability to make spectacular leaps out of the water.

Fishes
Life Cycles

Fishes are mostly active, mobile animals, and some species, particularly among those living in the open sea, may range throughout a vast area. In order to breed it is necessary for both sexes to locate each other at the right time and the right place.

In species such as salmon, which do not court or pair, the coordination necessary for successful breeding may be achieved by seasonal migration to a specific locality (▷ photo and p. 170). In such fishes the seasonal maturation of the gonads (the testes and ovaries) ensures that both sexes are ready to breed on arrival at the spawning site (▷ p. 185).

Male and female fishes

The sex of fishes is often less strictly demarcated than in other vertebrates. In certain species, including some of the lantern fishes, each fish possesses both male and female reproductive organs. In others, such as wrasses, a particular female individual in a shoal (often the largest) will change sex to replace a male that has died.

The sexes of some species look very different throughout their lives. Perhaps the most extreme examples are seen in certain families of deep-sea fishes, in which the males are for most of their lives small, degenerate, parasitic forms that remain attached to the much larger and very different-looking females. In this way they overcome the problem of the sexes failing to encounter each other in the dark ocean depths.

In species in which the sexes look very similar, changes in the body of one sex may take place during the breeding season. The male salmon, for example, develops enlarged hooked jaws. This aids synchronization of breeding condition and helps fishes in the correct identification of members of their own species. Elaborate courtship behaviour, distinc-

Salmon struggling up a waterfall to reach their spawning grounds. Salmon breed in fresh water, but the young migrate to the sea to feed and grow, returning to their native rivers after several years to breed.

tive markings, and changes in colour play a similar role (▷ box).

Fertilizing the eggs

Having either located an appropriate member of the opposite sex or arrived at the spawning site, fishes must then ensure that the eggs are fertilized. In most fishes this takes place externally – the males cover the eggs with sperm after they have been laid in the water (▷ p. 184). A cod may lay six million buoyant eggs, which float to the surface. Many of these are not fertilized or are eaten by other fishes, but a proportion survive to hatch because of their large number and wide dispersal. In some species males concentrate on fertilizing the eggs of a few or just one member of the opposite sex. The eggs of a female stickleback, for example, are usually fertilized by an individual male, who builds a nest in which the female is induced to lay her eggs by an elaborate courtship display (▷ p. 149). Some fishes, including sharks, rays, guppies, swordtails and the coelacanth, mate in pairs and copulate, the eggs being fertilized within the body of the female.

In courting and pairing species, the probability of an individual egg being fertilized is very much greater than in mass spawners such as cod, so these species generally produce fewer, larger eggs. The reproductive strategy of these fishes is thus broadly similar to higher vertebrates such as birds and mammals (▷ p. 185).

Protecting the fertilized eggs

Because eggs contain the food resources required by the developing embryo, they represent a nutritious source of food for predators. Another strategy that allows fewer eggs to be laid, with a better individual chance of survival, is the protection of the eggs after fertilization. In the cod such protection is negligible, but species such as the freshwater stickleback and many African cichlids build a nest in which the eggs are concealed and may also be guarded by one or both sexes. In the stickleback and many other nest-building species, it is the male that guards the eggs and hatchlings. The fertilized eggs of some copulating species, such as sharks and rays, are laid in horny protective capsules. Familiar examples are the 'mermaid's purses' of dogfishes, frequently found on beaches.

Probably the most advanced means of ensuring the survival of the fertilized eggs is seen in those species – including several sharks and tooth-carps – in which the internally fertilized eggs are retained within the female's body until they hatch. Some fishes – the so-called 'mouth-brooders' – incubate their eggs within their mouths for protection, and a number of species retain the young in the mouth even after they have hatched. Some cichlids do this, and the young, when eventually released, stay close to the female and at any sign of danger are taken into her mouth again until the danger has passed. Other eggs are laid within the body cavities of aquatic invertebrates or even out of the water, as in the case of the South American splashing tetra, which

COURTSHIP, COLOUR AND VISION

A tropical angelfish (*Pygoplites diacanthus*). As well as helping individuals to recognize others of their own sex, the striking coloration of this species may help to distinguish an individual that has reached breeding condition, since the pattern of markings changes with age.

In courting and pairing species, changes in coloration may take place in the body of one sex during the breeding season. Male sticklebacks, for example, develop a bright red belly. In others – notably tropical coral-reef species – astonishingly bright and highly distinctive colour patterns are believed to play a crucial role in species recognition, an essential precursor to successful breeding.

Such facts have been cited as evidence in the long debate as to whether or not fishes have colour vision (▷ box, p. 189). Experiments have now shown that, while many fishes have a developed sense of colour vision, others are only able to distinguish the brightness of an object, which varies according to its hue. With increasing depth, however, the significance of colour diminishes. Fishes living in the twilight zone of the sea, below the sunlit surface waters but not in total darkness (200–800 m / 660–2600 ft), have large eyes to gather what little light is available, predominantly at the blue end of the spectrum. Some fishes have a special reflective layer behind the eye (the tapetum) to maximize light collection, and this can be covered over to prevent the fish from being dazzled in brighter light. In the deeper waters below 800 m (2600 ft), where no light penetrates, fishes tend to have tiny degenerate eyes, as do certain subterranean cave fishes, some of which have lost their eyes altogether.

leaps onto overhanging vegetation to lay its eggs.

Young and adult forms

Young fishes are sometimes very different from the adults and may go through one or more changes in form (metamorphosis) before reaching the adult state (▷ p. 187). For example, the larvae of North American and European eels are so unlike the adults that they were at first thought to be a different kind of fish altogether (▷ p. 170). Another example is seen in the flatfishes. In early life they swim upright like most other fishes, but during development they rotate the body to present one flank to the sea bed, thus effectively lying on their sides (▷ photo, p. 76).   OC

The World of Fishes

Like other animals, fishes must eat, reach maturity and reproduce before they die. To this end they have evolved a spectacular array of predatory and defensive systems. These are a reflection of the diversity of the world's aquatic habitats and of the complexity of the communities that live within them.

Many aspects of a fish's physical form and behaviour are related to its method of obtaining food. Often the most obvious and important specializations involve the mouth and jaws. For instance, in Lake Victoria alone a single genus of cichlid fishes has diversified to capitalize on the various food sources available in the lake. Many of the hundreds of species that have evolved are distinguishable only by their specialized teeth and jaws, which are adapted to feed on a wide variety of food, including snails, insects, other fishes – and even, in some cases, just the eyes or scales of other fishes.

Feeding adaptations

In the most primitive fishes alive today, the hagfishes and lampreys, there are no true jaws at all. These creatures live as parasites on other fishes. The lampreys have sucker-like mouths in which there are rings of sharp, horny teeth. Attaching themselves to the bodies of other fishes with the sucker-like mouth, they rasp away at the flesh of their victim and drink the blood. Hagfishes are similar in habit, but they have teeth on the tongue, which they use to bore into a fish's body.

Sharks, rays and chimaeras have jaws made of cartilage (like the rest of their skeleton), and these are often armed with large teeth. This generally makes them highly efficient predators, although the largest of the sharks are filter-feeders (▷ below). The toothed sharks have rows of replacement teeth behind the ones in use. These move up to the front when the functional teeth get broken off or worn down. A few sharks, instead of having sharp blade-like teeth, have several rows united into a flat crushing plate – a feature that is also found in rays.

As well as having teeth in the jaws, bony fishes may have teeth on the tongue, on the roof of the mouth, on the bones supporting the gills, or in patches in the lining of the mouth. The pike, for example, has sharp, dagger-like teeth on most of the bones in the mouth. They are backward-pointing and hinged to prevent prey fishes from escaping once caught. The carps, on the other hand, have no teeth in the mouth, but robust crushing teeth in the throat for grinding up the plants on which they feed. Other bony fishes, including some cichlids, have a closely spaced comb-like row of teeth at the front of the jaws for scraping algae from rocks. The teeth of parrotfishes are fused into curved blades that form a strong beak. This is used to break off pieces of coral, which are crushed with grinding teeth in the throat, and then swallowed in order to digest the living polyps inside the coral.

The majority of fishes are carnivorous rather than herbivorous. The carnivorous species have bigger stomachs and smaller intestines than the herbivores; the latter have long coiled intestines, since plants take longer to digest than flesh.

Finding food

Many strategies have been adopted by fishes for locating and capturing living prey. Those that lie in wait for food organisms that venture too close include the 'angler fishes', a name confusingly given to several different groups of deep-sea fishes (▷ illustration, p. 221) as well as to the shallow-sea monkfishes. Both these and the frogfishes lure their prey by presenting a false 'bait'. Their method relies partly on visual concealment. Monkfishes, for example, have branched fleshy flaps that help to break up the outline of their body as they lie motionless in the mud, and their skin is mud-coloured. The first ray of the dorsal fin is elongated to form a long, thin rod with a little fleshy flap (the bait) at the tip. The monkfish waves the rod, causing the bait to move about in front of its huge, wide jaws. When small fishes approach to investigate, the massive jaws of the monkfish spring open and the small fish, together with a considerable amount of mud and water, are sucked into the cavernous mouth, which is full of sharp backward-facing teeth. The pike relies entirely on concealment and speed. It is well camouflaged and waits motionless in the cover of waterweeds until a suitable fish swims by. The pike then makes a lightning dash to seize its victim.

Some South American piranhas (freshwater fishes) are seemingly attracted in shoals to even small quantities of blood in the water. They are equipped with sharp cutting teeth, and there are many stories of humans and even horses being rapidly stripped to the bone by these fishes. Such stories are largely exaggerated, but a shoal of piranhas can certainly make short work of a relatively small swimming

---

MAN-EATERS?

None of the venomous or poisonous species of fish (▷ text) excites quite as much fear and curiosity as the sharks. These creatures can and do attack humans, sometimes with fatal results, but the relative infrequency of such attacks clearly indicates that we are not a normal item of food for any shark species, not even the dreaded great white shark. Fatalities are most often due to the shark biting the arms or legs and severing a major blood vessel. Barracudas and moray eels have also been implicated in attacks on divers, but such incidents are usually provoked and are too rare to qualify these fishes as dangerous to humans. A small number of apparently aggressive attacks on divers by stingrays has also been reported.

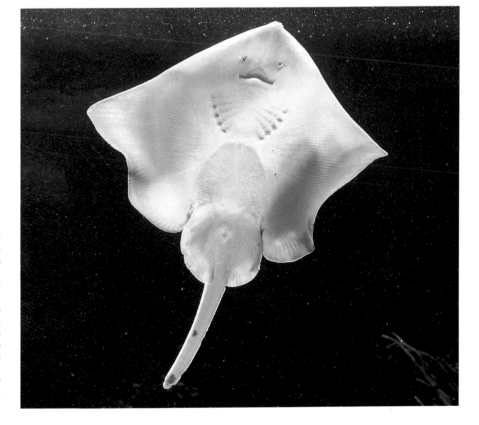

A skate's flattened body enables it to hide on the sea bed, where it lies in wait for its prey. These include crabs and mud-dwelling invertebrates, for which the skate's flat, crushing teeth are well adapted. Skates may also rise from the bottom to catch active prey such as herring, which they swim over and then drop down on from above.

The sand dab and other flatfishes can alter their coloration and markings so as to become almost invisible on the sea bed. In experiments in which flatfishes are placed on spotted or chequered surfaces, they have even attempted to assume these patterns in order to remain camouflaged.

mammal such as a capybara. The bluefish is a predatory marine fish that occasionally attacks fish shoals in large groups and kills many more than it can eat – a rare phenomenon in nature. The archerfish is remarkable for its ability to knock insects from vegetation overhanging the water by shooting a jet of water from its mouth. In order to do this these fishes must compensate for refraction at the surface of the water; they appear to be extremely accurate, usually hitting their target with the first shot (▷ box, p. 151).

A few fishes, including some flatfishes, burrow into the mud in search of prey, while others, such as the garden eels, burrow for the sake of concealment. With the front half of their bodies protruding, they wait for passing food organisms. The appearance of a whole colony of these fishes swaying in the current accounts for their common name. At the first sign of danger they all simultaneously vanish back into their burrows.

The largest sharks – the whale shark, the basking shark and the megamouth shark – are filter-feeders. They swim along with their mouths open, straining small planktonic animals from the water with their special sieve-like gills.

Many fishes undertake seasonal migrations between their breeding and feeding sites (▷ p. 170). Such movements of fish populations are of great importance to fishermen, who rely on a knowledge of them to make seasonal harvests. Although many factors may be involved, including water temperature, salinity, ocean currents, sunlight and bottom or coastal topography, food availability often plays a major part. This is also true of the daily vertical migrations made by some fishes between the surface waters and deeper waters.

Fish interactions

Many fishes form associations with other species, sometimes harmonious and sometimes competitive. Pilot fishes, for instance, follow large sharks, picking up

scraps when the shark feeds. They may also be less likely to be approached by a predator while in the company of such a large fish. Similar benefits – as well as that of effortless transport – may be enjoyed by remoras, which attach themselves to large fishes, turtles and even boats by means of a suction device on the top of the head. Some fishermen exploit this ability by attaching a line to a remora's tail and setting it off to swim after a large turtle. Once the remora has latched onto the turtle, the fisherman can haul it back to the boat. The so-called cleaner fishes form an unusual but important association with larger fishes, removing parasites and debris from their bodies (▷ box, p. 169).

Some fishes live among the stinging tentacles of sea anemones or jellyfishes, and are apparently immune to their stings (▷ p. 169). Others hide away in the body cavities of other animals, including starfishes, sponges and sea cucumbers. This type of relationship usually does no more than provide protective shelter for the fish, but some of the internal dwellers, not content with eating the food captured by their host, begin eating its internal organs and are thus truly parasitic. The South American candirú is a small, thin catfish with a pointed head that lives a parasitic life within the gill chambers of larger fishes, drinking blood from their gills. These little fishes are greatly feared for their supposed tendency to enter the urogenital openings of both men and women, where they become intractably lodged as they erect the spines on their gill covers, necessitating surgical removal. Apparently, humans fall victim when they urinate in the water, the stream of urine being mistaken by the candirú for the currents of water from the gills of a fish.

Defence mechanisms

The largest fish, the whale shark, reaches a length of 15·2 m (50 ft) and a maximum weight of 20 tonnes (tons). Once it has reached maximum size, this giant must be relatively secure from any potential predators, as must the basking shark at 10·4 m (33 ft) and the ocean sunfish at 4 m (13 ft) – and indeed any fish that attains more than a metre or so in length. In general, it is noticeable that relatively large fishes tend not to be equipped with extensive armour or elaborate defence mechanisms.

The pufferfishes and porcupine fishes are generally rather small, averaging about

---

FISHES AS FOOD

Fishes have always formed a staple food item in the diet of mankind. Most fish bodies contain a high proportion of muscle tissue, which is made up of approximately 65–80% water, 16–23% protein, and a variable proportion of fats (unsaturated, unlike that of mammals, and therefore healthier as a dietary constituent). Fish flesh also has a high vitamin content. Most of the familiar food fishes are marine species – even trout and salmon spend part of their lives at sea – although in some parts of Europe freshwater carp and other coarse fishes are eaten regularly. In tropical countries, however, where there is a greater abundance of freshwater fishes, these form a greater part of the diet in inland areas. Until recently only a small fraction of the available North Atlantic sea fishes were regularly marketed, and there seemed to be a prejudice against eating a number of quite nutritious and palatable species. In the last decade or two some of this conservatism has broken down and previously unsaleable varieties such as monkfish have become both popular and expensive, as have a number of unfamiliar imported species. Other food items provided by fishes include caviar, which consists of the roe (eggs) of sturgeons, and whitebait, which is a name given to young herring or sprats but which often includes the young of a number of diverse sea fishes.

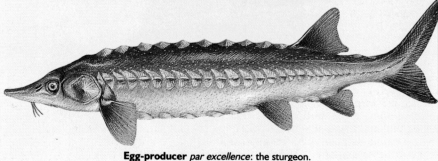

Egg-producer *par excellence*: the sturgeon.

12 cm (4¾ in), but they attempt to reduce their vulnerability in this respect by inflating their bodies with air or water so as to double or treble their size. In addition many of them are covered in spines, formed from modified scales; when the body is inflated, these stick out at all angles, making the fish a very difficult target for a predator. The boxfishes, seahorses and pipefishes have thickened scales interlocking to form a rigid shell around the body, with only the tail and other fins protruding for purposes of locomotion. Defensive spines are also found in the fins of some fishes. The three-spined stickleback, for example, can lock its anal, pelvic and dorsal spines in an erect position, making it very difficult for a predator to swallow. The catfishes are capable of the same defensive posture and may have very powerful barbed spines. The triggerfishes have a locking mechanism for their short, stout first dorsal spine. The mechanism can be released by lowering the relatively tiny spine behind it (hence 'trigger'). One of these fishes has been filmed in a coral crevice with its spine locked erect while a predator tries in vain to pull its armoured body free. The long, stout horn on the forehead of unicorn fishes is presumed to be a defensive mechanism.

Defensive camouflage is widely adopted in the world of fishes, particularly the form known as countershading (▷ p. 166). A fish's back tends to be darker than its belly, so that each aspect is camouflaged when seen, respectively, from above (against deeper, darker water) and from below (against the light penetrating the surface). The upside-down catfish of Africa habitually swims on its back and its countershading is correspondingly reversed. Bottom-living fishes are even more adept at camouflage. The flatfishes, for example, are able to vary their coloration and markings to match the sea bed where they lie (▷ photo). The leaf fish looks like a dead leaf, its chin barbel resembling the leaf stalk, while the leafy sea dragon has elaborate frond-like protrusions extending from the body so that it resembles a clump of seaweed. Yet another tactic involves mimicking a venomous or noxious animal (▷ box, p. 167). The snake eel, for example, has a shape and bright-banded coloration almost exactly the same as those of a venomous sea snake. Many fishes, including the glass catfishes, are transparent, which makes them difficult for predators to see.

Some cyprinid fishes (related to the carp) are capable of the so-called 'fright reaction'. If a fish is attacked or injured, it releases a chemical substance into the water. When individuals of the same species in the vicinity detect this substance, they automatically flee. Some fishes, such as the electric eel, electric catfish and torpedo ray, possess electric organs sufficiently powerful to stun other fishes; this capacity may be used in catching prey, but its main function appears to be to deter predators. Several other groups of fishes produce much weaker

discharges almost continuously, but these probably have a sensory function (▷ box, p. 191). A cooperative response to danger is seen in the shoaling or schooling habit (▷ box, p. 156).

Poisonous fishes

A number of fishes have developed toxic chemicals in their flesh, or spines that inject venom into their attackers. Both tactics are essentially defensive: fishes do not use their venom to capture prey or establish territories. Like their terrestrial counterparts, poisonous fishes are often conspicuously marked in order to provide a warning to would-be predators (▷ box, p. 167). There would be little point in toxic defences if enemies could only discover them after it was too late, so enemies must learn to associate the warning signals with danger. Some individual sacrifices are undoubtedly made in this learning process. Despite this general rule, the most venomous of all fishes, the stonefishes, are highly camouflaged, and their venomous spines seem to be a second line of defence. If trodden on, a stonefish may inject venom into the foot of an unwary bather, occasionally causing death through heart failure. The weeverfishes of European waters sometimes sting bathers in the same way, and although this is excruciatingly painful, it is not usually fatal. Even though weevers lie concealed in the sand, their venomous fin spines are advertised by a bold black patch on the fin. This must be an effective warning to other fishes, and it is said that harmless soles imitate this signal with a black patch on their pectoral fins. Unfortunately the warning is

The South American arapaima is one of the largest freshwater fishes (although probably not the largest, as is often claimed), an average specimen measuring around 2 m (6½ ft) in length. As the young hatch out from the eggs, which are laid in an excavated burrow, they swarm around the male's head, while the female warns off intruders. Highly prized as a food fish, the arapaima's numbers are becoming depleted.

not likely to be seen by humans. However, one could hardly miss the warning signals of a lionfish. The large, delicate, fan-like fins of these fishes, as well as the body, are covered in a beautiful pattern of vivid red and white stripes and mottled patches. This is undoubtedly to advertise the long venomous spines in the fins, and it has also made these fishes very popular in aquariums.

Among the fishes with poisonous flesh, the pufferfishes appear to cause the most violent and serious symptoms in man. Numbness leads to nausea and paralysis, and is sometimes followed by coma and death. Nevertheless, specially trained Japanese chefs still prepare these fishes, which they call 'fugu', in a way that is said to render them harmless, although many people have died from eating them. Some fishes that are normally quite edible become dangerously poisonous sporadically. This phenomenon, known as ciguatera poisoning, is thought to be due to the fishes feeding on poisonous algae that bloom irregularly in tropical waters.   OC

The surgeon fishes are so named because of the sharp, curved, blade-like spines on either side of the tail stem. These can be raised or lowered at will so as to stick out on each side; when not in use, they lie concealed within grooves in the body. A flick of the tail with the blades erect can severely injure a chasing predator.

Amphibians
Evolution and Classification

The amphibians, which include modern frogs, toads and salamanders, played a central role in the evolution of life on earth, since they were the first of the vertebrates to leave the water to live at least part of their lives on land. This meant that they and those that followed them – reptiles, birds and mammals – were able to colonize a huge array of previously unexploited terrestrial habitats. While modern amphibians are less diverse than their predecessors, they nevertheless provide some valuable clues to the kind of life their ancestors may have lived.

SEE ALSO

- THE AGES OF THE EARTH p. 4
- AMPHIBIANS: FORM AND STRUCTURE p. 79
- THE WORLD OF AMPHIBIANS p. 80
- REPTILES: EVOLUTION AND CLASSIFICATION p. 82

There is still no general agreement over the ancestry of these early amphibians. The traditional view (still held by many) is that they evolved from the same ancestral stock as the lobe-finned fishes, the group represented today by a single species of coelacanth (▷ p. 70); these fishes have fins with bony elements like the limbs of land vertebrates, not cartilaginous fin rays as in other fishes, and move their fins alternately, as if walking on land. More recently, however, an alternative theory has been proposed, putting forward the lungfishes (▷ p. 73) as a possible ancestral stock; these also display the distinctive 'walking' fin movements and share various structural features with amphibians (most notably, air-breathing lungs).

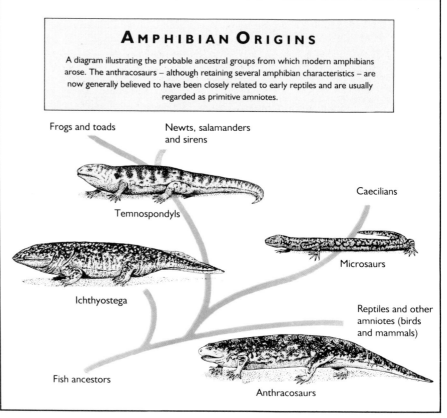

AMPHIBIAN ORIGINS

A diagram illustrating the probable ancestral groups from which modern amphibians arose. The anthracosaurs – although retaining several amphibian characteristics – are now generally believed to have been closely related to early reptiles and are usually regarded as primitive amniotes.

Frogs and toads

Newts, salamanders and sirens

Caecilians

Temnospondyls

Microsaurs

Ichthyostega

Reptiles and other amniotes (birds and mammals)

Fish ancestors

Anthracosaurs

AMPHIBIANS AND MAN

Amphibians have had a greater impact on our lives than might at first be apparent. They figure strongly in the folklore of many cultures. In Western cultures toads have long been seen as witches' familiars and no witch's brew would be complete without its eye of newt. Frogs appear in many forms of native art, from American Indian totem poles to Australian Aboriginal cave paintings. In many countries the first frog calls, such as that of the spring peeper of eastern North America, signal the coming of spring.

Frogs in particular have been the mainstay of many biology courses as a typical vertebrate for dissection. They have also been used extensively in medical research, notably into the physiology of nerve, muscle and heart function. Before the advent of more sophisticated techniques, frogs were used in pregnancy testing – injection of urine from a pregnant woman contains sufficient hormone to induce spawning in a gravid (egg-laden) female frog or toad. Toad skin-gland secretions were a source of Chen'Su or Senso, used in China for the regulation of heart beat before the discovery of digitalis.

Amphibians are also useful as sensitive indicators of the quality of fresh water (especially still groundwaters) and of changing environmental conditions. The recent and sudden disappearance – and possibly extinction – of some species has suggested that there may be international changes in groundwater (water table) levels and droughts associated with changing climatic conditions.

From water to land

What factors brought about the change to a land-based existence? It is now thought likely that the fish ancestors of the amphibians were drawn onto the land by the rich food sources – insects and other invertebrates – to be found there, for which there was at that time no competition (▷ p. 82). By coming out of the water these ancestral forms were themselves free from all predators, and therefore enjoyed ideal circumstances in which to survive and prosper.

The earliest known land vertebrate or *tetrapod* is *Ichthyostega*, which dates from the end of the Devonian period, some 370 million years ago. This creature had four well-formed limbs and a robust rib cage, which was needed to protect the internal organs under the full force of gravity. *Ichthyostega* is not the direct ancestor of modern amphibians (▷ illustration), but it closely resembled the vertebrates that made the first steps onto land.

Modern amphibians

The modern amphibians belong to three orders: frogs and toads to the order Anura; newts, salamanders and sirens to the Urodela or Caudata (the sirens sometimes being assigned to a separate order); and caecilians to the Gymnophiona. The members of each of these groups are highly distinct in body form, being adapted to life in different ecological situations. Typical frogs and toads live on or near the ground and take fast-moving prey such as insects and spiders, while newts and salamanders live in similar habitats but prefer less open situations and feed on slower-moving invertebrates such as worms, slugs and snails. Caecil-

ians are primarily burrowers and probably take earthworms and other soil-dwelling invertebrates.

The frogs and toads are generally believed to have arisen from a highly diverse group of amphibians known as the *temnospondyls*, which are first known from the early Carboniferous, some 350 million years ago. Some of these retained external gills like modern axolotls (▷ p. 81), but most were large, some growing up to 4 m (13 ft) in length, and similar in appearance to crocodiles or salamanders. It has been suggested that the temnospondyls may also be the ancestors of the newts, salamanders and sirens, and perhaps the caecilians too. However, some authorities consider it more probable that the latter are related to the *microsaurs*, a group of mainly small amphibians that occurred 300–250 million years ago. Some of these were specialized as burrowers and may be closely related to modern caecilians.

Although there remains some doubt over the immediate ancestors of modern amphibians, it is probable that they all originally evolved from a slow-moving ancestor that captured its food around the margins of ponds, a form from which the newts and salamanders in particular have little changed. Frogs and toads may have initially taken to lunging at faster-moving prey, developing larger, keener eyes, longer hind limbs to assist the forward thrust of the body, and a large head with a broader mouth and a wide gape to increase the likelihood of capturing prey. Caecilians became efficient burrowers, developing streamlined heads and worm-like bodies, while their eyes became reduced in size and function.   BTC

Amphibians
Form and Structure

The basic body plan of modern amphibians reflects their ancestry from the forms that made the transition from an aquatic to a terrestrial or semi-terrestrial life style. This shift was made possible by the solution of a number of major problems.

In order to spend at least part of their lives on land, amphibians must minimize the danger of drying out when exposed to air. At the same time, they must be able to obtain and use gaseous oxygen from the air for respiration. Finally, they have had to develop means of moving on land under the full force of gravity, without the buoyancy provided by water.

Anatomy

Amphibians are a diverse group, but most species share several common features. Amphibian skin is naked and kept moist by mucus secreted from special glands embedded in the skin; it is also permeable and capable of some control over water uptake and loss. The tongue is generally long and can be extended in capturing and controlling prey. Amphibians have simple sac-like lungs, which are well supplied with blood vessels, permitting gas exchange while limiting water loss. They are usually paired, but in caecilians the left lung is greatly reduced, while in lungless salamanders the lungs are lost altogether – they breathe through the skin and the mouth cavity, which is richly supplied with blood vessels. In addition to true eyelids, all amphibians except caecilians have a third eyelid or 'nictitating membrane' to keep the corneal surface clean and moist. The heart has three chambers, two atrial and one ventricular (▷ box, p. 181), though the detailed structure may vary in different groups.

The body form of each of the three groups of amphibians is adapted for prey capture in differing ecological situations (▷ opposite). Frogs and toads, which capture prey by 'sit-and-wait' tactics or by a rapid, short-distance pursuit followed by leaping or lunging with mouth agape, have squat bodies, with a short spine consisting of only six to ten vertebrae. The vertebrae have transverse processes (projections) but generally lack ribs. The hindmost (sacral) vertebra forms a hump-like projection in the middle of the back, and articulation at this point acts as a pivot in leaping. Adult frogs and toads lack tails; this may be important in permitting unencumbered leaping or may reflect an earlier adaptation to crevice- or burrow-dwelling. Newts, salamanders and sirens, which take slower-moving prey and walk or scramble, are more lizard-like in body form; they have 16 to 80 or more vertebrae, a relatively undeveloped sacral vertebra, and a tail complete with vertebrae. Caecilians may have 230 or more vertebrae and are limbless, worm-like amphibians, which are aquatic or burrow in soft earth. They may take earthworms or soil-living invertebrates and can lunge or strike at their prey with surprising speed.

Locomotion

In amphibians swimming is important in the larval stage and again in the adult, when it returns to water to breed. Terrestrial locomotion is important in the remainder of the adult's life – for obtaining food, finding a mate, and escaping predators. Frog and salamander tadpoles swim by lashing the tail from side to side, while aquatic caecilians, newts and salamanders swim in a fish-like manner with an S-shaped wave passing down the body from the head. Most frogs and toads swim by simultaneously thrusting the hind limbs against the water, but more primitive forms, such as the tailed frog of North America, have a rather less efficient method, using the hind limbs alternately.

Caecilians move on or beneath the ground by S-shaped undulations of the body. Frogs and toads burrow in one of two ways: midwife toads are amongst those that burrow head-first, digging with their fore limbs, while the spadefoot toads have a spade-like outgrowth on the foot and burrow into soft earth and sand by shuffling their feet. Most frogs and toads have well-developed hind limbs that can be folded and used to hop or leap, but some, such as the very short-limbed African rain frogs, can only walk or run in a mouse-like scramble. When combined with extensive webbing between the fingers and toes, as in some species of tree frog, leaping can be extended into controlled gliding or parascending. Terrestrial newts and salamanders move by lateral undulation of the body, moving diagonally opposite feet as the body bends.

Amphibian physiology

Amphibians are 'cold-blooded' or poikilothermic – they rely on the external environment to provide warmth to boost their metabolic activity, and their body temperature approximates to that of their surroundings. They can increase their body temperature by basking in the sun or reduce it by seeking shade or entering water.

Respiration takes place through the skin and the lungs, or the skin and the gills in tadpoles and permanently larval (*neotenic*) species such as axolotls (▷ p. 81). The skin is particularly important in pond-hibernating species and in lungless salamanders (▷ above).

There is a strong reliance upon water in amphibians, although some desert species are highly efficient at conserving water resources. Many species found in arid regions seek shelter during the hottest part of the day or may aestivate over several months (▷ p. 175): they drastically reduce water loss by forming a cocoon covering all the body openings except for the nostrils.

Control of body water is aided by uptake of water through the skin on certain parts of the body. Many toads, particularly in drier areas, can take up to 70% of the water that passes through the skin through the pelvic patch or seat – an area of baggy skin on the undersurface of the thighs, which is in contact with the ground when the toad is in a sitting position. Water is also important to amphibians for breeding (▷ p. 80).

Reliance on water can be a problem in cold conditions. Freezing of body water can be fatal, yet at least four North American species – the spring peeper, the wood frog, the chorus frog and the grey tree frog – can withstand freezing. The first three do so by increasing the level of glucose in the body (the grey tree frog uses glycerol), which acts as a kind of antifreeze. Ice may form in the blood and the spaces between cells, but does not do so within the cells. As long as no more than 50–60% of the body water freezes, the frog will survive. BTC

MODERN AMPHIBIANS

Newts, salamanders and sirens (order Urodela): Red-backed salamander (*Plethodon cinereus*)

Caecilians (order Gymnophiona): South American caecilian (*Siphonops annulatus*)

Frogs and toads (order Anura): American bullfrog (*Rana catesbeiana*)

The World of Amphibians

Amphibians, like other animals, perceive and respond to the world around them through the medium of their senses. Their sensory responses to food, enemies and potential mates are the key factors in their survival.

As 'cold-blooded' (poikilothermic) animals, amphibians may be either warm and active or cold and relatively inactive. If active, an amphibian needs to monitor its environment constantly and respond rapidly to sudden adverse or favourable changes, such as the appearance of a predator or potential prey. Another aspect of active behaviour is the need to locate a safe, stable haven, in order to minimize its vulnerability during its periods of inactivity.

Amphibian senses

An amphibian's perception of the world is somewhat different to our own. Sight, hearing and touch are generally the most highly developed senses in an amphibian, but sensitivity to temperature change, water loss, polarized light, and the earth's magnetic field may also be important.

Sight is probably the most important sense in frogs, toads, newts and salamanders, but less so in burrowing species and least of all in caecilians, which have vestigial eyes. Arboreal frogs have particularly large eyes set high on the top and sides of the head, giving good all-round and binocular vision. The visual cortex in a frog's brain (▷ p. 183) contains special groups of cells receptive to moving edges and to objects seen by the eye to move left to right and right to left, so enabling it to take flying and other fast-moving insect food and to be vigilant against predators when in exposed situations. Colour vision is important in mate recognition and courtship in newts and frogs, especially in brightly coloured species.

The sense of hearing is most developed in frogs and toads, and underlies a sophisticated vocal system used in locating a mate of the same species. In at least one group, the North American cricket frogs, there are call dialects, allowing females to distinguish males from different geographical regions. Touch is important in all amphibians, but especially in aquatic and burrowing species. Vibrations through the ground or through water provide information on the proximity of food, members of the same species (including potential mates), and enemies. In tadpoles and aquatic amphibians there is a lateral-line system like that in fishes, which is sensitive to water-borne vibrations (▷ p. 190).

A sense of smell is highly developed in newts and salamanders, and may also be in caecilians, frogs and toads. Newts react strongly – often violently – when presented with live food and may bite savagely at others competing for the same food item. Smell is also important in newt courtship, when the female is able to pick up chemical signals (pheromones) from the male.

Life cycles

Producing the next generation requires a great deal of mutual cooperation between male and female amphibians. The two sexes must attain breeding condition, arrive at the breeding site, and select a suitable mate of the same species; the male has to fertilize the eggs produced by the female, which then has to find a suitable egg-laying site. All this has to be achieved as quickly as possible, since amphibians are particularly vulnerable to predation during this time.

Attainment of breeding condition and migration to the breeding site are initiated by climatic changes such as changes in day length and temperature (▷ p. 174), which produce a hormonal response in adult amphibians. Many frogs, toads and newts show great fidelity to particular breeding sites, often migrating over several kilometres and passing apparently suitable alternatives to reach their 'home' pond (▷ p. 170). Vision and smell are important in locating a breeding pond, but amphibians may also be able to find their way on a cloudy day by using the earth's magnetic field or polarized light from the sun.

Males generally reach the breeding site first, whereupon they enter a pre-spawning phase. They are very active and yet may not eat; they are probably learning site landmarks and assessing the best territories. In frogs and toads no calling is heard during this period, which may last as long as two weeks in European common toads. At the end of the pre-spawning period the males space themselves out and start to defend individual territories. Frogs and toads have call battles for territories, the largest territories generally going to the most vociferous males. Actual fighting may take place – poison-dart frogs wrestle with one another and gladiator frogs live up to their name, using their sharp thumb spines to great effect, often seriously wounding their opponents.

When the females arrive at the breeding site, males give their proper mating or advertisement call. Females respond positively to the call of a male of the same species, and may also use it as a sign of a male's quality (his size or fitness as a mate). Although voiceless, male newts have an elaborate courtship display involving posturing and releasing pheromones, which they waft towards the females with their tails. The females may assess male fitness by the elaborateness of the courtship dance. Nothing is known of caecilian courtship behaviour except that there is a form of pre-mating display in aquatic species.

In caecilians fertilization of the egg is internal (▷ p. 184). The male has a pseudo-penis comprising the hind end of the genital opening (cloaca), which is turned outwards and inserted into the female's vent. In newts and salamanders fertilization is also internal, but the male produces a sperm package or 'spermatophore', which he deposits on the ground or on the bottom of a pond. He then guides the female over the spermatophore, which she takes up and stores internally. Her eggs (ova) may be subsequently fertilized as they leave the cloaca, while in some species they are retained in the female's body, where they develop into late-stage

Common toads (right) capture insects and other invertebrates with their long, sticky tongues, which flick out and in again within the space of one tenth of a second. They take only moving prey, which is swallowed by closure and downward movement of the eyes into the mouth cavity.

LIFE CYCLE OF A TYPICAL FROG

1. In a 'typical' frog the fertilized eggs, or spawn, initially have a spherical yolk, which assumes an oval shape after about a week. After ten days the position of the head, body and tail are apparent, and external gills soon appear. At about two weeks the tadpole hatches out from the egg mass and attaches itself to waterweed or some convenient surface by means of an adhesive organ; there is as yet no mouth – it feeds off its yolk supply.

2. Within a few days a mouth is formed and the tadpole feeds on algae, plant material and pond microorganisms. Soon after the mouth appears, internal gills are formed and the external ones are absorbed.

3. At about five weeks hind-limb buds appear, at seven weeks toes are discernible, and by ten weeks proper feet are formed.

4. By twelve weeks the major changes of metamorphosis commence: the stail is reabsorbed, providing a useful nutritive source for the remodelling of the body structure; the tadpole mouth disappears and a true mouth forms and widens; the eyes grow and eyelids develop; and the forelimbs are released. Internally there are major changes in the nervous, respiratory, circulatory and digestive systems.

5. The frog is sexually mature by about its third year, and returns to its native pond to breed.

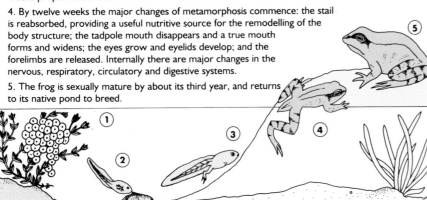

larvae or small juveniles. In most frogs and toads fertilization is external; the eggs are expelled from the female's vent and immediately fertilized by the male.

Growth and development

Amphibian eggs are relatively large and yolky (though never as large as even the smallest bird's egg); like fish eggs, they have gelatinous outer capsules but lack the protective membranes seen in higher vertebrates (⊳ p. 86). However, extra capsules may be found in the eggs of those species that lay their eggs out of water and show partial or complete direct development, in which the embryo leaves the egg as a late-stage tadpole or mini-adult.

The change from the larval to the adult stage (metamorphosis) is fairly abrupt and involves a major remodelling of body form. The change is far more drastic in frogs than in salamanders and caecilians, and is more obvious in frogs with a free-living tadpole (⊳ illustration) than in those species in which all changes occur within the egg capsule or in which the eggs are retained within the female, which gives birth to late-stage tadpoles or live young. In newts and salamanders the tadpole is more like the adult; it is carnivorous from the beginning and the fore limbs are the first to erupt.

Caecilians either lay eggs and have a free-living larval stage (similar to the adult except for the presence of large external gills) or give birth to live young.

Reproductive variations

Frogs and toads are the most innovative group of amphibians in terms of diversity of breeding behaviour. In midwife toads, the male carries one or more rosary-like egg strings entwined around its back legs, and the eggs hatch into late-stage tadpoles. There are a number of frogs that have egg pouches. Among these are Goeldi's frog, which carries its eggs in a depression on its back, the 'true' marsupial frogs of South America and Australia, and the two species of Darwin's frog, the males of which carry the young in the vocal sac until they emerge as either late-stage tadpoles or fully formed froglets. Perhaps the most bizarre form of parental care is gastric brooding, found in two Australian frog species. The female takes the young into her stomach, which suspends its digestive function to become a pseudo-uterus. Sadly, both species may now be extinct – one has not been seen since 1981, the other since 1985.

Some newts and salamanders have the ability to breed while still in the tadpole phase, a phenomenon known as *neoteny*. The best known of these sexually mature larval forms is the Mexican axolotl (a species of salamander). Neotenic newts are also known and are more common at high altitudes. The European fire salamander tends to give birth to live young at high altitudes.

Feeding and diet

Most amphibians are opportunistic feeders, taking almost any live invertebrate of suitable size. Larger species may also take fish, other amphibians, lizards, mice and small birds. Amphibians usually feed on what is most freely available in their environment – newt larvae, for instance, often eat frog tadpoles, and termites and ants are the favourite food of many tropical frogs and toads. There are some specialist feeders, such as the Mexican burrowing toad, which feeds exclusively on termites, and species of the Ethiopian genus *Tornierella*, which are snail specialists. There is even a fruit-eating frog – a small red tree frog (*Hyla truncata*), which matches the pink-red berries that seem to be its staple diet.

Defence strategies

Seen from a predator's viewpoint, amphibians – especially frogs – are a highly desirable food item. They provide a nutritious high-protein meal, with no fur, feathers or scales to deal with and relatively few bones. Thus frogs and their tadpoles are taken by a wide variety of mammals, birds, reptiles (especially snakes), and even insects such as dragonfly larvae and water scorpions. Given their widespread desirability as food, it is hardly surprising that amphibians have evolved a range of defensive capabilities.

The defence strategies employed by amphibians are principally non-aggressive – they have no venom-injection system like the fangs of snakes or the stings of insects. They tend to avoid confrontation by escape if possible, but if pressed are quite capable of effective defence. In some species the first line of defence is bluffing. Many species inflate themselves with air. For instance, the South African rain frogs – already rotund – become almost spherical, while European common toads enhance the effect by standing with stiff limbs and raising themselves off the ground. Other forms of dramatic posturing are seen in salamanders, which throw back the head and expose their brightly coloured undersurface. Such bluffing may be backed up by positive action if a predator persists. Toads and poison-dart frogs activate poison-producing glands in the skin, while various species present bony spines that penetrate the skin surface – the finger and toe tips of the West African hairy frog, hooks at the corner of the jaws of the crocodile salamanders, and the rib tips of the European sharp-ribbed salamander. Poisonous secretions are also found in tree frogs, many salamanders and newts, and caecilians.

Most amphibians are well camouflaged, with their body pattern and coloration closely matching that of their surroundings, thereby reducing the number of occasions when more strenuous defensive action is needed. Strikingly effective camouflage is seen in some species, such as the Australian waterfall and green-eyed frogs, the leaf-mimicking Asian horned toad, and the Central and South American glass frogs, which are virtually transparent in water.　　　　BTC

SEE ALSO

● AMPHIBIANS: EVOLUTION AND CLASSIFICATION p. 78
● AMPHIBIANS: FORM AND STRUCTURE p. 79
● SEX AND REPRODUCTION p. 184
● GROWTH AND DEVELOPMENT p. 186
● SENSES AND PERCEPTION p. 188

The slimy salamander gets its name from the sticky substance that is secreted from its skin when the animal is alarmed. This slime blocks the respiratory passages of a would-be predator, causing considerable distress.

Reptiles
Evolution and Classification

Although the first vertebrates to emerge from water to walk on land were early amphibians (⊳ p. 78), these animals never succeeded in making a complete break from water, so there was always a strict limit to the range of terrestrial habitats that they could exploit. The full conquest of the earth was left to the reptiles, the group that includes the dinosaurs and the living crocodiles, turtles, lizards and snakes.

The ability of reptiles to live away from open bodies of water depends on a number of features lacking in amphibians. Amongst the most important of these are the development of highly efficient mechanisms for conserving water within the body (⊳ p. 88) and the ability to reproduce by means of eggs that are not reliant on an external watery environment (⊳ box, p. 86).

The first reptiles

The first known reptiles, dating back some 340 million years to the early Carboniferous period, were small animals about 20 cm (8 in) long. The warm, moist conditions that prevailed at this time led to the formation of luxuriantly vegetated swamps (the origin of today's coal measures), which must have provided a wealth of new habitats ripe for exploitation. Initially these swamps were dominated by amphibians, but as conditions became hotter and drier towards the end of the period, around 300 million years ago, the swamps began to disappear and the reptiles – not being reliant on water – found themselves at an enormous advantage. From this time the true conquest of the land began.

On dry land it was necessary to hold and despatch struggling prey, and an important development in reptiles was an improvement in the mechanical strength of the skull. To allow the musculature of the jaws and neck to be increased, the reptilian skull generally became higher and narrower, and in various groups of reptiles other modifications to the skull appeared (it is on the basis of these that the major reptilian groups are defined). In the primitive reptile skull there were holes only for the eyes and nostrils; the resulting structure was strong but heavy, and could accommodate only limited musculature. This condition is seen today only in the turtles and the tortoises (the *anapsids*). In another group, the *synapsids* or mammal-like reptiles, an additional opening appeared in the side of the skull, in the lower cheek region, while in the *diapsids* this opening is supplemented by a second, higher opening. This last group includes most living reptiles – the crocodiles, lizards and snakes – as well as the extinct dinosaurs. Various other extinct reptiles, including ichthyosaurs and plesiosaurs, had only the upper opening and are known as *parapsids*, but it is thought that this group in fact comprises a number of diapsid lines that independently lost their lower openings.

Initially, the dominant group of reptiles was the synapsids (the mammal-like reptiles), the lineage that gave rise to the mammals and eventually ourselves (⊳ p. 98). At this time the main diapsid line, the eosuchians, consisted of small insect-eaters that were very similar in appearance to the living tuatara of New Zealand (⊳ p. 85). These, however, formed the stock from which arose the groups that came to dominate the planet – the dinosaurs on land, the plesiosaurs and mosasaurs in the seas, and the pterosaurs and later the birds in the air.

The coming of dinosaurs

The dinosaurs, the most advanced reptiles of all time, first appeared about 230 million years ago and were to dominate the earth for over 160 million years. Unlike

PTERODACTYL

THECODONT
(Protosuchus)

THECODONT
(Shansisuchus)

ORNITHOPOD
(Lesothosaur)

COELUROSAUR
(Compsognathus)

living reptiles – which either crawl, or walk with their limbs extended out to the sides – dinosaurs walked with their limbs directly under their bodies, just like modern mammals and birds. However, like modern reptiles, most of the dinosaurs are thought to have been 'cold-blooded' (poikilothermic; ⇨ p. 87). Many dinosaurs were of gigantic size, some weighing up to 100 tonnes (tons). Nearly 1000 species have been identified, and although the word 'dinosaur' is from the Greek meaning 'terrible lizard', there were herbivores as well as carnivores.

The ancestors of the dinosaurs are found among the early *thecodonts* ('socket teeth') that lived in the early Triassic period 240 million years ago. Primitive thecodonts were somewhat like modern crocodiles (which also evolved from them), with thick, flattened tails for swimming and longer, more powerful hind limbs for lurching at their prey. When they emerged on to dry land, the difference in the length of the limbs meant that the hind limbs had the potential to move at greater speed than the fore. This problem was overcome by the first dinosaurs lifting their forelimbs and running on their hind limbs as bipeds, with the heavy muscular tail acting as a counterbalance.

Triassic dinosaurs

The first true dinosaurs appeared on the single supercontinent Pangaea (⇨ p. 8) about 230 million years ago. From the very beginning, two different orders of dinosaur can be distinguished, the Saurischia and the Ornithischia. The *saurischian* ('lizard-hipped') dinosaurs had a pelvic girdle structured like that of modern lizards, whereas the pelvic girdle of *ornithischian* ('bird-hipped') dinosaurs was similar to that of birds.

The first ornithischians were *ornithopods* ('bird-footed'); these were initially small bipedal herbivores about 1 m (40 in) long. Like some modern crocodiles, they seem to have aestivated – they spent the hot, dry summers dormant in their burrows (⇨ p. 175). The first saurischians were the *theropods* ('beast-footed'), a group of bipedal carnivores about 2 m (6½ ft) in length. There were two contrasting kinds: the heavily built, large-headed *carnosaurs*, and the lightly built *coelurosaurs*, which had small heads and long necks. It was from the latter group that the birds are thought to have evolved (⇨ p. 90).

Jurassic dinosaurs

During the Jurassic period, 205–135 million years ago, giant *sauropods* – a group related to the theropods – were the dominant herbivores. These included several massive quadrupedal types such as the brontosaur and the diplodocus, while the largest was the brachiosaur, 25 m (80 ft) long and weighing 100 tonnes (tons). These giants spent much time browsing in lakes and rivers, where the water helped to support their great weight. At the same time the carnosaurs developed into huge, ponderous 12 m (40 ft) scavengers. In dramatic contrast, the lightly built coelurosaurs remained small, active hunters – one group was only 60 cm (2 ft) long.

A significant development in the ornithopods and other ornithischians during this period was the evolution of muscular cheeks, grinding teeth and a secondary palate separating the nasal passages from the main cavity of the mouth, which allowed them to breathe and chew food at the same time. While many ornithopods remained little changed as small herbivores to the very end of the age of the dinosaurs, the Jurassic is nevertheless marked by the appearance of large ornithopods, such as the herbivorous iguanodon. These dinosaurs are thought to have lived in herds of up to 30 individuals, which presumably afforded protection against predators. An alternative method of defence was seen in the stegosaurs ('plated reptiles'), which had two pairs of sharp spikes at the end of the tail. They also had a double row of vertical bony plates or spines running down the back and tail, but these are thought to have been primarily for display.

Cretaceous dinosaurs

A more solidly armoured type of ornithischian developed in the Cretaceous period (135–65 million years ago). These were the ankylosaurs ('fused lizards'), so called because thick bony plates covering the back and tail were often fused into a solid sheet over the pelvic region. ⇨

SEE ALSO

- THE PLANET'S CHANGING FACE p. 8
- REPTILES: FORM AND STRUCTURE p. 85
- REPTILES: LIFE CYCLES p. 86
- THE WORLD OF REPTILES p. 87
- THE COMING OF MAMMALS p. 98
- HOW ANIMALS MOVE p. 176

(205–135 million years ago)   **CRETACEOUS PERIOD** (135–65 million years ago)

EGOSAUR · DIPLODOCUS · ANKYLOSAUR · TYRANNOSAUR · CLAWED DINOSAUR (Deinonychus) · TRICERATOPS

THE EXTINCTION OF THE DINOSAURS

One of the greatest unsolved problems relating to the dinosaurs is why they suddenly vanished 65 million years ago. Numerous theories have been put forward, from changes in plant life that made the dinosaurs constipated, to small mammals eating up their eggs.

One of the most popular theories is that the earth was struck by a large meteorite, 15 km (9 mi) in diameter. This idea is based on the discovery of a thin layer of clay with a high concentration of iridium, a rare metal normally only found in such concentrations in meteorites. If such an object had struck the earth, it would have thrown up a cloud of dust that could have obscured the sun for several years. This would explain the disappearance of large land-dwellers while smaller animals survived.

The major difficulty with this theory is that the dinosaurs seem to have gone into decline 5 million years before their final extinction. Other forms of life also became extinct at around this time, but these extinctions were not only separated by tens of thousands of years but were highly selective, some groups of plants and animals apparently remaining quite unaffected.

For a theory of extinction to carry conviction, it must take into account the exact timing of extinctions and their curious selectivity. To date no theory has been able to do this, and the disappearance of the dinosaurs remains a mystery.

From the small bipedal herbivorous ornithopods there eventually evolved such giants as the three-horned triceratops, with its huge bony frill extending over the vulnerable neck region. Although the horns appear to be defensive structures, it is now believed that they were mainly used for trials of strength with rivals of the same species. Other descendants of the small ornithopods included the pachycephalosaurs ('thick-headed reptiles'), the tops of whose skulls were to up to 30 cm (12 in) thick. These dinosaurs were perfectly adapted to head-butting in the manner of sheep and goats, and may have lived in herds with a dominant male presiding.

The most successful of the herbivorous dinosaurs at this time were the hadrosaurs or duck-billed dinosaurs, which evolved from the large ornithopods. They also lived in herds, and maintained communal 'dinosaur nurseries', with adults caring for the young in much the same way as crocodiles do today (▷ p. 86).

As the herbivores became more advanced, so too did the carnivores. The best known of these was the fearsome-looking tyrannosaur, a bipedal scavenger 6.5 m (21 ft) tall, but by far the most formidable carnivores – although only 3 m (10 ft) long – were the deinonychosaurs or clawed dinosaurs. Hunting in packs and capable of leaping onto their prey, these bipedal predators bore a sickle-like claw on the hind foot for ripping open their victims.

Extinct marine and aquatic reptiles

As well as dominating land habitats, several groups of reptiles (none of which were dinosaurs) returned to the water. Amongst the first of these were the ichthyosaurs ('fish-lizards'), which appeared at the start of the Triassic period. These looked rather like sharks, with a triangular dorsal fin and an enlarged vertical tail fin, and they fed on fish and squid. Some species grew to lengths of up to 12 m (40 ft). Ichthyosaurs never left the water even to breed, but produced eggs that hatched inside the body and gave birth to live young.

During the Triassic the nothosaurs, which had long snouts and webbed feet, were common marine fish-eaters. In the succeeding Jurassic period, the seas were dominated by plesiosaurs, some of which were distinguished by their small heads and enormously elongated necks. They swam by means of four paddles, which in fact acted rather as hydrofoils, enabling them to 'fly' underwater in a fashion similar to penguins and sea lions. The long-necked plesiosaurs were fish-eaters, while the short-necked species (also known as pliosaurs) fed on squid. During the Cretaceous period a group of monitor lizards evolved into ocean-going 'sea serpents' known as mosasaurs.

At the end of the Triassic the emergence of *Terrestrisuchus* ('land crocodile') marked the beginning of the crocodiles. Although at first entirely terrestrial, members of the group returned to the water to occupy the ecological niche of semi-aquatic carnivore and scavenger, developing long snouts with numerous fish-trap teeth. As the typical crocodilian specializations evolved, these reptiles took on the appearance of the primitive thecodonts that had occupied this same ecological niche at the beginning of the Triassic (▷ above).

Gliding and flying reptiles

The first vertebrates to take to the air are found amongst the reptiles. The earliest of these, dating back to Permian times (300–250 million years ago) and similar in appearance to the living flying dragon of the Far East, were gliding lizards in which the 'flight' membrane was supported by elongated ribs. In addition to various types of glider, some reptiles adopted a parachuting method of moving through the air. One such was *Longisquama* ('long-scale') of the Triassic, which developed enormously elongated scales running in pairs along the back.

True flight was achieved towards the end of the Triassic by the pterosaurs, which had full flight membranes extending from the wrists to the tail (▷ illustration, p. 176). While early forms had long bony tails and numerous teeth, later forms reduced their weight (like the birds) by developing air-filled bones and by losing their teeth and tails.   BH

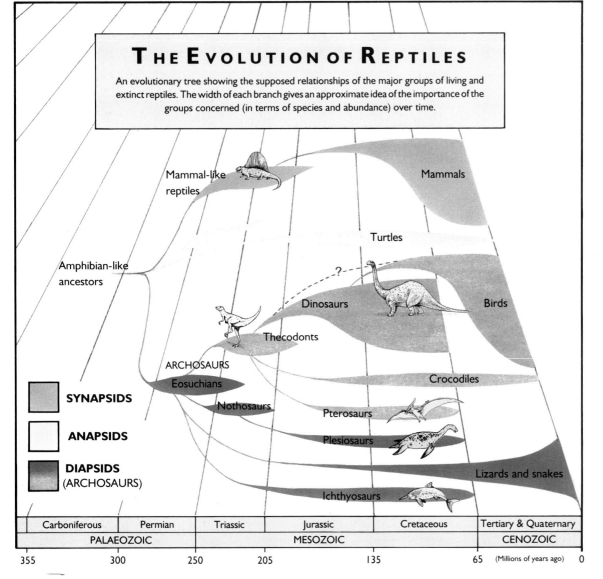

THE EVOLUTION OF REPTILES

An evolutionary tree showing the supposed relationships of the major groups of living and extinct reptiles. The width of each branch gives an approximate idea of the importance of the groups concerned (in terms of species and abundance) over time.

Mammal-like reptiles

Mammals

Amphibian-like ancestors

Turtles

Dinosaurs

Birds

Thecodonts

ARCHOSAURS

Eosuchians

Crocodiles

SYNAPSIDS

Nothosaurs

Pterosaurs

ANAPSIDS

Plesiosaurs

DIAPSIDS
(ARCHOSAURS)

Lizards and snakes

Ichthyosaurs

Carboniferous	Permian	Triassic	Jurassic	Cretaceous	Tertiary & Quaternary
PALAEOZOIC			MESOZOIC		CENOZOIC

355   300   250   205   135   65   (Millions of years ago)   0

Reptiles
Form and Structure

The reptiles are such a diverse group that it is impossible to characterize a single reptilian body plan – consider the differences between a turtle, a lizard, a crocodile, and a dinosaur. Generalized reptiles existed early on in the evolution of the group, but the modern forms have diverged enormously from the original body pattern because of their remarkably diverse specializations.

Early reptiles were small animals, about 20 cm (8 in) long, with long bodies, short sprawling limbs, and small skulls. In comparison with amphibians, the main changes were in the strengthening of the limbs for walking on land, the development of a distinctive neck, and a reduction in the relative size of the head. Major modifications also took place in the skull (▷ p. 82).

Turtles

The most characteristic feature of turtles, tortoises and terrapins (order Chelonia) is the 'shell', properly termed the *carapace* (the upper part) and the *plastron* (the flatter part beneath the belly). Despite the impression given in cartoons, turtles are unable to climb out of their shells – the carapace and plastron are fused to the ribs, the vertebrae, and the shoulder and hip girdles. The carapace and plastron are made up of between 20 and 30 bony plates, which are covered by keratin, or horn. The coloured patterns of many turtleshells reside in the keratin

covering, not in the bone itself.

How does a turtle grow if it is enclosed in a bony shell? All the separate elements of the shell are present in hatchling turtles and they all grow independently. During growth, each element lays down rings of new bone around its periphery. This means that each bony plate keeps the same shape during growth, and the overall shape of the carapace and plastron is not distorted. The number of growth rings around each bony plate is thus an indication of the age of a turtle, just as tree rings can age a tree. However, the number of rings is not exactly equal to a turtle's age in years, since the number laid down each year is variable, depending on the availability of food and weather conditions.

The turtle skull is a bony box, with two large eye sockets. Turtles have no teeth, which were lost early in their evolution; instead, the sharp edges of the jaws are lined with a bony beak, with which they can handle most plant and animal foods. Their limbs are short and strong, and may be adapted for walking on land or for swimming. In the latter case, the foot is expanded into a broad paddle.

Turtles can pull their limbs and the short tail inside the shell for protection. They can also retract the head, and this is done in one of two ways. The pleurodires withdraw their heads by making a sideways bend in the neck, while the cryptodires (the majority of living forms) make a vertical bend.

Crocodiles

The crocodilians (order Crocodilia) are a small group, consisting of only eight genera of crocodiles, alligators and gavials (or gharials). They all share a body armoured with bony plates set in the skin of the back, a long deep-sided tail used in swimming, short limbs, and a long-snouted skull.

Crocodilians are adapted for life in water, although they can also move about on land. The skull is much modified for capturing prey, the long snout being lined with many pointed teeth used for spearing fish or grasping flesh. In addition, crocodiles have a secondary palate – a bony division between the nasal capsule and the mouth that allows crocodiles to breathe underwater, even when the mouth is full of water. (The secondary palate of mammals is an entirely independent acquisition; ▷ p. 98.)

Lizards, snakes and the tuatara

Lizards and snakes, together with the amphisbaenians (▷ below), form the reptilian order Squamata. They are the most successful modern reptiles, each of the two major groups containing several hundred species. Although the largest living lizard – the Komodo monitor or dragon (▷ p. 89) – grows to a length of 3 m (10 ft), most species are small, and some tiny geckos are only a few centimetres long. Modern lizards mainly feed on insects and small prey, but larger species are often herbivores.

The principal modifications in lizards involve the skull, which is made up of several separate mobile elements (a form of modification known as *cranial kinesis*). There is a joint on top of the snout, one at the back of the palate, and a pair just above the jaw joint, at either end of the quadrate bone. The snout–palate joints allow the snout to tilt up and down, while the quadrate joints allow the jaws to gape even wider. The kinetic skull of lizards improves their ability to grasp food items.

Snakes have two main characteristics: the loss of their limbs and extreme cranial kinesis. Evolving from lizard ancestors 120 million years ago, snakes have become highly successful hunters, killing their prey by suffocation, by biting, or by venom (▷ p. 89). The smallest snakes are 20 cm (8 in) long, and feed on termites, while the largest constrictors may reach 10 m (33 ft) in length. The skull in all forms is an extremely loose structure, with numerous extra joints that allow a snake to dislocate its skull in order to swallow huge prey animals that may be several times the normal diameter of the mouth. Lacking limbs, snakes have developed a number of highly characteristic modes of movement (▷ p. 176).

The related amphisbaenians are small, limbless burrowing lizards that live in tropical countries. Their skulls are reduced to small battering rams for burrowing purposes, and they are blind.

The tuatara of New Zealand (order Rhynchocephalia) is a medium-sized lizard-like animal with a primitive non-kinetic skull. It feeds on invertebrates such as snails and worms, and occasionally on small lizards and young sea birds. Its closest relatives are known from the Triassic and Jurassic, some 220–150 million years ago early in the age of the dinosaurs, and the single living species is the last remnant of this formerly more diverse group. MJB

LIVING REPTILES

Crocodiles, alligators and the gavial (order Crocodilia): saltwater or estuarine crocodile (*Crocodylus porosus*)

Snakes (order Squamata, suborder Serpentes): king cobra (*Ophiophagus hannah*)

Lizards (order Squamata, suborder Sauria): sand lizard (*Lacerta agilis*)

Turtles, tortoises and terrapins (order Chelonia): green turtle (*Chelonia mydas*)

Reptiles
Life Cycles

The behaviour of reptiles is generally far less complex than that of mammals and birds. For many species, mating and reproduction involve very little in the way of courtship display or parental care.

This is not to suggest that such behaviour is altogether lacking, however. The males of many reptilian species engage in ritualistic combat during the breeding season, while many lizards and snakes (usually the males) become territorial.

Turtles

Many turtles and tortoises engage in a rather crude kind of display and combat before mating. For instance, in the red-legged and South American forest tortoises the males blunder about the forest floor, challenging any object that they think may be a rival male. If the object challenged is a male of the wrong species, the tortoise wanders off. If the challenge is answered, then the tortoises fight by banging their shells together in an attempt to overturn their would-be rival. The successful male then continues on his way, while the loser tries to turn himself the right way up again. If the object that is challenged is neither a male of the same species nor a male of another species, the tortoise moves closer for further investigation. Frequently he finds that the object of his affections is a rock or a log, and he retires. However, if it is a female, he sniffs around in order to determine that *she* belongs to the right species, and if she is, he proceeds to court her prior to mating (▷ photo).

All turtles lay eggs. Using her hind legs, the female digs out a nest into which she

The mating of land tortoises, such as these Galápagos giant tortoises, is an awkward affair. The sexual openings, located beneath the tail, must be brought into contact for the transfer of sperm. This involves the male balancing on top of the female, and of course he frequently falls off. The procedure is a little easier for freshwater and marine turtles, since they can mate while in water.

lays her clutch of eggs, which may range from four to six eggs in small species, to more than a hundred in the large sea turtles. Egg-laying in sea turtles often takes place after long migrations of hundreds of kilometres (▷ p. 170). Turtles can modify their laying pattern to suit local conditions. The painted turtle of North America lays two clutches of about eleven eggs each year in the northern USA, where the activity season is short, but four or five clutches of four eggs each in Louisiana, where the climate is warmer. The same number of young are produced, but the greater number of nests reduces the impact of predation. Nest predation is indeed a serious problem for all turtles, whether land or sea, and most of the eggs or young are eaten in the nest by predatory mammals, birds or crabs.

Turtles show no parental care, so far as can be determined. The eggs are simply laid and then abandoned. However, hatchlings may interact with one another in important ways, and this has been studied in sea turtles in particular. Large clutches of eggs may be buried up to 50 cm (20 in) deep in the sand, in order to protect the nest from predators. As the hatchlings start to emerge, they thrash about, and this seems to stimulate all the other eggs to hatch. The hatchlings then move upwards through the sand en masse, resting from time to time to muster their energy. They then suddenly burst forth at the surface and head for the sea as a horde. Predators such as crabs, foxes and raccoons are waiting for them, but the sheer numbers of struggling young generally mean that a few reach the sea and relative safety.

Crocodilians

Until recently it was commonly assumed that crocodilians were brutish parents that paid no attention to their young – other than possibly eating them. However, it is now known that all species build elaborate nests for their eggs, specifically designed to prevent flooding, and that the females of some (and possibly all) species guard their nests against predators. A fascinating (and unexplained) feature of crocodilians is that the sex of their young is determined by small changes in temperature in the period prior to hatching.

When the young begin to squeak in the nest, the female may scrape away earth and sticks to help them out. In some species, both parents may also assist them by gently removing the eggshells. Female American alligators (and perhaps other species) then pick up the newly hatched young in their mouths and carry them to the water. The hatchlings are capable of hunting for themselves from the start, but may remain close to their parents for up to two years. Parents will also respond to distress calls from their young.

Lizards and snakes

Snakes have very poor sight, so it is not likely that they can use their skin patterns and colours to identify members of their own species. They probably recognize mates and enemies mainly by smell. Males engage in ritualized fighting. In a

THE REPTILIAN EGG

The reptilian egg provides everything necessary for the early development of the young. As with all living things, the initial stages of development have to take place in a watery environment; hence the reptilian embryo is enclosed in its own private 'pond' by a membrane called the *amnion* (▷ pp. 186–7). On account of this structure, reptiles (together with their evolutionary descendants, the mammals and birds, which share this feature) are referred to as *amniotes*. A yolk enclosed in the yolk sac provides food for the developing embryo, while a further membrane (the *allantois*) acts as a combined respiratory and excretory organ, for gas exchange and the disposal of wastes. The *albumen* (egg white) provides extra cushioning and also serves as a reservoir of water and proteins. The entire structure is surrounded by a protective membrane (the *chorion*) and by the shell. In lizards and snakes the shell is leathery, but in turtles and crocodiles (and birds) it is hard and made of calcium carbonate, the calcium from which contributes to the formation of the internal bony skeleton. The ability to reproduce by means of such an egg has enabled the reptiles to live away from open bodies of water and to make a spectacular conquest of the land.

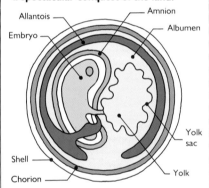

bizarre analogue of arm wrestling, a pair of snakes rear up and intertwine their bodies, the snake that forces the other to the ground being the 'winner'. This ritual involves much swaying around, and for this reason used to be interpreted as dancing. Its purpose appears to be to test strength, and hence fitness, prior to mating, and injury is avoided since the males do not bite each other.

In contrast, lizards have better eyesight than snakes and often use visual cues in pre-mating performances. For example, some species can expand their neck pouches to give a vivid display of colour to rivals. While most lizards and snakes lay eggs in the same way as other reptiles, a number of species, particularly in cooler areas, retain the eggs inside their bodies and produce live young. The mother can bask in the sun in order to maintain a warm body temperature, and so assist the incubation of the eggs. This avoids the risk of eggs in a nest becoming too cold and dying. It probably also gives the young some added protection from predators.

MJB

The World of Reptiles

A Gaboon viper. In the vipers and pit vipers the poison fangs are located at the front of the jaw and rotate into the striking position, while in other front-fanged snakes, such as cobras, mambas, coral snakes and sea snakes, the fangs are fixed in position. Striking at prey activates the release of venom from a pair of glands in the snout, and the venom passes down through tubes in each of the hollow fangs.

Reptiles are 'cold-blooded' animals, just like fishes and amphibians. This has often been misunderstood to mean that they are cold and slimy to the touch, and in some sense 'inferior' to birds and mammals. The very word 'reptile', meaning 'creeping' animal, reflects the common dislike of reptiles. Snakes in particular, from the Serpent in the Garden of Eden onwards, have been indiscriminately tarnished with a reputation for being both venomous and devious. In other cultures, lizards are thought to be poisonous.

Most of these ideas are based on misconceptions or partial truths. Although still current as recently as about 1950, the various myths about reptiles have since been dispelled by the large amount of research conducted both in the field and in the laboratory.

Controlling body temperature

Research in the 1940s demonstrated that reptiles have sophisticated systems of

thermoregulation – they can control their body temperature very precisely, principally by behavioural means, and each species has a preferred body temperature. Within the normal activity range of temperatures, which may be as small as 4 °C (7 °F) or as great as 10 °C (18 °F), the animal behaves normally. Beyond this is the voluntary range, which is defined by the lowest and highest temperatures at which the reptile can be active. Beyond these are the critical minimum and maximum temperatures – so cold or so hot that the animal becomes immobile and cannot take action to improve its situation. The lethal minimum and maximum temperatures are even more extreme, and mark the physical limits at which death occurs. Normally reptiles keep well within their normal activity ranges, passing into the cooler end of the voluntary range at night. They adjust their body temperature simply by moving between shade and light, depending upon whether they need to lose heat or take it in.

Reptiles are said to be *poikilotherms* (or *ectotherms*) – that is, they have variable body temperatures and control their temperatures by external means (⊳ box). Birds and mammals, on the other hand, are *homoiotherms* (or *endotherms*) – they have constant body temperatures and use internal temperature-control mechanisms. These terms are roughly equivalent to what people mean when they say 'cold-blooded' and 'warm-blooded' respectively, but the vernacular forms are somewhat misleading. A lizard living beside a rat might well have warmer blood at midday.

A normal daily regime for a desert lizard would be to creep sluggishly from its night-time hiding place. Its body temperature is too low to allow normal activity, but just enough to allow it to begin basking. Commonly performed on a bare rock, basking is a major feature of the morning for a reptile, and it is by means of this that its body temperature is raised into the normal activity range. Heat comes from several sources: direct radiation from the sun above, reflected radiation as the sun's rays bounce off neighbouring rocks, convection from the warm air around, and conduction from the heated rocks beneath.

Once its body temperature has reached the normal activity range, the reptile

darts about in search of food, while remaining ever vigilant for predators. In a desert, however, air temperatures continue to rise, and many lizards in such areas run a serious risk of overheating. The reptile may now place itself high in bushes exposed to the wind, so that it loses heat by convection. When temperatures become very high towards midday, the reptile may hide in the shade, and lose heat by convection to the air, by evaporation of fluids, and by conduction to the earth, which is relatively cool and moist.

Lizards have a large and subtle repertoire of behavioural techniques for controlling their body temperatures. They adopt a variety of postures, standing broadside to the sun to take in heat or facing the sun in order to minimize heat uptake. Many species can alter their body contours in such a way as to control heat transfer: they can expand their ribs sideways and spread their body out like a pancake on the rocks for maximum heat uptake. Some species even alter their colour from light to dark in order to increase the rate of heat absorption.

The achievements of reptiles in temperature control mean that they can live successfully in hot and cold parts of the world. Certain lizards are able to remain active in cool mountainous temperate-zone areas, maintaining their body temperatures 30 °C (54 °F) above the near-freezing air temperatures. But is this achieved simply by basking and seeking shade as appropriate?

Research in the 1960s showed that reptiles even have internal mechanisms for temperature control. Lizards, snakes and turtles can heat up faster than they cool down by controlling their blood flow during basking and cooling phases. While basking, the surface blood vessels open up, so maximizing the uptake of heat. In the late afternoon, when air temperatures begin to fall, these surface blood vessels are closed off in order to retain the heat as long as possible in the core of the body. These responses are of course automatic and are not carried out consciously.

Water conservation

Lizards and snakes are generally better at conserving water than mammals and other reptiles, and this makes them more tolerant of arid conditions. It is often said

---

WHY BE A REPTILE?

If poikilothermy or 'cold-bloodedness' is so typical of reptiles and so 'inferior' to the internal (homoiothermic) mechanisms used by birds and mammals, why be a reptile? Various new studies (⊳ text) show that the story is rather more complicated than had once been assumed. Reptiles are not simply passive vessels battered one way and another by changes in air temperature. They can maintain their body temperatures within fixed ranges by means of behavioural and even physiological mechanisms.

Other recent findings further complicate the simple older view. Some reptiles may even be able to generate heat internally, as mammals and birds do. Certainly, some incubating snakes and large sea turtles appear to maintain constant body temperatures much higher than their surroundings, possibly by means of muscular activity or even by a form of shivering.

In addition, there is a size-related phenomenon that actually bridges the gap between poikilothermic and homoiothermic animals. As reptiles attain greater body size, their temperature fluctuations are damped, since the bulk of the body acts as a form of insulation against heat loss and gain. In other words, while a small reptile's body temperature may follow fluctuations in air temperature fairly closely, larger species can escape this simply by being large. Thus large adult alligators have been found to show much smaller ranges in body temperature than juveniles. For this reason it has been suggested that large dinosaurs could have had constant body temperatures while being poikilothermic.

The main advantage of reptilian poikilothermy is that reptiles require less food than mammals of the same size. The difference is substantial: as much as ten times less. Put another way, mammals and birds use about 90% of the food they eat in order to maintain a constant body temperature by physiological means. So why not be a reptile, and eat less? Reptiles can live in areas of sparse food supply that would be uninhabitable to a mammal. While a snake or a crocodile only needs a meal every few weeks, a mammal of equivalent size must spend most of its time foraging for food and so lay itself open to predation and other dangers.

that these reptiles conserve water by cutting down evaporation through their scaly skin, but they actually lose water in this way nearly as fast as mammals do. Their tolerance of arid conditions is in fact principally due to their low metabolic rates, their salt glands, and excretion of uric acid.

A low metabolic rate is an aspect of poikilothermy. The metabolic rate is the rate of cellular activity (▷ p. 21), and is measured as the rate at which oxygen or food is used up. Hence homoiotherms broadly have much higher metabolic rates than poikilotherms. Since metabolic rate is proportional to the rate of oxygen use, it is also proportional to the rate of respiration; and the more an animal breathes, the more water it loses (▷ p. 180).

Some species of lizard, snake and turtle (as well as many birds) have salt glands on the side of the snout from which rich brine solutions are shed. This is a water-saving device since it cuts down dramatically on the amount of urine that has to be produced. Unwanted salts build up in the blood streams of all animals, and are generally passed out of the body in dilute form in the urine (▷ p. 179). By disposing of various salts in a nearly solid form from the salt glands, certain reptiles are able to maintain the correct salt balance in their bodies while saving water at the same time.

The final water-saving adaptation in lizards and snakes (as well as birds) is the production of near-solid urine. Unwanted salts from the blood are formed into uric acid crystals containing sodium, potassium and ammonium salts. These crystals form the whitish mass typically seen in bird and reptile droppings. Mammals and many amphibians and fishes produce urine in the form of urea, which requires a great deal of water for its disposal.

Reptilian senses

With the exception of lizards, most reptiles appear to have a rather poor sense of sight. Turtles and crocodilians rely mainly on their senses of smell and hearing in order to detect enemies, prey and mates. Although many lizards also depend on detecting scent and sound, others, such as iguanas and agamid lizards, have relatively good vision. Indeed, many of these lizards have bright and varied col-

oration and perform complex displays before mating (▷ p. 86), all of which indicate a good sense of sight.

The sensory systems of snakes seem to be different from those of other reptiles in many ways. It is thought that snakes evolved from lizards over 100 million years ago, and passed through a burrowing phase during which the normal lizard eye became reduced. The eyes later re-evolved for use above ground, but in a quite different form from the original lizard pattern. Snakes lack external ears (probably another sense organ lost during the burrowing phase of evolution) and are presumably deaf to each other, to the calls of their prey, and to the music of the snake charmer. The latter achieves his effects by inducing the snake to watch the hypnotic side-to-side movements of his pipe.

Snakes sense their prey by vibrations on the ground, which they detect through their whole body, by smell, and by sight if it is close. Rattlesnakes and other pit vipers also have a heat-sensitive organ located in a pit between the nostril and the eye (▷ box, p. 191).

A Jackson's three-horned chameleon (left) catching a fly. The swivelling 'turrets' that house a chameleon's eyes can move independently to give a three-dimensional image, essential for striking at distant prey. The chameleon then shoots out its long sticky tongue to catch the unsuspecting insect.

A crocodile attacking a wildebeest (below left) on the Mara River, East Africa. Crocodilians often grab an animal's leg at the edge of a river and pull it under until it drowns. They may then proceed to eat it straightaway or stow the carcass in an underwater 'larder' for later enjoyment. Sometimes a crocodile throws its whole body into violent twisting motions as it tries to tear a piece of flesh from its prey.

An African rock python (below) making a meal of a Thomson's gazelle. Snakes have numerous joints in the top and back of the skull and in the palate. At full stretch, a snake can typically swallow prey animals several times its own body diameter, an ability that may be at the root of the success of snakes as a group. A snake attempts to orientate its prey so that it can be swallowed head-first. Once swallowing begins, the prey animal cannot escape, since a snake's teeth curve backwards, so preventing it from slipping out.

Diet and feeding

Crocodilians are all carnivores, feeding either by snapping at fishes underwater or by seizing larger land animals at a river's edge (▷ photo).

Different species of turtle and tortoise feed variously on animal and plant material. Many pond turtles take small fishes and invertebrates, which they stalk in the murky waters. Sea turtles mainly feed on seaweed, jellyfishes and possibly small fishes, while tortoises eat either insects or plants. Turtles generally take advantage of easy food supplies: during locust plagues, North American box turtles stuff themselves so much that they cannot withdraw their fat legs into their shells!

Lizards include a variety of herbivores and carnivores, many of which (including the chameleons; ▷ photo) are highly adapted to exploit a particular food source. Large iguanids feed on plants, and the marine iguana of the Galápagos is unique in diving for seaweed. The varanids (monitor lizards) are voracious carnivores, the largest of which – the Komodo dragon – reaches a length of 3 m (10 ft) and has been reported to kill adult water buffalo and humans. Generally, however, it feeds on deer and wild goat, which it waits for beside their regular trails and captures by pouncing from cover.

All snakes are carnivorous, although the majority are small animals that feed on insects and worms. Venom and constriction are confined to only a few snake groups. The latter technique is used by the New World boa constrictors and anacondas and the pythons of the Old World. A constrictor coils around a prey animal, tightening its coils as the animal breathes out. The prey is thus progressively prevented from breathing, and is finally swallowed. While lizards have relatively loosely constructed kinetic skulls, a snake's skull is highly modified for swallowing large prey whole (▷ photo and p. 85).

Several groups of advanced snakes subdue their prey by venom, which is delivered by two modified teeth (fangs) in the upper jaw. In poisonous colubrids, such as the boomslang, the fangs are near the back of the jaw, and the prey is held tight while the venom is delivered. Other venomous snakes have fangs at the front of the jaw (▷ illustration, p. 87).

Snake venom is a complex cocktail of enzymes and toxins, its composition varying from species to species and even within a species. The different components have different effects, ranging from tissue destruction and haemorrhage to paralysis of the respiratory system. In many snakes the instantaneous effect of their venom is crucial, since if a prey animal were not stopped dead in its tracks, it would blunder away to die elsewhere and would not be found. Other snakes, however, including most of the vipers, have a 'play-safe' strategy – they avoid the risk of injury during the death throes of their prey by letting it wander off to die and later track down the corpse by its scent.

Certain cobras can project their venom 2 m (6½ ft) or more, and aim for the eyes. The venom causes instant blindness and can lead to death, even when delivered at a distance.

Defence

The commonest mode of defence for reptiles is to lie low in unobtrusive hiding places. The fact that reptiles have low metabolic rates means that they do not have to move about in search of food as much as birds and mammals do. Hence crocodiles and turtles spend a great deal of time immobile, hard to distinguish from logs or rocks. Many lizards and snakes that move about among leaves on the ground or in trees have camouflage patterns of mottled brown or leaf-green. Chameleons can even alter their colours to match their background by enlarging or constricting special colour-bearing cells within the skin.

If they are caught, many lizards save themselves by shedding their tails (a stratagem known as *autotomy*). Specialized fracture planes in the vertebrae of the tail allow the lizard to cast off the end of the tail and run away, leaving the perplexed predator with a twitching worm-like remnant. A new tail will grow in its place, albeit rather less impressive than the original.

The Chinese alligator. A victim of hunting, pollution and disappearing habitat, this species will probably be extinct in the wild before the end of the century.

Other reptiles use more direct means of defence. Snake venom may have evolved as a defensive attribute and is still used for this purpose. Poisonous snakes will bite predators that come too close, either to give them a painful warning or to kill them. The defensive aspect is heightened in many snakes by warning behaviour. For example, rattlesnakes have their 'rattles' precisely for this reason. The rattle is made from dead scales, which brush against each other when the tail is vibrated. A warning from this is usually enough to send another animal on its way without the need for venom. Cobras rear up and expand the ribs in the neck region to give a clear warning to mongooses and other predators.

Many poisonous snakes also display bright and intricate warning colours (▷ p. 166). The North American coral snakes, for instance, have distinct hoops of red, black and yellow around their bodies to warn predatory birds that they are poisonous. The success of this warning coloration has led to extensive mimicry by many small non-venomous snakes that live side by side with coral snakes. The mimicry is so effective that it has led to immense confusion when zoologists try to sort out their true affinities!

MJB

SEE ALSO

- REPTILES: EVOLUTION AND CLASSIFICATION p. 82
- REPTILES: FORM AND STRUCTURE p. 85
- REPTILES: LIFE CYCLES p. 86
- PREDATOR AND PREY p. 164
- DEFENCE, DISGUISE AND DECEIT p. 166
- ANIMAL PHYSIOLOGY pp. 178–83
- SENSES AND PERCEPTION p. 188

Birds
Evolution and Classification

The enormous popular interest in birds has been matched in recent years by the proliferation of ornithological societies and the growth of birdwatching as a recreational activity. Unfortunately, the expertise gained by such pursuits (which is often very considerable) rarely extends to any real insight into the evolutionary origins of birds.

Even professional scientists, intent on tracing human origins, have tended to focus their attention on the vertebrate groups – fishes, amphibians, reptiles and mammals – that are thought to lead to the 'ultimate' mammal, man. The evolution of birds has sometimes been regarded as little more than an interesting diversion from that line; yet it is a fascinating and complex story in itself.

The evolution of birds

It is now generally agreed that birds evolved from one of the small, lightly built, flesh-eating coelurosaurian dinosaurs (⊳ p. 83). Unfortunately, no unmistakable fossil intermediates have been found, so we can only guess how and when the first bird-like forms diverged from their dinosaurian ancestors. Some experts believe that 'pre-avis' – the form intermediate between dinosaurs and birds – was an arboreal animal that gradually developed the power of flight through various gliding phases in order to facilitate faster movements between the branches of trees. Others suggest that it was a ground-dwelling runner and leaper that slowly increased its leaps with the aid of elongated fore limbs.

Whichever scenario is preferred, exactly when these evolutionary changes started cannot be pinpointed, but clearly it must be well before the first recognizable bird appears in the geological record. The remains of this creature – *Archaeopteryx* – date back to the latter half of the Jurassic period, about 154 million years; this chicken-sized animal represents a fascinating blend of reptilian and avian features, and for this reason has been hailed as a 'missing link' between the reptiles and the birds (⊳ box). Unfortunately, the primitiveness of *Archaeopteryx*'s skeleton provides little indication of how it may be related to any other known bird. The general consensus is that it represents a relatively early branch that diverged from the main avian line and apparently left no relatives. However, a number of small bones dating from the late Triassic have recently been claimed as avian; if the identification is correct, it would appear that the first birds evolved about 220 million years ago and that *Archaeopteryx* is well along the line of bird evolution.

We have to wait until the next geological period, the Cretaceous, some 30 million years after the appearance of *Archaeopteryx*, before more fossil birds appear. By this time, birds had spread around the globe – remains are known from as far afield as Australia, Asia and Europe – and most of the primitive skeletal characteristics observable in *Archaeopteryx* had been lost. In *Ambiortus* and *Sinornis*, for instance, the wrist bones are fused almost exactly as in modern birds, and the breastbone has become broader and keeled, thus providing a firm anchorage for the enlarged pectoral muscles necessary for powered flight (⊳ diagram, p. 92). During this same period, a number of birds, including *Enaliornis*, an early hesperornithid resembling a modern diver or grebe (⊳ opposite), became adapted to cope with a fully aquatic environment.

As the Cretaceous period progressed, other aquatic species evolved, including the spectacular giant, flightless, diver-like *Hesperornis*, and the aerial, tern-like *Ichthyornis* (both belonging to the Odontornithes; ⊳ opposite). However, these two forms still retained true teeth similar to those of *Archaeopteryx*. It was not until 70 million years ago, at the very end of the 'age of the reptiles', that birds attained a skeleton that was in all respects modern in character.

Whatever caused the extinction of the dinosaurs, it did not interfere with the successful evolution of birds (or indeed of other important groups such as mammals and flowering plants). Thus by the beginning of the Tertiary period, 65 million years ago, birds had become relatively common and had apparently adapted themselves to ecological niches that they had not previously explored. Although it is currently believed that the original lineage of most modern bird groups goes back to the Cretaceous, it is not until the early Tertiary that fossilized bones attributable to these groups can be recognized. For example, birds identifiable as penguins, cormorants, shorebirds, gamebirds, owls, nightjars, swifts and rollers made an appearance within the first 40 million years of this period. Needless to say, some spectacular forms failed and became extinct during this time. These include the 2 m (6½ ft) tall flightless *Diatryma* from America and Europe, and the albatross/pelican-like bony-toothed seabirds, which ranged the oceans until about 3 million years ago. The largest example of this marine group (*Pelagornis*) probably had the longest wings of any bird, spanning in excess of 5 m (16½ ft).

Throughout the Miocene epoch, birds inseparable from modern genera began to appear, and by the end of the Pliocene, 1·6 million years ago, birds indistinguishable from our present-day species were in existence. It was, however, in the succeeding Pleistocene that the modern distribution of the birds we know today was largely established.

Bird classification

For most of the last 60 or so years the class Aves, to which all birds belong, was arranged in various modifications of a classification that recognized 28 orders of modern birds. Within the last few years, however, a new and hopefully more objective reclassification based on DNA analysis has been proposed; this recognizes only 23 orders, five being merged with others (⊳ opposite). Nevertheless, any classification must be flexible, since the component parts are constantly under review and liable to be altered as new evidence of relationships becomes available. The principal criticisms of the new system have so far concerned its timescale, since some authorities believe that the main adaptive radiation of birds should occur just after the Cretaceous/Tertiary boundary, following the disappearance of the dinosaurs. Be that as it may, the new arrangement still provides a valuable framework for re-evaluating bird relationships.   CW/CH

ARCHAEOPTERYX

A major deposit of exceptionally fine-grained 'lithographic' limestone (formerly used in the printing process known as lithography) was laid down by an inland sea that covered an area of what is now southern Germany during the latter half of the Jurassic period, about 154 million years ago. In 1861 workmen quarrying stone from this deposit split a slab of rock to find an unfamiliar fossil. It was sent to the German palaeontologist Hermann von Meyer, who identified the remains as a bird's feather and appropriately named it *Archaeopteryx* ('ancient wing') *lithographica*. In the following year a partial skeleton, with wing- and tail-feather impressions, was excavated from the same region. This specimen was brought to Friederich Haberlein, a local medical practitioner and amateur palaeontologist, who accepted it in lieu of medical expenses. It was clear that this fossil did not represent the remains of an ordinary bird and it became the focal point of much scientific discussion. Haberlein, requiring money for a dowry, eventually sold the fossil (together with 1700 other choice specimens) for £700, to what is now the British Museum (Natural History) in London.

Archaeopteryx was described in detail by the Curator of the Museum, Richard Owen. He noted that the feathers are structurally indistinguishable from those of living birds (⊳ diagram, p. 92). The wing plumes are differentiated into primaries and secondaries, as in modern birds, while the primaries (the main flight feathers) have a profile that is found only in birds capable of flight. The tail feathers, however, are arranged differently from those of living species, for they appear as pairs, each of which is associated with an individual tail vertebra. In all modern forms the bony tail has been reduced to a pygostyle (parson's nose) from which all of these feathers emanate. Possibly the most bird-like skeletal feature in Archaeopteryx is the well-developed furcula (wishbone), which would have been expected in an animal capable of some form of flight.

At the same time, the skeleton retains many characteristics that are regarded as fundamentally reptilian. In addition to the long bony tail, there are true teeth in the jaw, the wrist bones are not fused as in modern birds and seem to keep functional fingers, and the breastbone (sternum) is not broadened and keeled. Indeed, with the exception of the feathers, there is no single feature in Archaeopteryx that has not been recognized among the remains of certain dinosaurs. So it is very fortunate that feather impressions were preserved, for without them it would have been difficult to separate the remains of this early bird from those of a small dinosaur.

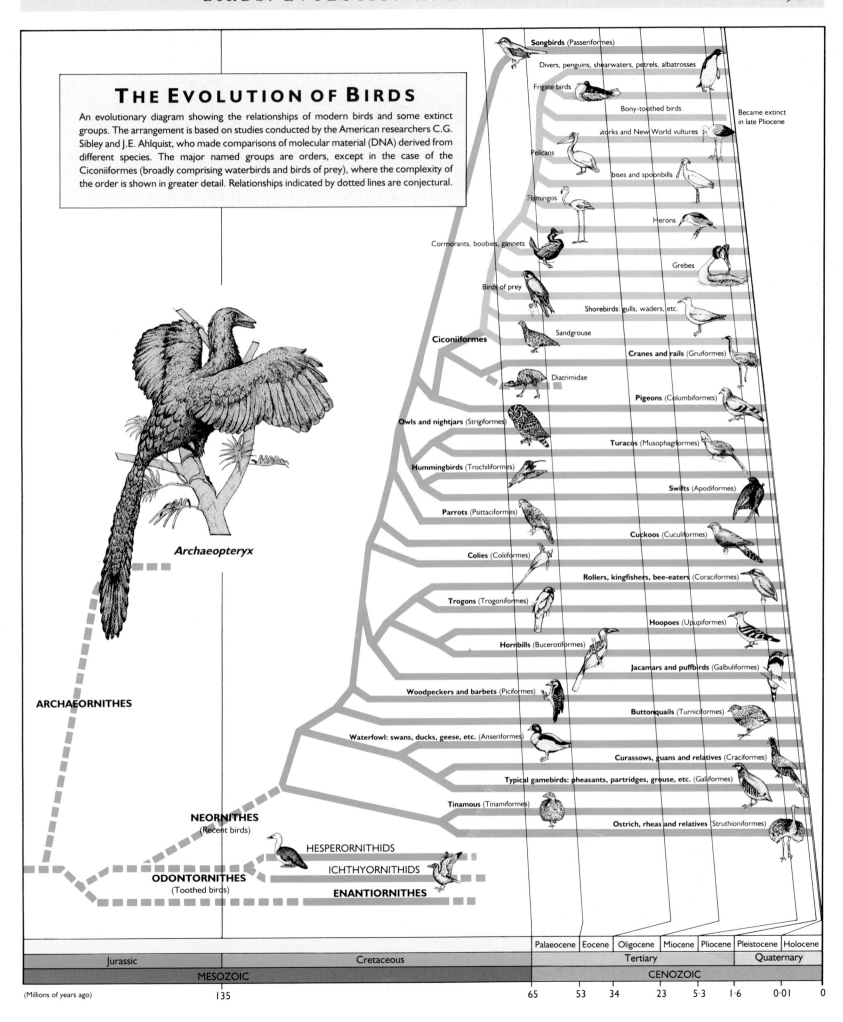

THE EVOLUTION OF BIRDS

An evolutionary diagram showing the relationships of modern birds and some extinct groups. The arrangement is based on studies conducted by the American researchers C.G. Sibley and J.E. Ahlquist, who made comparisons of molecular material (DNA) derived from different species. The major named groups are orders, except in the case of the Ciconiiformes (broadly comprising waterbirds and birds of prey), where the complexity of the order is shown in greater detail. Relationships indicated by dotted lines are conjectural.

Archaeopteryx

ARCHAEORNITHES

NEORNITHES
(Recent birds)

ODONTORNITHES
(Toothed birds)

HESPERORNITHIDS

ICHTHYORNITHIDS

ENANTIORNITHES

Songbirds (Passeriformes)

Divers, penguins, shearwaters, petrels, albatrosses

Frigate birds

Bony-toothed birds

Became extinct in late Pliocene

Storks and New World vultures

Pelicans

Ibises and spoonbills

Flamingos

Herons

Cormorants, boobies, gannets

Grebes

Birds of prey

Shorebirds: gulls, waders, etc.

Ciconiiformes

Sandgrouse

Diatrimidae

Cranes and rails (Gruiformes)

Pigeons (Columbiformes)

Owls and nightjars (Strigiformes)

Turacos (Musophagiformes)

Hummingbirds (Trochiliformes)

Swifts (Apodiformes)

Parrots (Psittaciformes)

Cuckoos (Cuculiformes)

Colies (Coliiformes)

Rollers, kingfishers, bee-eaters (Coraciformes)

Trogons (Trogoniformes)

Hoopoes (Upupiformes)

Hornbills (Bucerotiformes)

Jacamars and puffbirds (Galbuliformes)

Woodpeckers and barbets (Piciformes)

Buttonquails (Turniciformes)

Waterfowl: swans, ducks, geese, etc. (Anseriformes)

Curassows, guans and relatives (Craciformes)

Typical gamebirds: pheasants, partridges, grouse, etc. (Galliformes)

Tinamous (Tinamiformes)

Ostrich, rheas and relatives (Struthioniformes)

Palaeocene	Eocene	Oligocene	Miocene	Pliocene	Pleistocene	Holocene

Jurassic	Cretaceous	Tertiary	Quaternary

MESOZOIC	CENOZOIC

(Millions of years ago)   135   65   53   34   23   5·3   1·6   0·01   0

Birds
Form and Structure

The great majority of the 8600 or so species of bird alive today are able to fly. The stresses imposed by flight are very great, and a high degree of structural adaptation has been necessary to make it possible. For this reason birds that are capable of flight are remarkably uniform in those aspects of their structure that are associated with flying, which include the wings and feathers, and various modifications related to weight reduction.

SEE ALSO

● HAWAIIAN HONEYCREEPERS (BOX) p. 17
● BIRDS pp. 90, 94–7
● ANIMAL PHYSIOLOGY pp. 178–91

The three features that most conspicuously distinguish birds from other animals are a beak, a pair of wings, and feathers. Wings are (or were) found in the other animal groups that have achieved flight (insects, bats and pterosaurs), and certain fishes and turtles, for instance, have beaks (or at least structures that closely resemble them). Feathers, however, are a uniquely avian characteristic and are seen in no other animal, living or extinct.

Wings and flight

Birds' wings are highly modified fore limbs: the digits are reduced in size, and the wrist bones are elongated and fused, to provide the supporting structure for the flight feathers (▷ diagram). The wings are attached to the skeleton by mobile shoulder-joints and by the *furcula* (wishbone), the latter increasing the spring of the wing beat and helping to damp the unequal and jarring stresses imposed when a bird wheels in the air or changes speed suddenly. The great power needed for flight is provided by two pairs of massive pectoral muscles anchored to the large, keeled *sternum* or *breastbone*. These muscles work in the same direction, but one set pulls directly on the wings to bring them down and in, while the other is carried over the shoulder by a tendon-pulley arrangement to pull the wings up and back. This system works for birds weighing anything between 2 g/14 oz or less (the bee hummingbird of Cuba) and 18 kg/40 lb (various species of bustard).

A bird's wing has a characteristic profile (the so-called *aerofoil section*, similar to that of an aeroplane wing), with a convex upper surface and a concave lower surface. Air moving over the wing has to travel further, and thus faster, over the upper surface than the lower; this causes a reduction in air pressure above the wing, and hence creates lift.

Although this wing profile is shared by all flying birds, there is very great variety in styles of flight. Most of the larger flying birds rely principally on gliding or soaring. Large land birds such as vultures, buzzards and eagles have broad wings with splayed feathers, which allow them to 'float' on columns of warm rising air. Albatrosses and other large sea birds use long, thin wings (spanning over 3 m/10 ft in the wandering albatross) to glide on strong ocean winds, typically accelerating rapidly downwind and then turning and climbing steeply against the wind. In contrast, the majority of smaller birds flap their wings to fly. Lift and forward thrust are created simultaneously on the downstroke by twisting the wing (especially the wing tip) as it cleaves through the air; on the upstroke the wing is twisted the other way to reduce drag. Hummingbirds hover by means of a unique figure-of-eight 'stirring' motion of the wings, at a rate of 50 or more beats a second; by rotating their wings, first one way, then the other, they produce lift on both the up- and the downstroke.

To reduce drag and wind resistance as they move through the air, most flying birds have very streamlined bodies. They also show various modifications related to weight reduction. All modern birds have lost their teeth, and – except in the flightless ratites (▷ box, p. 97) – many of the larger bones are hollow, with a sponge-like internal strutting that is both strong and light.

In contrast to freshwater birds, which typically swim by means of webbed feet (▷ below), birds adapted to swimming and diving in the sea often propel themselves through the water by means of blade-like wings. Auks and penguins swim in this way, and the latter have lost the power of flight altogether (as did the now-extinct giant auk).

Feathers

Feathers are made of the protein keratin, the same material as in the hair and nails of mammals. Closest to the skin is a layer of soft, fine feathers (*down*), which is the principal insulating cover. Above this are the *contour feathers*, which serve to streamline the bird's body and act as a secondary heat-trap; these (like mammalian hair) can be erected to increase the thickness of the insulating layer and thus to regulate body temperature. The *tail feathers* and the *flight feathers* on the

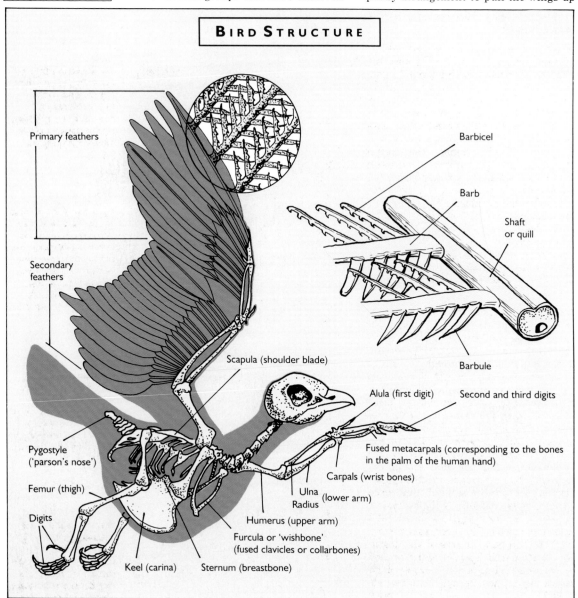

BIRD STRUCTURE

Primary feathers

Secondary feathers

Scapula (shoulder blade)

Pygostyle ('parson's nose')

Femur (thigh)

Digits

Keel (carina)   Sternum (breastbone)

Furcula or 'wishbone' (fused clavicles or collarbones)

Humerus (upper arm)

Radius
Ulna (lower arm)

Carpals (wrist bones)

Fused metacarpals (corresponding to the bones in the palm of the human hand)

Alula (first digit)   Second and third digits

Barbicel

Barb

Shaft or quill

Barbule

wings have the strongest shafts or *quills*, in order to withstand the tremendous stresses of flight and steering. With the exception of down, birds' feathers have a complex microstructure that allows the vanes on each side of the quill to 'knit' together to produce a smooth surface, important in minimizing air resistance during flight (▷ diagram).

Feathers may be adapted to uses other than flight and body insulation. The African sandgrouse's breast feathers have split and frazzled barbs. Nesting in the arid Namib Desert of southwestern Africa, this model parent flies off to the nearest spring or river (which may be a hundred or more kilometres distant), soaks its breast in water, and then flies back to its nest to give its chicks a drink. Birds of the heron family have downy 'powder feathers' in patches on the breast and rump. These constantly disintegrate to release keratin as a powder for use in preening and cleaning fish oil and grease from the bird's body. The eiderduck has extremely soft down on its breast, which it plucks off to insulate its nest against Arctic cold and winds (hence the use of these feathers to fill bed-quilts, the original 'eiderdowns').

As feathers are a bird's prime asset, it spends much time and effort *preening* – carefully cleaning and arranging its plumage with the bill. Next to a bird's tail is a special gland (the *preen* or *uropygeal gland*) that secretes skin oil used in preening. This oil is particularly important for sea and water birds, since it waterproofs the surface feathers.

Beaks, feet and claws

Like its feathers, a bird's beak or bill is composed of hardened keratin, but it is not dead tissue. Over the jaws there is a beak-generating organ (rather like the quick of a nail), which produces the beak's epidermal tissues and provides internal nutrition to the beak material. Nerves and fine blood vessels are present throughout the beak. In the absence of teeth, the beak has become the principal means by which most birds obtain food. As such it has been subjected to intense evolutionary pressure and become highly adapted to meet the needs of a wide variety of different feeding strategies (▷ illustration).

While the fore limbs of birds are modified into wings, the hind limbs – the legs – are usually covered in scales and armed with curving nails, the *claws*. Ground birds use their claws to dig and gather food, while perching birds use them to grip branches. Three digits face forwards for propulsion, and one faces backwards as a support, but the latter is not always fully developed. Owls and tree-living birds such as woodpeckers and trogons are *zygodactylous* – the third toe also faces backwards, increasing the strength of their grip. In owls, all the toes are equally well developed, with very sharp talons for grappling with prey.

Unlike sea birds such as penguins, which use their wings as flippers, propulsion in water birds is generally achieved by kicking with powerful legs. Ducks, swans and loons (divers) have webbed feet, while grebes and finfoots have fringed paddle-like toes.

Bird physiology

Although traces of their reptilian ancestry are still to be seen in the internal structures of birds, the similarities are perhaps less impressive than the differences. The most significant of these is that birds – like mammals but unlike 'cold-blooded' reptiles (▷ p. 87) – are warm-blooded or homoiothermic: they maintain their body temperature at a constant level, usually well above that of their surroundings, by means of internal (metabolic) mechanisms. In order to provide energy for their powered flight, birds need to maintain an extremely high metabolic rate, and their body temperatures (averaging about 40 °C/104 °F) are uniformly higher than those of mammals. In order to pump a constant supply of energy-rich blood to the flight muscles, birds also have large hearts that work at prodigious rates – up to around 600 beats per minute in hummingbirds.

A high metabolic rate also demands an efficient respiratory system, to keep the blood well supplied with oxygen. In contrast to other vertebrate lung systems, which involve a tidal (in-and-out) movement of air to and from the lungs, birds have a unique arrangement of air sacs attached to the lungs that allow a one-way flow of air through the system (▷ p. 181). At the base of the trachea (windpipe) is a bird's singing organ, the *syrinx*, which is particularly well developed in the songbirds. As air vibrates the membranes of the syrinx, its muscles alter their tension, allowing tones of different pitch and timbre to be produced. Thanks to this organ, birds are able to produce an extraordinarily rich repertoire of songs and other vocalizations (▷ p. 152).

Much of a bird's digestive and excretory system follows the reptilian pattern (although the digestive processes themselves are much more rapid). Waste products are voided through a single chamber (the cloaca), which also doubles as the genital opening (▷ diagram, p. 185), and – like lizards and snakes – birds produce near-solid urine, in the form of uric acid (▷ p. 88). Several ocean-going birds, including gannets and albatrosses, exrete excess salt (mainly from sea water taken in with the food) not through the kidneys, but through modified tear glands, ejecting the salt-rich fluid from their nostrils or mouth. A similar system is found in some turtles, lizards and snakes (▷ p. 88). However, two characteristically avian features of the digestive system are the gizzard, which is used to fragment food, and the crop, which serves as a food-storage chamber (▷ p. 178). ML-E

BEAKS

Toco toucan. Toucans have huge beaks, almost as long as their bodies. The beak is used to pluck fruit from branch tips, and may be important in display.

Caribbean flamingo. Towering on their long thin legs, flamingos hold their beaks upside down to sieve water fleas and small fish from shallow lakes and ponds.

Rufous hummingbird. Hummingbirds use their bills to reach to the bottom of tubular flowers, from which they suck nectar through their tube-shaped tongues.

Gold-and-blue macaw. Parrots have powerful beaks, enabling them to lever themselves up trees and to break open the hardest nuts; however, they can also be used with great delicacy in grooming and courtship.

Goshawk. Birds of prey have sharp beaks ideally suited to slashing and dismembering their prey. Their nostrils are protected from clogging with blood by tufts of feather.

Black skimmer. Skimmers feed on the wing, with the lower jaw cutting the surface water and the upper jaw ready to snap down on any aquatic life that is channelled in.

Birds
Life Cycles

Many animals have evolved complex patterns of breeding behaviour in order to create the best possible conditions in which to mate and rear their young, but none has done so more spectacularly than birds. Birds – like humans but unlike many other mammals – generally have a rather poor sense of smell, so scent plays a minor role in their reproductive behaviour. Instead, they rely on a stunning repertoire of songs, dancing displays, and the parading of brightly coloured plumage to attract mates and outshine rivals.

The various activities associated with breeding – defending a territory, producing a clutch of eggs, finding food for the young – represent a great investment in time and energy for birds (▷ box, p. 159). Food must be plentiful at this time, so in temperate regions the breeding season is usually restricted to spring and summer.

Establishing a territory

In order to secure essential resources such as food, nesting material and a nesting site, male birds typically become territorial in the breeding season, actively defending an area (often the same area each year) against rivals (▷ p. 156). For colonial sea birds, which have the food resources of the sea at their disposal, such a territory may be no more than a square metre or so, while large birds of prey may preside over an area of several square kilometres.

Many male birds stake their claim to a territory and advertise their presence by means of song (▷ p. 152). The size of a territory will depend partly on the local population density, but bigger and better sites are taken by older, more experienced birds. Continued singing and threat displays are usually enough to keep rivals from trespassing, but active pursuit and fighting may occasionally occur. Song may also be used to attract a mate, a singer with a large and varied repertoire generally being more attractive to females than an inexperienced juvenile.

Courtship and display

Many birds use some form of visual display to attract the attention of females. Such displays are the opportunity for males to demonstrate that they are more eligible suitors than their rivals, but it is the female who makes the final selection.

Female birds often appear to be rather capricious in their preferences. Hen swallows and pheasants, for instance, select their mates on the basis of the size of their tails and leg spurs respectively. Because better-endowed males get to breed more frequently, their preferred characteristics tend to be perpetuated in succeeding generations and to become ever more exaggerated. It is due to this 'sexual selection' (▷ box, p. 159) that such extravagances as the peacock's tail and the magnificent plumage of the birds of paradise are thought to have evolved. Not content with the extraordinary plumes on the wings, tail or head, female birds of paradise also require the males to perform astonishing gymnastic displays. In the case of the blue bird of paradise, for instance, the male hangs upside down from a prominent branch, vibrates his feathers, and may bounce up and down to impress a female.

Birds such as the black grouse, the ruff and the prairie chicken mate in communal display grounds called *leks*, while others, including the Australasian bowerbirds, build special structures to attract females (▷ photos). Weavers (▷ p. 161) are assessed on the basis of their nests, and as expertise comes with practice, experienced males are more successful in attracting females.

Courtship displays have other functions besides impressing females. When a female and a male come together, the latter is normally more willing to mate and must induce a similar state of readiness in the female. Indeed, the female's natural response at the approach of a male may be to flee or attack, so an important aspect of the courtship ritual is to inform the female of the male's intention to mate with her. In monogamous species, where both parents are required to build the nest and rear the young, courtship displays may be important in establishing and reinforcing pair-bonds. In grebes, for instance, pair-bonding is achieved by a variety of elaborate and highly ritualized displays (▷ box, p. 155).

Mating and egg-laying

The great majority of male birds do not have an intromittent organ that can be inserted into the female's genital opening (or cloaca), so sperm is transferred by the male balancing on the female's back and bringing his cloaca into contact with hers (▷ box, p. 185). The sperm is stored within the female's reproductive tract and shared out amongst the ova (eggs),

Diverse ways to impress a female: The great grey bowerbird (above) and the other members of the Australasian bowerbird family construct special 'bowers' or courting grounds from twigs and branches, which they decorate with flowers, leaves, pebbles, shells and other brightly coloured objects. Despite its appearance, a bower is not a nest: a female attracted to a bower to mate subsequently builds a nest of her own and cares for the brood unaided.

Male ruffs (left) congregate at a communal display ground called a lek, where they engage in fierce competition for small mating territories; females apparently prefer males holding territories near the centre of the lek. Like the bowerbirds, the male ruff takes no further interest in his posterity once he has mated, leaving the task of incubation and rearing entirely to the female.

which mature in the subsequent laying period.

A bird's egg essentially follows the reptilian model (⊳ p. 86), with a nutritive yolk and various associated membranes. After the hard shell of calcium carbonate has been deposited, characteristic colour markings are added by pigment glands in the lower oviduct. The main purpose of these markings is camouflage, but where there is considerable variation in colouring, as in some colonial nesters, they may also help parents to identify their eggs. Eggs are laid one at a time at regular intervals until the clutch is complete. Clutch size varies greatly from species to species and may vary even within a species, depending on the food resources available in a given area.

Nesting

Most birds build a nest in which to lay their eggs. This helps to keep the eggs warm and dry, and also offers some protection against predators. In species in which the young are entirely dependent on their parents for food and protection (⊳ below), the nest must also serve as a sanctuary for the hatchlings, and it is typically such birds that build the more elaborate nests. Nest-building ability is partly instinctive, but greater expertise is acquired with experience. The first attempts of accomplished builders such as weavers (⊳ p. 161) are often ramshackle affairs, but their technique improves in subsequent years.

The type of nest built depends not only on a bird's particular requirements but also on the kinds of site and building material available. In the emperor penguin the 'nest' is no more than a flap of skin between the feet and the warm, feathered belly (⊳ photo), while gannets, cormorants and other sea birds that nest on cliffs, where they can rely on a high degree of natural protection, often throw together simple, untidy nests of marine debris.

Most birds gain added security against predators and the elements by nesting clear of the ground. Such nests are usually set in a forked branch and may be cup- or plate-shaped, depending on the size of bird. They are made by interweaving flexible stems or twigs, and are often lined with feathery down, moss, soft leaves and other materials. Birds such as woodpeckers and hornbills gain even greater protection by nesting not in the branches of a tree but within the body of the tree itself (⊳ p. 161).

Many birds use mud to reinforce their nests, but a few have made a speciality of exploiting this material. The red hornero lives in flat, open areas of South America, where hiding a nest is impossible. As soon as rain arrives, both birds of a pair collect lumps of wet clay and build globular nests on posts, branches or any other firm support. The finished structure looks like a rustic mud oven, hence the hornero's common name 'ovenbird'. In Europe an equally characteristic sight is the mud nests of the house martin and the barn swallow, which are often seen fixed under the eaves of houses or on top of roof beams.

Exotic nesting materials are employed by some birds. The female white-eared hummingbird of Mexico, for instance, builds a framework for her nest from spider silk fixed between forks in adjacent branches and then interlaces delicate fibrous materials such as animal hairs, wool and lichens to create a strong but light cup suspended in mid-air. The cave swiftlet of the Indonesian archipelago uses only saliva to build a cup nest, which is stuck firmly to the roofs and walls of sea caves. These nests are the main ingredient of Chinese bird's-nest soup, a factor contributing to the rarity of the bird.

Incubation and hatching

The warmth of the nest is not sufficient in itself to bring the eggs safely to hatching. Birds must actively incubate their eggs, usually by sitting on them and turning them at regular intervals. In many species both parents are required for brooding, one bird incubating while the other searches for food. Often incubation begins only when the clutch is complete, so that all the eggs hatch simultaneously. Various species, however, including many birds of prey, begin to incubate as soon as the first egg is laid, so that the larger, first-hatched chick has a better chance of surviving if food is in short supply.

A novel method of incubation is used by the moundbirds or megapodes of Australasia. The male birds build enormous incubators, up to 5 m (16 ft) high and 12 m (40 ft) wide in the case of the common scrubfowl. The mallee fowl, inhabiting the dry scrub of southern Australia, digs a pit and laboriously collects plant material (rather sparse in his environment), which he throws in and piles up. As soon as his compost heap has been wetted by the autumn rains, he covers it over with more vegetation and soil. The brush turkey of eastern Australia lives in forested habitats and so has the advantage of a good supply of leaves, which he gathers together in huge heaps. Both species have temperature-sensitive tongues, which can detect fluctuations from the ideal incubation range of 33–35 °C (91–95 °F). Each day the male tests the interior of the pile with his beak. If it is too warm, he scratches off the surface; if too cool, more cover or vegetation is added. Females are attracted to the mound, mate and lay eggs in tunnels scattered round the mounds, and then leave. As they hatch, the chicks dig their way out of the mound and are not cared for after they emerge.

Some chicks 'peep' to indicate that they are too cold or hot inside the egg, stopping only when the parent has manoeuvred them to a position that removes the unwanted stimulus. In many species a characteristic form of peeping may also occur as the time for hatching approaches. The chick has a hard knob on its beak (the so-called 'egg tooth'), which it taps against the inside of the shell until it cracks, after which a tremendous amount of effort leads to complete hatching.

Parental care

The amount of parental care required by the hatchlings depends greatly on their condition at hatching. In *nidicolous birds*, such as birds of prey, swifts, woodpeckers and songbirds, the newly hatched chicks are naked, blind and defenceless; they are entirely dependent on their parents for food and protection, and must remain in the nest for several weeks until fledged. An unstinting supply of food is necessary, and – especially for smaller birds – great efforts have to be made to satisfy the demands of the begging chicks.

In *nidifugous birds*, such as gamebirds, wildfowl, grebes and divers, the chicks hatch in a relatively advanced state – their eyes are open and they have a covering of fluffy down. The eggs hatch at about the same time, and the chicks are mobile within hours, leaving the nest for good soon after hatching. Although the young are capable of foraging for themselves, much vigilance is required from the parent or parents, to keep them together and clear of danger. Bonding between parent and chick is particularly important in these species, since mutual recognition and a drive to stay together is vital for the survival of the chicks. Konrad Lorenz's famous work with greylag geese showed that bonding takes place with whatever the chick sees immediately after hatching, by means of a process known as *imprinting*. Thus it is possible to induce a chick to follow a human, a dog, or even a balloon, provided that it is the first object encountered.

A number of birds, including honeyguides, widow birds and various species of cuckoo, have abrogated all parental responsibility. Not only do they not look after the hatchlings, but they rely on other birds to incubate their eggs (⊳ p. 168).   ML-E

Emperor penguins breed and raise their young in the freezing environment of the Antarctic. As soon as it is laid, the single egg is entrusted to the male, who, huddling together with thousands of his fellows, cradles it in a cosy nook between his feet and his warm, feathered belly. Feeding out at sea for the whole two-month incubation period, the female finally returns to relieve her mate (now a shadow of his former robust self), allowing him to head off in search of food.

SEE ALSO

● BIRDS: EVOLUTION AND CLASSIFICATION p. 90
● BIRDS: FORM AND STRUCTURE p. 92
● THE WORLD OF BIRDS p. 96
● ANIMAL COMMUNICATION p. 152
● TERRITORY, MATING, SOCIAL ORGANIZATION p. 156
● ANIMAL BUILDERS p. 160
● THE RHYTHMS OF LIFE p. 174
● SEX AND REPRODUCTION p. 184
● GROWTH AND DEVELOPMENT p. 186

The World of Birds

A bird's view of the world is perhaps not so very different from our own, in so far as the senses on which birds primarily depend – sight and hearing – are also those that are best developed in humans. Birds communicate with one another through a range of visual and auditory signals – through a language of brightly coloured plumage and a repertoire of songs and other noises. Irrespective of the 'meaning' of such signals to other birds, the signals themselves are immediately appealing and attractive to us, and this is probably one of the main reasons for the widespread popular affection for birds.

The power of flight, in particular, has made birds among the most ubiquitous of animals – they are to be found in virtually every habitat in every continent; and their continent-spanning migrations have undoubtedly made them the best-travelled (▷ pp. 170–1). Many birds live in close contact with humans, often nesting in buildings and feeding in streets and gardens, so they are also among the most conspicuous of wild animals.

BIRD COLOURS

The magnificent array of colours seen in birds is produced by a combination of structural effects and actual pigments in the feathers themselves. Light may be absorbed or dispersed by feathers, resulting in unshiny dark or light areas, or may be diffracted by the feather filaments, producing the marvellous iridescence and metallic sheens seen, for example, in starlings and peacocks. Most reds and oranges result from the presence of rhodopsin (a red pigment also found in the retina of the eye) and carotenoid pigments related to vitamin A, all of which are derived from foodstuffs. Melanin, the dark pigment in mammalian hair and skin, is responsible for yellows, browns, greens and blues. Turacos have brilliant red and green colours different from those of all other birds. These colours are based on copper-containing pigments related to haemoglobin, the oxygen-carrying protein in red blood cells. The pink colour of flamingos comes from the small aquatic crustaceans on which they feed.

The resplendent colours of the sword-billed hummingbird.

Like the fur of many mammals, the plumage of some birds undergoes seasonal colour change. The feathers of the ptarmigan, for instance, which lives in snowy winter conditions, become white in winter, so that the bird remains well camouflaged throughout the year. Just as in mammals, such changes are controlled by the pituitary gland (▷ p. 174).

Feeding habits

Although the form of a bird's beak is usually the best clue to its feeding habit (▷ p. 93), each species shows a number of other adaptations to a particular method of finding food. The structure of the feet, the relative development of each of the senses, the form of the wings (dictating a particular pattern of flight) – all are important in equipping a bird for effective foraging or hunting.

Most ocean birds are fish-eaters, and their bills are typically sharp and strong, with serrated edges and often hooked tips to hold slippery fish. Pelicans dive into the sea and catch fish with their long bills, storing them in a pouch under the bill. Storm petrels of the southern hemisphere skip at the surface of the sea, collecting small fish, plankton and crustaceans. Groups of gannets dive from heights of up to 30 m (100 ft) to take shoaling fish. Ocean birds follow their food wherever it is most abundant, often to the extremes of Arctic and Antarctic oceans.

Coastal birds such as fulmars and auks live on rocky cliffs. Others, including waders, terns and some gulls, prefer flatter ground, while puffins live in burrows. All either catch fish by diving and swimming or gather invertebrates from the intertidal zone (▷ pp. 220–1). The Atlantic puffin has a stout bill striped yellow and red with which it catches several fish at a time, diving underwater and swimming strongly with its wings. Waders such as the oyster-catcher, redshank and avocet, all birds of Atlantic beaches, probe with their beaks for invertebrates such as crustaceans, shellfish and worms on intertidal mud flats and shores. The distinctive bill form of each species means that the many different waders exploit different food resources and competition between species is kept to a minimum.

Some sea birds are scavengers, such as the snowy sheathbill of the South Atlantic and Antarctica, which eats seaweed, carrion and young seals. Although feeding on fish at other times, its counterpart in the North Atlantic, the great skua, harries nesting and feeding birds during its own nesting season to steal their eggs, chicks and food (▷ photo, p. 164). The frigate bird of the Galápagos Islands has a long hooked beak with which it scoops up fish, squid and jellyfish. It also terrorizes other birds into regurgitating their food, which it then steals. Many gulls have adapted to feeding on human refuse, and are steadily moving inland in some areas.

Most water birds – ducks, geese, divers and others – dabble and dive for plants, insects, molluscs and crustaceans. Some, such as the mergansers and red-throated divers of North Atlantic and Pacific waters, dive for fish.

Birds of prey are almost exclusively carnivorous, feeding on birds, reptiles, mammals, and sometimes fish. They have

Yellow-billed oxpeckers. The two species of oxpecker, African relatives of the starling, ride about on large herbivorous mammals such as cattle, antelopes and giraffes, feeding on ticks and other skin parasites. The association is usually symbiotic – both parties stand to gain by the relationship (▷ p. 169) – but sometimes it borders on the parasitic: on occasion these birds abuse their privilege, pecking at surface wounds and eating pieces of flesh.

strong hooked beaks and well-developed musculature for holding and tearing prey, and their toes, arranged three forward and one back, are armed with long, sharp, curving claws (*talons*). The osprey is widely distributed throughout the world and eats fish, diving at them feet-first and gripping them with its claws and spiked soles. The American bald eagle is also a fish-eater, and often follows salmon on their spawning runs to catch exhausted fish. Most hawks and eagles perch or soar until they sight prey on the ground, and then drop from a height to stun and kill it. Others, such as the peregrine falcon, hobby and goshawk of the northern hemisphere, are aggressive and acrobatic hunters, chasing and catching birds on the wing.

Old and New World vultures are carrion-eaters. Most have featherless heads and necks, and sharp beaks to cope with dead animals. However, they belong to different families, and their similarities are therefore examples of convergent evolution (▷ p. 17). The small Egyptian vulture also eats animal droppings, preferring lion faeces, which are high in undigested protein; it sometimes feeds on ostrich eggs, using stones to crack them open (▷ box, p. 151).

Seed- and nut-eaters usually have short, robust bills. They may be sharply curved, and in species such as the crossbill the tips of the bill overlap, so allowing the bird to remove the seeds of pine cones. Parrots make great use of their tongues, rolling around nuts and seeds in their beaks to test their shape and find the best place to crack them open.

Hummingbirds are nectar-sippers (▷ illustrations, p. 73 and this page), and may be important pollinators of various long-tubed flowers (▷ p. 42). They also eat insects, catching them in flight or on flowers. Except when nesting, swifts spend all their life in the air, and feed entirely on insects, which they catch in flight.

Bird senses

For most birds vision is the key sense (▷ p. 189), and their eyes are typically large and well developed. Hunting birds such as hawks, eagles and owls have eyes set towards the front of the head, so providing good forward and binocular vision, which is vital in pinpointing prey and accurately judging distances when striking. Hawks have a double fovea (part of the retina) in which there are only cones – nerve cells sensitive to fine detail in bright light – while vultures have a higher magnification in the centre of their field of view, enabling them to recognize prey even if it is not moving. Nocturnal birds such as owls, frogmouths and nightjars have large eyes with a high proportion of rods – retinal cells sensitive to movement in dim light. For many birds, particularly those that feed on the ground, vigilance against predators is crucial to their survival, so they typically have eyes set at the sides of the head, giving limited binocular vision but good all-round vision.

A different problem faces birds that spend at least part of their time underwater. Dippers and divers have flexible lenses, allowing them to adjust to the different refractive indices of air and water. Penguins, on the other hand, have flat corneas (the transparent part at the front of the eye) ideal for underwater vision but making them very short-sighted on land.

Many sea birds are attracted to red, possibly because oil droplets in their retinal cones absorb green and blue light. Nestlings peck at red spots on the beaks of adult birds, stimulating regurgitation of food (▷ pp. 148–9). Penguins, which make nests out of pebbles, habitually steal the 'best' ones from their neighbour's nest, and a pile of red pebbles put at one edge of a penguin rookery will rapidly migrate towards the other side.

The sense of hearing is also well developed in most birds. Owls have left and right earflaps in slightly different positions on their heads, allowing them to fix the position of noises more accurately. They also have large eardrums to amplify tiny sounds. Their feathers are very soft, resulting in almost soundless flight – an adaptation that fishing owls lack, not needing silence to catch their prey. The South American oilbird relies on its hearing, although not in order to detect prey – it feeds on seeds and fruits. When foraging in the tropical forest by night or when flying in the dark caves in which they live, these birds navigate by listening to the echoes of their vocal clicks, using a system of echolocation similar to that of bats (▷ box, p. 119).

Smell is generally poorly developed in birds, but a notable exception is the kiwi of New Zealand. It has very small eyes and detects earthworms and grubs by smell, having nostrils at the very tip of its long flexible beak. The part of the kiwi's brain associated with smell is far better developed than in other birds. The kiwi, like many other birds, also has sensitive bristles (vibrissae) at the base of its bill by which it can detect slight vibrations as it probes for food in the leaf litter.   ML-E

BIRDS AND MAN

The flesh and eggs of birds have formed a staple part of the human diet since prehistoric times. While many wild species are locally important as food birds, domesticated forms of the red jungle fowl (the domestic fowl or chicken), the turkey, the greylag goose and the mallard duck are farmed commercially on a worldwide basis. The shooting of gamebirds such as grouse, pheasants and partridges may also provide food, but in these cases the primary motivation is sport rather than food provision (▷ photo, p. 213, and p. 229). In spite of growing opposition, large numbers of migratory birds such as larks and thrushes are trapped and killed for food each year as they fly over southern Europe. Birds' feathers have also been valued for their aesthetic and insulating qualities: the soft down of eiderducks is used as filling for duvets or 'eiderdowns', while the decorative plumes of birds such as the ostrich, peacock and great white egret have at various times been sought-after as fashion accessories.

Hunting for food or sport has led to the extinction of some bird species, such as the dodo, the giant auk, the American passenger pigeon and the giant moas of New Zealand, but this is not the only damage caused by human activities, nor the most severe. Numerous tropical species, including curassows and many members of the parrot family, have suffered as a result of rainforest clearance, while wetland drainage has spelled disaster, locally at least, for populations of marshland birds such as bitterns, spoonbills, godwits and avocets. Many birds have been deliberately introduced to areas outside their natural range, and a few, such as the house sparrow and the starling, have made spectacular conquests, but more often native bird populations have suffered as a result of introductions, especially of mammalian predators such as dogs, cats, weasels and mongooses. Pictures of oil-clogged guillemots or razorbills have brought public awareness of the effect of oil pollution on sea and coastal birds, but more insidious and probably greater damage is done by other forms of chemical pollution, as when agricultural pesticides and toxic industrial wastes become concentrated in bird food chains (▷ p. 227).

The negative impact of birds on man is negligible by comparison. Very few birds are significant as vectors (carriers) of disease, an exception being the feral pigeon, which has found urban life very much to its liking and can transmit diseases such as psittacosis to humans through its droppings. Much more important is the role of birds as crop pests. Perhaps only the red-billed quelea of Africa can be said to reach true plague proportions (▷ illustration, p. 230), but the combined depredations of seed-eaters such as sparrows and bullfinches are considerable. Geese, particularly the greylag and the Canada goose, may also be very destructive to newly planted grass and crops such as carrots. A few species, including herons and kingfishers, have been implicated in depletion of freshwater fish stocks, but in fact their impact is probably slight. The presence of flocks of birds in the vicinity of airports is the cause of some concern, and a number of serious accidents have been caused by bird strikes; large amounts of time and money have been invested in efforts to minimize this problem.

Spix's macaw: like many exotic species, a victim of both habitat destruction and the depredations of collectors. The known wild population of this species has been reduced to a solitary male, so unless a captive breeding programme can be set up among the handful of specimens held in captivity, this bird is destined to extinction in the wild.

SEE ALSO

- BIRDS pp. 90–5
- ANIMAL COMMUNICATION p. 152
- TERRITORY, MATING, SOCIAL ORGANIZATION p. 156
- ANIMAL BUILDERS p. 160
- PARASITISM AND SYMBIOSIS p. 168
- MIGRATION p. 170
- THE RISE AND RISE OF THE COLLARED DOVE (BOX) p. 173
- SENSES AND PERCEPTION p. 188

FLIGHTLESS BIRDS

The 10 species of ratite, including cassowaries, rheas, the emu and the ostrich, are the most primitive living birds, and are the remnants of a group once found worldwide. All the ratites are flightless and do not have the adaptations seen in flying birds: they have small wings, unbarbed downy feathers, flat breastbones, and often large, solid legbones. No skeletons or fossils of ancestors with fully formed wings have yet been found, and it is now thought likely that such birds were never capable of flying. The closest living relatives of the ratites, the tinamous of Central and South America, are capable of rather weak and clumsy flight, but are largely ground-dwelling.

Apart from the kiwi of New Zealand, 50 cm (20 in) tall, the ratites range in size from 1·5 (5 ft) to the largest living bird – the ostrich of Africa, towering above humans at up to 2·75 m (9 ft). A curious feature of the ratites is that the eggs are usually incubated by the male, and in the case of the ostrich several females lay their eggs in a communal nest. The egg of an ostrich is equal in volume to 25–40 chicken eggs, while the huge extinct elephant birds of Madagascar, which were also flightless, laid eggs that weighed around 10 kg (22 lb).

Some species of flying bird, such as the New Zealand rails and the kakapo, a nocturnal parrot, have become flightless, usually in island habitats in the absence of predators. The severe pressures faced by such species, both from humans and from introduced predators, is illustrated by the well-known fate of the dodo. This large flightless relative of the pigeons lived in security on Mauritius until the island became a port of call for passing sailors; by the end of the 17th century it had already been hunted to extinction. Some other birds, including penguins and the recently extinct giant auk, presumably lost the power of flight in the process of adapting to life in water (▷ illustration, p. 17).

The Coming of Mammals

As the dominant species of terrestrial mammal alive today, we naturally take a special interest in our fellow mammals. In terms of numbers alone (whether of species or of individuals), the class Mammalia, with fewer than 5000 species in total, is not particularly significant, even amongst the vertebrates – there are roughly four species of fish to every mammalian species. But many of the larger marine animals and most of the larger land animals are mammals, and as such their importance far outstrips their numbers.

In terms of geological time, the significance of mammals as a group is a relatively recent phenomenon. Although the earliest mammals appeared over 200 million years ago, for the first two thirds of their history they were overshadowed by the mighty dinosaurs and other reptiles. It is only during the last 65 million years that mammals have come increasingly to the fore. During this time a spectacular array of mammals has evolved, the extraordinary diversity of which is only partly mirrored in the species alive today.

The ancestors of the mammals

The first reptiles that dominated the land over 300 million years ago were the synapsids or mammal-like reptiles (▷ p. 82). Their sprawling gait and posture were typically reptilian, but the detailed structure of their skulls suggests that they formed the ancestral stock from which the mammals arose. All early synapsids were carnivores, the smallest ones feeding on insects and worms, the larger ones on smaller reptiles, but by about 250 million years ago one major group, the *dicynodonts* ('two dog tooths'), had become herbivores.

By the Triassic period, 235 million years ago, synapsids such as the mainly carnivorous *cynodonts* ('dog tooths') were much more mammalian in appearance. The structure of their skeletons had altered to allow a more upright stance and gait, and their skulls had become modified to accommodate teeth of a more mammalian pattern and a secondary palate separating the food and air passages (▷ box). It cannot be known for certain when milk production and suckling first developed, but there is some indirect evidence that it was among the more advanced cynodonts. In one group there appears to have been a dramatic change in the pattern of reptilian tooth replacement, and since there is a clear disadvantage for the mother if a suckling offspring has a fully functional set of teeth, this may be related to the onset of suckling. It is also possible

SEE ALSO

● THE PLANET'S CHANGING FACE p. 8
● EVOLUTION pp. 12–17
● THE CLASSIFICATION OF LIFE p. 18
● MAMMAL GROUPS pp. 102–45

WHAT MAKES A MAMMAL A MAMMAL?

When we look at a tiger or a horse, we may have no difficulty in saying that it is a mammal, but the situation is not always so simple. A shrew may look like a reasonably typical mammal, but closer inspection will reveal that – like reptiles and birds – it has a common chamber (the cloaca) into which the alimentary canal and the urinary and genital ducts open. Still more perplexing are the duck-billed platypus and the echidnas, which lay eggs – not at all, one would have thought, a mammalian characteristic. So how do we recognize a mammal? In other words, what features make a mammal a mammal?

The name 'mammal', derived from the Latin *mamma* meaning a 'breast', points to one of the most fundamental features of the group – the production of milk from mammary glands in the female. Milk is a nutritious fluid containing varying amounts of protein and fat, and its production is not found in any other group of animals. However, the possession of external structures (nipples or teats) with which to suckle the young is not in itself definitive of mammals, since the monotremes (echidnas and the platypus) do not have such structures.

Unlike reptiles, mammals are homoiothermic ('warm-blooded') – they maintain a constant internal temperature by means of a high metabolic rate. This is not unique to mammals – birds are also homoiothermic and use essentially similar internal mechanisms – but their external 'lagging' is: the insulating covering of hair or fur is peculiar to mammals, just as feathers are to birds.

A unique feature of all mammals is the possession of three sound-conducting bones (or ossicles) in the middle ear – the stapes, incus and malleus (▷ p. 191). Two of these (the incus and malleus) are thought to be derived from part of the jaw hinge in reptiles, but in mammals the bone of the lower jaw (the dentary or tooth-bearing bone) is single and articulates directly with the rest of the skull.

As one of the principal means by which mammals gather their food, the teeth have been greatly modified from those of reptiles. They have increased in size and been reduced in number, and they have also become specialized in different parts of the jaws for different functions. At the front, *incisors* have developed for biting and *canines* for stabbing, while at the back cusped teeth (*molars* and *premolars*) are

adapted for chewing and grinding. The general lengthening of the crowns of the teeth (the parts above the gums) required more solid anchorage, so the roots became longer and sockets for the teeth developed in the jawbones. The consequent deepening of the upper jawbone (maxilla) resulted in the roof of the mouth becoming arched, so that there was enough space for air to pass even when the mouth was filled. It is thought that a flap of skin developed across this space and that a bony sheet subsequently spread from the maxilla, so forming a secondary palate separating the food and air passages. This feature (which evolved independently in reptiles) allows the young to suckle and breathe at the same time, and also allows food to be processed in the mouth. The pattern of tooth replacement in mammals, with a set of milk teeth followed by a permanent set of teeth in the adult, is also unique to mammals; as with the development of the secondary palate, this feature is believed to be associated with suckling (▷ text).

The way in which the head articulates with the rest of the body is another feature that separates mammals from all other animals. The joint system of the neck, involving a pivotal joint (▷ p. 177) between the first and second vertebrae (the atlas and the axis), allows the head to be turned from side to side, but not to the same degree as birds, which have a system that allows rotation of the head through 180° if necessary.

Among the most significant characteristics of mammals – though the hardest to quantify – is their intelligence. The cerebral cortex (the part of the brain concerned with intelligent behaviour) is better developed in higher mammals than in any other group of vertebrates (▷ p. 183).

Mammals are therefore distinguished from other animal groups by a collection of characteristic features, although not all of them are found in all mammals. In the evolution of mammals from their reptilian ancestors, these various features were presumably acquired one after another, but – given the incomplete state of the fossil record – it is not possible to tell which came first. It is only by the simultaneous possession of a number of such characteristics that a mammal can be recognized, and there is therefore no single feature that can be said to define the group.

that the shift to homoiothermy ('warm-bloodedness') also occurred at around this time. Among later cynodonts there are pits in the upper jawbone, which may indicate the possession of whiskers and hence a covering of insulative fur or hair, a feature probably associated with warm-bloodedness. The presence of a secondary palate would have permitted the young to suckle and breathe at the same time, and this may be linked to the need for rapid feeding in warm-blooded animals with a high metabolic rate.

In one of the most striking faunal turnabouts in the history of life on earth, the synapsids began to go into a serious decline. By the Jurassic period, 185 million years ago, only a few herbivores had managed to survive. In their place came the dinosaurs, which had emerged about 230 million years ago and which were to dominate life on land for the next 160 million years (▷ p. 82).

The first mammals

It was during the later Triassic, about 220

million years ago, that the first true mammals appeared. Throughout the age of the dinosaurs, the mammals remained small, and were shrew- or rat-like in appearance and life style. They probably managed to subsist side by side with the dinosaurs because – unlike the latter, which were cold-blooded – they could adapt to nocturnal activity. By the end of the Cretaceous, 65 million years ago, the first primate, *Purgatorius*, had made its appearance (⊳ p. 136).

When the dinosaurs became extinct at this time and the ecological niches they had occupied became vacant, the mammals began to extend their range and to adapt themselves to a wider range of habitats. Despite the disappearance of the dinosaurs, however, the mammals did not lack competition. Birds – the other group of warm-blooded animals – also began to radiate dramatically during this period (⊳ p. 90), and for a time giant flightless birds such as the *Diatryma* dominated terrestrial habitats. Nevertheless, even in the face of this competition, the mammals were able to establish themselves.

The age of the mammals begins

When the dinosaurs began to spread over the planet, all the continents were united in the single landmass of Pangaea; in contrast, at the beginning of the age of mammals 65 million years ago, the continents were already largely separated (⊳ p. 8). Although not closely related, similar-looking mammals evolved independently on the different continents to fill the same kinds of niche. In Australia, which was isolated throughout much of the Tertiary, there evolved marsupial cats and dogs, marsupial flying squirrels, and even marsupial moles (⊳ p. 103). On the grasslands of South America there were animals resembling modern camels, horses, rhinoceroses and elephants, and

even a large marsupial cat; there were also unique creatures, such as the giant ground sloths and armoured glyptodonts (closely related to the living armadillos), which survived into historical times (⊳ illustration).

A major event in the history of the mammals occurred during the Miocene epoch, 23 million years ago, when Africa collided with Eurasia at either end of what is now the Mediterranean. This new land connection allowed the mammals of Europe and Asia to invade Africa, and by the same token the unique African mammals, the elephants and apes, to migrate north to conquer Eurasia.

The spread of grasslands

Fundamental changes in mammalian life came about with the gradual replacement of forests and woodlands by grasslands during the late Oligocene and early Miocene epochs, 28–18 million years ago. Perissodactyls were initially the dominant group of hoofed mammals, living as browsing forest-dwellers and probably reaching their zenith about 40 million years ago. As the grasslands spread, a few perissodactyls, such as the horses, successfully turned to grazing, but increasingly they came to lose out in competition with grazing artiodactyls such as cattle and antelopes (⊳ p. 112). In the more open conditions of the developing grasslands fleetness of foot was required to escape predation, and both horses and most artiodactyls became long-legged and fast-running. As well as being able to move quickly, such grazers were able to observe the approach of predators at greater distances than had been possible in forested conditions. This had a dramatic impact on the main groups of carnivores, many of which subsequently became extinct.

In this new environment there was therefore a premium on intelligence, and it was

The honey possum, an Australian marsupial that feeds on nectar. Although the isolation of Australia and New Guinea has allowed a wide range of marsupials to evolve and survive there, elsewhere the placental mammals have ousted virtually all marsupials from their ecological niches – the only major exceptions being the omnivorous opossums of the Americas.

on the grasslands that the modern group of carnivores, including cats and dogs, came to the fore. These predators are generally unable to catch their prey simply by chasing it – they do not have

SOUTH AMERICAN MAMMALS
65–23 million years ago

GIANT GROUND SLOTH
(*Megatherium*)

LITOPTERN
(*Theosodon*)

MARSUPIAL CAT
(*Thylacasmilus*)

GLYPTODONT
(*Daedicurus*)

the speed capability. Instead they use complex stratagems: dogs often hunt in packs, while cats, although typically solitary, use cunning and stealth to stalk their victims (▷ p. 165).

Another consequence of the great reduction of forests was that the tree-dwelling primates found themselves progressively forced onto the open grasslands. However, because of their intelligence and ability to work together in teams, the savannah-living apes of the Miocene in Africa and Eurasia were able not only to survive but to spread over vast areas.

The present-day game reserves of East Africa, with their great variety of specialized mammals, represent the type of ecosystem that first arose in the Miocene grasslands. This diversity came to an end in most parts of the world with the ice ages of the Pleistocene epoch, 1·6 million to 10 000 years ago (▷ p. 10).

The epoch of the ice ages

In the wake of dramatic climatic change, there follow a wide range of environmental conditions and a succession of different plant types. In such conditions it is difficult for an animal to specialize on one type of food; the key to survival and success is to be adaptable, and so it was the generalists rather than the specialists that dominated life in the Pleistocene.

Cattle (in the form of bison in North America), sheep and pigs were the key herbivores everywhere, and the most adaptable of the carnivores – the cats and dogs – held their own by virtue of their superior intelligence. But it was the rats and mice that became the most numerous and successful of all the mammals. One other highly adaptable mammal, with an ability to use tools and harness fire, also began to emerge: man.

When the ice caps spread from the polar regions into central Europe, the permafrost and tundra reached as far south as the Alps and the Himalaya. The great plains of Eurasia and North America were inhabited by woolly mammoths and woolly rhinoceroses (▷ illustration), which grazed on the relatively rich plant life that the tundra supported at this period. These large mammals were preyed upon by carnivores such as sabre-toothed cats, with their long stabbing canine teeth. Caves became important as shelters for many animals, including humans and the large cave bears that they hunted. Indeed, much of our detailed knowledge of the appearance of ice-age mammals comes from the figurines of Cro-Magnon man of 30 000 years ago, and from later cave paintings.

Why are most of these great mammals of the Pleistocene no longer around? One theory suggests that they were victims of man's improving technology and were simply hunted to extinction: these extinctions seem to correlate with the advent of the hand-axe culture of early man and his use of fire. With the end of the last ice age 10 000 years ago, the plant-rich tundra disappeared, and this may also have contributed to the extinction of so many species.

The classification of mammals

All mammals, together making up the class Mammalia, suckle their young with milk produced by mammary glands in the female (▷ box, p. 98), but in other respects they differ in the way in which they reproduce. It is on the basis of these differences that the three major groups of living mammals are defined.

The smallest group, the monotremes, includes a single order with just three species, all of which are confined to Australasia – the duck-billed platypus and the echidnas (▷ p. 102). These are distinguished from all other mammals by the fact that they lay eggs, and are known as prototherians (subclass Prototheria).

Although now generally accepted as true mammals, they retain a number of primitive features. Even though they are furry and feed their young on milk, they lay eggs that are structurally very similar to those of a bird, and – like birds – they incubate them in a nest or burrow. The prototherians do not have breasts or teats but have special glands that ooze milk, which the young lap up from the fur, rather than suckling in the normal way.

All other mammals bear live young and are known as therians (sometimes considered as the subclass Theria), but reproductive differences within this group allow further division into two subgroups (subclasses or infraclasses). The first of these, Metatheria, again includes just a single order, which comprises the marsupial ('pouched') mammals (▷ p. 103). These too are best known from Australasia, but they have a significant representation in South America, principally through the opossum family. Although marsupials give birth to live young, the young are born at a very early stage of development. The tiny offspring typically makes its way from the birth canal to the pouch, where it attaches itself to a nipple and remains for a protracted period until capable of feeding itself.

The great majority of living mammals belong to the second subgroup of therians, the eutherians (subclass or infraclass Eutheria) or placental mammals (▷ pp. 106–45). The name of the group is derived from the placenta, a special organ by which the embryo is fed directly from the mother's blood supply and by which waste products are removed (▷ p. 186). The length of time that the young grow within the mother (the gestation period) varies considerably from species to species, as does the degree of development at birth, but no placental mammal produces offspring as undeveloped as that of a marsupial.   BH

ICE AGE MAMMALS
1.6 million–10 000 years ago

WOOLLY MAMMOTH
(Mammuthus primigenius)

SABRE-TOOTHED CAT
(Machairodus latidens)

WOOLLY RHINO
(Coelodonta tichorhinus)

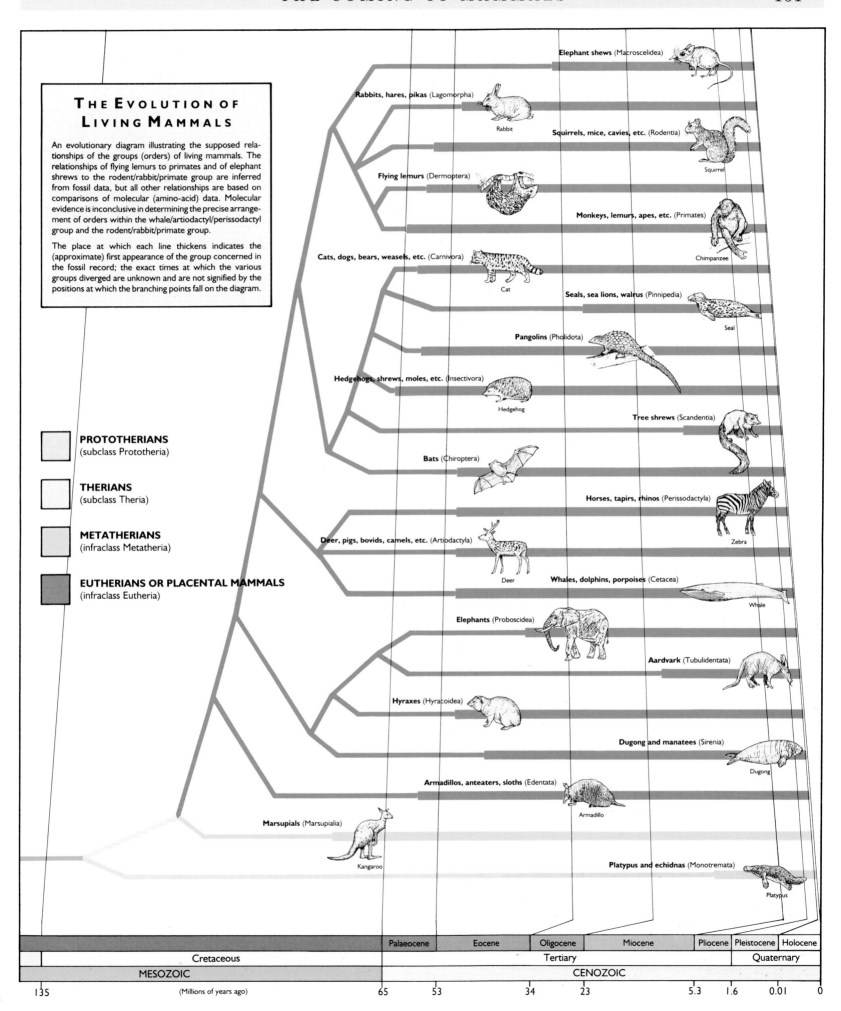

THE EVOLUTION OF LIVING MAMMALS

An evolutionary diagram illustrating the supposed relationships of the groups (orders) of living mammals. The relationships of flying lemurs to primates and of elephant shrews to the rodent/rabbit/primate group are inferred from fossil data, but all other relationships are based on comparisons of molecular (amino-acid) data. Molecular evidence is inconclusive in determining the precise arrangement of orders within the whale/artiodactyl/perissodactyl group and the rodent/rabbit/primate group.

The place at which each line thickens indicates the (approximate) first appearance of the group concerned in the fossil record; the exact times at which the various groups diverged are unknown and are not signified by the positions at which the branching points fall on the diagram.

PROTOTHERIANS
(subclass Prototheria)

THERIANS
(subclass Theria)

METATHERIANS
(infraclass Metatheria)

EUTHERIANS OR PLACENTAL MAMMALS
(infraclass Eutheria)

Elephant shews (Macroscelidea)

Rabbits, hares, pikas (Lagomorpha)

Rabbit

Squirrels, mice, cavies, etc. (Rodentia)

Squirrel

Flying lemurs (Dermoptera)

Monkeys, lemurs, apes, etc. (Primates)

Chimpanzee

Cats, dogs, bears, weasels, etc. (Carnivora)

Cat

Seals, sea lions, walrus (Pinnipedia)

Seal

Pangolins (Pholidota)

Hedgehogs, shrews, moles, etc. (Insectivora)

Hedgehog

Tree shrews (Scandentia)

Bats (Chiroptera)

Horses, tapirs, rhinos (Perissodactyla)

Zebra

Deer, pigs, bovids, camels, etc. (Artiodactyla)

Deer

Whales, dolphins, porpoises (Cetacea)

Whale

Elephants (Proboscidea)

Aardvark (Tubulidentata)

Hyraxes (Hyracoidea)

Dugong and manatees (Sirenia)

Dugong

Armadillos, anteaters, sloths (Edentata)

Armadillo

Marsupials (Marsupialia)

Kangaroo

Platypus and echidnas (Monotremata)

Platypus

	Palaeocene	Eocene	Oligocene	Miocene	Pliocene	Pleistocene	Holocene
Cretaceous			Tertiary			Quaternary	
MESOZOIC			CENOZOIC				

135   65   53   34   23   5.3   1.6   0.01   0

(Millions of years ago)

Echidnas and the Platypus

There are just three living species of monotreme or 'egg-laying mammal' – the duck-billed platypus and two species of echidna or spiny anteater – none of which is found outside Australasia. Although they lay eggs and are thought of as 'primitive' (by association with the egg-laying reptiles and lower vertebrates), monotremes in fact possess all the other features that characterize mammals. Not only do they have fur and suckle their young but they have various structural features, such as those associated with the middle ear and the lower jaw, that are uniquely mammalian (▷ p. 98). Their body temperatures are rather variable, but they are warm-blooded (homoiothermic) and can maintain their temperature well above that of their surroundings.

SEE ALSO

● THE COMING OF MAMMALS
p. 98
● SIXTH SENSES (BOX)
p. 191

The name 'monotreme' (meaning 'single-holed') is derived from the supposedly primitive feature that the ducts of the genital and excretory systems and the intestine open into a common chamber (the cloaca), and so share a single opening to the exterior (▷ p. 185). In spite of this and of the fact that they lay eggs, their reproductive system is in most respects typically mammalian. The young spend only a brief period of development in the egg, which is soft-shelled and hatches after some 10 days. The newly hatched young are grossly underdeveloped and are entirely dependent on their mother and her milk.

The duck-billed platypus

The platypus of eastern Australia and Tasmania is a small semi-aquatic mammal that forages underwater for freshwater invertebrates. Males are larger than females, and can reach 60 cm (24 in) in length. Its fur is dense and water-repellent, and it has short limbs with fully webbed fore feet to provide propulsion and partially webbed hind feet for steering. The tail is broad and flattened, and acts as a rudder. The platypus's strangest feature is the broad, duck-like bill. Pliable and sensitive to touch and to electrical fields (▷ p. 191), this is used for locating food and navigating underwater, since both ears and eyes are closed during dives. The nostrils are placed on top of the bill, near the tip. Behind the bill are two cheek pouches, which contain horny ridges and are used to store and masticate food as it is being collected. Platypuses are born with teeth, but these are lost in youth.

Female platypuses (unlike echidnas) do not have an egg pouch; instead, between one and three (usually two) eggs are laid in a nesting burrow where the young are suckled for three to four months after hatching. The mother's mammary glands do not have teats, so the youngsters lap up the milk as it oozes out onto the surrounding fur.

The platypus is one of the very few venomous mammals (there are a few venomous insectivores; ▷ p. 120). Males have horny poison spurs projecting from the back of the ankle of each hind foot. The hollow spur is attached by a duct to a venom gland in the thigh, and can be erected from a fold of skin. A firm jab of the hind limb causes a flow of poison that can kill a dog and cause considerable pain to a human. The enlarged size of the venom glands in the breeding season, coupled with increased use of the spurs in aggressive encounters between rival males, suggests that they function in disputes over access to females as well as providing defence against predators.

Echidnas

The short-beaked echidna is the smaller of the two species, growing to around 40 cm (16 in) in total length. It is also the more common, occurring in New Guinea, Australia and Tasmania; the distribution of the larger long-nosed species is uncertain outside New Guinea. Both species have large, powerful claws for digging and long, narrow snouts used to probe for food. The nostrils are located at the tip of the snout and are used to sniff out hidden prey. The short-beaked echidna specializes in a diet of ants and termites, while the long-beaked species feeds on various insects and earthworms. Within the small mouth there is a long, thin, flexible tongue, which is covered in sticky saliva and can be protruded to catch prey. Although lacking teeth, echidnas have pads of horny spines on the palate and on the back of the tongue with which they grind up their food.

Female echidnas possess a rudimentary pouch, which becomes enlarged during the breeding season and into which a single egg is placed. The mammary glands at the front of the pouch lack teats (as in the platypus), so milk is likewise licked up from the fur around the openings of the glands. Like the platypus, echidnas possess ankle spurs, but these cannot be erected and the venom gland is non-functional. However, they can curl up into a ball when threatened, and their thick covering of fur and large spines (which are modified hairs) provide excellent protection.

Origins of the monotremes

Monotremes are generally assumed to be the most primitive of living mammals, but their origins are in fact uncertain and their fossils are known only from relatively recent times, so providing few clues to their history. The modern platypus, for example, is very similar to fossils found in rocks of the mid-Miocene epoch, some 10 million years old. It has been suggested that the present-day monotremes are the descendants of a long-extinct and possibly much more widespread group of mammals, but no fossil monotremes have been found outside the Australasian region. It is possible that they are old, specialized derivatives of very ancient mammals, now extinct, that flourished in North America from the late Cretaceous period and through the Eocene epoch, 65 to 34 million years ago, and that their survival today is due to their isolation in the Australasian region.   JSC

The short-beaked echidna is an expert at excavating ant and termite nests, the inhabitants of which it then proceeds to sweep up with its long sticky tongue. These animals are also well known for their skill at burrowing: if danger looms, they dig vertically down, all but the tips of their spines disappearing from view in a matter of seconds.

STRANGER THAN FICTION

When the first platypus specimen was brought to Britain in the late 18th century, it was thought to be so bizarre that it must be a hoax – that the skin of a mammal had been skilfully stitched to a duck's beak. Later, as more complete specimens arrived, the platypus was accepted as genuine but not as a mammal. How could a creature with a reproductive system similar to birds and reptiles and which appeared to lay eggs rather than bear live young possibly be a mammal?

The discovery that the platypus possessed milk glands – an exclusively mammalian feature – helped to resolve the controversy. Even so, with its egg-laying habit and its reptilian skeletal features, it has been widely thought of as a very primitive mammal. Like reptiles, it was believed to be unable to regulate its body temperature internally in the way that other mammals do. However, recent studies show that even this is not the case, and its status as a mammal is no longer in doubt.

Marsupials

The marsupials include the possums and opossums, bandicoots, wombats, kangaroos and wallabies, the numbat and the koala. Altogether there are some 266 species belonging to 18 different families. Although the marsupials were formerly widespread and even today extend into most parts of Australasia and into Central and North America, the richest diversity of species is found in Australia (containing 45% of species) and South America (30% of species).

Marsupials range from tiny insectivorous dasyures of 2 g (1/14 oz) to the large grazing kangaroos weighing up to 90 kg (200 lb). They represent a whole spectrum of mammal types, including herbivores, carnivores, omnivores and even nectar-feeders, and occupy many habitats from desert and open grassland to hot, lush tropical forest. Most are ground-dwelling, but some are specially adapted for life in the trees, underground or in water.

General features

Although immensely varied in form, marsupials are united as a group and distinguished from all other mammals by various features of their reproduction. In its form and development the marsupial egg more closely resembles that of reptiles and birds than the eggs of placental mammals. In contrast to placentals (▷ p. 101), whose young generally grow to a relatively advanced stage within the female prior to birth, marsupial young are born in an undeveloped state and even after birth grow slowly. The new-born young of a 30 kg (66 lb) female eastern grey kangaroo weighs only 0.8 g (1/30 oz). After 300 days it weighs about 5 kg (11 lb), but it continues to suckle until about 18 months of age.

Despite being small and blind, the new-born marsupial has to make its own way unassisted from the birth canal to the nipples. This is achieved with the aid of the fore limbs, which (with the head) are better developed than the other parts of the body, although exactly how it finds its way is not known. Once it closes its mouth around a nipple, the end of the nipple enlarges and the youngster remains firmly attached there for between one and two months.

In many marsupials the youngster is protected by a pouch of hair-covered skin (the *marsupium*), which covers the nipple area. Although the name of the group is derived from the marsupium, this feature is not in fact definitive of the group, as some small marsupials have no pouch. The pouch varies considerably in form, from a mere fold of skin on either side of the nipple area in mouse opossums to a deep pouch completely enclosing the teats in the arboreal climbers and leapers. Some pouches open forwards (as in kangaroos and possums), while those of burrowers such as bandicoots open backwards. In many marsupials the pouch develops only during the breeding season, while in some species both females and males possess a pouch.

A further peculiarity of the reproductive system of marsupials (and another feature that sets them apart from placental mammals) is their lack of a true placenta by which nutrients pass from the mother's body to the embryo (▷ pp. 186–7). Instead, the embryo is nourished by nutrients transferred from within the uterus via a yolk-sac placenta. The female reproductive tract is forked, with a double uterus and vagina, as is the male's penis in many species.

Other characteristics that distinguish marsupials from placental mammals are the larger number of teeth (usually 40 to 50) and certain features of the skull, including a relatively small brain case. Although marsupials have often been considered less advanced than placentals and hence rather primitive, studies show that their problem-solving abilities (▷ p. 150) are just as good as those of many placentals. The peculiar reproductive system, though sometimes thought of as primitive and inefficient, has several advantages. Few resources are invested in developing embryos during gestation; instead, the major commitment comes after birth and during lactation. If her offspring should die, a female can reproduce again soon afterwards, without having wasted time and energy in producing a large offspring after a prolonged gestation.

Evolutionary origins

The origins and present-day occurrence of the marsupials are inextricably linked with the changes in the configuration and distribution of the world's landmasses over the course of geological time (▷ pp. 8–9). The earliest known fossils of these mammals are some 75 million years old, dating from the Cretaceous period of North and South America, and there is evidence to suggest that they diverged from the other major group of mammals, the placentals, at least 100 million years ago. They are known from more recent fossils found in Europe, of Eocene age (53–34 million years old); the earliest fossil marsupials yet found in Australia date from the Miocene epoch, some 23 million years ago, by which time most of the modern families had already become established. Since none of the earliest fossil marsupials has been found in Europe, Africa or Asia, it is probable that they originated in North America 75 to 100 million years ago and from there spread to Australia via South America and Antarctica, since these three southern continents were united as a single landmass until about 45 million years ago. At this time Australia separated from South America and Antarctica, which remained united for a further 15 million years. Fossil marsupials have been found in Antarctica dating back to the period when Australia broke away, a time when the climate there was far more equable and was able to support forests.

As placental mammals became increasingly more numerous and diverse, they appear to have ousted the marsupials from many of their original ranges. They became extinct in North America 20 to 25 million years ago, and also in Europe. South America was isolated from North America at this time, and here – in the absence of placentals – the marsupials diversified to fill a variety of ecological niches, as herbivores, carnivores and omnivores (▷ pp. 194–5). However, when a land bridge was re-established between the two Americas some 3.5 million years ago, placentals were able to spread southwards, and many of the marsupials, including the larger carnivores (which

15-day-old opossums fastened to the teats in their mother's pouch. In contrast to placental mammals, marsupials are born in a highly undeveloped state. Blind and unassisted, the new-born marsupial must undertake an arduous journey from the birth canal to find an available nipple.

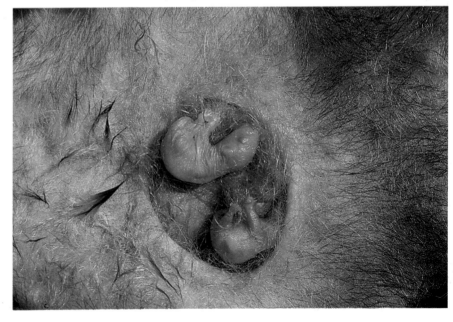

A female koala with young. Despite their docile appearance, koalas are frequently aggressive to one another and will defend themselves vigorously against dogs and other intruders.

resembled hyenas and sabre-toothed cats) subsequently became extinct. Today, only the omnivorous marsupials remain in South America, although some – notably the opossums – were successful enough to move northwards and re-colonize much of North America.

The long isolation of Australia and the absence of placental mammals permitted marsupials to persist and diversify there. As changes in climate and vegetation occurred, some of the larger forms became extinct. These included giant herbivores such as the 3 m (10 ft) tall browsing kangaroo *Procoptodon*, and the 2·1 m (7 ft) long rhinoceros-like *Diprotodon* (the most massive known marsupial), as well as large carnivores such as the marsupial lion *Thylacoleo*. Nevertheless, Australia still contains an enormous variety of marsupials, which, like the placentals, have evolved to fill many different ecological niches. The result is a remarkable similarity in form and function between the isolated marsupials in Australia and their placental counterparts elsewhere in the world. Examples of such convergent evolution (▷ p. 14) include the marsupial mole (which closely resembles the placental moles), the marsupial sugar glider with its 'flight' membrane (which resembles flying squirrels), and the thylacine or Tasmanian wolf (the counterpart of the placental wolves; ▷ box).

Opossums

In the Americas the marsupials are represented by a highly diverse and successful group, the opossums (family Didelphidae). There are about 70 species of these, ranging in size from the tiny Formosan mouse opossum with a head-and-body length of less than 7 cm (2¾ in) to the cat-sized Virginia opossum. Most are good climbers with fore and hind feet adapted for grasping, in which they are assisted by an opposable big toe on each hind foot and often by a prehensile tail. Opossums are found throughout most of South and Central America, and in eastern North America as far north as Ontario. Many are unspecialized, rather rat-like species, which feed on fruit, insects, small vertebrates, carrion and even garbage. Others are more specialized, including the yapok or water opossum, which has webbed feet and is adapted for a semi-aquatic life; the female's pouch opens backwards and can be closed with the aid of sphincter muscles to form a watertight chamber.

Many opossums are prodigious breeders. The gestation period of about 13 days is one of the shortest found in any mammal. Up to three litters may be born in a season, each containing many young. The Virginia opossum usually produces around 21 tiny young per litter; it normally has 13 nipples but not all are functional, so many of the young cannot be suckled and soon die. When the pouch becomes too small to accommodate the growing youngsters, some come out and ride on their mother's back. The high reproductive rate is offset by high mortality: few opossums survive to their third birthday.

Kangaroos and wallabies

Mention of marsupials often conjures up the image of kangaroos. The descriptions of these extraordinary beasts made in 1770 by the naturalists on Captain Cook's expedition to the South Seas aroused great interest back in Europe. There are over 50 species of kangaroo and wallaby in Australia, together forming the family Macropodidae (meaning 'big-footed'). They have a characteristic body shape, with short fore limbs held clear of the ground and greatly enlarged, muscular hind limbs adapted for hopping. They are capable of huge leaps of 6 m (20 ft) or more, providing a highly energy-efficient mode of locomotion (▷ p. 177). The large tail is used as a counterbalance when hopping, and as a prop to support the body when the animal is stationary or moving slowly. The hind feet have two large clawed toes and two very small ones, which are bound together and used in grooming. Tree kangaroos have rather shorter hind limbs with broader feet, and are able to climb with the aid of their strong claws.

Kangaroos and wallabies are adapted for a herbivorous diet. Most species have a large, sac-like stomach similar to that found in ruminant placental mammals (▷ p. 112). They have a large gap (diastema) between the incisors and the cheek teeth, typical of placental herbivores, and grinding cheek teeth. Also characteristic are the two large, forward-pointing incisors in the lower jaw, which move from side to side against six upper incisors. Many species are grazers, occupying the same ecological niche as sheep, cattle, deer and antelope elsewhere in the world. The largest is the red kangaroo of arid central Australia, which grows to a height of 1.5 m (5 ft) at the shoulder.

SEE ALSO

● THE PLANET'S CHANGING FACE p. 8
● THE COMING OF MAMMALS p. 98

The koala

Despite their winsome looks and their slow, dull demeanour, koalas are in fact highly specialized marsupials, adapted for an arboreal life and a diet of leaves. They have rounded, robust bodies clothed in soft, dense fur and a mere stump of a tail. The feet are large and each toe possesses a stout claw to assist climbing. The first and second toes of the fore feet are opposable to the remaining three, allowing them to grip branches as they climb and to grasp their food. They are agile climbers but clumsy when on the ground, so they rarely venture out of the trees.

Koalas feed on various tree species, but the bulk of their diet consists of leaves of only two to four species of eucalyptus. Besides being highly fibrous with a rather low protein content, eucalyptus leaves contain high concentrations of aromatic oils and other chemical compounds that are toxic to many herbivores. Koalas are adapted to overcome these problems: their high-crowned cheek teeth thoroughly grind up the leaves before they are swallowed, and they possess a huge caecum (appendix) – the largest of any mammal in proportion to its body size – which contains bacteria that assist digestion. The undesirable leaf compounds are detoxified in the liver and excreted. The low quality of their diet can only support a leisurely, energy-conserving life style, so koalas sleep for 18 hours a day.

European settlers valued koalas for their soft pelts. With large-scale hunting, followed by forest clearance, these animals seemed doomed to extinction. However, bans on hunting and a programme of careful management of their populations and habitats have rescued the koala, and it is once again widely distributed in eastern Australia.

Carnivorous marsupials

In addition to the primarily vegetarian marsupials, there are many carnivorous representatives, the majority of which belong to the family Dasyuridae of Australia and New Guinea. There are nearly 50 species of dasyurid, most of which are small and rather mouse-like, with long tails and snouts, and large eyes and ears. The smallest are the shrew-like ningauis, which weigh as little as 2 g (¹⁄₁₄ oz); the largest are the weasel-like quolls, the marsupial cat, and the Tasmanian devil. Dasyurids feed primarily on invertebrates such as insects, spiders and earthworms, but they will also take lizards and other small vertebrates as well as fruit and other vegetation. Despite its size and its reputation for ferocity, the Tasmanian devil is thought to live mostly on carrion; its large, powerful jaws and teeth allow it to rip up carcasses and crush bones. The cat-sized quoll or tiger cat of Tasmania is a more active predator. It is a good climber and preys on small wallabies, sugar gliders and reptiles.

The majority of dasyurids are solitary, but to assist location and attraction of mates many species have special mating calls. During the breeding season males tend to call at night, and females call when they are receptive to the attentions of males. While many of the smaller species may have two or more litters a year, each containing up to 12 young, the larger ones, such as the Tasmanian devil, take much longer to raise their young and produce only one litter a year of two to four young. The young are suckled for between eight and nine months, and take around two years to become sexually mature.   JSC

The spotted-tailed quoll or tiger cat, one of the larger marsupial carnivores, weighing about 1·5 kg (3¹⁄₃ lb). Although adept climbers, quolls spend much of the time on the forest floor, where they hunt for prey such as small wallabies, sugar gliders and reptiles.

THE THYLACINE – DEAD OR ALIVE?

The largest of the modern marsupial carnivores was the thylacine, also known as the Tasmanian wolf or tiger. Although fossil remains of this animal indicate a formerly wide distribution in Australia and New Guinea, throughout historical times it has been known only in Tasmania. Today it is thought to be extinct, but reported sightings of it, even in recent years, have kept up hopes that it might still be alive in remote parts of this rugged island.

The thylacine was wolf-like in size and appearance, although the body (like that of hyenas) sloped downwards away from the shoulders. Unlike dog-like carnivores, however, it possessed a long, stiff tail, and a sandy-coloured coat with dark stripes across the back. Because of its unique features it was classified in a family of its own, the Thylacinidae. It probably lived on wallabies, possums and other vertebrate prey, but with the arrival of European settlers it quickly gained a reputation as a sheep-killer, and bounties were offered for its destruction. Between 1888 and 1909 this led to the death of at least 2268 thylacines. Its numbers declined drastically not only as a result of hunting but also, it is thought, on account of an epidemic disease. The last one to be seen alive was captured in western Tasmania in 1933 and died in Hobart Zoo in 1936. Sadly, despite several organized searches for the animal, no substantiated sightings have been made, and no positive evidence exists of its survival today.

Anteaters, Armadillos and Sloths

The anteaters, armadillos and sloths together form the order Edentata. The scientific name means 'toothless', although in fact only the anteaters are entirely lacking in teeth. The species of edentate mammal alive today are the remnants of a once much larger group of South American mammals, most of which became extinct in prehistoric and historic times.

Armadillos and sloths lack true incisors and canines, but they do possess simple, peg-like cheek teeth (molars and pre-molars). These teeth lack an enamel covering and are rootless, growing throughout the animal's life. The anteaters feed almost exclusively on ants and termites, while armadillos are omnivores, with a diet of insects, carrion and tubers, and sloths are truly vegetarian.

SEE ALSO

● THE PLANET'S CHANGING FACE p. 8
● THE COMING OF MAMMALS p. 98
● NUTRITION AND FOOD SELECTION p. 162

General features

This group of mammals is distinguished from all others not by toothlessness but by the strange and unique additional articulations between the vertebrae at the base end of the spine. These bear bony projections that give extra reinforcement, particularly when digging. Armadillos are especially adept and active burrowers.

In addition, edentates have rather simple skulls, and they possess a double rather than a single inferior vena cava (the vein that returns blood to the heart from the posterior part of the body; ▷ p. 181). Female edentates have a divided uterus, rather similar to that of marsupials, and a common urinary and genital duct. All edentates have rather low body temperatures and metabolic rates. For this reason, and as a result of the low quality of their plant diet, sloths are particularly economic with their use of energy: they move slowly, travel little, and occupy small home ranges.

Origins

The edentates living today are merely the relicts of what was formerly a highly diverse and bizarre group of New World mammals that had arisen by the early Tertiary period, some 60 million years ago (▷ p. 98). Throughout most of the Tertiary, North and South America were not connected, remaining separate until the land bridge between the two was re-established some 3·5 million years ago (▷ p. 9) During the early part of the period South America was linked to Australia and Antarctica, but both of these had broken away by 30 million years ago, so for most of their history the edentates shared the continent only with early marsupials and a few other primitive mammals.

This state of isolation seems to have provided the opportunity not only for a great diversification of species but also for the evolution of spectacular giant forms (▷ illustration, pp. 98–9). Three families of giant sloths, including some species the size of today's elephants, had appeared by the early Oligocene epoch, some 34 million years ago. Contemporary with them were four families of highly armoured armadillo-like animals, the largest of which (*Glyptodon*) reached 5 m (16½ ft) in length and possessed a huge, rigid, bony shell on its back. Some of these huge mammals survived into historic times and are mentioned in the legends of the Patagonian Indians. Today, however, it is only their smaller relatives that persist. The enormous extinct edentates were probably slow, unspecialized herbivores, which were quickly ousted by smaller, more active mammals moving in from the north as North and South America rejoined. Those that survived were more specialized ant-eating, leaf-eating or burrowing forms, occupying particular niches for which they were better adapted than the northern invaders.

Anteaters

The range of the four species of anteater and tamandua (family Myrmecophagidae) extends from southern Mexico as far south as northern Argentina. All species

The giant anteater is the largest of the anteaters, growing to an overall length in excess of 2 m (6½ ft). An expert predator of ants and termites, this animal can extend its narrow, sticky tongue some 60 cm (2 ft) into the insects' nests and flick it in and out at an astonishing rate of up to 150 times a minute. This is necessary to satisfy its appetite for around 30 000 insects a day.

A three-toed sloth making a rare visit to the ground. Although surprisingly adept at swimming, sloths are seriously out of their element on the ground, having to drag themselves along with their clawed limbs.

lack teeth and specialize in a diet of ants and termites. The surface of their long, narrow tongue has minute spiny projections and is covered in sticky saliva, which traps their insect prey. The middle toes of the fore feet are armed with particularly long, sharp claws, which are used to tear open the nests of ants and termites, and are also useful in defence. As a result, anteaters walk on their knuckles and the sides of their fore feet, with the claws curled inwards. This produces a characteristic, awkward-looking gait, but they are strong, swift movers when the need arises.

The giant anteater, with its extremely long narrow snout, thick fur and bushy flag of a tail, is the most striking species (▷ illustration). A solitary animal occurring in a range of habitats from open grassland to tropical forest, it has a particularly acute sense of smell to assist in locating the large ground-dwelling ants and termites on which it feeds. Within its home range it visits the insect colonies in turn, feeding briefly on each to avoid overexploiting its prey. Although capable of independent movement soon after birth, the single youngster produced by the giant anteater rides on its mother's back for up to a year, and does not become totally self-reliant until fully grown at two years of age. The tamanduas are smaller, arboreal anteaters with prehensile tails; these are naked on the underside to assist climbing.

Sloths

There are two groups of living sloths – three species of three-toed sloth (family Bradypodidae) and two species of two-toed sloth (family Megalonychidae). All are adapted for life in the trees of the tropical rainforests of Central and South America, where they feed on a diet of leaves. They have rounded heads with rather flattened faces, and their feet are highly modified, with five toes on the hind feet and either two or three on the fore feet, each possessing a huge, hook-like claw used to grip branches. Unlike other arboreal mammals, they hang from the branches, and they rarely come to the ground. Sloths (like many other herbivores; ▷ p. 112) have large, sac-like stomachs containing bacteria that assist digestion. Nevertheless, digestion is very slow and it may take a month or more for food to pass from the stomach to the intestine.

Sloths have thick coats made up of a short, dense undercoat overlaid with longer, coarser hairs. The coat often assumes a greenish hue caused by blue-green algae that grow in grooves running the length of the hairs, and this provides excellent camouflage. The three-toed sloths are very unusual in having eight or nine neck vertebrae instead of the usual seven found in most other mammals (even giraffes); this gives extra flexibility in the movement of the head. Female sloths produce a single youngster at a time, which is carried for six to nine months by its mother.

Armadillos

The 20 species of armadillo (family Dasypodidae) occur throughout much of South and Central America, and even extend into the southern USA and some of the West Indies. They have a highly distinctive appearance, with bony armour shielding both the head and the body. This protective covering is arranged as a series of hard, rigid plates covered by horn and connected to the underlying flexible skin. Even the tail and upper surfaces of the limbs are armoured, but not the belly, which is soft and hairy. In addition to this defence, some species, including the three-banded armadillo, can roll into a ball when threatened. Armadillos are fast burrowers, loosening the soil with the fore limbs – which are particularly strong and bear stout claws – and then kicking it away with the hind legs.

Armadillos are adaptable animals with an omnivorous diet, which has allowed them to exploit a wide range of habitats. They feed on a variety of invertebrates as well as some plant matter. The common long-nosed armadillo has successfully spread into and colonized much of the southeastern USA, and can be a pest in arable land, where it causes damage to crops by digging and feeding. Armadillos have an unusual reproductive system: they exhibit delayed implantation (▷ box, p. 185), which, coupled with a gestation period of around four months, allows young to be produced in spring when food is plentiful, although mating may have occurred many months previously. The long-nosed armadillo normally produces quadruplets, all of the same sex, which have developed from a single fertilized egg and are thus identical.   JSC

PANGOLINS – THE SCALED MAMMALS

The Old World counterparts of the ant-eating edentates are the seven species of pangolin or scaly anteater (order Pholidota), which are found in Africa, India, China and Southeast Asia. They too lack teeth, have narrow snouts and long, sticky, protrusible tongues, and specialize on a diet of ants and termites. The tongue of the giant pangolin is some 70 cm (28 in) in length and is attached at its base, via a sheath, to the pelvis; it is lubricated with saliva produced by a huge gland located in the chest. Once swallowed, ants and termites are ground up in the stomach, which has a horny lining.

Pangolins have closeable nostrils and thick eyelids to protect them from insect attack as they feed. They also have powerful feet with long, curved claws with which they gain access to termite and ant colonies. Like the New World anteaters, they walk with a shuffling gait on the outsides of their fore feet with the claws curled inwards, but they can move more rapidly by running on their hind legs alone, using the long tail as a counterbalance.

Pangolins are unique among mammals in having an armour of horny, overlapping scales, which grow from the underlying skin and protect the whole body except the underside. The scales can be shed and replaced individually. As additional defence, these animals can curl into a ball when threatened. Several species are good climbers, and two of the African species are truly arboreal, possessing prehensile tails with which they can hold onto branches.

A tree pangolin with young

Elephants and Hyraxes

The earliest ungulates, or hoofed mammals, arose in the Palaeocene epoch some 65 million years ago, with the appearance of a group of unspecialized mammals called the condylarths. These were the ancestors of numerous large herbivorous mammals, including the modern-day odd-toed and even-toed ungulates (▷ pp. 110–16). They also gave rise to a group of primitive ungulates in Africa, which, by the Eocene (53 million years ago), had separated into four distinct orders: the proboscideans or trunked mammals (including the elephants), the hyraxes, the sirenians or sea cows (▷ p. 142), and the group represented today only by the aardvark.

The primitive ungulates may look very different, but they share a number of anatomical and biochemical features. Elephants and hyraxes both have toes bearing short, flattened nails rather than well-developed hooves, and both have grinding cheek teeth.

Elephants – the land giants

Elephants are the largest living land mammals, and – with the exception of man – the longest lived. The largest yet found was an African elephant weighing 10 tonnes (tons) and measuring 4 m (13 ft) at the shoulder. The massive body is supported by column-like legs and rounded feet, with three to five toes bound around a pad of elastic tissue that absorbs the impact as the animal strides along. The skin is only sparsely haired, and lacks the moisturizing glands of other mammals. Frequent bathing with water or dust is needed to keep it in good condition and free from parasites. The ears are large and help to radiate heat from the body in the hot climate.

The elephant's most characteristic feature is the long, flexible, muscular trunk, which is really an elongation of the nose and upper lip, with the nostrils at the end. The tip of the trunk is sensitive and dextrous, and is used to grasp foliage and other small objects. As well as for food gathering, the elephant uses its trunk while bathing and for drinking, drawing up water through the trunk and blowing it out into the mouth. The trunk is also employed in caressing, greeting and threatening other elephants, and for amplifying calls, including trumpeting.

The arrangement of an elephant's teeth is peculiar and characteristic. They possess no canines or lower incisors. The tusks are much enlarged upper incisors that grow throughout life in both sexes. Composed of hard dentine (ivory), tusks can reach 3.5 m (11½ ft) in length and together weigh 110 kg (242 lb) or more. They enable an elephant to dig up roots and prise bark off trees, and they are important as weapons and for display in social encounters. The cheek teeth are massive and highly ridged for grinding up coarse vegetation. They form in the back of the jaw and move forward as they grow, pushing out the front ones as they wear out. Elephants are not ruminants, but – like horses – they have a large caecum (intestinal pouch) where bacterial fermentation assists the digestion of food. They feed on grasses, leaves, roots, bark, flowers and fruit, and have prodigious appetites, eating up to 150 kg (330 lb) of food a day. They may also drink in excess of 100 litres (176 pints) of water daily.

Elephant society

Like many other hoofed mammals, elephants are gregarious and herd-dwelling. Male and female African elephants tend to live apart. Females live in closely knit cooperative groups of two or three related adults and their young, headed by the oldest female, or matriarch. As young

Rock hyraxes on Mount Kenya, East Africa. Being unable to regulate their body temperature effectively, hyraxes are often seen basking in the sun and huddling together in groups, particularly in the early morning after the cold of night.

females mature, they gradually move off to form new family groups. Adult males live alone or in small, more transient bachelor groups. Sexual maturity comes at about 10 years of age. Females produce a single calf every three to four years after a gestation period of about 22 months. Elephants can live up to 80 years, but few reach the age of 30 in the wild. They communicate not only by vocalizations but also by touch, smell and visual postures. Besides the trumpeting calls made when they are angry, aroused or alarmed, elephants produce low-frequency rumbling sounds by means of the larynx (previously mistaken for stomach noises), which can carry up to a kilometre. These are probably used to maintain contact between members of a group.

The African elephant inhabits grasslands, woodlands and forests, its range once covering much of Africa south of the Sahara. Today, however, most of them are confined to reserves. The Asian or Indian elephant is a less massive species with smaller ears and tusks (the latter rudimentary in the female), hairier skin, and a prominently domed head. It once ranged from the Middle East through southern Asia south of the Himalaya to northern China. It is primarily a forest-dweller, and today it has a restricted distribution in

---

ELEPHANT ORIGINS

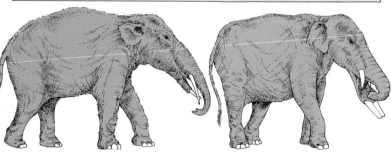

Deinotherium (below). Splintering off from the main proboscidean stock some 40 million years ago, deinotheres developed strange, downward-pointing tusks in the lower jaw, which were probably used to strip bark off trees. Deinothere species occurred widely in Africa and Eurasia, some of them surviving into Pleistocene times (1·6 million to 10 000 years ago).

Gomphotherium. Gomphotheres were browsing mastodons that first appeared in the early Oligocene, some 34 million years ago. Through the Miocene (23–5·3 million years ago) they became widely established as the dominant large mammals. Their subsequent history is one of gradual displacement by true elephants, but remnants of the group survived into the present era, a South American species becoming extinct as recently as AD 200–400.

Amebelodon. The later evolution of the gomphotheres is marked in some forms by an extreme elongation of the lower jaw into shovel-like tusks, which were probably used to plough up plant roots or water plants. An example is *Amebelodon*, which flourished in North America in the late Miocene, around 10 million years ago.

Woolly mammoth (*Mammuthus primigenius*). Close relatives of modern elephants, mammoths spread through Africa, Eurasia and America in the Pleistocene, 1·6 million years ago. The woolly mammoth was one of the smaller mammoths, specially adapted to endure the freezing conditions of the subarctic tundra of northern Eurasia and North America. It became extinct about 10 000 years ago, its demise hastened by climatic changes and human hunting.

parts of the Indian subcontinent, Southeast Asia and southern China; numbers remaining in the wild are low.

Elephants and man

The Asian elephant has had a special relationship with man for thousands of years. There are records of tamed Asian elephants being used as beasts of burden at least 5500 years ago in the Indus Valley, and they have been widely used in agriculture and forestry ever since. Nevertheless, they have never been truly domesticated as a species, since they are difficult to breed in captivity. During the rutting season, the males of both species go into 'musth': for two to three months the body produces high levels of male hormone, accompanied by glandular secretions, and they exhibit increased sexual activity. In captivity they become aggressive and uncontrollable. Most captive, trained elephants are still obtained as youngsters from the wild. Tame elephants continue to be used on ceremonial occasions and as marks of wealth and status. They have featured widely in legends and religions, particularly in Far Eastern cultures. African elephants have also been tamed and used as working beasts, but less widely. The most notable example was the Carthaginian leader Hannibal, who used these animals in his wars against the Romans over 2000 years ago.

Given the special relationship that man has cultivated with elephants, it is ironic that he is bringing them to the brink of extinction, principally by over-hunting for their ivory tusks. The demand for ivory and the high price it commands encourage poaching; despite extensive conservation measures, this has caused a major decline in numbers, and both species are highly endangered. When confined in national parks and reserves of limited area, elephants create another kind of problem. Needing to satisfy their enormous food requirements, these large mammals cause widespread habitat destruction, pulling down and uprooting trees and shrubs. As their populations

increase, so does the scale of the problem, particularly in the semi-arid, drought-prone areas where most African elephants are found. If uncontrolled, this behaviour may lead to food depletion and slow death by starvation, so a policy of culling has been introduced in some areas.

Hyraxes

Hyraxes are small, rather rodent-like mammals about the size of a rabbit. They have robust bodies covered in thick brown fur, and a short stump of a tail. There is a prominent scent gland on the back, probably used in communication. The legs are short and sturdy, and the front feet have four toes with flattened nails like small hooves. The hind feet have three toes, one of which possesses a long, curved claw. The feet have naked pads that are kept moist by a sweaty secretion and special muscles that cause the middle of the sole to retract. Acting like suction cups, the feet give remarkable grip – an especially useful adaptation for an animal that spends much time climbing and jumping among rocks and bushes.

Hyraxes have a short snout and a cleft upper lip from which protrudes a single pair of elongated, sharp incisors that grow continuously. There is a gap between the incisors and the cheek teeth, as in other herbivores. They do not ruminate, but digestion is assisted by bacterial action in the gut. Hyraxes manage to survive in arid regions where little drinking water is available and where food is relatively sparse. As an adaptation to their low-energy diet, they have low metabolic rates; a consequence of this is that they are unable to regulate their body temperature as effectively as most mammals, and use behavioural means to boost their body heat (▷ photo).

Rock hyraxes or dassies have a wide distribution in Africa and the Middle East. They are particularly associated with rocky outcrops, which they use as refuges, venturing out to feed on grasses and other plants nearby. They are highly

▷ box

THE AARDVARK

The aardvark is one of the most bizarre of all mammals. Pig-like in size and appearance, it has a long, tubular snout and long ears. The stocky body is sparsely covered in greyish bristly hair, and the tail is strong and muscular. The legs are rather short and there are four to five toes that bear large, blunt claws used for digging. The aardvark feeds mostly on ants and termites, which are caught with its long, sticky tongue. It has peg-like cheek teeth and no front teeth. Aardvarks are solitary and nocturnal, inhabiting grassland, scrub and open woodland in Africa south of the Sahara. They are rarely seen, but their presence is betrayed by their digging activities. They are prodigious burrowers, excavating termite mounds while foraging, and creating extensive burrows for shelter.

Although superficially resembling other anteaters, the aardvark is not related to them, any similarity being due to convergent evolution (▷ p. 16). Virtually a living fossil, it is the sole survivor of an evolutionary line that branched off over 50 million years ago from a group of very early ungulates that also gave rise to the elephants, the hyraxes, and the sirenians or sea cows. Indeed, on the basis of molecular evidence, the aardvark is – paradoxically enough – the closest living relative of the elephants (▷ pp. 98–101). Even so, the relationship is distant at best, and the aardvark is assigned to an order of its own, the Tubulidentata.

gregarious and live in colonies of up to a hundred or more individuals, although the basic social unit comprises a territorial male with several females and their young. Tree hyraxes are found in forested and wooded areas, and are good climbers, feeding mostly in the trees. They tend to be more solitary and occur in twos and threes.

Improbable as it may seem from their appearance, hyraxes (together with the aardvark; ▷ box) are the elephant's closest living relatives. Like the elephant, the hyrax is a primitive ungulate whose forebears separated from the main ungulate stock into a distinct order, the Hyracoidea. The eleven small species of hyrax alive today are but a tiny sample of a once diverse group of mammals living in Africa and the eastern Mediterranean, some of which were as large as pigs. Their decline coincided with the increase in artiodactyl species such as cattle and antelope during the Miocene epoch, 23 million years ago. JSC

SEE ALSO

● THE COMING OF MAMMALS pp. 98–101
● TROPICAL GRASSLANDS p. 202

The two species of elephant alive today are but a tiny remnant of the order Proboscidea, a group that was formerly highly successful, with perhaps as many as 350 species found in every continent except Australia and Antarctica. The origins of the proboscideans, or trunked mammals, can be traced back to the early Eocene, some 53 million years ago. The earliest forms were small herbivores, rather similar to tapirs in appearance, but the subsequent history of the group over the next 50 million years is one of spectacular diversification to a wide range of habitats from tundra to tropical forest. This history is chiefly characterized by the development of the features most conspicuous in the living elephants – great body size and large tusks and trunks – but during the long evolution of the group there were some fascinating and bizarre variations on this basic theme.

African elephant (*Loxodonta africana*)

Indian elephant (*Elephas maximus*)

Horses, Rhinoceroses and Tapirs

The order Perissodactyla contains 16 species of odd-toed ungulate or perissodactyl in three families – the horses, asses and zebras, the tapirs, and the rhinoceroses. All are strictly herbivorous, and – with the exception of the forest-dwelling tapirs – they are typically adapted for fast running on open plains, where they feed mostly on grasses.

Today, the natural range of the horses, asses and zebras and the rhinoceroses is restricted to Africa and Asia, although domesticated horses and asses have been introduced to other areas where they have established feral populations. Tapirs are found only in Southeast Asia and in South and Central America. The members of the horse family, in particular, are gregarious, living in small social groups or larger herds. Horses and asses have been domesticated since ancient times and have long been important as beasts of burden and as a means of transport.

General features

In all perissodactyls, the weight of the body is borne on the central toes, with the main axis of the limb passing through the central (third) toe, which is the longest. In the horses, only the single central toe on each foot is functional. In rhinoceroses three hoofed toes are present on all feet, while in the tapirs four functional toes have been retained on the fore feet and three on the hind feet. All other toes in these mammals have been lost or reduced to mere vestiges. Of the living perissodactyls, only rhinoceroses possess horns, which are composed of compacted hairs and mounted on small bony prominences on the snout.

Unlike the artiodactyls (▷ box, p. 112), perissodactyls never ruminate, and they have smaller, simpler stomachs. Although

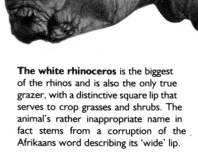

The white rhinoceros is the biggest of the rhinos and is also the only true grazer, with a distinctive square lip that serves to crop grasses and shrubs. The animal's rather inappropriate name in fact stems from a corruption of the Afrikaans word describing its 'wide' lip.

their digestive processes are assisted by bacterial fermentation, this does not occur in the stomach but in the enlarged caecum (appendix) and in the hindgut. Since food is not retained in the digestive system for as long as it is in artiodactyls, the digestive process is not quite so efficient.

In comparison to artiodactyls, perissodactyls do not rely so heavily on odour as a means of communication and do not possess the same array of scent glands. However, they have a greater diversity of vocalizations and facial gestures, which play an important role in their social life.

The species of perissodactyl alive today represent just a fraction of this once widespread and flourishing group of mammals, which probably evolved in North America in the late Palaeocene epoch some 55 million years ago, and reached its zenith 40 million years ago in the middle of the Eocene, at which time they were the dominant hoofed mammals. Their decline was probably due to a change in climate from the warm, lush conditions suitable for a diversity of forest-dwelling, browsing forms, to drier, more open conditions that required special adaptations in order to graze and escape predators.

Rhinoceroses

The five species of rhinoceros (family Rhinocerotidae) are massively built animals, with large bodies, short stocky legs, and huge heads. The simple horns on the top of the snout (one or two, depending on the species) are fibrous, being composed of densely compacted hair rather than horn as in cattle. Their skin is virtually hairless. Their eyes are relatively small and situated far out on the sides of the head, and they have rather poor vision. Their sense of smell is excellent, however. They are essentially solitary animals, although the white rhino is more social.

The biggest of the living species is the white rhinoceros of Africa, weighing over 3·5 tonnes (tons) and reaching 2 m (6½ ft)

HOOFED MAMMALS

During the course of evolution some mammals have replaced claws with hooves as an adaptation to both speed and endurance in running. Known collectively as *ungulates*, these animals, which include horses, pigs, camels, deer and cattle, are distinguished on this basis from all other mammals. Ungulates are typically large, terrestrial herbivores, which walk and run on the tips of their toes with the heel raised off the ground. The toes are reduced in number, and the bone at the tip of each toe is broadened and covered by a tough, protective hoof of keratin (the same protein that is found in hair and nails). A fleshy pad inside the foot absorbs the shock as it strikes the ground.

The living species of hoofed mammal are divided into two orders based on the number of toes they possess: the even-toed ungulates or artiodactyls (▷ pp. 112–17) and the odd-toed ungulates or perissodactyls. Both orders diverged from a common hoofed ancestor in the early Palaeocene epoch, some 60 million years ago.

Despite the variation in body form of ungulates, from the elegant horses and antelope to the massive hippopotamuses and rhinoceroses, all species have barrel-shaped bodies with large stomachs for holding food, long limbs of approximately equal length, and a long head held high in a horizontal position on the neck. Their skin is thick and covered by stiff hairs rather than soft fur. Ungulates have good senses of smell and hearing, and their ears can be rotated to locate the direction of sounds. Their large eyes give good day and night sight, and they are widely spaced on the side of the head, so providing a wide field of view and relatively good binocular vision.

Ungulates have undergone various adaptations to equip them for a diet of grasses, herbs and foliage. Unable to manipulate food with their feet, they must pluck it directly from the plant. To assist them in this, they have mobile, sensitive lips and long, mobile tongues, and their incisor teeth are adapted for cropping. Unlike fruits and seeds, the leaves and stems on which ungulates mostly feed are tough, fibrous and hard to digest. Their cheek teeth are large with ridged grinding surfaces, which fragment the food before it enters the digestive system. To assist the grinding action, the musculature and the flattened articulating surfaces of the jaws permit side-to-side movements of the lower jaw, in contrast to the more vertical movements typical of other mammals. There is also a gap between the incisors and the cheek teeth (the diastema), which facilitates chewing of long grasses and stems.

Although the actual digestive process differs between perissodactyls and artiodactyls, all ungulates have long intestines with a large stomach and caecum (appendix) to enable maximum digestion of food, usually assisted by bacterial fermentation (▷ pp. 112 and 169). One consequence of the resistance of vegetation to digestion is that large quantities must be eaten, and these herbivores spend much of their waking lives feeding.

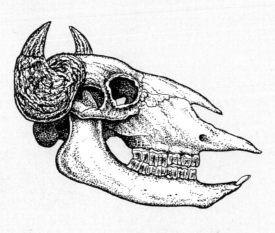

The skull of a bison

high at the shoulder. It is a grazer and has square lips for cropping grasses and herbs. The other African species, the black rhinoceros, is primarily a browser, and has an upper lip that tapers to a point and is prehensile, allowing it to grasp and pluck leaves and twigs from shrubs. Both species possess two horns on the snout. These animals were once widespread in savannah and woodland south of the Sahara, but uncontrolled hunting has drastically reduced their numbers, so that only small, scattered populations are found today, mostly confined to national parks and reserves. Both species are highly endangered.

The three Asian species of rhinoceros are also browsers with tapered upper lips. The largest of these – the Indian rhinoceros – is similar in height to the white rhino but less massively built. Its skin is characteristically folded at the joints, giving an armour-plated appearance, and it has a single horn. Similar to this species, but smaller, is the Javan rhinoceros, which is one of the world's most endangered mammals (▷ box). The smallest of the rhinoceroses is the Sumatran species, which is also a browser. Unlike the other species, its skin has a covering of bristly hairs. It is often found in pairs. Like the Javan rhino, its range has been greatly reduced and it too is an endangered species.

Horses, asses and zebras

The seven species of horse, ass and zebra (family Equidae) are fast-running ungulates with long heads and necks, and slender limbs. They have long tails and a mane of stiff hair covering the neck. Horses and asses are uniform brownish or greyish in colour, but the zebra's coat is spectacularly striped. They are all typically inhabitants of grasslands and semi-arid lands, where they live in herds and feed on grasses.

The small, stocky, dun-coloured Przewalski's horse of Mongolia and Sinkiang (western China) is the only true wild horse still in existence. Although none has been sighted in the wild since 1968, it is common in zoos. The ancestor of the modern domestic horse is thought to be the grey, pony-like tarpan, which once occurred on the steppes of southeastern Europe but is now extinct in the wild. The African wild ass (ancestor of the donkey) and the Asian wild ass still maintain small, isolated populations but are nearing extinction. South of the Sahara, four species of zebra were once widespread. The common zebra (Burchell's zebra) is still quite common in eastern and southern Africa, but the mountain zebra and Grevy's zebra have been greatly reduced in number. The quagga (the southernmost species), which had stripes only on its head, neck and forequarters, was driven to extinction in the 1880s.

The earliest horse-like mammal, *Hyracotherium*, evolved in the early Eocene around 50 million years ago. It was a small, dog-sized forest-dweller, which browsed on foliage and had simple teeth,

which lacked the enamel grinding ridges of modern horses; it had four toes on the front feet and three on the hind, and had not developed hard hooves. When conditions became drier in the Miocene epoch (23–5·3 million years ago) and the forests gave way to large expanses of grassland, the equids became increasingly specialized and many new forms appeared. They increased in size, their teeth were modified, and they became more swift of foot, the number of toes being reduced still further. Since the middle Oligocene 30 million years ago, the middle toe alone has supported the body, with extra support provided by ligaments.

Tapirs

Tapirs (family Tapiridae) are amongst the most primitive of large mammals. There are four species, three occurring in South and Central America, and one in Southeast Asia. As forest-dwellers, they are sturdily built to push their way through the dense undergrowth. Their most distinctive feature is the short, sensitive, flexible trunk or proboscis, which is formed by an extension of the nose and upper lip. This is used to probe and sniff about in the forest and to seize hold of vegetation and bring it towards the mouth. They browse and graze on shoots, leaves, buds, fruits and grasses.

Tapirs are good swimmers and are never found far from water. If alarmed, they often retreat into the water and may

Brazilian tapir

Malaysian tapir

remain submerged for several minutes. Solitary, nocturnal animals, tapirs are shy and difficult to study. The largest species – the Malayan tapir – reaches 2.5 m (8¼ ft) in body length and has striking black and white coloration, providing excellent camouflage in its shady forest habitats. The other species are uniform in colour as adults, although young tapirs are striped and blotched.   JSC

SEE ALSO
● THE COMING OF MAMMALS p. 98
● ARTIODACTYLS pp. 112–17
● BODY LANGUAGE (BOX) p. 154

RHINOS UNDER THREAT

Although habitat destruction in many parts of their natural ranges has played an important part in decimating rhino populations, the prime threat to their existence is hunting for their horn, which is highly valued in Asia. For centuries it has been reputed to have useful medical and aphrodisiac properties, for which there is no evidence whatsoever. This belief has led to a great demand for powdered horn and to an increase in its commercial value. Unfortunately, the rarer the product, the higher the price it commands; this gives an added incentive to rhino poachers, who persevere in spite of conservation measures.

The Javan rhino, one of the world's most endangered mammals. Once widespread in Southeast Asia, it is now reduced to a few isolated individuals in mainland Asia and to a small population in a reserve in Java. Given the wholesale destruction of its habitat, this animal seems doomed to extinction.

Pigs, Peccaries and Hippopotamuses

Together with peccaries and hippopotamuses, pigs are the only artiodactyls not to ruminate as part of their digestive process (▷ box). In pigs, the stomach is relatively simple, with two chambers, while hippopotamuses have a rather more complex three-chambered stomach.

In the case of pigs and peccaries at least, the relatively unmodified stomach may reflect their somewhat catholic tastes. In contrast to other artiodactyls, which are strictly herbivorous, these animals will feed on a variety of plant and animal matter, depending on availability.

Pigs

The eight species of pig or hog belong to the family Suidae, which is widespread throughout the Old World. Pigs are stockily built artiodactyls with four toes on each foot. They have short necks and long heads, and their thick skins are sparsely covered by coarse bristly hairs. In males there is sometimes a bristly crest or mane of hairs along the spine. While the adults are uniform in colour, the young of many species are striped. A pig's most distinctive feature is its long, muscular, flexible snout, which is strengthened by a bone and terminates in a tough, cartilaginous disc bearing the nostrils. It is used to

The wild boar, ancestor of the domesticated pig. Although most species of pig are uniformly coloured as adults, their young are often striped. This enhances their protective camouflage in the dappled light of their woodland and forest habitats.

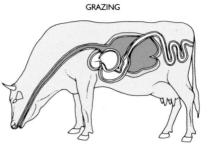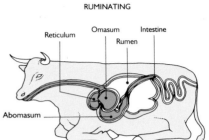

ARTIODACTYLS

GRAZING

RUMINATING

Reticulum   Omasum   Intestine
                      Rumen
Abomasum

The order Artiodactyla – the even-toed ungulates – is the most diverse group of large mammals alive today, with some 200 species whose natural range covers every continent except Australasia and Antarctica. Artiodactyls include such familiar animals as pigs, cattle, sheep, deer and antelope. They are far more numerous than the perissodactyl ungulates (▷ pp. 110–11), and – as the dominant herbivores in many terrestrial habitats – they have had a major role in shaping the environments and the plant communities in which they live. These animals have been of immense importance to mankind, who at first hunted them for food and skins, and later domesticated them. Most of the domesticated animals upon which human civilization has relied have been drawn from the ranks of the artiodactyls, and they have been transported by man to all parts of the globe.

Like the perissodactyls, the artiodactyls are typically big mammals with large bellies to carry their bulky food of grasses, herbs and browse. They have large heads with long jaws and a battery of grinding teeth, and long, thin legs built for speed and endurance. There are either two or four weight-bearing toes on each foot, with the axis of the limb passing between the third and fourth toes. This forms the typical semicircular 'cloven' hoof. The remaining toes are much reduced in size or merely vestigial, and do not touch the ground. The first toe has been lost altogether in all species.

A unique feature of the artiodactyls is the reduction in number or the complete loss of the upper incisor teeth (▷ illustration, p. 110). In their place is a tough, horny pad against which the lower incisors work. In many species the canines have also been lost or reduced, although in some, such as the pigs, they have become large and tusk-like. Many species bear horns or antlers on the head.

Another characteristic peculiar to the artiodactyls is the habit of ruminating or chewing the cud. With the exception of pigs, peccaries and hippopotamuses, all artiodactyls are ruminants. This is their solution to the problem of maximizing digestion of the hard, fibrous vegetation

on which they feed. Associated with this process is a complex, multi-chambered stomach (four-chambered except in the camels) incorporating a large fermentation chamber (the rumen), which contains bacteria that help to break down the food. As food is cropped, it is swallowed, with little chewing, and stored in the rumen. In this way the large quantities of food required by these mammals can be eaten quickly and digested at leisure and in safety. While in the rumen, the food is moistened, softened and subjected to bacterial action. The partially digested plant material, known as 'cud', is then regurgitated in small quantities into the mouth for more thorough mechanical fragmentation by the grinding cheek teeth, and mixed with saliva. It is swallowed for a second time, bypassing the rumen, and progresses to the other chambers of the stomach, where it is subjected to further action by acids and digestive enzymes. Rumination and digestion are accompanied by the release of gases, which are expelled by belching and breaking wind. The whole process of digestion of a particular food item and its passage through the intestine for absorption of nutrients takes several days in these mammals, compared with just a few hours in carnivores feeding on more readily digested meat.

Because artiodactyls are large, with robust bones that resist decay, they have one of the best fossil records of all mammal groups, going back to the early part of the Eocene epoch, 50 million years ago. Whereas the perissodactyls evolved and flourished mostly in North America, the artiodactyls probably arose in the Old World. The first representatives of this order had four distinct toes on each foot, a full set of teeth, no horns or antlers, and a simple non-ruminating stomach. Only later did reductions in toes and teeth occur, and the complex many-chambered stomach develop.

Like the perissodactyls, the artiodactyls are typically gregarious, but – in contrast to perissodactyls – they have a range of scent glands on their feet and faces, and communication by scent is an integral part of their social life.

turn over the soil and sniff out the roots, seeds, fruits and fungi on which they mostly feed. Pigs are primarily woodland- or forest-dwellers, and will feed on a range of plant material, invertebrates and even carrion found on the forest floor. Wild pigs have well-developed canines, which

in males form elongated, curved tusks. These sharp, pointed tusks grow throughout life, and can be lethal weapons.

Pigs generally live in family groups (known as 'sounders') of four to six individuals, although wart hogs and wild boar often occur in much larger groups. They are highly vocal, communicating with a large repertoire of squeaks, grunts and chirrups. Pigs have notably high reproductive rates: males produce copious quantities of semen, and females may have two or more litters a year, each with six or more young. Mankind has capitalized on this characteristic by selectively breeding domestic pigs, which commonly have much larger litters. They have also become much larger in body size than their wild counterpart and forebear, the wild boar, and have developed a curly tail and lost most of their covering of bristles.

The wild boar is still widespread in continental Europe. Once common in Britain, it was hunted to extinction in the 17th century. It also occurs in North Africa, but south of the Sahara it is replaced by three other species. The most familiar of these is the wart hog, so called because of the wart-like growths on its face. This species is still common in much of Africa, where it inhabits wooded savannah. The African bush pig, or red river hog, is found on the African mainland and on Madagascar and the Comoro Islands, where it may have been introduced by man. The giant forest hog occurs in central Africa and also in parts of eastern and western Africa, but is less well known – indeed, it was not scientifically described until 1904.

The smallest of the pigs is the pygmy hog of the foothills of the Himalaya. Only 50 cm (20 in) long as an adult, it is critically endangered as a result of habitat destruction. The babirusa, found on Sulawesi and certain other Indonesian islands, is also a threatened species. In contrast, domestic pigs have been introduced to many areas, where they have become wild and increased in number. They themselves can then become the cause of habitat destruction through their rooting up of vegetation and soil, posing a threat to endemic (particularly island) species.

Peccaries

The three species of peccary (family Tayassuidae) are confined to Central and South America, and occupy the same niches as pigs in the Old World. Peccaries are similar to pigs in appearance and habits, but are rather smaller. Only two toes on each foot touch the ground, and their canines are well developed into tusks, but these are small compared with those of pigs and curve downwards rather than upwards. Peccaries have a scent gland on the rump, just in front of the tail, which secretes a musky odour when the animal is excited. They are highly gregarious, moving around in large groups that may number up to 100 individuals, all of which may turn en masse to ward off a would-be predator. The Chacoan peccary was known only from fossil remains before live specimens were found in the early 1970s in the Gran Chaco of Bolivia, Paraguay and northern Argentina.

Hippopotamuses

There are only two species of hippopotamus (family Hippopotamidae), both restricted to Africa. Weighing up to 4.5 tonnes (tons), the common hippopotamus is one of the heaviest of terrestrial mammals, second only to the elephants, and similar in size to the white rhino. The common hippopotamus is amphibious, spending most of the heat of the day resting partially submerged in rivers or lakes and emerging at night to graze on grasses and herbs. The eyes are set high on the head, remaining above water when the animal is submerged; the nostrils are on top of the snout and are closeable. The skin is grey in colour and almost hairless, and contains pores that secrete a thick, oily, pinkish substance. The colour of this secretion gave rise to the belief that hippos 'sweat blood', but its function is in fact to protect and lubricate the skin.

The ancestors of the hippopotamuses spread right across Africa and Eurasia, and the common hippopotamus itself was once widespread in Africa, extending to the Nile Delta, but its range has contracted greatly as a result of overhunting and habitat destruction. However, they are still locally abundant, and may be seen in large numbers along stretches of the Zambezi river, among other places. The pygmy hippopotamus is found only in the forests of West Africa, and is much smaller, weighing a mere 225 kg (500 lb). Unlike its larger relative, it is not aquatic, and is less social. JSC

SEE ALSO

● THE COMING OF MAMMALS p. 98
● HOOFED ANIMALS (BOX) p. 110
● OTHER ARTIODACTYLS pp. 114–17

Hippopotamuses are amphibious animals that spend most of the hot day partially submerged to avoid excessive water loss, only emerging at night to graze on grasses and herbs. They are excellent swimmers and can even walk on the bottom of rivers and lakes.

Cattle, Sheep, Antelopes and their Relatives

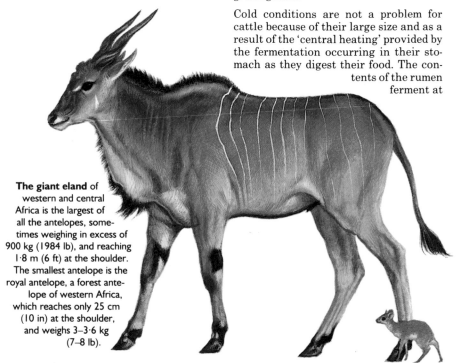

Cattle, sheep, goats, antelopes and duikers form a large and highly diverse group of artiodactyls (▷ p. 112), amounting to some 128 species together making up the family Bovidae. Bovids have a wide distribution over Eurasia and North America, but they are best represented in Africa and the Middle East, where there is a wealth of species exhibiting different dietary specializations and adaptations to various habitats. All are ruminants, possessing a four-chambered stomach, the culmination of artiodactyl evolution (▷ box, p. 112).

African buffaloes (above right) congregating at a watering hole in Kenya. These large, powerful animals have successfully adapted to a range of habitats from forest to semi-desert. However, they are rarely found far from a good water source, since they drink copiously and enjoy resting in water or mud.

Amongst the bovids are many of man's most important domestic animals (▷ box). Although of inestimable economic value to mankind over many thousands of years, the numerous domesticated forms of cattle, sheep and goats have frequently had a damaging impact on wild species and on the balance of natural ecosystems.

General features

Bovids show considerable variation in size and form, from the small, sleek duikers to the massive bisons and buffalo. They are adapted for life in a wide range of habitats, including hot semi-arid deserts, tropical and temperate grasslands, and cold arctic tundra. Some, such as goats and sheep, inhabit rocky, mountainous terrain. Only two toes on each foot are functional and used in walking; the others are either reduced or absent. All adult male (and some female) bovids carry horns, which consist of a bony core attached to the skull and a hard outer covering of horny material. These are simple, unbranched structures and are never shed, but they vary greatly in shape and size, from the long corkscrews of the greater kudu to the ramshorn spirals of the bighorn sheep. All bovids are strictly herbivorous, whether as grazers of open grassland or as browsers of woodland shrubs. They feed by grasping herbage with their long, prehensile tongues and cutting it off with the lower incisors. Most species are social and many live in large herds. Within this diverse family five major groups or subfamilies are recognized.

Cattle

The subfamily Bovinae includes cattle, bison, buffalo and most of the other larger bovids, together with the spiral-horned antelopes such as eland, bongo and bushbuck. Many species of cattle evolved in the open plains of Eurasia during the Pliocene epoch (5·3–1·6 million years ago), but with changing climate and man's increasing impact as a hunter, many became extinct, including the aurochs, the ancestor of most domestic cattle (▷ box). Today, species of cattle such as gaur, banteng and kouprey are rare in the wild, although they too gave rise to domesticated forms. In the Far East the water buffalo and (to a lesser extent) the yak are also important as domestic animals, principally as beasts of burden but also as sources of milk, meat, leather and other products.

Cattle have good senses of hearing and sight. Although they have acute vision, domestic cattle (and Spanish fighting bulls) cannot distinguish red, contrary to popular belief. The sense of smell is well developed and important as a means of communication. Forest species such as gaur and banteng live in small groups. American and European bison in more open habitats live in larger groups of 10 to 20 individuals, while the African buffalo may live in herds of a hundred or more. Some species were once very numerous: 40 to 60 million bison roamed North America even in recent times (▷ p. 232), and as late as 1900 wild yak occurred in large herds in the Himalaya. All have increasingly been ousted by their domestic relatives, with which they compete for grazing land.

Cold conditions are not a problem for cattle because of their large size and as a result of the 'central heating' provided by the fermentation occurring in their stomach as they digest their food. The contents of the rumen ferment at

DOMESTICATED BOVIDS

Most breeds of domestic cattle are descended from the wild aurochs (*Bos primigenius* or *taurus*). The aurochs was a massive horned artiodactyl that once roamed over Europe and Asia, but by the Middle Ages it had been driven to extinction in most areas, the last aurochs being killed in Poland in 1627. It was domesticated about 8000 years ago, but some breeds, such as Scottish Highland cattle, Spanish fighting cattle and Chillingham Park cattle, still resemble their wild ancestor in build and behaviour. Zebu cattle, with their characteristic hump, are also forms of *Bos primigenius*. They originated in India, and are particularly valued in tropical regions for their ability to resist heat, ticks and parasitic insects.

Evidence suggests that sheep are descended from a wild species similar to the mouflon and were domesticated about 10 500 years ago in the eastern Mediterranean. With selective breeding the characteristic thick woolly fleece has developed, the like of which is not found in any wild species. Today there are over 800 breeds of domestic sheep worldwide.

Goats were domesticated in southwestern Asia about 8500 years ago, probably from the Cretan wild goat, which had a wide distribution in the Near and Middle East. They play a major role in the rural economy of many Third World countries. However, with their ability to exploit all kinds of plant foods, domestic and feral goats are extremely destructive, and their large numbers and widespread occurrence pose a threat to many fragile habitats.

The giant eland of western and central Africa is the largest of all the antelopes, sometimes weighing in excess of 900 kg (1984 lb), and reaching 1·8 m (6 ft) at the shoulder. The smallest antelope is the royal antelope, a forest antelope of western Africa, which reaches only 25 cm (10 in) at the shoulder, and weighs 3–3·6 kg (7–8 lb).

40 °C (104 °F), so cattle rarely have to shiver in order to generate extra body heat, even at temperatures as low as −14 °C (7 °F). Wild cattle of cold climates also have extra thick coats with long fringes of hair, in contrast to tropical species, which have short hair and fleshy dewlaps to assist heat loss.

Sheep and goats

The sheep and goats, together with their relatives the tahrs, takins, serows and gorals, belong to the subfamily Caprinae. As a group, they show remarkable tolerance of extreme climatic conditions – barbary sheep, for instance, are adapted to life in deserts, while the musk ox subsists in arctic tundra. Other species, such as wild goats, chamois, ibex and tahr, are able to cope with very difficult terrain, inhabiting mountainous areas where they exploit vegetation high up on cliffs. All species possess a coat of wool or long hair, and are particularly conspicuous for their long, curving horns, which are best developed in males and used in defending resources, and fighting and competing for females. Their ability to thrive in difficult conditions has made them particularly valuable as domestic animals, providing meat, milk, wool and skins.

Antelopes

The majority of antelope-like bovids – with the exception of the spiral-horned antelopes (▷ above) and the duikers (▷ below) – belong to one of two subfamilies: the Hippotraginae (the grazing antelopes), including oryxes, kobs, waterbuck and wildebeest; and the Antilopinae, which includes the smaller antelopes such as gazelles and impala. These subfamilies contain many species, all of which are elegant, long-legged animals typically adapted for fast running. Most are grazers or browsers inhabiting grasslands and woodlands in Africa, the Middle East, India and China. Some, such as oryxes and addax, are adapted for life in arid lands and can survive for long periods without drinking. Others are adapted for life in flooded grasslands; these include waterbuck and lechwes, which are good swimmers and will feed on grasses and reeds while partially submerged at lake sides. The sitatunga has long, splayed hooves enabling it to move easily over swampy ground.

The open grasslands of eastern and southern Africa are still home to large herds of wildebeest or gnu. These prefer to graze on short grasses, and in areas of seasonal shortages in food and water they and many other species of antelope are migratory (▷ p. 171). Another very familiar species of African antelope is the golden-brown impala, which also occurs in herds, sometimes numbering a hundred or more. They feed on various plants and are most numerous in open woodlands, where a mixture of vegetation is available. One of the rarest antelopes is the Arabian oryx of the arid Arabian peninsula. This handsome beast, with its white body, dark legs and facial markings, and long, straight, tapering horns, had been hunted to extinction by 1972. However, it has been given a second lease of life by a captive breeding programme followed by a successful release into the wild in Oman, and careful protection measures are now in force.

Duikers (subfamily Cephalophinae) are small antelope-like bovids inhabiting forests and thickets in Africa. Unlike the majority of bovids, they are not gregarious but live alone or in pairs; they tend to be monogamous and may mate for life. Beneath each eye they have a particularly large scent gland, which differs from those of other antelopes. It is used to mark objects and other animals, but its precise function remains unknown.   JSC

SEE ALSO

- THE COMING OF MAMMALS p. 98
- HOOFED ANIMALS (BOX) p. 110
- ARTIODACTYLS (BOX) p. 112
- OTHER ARTIODACTYLS pp. 112 and 116
- TERRITORY, MATING, SOCIAL ORGANIZATION p. 156
- MIGRATION p. 170

CAMELS AND THEIR RELATIVES

The family Camelidae is today represented by just six species of artiodactyl (▷ p. 112), including the camels of the Old World and the guanacos and llamas of South America. All have long, thin necks, small heads, slender snouts and cleft upper lips, and there are only two toes on each foot. All species are adapted to life in arid or semi-arid plains, grasslands and deserts.

Camelids differ from other ruminants, such as deer, sheep and cattle, in that the stomach has three chambers rather than four. They are also unusual in that the digestive glands are located in special sacs in the rumen of the stomach. Another extraordinary feature (distinguishing camelids from all other mammals) is that their red blood cells are oval, not round, so preventing their blood from thickening with a rise in temperature.

The Old World camels are the tallest of the artiodactyls. The one-humped Arabian camel or dromedary reaches 2·4 m (8 ft) at the shoulder, marginally bigger than the two-humped Bactrian camel. The dromedary is a native of North Africa and the Middle East, but it survives today only in its domesticated form. Feral populations descended from introduced dromedaries also occur in Australia. The Bactrian camel of Central Asia has also been domesticated, and wild populations are now found only in the remote deserts of Mongolia and western China.

Both species of camel are highly adapted to desert conditions. The two toes of each foot are joined by a web of skin with a tough pad beneath (as in all camelids), allowing the foot to splay out and preventing the animal from sinking into soft sand. The large eyes are protected from wind-blown sand by long, thick lashes, and the nostrils are closeable. Contrary to popular belief, the humps are not used to store water but are composed of fatty tissue, providing an energy store that can be used in times of food shortage. Camels are able to go for long periods without drinking (a week or so when working, several months at other times), and have a number of adaptations to help conserve water. They have the ability to accumulate heat slowly within the body during the day as the environmental temperature rises, and to dissipate it in the cool of the night. Their thick coat restricts water loss through evaporation, and they have relatively few sweat glands and do not begin to sweat until the temperature outside reaches 40 °C (104 °F). Highly efficient kidneys ensure that the water content of the urine is kept to a minimum, and camels can survive a water loss of some 40% of their body weight without harm. They are also able to drink prodigious quantities of water quickly when the opportunity arises – over 100 litres (26 US or 22 imperial gallons) in a matter of minutes. In other mammals this would create serious physiological problems amounting to water intoxication.

In South America the camel family is represented by four species: the wild guanaco and vicuna, and the llama and alpaca, which are domesticated descendants of the guanaco but which are now usually given the status of separate species. They are relatively small, slender-legged camel-like mammals that feed on grass; their hair is soft and woolly, and is highly valued. Guanacos and vicunas live in small herds of one male and several females, plus their young. Other males form separate bachelor groups. Both species are adapted to the high altitudes and difficult, rocky terrain of the Andes. The domesticated llamas have been used in this area as pack animals and as providers of wool, meat and milk for some 4500 years, and were important in the economy of the Incas.

Guanacos in the Paine National Park, Chile.

Deer, Giraffes and the Pronghorn

There are some 38 species of deer belonging to the family Cervidae, and they have a widespread distribution in many habitats across Eurasia, North Africa, North and South America, and many continental islands. They include such species as red, fallow, sika, roe and mule deer, wapiti, moose, and caribou or reindeer. The range of many species has been extended by man, who has introduced them to many areas; exotic species such as muntjac and sika deer, for instance, have been introduced to Britain, and red deer to New Zealand.

The giant deer *Megaloceros* with a modern red deer for scale. Although *Megaloceros* – the Irish elk – died out in Ireland 11 000 years ago, it survived in Central Europe until about 500 BC. Remains of this animal found in peat bogs in Ireland indicate that its antlers may have had spans of 3·7 m (12 ft) or more.

The largest species of deer is the moose or elk of northern Eurasia and North America, which can exceed 2·5 m (8¼ ft) in length. In the past, however, much larger species occurred, such as the giant deer *Megaloceros* (▷ illustration).

Deer: general features

Compared with many other hoofed mammals, deer are finely built, with long, narrow legs and slim bodies. They are ruminant artiodactyls with a four-chambered stomach (▷ box, p. 112).

Their weight is borne on the two central toes of each foot, the others being reduced or lost. Adults are usually brown in colour, but the young of many species are spotted with white. Deer are typically social mammals living in well-defined groups.

The most distinctive characteristic of deer is the branched antlers. Except for reindeer or caribou, these are borne only by males. They have a bony core and are supported on permanent skin-covered projections (*pedicels*), and – unlike the horns of other ruminants – they are shed annually. In temperate regions antlers begin to grow in early summer. While growing they are soft and tender, and are enveloped by a thin skin, which is well supplied with blood vessels and covered by short, fine, velvety hair. In mid-summer, when they have reached their maximum size, the blood supply is cut off and the skin covering dries up, loosens, and is rubbed off, leaving the hard, bony core. Acting as advertisements of sexual prowess, the antlers are used during the rutting (breeding) season by sparring males as they compete for females (▷ photo). The larger the male and the bigger his antlers, the more able he is to attract and defend a harem of females with which to mate. Following the rut, the antlers are cast. Year by year they become progressively larger, developing more branches and points with age until a maximum for each species is reached. A so-called 'royal' red deer stag has twelve points. Tropical deer may not have a fixed breeding season and so may not shed their antlers at any particular time of year. Chinese water deer do not possess antlers; instead, the canines of the males are elongated and tusk-like.

Red deer stags fighting for sexual favours. In this species the sexes live apart for most of the year – males in small bachelor herds, females in separate groups with their young. However, during the rut (breeding season) males become highly aggressive and gather harems of females, which they vigorously defend against other males.

Most deer have well-developed facial glands situated in a depression in front of each eye, and these exude a scent that is rubbed onto vegetation. They also have scent glands on the legs and between the toes, which help to mark out territories and to promote group cohesion.

Musk deer

The three species of musk deer, belonging to the family Moschidae, are like rather primitive deer. They are ruminants, and reach a height of about 60 cm (24 in) to the shoulder. Musk deer lack antlers, but males have long, pointed upper canine teeth that project down below the lips. These shy, solitary animals are mostly inhabitants of forests at high altitudes in central and eastern Asia, and little is known of their habits.

The name 'musk deer' is derived from scent glands on the abdomen of mature males; these produce a brownish, waxy secretion (musk), which is stored in a pouch or pod on the belly near the tail. This substance is much in demand for the manufacture of perfume and soap in the West, and for medicaments in the Far East, where it commands high prices. During the 1970s in Nepal, musk was worth more than its own weight in gold. Consequently, musk deer have been grossly overhunted and are rare in many parts of their range. However, the musk can be extracted quite easily from the pouch while the animals are still alive, so captive breeding and farming of these deer is taking place in China in an attempt to satisfy both the market and the conservationists.

Chevrotains

The four species of chevrotain or mouse deer (family Tragulidae) are small, graceful, deer-like ruminants. Mostly solitary, they are found near water in forests and swamps in western and central Africa, and in Southeast Asia. Ranging from rabbit-size to a height of a mere 35 cm (14 in) at the shoulder, they have small heads, relatively large bodies, and long, pencil-thin legs. Each foot has four well-developed toes.

Chevrotains are intermediate in form between pigs and deer, and are generally regarded as the most primitive ruminants. Like most other ruminants, they have a four-chambered stomach, but the third chamber is poorly developed. The upper incisors are missing (a ruminant feature), but they have no horns or antlers. Instead, the males have enlarged, continually growing canines (like pigs), which are narrow and pointed, and project below the lips like small tusks. They also possess no scent glands beneath the eyes or on the feet, unlike deer and other ruminants.

Giraffes

The family Giraffidae contains only two living species, both now confined to sub-Saharan Africa. One of these – the familiar giraffe – is the tallest land animal, mature males growing to a height of around 5.5 m (18 ft). Its long neck possesses no more than the usual seven vertebrae found in most mammals, but each is greatly elongated and articulates with a ball-and-socket joint, so allowing great flexibility. A series of valves within the blood vessels in the neck helps to regulate the blood supply to the head when the animal is holding its head up to browse or down to drink. So long are the legs that giraffes have to splay their fore legs and bend their knees in order to drink at ground level. They also have long, extensible tongues and long, sensitive lips with which they can delicately pick leaves from the trees and shrubs on which they browse.

A giraffe's feet are robust and possess two hoofed toes. The body slopes downwards from the shoulders, and there is a long tail ending in a tuft of long hairs, which serves as an effective fly-whisk. The stomach is four-chambered, as in other ruminants. The horns of giraffes are unique. Present in both sexes, they have a bony core fused to the skull, and are covered by skin and hair; they grow slowly throughout the animal's life and are never cast. As well as the paired horns on the top of the head, there is also a small horn in the middle of the forehead. The colouring of giraffes (usually yellowish buff) and the size and shape of the characteristic brown blotches are highly variable, and several subspecies and distinct geographical races are distinguished.

Giraffes are social mammals, living in loosely associated groups or small herds in open woodland and wooded grassland in eastern and southern Africa. Once common in North Africa, they were captured by the Romans and transported back to Rome for exhibition in the fights between men and wild beasts in the Coliseum. Today, large populations are found only in Tanzania and surrounding areas.

The only other member of the Giraffidae is the much smaller okapi (▷ illustration), which reaches a height of 1·6 m (5½ ft) at the shoulder. The okapi is a solitary inhabitant of dense tropical forests in central Africa whose habits are little known. 'Okapi' is the name given to it by the pygmies who traditionally hunted it for food. It resembles the giraffe in structure, but it has a much shorter neck, and its coat is dark and lustrous with pale stripes on the legs and hindquarters. It too is a browser, extending its protrusible tongue to reach foliage. So long is the animal's tongue – over 50 cm (20 in) – that it can extend it to lick and clean its own eyes. Unlike giraffes, only the males have horns.

The pronghorn

There are no true antelopes in the New World, but in North America the pronghorn is their ecological equivalent. This grazing artiodactyl combines the characteristics of antelope and deer, but is distinct enough to be assigned to its own family (Antilocapridae). The single species alive today is the sole survivor of a once more widespread and diverse group of ruminants. It is found in western USA and Canada, and parts of Mexico, where it inhabits open grassland and dry scrubland. It feeds mostly on grasses, and can survive long periods without drinking. Its upper body is reddish brown, with white below and bands of white on the neck. Males and most females possess short, forked horns; those of the male are longer, with forward-pointing prongs below backward-pointing tips. Like those of cattle, these consist of a bony core covered by a horny casing, but as in deer they are shed annually after the breeding season. Pronghorns possess two toes on each foot, and unlike all other artiodactyls all vestiges of the other toes have disappeared. With their long, slender legs they are extremely swift runners, achieving speeds up to 87 km/h (54 mph), and their long pointed hooves are cushioned against the shock of their enormous 8 m (26 ft) strides.

For much of the year, pronghorns live in loosely associated herds of both sexes and all ages. During the rut, however, males actively defend territories and groups of females against competing males. Pronghorns suffer from insatiable curiosity, which has contributed to their decline following the arrival of Europeans in North America. They will travel considerable distances to inspect moving objects, including the handkerchiefs on poles that were used to bring them within firing range. Overhunting for meat and sport, coupled with loss of habitat, reduced their numbers drastically, but conservation measures have allowed a recovery of their populations. Pronghorn herds will never return to their original thousands, but their survival as a species now seems secure. JSC

SEE ALSO
- THE COMING OF MAMMALS p. 98
- HOOFED ANIMALS (BOX) p. 110
- ARTIODACTYLS (BOX) p. 112
- OTHER ARTIODACTYLS pp. 112–15
- TERRITORY, MATING, SOCIAL ORGANIZATION p. 156

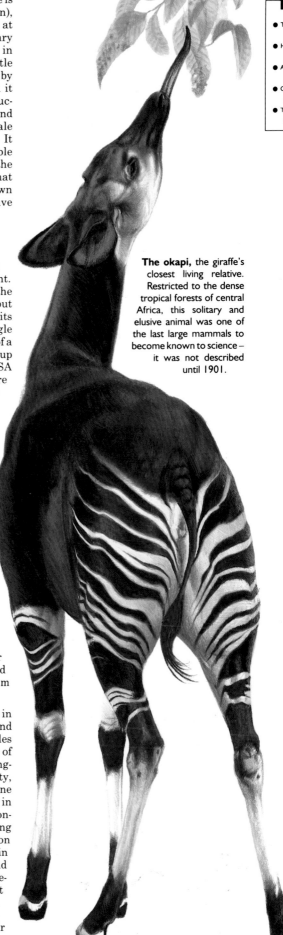

The okapi, the giraffe's closest living relative. Restricted to the dense tropical forests of central Africa, this solitary and elusive animal was one of the last large mammals to become known to science – it was not described until 1901.

Bats

Although certain other mammals are capable of gliding – sometimes for considerable distances – bats are the only mammals capable of true flight. The flying membrane of bats consists of skin stretched between the four extremely elongated fingers of each hand; only the thumb remains free, and is used for grooming. The elastic membrane is attached to the bat's ankles, and in many species it is also connected to the tail. While many species of bat have light fluttery flight, others are powerful fliers capable of covering great distances. Migratory species, such as the European mouse-eared bat, regularly fly 400 km (250 mi) or more.

Bats occur all over the world except the colder regions, above the tree line, and on some remote oceanic islands. They are extremely numerous and diverse, second only to rodents in number of species: nearly a thousand species have been described, and new ones are regularly discovered, particularly in the tropical forests. Because all bats fly at night, they are mostly rather drab-coloured – usually various shades of brown. A few are more striking, however, such as the white bat of Central America and the Rodriguez fruit bat of Rodriguez Island (near Mauritius in the Indian Ocean), which has golden fur. Bats vary in size from one of the smallest known mammals, Kitti's hog-nosed bat of Thailand, which is about the size of a bumblebee and weighs around 2 g (¹⁄₁₄ oz), to the fruit bats, which may weigh over 1·5 kg (3⅓ lb) and have wing spans in excess of 1·5 m (5 ft).

Classification and origins

The order Chiroptera, to which all bats belong, is divided into two unequal groups or suborders: the Megachiroptera consists of a single family (Pteropodidae) of about 170 species of fruit bat, or flying fox; the Microchiroptera includes all other species of bat – around 800 species in 18 distinct families.

The evolutionary origins of bats are obscure. The earliest fossils, dating back some 50 million years to the Eocene epoch, show that the adaptations necessary for flight had already evolved by this time. These fossil forms are essentially similar to modern bats, so it may be assumed that the group itself originally emerged considerably earlier. It has been suggested that bats may have evolved from insectivore-like ancestors that progressively developed a gliding ability.

Fruit bats

The fruit bats, or flying foxes, are found in the tropical and subtropical regions of Australia and the Old World. As their name suggests, fruit bats mostly eat fruit, but some feed on nectar as well. Their alternative name is also apt, as most fruit bats have dog-like faces, and many have the rusty coloration of the red fox. Most fruit bats have large eyes and rely prin-

The northern blossom bat of northern Australia, New Guinea and the adjacent islands is one of the smaller fruit bats. The bats of this genus, *Macroglossus* ('big-tongued'), emerge at night to feed on nectar and pollen, which they extract with their long, extensible tongues. They are extremely important as pollinators of the plants on which they feed (▷ p. 42).

cipally on their sight, and are therefore usually active at dusk or dawn, flying up to 70 km (44 mi) in search of fruit. A few are nocturnal, however, using echolocation for navigation (▷ box).

ECHOLOCATION

With the exception of most species of fruit bat, the majority of bats are dependent for navigation on echolocation – an extremely sophisticated form of sonar. Echolocation involves the emission of high-frequency calls or clicks (mostly above the range of human hearing) and detecting the echo that returns from intervening objects. Depend-ing on the nature of the sounds emitted, bats can obtain information on the distance, direction and relative velocity of an object, and also on its shape and texture. Microchiropteran bats produce calls in the larynx and emit them through the mouth or through the nose, where they may be focused by elaborate nose leaves, as in the horseshoe and leaf-nosed bats. The few fruit bats that echolocate use sounds produced by clicking the tongue against the side of the mouth.

Since echolocating bats feed mainly on insects, particularly night-flying moths, an intense evolutionary 'arms race' has been waged between hunter and quarry. Because certain moths have developed evasive strategies when they hear bats approaching (▷ p. 166), some bats have huge ears and emit very low-volume sounds. In an extraordinary analogue of the warning coloration seen in many animals (▷ box, p. 167), some moths produce ultrasonic clicks that warn echolocating bats that they are distasteful and do not make good meals.

Orientation by means of echolocation is not limited to bats. A similar system is used by dolphins and many toothed whales (▷ p. 144), and by birds such as cave swiftlets and oilbirds (▷ p. 94). Among mammals, various species of shrew and tenrec have been shown to use a rather crude form of echolocation to detect relatively large objects in darkness.

The fringe-lipped bat, a member of the New World leaf-nosed bat family, showing the large ears and elaborate nose structure typical of echolocating bats. This species is believed to be able to discern the social calls of the pond-dwelling frogs on which it preys.

Fruit bats often live in the tops of trees in communal roosts. They frequently occur in huge numbers – a single colony sometimes numbering over a million members. Some of the smaller fruit bats roost in the roofs of caves, in colonies of up to 9 million members. They often move in flocks of a thousand or more to suitable feeding sites, as figs or other fruit ripen on a particular tree. These feeding movements play an important part in the dispersal of the seeds of trees in tropical forests (▷ p. 43). Fruit bats sometimes feed in citrus and other plantations, but since they normally eat fruit that is ripe or over-ripe, and most commercial harvesting is of under-ripe fruit, they actually cause little damage. Despite this, they have been extensively persecuted by fruit farmers.

Feeding techniques

Bats have evolved a wide variety of feeding techniques. Most are insectivorous, often consuming huge numbers of tiny insects such as midges and mosquitoes in the course of a single night. In the tropics many species feed on nectar and pollen, and are often extremely important pollinators for night-flowering trees, cacti and other plants (▷ photo). Flower-feeding bats tend to have long, extensible tongues or pointed snouts in order to reach into flowers.

The three species of vampire bat, confined to the New World, are the most specialized of all bats, feeding entirely on the blood of warm-blooded vertebrates. Their front teeth are modified into two triangular razors, which they use to make a small incision, rarely felt by the victim. They then lap up the blood. The bat's saliva contains anticoagulants to prevent the blood from clotting, and the victim may continue to bleed after the bat has finished feeding. Although vampires are a problem in some parts of the New World, particularly in cattle-rearing areas and where rabies is common, the dangers are often greatly exaggerated.

A few species of bat are carnivorous, the largest being the Australian giant false vampire, which feeds on mice, small marsupials, birds and even other bats. Other species of false vampire occurring in Central and South America feed on opossums, lizards and other bats. There are also a few species that regularly catch fish.

Bat roosts

Many species of bat roost in caves or crevices, and some of the larger roosts are among the biggest known concentrations of any one mammal (▷ photo). A single colony of free-tailed bats in Eagle Creek, Arizona, once contained some 50 million members. This number subsequently fell to about 600 000, probably as a result of the widespread and often indiscriminate use of insecticides in North America in the 1960s.

While roosting, most bats hang upside down by their feet, and may even give birth in this posture. The dung, or *guano*, of cave-dwelling bats has been mined in

many parts of the world as a rich source of agricultural fertilizer.

In certain tropical caves, such as those in Borneo, several million bats of a number of different species live together. Complete ecosystems have evolved that are dependent on these bats. A wide range of invertebrates feed on the bats' dung and food remains, while snakes, insectivores and other vertebrates prey on the invertebrates and the bats themselves.

Hibernation

Most species of bat living in temperate regions either migrate in winter to areas where food is available or hibernate (▷ p. 175). During hibernation bats use about a tenth of the oxygen needed when active, and can rely on energy stored as fat. Most bats allow their body temperature to drop and their metabolism to slow

down both when sleeping and when hibernating.

Declining populations

In parts of the world where there is intensive farming, especially in Europe and North America, bat populations have undergone catastrophic declines in the past few decades. Damage to habitat has been a significant factor, but the widespread use of persistent pesticides in agriculture and of highly toxic insecticides for treating building timbers has also played a major part. Although most bats produce only one offspring a year, this low rate of reproduction is normally balanced by the fact that they are long-lived animals, with even small bats known to live 20 years or more in the wild. As a result, declines in numbers can take a long time to be noticed – and even longer to be reversed.　　　　JAB

Female Mexican free-tailed bats congregate in vast summer nursery roosts in 13 caves in Texas, USA. At one time their numbers were estimated at around 100 million individuals, but their population crashed catastrophically in the 1960s to perhaps 2 to 3 million. This decline was attributed to the indiscriminate use of DDT. This insecticide is now barred in the USA, but it is still used in Mexico, where these bats spend the winter.

SEE ALSO

● THE COMING OF MAMMALS p. 98
● DEFENCE, DISGUISE AND DECEIT p. 166
● HOW ANIMALS MOVE p. 176

Insectivores

The insectivores, which include such familiar animals as hedgehogs, shrews and moles, form a diverse and highly successful group of mammals. The order to which they belong (Insectivora) contains around 345 species, and with the exception of Australasia and Antarctica there are representatives on every continent, where they occupy all kinds of terrestrial and semi-aquatic habitats.

The insectivores can be conveniently divided into three major groups: the hedgehog types (Erinaceomorpha), the shrew types (Soricomorpha), and the tenrec types (Tenrecomorpha). For many years the tree shrews (⊳ p. 136), the colugos or flying lemurs (⊳ p. 137), and the elephant shrews (⊳ box) were included in the Insectivora. However, they are all so different from the members of this group and from each other that they have each been assigned to their own order.

General features

As a group, insectivores differ considerably in size and appearance, but typically they have long, narrow snouts, which are often highly mobile, and relatively small eyes and ears. Their limbs are short, and they walk with the soles and heels of their feet flat on the ground (plantigrade). Most species are small, active insect-eaters, although they will feed on a wide variety of invertebrates together with the occasional small vertebrate and some plant material. Their teeth are rather uniform, with the cheek teeth possessing sharp cusps to help crush the hard exoskeletons of insects. The fur of many species is short, dense and velvety. All species have a well-developed sense of smell.

Insectivores are considered to be the most primitive of the placental mammals. For example, like reptiles and lower vertebrates, they retain a common chamber (the cloaca) via which the genital, urinary and faecal systems reach the exterior. As well as having certain primitive features of the skull and skeleton, they have relatively small brains and their testes remain in the abdomen and do not descend into a scrotal sac. Nevertheless, superimposed upon these primitive features are many specializations that have equipped insectivores for many different modes of life, on and under the ground, in water and in the trees. The solenodons, the Eurasian water shrews and the short-tailed shrew are unique among mammals in that they have a poisonous bite, which helps to immobilize their prey.

Hedgehogs and moonrats

The 12 species of hedgehog and the 5 species of moonrat together make up the family Erinaceidae. The hedgehogs are distributed through Europe, Asia and Africa, and because of their relatively large size, their tameness and their frequent occurrence in gardens, parks and golf courses, they are familiar and popular animals. Like other insectivores they have long, mobile snouts, but their eyes and ears are comparatively large. Their most distinctive feature is the spines, which are in fact modified hairs and cover the animal's back and sides. The number of spines varies from species to species; the European hedgehog, for instance, carries about 5000 in total. On the lower flanks and belly, the spines give way progressively to stiff hairs and soft fur. Within the spines air-filled chambers separated by thin plates provide a strong but light structure. Near the base of each spine there is a flexible section that allows it to bend rather than break on impact. Muscles at the base of the spines pull them erect when the hedgehog is threatened, while massive muscles covering and encircling the back, just beneath the loose and flexible skin, allow the body to be curled up into a tight ball. Young hedgehogs are born naked: the spines are visible beneath the skin but do not break through for several hours after birth.

Hedgehogs feed on a wide range of invertebrates, including earthworms, beetles and slugs. They also scavenge for the remains of vertebrates, and will eat some vegetable matter, such as berries and fruits. Under adverse conditions, such as those experienced during northern temperate winters, hedgehogs are able to hibernate (⊳ p. 175). The animal's body temperature falls towards that of its surroundings to minimize the energy used in producing heat, and it can survive for several months on its stored body fat.

Moonrats or gymnures resemble spineless hedgehogs and are found in China and Southeast Asia. They inhabit forests and areas with thick cover, and feed on invertebrates and small vertebrates, which are sniffed out using their long, probing snouts.

Shrews, moles, desmans and solenodons

Despite their small size – most species weigh only 2–15 g (1/14–1/2 oz) – shrews have gained a reputation for their ferocity, belligerence and a voracious appetite. The family to which they belong (Soricidae) is the most numerous, diverse and widespread of all insectivore groups, with nearly 270 species occupying every terrestrial and semi-terrestrial habitat in all continents except Australasia and Antarctica. They have adapted to all conditions, from the hot, dry deserts of Arizona to the cold Arctic tundra.

Shrews have characteristically long, pointed snouts, which are used to probe for invertebrate prey amongst soil and leaf litter, and their fur is short and velvety. Although some species from more tropical climates are capable of short periods of torpor, most remain active day and night all through the year, and none hibernates. The Etruscan shrew, weighing about 2 g (1/14 oz), is amongst the smallest of all mammals. Because of the problem of maintaining a constant temperature in so small a body with such a large surface-to-volume ratio, the metabo-

A European hedgehog suckling its 19-day-old young. When born, young hedgehogs are naked, but the spines break through within a few hours.

A European mole making a rare visit to the surface.

lic rates of shrews are high, particularly in those species inhabiting cold climates. Consequently they must eat a great deal to satisfy their high energy requirements: a pygmy shrew weighing 3 g (⅒ oz) must eat nearly one and a half times its own body weight daily, and requires a meal every 3 to 4 hours if it is to survive.

Shrews are mostly solitary creatures, exhibiting territorial behaviour. They are extremely aggressive towards their fellows, producing loud vocalizations when they meet and sometimes fighting. Their scent glands, including those on their flanks, are an important aspect of their social behaviour and give shrews a characteristic musky odour, which most mammalian predators, including cats, find distasteful. Several species have become adapted for aquatic foraging in streams and possess thick well-waterproofed fur. Some merely have stiff hairs fringing the feet, to increase their surface area and assist propulsion through water, but others have webbed or partially webbed feet.

In spite of their very different ways of life, the subterranean moles and the semi-aquatic desmans (family Talpidae) are very similar in appearance. Both have cylindrical bodies with elongated, mobile snouts, which extend out below the lower lip. The tip of the snout of the star-nosed mole is furnished with fleshy tentacles to increase its value as a touch-sensitive organ used to locate prey. The eyes of moles are very small and may be covered by skin in some species, and they have no external ears. Moles have short tails, but those of desmans are long and flattened, and are fringed with stiff hairs to act as rudders while swimming. A mole's fore limbs are highly modified for digging – they are short and broad with strong claws, and turn outwards. The limbs of desmans are longer, the hind feet being fully webbed and the fore feet partially so, to assist propulsion in water. Their nostrils and ears can be closed by valves, and their fur is particularly water-repellent.

The 27 species of mole spend most of their solitary lives beneath ground in extensive burrow systems, where they feed largely on earthworms found as they patrol their tunnels. Surplus prey are immobilized by a bite and then stored for later consumption. Desmans live beside streams and feed exclusively on aquatic invertebrates such as freshwater shrimps and caddis-fly larvae.

The two species of desman are today confined to the Pyrenees of southern Europe and parts of Russia, but moles have a much wider distribution in Europe, Asia and North America. Moles are regarded as pests, and have been trapped and killed in large numbers over the centuries. Their skins were formerly used to trim garments and hats. Desmans were highly prized for their lustrous fur, which led to gross overhunting, and for this reason Russian desmans are still endangered. Pyrenean desmans are more at risk from habitat destruction.

Of the two remaining species of solenodon alive today (family Solenodontidae), one is restricted to Cuba and the other to Hispaniola in the Caribbean. They are rather shrew-like in appearance, with very long, highly flexible snouts used to investigate crevices, and large claws to dig out prey from the soil and from beneath bark and stones. They feed largely on invertebrates and occasionally amphibians, reptiles and small birds. They are amongst the largest of insectivores, weighing up to 1 kg (2·2 lb). Before the arrival of European colonists in the 16th century, solenodons were one of the dominant carnivore groups in the West Indies. Today, both species are extremely rare and threatened with imminent extinction as a result of habitat destruction and predation by carnivores introduced by man. The threat is exacerbated by their low reproductive rate: although they have a relatively long life span (six or more years), they produce only one or two young per litter and these remain with their mother for several months.

Tenrecs and golden moles

The family Tenrecidae includes 31 species of tenrec, which are restricted to Madagascar and the Comoro Islands, and 3 species of otter shrew – including the giant otter shrew with a body-and-tail length of up to 64 cm (over 2 ft) – found in western and central Africa. Tenrecs were amongst the first mammals to colonize Madagascar (some 60 million years ago), and they have evolved there to fill a wide range of ecological niches as insectivores, some being truly terrestrial, others semi-aquatic. They are largely nocturnal and their eyesight is rather poor, but they have well-developed senses of smell, hearing and touch. Many species occupying terrestrial habitats are small and shrew-like, with long tails. Rice tenrecs are burrowers resembling moles, with enlarged fore feet to assist digging, reduced eyes and ears, and very short, dense fur. Aquatic tenrecs, including the otter shrews, live beside rivers, lakes and swamps and are good swimmers. Several have webbed or partially webbed feet and long tails used as rudders. They forage for aquatic prey such as crustaceans, molluscs, fish and amphibians. Yet others resemble hedgehogs, with no tail (or a very short one) and varying degrees of spininess. Some can roll into a ball when threatened.

The body temperatures of tenrecs are relatively low and variable, and many are able to enter torpor for periods ranging from a few days to several months when conditions are unfavourable for foraging. Many species have large litters: common tenrecs produce an incredible average of 20 young per litter, and to feed them the females have up to 29 nipples – the most found in any mammal.

The 18 species of golden mole (family Chrysochloridae) of southern and central Africa are adapted for a subterranean burrowing life. They have robust, compact bodies with short limbs and no visible tail, and their fore legs are equipped with powerful claws for digging. Their eyes are hidden by hairy skin, the ear openings are covered by fur, and their nostrils are protected by a leathery pad on the snout, which is also used in digging. Golden moles have very water-repellent fur; it is bronze-yellow in colour (with an iridescent sheen in some species), which accounts for the common name of the group. They live in extensive burrow systems, where they feed on earthworms and other invertebrates. JSC

SEE ALSO
● THE COMING OF MAMMALS p. 98
● TREE SHREWS (BOX) p. 136

ELEPHANT SHREWS

Elephant shrews superficially resemble the true shrews, and indeed used to be classified with them in the order Insectivora. However, they are now recognized as being unique and have been assigned to their own order, the Macroscelidea, comprising one family, the Macroscelididae. Elephant shrews are very distinctive in appearance, with extremely long mobile snouts and long legs. Their hind legs are longer than the fore legs, enabling them to bound along with remarkable speed and agility. They have long tails and large eyes, and their cheek teeth are high-crowned like those of herbivores. Varying from mouse-sized to rat-sized, they feed on invertebrates together with some vegetable matter such as fruits and seeds. They are widespread in Africa, with some 15 species, but occur nowhere else in the world.

Carnivores
General Features

Carnivores – mammals of the order Carnivora – have long been the focus of special interest for mankind. Some have been ruthlessly persecuted on account of their supposed predatory impact or hunted to the point of extinction for their luxuriant furs, while others – domesticated cats and dogs – have been lovingly invited into our homes.

Hesperocyon, an early carnivore from the Oligocene epoch of North America (34–23 million years ago). Although somewhat similar to a civet in appearance, the structure of the teeth and inner ear show that this animal was in fact more closely related to modern dogs.

SEE ALSO

● THE COMING OF MAMMALS p. 98
● CARNIVORE GROUPS pp. 123–31
● PREDATOR AND PREY p. 164

Although the name 'carnivore' is applied generally to any carnivorous, or flesh-eating, animal (⊳ p. 194), the term is used specifically to describe the group of mammals belonging to the order Carnivora. This order contains a diversity of mammals, ranging in size from small weasels to the huge polar bear. Although most are flesh-eating, many are omnivorous or even (as in the case of the giant panda) vegetarian.

There are some 230 species of carnivore in seven families, divided into two major groupings or suborders on the basis of their evolutionary relationships (⊳ pp. 123–31). One group (Aeluroidea) contains the cats, the civets and their relatives, and the hyenas. The other (Arctoidea) contains the dogs and their relatives, the bears, the raccoons and their relatives, and the weasels and their relatives. Despite their diversity of form, members of this large order are united by the structure of their skulls and, more particularly, their teeth. Many of the adaptations of carnivores are associated with the catching and killing of prey.

Distinguishing features

Most characteristic of carnivores are the canine teeth, which are usually large, curved and dagger-like, and used for stabbing and tearing (⊳ illustration). All carnivores have small, pointed incisors adapted for holding prey and nipping flesh from bones. Of the cheek teeth, the premolars are usually adapted for cutting, while the molars have sharp, pointed cusps for gripping and crushing. A typical feature is the modification of the last upper premolar and the first lower molar (the *carnassials*), which usually have sharp cutting edges and work perfectly together with a scissor-like shearing action; these can cut through the tough hide, sinews and muscles of prey. The carnassials are best developed in the more carnivorous representatives, such as members of the cat and dog families, and least developed in the more omnivorous species. In vegetarian members of the order, such as the giant panda, the sharp cusps of the cheek teeth are replaced by smoother grinding surfaces.

The skulls of carnivores are heavy, with strong musculature to provide the power to split bones and cut flesh, and to drag or carry prey. The temporalis muscle, which runs from the side of the skull to the process (projection) on the lower jaw, provides the power for stabbing with the canine teeth. In the more carnivorous species, the skull has crests along the top and back to provide a larger surface for the attachment of this muscle. The masseter or chewing muscle runs from the arch on the side of the skull to the lower jaw, and provides the force necessary for biting and cutting. The upper and lower jaws articulate in such a way as to permit only vertical, not side-to-side, movement.

Carnivores have four or five clawed toes on each limb, and the first toe is often reduced in size or absent; in some species, including most cats, the claws are *retractile* – they can be withdrawn into sheaths. Some carnivores are *plantigrade*, walking with their soles and heels flat on the ground, while others, such as cats and dogs, walk only on their toes (*digitigrade*). The collar bone is reduced in carnivores, so permitting a longer stride when running on all fours.

The efficiency of carnivores as killers depends on complex repertoires of beha-

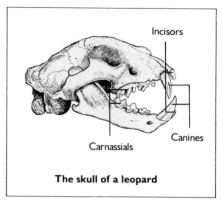

Incisors

Canines

Carnassials

The skull of a leopard

viour, including great stealth and a range of attack strategies by which they can secure their prey without incurring injury to themselves. Mongooses and weasels, amongst others, bite into the back of the head, so cracking the braincase and disabling the prey. Others, such as cats, bite at the neck, so separating the vertebrae and damaging the spinal cord. Prey may also be violently shaken to dislocate the vertebrae. Large prey may be throttled by gripping the throat or nose.

The senses of smell, hearing and touch are well developed in carnivores (⊳ pp. 188–91). Scent is particularly important in their social life, and they are notoriously smelly creatures. Both urine and faeces are scented and deposited as territorial markers. The cocktail of scents produced provides details not only of an individual's identity, but also of its sex, breeding condition and social status. Anal scent glands produce pungent scent, which is used for defence in some species such as skunks.

Distribution

The natural range of carnivores is very wide, extending to all continents except Australasia and Antarctica. They occupy many habitats, from tropical rainforest (e.g. civets and ocelots) to hot deserts (fennec foxes) and the high Arctic (polar bears). Aquatic habitats are also used, by the river and marine otters. Their natural ranges have been increased through introductions by man: the dingo was introduced to Australia in prehistoric times, to be followed more recently by the red fox; the stoat was taken to New Zealand, and Indian mongooses were released in the Caribbean.

Origins and evolution

The Carnivora evolved from primitive, insectivorous mammals and were represented initially by rather unspecialized, arboreal meat-eaters, which nonetheless possessed the characteristic carnivore feature of carnassial teeth. During the early Palaeocene epoch (65 million years ago), possibly in response to increasing diversity of prey types, there was a great divergence of these meat-eating mammals, from which the modern families of carnivores arose. During their evolution some spectacular forms have occurred, particularly in the cat family (⊳ pp. 98–101).   JSC

CARNIVORES AND MAN

Mankind has a love–hate relationship with carnivores. We cherish the domestic dog and cat, which we have taken with us to all parts of the globe. Simultaneously, we have hunted many almost to extinction for their highly valued furs or for their predatory impact (real or imagined) on our domestic stock. The dog was one of the first animals to be domesticated, probably from the wolf, and there is evidence of a close relationship between man and dog going back over 14 000 years. It may have been encouraged in hunter-gatherer societies as a hunter or guard, or merely as a companion. Cats were domesticated much more recently, no more than 4000 years ago, and for this reason are generally much more similar to their wild ancestors than is the case with dogs.

Civets
and their Relatives

The civets and their relatives – genets, mongooses, linsangs and fossas – amount to some 66 species of carnivore that together make up the family Viverridae. Although they have been introduced elsewhere (▷ below), their natural range is restricted to the Old World, where they occur widely in southern Europe, the Middle East, southern Asia and Africa; they are also the only carnivore group to have colonized Madagascar. Viverrids are typically small mammals with long, lithe bodies, short legs, long, bushy tails and pointed snouts. Many are patterned with dark stripes or spots.

SEE ALSO

- THE COMING OF MAMMALS p. 98
- CARNIVORES: GENERAL FEATURES p. 122
- OTHER CARNIVORE GROUPS pp. 124–31

Viverrids closely resemble ancestral carnivores, particularly with respect to their relatively unspecialized teeth and skulls. They are generally omnivorous and opportunistic, feeding on a wide range of items including small mammals, birds and their eggs, reptiles, insects and fruit.

General features

Viverrids are generally quick, active mammals, typically inhabiting forests and areas of dense cover. Most have five clawed toes on each foot, and in some species the claws are retractile – they can be withdrawn into sheaths. Some walk on the soles of their feet (plantigrade), but many are digitigrade, running on their toes like dogs and cats. Although the majority are ground-dwelling, many are good climbers. The binturong of Southeast Asia is a tree-dweller, and is one of only two carnivores to have a prehensile tail (the other being the kinkajou; ▷ p. 130). Others are good swimmers, and the Asian otter civet and the Congo water civet are semi-aquatic.

Viverrids have long, flattened skulls and well-developed canines and carnassial teeth (▷ opposite). They possess acute senses of sight, smell and hearing, and catch their prey by stalking and pouncing. Many are solitary or live in pairs or small families, but a few are more social and live in larger groups (▷ box).

As with other carnivores, viverrids are smelly creatures. They have scent glands in the anal area, which are associated with the genitalia. These play an important part in their social life: scent is rubbed onto objects in their home range to communicate information about the owner. So persistent is the musky odour from these glands that it is collected from several species of civet for use as a base in the perfumery trade, where it is known as civet. It is also used for medical purposes in treating skin disorders, and as an aphrodisiac (although there is no evidence that it has any such effect).

Civets and genets

Civets and genets are cat-like carnivores, mostly solitary in habit and hunting by night. They mainly inhabit forested areas and are good climbers. They are amongst the more omnivorous of the carnivores, feeding on fruits and seeds as well as small vertebrates and invertebrates. Palm civets are mostly arboreal, assisted in climbing by their curved retractile or semi-retractile claws. The fruit-eating habit of palm civets may result in their becoming pests in plantations, and they are particularly noted as banana thieves. They also rob fermenting palm sap (toddy) from the vessels attached to trees to collect it for human consumption, and for this reason are sometimes called 'toddy cats'.

Like other carnivores, civets and genets have well-developed scent glands associated with the genitalia, which play an important part in communication. In some species, such as the binturong, scent is simply spread as an individual brushes against vegetation or objects in its path. Others actively rub the glands on prominent objects as they squat against them. The Indian civet carries scent in a large muscular pouch by the anus, which can be turned inside out and pressed against objects. Genets often mark objects out of immediate reach of their glands by performing a handstand.

Genets are small, attractive carnivores with dark spots or stripes on the body and dark rings around the tail. They are arboreal, but will also hunt on the ground for the small mammals, lizards, frogs and insects on which they mostly feed. Genets are readily tamed, and during the Middle Ages in Europe – before they lost their place to the domestic cat – they were kept in houses to control rats and mice.

Mongooses

Mongooses are widespread and successful viverrids occurring in Africa, the Middle East, India and the Far East, where they occupy various habitats from desert to savannah and forest. They are small, agile, active mammals with long faces and bodies, short legs, small rounded ears, and long bushy tails. They have a large anal scent sac and use odour to mark their home ranges. Most species are solitary, but some, such as the slender-tailed meerkat and the banded and dwarf mongooses (▷ box), are social. Contrary to popular belief, mongooses do not live entirely on snakes, nor are they immune to snake venom; instead, they rely on their skill, agility and thick fur to avoid being bitten. They are opportunists, feeding on a range of small vertebrates, invertebrates, and even fruits. Those species that feed principally on vertebrates have well-developed carnassial teeth.

In the late 19th century the Indian mongoose was introduced into several West Indian and Hawaiian islands to control rats in sugar-cane plantations. Although it successfully preyed on the rodents, it also attacked chickens and indigenous birds and reptiles, and so became a pest itself on some islands.   JSC

THE DWARF MONGOOSE

The diminutive dwarf mongooses are unusually gregarious for carnivores. Occupying old termite mounds for shelter, they live in social groups that normally contain between 8 and 12 members but which may number up to 40. Their social system is run on matriarchal lines: females are dominant to males and a strict hierarchy is observed. The group is led by a mature female, and second in rank is her mate. Together they form a monogamous pair and are the only ones to breed. Below them rank younger females and males, which include both offspring and immigrants from neighbouring groups. Within the group there are many examples of co-operative behaviour: some individuals act as sentinels, scanning for predators; others act as baby-sitters, feeding, cleaning, and keeping watch over the young.

Cats

Of all the carnivores, it is the cats that conjure up the image of the consummate predator. These lithe, muscular creatures are indeed the most highly adapted for a hunting, meat-eating life, and they live almost exclusively on vertebrate prey.

Tigers (right) are unusual amongst cats for their readiness to take to water. This allows them to keep cool in the heat of their predominantly tropical habitats, but they are also good swimmers, often fording rivers and streams, and even chasing prey at the edges of lakes and rivers.

The family Felidae contains 34 species, and comprises the big cats, such as lions, tigers and leopards, and also many species of small cat, including the lynx, bobcat, ocelot and wild cat. Although some – notably lions – are social and live in groups, most are solitary and often inhabit forests, where their habits are difficult to study.

Many cats have highly distinctive coats of yellowish-brown, patterned with dark stripes, spots or blotches. Their coats are kept glossy by frequent grooming with tongue and paws. So valued are their patterned skins that hunting by man threatens many cat species with extinction.

General features

Unlike other carnivores, the skull and jaws of cats are rather short. The cheek teeth or carnassials have sharp cutting edges (⇨ p. 122), and the canines are large and used for gripping and killing prey, usually with a single bite to the throat or neck. The jaws have a hinge-like articulation, and movement is restricted to the vertical plane. In combination with the large chewing muscle (the masseter), this provides an immensely strong grip and allows struggling prey to be held tight. The surface of the tongue is covered with sharp papillae (pimple-like projections), which help to rasp flesh from a carcass.

Cats feed on a variety of small to large vertebrates, mostly mammals and birds. They are good climbers and very agile, with an uncanny ability to land on their feet, even after an awkward jump or fall. Their reflexes are quick, and sharpened by their good sense of balance and orientation. Cats walk on their toes (digitigrade), and their feet are well haired around the pads, allowing silent stalking of prey. All cats have sharp claws to grip prey, and in all species except the cheetah these can be retracted into sheaths, so protecting them from wear and keeping them sharp.

One of the most important attributes of cats is their eyesight. They have large eyes set at the front of the head, giving good binocular vision and visual acuity – essential for judging distances when running and pouncing. Cats see well in daylight, but their eyes are also highly adapted for night vision. Like many nocturnal vertebrates (as well as fishes), their sight is enhanced by a reflecting

A pride of lions (right) in the Ngorongoro Crater, northern Tanzania. Lions are the most social of all the cats, living in groups that consist of up to 6 males and 12 related females, plus their young.

layer in the eye (the tapetum), from which light is bounced back to stimulate the receptor cells in the retina (⇨ p. 189). The presence of the tapetum accounts for the shine of a cat's eyes when they are caught by torchlight at night. Cats also have an acute sense of hearing, with large mobile ears and an ability to detect high-frequency sounds. Their senses of taste and smell are well developed, although not to the same degree as in dogs. The long, stiff facial whiskers are sensitive to touch and particularly useful when a cat is active at night.

The ability to purr is a characteristic of the small cats of the genus *Felis*, and is probably used as a contact noise between mother and young. Although lacking this ability, the big cats of the genus *Panthera* are able to advertise their presence by roaring: they possess a pliable cartilage instead of rigid bone at the base of the tongue, which allows greater movement.

The earliest cats evolved at the beginning of the Oligocene epoch, some 34 million years ago (⇨ pp. 98–101). Many of the extinct forms were very large, including the fearsome sabre-toothed cats.

Tigers

The tiger is the largest of the cats alive today, with a head-and-body length of 2.5 m (8 ft) or more and weighing up to 260 kg (573 lb). It occurs in a range of forested habitats, where its distinctive coat colouring of dark stripes on a pale, golden background provides excellent camouflage. Although once widespread over

much of Asia, the tiger is now restricted to parts of India, Nepal, Southeast Asia, Indonesia and China, where it is mostly confined to reserves. Its numbers are dwindling as a result of habitat destruction and human persecution.

Tigers generally employ a stalk-and-ambush strategy to catch and kill large prey such as deer. The hind limbs are longer than the fore limbs, so enhancing the animal's ability to jump and pounce. The fore limbs are particularly well muscled, and the paws possess sharp, retractile claws, so that struggling prey can be seized and held. Tigers occasionally attack humans, and the causes of this 'man-eating' behaviour have been much discussed. However, the likely reason is simply that humans fall within the natural size range of a tiger's prey, and make fairly easy prey (being unable to run fast), particularly for old or injured tigers, which most often become man-eaters.

Tigers are essentially nocturnal, solitary animals. They occupy large home ranges. Those of females in southern Nepal measure about 20 km^2 (7¾ sq mi) with little overlap between neighbours. Males have even larger home ranges, which incorporate those of several females. Tigers exhibit territorial behaviour to deter intruders, including spraying trees and bushes with urine mixed with scent from anal glands, deposition of scented faeces in strategic places, and scratching marks on trees.

Lions

Like the tiger, the lion was once much more widely distributed than it is today.

Even within historic times, lions were found in southern Europe and in the Middle East through to India, as well as in most of Africa. Today they are confined mostly to sub-Saharan Africa, where they inhabit open savannah. Unlike tigers, these large cats live in groups – indeed, they are the most social of all cats. Their social organization is centred on the pride, which comprises between 4 and 12 related adult females and their offspring, plus 1 to 6 males. Within the pride, individuals associate in smaller groups of three or more.

Lions feed on large herbivores such as wildebeest, zebra and antelope, most of the hunting being carried out by females on a cooperative basis. Their prey are large and swift-running, so stealth is used to approach as close as possible before making a short dash and a grab to bring down the prey. All members of the group then share the feast. Male lions are clearly distinguishable from females by their thick, dark mane around the neck and shoulders.

Leopards

Despite hunting for its exceptionally fine spotted coat, the leopard is still the most widely distributed of all the cats, being found over most of sub-Saharan Africa, parts of North Africa and the Middle East, and much of southern Asia and the Far East. This relative success can be attributed to the leopard's adaptability: it is able to catch a wide range of small prey, such as birds and small mammals, as well as larger ones. It is a solitary cat, which hunts at night, and is a good climber, often dragging its prey up into a tree to protect it from other carnivores and scavengers.

The leopard shows considerable variation in coat colour, with a black form (the so-called black panther) being quite common, particularly in Asia. In the New World, the only large cat of the genus *Panthera* (the big cats) is the jaguar, a forest-dweller from South America, which resembles the leopard.

Cheetahs

With a reputed ability to sprint at up to 96 km/h (60 mph), the cheetah is generally believed to be the fastest land animal (⇨ p. 177). Although still widespread over much of the African savannah, few of these animals remain in the Middle East and southern Asia, where they once ranged freely. Females are essentially solitary, but males often live in permanent groups.

Unlike the thickset tiger and lion, the cheetah is slim and lithe, with relatively long legs adapted for speed. Its prey of fast-running antelope is stalked and then pursued in an explosive burst of speed. As with other large cats, once the prey is caught, it is suffocated with a bite to the throat. In the Middle Ages cheetahs were kept as hunting animals by Arabs, Abyssinians and Mogul emperors. Like that of other spotted cats, the cheetah's fur is highly prized, and – despite attempts to

protect them – these graceful animals are an endangered species today.

Small cats

In addition to the large cats, there are many species of small cat of the genus *Felis*, which are mostly solitary and whose habits are not well known. They include such species as the lynx, the bobcat, the ocelot and the wild cat. The domestic cat is thought to be descended from the African subspecies of the wild cat, which was probably domesticated in ancient Egypt around 2000 BC.

The small cats often live in forested habitats, where they employ stealth followed by a pounce to secure their prey of small mammals, birds, and occasionally reptiles and amphibians. The largest member of the genus *Felis* is the puma, cougar, or mountain lion, which has a wide distribution in both South and North America. Although its numbers have declined as a result of hunting and habitat destruction, small populations still survive in remoter parts of the USA and Canada. Many of the small spotted cats have been severely endangered by the fur trade, which reached a peak in the 1960s and 1970s. For example, 30 000 margay skins were sold in 1970, and even in 1983 250 000 spotted-cat skins were traded worldwide.   JSC

HYENAS – THE PERFECT SCAVENGERS

Although they superficially resemble dogs, the hyenas and aardwolf of Africa, the Middle East and India are in fact more closely related to civets and cats, and are placed in a family of their own (the Hyaenidae). Hyenas have large ears and eyes, a long bushy tail, and a back that slopes characteristically down to the hindquarters. Their most important features are the powerful jaws and large bone-crushing teeth, which ideally suit them to a scavenging life. This is best seen in the spotted hyena, which is able to crush large bones and bite through tough hide. The brown and striped hyenas are rather smaller and less massively built, and feed on insects, eggs and fruits as well as carrion and larger prey. The aardwolf has weak jaws and much smaller teeth, and feeds mostly on termites.

Not only can hyenas eat food that other carnivores cannot readily ingest but they also have a remarkable digestive system by which they are able to break down the organic matter of bone and other hard tissues. The remaining indigestible components are eliminated by regurgitation. Spotted hyenas are also active predators. They live in groups with a complex social organization and communicate by vocalizations and scent. They cooperate in hunting prey such as wildebeest, which are too large for an individual to tackle unassisted.

SEE ALSO

- THE COMING OF MAMMALS p. 98
- CARNIVORES: GENERAL FEATURES p. 122
- OTHER CARNIVORE GROUPS pp. 123, 126–31
- PREDATOR AND PREY p. 164
- THE LYNX AND THE SNOWSHOE HARE (BOX) p. 172

A European lynx with its prey. As well as taking small mammals such as hares, lynxes prey on birds and occasionally young deer. Generally solitary and nocturnal, these small cats are found mainly in forested habitats in Europe, Asia and northern North America.

Dogs and their Relatives

The fennec fox, the smallest of all the foxes, lives in the deserts of North Africa and Arabia. Mainly active at night, it escapes the daytime heat by denning up in its underground burrow. The animal's very large ears assist in dissipating heat and locating prey.

Amongst the 35 members of the family Canidae are wolves, foxes, jackals, wild dogs, the coyote and the domestic dog. They have a worldwide distribution, being found in every continent except Antarctica. The dingo or wild dog of Australia was probably introduced there by man in prehistoric times, but is now completely wild. A wide range of habitats has been exploited by members of this family (canids): fennec foxes occur in hot deserts, hunting dogs and jackals in savannah grassland, arctic foxes in the freezing northern tundra.

All canids are adapted for fast, long-distance running in pursuit of their prey. Unlike most other carnivore families, the majority of canids are highly social and live in groups.

General features

In contrast to several other carnivore families, canids are all remarkably similar in appearance and habit. They have slim, muscular, deep-chested bodies supported by long, slender limbs, bushy tails, long muzzles, and large, erect ears. Their teeth are typical of the carnivores (▷ p. 122) – well-developed carnassials, powerful crushing molars, elongated dagger-like canines, and small, pointed incisors. Canids have keen sight and hearing, and the sense of smell is particu-

larly well developed (▷ pp. 188–9), being important both in foraging and in social communication. They also communicate by an elaborate system of vocalizations and visual gestures.

Most have five toes on the fore feet but only four on the hind feet, each bearing a robust, blunt claw that cannot be retracted. The first toe of the front foot is small and vestigial, as is the first toe of the hind foot in domestic dogs (the so-called dew claw). Canids mostly have a digitigrade gait, walking or running on their toes. They have a variety of locomotory gaits, ranging from ambling to trotting and galloping. They are entirely terrestrial, although they can swim and the North American grey fox can climb trees. Most species are uniformly coloured or speckled, but the African hunting dog is blotched and one species of jackal is striped.

Although canids will eat fruits, nuts, invertebrates, eggs and carrion, they are largely predatory, feeding mostly on mammals and birds. Generally long-limbed and with great stamina, they are able to chase

prey for considerable distances, but they also hunt by stalking or pouncing. The larger social species, such as wolves and hunting dogs, can kill prey larger than themselves by hunting cooperatively. Smaller canids feed on birds, rodents and other small vertebrates, and hunt singly or in pairs. All species occupy regular home ranges, all or part of which may be maintained as a territory with the aid of scent-marking.

Dogs and dog-like carnivores originated in North America during the late Eocene, some 40 million years ago. The ancient forms were rather mongoose- or civet-like in appearance (▷ illustration, p. 122).

Wolves

There are two species of wolf alive today. The grey or timber wolf has the greatest natural range of any land mammal apart from man, extending across North America, Europe, Asia and the Middle East. The other species is the smaller red wolf of southeast USA, which is now thought to be extinct in the wild. There are several subspecies of the grey wolf, which vary in size, coloration and distribution. Once they were much more numerous and widespread, but today they are mostly inhabitants of remote wilderness areas, where they do not conflict with man's interests. Despite widespread persecution in Europe, grey wolves still maintain small populations in the forests of eastern Europe and in mountainous areas in the Mediterranean region. Active conservation measures coupled with education to encourage interest in and support for wolves is helping their survival.

Wolves have entered folklore as vicious and evil creatures, which cause havoc to domestic stock, attack humans, and even steal babies. They will indeed kill domestic animals given the opportunity, but are generally wary of man and rarely attack. Reports of their viciousness are greatly exaggerated. Wolves are social, group-living predators, which feed mostly on large mammals such as deer, moose and caribou. These prey are considerably larger than a wolf, and so are hunted by cooperative packs. Wolves will also eat small vertebrates, carrion and some vegetable matter, and will scavenge for food. Packs occupy large home ranges – up to 1000 km² (386 sq mi) – and in the far north they are nomadic, following the migrations of their prey (▷ p. 171). In forests they maintain more discrete territories by means of scent-marking and vocalizations, including the characteristic mournful howls, which promote group identity and keep neighbouring packs away.

Wolf packs generally comprise between 7 and 20 individuals and are strictly hierarchical. At the head of the pack are a breeding male and female, which usually mate for life and whose young are raised with the help of other pack members. The social hierarchy is maintained by an elaborate system of gestures and postures indicating threat and submission (▷

Two dingoes attempting to get the better of a monitor lizard. Although no members of the dog family are indigenous to Australia, the dingo – a descendant of domesticated dogs – is now completely wild, having been introduced by the Aboriginal settlers of Australia more than 10 000 years ago.

THE AFRICAN HUNTING DOG

The African hunting dog or wild dog is a formidable hunter, and – unlike most other canids – is exclusively carnivorous. Its distinctive coat of black, yellow and white blotches (varying according to the individual) is short and rather thin. The long tail is not brush-like as in other canids, and usually has a white tip. The large rounded ears give good hearing, while the long, thin legs are well adapted to fast running. In accordance with its carnivorous diet, the carnassial teeth are well developed.

Hunting dogs mostly inhabit savannah and open woodland in central, southern and eastern Africa, where they live in social groups. The average pack size is about 8 adults, but ranges from 2 to 20 individuals, plus their pups. They have large home ranges and travel long distances – up to 50 km (30 mi) a day – in search of prey. Antelope such as gazelle, impala and wildebeest make up the bulk of their prey, although they will also eat smaller mammals. Their hunting success, particularly when pursuing large and swift prey, is based on their cooperative behaviour, which allows them to head off their chosen

prey. Usually a young, weak or isolated antelope is selected and then pursued relentlessly in a straight chase. These dogs are capable of running at speed – around 50 km/h (30 mph) – for 5 km (3 mi) or more, and so can tire their prey before finally bringing it down. All members of the group then share the feast, and adults (not only parents) will regurgitate food for the younger pups.

Like wolves, hunting dogs are notably hierarchical, but adult males and females establish separate hierarchies. The dominant male and female are usually the only ones in the group to breed, and they inhibit reproduction in the others by physiological and behavioural means. There is intense rivalry between females for the breeding position. Unusually for social mammals, the stable core of the group comprises several related males, plus one or more (usually two) females, who are related to each other but not to the males. It is females rather than males that leave their natal pack, giving rise to the unusual ratio of two males to every female in most groups.

box p. 154). Much of this posturing can still be seen in our domestic dogs, which are descended from the grey wolf.

Jackals

Jackals are small, fox-like canids with an interesting social life based on co-operation. They are found in Africa, through the Middle East to southern Asia and in parts of southeastern Europe, where they are mostly inhabitants of dry, open areas such as savannah and semi-desert. They are yellowish to russet-brown in colour, and the silver-backed species has a distinctive brindled black-and-white back.

Although jackals are often portrayed as cowardly scavengers, this is far from being the case. In fact they feed on small vertebrates, invertebrates and fruits as well as on carrion. Golden and silver-backed jackals can catch larger prey such as young gazelles by hunting cooperatively. They will then share the prey and even regurgitate food for the pups or nursing females back at the den. Jackals live in monogamous pairs or family groups. In the silver-backed jackal, some youngsters remain with the parents and

act as helpers in rearing and guarding subsequent litters. Such behaviour has been shown to significantly increase pup survival. In return, the helpers benefit from the relative security of living as part of a group, and may inherit the territory when their parents die.

Foxes

Foxes are small canids with slender muzzles, large pointed ears, and long bushy tails. There are 21 species distributed over all continents except Antarctica (the red fox has been introduced into Australia). They inhabit areas as diverse as freezing arctic tundra, hot deserts and tall grasslands. With an age-old reputation for cunning and wiliness, foxes are successful and adaptable creatures. The red fox, in particular, is now a common urban mammal found in many towns and cities, where it dens up in parks, gardens and areas of rough land (▷ p. 212). They are opportunists, eating whatever is to hand, whether it be rabbits, small rodents, birds, invertebrates, fruit or carrion. Field voles are the favoured prey of red foxes. Excess food is often cached and later retrieved.

Although foxes are not pack-hunting animals, tending instead to forage alone, they do not live a solitary existence either. They often live in monogamous pairs, or even in small, loosely associated groups, usually comprising one male and several females. Cooperative behaviour in the form of helpers at the den has been observed in groups of arctic and red foxes. Scent-marking is an important part of their social life, and both scented urine and faeces are deposited in strategic places around the territory. As well as a pair of scent sacs on either side of the anus, foxes have a small gland in the skin on the underside of the tail near its base, and there are other scent glands between the toes. They also communicate by sound and produce a range of vocalizations, including howls, barks, yaps and screams.

Foxes have been widely hunted for their skins, particularly the red, grey and arctic species. They are also hunted for sport and on account of their occasional predation of game birds and of domestic stock such as chickens and lambs. Foxes suffer from and can transmit rabies, and so have been subject to widespread killing in attempts to control their numbers. JSC

SEE ALSO

● THE COMING OF MAMMALS p. 98
● CARNIVORES: GENERAL FEATURES p. 122
● OTHER CARNIVORE GROUPS pp. 123–5, 128–31
● BODY LANGUAGE (BOX) p. 154

Bears

Amongst the eight species that make up the family Ursidae are the largest of all the carnivores – the polar and grizzly bears. All bears are similar and distinctive in appearance: they have large heads with relatively small, rounded ears, dog-like faces with long snouts, and heavily built bodies with powerful limbs and a short tail. Their coats are dense and usually uniformly brown, black or white in colour.

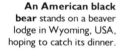

SEE ALSO

● THE COMING OF MAMMALS
  p. 98
● CARNIVORES: GENERAL
  FEATURES p. 122
● OTHER CARNIVORE GROUPS
  pp. 123–7, 130–1

Bears are widespread in the northern hemisphere, including America and Asia (and a few surviving in Europe), but are absent from Africa. One species, the spectacled bear, lives in northern South America. They occupy habitats as diverse as tropical forest and arctic tundra.

General features

Bears have massive skulls and quite unspecialized teeth compared with other carnivores (⇨ p. 122). The canines are enlarged, but the incisors are simple, the carnassials undeveloped, and the cheek teeth large and broad with rather flattened, rounded cusps. In contrast to other carnivores, their lips are not tightly tethered to the gums, but are quite free and hence very mobile and flexible. Bears have five toes on each foot, each bearing a large, curved claw used in digging and tearing. They have a plantigrade gait, with the sole and heel on the ground, and they walk in a shuffling manner on all fours. They are also able to walk for short distances on their hind legs, and most species can climb well. Their eyesight and hearing are rather poor, but they have a highly developed sense of smell.

Typically omnivorous and opportunistic, bears feed mostly on fruits, leaves, nuts, shoots and honey, and occasionally on small mammals and fish. They will also scavenge for food. The most carnivorous is the polar bear, which feeds almost exclusively on fish and seals. In winter bears usually hibernate (⇨ p. 175): they become very fat during the autumn, and as the weather becomes colder, they retreat into a den, where they sleep through the winter, living off their stored body fat. The heart rate drops markedly, but their body temperature is not greatly reduced and they are easily aroused from their winter sleep and may wake during mild spells. Hibernation is useful where food, particularly vegetation, is short in winter. Bears are solitary creatures with large home ranges, and their population densities are low.

Bears are relatively recently evolved carnivores. They first appeared in Europe sometime in the Oligocene epoch, some 30 million years ago, as small, somewhat dog-like carnivores. Subsequently much larger and more massive forms evolved, resembling those of today.

The brown or grizzly bear

Brown bears are large, powerful creatures, which have one of the widest distributions of any mammal, occurring through central and northern Eurasia and North America. Large populations remain only in remoter areas, away from human interference. Although historically much more widespread in Europe, they still maintain isolated populations in many parts, including Italy, Spain and Poland. Individuals can weigh as much as 780 kg (1720 lb), but are generally considerably lighter. European brown bears are much smaller than their American counterparts, and usually weigh less than 250 kg (550 lb). Most are dark brown, but paler and darker forms are common. In North America the long hairs on the back and shoulders have whitish tips, giving a grizzled appearance and accounting for their common name.

Brown bears often occupy open areas where they have access to cover, including tundra, coastal habitats and mountains. In Russia and in southern Europe they mostly live in forests. They are omnivorous, and much of their diet consists of vegetation, including tubers, which are dug up with their large claws, and berries. Insects, small mammals, fish and carrion are also eaten, and large mammals, such as young caribou and even domestic stock, may be taken if available. During the summer, when salmon are moving upstream to spawn, these bears gather at riversides to catch them. Most of the time, however, they are solitary.

In winter brown bears hibernate. Each makes a den in the lea of a rock or amongst tree roots, and lays down a bed of dry vegetation. One or two cubs are born in the den during the winter and remain with the mother for at least a year, often longer. They do not become mature until the age of 10 and can live for at least 25 years in the wild. Their numbers have been severely reduced by hunting and habitat destruction, and they are endangered in many areas. Although only occasionally a threat to humans, these large bears can become a nuisance as scavengers and hunters around settlements.

Polar bears

Rivalled only by the Kodiak bear (a subspecies of brown bear; ⇨ illustration), the polar bear is one of the largest of all carnivores, with a head-and-body length of up to 2·5 m (8¼ ft) and a height at the shoulder of 1·6 m (5¼ ft). Mature males can weigh up to a massive 800 kg (1764 lb). The coat is dense, shaggy, and white or

An American black bear stands on a beaver lodge in Wyoming, USA, hoping to catch its dinner.

Polar bears, perfectly adapted to a semi-aquatic life in the hostile climate of the freezing Arctic wilderness.

yellowish in colour, providing camouflage, warmth and water-repellence in the pack-ice environment in which they live. A layer of fat beneath the skin provides extra insulation. Polar bears swim and dive well, the large, broad feet functioning as oars when swimming. Their soles are well furred to protect their feet on the ice, while their large, curved claws assist grip on the ice as well as on their prey. Compared with other bears, their heads are rather small and their necks long, so they can readily swim with the head above the water.

Polar bears occur in coastal regions around the North Pole, including Greenland, northern Canada, northern Siberia and the Arctic islands. They extend as far north as 88°, and often follow the seasonal distribution of pack ice. Although polar bears prey mostly on seals, they occasionally feed on walruses, small mammals, birds, eggs and even vegetation. They will also scavenge for food, including around rubbish tips at human settlements. Seals are usually captured by waiting at a breathing hole in the ice or at the edge of open water, and seizing them as they surface.

Polar bears have large home ranges and may travel 20 km (12½ mi) or more in a day while foraging. They are solitary for the most part, although they may congregate around good food sources. A popular place is the rubbish dumps at Churchill on the southern shore of Hudson Bay in Canada, where they can easily be observed.

Only females, particularly if pregnant, hibernate for lengthy periods in winter. They move inland and excavate dens in the snow, where they bear their young in litters of up to four. Polar bear milk has a very high fat content, so the cubs grow quickly and are ready to emerge from the den in the spring.

These giant animals have no predators apart from man. They have been killed for their fat and fur by local peoples for centuries, but hunting is now restricted over much of their range.

The sloth bear

The sloth bear of the Indian peninsula and Sri Lanka is more specialized than other bears, being adapted for an arboreal life and feeding on termites and ants. It is typically large, black and shaggy, with a pale V-shaped mark on the chest, and weighing between 55 and 145 kg (120–320 lb). It has long, curved claws, enabling it to hang beneath branches in a sloth-like manner.

Particularly remarkable are the modifications of the sloth bear's feeding apparatus. It has a long, mobile snout with naked, prehensile lips and a long tongue. The central pair of upper incisors are missing, creating a gap between the teeth. Termite nests are dug up with the claws, the dust blown off, and the insects noisily sucked up through the gap in the front teeth with the aid of the lips and tongue. These bears will also feed on other insects, birds, eggs, carrion, flowers and fruit.

The giant panda

With its distinctive black and white fur and winsome features, the giant panda is one of the best known and loved of all mammals, but it is also one of the world's rarest. It is found only in central China, in montane forests where dense thickets of bamboo grow. It subsists almost entirely on bamboo shoots and roots, although it will occasionally eat other plant material and even small vertebrates.

The least carnivorous of all the members of the order Carnivora, the panda has several adaptations to its special vegetarian diet. It has particularly large chewing muscles, and bigger, broader cheek teeth with more rounded cusps than other bears. On the fore feet, one of the bones of the wrist is modified into the so-called 'sixth finger' – it is enlarged, elongated and thumb-like, allowing the animal to grip bamboo stems while sitting upright in its customary feeding position. Despite these modifications, pandas are not very efficient at digesting bamboo, which is hard and fibrous, and this may account for their leisurely, energy-conserving way of life. They are essentially solitary, and rear only one or two cubs at a time. Sexual maturity is not reached until four years of age or more.

The slow reproductive rate of pandas may contribute to their rarity – there are thought to be fewer than a thousand in the wild today. They are protected, but local habitat destruction and occasional poaching tend to isolate populations and reduce their numbers. A major factor leading to mortality in pandas is the natural die-back of bamboo that follows flowering and seeding. This event occurs only about once every 100 years, and last happened in the 1970s. Restriction of panda populations resulting from agriculture and other human activities means that they cannot easily move to new areas where bamboo is flourishing. Although pandas can be maintained in captivity, they are notoriously difficult to breed.

Because of its unusual anatomical features, the taxonomic position of the giant panda has been somewhat controversial. In the past it has been variously allied with the raccoons, or with the red panda in a family separate to themselves. Today it is more often viewed as a legitimate member of the bear family. JSC

BEAR FEROCITY – FACT OR FALLACY?

The Kodiak bear of Alaska – usually considered to be the largest living member of the Carnivora, measuring around 2·5 m (8 ¼ ft) from nose to tail.

Bears have a reputation for ferocity and aggression, and this has sometimes led to their persecution. Although they look cuddly, they are large and powerful, and bold when threatened. Most incidents of aggressive behaviour towards humans involve female bears accompanied by their cubs, which they will vigorously defend. Bears do not see or hear very well, and attacks can result when a bear is frightened or alarmed. If they become habituated to the presence of humans, bears can become a nuisance and a potential danger, especially around settlements or picnic areas in parks where they scavenge for food. They also come into conflict with humans as their natural habitats are cleared for agriculture, and they may then attack crops and occasionally domestic animals.

Raccoons
and their Relatives

The 16 species that make up the raccoon family (Procyonidae) include not only the raccoons proper but also the coatis, ringtails, olingos, kinkajou and red panda. All except the red panda are found only in the New World. They are thought to be related to the bears, and are more dog-like than cat-like in appearance, with rather foxy faces.

SEE ALSO

● THE COMING OF MAMMALS
p. 98
● CARNIVORES: GENERAL
FEATURES p. 122
● OTHER CARNIVORE GROUPS
pp. 123–9, 131

Procyonids are small carnivores and typically have long bodies with long tails. Although mostly uniformly brownish, reddish or greyish in colour, several species have dark rings around the tail and distinctive dark and pale markings on the face (the so-called face mask). The members of this family are omnivorous rather than truly carnivorous, and are relatively unspecialized members of the Carnivora. Most are ground-dwelling, but all are good climbers and some, such as the kinkajou and red panda, are primarily arboreal.

General features

Procyonids have five toes on each foot, bearing short curved claws that can never be fully retracted. They walk on the soles of their feet (plantigrade) in a bear-like manner. The face is generally short and broad with a pointed snout, and the coatis have particularly long, flexible noses. The short ears are usually rounded and well furred.

The teeth of procyonids are not specialized in the same way as most other carnivores (▷ p. 122). Although the canines are elongated, only in the ringtails are the carnassials developed. The premolars are small and sharp, the molars rather broad and rounded; the incisors are also unspecialized. Procyonids generally feed on invertebrates and small vertebrates, but much of their diet consists of fruit and other vegetable matter. They are mostly

The ring-tailed cat, or cacomistle, is not a cat at all, but one of the smaller members of the raccoon family. It measures up to 1 m (39 in) in length, around half of which is accounted for by the long bushy tail. Occurring in the southern USA and into Mexico, ring-tailed cats are largely nocturnal and tree-dwelling, feeding on a wide range of small animals, nuts and fruit.

THE RED PANDA

The red or lesser panda is the only procyonid with a natural distribution outside the New World. Its range is restricted to southern China across to the Himalaya, where it lives in mountain forests and bamboo thickets at high altitudes, between 1800 and 4000 m (5900–13 100 ft). It has luxuriant russet-red fur and a long, bushy tail, and its face is rather fox-like, with large ears and a mask of white and dark-coloured fur. The feet are hairy and the claws can be partially retracted.

Like its more illustrious namesake, the giant panda (▷ p. 129), the red panda has an extra thumb-like structure on each fore foot, formed by an enlargement of one of the wrist bones, which assists it in grasping objects. In the red panda, however, this unusual structure is only rather rudimentary. Red pandas are good climbers, but also spend a large part of their time on the ground. They frequently spend much of the day curled up on a branch sleeping, with the tail over the head, and for this reason have earned the alternative name 'cat bear'. Although mainly vegetarian, feeding on bamboo sprouts, grasses, roots and fruits, they will also take insects, birds, eggs and small vertebrates. The numbers of red pandas are declining as a result of the destruction of the forests in which they live.

solitary mammals, although female coatis live in social groups.

Raccoons

Raccoons are notorious in the Americas for their inquisitive nature and mischievous habits, which include raiding crops and rubbish tips. They can use their fore paws with extraordinary dexterity, adeptly picking up food items and even scooping out fish and crayfish from streams and ponds. The common raccoon is an adaptable creature and often lives in close proximity to man, both in rural and urban situations (▷ p. 213). It is the commonest species of raccoon, with a wide distribution in Central America and the USA, and in recent years it has successfully extended its range northwards into southern Canada. Like other procyonids, it does not hibernate, but in winter in the colder parts of its range it develops a thick coat and becomes much less active, spending most of the time in its nest.

Raccoon fur is valued commercially, and trapping and hunting of raccoons is an annual event in North America. Because of their value as fur-bearers, raccoons were formerly introduced to various parts of Europe and Russia, and wild populations still persist in some areas. Like a number of other carnivores, the common raccoon is known to transmit rabies, which has led to its persecution, especially in the southeastern USA.

Coatis

Coatis are conspicuous for their long, slightly upturned and flexible snouts, which they use to probe for prey, and their long, black-ringed tails. They are quick, agile mammals, and good climbers. The fore feet have long claws used in climbing and for rooting out food. Their prey includes insects and other invertebrates, birds, eggs, and the occasional small vertebrate, but they also eat fruit.

Coatis inhabit forests and scrublands in South and Central America, and are mostly active by day. Adult males are solitary, but females live in social groups of 5 to 12 individuals plus their offspring. Within the group, cooperative behaviour takes a number of forms – watching for predators and chasing them away, mutual grooming, and even suckling each other's young. Lone males are allowed to join the group only during the breeding season. Females build platform nests in trees, where they bear litters of three to five young. As soon as the youngsters are old enough – at five or six weeks – they are brought down from the nest to join the rest of the group.

Kinkajous and olingos

The kinkajou and the olingo are similar in appearance and habits, with long bodies and tails, and short legs, and both inhabit tropical forests in South America. Kinkajous, however, have prehensile tails, and long protrusible tongues, possibly used to reach for honey and nectar. They feed mainly on fruit, while olingos are more omnivorous, also taking invertebrates and small vertebrates. Unlike most carnivores, the kinkajou lacks anal scent sacs; it has scent glands on its chest and belly instead.   JSC

Weasels and their Relatives

The weasels, badgers, skunks, otters and their relatives together form the family Mustelidae; with 67 species, it is the largest of all carnivore families. Mustelids are widely distributed throughout the world, being absent only from Australasia, Madagascar and Antarctica. Many, like the weasels (the smallest of all carnivores), are small, lithe creatures with long bodies and short legs, but others, such as badgers, wolverines and skunks, are considerably larger and stockier.

SEE ALSO

● THE COMING OF MAMMALS p. 98
● CARNIVORES: GENERAL FEATURES p. 122
● OTHER CARNIVORE GROUPS pp. 123–30

Although diverse in appearance and habits, mustelids are united by their possession of well-developed carnassial teeth and anal scent glands (⊳ p. 122). They occupy many habitats, freshwater and marine as well as terrestrial, and are common and important predators, particularly in northern temperate regions.

General features

Mustelids are often uniformly brownish or greyish in colour, but some, including the badgers and skunks, have distinctive spots or stripes. In northern regions, weasels and stoats often turn white in winter (⊳ p. 174). All species have five toes on each foot, which bear claws that cannot be retracted. Mustelids mostly walk on their toes (digitigrade); small species run in a scampering, bounding manner, while larger ones have a more shuffling gait. Despite the carnassials and the elongated canines, the teeth are relatively unspecialized. While the weasels, stoats, polecats and otters are carnivorous, feeding mostly on small vertebrates, other mustelids are more omnivorous, with a mixed diet of invertebrates, fruits and seeds. Their senses of sight and hearing are well developed, but they mostly hunt by scent.

All species have well-developed anal scent glands, which may be used in defence as well as in social communication. These glands produce an oily and pungent scent, which is stored in a sac opening into the rectum via a sphincter. When threatened, the sphincter is opened and the scent sprayed out through the anus towards the intruder.

Most mustelids are solitary, and the sexes only come together briefly during the mating season. Mating is unusually prolonged and vigorous, since females produce eggs only following such active stimulation (a phenomenon known as induced ovulation). To assist the males in this, the penis is supported by a bone (the baculum), a feature characteristic of the order Carnivora.

Many mustelids, including the stoat (ermine), marten, mink and sable, have long, dense, luxuriant winter coats, which are highly valued. Their skins have been traded for centuries, contributing significantly to the local economy in northern lands. High demand for skins has often led to over-trapping – it takes about 300 ermine skins to make a coat, for example. With these valuable species under threat, controls on hunting have been introduced in many areas. A more sustainable supply has been produced by farming species such as mink, but changes in fashion and increasing conservation awareness have reduced demand in recent years.

Weasels

Weasels, stoats and polecats are fierce predators, and – despite their relatively small size – can kill prey such as rabbits that are much larger than themselves. They also eat small rodents, birds, lizards and insects. They have a widespread distribution in the Americas, Africa and Eurasia. The striped weasel, marbled polecat and zorilla have distinctive black and white (or black and yellowish) markings on their heads and bodies. Weasels are the smallest of the mustelids – indeed, of all carnivores – and have particularly long, lithe bodies, long necks and short limbs.

Badgers

Badgers are stocky mustelids with powerful jaws, well-developed carnassials, and elongated snouts used for rooting out food. They have large claws used for digging, and excavate extensive burrows (known as setts). Most species are nocturnal, with a good sense of smell but rather poor eyesight. They eat small vertebrates, invertebrates and fruit, while the European badger feeds mainly on earthworms. Unlike many other mustelids, the latter lives in social groups of about 12 members, which share the same sett but forage individually. The coarse,

stiff hairs of badgers were formerly used in making shaving and paint brushes.

Otters

Otters are the only truly aquatic mustelids. They have sinuous bodies, thick tails and short limbs, and the feet are webbed in most species. The ears and nostrils can be closed during swimming. The fur has a short, dense undercoat with overlying long hairs, providing good water-repellence and insulation. The whiskers on the muzzle are long and touch-sensitive, so assisting the animal in searching for prey underwater. Most otters feed on fish, frogs, crayfish, crabs and other large aquatic invertebrates.

The sea otter of the north Pacific is the largest of the mustelids and is exclusively marine, rarely coming ashore even to produce its young. It feeds mostly on sea urchins, clams and abalones, which it pulls off rocks with its fore feet and crushes with its rounded molar teeth. It is also one of the few tool-using mammals (⊳ p. 150), using stones carried from the sea bed to knock shellfish off rocks and to break open their hard shells. The stone is placed on the chest of the floating otter and used like an anvil.

Skunks

The New World skunks are notorious for their ability to squirt evil-smelling liquid from their anal scent sacs at an intruder, a trait developed by young skunks as little as a month old. Warning is usually given beforehand by stamping the front feet, raising the tail, and walking in a stifflegged manner. The bold black and white stripes of skunks also act as a warning to would-be predators. If the intruder persists, scent is aimed at its face, and – besides its pungent, sulphurous odour, which persists for days – it causes severe irritation to the eyes. Skunks have an accurate range of some 2 m (6½ ft), although the scent will carry a good 4 m (13 ft) or more.   JSC

The common or European polecat (right), a solitary, nocturnal animal found in woodland habitats in Eurasia and North Africa. Hunting mainly on the ground, polecats feed on a wide range of small animals. Like other mustelids, they have well-developed anal scent glands. Used both for marking territory and for defence, the secretions from these glands are notoriously foul-smelling.

Rabbits, Hares and Pikas

The familiar hares and rabbits (family Leporidae) and the smaller, rather rodent-like pikas (family Ochotonidae) together form the order of 'hare-shaped' mammals, the Lagomorpha. Rabbits and hares have long, mobile ears, large eyes, and a short 'scut' of a tail; their hind limbs are exceptionally long, enabling them to bound swiftly over open ground. In contrast, pikas have short legs, both front and back, and are better adapted for clambering in the rocky habitats in which they are commonly found. They have no tail and much shorter and more rounded ears than hares and rabbits.

Arctic hares (top right) grazing in the bleak landscape of Ellesmere Island, a Canadian territory in the Arctic Ocean.

Rabbits and hares occupy a wide range of habitats worldwide, from the cold Arctic tundra of the snowshoe and arctic hares to the hot semi-desert of the black-tailed jack rabbit, the tropical montane forests of the Sumatran hare, and the warm swamps of the North American swamp rabbit. Pikas are more restricted in habitat, and are found in alpine areas of Asia and North America, and on the Asian steppes. The large-eared pika of Nepal dwells at some of the highest altitudes of any mammal, occurring at up to 6100 m (20 000 ft).

General features

Despite their rather dissimilar looks, all lagomorphs share certain characteristic features. They have long, exceptionally soft fur, and their feet are fully furred (in contrast to rodents, which have naked feet). Their eyes are set high up on the sides of the head, so providing a wide field of view, and their nostrils are narrow and slit-like, with a fold of skin that allows them to be opened and closed. Although, like rodents, lagomorphs have continu-

SEE ALSO

- THE COMING OF MAMMALS p. 98
- RODENTS p. 133
- POPULATION DYNAMICS p. 172
- NERVOUS SYSTEMS p. 182
- SEX AND REPRODUCTION p. 184
- PESTS p. 230

RABBITS AND MAN

Rabbits have been important to mankind as sources of meat, fur and sport for many centuries. European rabbits were kept in walled enclosures by the Romans, and in managed warrens during the Middle Ages. Over the years they have been transported to many areas outside their original range in southern Europe. Many populations today, including those in Britain and Australia, are descendants of introduced animals, often escapees. In these new areas they lack their full complement of natural predators, and – with their highly successful reproductive system – have rapidly expanded their populations and gained pest status on account of their extensive and destructive grazing on pasturelands and crops. Trapping, shooting and poisoning have been widely used in attempts to reduce their numbers, but the most successful control measure was the introduction of the virus causing myxomatosis. This virus is quite harmless to its natural host, the South American forest rabbit, but it proved virulent and fatal to the European species. Following its introduction to Australia in 1951/2 and subsequently to Britain and elsewhere in Europe, rabbit populations were devastated. Now they have developed a degree of immunity to the disease, which has allowed their numbers to increase once more.

ously growing gnawing incisors suited to their vegetarian diet, they also have an additional pair of small, peg-like incisors behind the large set in the upper jaw. The relatively long, flexible necks of the lagomorphs allows them to turn their heads to a greater degree than rodents. Despite these differences, studies of their RNA have shown that lagomorphs are more closely related to rodents than to any other group of mammals. The oldest known lagomorphs date back some 50 million years, to the early Eocene epoch of eastern Asia.

The digestive system of lagomorphs is highly adapted to deal with the bulky grasses, herbs and other vegetation on which they feed (▷ p. 179). They have a large caecum (appendix) containing bacteria that assist in the digestion of cellulose, the main constituent of plant cell walls, and – like rodents – they practise *refection* or *coprophagy*: by swallowing and redigesting the soft, black pellets produced after the food's first passage through the gut, more of the valuable nutrients can be extracted.

Rabbits and hares

There are some 44 species of rabbit and hare. Rabbits are typically burrowers, while hares live above ground in more open habitats and are adapted for high-speed escape from predators – in full flight they can reach up to 80 km/h (50 mph). The large ears of rabbits and hares aid the detection of predators, especially when the head is lowered while grazing, and also provide a large surface area for losing heat in warmer climates. In far northern regions the winter coat of certain species is pale in colour (often white), so providing them with effective camouflage. Rabbits are born blind and naked, while hares are born fully furred and with their eyes open.

Rabbits reach sexual maturity at an early age (three months in the female European rabbit), and usually produce large litters after short gestation periods. Furthermore, females are able to conceive again

immediately after giving birth. These features contribute to the high reproductive rates so typical of these mammals, but they also suffer high rates of mortality. They are important prey for many predators, such as foxes, lynx and buzzards, and they are vulnerable to disease and to adverse climatic conditions and food shortages. Both rabbits and hares are subject to marked fluctuations in population size, as the example of the snowshoe hare of North America shows (▷ p. 172).

Pikas

There are 14 species of pika. Most of them inhabit the rocky sides of mountains, retreating into crevices for protection, but others, including the Daurian pika, construct burrows in the Asian steppe. Pikas are mainly solitary and territorial. They feed on grasses, sedges and even mosses and lichens and, like other lagomorphs, practise coprophagy (▷ above). In winter they tunnel through the snow to feed on the bark of trees and shrubs. Although they do not hibernate, they collect stores of plants during the summer for use as additional food in winter. Rock-dwelling pikas store their winter provisions in crevices, while steppe-dwellers construct piles of vegetation, sometimes up to 6 kg (13 lb) in weight, near their burrows.   JSC

A North American pika making hay to tide it over the winter in the Colorado Rockies, USA.

Rodents

Rodents are the most diverse, numerous and ubiquitous of all mammals, accounting for nearly 40% of all mammalian species. The order to which they belong (Rodentia) contains almost 1700 species distributed among some 30 different families. Many are very small, weighing less than 10 g (⅓ oz), but the majority are larger (up to 1 kg/2·2 lb), and the biggest – the capybara – grows to 66 kg (146 lb).

Rodents are highly adaptable and opportunistic mammals, and are found in all terrestrial habitats in every continent except Antarctica. They occupy such diverse regions as freezing Arctic tundra, where they live and breed under snow, and hot, dry deserts, where they can live their whole lives without drinking, deriving moisture from the seeds on which they feed. While most are strictly terrestrial and ground-living, many are arboreal and some are semi-aquatic. The success of rodents as a group can be attributed to their unselective, generalized feeding habits and to their enormous capacity to reproduce, which enables them to adapt rapidly to changing environmental conditions.

The fossil record of rodents is rather poor, but remains have been found in North America dating from the Palaeocene epoch, some 57 million years ago, by which time all their major features

had developed. The first rodents were squirrel-like, but during the succeeding Eocene epoch (53–34 million years ago) a major diversification took place.

The order Rodentia can be divided into three major groups, or suborders, on the basis of the arrangement of the jaw muscles and various associated features of the skull. The most significant feature from a taxonomical point of view is the masseter, or chewing muscle, which closes the lower jaw and pulls it forward, facilitating the gnawing action so characteristic of rodents. One suborder contains the squirrel-like rodents (Sciuromorpha), another the cavy-like rodents (Caviomorpha), and a third the mouse-like rodents (Myomorpha).

General features

Despite their diversity in appearance, all rodents are similar in basic design and share certain important features. The teeth of rodents are particularly characteristic, and include two pairs of chisel-like incisors that can gnaw through the toughest nut (the name 'rodent' is derived from the Latin verb meaning 'to gnaw'). These have open roots and – like those of rabbits and hares (▷ p. 132) – grow continuously throughout the animal's life. The upper pair grow over the lower pair and their tips abrade against each other. The teeth are constantly worn down by this action as well as by gnawing on hard food objects, and are thereby kept razor-sharp. Between the incisors and the cheek teeth there is a large gap (the diastema), which facilitates gnawing. The lips can be drawn into this gap, so sealing off the mouth from woody and indigestible fragments of husk

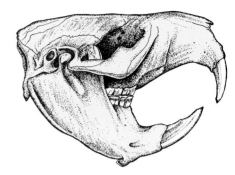

Skull of a beaver

sliced off by the incisors. The canine teeth are missing, but there are large cheek teeth (molars), which are covered in convoluted layers of enamel that create a complex and efficient grinding surface. The different patterns produced by these layers are a useful taxonomic aid, allowing many species to be distinguished from one other. Many rodents have cheek pouches, which may reach back to the shoulders and are used for storing and carrying food. These folds of skin are lined with fur and can be turned inside out for cleaning.

Rodents are primarily vegetarian, feeding mostly on seeds, fruits, leaves and roots, but they often also eat invertebrates such as beetles, grasshoppers and insect larvae. A few are carnivorous, including the Australian water rat, which feeds on small fish, frogs and molluscs. To assist digestion of tough plant material, rodents have a large caecum (appendix) containing bacteria that break down the cellulose of plant cell walls. To make full use of the available nutrients, rodents – like rabbits and hares – practise *refection* or *coprophagy*: the soft, moist faecal pellets produced after the first passage of food through the gut are eaten to allow further digestion. The hard, dry faeces resulting from this second passage through the gut are then discarded. Rodents have well-developed senses of smell and hearing. All species have long, touch-sensitive whiskers (known as *vibrissae*), and nocturnal rodents have large eyes.

Squirrel-like rodents

The squirrel-like rodents share a relatively unspecialized and primitive arrangement of the jaw muscles and associated features of the skull (although this does not necessarily indicate a close evolutionary relationship between the very diverse members of this group). In particular, the deep branch of the masseter muscle is very short and used only to close the jaw. The families in this group probably diverged from each other and from other rodents very early in their evolution. Besides the true squirrels (arboreal seed-eaters found in most parts of the world), the group includes many terrestrial squirrel-like species (marmots, ground squirrels and chipmunks) as well as kangaroo rats, pocket gophers, beavers and the unusual springhare.

Squirrels are amongst the most widespread of mammals, with some 270 species occupying every continent except Austra-

A marmot in Banff National Park, Canada. Their warning whistles are a familiar sound in the high valleys of the Rockies and European Alps, and in areas frequented by walkers they can become quite tame.

A southern flying squirrel about to launch into a glide. The furry membrane is clearly visible, folded up between the fore and hind limbs. Although capable of glides of 60 m (200 ft), this North American flying squirrel has nothing on its southern Asian counterpart, the red giant flying squirrel, which has been known to glide up to 450 m (1500 ft).

lasia and Antarctica. Squirrels, particularly tropical species, are also some of the most brightly coloured of mammals, with stripes and patches of black, white and russet fur. Most have large eyes and keen sight, with a remarkable capacity to judge distances between tree branches – a vital ability for arboreal climbers and leapers. They have short fore limbs and their toes bear sharp claws, which assist climbing in arboreal species. Their long hind limbs provide the propulsion necessary for leaping. Flying squirrels have a furry membrane (the patagium) along each side of the body, running from the hind limb to the fore limb, where it is attached to the wrist. By extending their limbs, these animals greatly increase the surface area of their body and are able to glide from tree to tree in a controlled manner, by using their tail as a rudder. When not in use, the membrane is tucked away so as

not to impede running and climbing. Ground squirrels, marmots and prairie dogs (▷ p. 161) form large colonies that inhabit underground burrow systems in treeless, often alpine, areas, from which they venture out to feed on grasses and herbs. In the northern parts of their range, they may be active for just three months in the brief summer, during which time they must breed and feed in readiness for their long winter hibernation (▷ p. 175). Pocket gophers are highly adapted to a subterranean life, with enormous incisors and powerful front claws to assist digging.

The beavers are among the largest of rodents, growing to lengths of 1·2 m (nearly 4 ft) excluding the tail, and are highly adapted to a semi-aquatic way of life. They have thick coats with an outer layer of long, water-repellent guard hairs and a soft, dense inner coat providing heat insulation. Their hind feet are large and webbed, and the tail is flexible and horizontally flattened into a paddle; these together give power and control to the swimming beaver. The nose and ears can be closed during dives, and the eyes are covered by a translucent membrane. So that it can gnaw and carry sticks underwater without drowning, a beaver can seal its mouth by closing its lips behind the incisors and blocking the throat with the back of its tongue. Beavers live in family groups, each with its own territory, and they cooperate to modify their environment in a way not seen in other mammals, moving mud and sediment from streams and gathering branches to construct lodges and dams (▷ illustration, p. 161). The European beaver was once widespread through Eurasia, but today only isolated populations survive in France, Germany, Scandinavia and central Russia. The American beaver was hunted almost to extinction during the booming fur trade of the 18th and 19th centuries, but has been re-established and now survives across most of its natural range in North America.

Cavy-like rodents

Although the cavy-like rodents include many of the larger species, they are typified as a group by the familiar guinea pig or cavy, a native of South America. The rodents of this group are diverse in appearance, ranging from the small cavies, degus and hutias to the chinchillas, pacas, maras and agoutis, and the large porcupines and the capybara. Nevertheless, they share sufficient features (particularly those relating to the arrangement of the chewing muscle and the skull) to constitute a natural, interrelated group. Unlike the arrangement in the squirrel-like rodents, the deep branch of the masseter muscle in these mammals extends forwards to the snout to facilitate gnawing. Cavy-like rodents are also characterized by the production of relatively small litters after long gestation periods, with the young well developed at birth. Unlike blind, naked new-born mice, cavies are born with their eyes open and a good covering of fur. Most species are terrestrial, but some porcupines are arboreal and the capybara (the world's largest rodent) is amphi-

---

OF MICE AND MEN

With their great reproductive powers, rodents have an inordinate capacity to increase their numbers. Around 200 000 house mice and 1200 common voles per hectare (81 000 and 485 per acre) were recorded during population peaks in Central Valley, California, in 1941/2. Rodents in such densities can cause extensive damage to natural vegetation by their feeding habits and soiling, but the extent of the damage they cause is more often assessed in relation to agricultural crops and stored foodstuffs. The Food and Agriculture Organization of the United Nations has estimated that the damage brought about by rodents worldwide amounts to 42·5 million tonnes (tons) of food a year. They consume the equivalent in food to the world's total annual production of cereals and potatoes.

Despite the large number of rodent species, only a few are of economic importance as pests. These include the brown rat, roof rat, cane rat, Nile rat, multimammate rat and house mouse, all of which damage stored foodstuffs by eating them and soiling them with their droppings. In Europe the common hamster and the root vole cause damage to crops in the fields by their burrowing and feeding activities, as do gerbils in Africa. Squirrels, including the grey squirrel, strip bark from trees, but the damage is generally localized and not of great economic significance. Rodents also cause structural

damage to buildings. Damage to electrical cables due to gnawing is the cause of numerous fires on farms, and they undermine sewers and roads by their burrowing.

Rodents are important transmitters of disease, and over 20 pathogens are carried by them. Among these is the bacterium that causes bubonic plague ('the Black Death'), which is transmitted to humans by the bite of the rat flea. In the first great European epidemic of the Black Death in the late 1340s, one third of the population died of plague in many areas, and in all some 25 million people died from the disease in Europe between the 14th and 17th centuries. A further 11 million people died in India during plague outbreaks between 1892 and 1918. Rodents also carry leptospirosis (Weil's disease), Lassa fever, rat-bit fever and murine typhus, all of which are potentially fatal.

Despite their bad reputation as pests and carriers of disease, rodents are also useful to mankind. Many continue to be important sources of meat for local peoples, including the capybara in South America and the cane rat in West Africa. Others, such as the beaver, coypu and chinchilla, have provided highly valued furs. The domesticated house mouse and brown rat are probably the most widely used animals in laboratory research and in the testing of drugs and other products.

bious and has webbed feet. Many, including cavies, are social and live in family groups.

Porcupines occur in both the New and the Old Worlds and are distinguished by their long protective spines and quills, which are in fact modified hairs, the former being solid and the latter hollow (like a feather). Even the young are born with quills. When threatened, porcupines erect and rattle their spines and quills, and may run sideways or backwards at the enemy. Quills can become detached and penetrate the intruder's skin on contact. Although not poisonous, they are difficult to remove and may produce septic wounds. The New World porcupines are excellent climbers, with large claws and naked soles to their feet to give extra grip, and some species have prehensile tails to hold onto branches. They feed on leaves, seeds, berries, bark and other vegetation.

The African mole rats spend their whole lives below ground in extensive burrow systems. They inhabit dry areas (but not deserts), where they feed on roots and tubers. Virtually sightless and lacking external ears, they have compact bodies and short limbs to facilitate movement through their burrows. They have extremely loose-fitting skin, which makes them very agile and allows them to turn round easily within the confines of the burrow, even though they can move along their tunnels backwards as well as forwards. Their enormous incisors and the long, sharp claws on their fore feet are used for excavating soil. While most species are solitary or live in small groups, studies of the naked mole rat have revealed a complex and remarkable social system (▷ p. 158).

Mouse-like rodents

More than a quarter of all mammal species belong to the suborder Myomorpha, the most familiar members of which are the house mouse and brown rat. Most are small, terrestrial, nocturnal seed-eaters, which are particularly notable for their high reproductive capacities. Although rather a diverse group, all members share the same arrangement of the chewing muscles and the same structure of the molar teeth. The deep and lateral branches of the masseter muscle are huge and extend forwards towards the snout, providing the most powerful gnawing action of all the rodents.

The majority of the mouse-like rodents belong to the mouse family (Muridae), which contains over 1000 species. Members of this family occur even in Australia and New Guinea, where there are no other native placental mammals. The group also includes the arboreal dormice, the athletic jumping mice, the desert-dwelling jerboas and gerbils, the voles and lemmings, and the Eurasian blind mole rats. Members occur all over the world, and some – notably the house mouse and the brown and black rats – have been introduced by man (often unwittingly) to areas far outside their original range.

Mouse-like rodents are typically short-lived but capable of prolific breeding from an early age. They have short oestrus cycles and gestation periods (▷ p. 185), and long breeding seasons. Female house mice are sexually mature by 2 months, produce up to eight young per litter after a gestation period of only three weeks, and can breed again three to four weeks later. However, they rarely live longer than two years (and usually much less in the wild). Many species of rat and mouse continue to have a major impact on mankind as pests of stored food products and as carriers of disease (▷ box).

Voles and lemmings are small rodents with blunt noses and short tails. In

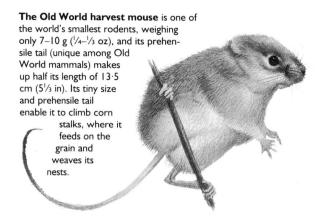

The Old World harvest mouse is one of the world's smallest rodents, weighing only 7–10 g (¼–⅓ oz), and its prehensile tail (unique among Old World mammals) makes up half its length of 13·5 cm (5⅓ in). Its tiny size and prehensile tail enable it to climb corn stalks, where it feeds on the grain and weaves its nests.

northern temperate regions their populations undergo marked cycles every three to four years, with high densities of up to 330 per hectare (133 per acre) crashing to levels of 50 per hectare (20 per acre) or less as their food supply dwindles and sociological problems and disease lead to death or emigration. Norway lemmings are well known for their mass emigrations, triggered by overcrowding in areas of high population density (▷ p. 172). Voles and lemmings are important food for owls, foxes and other predators, whose numbers also fluctuate with the abundance of their prey.

The Eurasian blind mole rats are highly adapted for a life below ground. Their eyes are completely hidden under the skin and they have no external ears or tails. Their incisor teeth are so large and protruding that these animals can use them for digging without opening their mouths. They possess a line of short, stiff, tactile hairs on each side of the head, which probably assist orientation in their dark world. They live in dry habitats, such as the steppes of southern Russia, and feed on underground bulbs and tubers, rarely coming to the surface.   JSC

SEE ALSO

● THE COMING OF MAMMALS p. 98
● NAKED MOLE RATS (BOX) p. 158
● ANIMAL BUILDERS p. 160
● POPULATION DYNAMICS p. 172
● DEVELOPMENT OF THE MOUSE EMBRYO (BOX) p. 186
● PESTS p. 230

A capybara with its young. Weighing up to 66 kg (146 lb) and growing up to 1·3 m (4 ft 3 in) in length, the capybara of South America is the largest living rodent. But even the modern capybara would have been dwarfed by its extinct relatives, some of which were the size of rhinos.

Primates General Features

Of all the mammals, primates have long been the focus of the most intensive study, not least because man himself is counted amongst their number. In addition to *Homo sapiens*, the order Primates contains some 180 species, including lemurs, lorises, monkeys and apes. These animals generally show a high degree of adaptability and complexity in their behaviour, and this has allowed them to exploit a wide variety of ecological niches. The majority of primates are adapted for life in trees, but many are ground-dwelling. Similarly, they have adopted a broad range of feeding habits: some species depend principally on leaves, fruits and other vegetable matter, while others are insectivorous and many have a varied diet, eating fruits and leaves but also hunting live prey.

SEE ALSO

- THE COMING OF MAMMALS p. 98
- PRIMATE GROUPS pp. 137–41
- LEARNING AND INTELLIGENCE p. 150
- BODY LANGUAGE (BOX) p. 154
- ANIMAL PHYSIOLOGY pp. 178–91

Carl Linnaeus, the founder of modern biological taxonomy (▷ p. 18), recognized that man was a primate and accorded the order Primates the highest rank in his classification of animals published in 1758. However, this view of man's position was not readily accepted until Darwin published his ideas on evolution a hundred years later (▷ p. 12).

The first primates appeared in North America and Europe in the Palaeocene epoch (65–53 million years ago). Early forms such as *Purgatorius* were small and bore a striking resemblance to the living tree shrews (▷ box). They are thought to have inhabited forests, where they fed primarily on insects. Their teeth were similar to those of rodents, with long gnawing incisors; this, as well as molecular evidence, suggests that primates and rodents may have diverged from a common ancestor (▷ p. 101).

Primate characteristics

Primates are distinguished as a group by a number of common characteristics. Each of their hands and feet have five digits, but in most species the curved claws found on the digits of other mammals are replaced by flattened nails, and often the thumb (and sometimes the big toe) can be opposed to the other digits, allowing these animals to grasp and manipulate small objects (▷ illustration). The snout of primates is typically short, and in comparison to most mammals the sense of smell is relatively unimportant. Instead, primates are much more reliant on their sight. Their eyes are characteristically large and forward-looking, so their visual fields overlap to give good binocular vision.

These characteristics were once thought to be associated with an arboreal way of life – the structure of the hands to allow grip around branches, the well-developed vision to facilitate leaping between branches. However, these features are not found in other arboreal animals such as squirrels and are now interpreted as adaptations for nocturnal, visually directed predation on fast-moving insects in the outlying branches of trees. Prehensile hands are common in mammals that forage for insects or vegetation in such locations, and binocular vision, while not essential for gauging distance during leaps, is important in accurately striking at fast-moving insects. The reduction in the sense of smell may simply be a consequence of the increased convergence of the eyes necessary for binocular vision.

Another highly significant feature of primates is their relatively large brain size, which gives them a great capacity for learning and adaptable behaviour. It also allows complex social interactions to develop. The primates are indeed the most social of all mammals, showing a range of social systems from semi-solitary animals and monogamous pairs to the complex hierarchical systems of baboons and chimpanzees. A variety of different communication sounds are produced, and in the higher primates facial expressions are important in social interactions (▷ p. 154), as are touch and scent.

Female primates produce one or two young, which remain with their mother for a considerable period – 3 years in monkeys, 3 to 5 years in apes, and 14 to 18 years in man. These long periods of dependence seem to be correlated with increased brain size and give many opportunities for learning the customs and techniques that underlie their complex social systems.

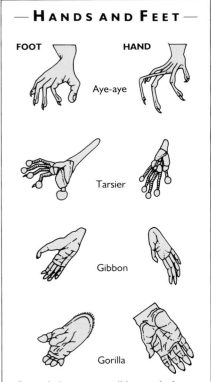

— HANDS AND FEET —

FOOT   HAND

Aye-aye

Tarsier

Gibbon

Gorilla

One of the most striking and characteristic features of primates is the high degree of modification in the hands and feet, which has allowed these animals to grasp and manipulate objects with greater versatility than any other group.

The aye-aye is unusual in having claws rather than nails on all its digits except the big toe; rather than gripping like most primates, it climbs by digging its claws into tree bark (▷ p. 137). Tarsiers have disc-like friction pads on their fingers and toes to enhance grip (▷ p. 138). Gibbons have extremely long fingers and deeply cleft thumbs and big toes that can be strongly opposed to the other digits (▷ p. 140). The gorilla's hand is adapted for 'knuckle walking', but considerable dexterity is still possible, as in all the great apes.

Classification of primates

Many different forms of primate classification have been proposed, and even today there is no general agreement on the matter. According to a widely (though not universally) accepted system, the order Primates is divided into two major groups or suborders. The lemurs and lorises (sometimes known as the *lower primates* or *prosimians*) together form the suborder Strepsirhini, while the monkeys, apes and man (variously called the *higher primates*, *simians* or *anthropoids*) make up the suborder Haplorhini.

The higher primates are themselves divided into two major groups, on the basis of certain structural features of the nose – the Platyrrhini (the New World monkeys) and the Catarrhini (the Old World monkeys, apes and man). Recent classifications also include the tarsiers as higher primates, although some authorities place them in a separate suborder, the Tarsiodea.

GS

— TREE SHREWS —

A Malayan tree shrew

The 18 species of tree shrew (family Tupaiidae) are small shrew- or squirrel-like animals that live in tropical forested habitats in Southeast Asia, where they feed on insects or fruit. There has been much debate over whether they should be classified as primates or insectivores (the group that includes the true shrews). Like insectivores, they have three incisors in the upper jaw, but the region of the brain associated with smell is smaller than in insectivores and the cerebral hemispheres are relatively large, and as in primates there is a complete bony bar running behind the eyes. Tree shrews resemble lemurs in many ways, but they differ in the details of their reproduction and behaviour. Females produce one to four young and apparently suckle them only once every 48 hours. On the basis of these various unusual features, tree shrews are now generally classified in an order of their own (Scandentia). Whatever their affinities, these curious animals bear a striking similarity to early fossil primates, and therefore offer a tantalizing clue to what the first primates may have looked like and perhaps how they behaved.

Lemurs, Lorises and their Relatives

The lower primates (suborder Strepsirhini) fall into two main groups – the lorises and the related bushbabies and pottos, and the lemurs. They are generally small, often nocturnal primates that inhabit the tropical forests of Africa (especially Madagascar) and southern Asia. These animals used to be considered primitive, but they are now recognized as highly specialized forms that have evolved considerably from their early primate ancestors. The lemurs, in particular, are fascinating animals, having evolved in isolation on Madagascar for most of the last 60 million years, since the time when the island became separated from the African mainland.

SEE ALSO

● THE COMING OF MAMMALS
  p. 98
● PRIMATES: GENERAL FEATURES
  p. 136
● OTHER PRIMATE GROUPS
  pp. 138–41

Most of the lower primates have long, bushy tails and long, narrow muzzles, the tip of which – the rhinarium – is naked and moist (in contrast to higher primates, which have a dry rhinarium). They generally have flat nails on all their digits except for the second toe, where a claw is retained for grooming. As a further aid to grooming, the lower incisor and canine teeth grow nearly horizontally to form a characteristic 'dental comb'.

Lemurs

There are five families of lemur with a total of 21 species. Most of these are found only on Madagascar, although some have been introduced by man to the nearby Comoro Islands. The 8 species of true lemur (family Lemuridae) are cat-size animals, growing to around 45 cm (18 in) plus a tail that is roughly the same length again. They are typically omnivorous in diet and are found in forested habitats, where they are active by day and highly adept at climbing and leaping from branch to branch. They are social animals, forming troops of between 10 and 24 individuals in which females are dominant over males. Each troop is territorial, marking branches within its territory with scent from a gland on the upper arm and defending it against neighbouring troops. A lemur's long tail is important in intimidating rivals. Dominant males wipe the tail over scent glands on the anus, and then hold it upright and wave it towards rivals. Such 'stink' battles are used to determine rank and ownership of territories.

Among the six species in the family Cheirogaleidae are the smallest of all primates – the dwarf lemurs, growing to a head-and-body length of around 19 cm (7½ in), and the mouse lemurs, which are even smaller at 13 cm (5 in). These diminutive creatures are nocturnal, and have large ears and large eyes with a reflective layer (the tapetum) behind the retina to improve their night vision. They feed on fruit, leaves, tree gums, nectar and pollen. During the dry season, from May to September, when food is scarce, dwarf lemurs hibernate.

The four species of indrid or woolly lemur (family Indriidae) are larger animals; the largest, the indri, reaches a length of 90 cm (3 ft) excluding a short tail. Although slow climbers, they make vertical leaps between trunks and boughs. They are active by day, feeding on shoots, flowers and fruit. All species are vocal, and one produces loud resonant calls early in the morning, probably to demarcate its territory. The related sifakas are social fruit-eaters, also active by day. Some species have a fold of skin that runs from the forearms to the torso and is fringed with long hairs. This is held out as the animal leaps, allowing it to glide from tree to tree.

The bizarre-looking aye-aye (the only species in the family Daubentoniidae) grows to a head-and-body length of around 40 cm (16 in), but its shaggy hair and long, bushy tail make it appear larger. It is an unusual creature in many ways: it is the only primate to have its nipples in the abdominal region rather than on the chest, and all its digits except the two big toes have claws, not nails. The middle finger is extremely long and thin, and is used to hook insect larvae out from under bark and to scoop out the flesh of coconuts through a hole gnawed in the shell by the single pair of long incisor teeth. Aye-ayes are nocturnal and appear to be solitary.

The sportive (or weasel-like) lemurs (two species in the family Lepilmuridae) get their name from their habit of raising their hands like a boxer when threatened. These shy, nocturnal creatures feed mainly on leaves, and they live singly in territories marked with urine and secretions from glands near the anus.

Lorises and galagos

The family Lorisidae comprises 10 species, including the slow-moving pottos and lorises as well as the agile bushbabies or galagos (although some authorities put the latter in a separate family). They occur from Africa to India and Southeast Asia, and are nocturnal, mainly arboreal animals with forward-facing eyes. Insects form the bulk of their diet, but seeds and flowers are also eaten and some bushbabies may feed on tree gums. They appear to be mainly solitary and occupy territories, which in bushbabies are marked with urine. Groups of related bushbaby females may remain together and mate with a single male, who oversees a larger territory for up to a year. Some reports suggest that pottos may be monogamous.

Lorises and pottos (including the angwantibo) are small animals, growing to no more than about 36 cm (14 in) excluding a very short tail. They climb branches hand-over-hand using fore and hind limbs of almost equal length. Their small round heads have large eyes but small ears. Galagos are squirrel-sized with large eyes and ears. They range in size from Demidoff's bushbaby, which is just 25 cm (10 in) in length, more than half of which is tail, to the greater galago, which grows to around 37 cm (14½ in) excluding a tail that is longer than the combined length of the head and body. In contrast to pottos and lorises, galagos are agile leapers. Their hind feet are long, and the digits and palms of their hands have ridged pads to improve their grip around branches. They produce about 10 different basic sounds used in communication. GS

Ring-tailed lemurs are active by day, sometimes climbing trees but spending much of the time on the ground. They usually produce a single young, which is well furred and open-eyed at birth but rides around clinging to its mother's fur in the early stages. It gradually becomes more independent and reaches sexual maturity at around 20 months.

New World Monkeys

The New World monkeys together make up one of the two major groups of higher primates (▷ p. 136). These monkeys, found in forested habitats throughout Central and South America, are themselves divided into two main groups: the family Callitrichidae contains the small marmosets and tamarins; the other, Cebidae, is a more diverse group containing a variety of species including capuchins, howler monkeys and spider monkeys.

The scientific name of the New World monkeys, Platyrrhini, means 'broad-nosed' and refers to the fact that their nostrils are widely spaced and usually face outwards rather than downwards. It is by this feature that they are most readily distinguished from the group that contains the Old World monkeys and the apes (the catarrhines). The divisions within the platyrrhines are defined principally on the basis of their dentition and the form of their digits: the marmosets and tamarins have 32 teeth and sharp, curved claws on all digits except the big toe, while the cebid monkeys have 36 teeth and nails on all digits. Goeldi's marmoset is unique in having 36 teeth and curved claws on most of its toes, so this small black marmoset is put in a family of its own (Callimiconidae). It is active by day, feeding on fruit and insects, and occasionally small vertebrates.

Marmosets and tamarins

The callitrichid primates – marmosets and tamarins – are found in forested habitats in central and eastern South America, especially in the Amazon basin. The 20 species are small, the largest growing to

Spider monkeys, as well as several other species of New World monkey, including capuchins and howlers, have fully prehensile ('grasping') tails, which serve as a fifth limb during climbing. This feature is not shared by any of the Old World primates.

TARSIERS

The tarsiers are something of a taxonomical conundrum. In general appearance, they are not unlike bushbabies with long, stringy tails, and like the majority of lower primates, they are nocturnal and have claws on their second and third digits. However, they have a number of features linking them with the higher primates, including a dry rhinarium, the absence of a dental comb, and the structure of the inner ear. Tarsiers have been variously classified with the lower primates in the Strepsirhini or in a suborder of their own, but are now generally included as the family Tarsiidae within the suborder Haplorhini (higher primates).

The three species of tarsier are all small; they grow to a body length of no more than 16 cm (6⅓ in) and have a naked, rat-like tail up to 27 cm (10½ in) long. They inhabit dense forest in Borneo, Sumatra and the Philippines. Their name is derived from the unique elongated, fused tarsal (ankle) bones in the hind limbs, which allow these animals to make spectacular leaps between trees. Their fingers and toes have disc-like pads for gripping. The enormous forward-facing eyes (equivalent to grapefruit-sized eyes in a human) and the remarkable ability of turning the head through 180° in each direction are also unique characteristics found in no other primate.

A western tarsier, Borneo.

---

no more than about 34 cm (13½ in), excluding a tail that is typically longer than the rest of the body. They weigh up to 1 kg (2·2 lb) and include the smallest of the higher primates, the pigmy marmoset, which weighs only 120 g (4¼ oz). Many species have tufts or ruffs of fur on or around the head; the common marmoset has white ear tufts, while the golden lion tamarin sports a striking silky golden mane.

Marmosets and tamarins, being relatively light, can reach the outermost branches of the forest trees. Most marmosets are agile leapers, while tamarins and pygmy marmosets move more vertically through the trees. Unlike many other primates, the thumb of these animals is not opposable to the fingers, so they cannot readily grip branches or manipulate objects. They are active by day, hunting for insects and other invertebrates, but many also feed on tree sap, which they reach by gnawing through the bark while clinging firmly with their claws. Sap is indeed a major constituent of the diet of marmosets, which mark their holes with urine and scent from a genital gland and defend them over an area of 1 hectare (2½ acres). For tamarins, sap is less important, and they use the sap exposed by other animals; their diet is generally more varied than that of marmosets and includes small vertebrates, flowers, fruit and nectar. This probably accounts for their larger territories of up to 5 hectares (12⅓ acres).

Callitrichids tend to live in small family groups, numbering up to 13 in marmosets, and between 6 and 25 in tamarins. Adults are monogamous, since females generally bear twins and need the help of their mate and older offspring to carry them. Reproduction in female offspring is suppressed by chemical signals from the mother.

The rainforest habitats of marmosets and tamarins are rapidly being destroyed; for instance, only 5% of the coastal rainforest of Brazil survives. The golden lion

tamarin nearly became extinct outside of captivity, but a captive breeding programme begun in the 1970s has led to the successful reintroduction of animals into the wild.

Cebid monkeys

Most of the 31 species of the family Cebidae are considerably larger than the tamarins and marmosets, the largest species (the muriqui or woolly spider monkey) weighing up to 15 kg (33 lb) and growing to a body length of around 72 cm (28 in). With the exception of the uakaris, which have short tails, all cebids have long tails, often longer than the rest of the body. The uakaris are also conspicuous by their naked faces and foreheads, which are often bright pink or red.

Cebid monkeys live from the floor to the canopy in tropical forests throughout South America, and except for the night monkey all are active by day. Howler monkeys feed mainly on leaves, but most species have a broader diet including leaves, fruit, seeds and small invertebrates. Capuchin, spider and howler monkeys have prehensile tails that can be wrapped around branches and so act as an extra limb during climbing. In the howler and spider monkeys the underside of the tail tip is naked and touch-sensitive, and in the latter the thumb is reduced so that the hand forms a hook.

The social structure varies among cebids. Night and titi monkeys are monogamous and live in small family groups. The latter show intense contact behaviour: pairs and sometimes whole families sit close together with tails entwined. Squirrel monkeys live in groups of up to 300, within which there are further divisions into single-sex subgroups. Cebids move within home ranges that may be small or large (up to 250 hectares / 618 acres), and these may be demarcated by loud calls. Howler monkeys have a special laryngeal resonating chamber, and their calls can be heard over distances of several kilometres.   GS

Old World Monkeys

The 70 or so species of Old World monkey are widespread in Africa, Asia and Indonesia, where they are found in a greater variety of habitats than their New World counterparts, ranging from tropical forest to open savannah. The single family to which they belong (Cercopithecidae) is divided into two major groups: the cercopithecine monkeys, which include macaques, baboons and vervets; and the colobine monkeys, which comprise langurs and colobus monkeys.

In contrast to the New World monkeys, the higher primates of the Old World (the catarrhines, comprising the cercopithecid monkeys and the apes) have nostrils that are close together and face downwards. Old World monkeys are generally larger than the New World species, and many have bare patches of thickened skin around the buttocks (the 'sitting pads'). These can be highly coloured and often swell up when the animals (particularly females) are reproductively receptive. Old World monkeys, like most of the New World species, are active during the daytime. Many species have long tails, but these are never fully prehensile; others have no tail at all.

Old World monkeys are generally highly social animals. Group sizes range from ten to several hundred, and both size and social structure often vary even within a species, depending on food, shelter and

SEE ALSO

- THE COMING OF MAMMALS p. 98
- PRIMATES: GENERAL FEATURES p. 136
- OTHER PRIMATE GROUPS pp. 137–8, 140
- LEARNING AND INTELLIGENCE p. 150
- BODY LANGUAGE (BOX) p. 154

SOCIAL BEHAVIOUR IN BABOONS

Baboons are largely ground-dwelling monkeys, and are highly social, living in groups of 13 to 185 individuals. Savannah baboons have a rich food supply and form groups of about 50 members, which include many males together with females and their young. The social organization within a group is somewhat loose. Hierarchies are formed among both males and females, while male–female pair bonds last only during mating. The dominant males mate more than subordinates; they also surround the females and young when the group is on the move, to protect them against any threat. In contrast, hamadryas baboons live in rocky terrain where food is scarce, and families of a male with several females and their young form tight social groups. Several family groups forage together as a 'clan' of around 20 baboons, and a number of clans travel together in 'bands' of around 60. At night several bands group together for safety and form a loose aggregation called a 'troop'.

Amongst baboons members of social groups spend much time grooming each other. Such grooming appears to reduce tension, particularly between males, and cements bonds between group members. All baboons groom in this way, but dominant animals receive most attention. Individuals often have preferred grooming partners. In male olive baboons, a phenomenon known as 'reciprocal altruism' occurs. If one of the grooming partners is in dispute with a third animal, the other will come to its aid and may even help it to take a mate from a dominant opponent. The assistant male in turn receives aid from its partner when necessary. Grooming appears to help to cement such reciprocal alliances.

A male proboscis monkey in Kalimantan, Borneo. Juveniles of this species have long but fairly rigid noses, which in females cease to grow at maturity. In males, however, the nose continues to grow unabated, eventually becoming huge and pendulous. When the male makes his loud, honking cry, his nose straightens out.

pressure from predators. All species take vegetable food, but many supplement this with the flesh of small animals.

Cercopithecine monkeys

The 45 species of cercopithecine monkey vary in size from the small talapoin monkey, which grows to around 40 cm (16 in), excluding a long tail, to the chacma baboon, which can exceed 1 m (39 in) in length and weigh over 30 kg (66 lb). The face is dog-like in the macaques and baboons, but more rounded and flatter in the vervets. Cheek pouches allow these animals to gather a meal of food and then digest it in safety.

Macaques and their relatives are widely distributed across Southeast Asia, extending north into China and Tibet. The most northerly species, the red-faced or Japanese macaque, often endures snow – it has long, dense fur and hugs its companions for warmth. The tailless barbary ape on the Rock of Gibraltar is the only non-human primate found in Europe and was probably introduced from populations in North Africa. It is in fact a macaque, not an ape, but unlike other macaques the males help care for the young. Java, or crab-eating, monkeys are agile swimmers and feed on crustaceans. The rhesus monkey, familiar as a laboratory animal, adapts well to urban life and is often found in large groups around markets and temples in the foothills of the Himalaya.

Baboons and mandrills, the largest cercopithecine monkeys, live in sub-Saharan Africa – baboons in open savannah, mandrills in forested habitats. They show a marked sexual dimorphism (⊳ p. 159), males being almost twice the size of females. Mandrills, with their brightly coloured red or blue snouts and white hair tufts and beards, are the most highly coloured of any mammal. Their diet consists of grass, seeds, nuts, buds, insects,

and occasionally small mammals. Baboons and the related mangabeys have a variety of facial and vocal signals. In male mangabeys resonating vocal sacs amplify their calls, which serve to space out adjacent groups.

The smaller vervets are ground-dwelling monkeys that feed on leaves and insects. They have different alarm calls to signify different predators; snakes, eagles and leopards each evoke a distinct call and produce the appropriate flight reaction in other members of the group. The related diana monkeys live in the forest canopy and are strikingly coloured, with black and white on the face and chest, and chestnut patches on the back and hind limbs.

Colobine monkeys

The 24 species of colobine monkey include the colobus monkeys of equatorial Africa and the langurs of eastern Asia. The largest species, the hanuman langur, grows to a body length of over 1 m (39 in) and weighs about 20 kg (44 lb). Their faces are typically rounded with a short snout. Leaves form a large proportion of their diet, and like ruminants (⊳ p. 112), they have large, complex stomachs containing bacteria that help to digest the cellulose of plant cell walls. These monkeys are highly adapted for life in the trees, and their thumb is reduced to a stump to give a hook-like hand.

The langurs live in mountains, rain forests and dry areas in Asia, and also forage for food in villages and towns. They include the proboscis monkey, which gets its name from the male's elongated nose (⊳ photo). Among the colobus monkeys of Africa is the conspicuously coloured black-and-white colobus, which has been hunted for its fine fur. It lives in small groups of 8 to 15 individuals with a dominant male, while the red colobus lives in larger groups of 12 to 80.   GS

Apes

Amongst the apes are to be found our closest living relatives in the animal kingdom, and for this reason as well as for their own intrinsic interest they have been the most intensively studied of all animal groups. There are 14 species of ape, or hominoid primate, and these are conventionally subdivided into the lesser apes (the gibbons), the great apes (the orang-utan, gorilla and chimpanzees), and man – although the exclusion of man from the great apes has more to do with vanity than biology.

Together with the Old World monkeys, the apes form the group of higher primates known as the catarrhines (▷ p. 136), which are distinguished from the monkeys of the New World on the basis of the structure of the nose. Apes differ from their fellow catarrhines, the Old World monkeys, in a number of ways. All apes lack a tail (which has been reduced to a small bone called the coccyx), and they have broad chests with shoulder blades set at the back rather than to the sides. The size and complexity of the brain (especially the cerebral cortex) – a feature that distinguishes higher primates as a whole – is particularly marked in the apes.

Gibbons

The nine species of gibbon (family Hylobatidae) are all found in the forests of Southeast Asia and Indonesia. In comparison to the other apes, they are slenderly built and relatively small. Most grow to a body size of about 50 cm (20 in), although the largest – the siamang – is around 90 cm (3 ft) and weighs up to 11 kg (24 lb). They have rounded faces with a short hairless muzzle. Of all primates, gibbons are perhaps the most highly adapted for swift movement through the forest canopy. They have very mobile shoulder joints, long fingers and toes, and extremely elongated arms. So equipped, they are able to move effortlessly through the trees, swinging hand over hand and occasionally leaping between more distant branches. Gibbons gather shoots and fruit, but they also eat insects, eggs and small vertebrates. Both sexes have large canine teeth.

Gibbons are monogamous and live in small family groups. They defend a territory of 10–40 hectares (25–100 acres), indicating their presence to rival groups with loud but tuneful calls that are amplified by large throat sacs. Both sexes sing, and males and females may sing duets, calling alternately.

Orang-utans

The orang-utan, the only species in the family Pongidae, is restricted to Borneo and Sumatra. Its name is of Malay origin, meaning 'man of the forest', and it is found in a range of

An orang-utan in the forests of Borneo. Although man's most distant relatives among the great apes, these animals are nevertheless highly intelligent and show great adaptability in behaviour. As ever-widening inroads are made into their habitats by logging and forest clearance for agriculture, these solitary creatures face a growing struggle for survival in the wild.

wooded habitats, from mangrove swamp and lowland wood to mountain forest. Orang-utans are relatively large primates: males may grow to 97 cm (38 in) in body length and weigh up to 90 kg (198 lb), while females are considerably smaller, weighing about 50 kg (110 lb). They have large heads with a protruding muzzle and are covered in reddish-brown hair.

Despite their size, orang-utans spend most of their time in the lower to middle branches of trees, where they climb by means of their long and powerful arms. They sleep in the trees, usually making a fresh nest each night, complete with both base and roof. Both the thumbs and the big toes are opposable to the other digits, so enabling these animals to manipulate their favoured food, fruit and nuts. Their large, powerful jaws can crack open hardshelled seeds, but they also eat shoots, leaves, ants, termites and tree bark when fruit and nuts are scarce. Rainwater is licked from leaves or scooped out of hollows in trees with cupped hands.

Orang-utans tend to be solitary. Males move within large overlapping territories, indicating their presence with roaring barks that build up to a loud crescendo. Throat sacs amplify their calls, which can carry over several kilometres. Territory-holders usually avoid

A mature male gorilla or 'silverback'. Gorillas live in social groups, usually presided over by a silverback, but there is very little aggressive behaviour within a group. Males rarely fight over females and subordinate males are allowed to mate, although the dominant males mate with females during their most fertile period.

each other, but if they do meet, fights can break out. Females live alone with their young or in 'mother groups' of two or three females and young, and move within smaller areas. When ready to mate, a female appears to find a male by following his calls. She remains with him for a few days or a few months at most, and then separates to rear the single young alone.

Gorillas

Together with chimpanzees and man, the single species of gorilla belongs to the family Hominidae. There are three distinct subspecies found in restricted areas in central Africa. Gorillas are the largest of the primates and are massively built: males measure around 170 cm (67 in) when standing and weigh about 150 kg (331 lb); females are considerably smaller than males, weighing 80 kg (176 lb) on average. The skull is heavy with prominent ridges over the eyes and the nose is flat with broad openings for the nostrils. Gorillas have smooth black skin covered in black hair, which tends to go greyish-white on the backs of fully mature males – the so-called 'silverbacks'.

Gorillas, especially older animals, spend much of their time on the ground. The young ones climb trees readily, and they and mature females may make nests in low branches. Movement on the ground is on all fours, the knuckles taking the weight at the front of the body. Gorillas are entirely vegetarian; they feed on herbs, particularly wild celery, shoots and berries, and spend much of their time foraging. This food also seems to supply all their water requirements.

Gorillas generally live in groups of 5 to 30 animals, consisting of one or more adult males, juveniles, and females with their young. There is a dominance hierarchy among males but less ranking among females. The dominant male is usually a

silverback, but very little aggression is seen in these animals (⊳ illustration). Groups move within large territories of 23–40 km² (9–15½ sq mi), but they rarely fight if another group is encountered. The silverbacked males first stare at their rivals, and if this fails to deter them, they display by uttering loud cries, throwing leaves in the air, beating their large chests, and finally running sideways, tearing up vegetation, and banging their hands on the ground. Only if this intimidation is not effective do the rivals come to blows. Sometimes a male from another group kills the infant of a female, who may then join the other group. The little grooming that does occur in gorillas is mainly between mothers and young.

Chimpanzees

The two species of chimpanzee are our closest living relatives, and are now often put in the same subfamily as man, the Homininae. They are similar in anatomical structure, in the shape of the brain, and also in the structure of molecules such as DNA and RNA (⊳ p. 22) and the blood protein haemoglobin. Like man, chimpanzees are said to express feelings such as joy, grief, fear and rage.

The common chimpanzee is a medium- to large-sized primate. Males are larger than females, weighing about 40 kg (88 lb) and growing to an average height of 1 m (39 in). They live in woodland and savannah in central Africa. Their highly individual faces have pronounced brow ridges and protruding lips and snout. Chimpanzees spend much of the time on the ground, where they walk bipedally or on all fours, moving on their knuckles like gorillas. They also climb and jump in the trees. The pygmy chimpanzee or bonobo is not – in spite of its name – much smaller than its common cousin; it is more arboreal, however, and is restricted to tropical rainforests around the Zaïre river and its tributaries.

A chimpanzee's diet is mixed: leaves, buds, seeds, eggs and small mammals are eaten regularly, but larger prey are also caught. In West Africa a complex cooperative hunting strategy has been observed. Small groups of adult males move through the forest until suitable prey such as colobus monkeys are found. A young male then acts as 'driver', keeping the monkeys moving forwards, while on either side of their path more chimpanzees act as 'blockers', preventing their quarry from escaping. 'Chasers' spring up into the trees after the monkeys. The most experienced chimpanzee of the group, the 'ambusher', climbs into a tree ahead of the monkeys, effectively blocking their way. As the monkeys turn back, they are caught by the chasers, which move in for the kill. The meat is then shared out among the hunting group by the dominant animals.

Social groups of chimpanzees range in size from 2 or 3 to 60 or so animals. The larger 'communities' are made up of several 'bands' – loose and changeable associations of males, females and their young that feed together. The only strong

---

HOW CLEVER ARE CHIMPANZEES?

After ourselves, chimpanzees are undoubtedly the most intelligent of the primates. They show an extraordinary capacity to modify their behaviour and to use objects for new purposes. One chimp in the Gombe National Park, Tanzania, learned that banging water cans together made his aggressive display more effective. Several of the great apes can solve puzzles, such as piling up boxes to reach food overhead or fitting sticks together to reach outside their cages, but the chimpanzees show the greatest adaptability in tool use and can even make their own tools (⊳ p. 150). They strip leaves from twigs to 'fish' for termites in termite mounds, and chew leaves to form a rough sponge for soaking up water in otherwise inaccessible tree hollows. Such behaviour requires learning, planning and insight.

Attempts to teach chimpanzees to speak have failed because they do not have the necessary vocal apparatus, but attempts to teach them sign language have been more successful (⊳ p. 155). The results of such experiments show that chimps can use abstract symbols to indicate objects and actions, and can use abstract terms such as 'like' and 'different'. But how far they indicate capacity for thought or language is still open to question.

A young chimpanzee using a stick to 'fish' for termites.

bonds are between a mother and her offspring, and these can last a lifetime.

Hierarchies occur amongst both males and females. In their fertile period females develop sexual swellings around the genital area. Early in the swelling period young males are allowed to mate, but only high-ranking males mate during the period of peak swelling. Older offspring help to care for the young, even if the mother should die. Females may leave their own group when adolescent and join neighbouring groups.

Chimpanzees have a wide repertoire of different sounds and facial expressions, such as pouting and grinning (⊳ p. 154). They may threaten each other with dramatic displays – waving branches, charging opponents and even hurling rocks at them. Tactile signals such as hugging, touching, kissing, embracing and grooming are also used, particularly as signs of appeasement.

Communities move within areas of up to 15 km² (6 sq mi). The boundaries are patrolled regularly by small groups of adults. When strangers are encountered, the patrol displays and may charge, hurl rocks and call loudly. Small groups or lone strangers, particularly females, are often viciously attacked and beaten, and young animals may be killed.   GS

SEE ALSO

● THE COMING OF MAMMALS p. 98
● PRIMATES: GENERAL FEATURES p. 136
● OTHER PRIMATE GROUPS pp. 137–9
● LEARNING AND INTELLIGENCE p. 150
● BODY LANGUAGE (BOX) p. 154

Marine Mammals
General Features

The mammals first evolved on land – all are air-breathing and warm-blooded (homoiothermic) – but at least three separate groups have returned to the seas, where their fish ancestors had lived hundreds of millions of years before, and have become adapted to life in water. The cetaceans (whales and dolphins) and the sirenians (manatees and the dugong) live their entire lives in water, and have lost the hind limbs and the ability to move on land. Seals and walruses (the pinnipeds) come ashore to breed and retain their hind limbs, but move awkwardly on land.

The pressures of adapting to life in water have led to a degree of uniformity in the appearance of marine mammals, but their origins are in fact very diverse. Cetaceans are thought to have evolved from the same ancestral stock as the ungulates or hoofed mammals (p. 110), diverging from their common ancestor at the beginning of the Tertiary period, some 65 million years ago. Early forms were not so well adapted for aquatic life as their living descendants, although the hind limbs appear to have become vestigial at an early stage. The sirenians are related to the proboscideans, which include modern elephants; all early fossil forms have been found in tropical marine deposits. The pinnipeds are believed to be very closely related to carnivorous mammals such as bears and weasels, and are indeed sometimes included in the order Carnivora.

Adaptations for life in water

A major problem for a warm-blooded animal in water is reducing loss of body heat. All marine mammals tend to be large in size, since a high surface-to-volume ratio reduces heat loss. Sirenians reach weights of 900 kg (1984 lb), while the largest seals – the elephant seals – can approach 4 tonnes (tons) in weight. The cetaceans boast the largest animals that have ever lived: blue whales can grow to lengths of 35 m (115 ft) and weigh up to an estimated 130 tonnes. Loss of heat is also reduced by an insulating layer of fat beneath the skin known as blubber. In whales this can be up to 1 m (3¼ ft) thick.

Like fishes, marine mammals have streamlined bodies to reduce drag as they move through water. The cigar or torpedo shape has proved to be most efficient and is seen in both fish and marine mammals. Movement is hindered by parts of the body that protrude, so external features such as limbs, ears and hair are reduced. In cetaceans and sirenians all external signs of the hind limbs have disappeared, the pelvic girdle is vestigial, and hair is virtually or totally absent. The seals and walruses need to move on land to breed and so retain hair for warmth and also have hind limbs, but these are reduced in length and are not very efficient on land.

The fore limbs of all marine mammals, and the hind limbs where present, are modified to form webbed paddles. The hands are long and flat, and in the cetaceans there are extra finger bones to give increased length. Only the parts of a limb below the elbow or knee protrude and movement is limited, generally to an up and down movement, with slight rotation of the paddles to assist manoeuvrability. In cetaceans and sirenians horizontally flattened tail flukes stiffened with cartilage provide the propulsion for swimming.

Many marine mammals can stay submerged for up to 30 minutes (although shorter dives are more common), and some for as long as two hours. They can close off the air passages and survive on low levels of oxygen by using anaerobic respiration in the muscles (box, opposite).   **GS**

ADAPTING TO LIFE IN WATER

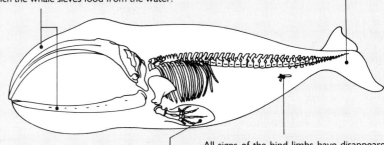

The massive jaws lack teeth and are arched, to provide space for the horny baleen plates with which the whale sieves food from the water.

The horizontal tail is a new structure made of cartilage and has no skeletal base.

The fore limbs are modified into paddles, with additional finger bones for extra length.

All signs of the hind limbs have disappeared except for small, isolated fragments of the pelvic girdle.

The pressures imposed by living, feeding and moving in water are very different from those experienced on land, and all groups of aquatic mammals have undergone great physical modification. The most extreme adaptation is seen in the baleen whales, such as the bowhead whale shown here.

SIRENIANS – THE MERMAIDS OF LEGEND?

The sirenians or sea cows are the only herbivorous marine mammals. Although they probably take in some animal matter concealed in their food, they are essentially grazers, feeding on seaweed and other marine plants in tropical and subtropical coastal waters. There are only two families in the order Sirenia: the family Trichechidae contains three species of manatee (the round-tailed sea cows), which live on either side of the Atlantic; and the Dugongidae (the fork-tailed sea cows), which today comprises a single species of dugong, found around the Indian Ocean, the Red Sea and the Western Pacific.

Manatees are large, growing to lengths of 2·5–3·5 m (8–11½ ft) and weighing between 200 and 600 kg (440–1320 lb). Their bodies are hairless or sparsely haired, and they have small external ears. They graze by means of a thick, muscular upper lip, which is deeply split and covered in short, stiff bristles. Only molar teeth are present (up to 40 of them), and these are constantly worn down by the animal's diet of tough, fibrous seaweed. Worn teeth are discarded and new ones grow from the back of the jaw.

The dugong broadly resembles manatees in overall appearance, but its tail is forked and the large upper lip is flexible and can be used to tear up the roots of sea plants. Dugongs grow to similar lengths to manatees, but may be heavier, reaching weights of up to 900 kg (1984 lb). They have five or six molars and the incisor teeth form small protruding tusks in the male. Mothers suckle their single young from teats just under the armpits while partly raised out of the water, and this may perhaps have given rise to the legend of mermaids.

Sea cows are endangered by hunting and other human activities. Off the Florida coast many manatees are killed or injured by powerboats, and their coastal habitat is being destroyed. Although protected by law, dugongs are threated by pollution – particularly by oil in the Persian Gulf – and they are frequently caught in fishing nets. A salutary reminder of the dangers facing these gentle creatures is given by the fate of their giant relative, Steller's sea cow. This was the largest of the sirenians, growing to lengths of 7 m (23 ft) and probably weighing around 4 tonnes (tons), and was the only species adapted to cold water. Within 21 years of its discovery in 1741 it had been hunted to extinction for its hide, blubber and meat.

West Indian manatees

Seals, Sea Lions and the Walrus

The mammalian order Pinnipedia comprises three families: the eared seals (Otariidae), which include fur seals and sea lions; the earless seals (Phocidae), which include the so-called 'true' seals and the enormous elephant seals; and the walrus, a single species in a family of its own (Odobenidae). These marine mammals are flesh-eating and closely related to land carnivores such as bears and weasels; indeed, they are often regarded as a suborder of the order Carnivora, rather than being given separate ordinal status.

The pinnipeds breed on land and typically gather in large groups at traditional sites called rookeries. In the eared seals and walruses there is marked sexual dimorphism (⊳ p. 159) – males are generally much larger than females; they often fight for possession of a territory and defend a harem of up to 50 females. With the exception of elephant seals, there is less social structure among earless seals and sexual dimorphism is less extreme.

When the females arrive at the breeding ground, they give birth to a single pup conceived the year before and then immediately come into oestrus (the period at which they are sexually receptive). In this way birth and mating are synchronized to the short time that the animals come ashore each year. The gestation period of most pinnipeds is only 9 to 10 months, but the fertilized egg remains dormant in the uterus for up to 3 months before implanting (a phenomenon known as 'delayed implantation'; ⊳ illustration, p. 185).

Earless seals

The 18 species of earless seal are the most highly adapted of the pinnipeds to life in water, and are widely distributed from Arctic to Antarctic waters. They have no external ears and can close the openings of the auditory canals as well as the nostrils. Their hind legs cannot be brought forwards, so when these animals move on land they have to shuffle along 'caterpillar fashion'. In the water the hind limbs are held backwards and moved laterally like the tail movements of a fish. The fore limbs are used for steering.

Most earless seals grow to around 1·2–1·8 m (4–6 ft) in length and weigh up to 100 kg (220 lb). However, the southern elephant seal of sub-Antarctic waters is the largest of all the pinnipeds (⊳ illustration). Males can measure up to 6 m (20 ft) and weigh 4 tonnes (tons), although females are about half this size. The name is

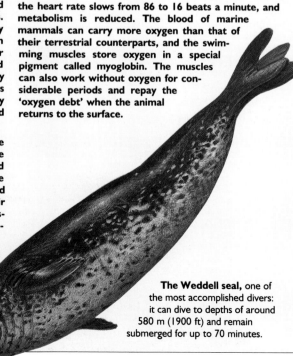

The southern elephant seal, the largest of the pinnipeds (⊳ text).

derived from the large trunk-like elongation of the male's snout, which can be up to 30 cm (12 in) long and is used to amplify territorial calls.

The diet of earless seals is mainly fish, crab and squid, but the leopard seal takes sea birds such as penguins, and the misnamed crabeater seal (⊳ illustration, p. 185) has sieve-like teeth to filter krill (small crustaceans) out of the water. Apart from the elephant seals, male earless seals tend not to form territories or to claim harems, but mate opportunistically with several females. Female earless seals do not feed during lactation. The young are suckled for only four to six weeks, but they grow at a rapid rate, doubling their birth weight in two weeks.

Eared seals

Unlike the true seals, the 14 species of eared seal (fur seals and sea lions) can bring the hind limbs forwards and so use all four limbs alternately on land. In water the front flippers are used to 'row' the animal through the water, while the hind limbs serve as a rudder. Eared seals eat mainly fish and squid, but some also take crustaceans and shellfish. The fur seals have very thick coats, with up to 50 000 hairs per cm² (300 000 per sq in); for this reason their skins have a high commercial value and have long attracted the damaging attentions of hunters. This thick hair is completely impenetrable to water, and together with the thick layer of fat beneath the skin, gives efficient insulation against cold water temperatures. Sea lions have thinner coats. The Californian sea lion is well known as a zoo and circus animal.

The walrus

The walrus is usually found in the shallow coastal waters of the Arctic. It is the second largest of the pinnipeds after the elephant seals. Males can reach a length of 4 m (13 ft) and weigh over 1·5 tonnes (tons). They feed on bottom-living invertebrates, such as shellfish, crabs, and sea urchins and other echinoderms, which they scoop up with their lips, but they may also eat fish, birds and even small seals. The snout is covered in an array of stiff but very sensitive bristles that assist in finding prey. The cheek teeth are reduced and flattened, but the upper canines form two long tusks. These are particularly large in males, which use them in territorial disputes with rivals. **GS**

DIVING

Marine mammals are not only able to stay submerged for long periods of time but also to dive to great depths. Bottlenose whales, for instance, will occasionally remain underwater for up to two hours, while sperm whales can descend to depths of 1000 m (3300 ft) or more. Dives are typically long and shallow, or short and steep. For humans, maximum submergence time is only a few minutes and return to the surface from deep dives brings with it the risk of 'the bends' – sickness caused by bubbles of nitrogen forming in the blood vessels and blocking the flow of blood to vital organs.

At the beginning of a dive marine mammals do not take a lungful of air, but exhale and collapse the lungs. The diaphragm lies at an oblique angle across the body, and as the animal dives and the external water pressure increases, the flexible chest wall collapses inwards and the diaphragm flattens, pushing the little remaining air from the lungs into the rigid-walled respiratory passages. Air (which is nearly 80% nitrogen) cannot therefore be forced into the blood under the increasing pressure, and so cannot form bubbles as the pressure is reduced on return to the surface. Expelling air also reduces buoyancy. These expert divers limit their oxygen requirements by reducing the flow of blood to all but the most vital organs, such as the brain and heart. 'Surplus' blood is stored in special sinuses. In seals the heart rate slows from 86 to 16 beats a minute, and metabolism is reduced. The blood of marine mammals can carry more oxygen than that of their terrestrial counterparts, and the swimming muscles store oxygen in a special pigment called myoglobin. The muscles can also work without oxygen for considerable periods and repay the 'oxygen debt' when the animal returns to the surface.

The Weddell seal, one of the most accomplished divers: it can dive to depths of around 580 m (1900 ft) and remain submerged for up to 70 minutes.

Whales

The order Cetacea is divided into two major groups. The larger of the two – the toothed whales (suborder Odontoceti) – includes dolphins, porpoises and the majority of the smaller whales. The other group – the baleen whales (suborder Mysticeti) – contains most of the true giants of the sea, including the blue whale, the largest animal ever known to have existed.

Whales spend their whole lives in water and are extremely efficient swimmers. Speeds of up to 36 km/h (22 mph) have been recorded for dolphins and up to 55 km/h (34 mph) for killer whales. Water resistance is decreased by streamlining and by fine grooves in the skin that keep the water moving over the body.

A school of humpbacks involved in cooperative feeding off the coast of Alaska. Some of the whales beneath the surface may blow a 'bubble net' around a mass of small fishes or crustaceans, which are afraid to swim through the screen of bubbles, and can thus be easily captured at or near the surface.

General features

Cetaceans feed underwater, and to prevent water entering the lungs, the windpipe can be sealed off from the throat by the larynx and the epiglottis. The nostrils are generally fused to form a single opening, the blow hole, which has a valve and lies on top of the head; it is therefore the first part to emerge when a whale 'breaches' or comes to the surface after a dive. The skull has two curious and unexplained features: it is asymmetrical – some of the bones on the left side are larger than those on the right – and some bones in the front of the skull are 'telescoped' over others. The neck is compressed and the neck vertebrae are often fused.

The eyes of cetaceans function equally well in and out of water, and the ears (although having no external structure) are very sensitive. Toothed whales can detect ultrasonic frequencies of up to 280 kilohertz (i.e. 280 000 vibrations or cycles per second); in contrast, humans can only hear up to pitches of 20 kilohertz. The bone containing the inner ear (the auditory bulla) is not attached to the skull as it is in other mammals, but is separated by cavities containing air and an oil-mucus foam. Sound is believed to reach the inner ear via the lower jaw and an adjacent fat body, a system that apparently ensures directional sensitivity underwater.

Female cetaceans produce a single calf every two to three years, after a gestation period ranging from 10 to 16 months. In the toothed whales the calf may be helped to the surface by the mother or attendant 'aunts' to take its first breath of air. Suckling occurs underwater. The milk is high in fat and protein, and is squirted into the calf's mouth by muscular contractions. The young grow rapidly: a blue whale calf can grow up to 9 m (30 ft) in seven months.

Cetaceans are gregarious to differing extents, and often live together in groups called 'schools'. There may be some social structure within the schools, and this may vary with the seasons.

Toothed whales

Toothed whales are found throughout the world, principally in oceans and seas but also in some lakes and river systems. They are thought to be more like the ancestral whales than the baleen whales.

Toothed whales feed on fish, cephalopods such as squid and octopus, and crustaceans. They may dive to great depths in search of prey, hunt in the midwaters, or scan the bottom of shallow rivers and coastal waters. Most species use echolocation (similar to the system used by

bats; ⇨ p. 118) to detect prey and surrounding obstacles. They emit clicks with frequencies of up to 265 kilohertz and at rates of up to 80 per second, the rate increasing as an object is approached. The sounds are thought to be produced by air being forced through valves near the blowhole. The sound may leave via the blowhole or through the 'melon' – a sac of oil in the forehead, which may act like an acoustic 'lens', focusing the sound to a narrow beam. Some toothed whales produce very loud sounds, which are thought to disorientate or even stun their prey.

The five species of river dolphin (family Platanistidae) probably most resemble the ancestral whales. They have long beaks with whiskers, and the neck vertebrae are not fused. These animals are relatively small, growing to lengths of 1·5–2·7 m (5–9 ft). They live in the turbid waters of the Ganges, Indus, Yangzte and Amazon rivers and in the coastal waters off Uruguay (the River Plate dolphin). Some authorities classify the animals from these different areas in separate families. The sense of sight is limited, and they are particularly reliant on echolocation. The eyes of the Ganges river dolphin lack lenses; it swims on its side and feels for the bottom with one flipper, picking up bottom-dwelling crabs and

The blue whale, the largest animal ever to inhabit the earth. The jaws alone of this astonishing creature are 6 m (19½ ft) long, and its heart is the size of a car.

fish. The Indus river dolphin hunts for catfish while swimming on its back.

Although there are as many as 18 species of beaked whale (family Ziphiidae), they have been little studied and their habits are not well known. They are medium-sized, up to 12 m (39 ft) in length, and feed mainly on squid. The jaws are extended to form a long beak, which bears one or a few pairs of teeth, one of which may protrude as tusks in males. The northern bottle-nose whale is a deep-diving form that can forage for up to two hours underwater, although shorter dives are more common (▷ box, p. 143).

The sperm whales (family Physeteridae) include the largest of the toothed whales and also some of the smallest. The two smaller species, the pygmy and dwarf sperm whales, grow to around 4 m (13 ft); little is known of their habits. The true sperm whale can reach lengths of up to 18·5 m (61 ft) and weigh 53 tonnes (tons). It hunts large squid, tuna and skate, commonly diving to depths of 1000 m (3300 ft). The huge head – accounting for a third of the total body length – is largely made up of the spermaceti organ, which contains up to 2000 litres (528 US / 440 imperial gallons) of fine oil. This organ may act as a reverberation chamber and increase the intensity of the clicks, or it may regulate buoyancy during dives. Spermaceti oil has been highly prized as a source of lubricant and as a base for cosmetics and margarine. Sperm whales are also the source of ambergris, a greyish-black substance found in the stomach and intestines and used in the perfume industry.

The family Monodontidae contains two species, the white whale or beluga and the narwhal. Belugas are 4–6 m (13–20 ft) in length, and in spite of the family name (meaning 'one-toothed'), they have 16 teeth in each jaw. They feed on fish, including commercially valuable species such as salmon and cod; fishermen have attempted to scare them off by replaying the calls of killer whales. Narwhals have

been called the 'unicorns of the sea', since males have a single spiral tusk in the upper jaw (in fact an enormously elongated incisor) which grows to around 2·7 m (9 ft). It is probably used in aggressive and sexual behaviour. These animals have long been hunted for their tusks, originally for their supposed magical and medicinal properties, now for their ivory. Females have no developed teeth. Narwhals feed on squid and bottom-dwelling fish such as flounders and skates.

Killer and pilot whales (family Globicephalidae; 6 species) are medium-sized whales, growing to around 6–8 m (20–26 ft), and have striking markings. Pilot whales live in nomadic schools, which may move through the water in single file. Killer whales live in family groups. They mainly feed on fish and molluscs, but may also cooperate to hunt and surround seals and other pinnipeds. Off the coast of Patagonia they swim near the steeply shelving coast and take young seal pups off the beach. Packs of killer whales will even attack baleen whales.

The 32 species of true dolphin (family Delphinidae) are up to 3 m (10 ft) in length and are found in coastal waters and in the open sea. They have many (up to 260) small pointed teeth. Bottlenose dolphins cooperate in driving shoals of fish towards the shore, and so sometimes force them into fishermen's nets. They communicate with one another by means of a variety of squeaks and cries.

Porpoises (family Phocoenidae; 6 species) are small, up to 1.5 m (5 ft), and live in bays and shallow waters in schools of up to 100 animals. Their short jaws contain up to 30 teeth. They feed on cuttlefish, squid and fish. The black finless porpoise does not have a dorsal fin, and like some other cetaceans, carries its young 'piggyback'.

Baleen whales

There are 10 species of baleen whale in three families. These large whales have no teeth; instead, there are sheets of horny material, called baleen or whalebone, which hang from the upper jaw. These are fringed with hairs on the inner edges and are used to filter plankton (mainly small shrimp-like crustaceans called krill) from sea water. Many of them feed in northern or southern seas but migrate to near the equator to breed.

Right whales (family Balaenidae; 3 species) are now rare. They are up to 20 m (66 ft) long, and their blubber can be 50 cm (20 in) thick, making them valuable to whalers – and hence the 'right' whales to hunt. The massive jaws of the bowhead are deeply arched (▷ illustration, p. 142), and bear 350 thin baleen plates up to 1·8 m (6 ft) long. These whales skim small crustaceans from the surface waters.

The family Eschrichtidae consists of a single species, the Californian grey whale, which inhabits coastal waters in the northern Pacific and is up to 12 m (39 ft) long. It feeds on bottom-dwelling crustaceans, using its snout to stir up

SAVING THE WHALES

Marine mammals have long been hunted for a range of products – meat, blubber (which provides oil), skins, spermaceti oil and whalebone (used in corsets among other things). Since the 17th century there has been a marked increase in hunting. The bowhead and right whales had been eliminated from the North Atlantic and Pacific by the 1850s, and the invention of the harpoon gun and the coming of modern steamships spelled disaster for many whale populations. Stocks of blue whales, fin whales and humpbacks, in particular, decreased dramatically during the 20th century, and these and other species are now endangered. The International Whaling Commission, founded in 1946, was initially set up to define hunting quotas, but it has become increasingly involved in conservation and protection. A general moratorium on whaling was agreed in 1985, but some countries continue to take large numbers of certain species (minke whales, for example) for so-called 'scientific' purposes. Smaller cetaceans such as dolphins and porpoises are also killed to prevent them taking fish, and – more recently – as alternatives to dwindling fish stocks.

Whales face other hazards, including fishing of their food resources and pollution. Sperm whales sometimes become entangled in underwater cables, and many thousands of dolphins and porpoises have been killed in the extensive drift nets and purse-seine nets used to catch tuna and other fish. Conservation measures involve bans not only on whaling but also on methods of fishing such as drift-netting. Educating the public to boycott unscrupulous companies and their products can also be effective.

Sometimes whales need saving from themselves. There has been considerable media interest in the phenomenon of whales becoming stranded on shallow beaches, and local conservation bodies have become involved in desperate attempts to refloat these marooned giants. It may be that such beaches fail to reflect echolocation signals, but recent work indicates that whales may use the earth's magnetic field for orientation and become confused when irregularities occur, for instance during magnetic storms. Damage to the inner ear from disease and parasites may also play a role.

The Yangtze river dolphin or baiji, one of the most endangered of all cetaceans. Its current population, estimated at around 300, is falling as a result of diminishing food supplies and pollution.

sediment, which it then sifts through short baleen plates.

The six species of rorqual (family Balaenopteridae) occur in all oceans and range in size from the minke at 10 m (33 ft) to the gigantic blue whale, which grows to a length of 35 m (115 ft) and weighs up to 130 tonnes (tons). Rorquals have deep furrows running down the throat, which can expand to increase the size of the mouth. Great mouthfuls of water are taken in and the krill is filtered off on baleen plates 1 m (39 in) long. The humpback whale sings a complex song of moans, squeaks and whistles lasting up to 30 minutes. The fin whale produces a very low-pitched song, at frequencies as low as 20 hertz, which may travel for hundreds of kilometres through the depths of the sea. GS

ANIMAL
BEHAVIOUR
AND
PHYSIOLOGY

Understanding Animal Behaviour

Human interest in the behaviour of animals must be as old as mankind itself. Indeed, for early man a good understanding of the ways of the animals that he hunted and on which he depended for food and clothing must have been crucial to his survival. However, although observation and description of animal life go back to the Greek philosopher Aristotle and before, the emergence of *ethology* – the study of animal behaviour – as a rigorously scientific discipline is a surprisingly recent phenomenon, and its distinctive techniques have only been fully developed within the last century.

In scientific usage, the term 'behaviour' embraces everything that animals do; animals are therefore behaving when they are active and when they are asleep. As we cannot ask animals why they are behaving in a particular way, students of animal behaviour have devised a methodology by which we can interpret their behaviour and make inferences about why they do what they do.

The pioneers of modern ethology

The recognition of ethology as a valid part of modern science is largely due to the pioneering work of three men in the present century – Konrad Lorenz, Karl von Frisch and Niko Tinbergen. These men observed animals in the field –

particularly birds, mammals and insects – and compiled detailed inventories of behaviour called 'ethograms'. They realized that many types of behaviour were performed in a similar or 'stereotyped' way by all members of a species, and as these 'fixed action patterns' were species-specific, they argued that they must have a genetic basis and so be 'innate'. Such behaviour patterns would therefore be handed down from generation to generation, and would be acted on by natural selection and evolve in just the same way as physical characteristics (⇨ pp. 12–17). It was found that in related species there were often subtle differences in behaviour, which was taken as further evidence that the evolution of the behaviour in question had occurred from a common ancestral pattern.

Central questions of ethology

Tinbergen pointed out that in trying to understand why animals behave as they do four distinct types of question can be asked. Questions of *causation* ask what internal and external factors cause an animal to act in a particular way. For instance, what is the effect of hormones or external stimuli? Questions of *function* ask what the behaviour is for. What is its immediate result and how does it benefit the animal? Questions of *development* focus on how the behaviour develops during the life of the animal, and on the interaction of genetic inheritance with environmental factors. Questions of *evolution* ask how the behaviour has evolved.

Some of these questions can be answered by observation of animals in their natural habitat, but others need some form of investigation and manipulation. Addressing questions of causation, for instance, often involves manipulating the external situation presented to animals or altering their internal state in order to see what features may be important.

All animals except humans are beyond direct questioning. So how do we begin to understand why they behave as they do? Why do vervet monkeys groom one another? And why do male vervets have such splendidly coloured appendages? Pioneering workers in the 20th century have forged a methodology that helps us to answer such questions.

Sign stimuli and motivation

One of the important findings of the early ethologists was that animals often respond to certain aspects of their environment and ignore or 'filter out' others. Soon after hatching, herring gull chicks peck at the parent's bill, which is yellow with a circular red patch near the tip; this behaviour results in the parent regurgitating food for the chicks. In his study of this behaviour pattern, Tinbergen set out to determine which aspects of the parent were important in eliciting the pecking response. He presented young chicks with various cardboard models and counted the number of pecks that each received,

---

GAMES THEORY

A leopard shelters in a cave as it is mobbed by baboons. If mobbing is dangerous, why not play safe and leave it to others? Games theory suggests an answer.

It is assumed that a behaviour pattern will only survive the pressures of natural selection if it is beneficial in increasing an animal's reproductive output. But as well as benefits, all behaviour has costs in terms of time, energy and risk of injury or predation. So the evolutionary gains must outweigh the costs. One way of investigating why animals behave as they do is to attempt to weigh up the costs and benefits of the various behavioural options open to them. The so-called 'optimality theory' predicts that through natural selection behaviour will evolve that gives maximum gain at minimum cost.

The behaviour of one animal, however, can affect the costs and benefits to another. The approach known as 'games theory' takes these various interactions into account. Behaviour is treated as if it were a game between members of a population in which there are several available behavioural options or 'strategies'. By estimating the costs and benefits of each stategy, it is possible to assess which may be the most successful. Often the answer is not simple. For example, in black-headed gull colonies a predator is driven off by mobbing – groups of gulls fly towards the intruder, shrieking raucously and dive-bombing it. This behaviour is beneficial as it drives the predator away from the nesting area, but it is also dangerous, as mobbing birds can be killed or

injured. Any black-headed gull that hangs back and is less daring than others is therefore at an advantage and should be more successful in producing offspring than 'daring mobbers'. But if the proportion of cautious mobbers in the population reached a point at which the proportion of daring mobbers was not sufficient to protect the colony, all birds would suffer loss of reproductive fitness. Games theory predicts that in any population of black-headed gulls the proportion of cautious mobbers cannot exceed a certain value (here calculated to be 45%). An 'evolutionarily stable strategy' – one that will be perpetuated in a population – is one that cannot be bettered in fitness terms by any other. In black-headed gull populations, neither daring mobbing nor cautious mobbing is stable on its own, so the population will tend to an equilibrium position of 45% cautious mobbers and 55% daring mobbers. These may be either different types of bird or birds at different stages of their life cycle, younger birds perhaps being more cautious than older ones.

ALTRUISM

A female pronghorn with twin fawns, Wyoming, USA.

As a consequence of natural selection (▷ p. 13), animals that succeed in competition with rivals of their own species – the 'fittest' animals – are thought to pass on the genes controlling their successful characteristics to their offspring. In other words, the fittest animals are those that reproduce most successfully and so leave more copies of their genes in succeeding generations. Behaviour patterns, therefore, should only survive in a population if ultimately they increase the reproductive output or 'reproductive fitness' of the gene-bearer. But there are many behaviour patterns that have long been a puzzle to ethologists, as they appear to *decrease* the reproductive success of the bearer. For example, the alarm calls given by birds and mammals such as ground squirrels, and the tail flashes of deer, rabbits and pronghorn antelope, all

of which warn of predators, can actually be harmful to the signalling animal, since it thereby draws attention to itself and is more vulnerable to attack. Some social animals even forego reproduction altogether in order to help others reproduce: in a bee colony, for example, only one female (the queen) reproduces; the other females (the workers) do not reproduce at all (▷ p. 63). An answer to the puzzle of such apparently selfless or 'altruistic' behaviour was provided by W.D. Hamilton, who pointed out that it is possible for an animal to increase the number of copies of its genes in succeeding generations other than by its own reproduction.

Related animals have genes in common. On average an animal will share 50% of its genes with either parent; in other words,

there is a 50:50 or 1-in-2 chance that any one gene present in an animal is also present in either one of its parents. Similarly there is a 1-in-2 chance that full siblings (brothers or sisters) hold a particular gene in common. This value of ½ or 0·5 is known as the 'coefficient of relatedness'. For grandparents and grandchildren the value is 0·25, and for cousins 0·125. So the closer the relationship, the more chance there is of animals having genes in common. An animal performing an altruistic act may, therefore, be increasing its genetic representation in the next generation, provided that it is a relative that it helps to reproduce successfully. On this argument altruistic behaviour should be more common between close relatives, and this is indeed generally the case; female ground squirrels, for instance, give alarm calls more readily in the presence of relatives than of non-relatives. In honeybees, males are haploid and females diploid (▷ p. 22), so females inherit a complete set of their father's genes and 50% of their mother's. On average female bees share 75% of their genes with their sisters. This high coefficient of relatedness (0·75) is greater than that between mother and daughter (0·5), so workers can increase their future genetic representation by helping their mother or a sister to reproduce. The process by which a gene is selected that causes its carrier to behave so as to reduce its own chances of reproducing while increasing those of a relative is known as *kin selection* (▷ p. 17).

When altruism occurs between non-relatives, as in the case of baboons (▷ p. 139), the benefactor and the beneficiary typically live in a closely knit social group, and the altruistic act is normally reciprocated. The assisted animal, therefore, cannot easily renege on its commitment, as it has to live with its partner and may require assistance again.

SEE ALSO
- EVOLUTION IN ACTION p. 16
- LEARNING AND INTELLIGENCE p. 150
- VARIOUS ASPECTS OF ANIMAL BEHAVIOUR pp. 152–73
- ANIMAL PHYSIOLOGY pp. 174–91

systematically altering the colour of the patch, the contrast of the patch (black through to grey) against a grey background, and the colour and length of the bill. He found that red was the most effective colour for eliciting pecking and that the contrast of the patch and the length of the bill were also important; the shape of the head, however, and the presence or absence of a body attached to the bill were apparently unimportant.

A feature or combination of features that elicits a response in an animal is called a *sign stimulus*. Such stimuli include any feature that an animal responds to, and may relate not only to food but also to other influences, such as nesting material and potential rivals or mates. If, as in the case of the gulls, the stimulus comes from another animal and is involved in social interactions, it is known as a *releasing stimulus*. Tinbergen also found that a long, thin bill received even more pecks than a realistic one and that a pencil with three red rings painted near the point was especially effective. Such 'supernormal' stimuli are rarely found in nature – a thin bill, for instance, will generally be of little use in feeding. However, the large size of a

cuckoo chick may act as a supernormal stimulus to its foster parents, eliciting feeding of the imposter in preference to their own chicks, if any have survived (▷ p. 168).

In many instances of social behaviour, the response of one animal to a releasing stimulus may itself be the releaser of a response by another. In his study of the courtship behaviour of the three-spined stickleback, Tinbergen showed that the appearance in the male's territory of a gravid (egg-laden) female – or even a model of a female – is the releaser for a characteristic 'zig-zag' darting dance by the male. This causes the female to adopt an upright courtship posture, to which the male responds by leading her to the nest he has made. Further interaction leads to the female spawning in the nest and to the male fertilizing the eggs. Clearly behaviour does not always occur in such a neat linear chain – actions are often repeated and the order of responses can change. Nevertheless, the idea of chains of reciprocal responses helps to explain how apparently complex behaviour patterns can build up through simpler mechanisms.

While in some cases animals appear to react almost 'automatically' to an appropriate sign stimulus, they do not always do so: outside the breeding season, male sticklebacks ignore models of females; gull chicks that have recently been fed do not respond to the parent's bill. The responsiveness of an animal to a particular sign stimulus is affected by internal factors such as blood-sugar or hormone levels, which bring about changes in the animal's internal state or *motivation*. Behaviour is therefore controlled by a combination of internal and external factors. A highly motivated animal – for instance, one that is very hungry – will respond to a sign stimulus, such as bitter-tasting food, that a fully sated animal will ignore.

The concepts of sign stimuli and motivation are useful in explaining what causes an animal to behave as it does at a particular moment in time. These ideas reveal only part of the story, however. A fuller account of a given behaviour pattern should also take into account both the evolutionary history of the animal concerned (in particular, how its behaviour has evolved) and the role played by learning (▷ p. 150).　　GS

Learning and Intelligence

Most animals are born with at least some of the responses appropriate to the various stimuli they encounter in their environment 'built in' to their nervous system. Such responses are performed correctly, or almost correctly, on the first occasion they are needed and appear to be little affected by the environment in which the animals develop. Female digger wasps dig a nest, bees construct combs, male sticklebacks mate, and female mice care for their young, all without previous experience. Such behaviour is *instinctive* and must be encoded in the animal's genetic material (▷ p. 22). It forms a large part of the behaviour of animals that live short lives in relatively stable environments and so have little time for mistakes. But purely instinctive behaviour is unvarying and cannot change in the short term to suit changing conditions.

SEE ALSO

- HOW CLEVER ARE CHIMPAN-ZEES? (BOX) p. 141
- UNDERSTANDING ANIMAL BEHAVIOUR p. 148
- VARIOUS ASPECTS OF ANIMAL BEHAVIOUR pp. 152–73
- ANIMAL PHYSIOLOGY pp. 174–91

Learning occurs when an animal shows a beneficial change of behaviour as a result of experience: squirrels find the most effective way to open nuts, birds learn to avoid unpleasant-tasting insects after one bad experience, rats learn to avoid food that makes them ill (▷ p. 163). Learning involves interaction with the environment and allows animals to modify their behaviour to suit changing or variable conditions. Such a process takes time, however, and requires a sufficient brain capacity to store and retrieve the relevant information. Animals that rely most on

A blue tit drinking milk from a bottle. Milk is not a normal part of a bird's diet – only mammals produce milk – but blue tits and great tits have taken advantage of the milkman's habit of leaving bottles sitting around on doorsteps. Some birds apparently recognize only certain types of bottle as bona fide 'prey'. Such a feeding habit is clearly not genetically determined; it is believed to spread through populations by a process known as cultural or social transmission.

learning generally have a relatively long life span and a period of parental or family care during which the brain can develop and mistakes can be made while the appropriate experience is acquired.

Nature versus nurture

Broadly speaking, up until the mid-1960s students of animal behaviour were divided into two opposed schools. The ethologists, mainly centred in Europe, considered that most behaviour was instinctive and genetically controlled, with little environmental influence. The psychologists, on the other hand, who were mainly American, believed that most behaviour was acquired through an animal interacting with its environment and so learning correct responses. These differences of opinion became enshrined in the so-called 'nature versus nurture' debate. It is now realized, however, that all behaviour has a genetic basis – even the ability to learn – and that animals differ in the extent to which they rely on genetically determined responses and those acquired through interaction with their environment. A good example of the interaction of genetic inheritance and learning is seen in the development of birdsong (▷ p. 153).

Habituation

A snail crawling across a board will retreat into its shell when the board is tapped. If the taps are repeated at regular intervals, the snail will retract less and emerge sooner on each occasion, until eventually it fails to respond altogether. It will, however, respond to a fresh stimulus, such as a bright light, so it is clear that the waning of the response is not due to fatigue. This waning of a response with repeated, unreinforced stimulation is known as *habituation*, and is the simplest form of learning. It enables animals to discriminate between familiar, regularly experienced stimuli and novel ones, and to respond only to the latter. In this way, an animal avoids burdening its sensory and nervous systems with unnecessary information.

Learning by association

Many animals can learn to associate two events that occur close together in time, provided that one of them is motivationally significant to the animal – for example, if it is an item of food or a painful stimulus. There are two types of such association learning – classical conditioning and instrumental (or trial-and-error) learning.

The study of *classical conditioning* was pioneered in the early 20th century by the Russian physiologist Ivan Pavlov. He found that if a hungry dog heard a bell before being presented with food on several occasions, it would eventually come to salivate in response to the bell alone. In classical conditioning a neutral stimulus (the conditional stimulus or CS), such as a bell or a light, precedes a motivationally significant or unconditional stimulus (US), such as food or an electric shock, which on its own elicits some response (the unconditional re-

MEMORY

Memory is the retention of a learned response or a perception over time. It involves learning, storage and retention of information. Mainly through research on humans, three basic types of memory are recognized: immediate, short-term and long-term memory. *Immediate memory* lasts a fraction of a second and represents the input stage of the memory process; it is very difficult to study in non-verbal animals. *Short-term memory* lasts several seconds or minutes and has limited capacity. For example, most people can remember an unfamiliar telephone number for a few minutes after dialling it, but cannot recall it after an hour or so. In animals short-term memory is studied by conditioning experiments (▷ text). Long-term memory is the retention of information for over 24 hours and it can last for many years. The amount of material stored appears to be unlimited, but recollection is not so accurate as in short-term memory, as details are often forgotten.

It is not known how memories are stored. One theory suggests that experiences and sensory information set up persistent electrical activity in so-called 'reverberating circuits' in the brain; another proposes that permanent changes occur in the biochemical processes or structures within special memory cells. Learning certainly involves an increase in the synthesis of nucleic acids (**DNA and RNA**, ▷ p. 22) in the brain, but it also appears to involve growth of nerve dendrites and synaptic connections (▷ p. 182). Animals do not retain all that they learn: memories that are not rehearsed may decay with time, or alternatively the establishment of new memories may obliterate or interfere with the storage and recall of earlier memories.

sponse or UR) – salivation or withdrawal. After several pairings, the CS alone comes to elicit the response, which is now called the conditional response (CR). Within limits, the strength of the CR increases with every additional pairing of the CS and US, but if the US is withheld for several trials, the response wanes through a process called *extinction*. The US therefore reinforces the response and is called a *reinforcer*. Classical conditioning allows an animal to respond in anticipation of some event or situation – the availability of food, perhaps, or the approach of a predator – by responding to something that reliably precedes it, such as the landmarks near a food source, the bright colour of a distasteful caterpillar, or the rustle of grass beneath the feet of an enemy.

Animals may also modify their behaviour as a result of the consequences of an action. A rat in a box fitted with a pedal will come across the pedal accidentally during its normal exploratory behaviour. If on pressing the pedal it receives a food reward, the rat soon learns to associate the pedal with the reward and will press it repeatedly. The motivationally significant event or reinforcer – here the food reward – is contingent upon some action or operation by the animal on its environment. This type of learning is therefore called *instrumental conditioning* or sometimes *trial-and-error learning*. Rats will learn to press a pedal if reinforced only intermittently – every 10 presses, say, or every 15 seconds – so long as reinforcement, whenever it is given, immediately follows a pedal press. As with classical conditioning, the strength of the response – for instance, the rate of pedal pressing – increases with training, and extinction occurs if the reinforcement is consistently withheld. Through instrumental conditioning animals learn to modify their behaviour to improve the outcome of their actions – for example, to learn to catch or manipulate food more effectively, to aim blows at a rival more accurately, or to carry infants more efficiently. In social animals, much of this type of learning occurs during play: kittens pouncing on balls of wool are perfecting their hunting technique, puppies and lion cubs play-fighting are learning the aggressive skills they will need later in life.

Social transmission

In a celebrated captive population of Japanese macaques one individual 'invented' the practice of washing sweet potatoes before eating them; within a few years the habit had been adopted by most of the other members of the population. A pattern of learned behaviour such as this – i.e. one that is found in one particular group within a species but not in others – is known as a *tradition*. Other examples include local dialects in birdsong; the traditional migration routes adopted by ducks, geese and caribou (reindeer); and the habit of opening milk bottles seen in blue tits. In all these cases the behaviour is passed on from one generation to the next through a non-genetic process called *cultural* or *social transmission*.

The behavioural mechanisms by which traditions are established and maintained in a population are difficult to ascertain, but they often do not require the complex learning processes of observation and imitation that are sometimes claimed for them. Young wild rats prefer food taken by their mothers, apparently by choosing flavours that were present in the milk, while adult rats select food taken by burrow mates, even if they have not seen them feeding; they detect the odour on the breath of the 'demonstrator' rat and choose food with that odour. In these cases apparent imitation of one rat's behaviour by another can be accounted for by a simple learning process. In some species animals are attracted to places where others have been feeding, and this probably accounts for the spread of the habit of milk-bottle opening among populations of blue tits. Other forms of social transmission, such as the use of sticks to 'fish' for termites seen in chimpanzees (▷ box, p. 141), probably do involve more complex learning processes, including imitation.　　GS

A sea otter bringing technology to bear on a clam.

An Egyptian vulture using a rock to crack open an ostrich's egg.

In the context of animal behaviour, tools are defined as external objects that are used as a functional extension of the body to achieve a short-term goal. The chewed leaves used by chimpanzees to soak up water are an extension of the hands, while the cactus spines with which woodpecker finches poke out insects from beneath bark are an extension of the beak. The stone picked up by an Egyptian vulture and dropped onto an ostrich egg and the stone used by a sea otter to break open clams and mussels on its chest are also tools by this definition, as they are used as extensions of the bill and paws respectively. However, according to the strict definition, the stone used as an anvil by a thrush to break open a snail shell would not count as a tool.

The use of tools is not necessarily a sign of intelligence. The solitary wasp *Ammophila* uses a small pebble in its mandibles to pound soil into its burrow, while the archerfish dislodges insect prey from twigs above the water of its native swamps by spitting a stream of water at it over considerable distances. Such actions may appear to demonstrate a degree of intelligence or insight, but they are in fact stereotyped, instinctive behaviour patterns that have evolved in a species and are characteristic of all members of the species. Other examples, however, particularly among primates, do indicate more flexible, intelligent behaviour. In the laboratory baboons, for instance, learn by trial and error to use appropriate objects to obtain food that is out of reach. Chimpanzees fashion the twigs that they use to extract termites from termite mounds and learn the technique by observation, imitation, and trial and error (▷ box, p. 141). Most tools are used to obtain food, but elephants and horses have been observed to use twigs to scratch themselves and chimpanzees hurl rocks at rivals.

Animal Communication

The roar of a stag and the faint whiff of a scent, the flash of a brightly coloured wing and the complex speech of humans – these are all means of communication. They are signals produced by one individual, the *signaller*, and they carry information about the state of the signaller to another individual, the *recipient*.

One of the most spectacular signals in the animal kingdom (right): the magnificent branching antlers of a reindeer stag. The size of the antlers and the number of branches increase each year, so indicating both the age of an individual and his competitive fitness.

A signal broadcasts information about the signaller's present state. The meaning of the signal to a recipient depends on the recipient's state and status. To a neighbouring male the territorial song of a bird reveals the presence of a rival, possibly to be challenged. To an unmated female the same song indicates a potential mate, while a mated female ignores it.

The majority of communication signals take place between members of the same species, but this is not invariably so. Warning coloration, for instance (⊳ below and p. 166), is directed at potential predators, while the distinctive coloration and display of cleanerfishes are intended to avoid misunderstandings with their clients (⊳ p. 169).

Functions of communication

A major function of communication is to bring the sexes together for reproduction. It is the means by which the species, sex, reproductive state (i.e. whether in breeding condition or not) and sometimes even the particular individuals concerned are identified. Courtship often involves complex displays in which one or both

BIRDSONG

The elaborate and often beautiful songs of birds are familiar sounds in both town and country, particularly in the spring when breeding begins. They are generally produced by males and serve both to defend a territory and to attract mates (⊳ pp. 94–5). Rival males tend to avoid areas occupied by singing territory-holders. In one study of great tits, males even avoided an area where great tit songs were played through loudspeakers. The deterrent effect was greater if a whole repertoire of songs was relayed rather than a single song; this may be one reason for the evolution of the complex variety of songs that some birds produce.

Another possible explanation for the complexity of birdsong is the behaviour of females. Male sedge warblers that sing complex and varied songs acquire mates more quickly than those singing simpler songs. In such cases the elaborateness of songs has been likened to an 'acoustic peacock's tail' developed to attract females (⊳ box, p. 159).

Songs are specific to a species and are used by birds to recognize members of their own species, but there may also be variation in

the form of local 'dialects' and even between individuals. A basic component of the song appears to be innate, but individuals have to learn details and the local dialect. Just after fledging, young chaffinches produce a rambling 'subsong', which is not very like the adult song, but the following spring, when the males first adopt a territory, this develops into a more species-like 'plastic song'. Gradually the full song develops. If chaffinches are removed from the nest at hatching and kept in isolation, when adult they can produce only a very abnormal song, which probably represents the innate component. If the birds are isolated in their first autumn, they can produce a fairly good but not perfect song the following spring. It appears that young birds have to learn to modify the innate component into the species-typical song during their first year, and that they perfect this by singing early in their second spring. It is also essential that they can hear themselves sing: birds deafened before they have developed the full song are unable to sing properly. The development of birdsong therefore involves an interaction of learning and innate behaviour (⊳ p. 150).

A bee-eater in full song.

partners posture and perhaps call to each other. Such displays enable partners to learn to recognize each other and to assess each other's suitability as mates (⊳ p. 156).

Communication is also important in spacing out animals, in marking territorial boundaries, and in establishing a position within a social hierarchy. Size or pitch of voice may indicate competitive status, as may the possession of weapons such as antlers or tusks. Among social animals, including many primates and dogs, positions of submission and dominance within a hierarchy are often signalled by particular body postures or facial expressions (⊳ box, p. 154).

Defence against attack by rivals or predators frequently involves intense or sudden signals. Mammals may let out loud roars or screams when threatened. Rabbits, deer and pronghorn use tail flashes to warn others of impending danger, while birds use various kinds of alarm call. Vervet monkeys use different alarm calls to indicate different predators, such as snakes, eagles and leopards (⊳ p. 139). Although such alarm signals often have features that reduce the likelihood of detection by a predator (birds and mammals, for instance, generally use low-pitched calls that are hard to locate), the signaller nevertheless places itself in some jeopardy; such apparently 'altruistic' behaviour has been explained, in evolutionary terms, by the process of kin selection (⊳ box, p. 149).

Unpleasant-tasting or poisonous animals such as wasps and snakes often indicate their unpalatability by adopting bright warning coloration (⊳ p. 166). Sometimes such noxious species mimic one another (many different species of bee and wasp, for instance, have distinctive black and yellow coloration), so that if a predator takes one species, it is likely to avoid all the others. At the same time many harmless species such as hoverflies 'cheat' predators by adopting warning coloration.

Information on food sources may be communicated through transmission of the odour or taste of the food from a successful forager to other members of the group. This method is used by both bees and rats. Bees also transmit information about food sources through complex dances (⊳ box).

The form of communication signal differs depending on the information to be conveyed, the distance over which it has to travel, and the habitat of the animals concerned.

Chemical signals

Chemical signals depend on the sense of smell and sometimes on the sense of taste (⊳ p. 188). Such signals can travel over long distances if carried on wind currents, but can only be perceived downwind. They may be used for long-term signalling, as when mammals such as hyenas and deer mark their territory, while the food trails left by ants are a form of short-term signalling.

A *pheromone* is a chemical signal or 'messenger' released by an organism that has

THE DANCE OF THE HONEYBEE

One of the most astonishing feats of communication in the whole of the animal kingdom is the 'dance language' of honeybees. When a worker bee finds a rich supply of food, she communicates a variety of information about it to her fellow workers by performing a dance inside the dark hive. If the food is within about 50 m (164 ft) of the hive, she performs a *round dance* – she dances in a circle, first one way, then the other. Other workers follow the dance, keeping in touch with the dancer by means of their antennae. They pick up the odour of the flowers she has visited by her scent and their taste from small amounts of food that she regurgitates. The recruits then fly out of the hive and search for the right smell within a 50-metre radius.

If the food is more distant, the returning worker performs a *waggle dance*, in the form of a figure-of-eight with a straightened central bar. During the straight part of the dance, she waggles her abdomen to and fro, producing bursts of sound. The further away from the hive the food is, the slower the rate at which the bee dances; the richer the food supply, the more intense the waggle and the sound produced. The angle at which the straight part of the dance is performed indicates the direction of the food source. The top of the comb is equivalent to the position of the sun outside, and the dance is performed at the same angle to this vertical position as the food source is to the sun. This symbolic dance can therefore code for the direction, distance and abundance of the food source, and – as in the round dance – odour cues indicate its nature.

The significance of the honeybee dances was discovered by the Austrian scientist Karl von Frisch. Putting out a dish of scented sugar water and marking the bees that came to it, he watched them as they returned to his specially designed observation hive. At first he thought that the dances merely served to pass on the odour of the food source to recruits. But then he set out other dishes of food either in a straight line with the first but at different distances from the hive, or at the same distance from the hive but in a fan-shaped array around it. New recruits went to the position of the original dish rather than to the nearest one, showing that they knew its exact location. In conducting his experiments he also discovered that honeybees can navigate even on a cloudy day, when the sun is obscured, by using polarized light. At first von Frisch's results were disputed and his methods criticized, but recently more rigorous experiments have confirmed his theory.

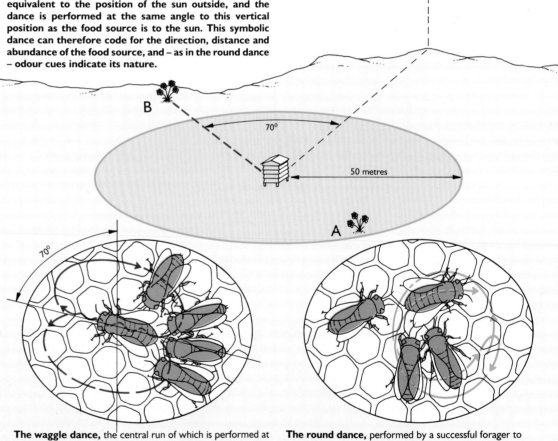

The waggle dance, the central run of which is performed at an angle of 70° to the top of the comb, to indicate that a more distant food source [B] is to be found by flying from the hive at this angle relative to the position of the sun.

The round dance, performed by a successful forager to indicate that a food source [A] is nearby (within a radius of about 50 m/164 ft), but not specifying its direction.

a specific effect (behavioural or physiological) on another member of the same species. Such signals are widely used throughout the animal kingdom. In social insects such as ants, termites and bees, many or most of the functions of the colony are mediated by means of pheromones. In termite colonies, for instance, the queen produces a pheromone that is thought to prevent the ovaries of other

females from developing (⊳ pp. 63 and 156), while an alarm pheromone produced by disturbed honeybees attracts and excites other workers. Amongst mammals pheromones are often important in reproduction. A wild boar produces a pheromone that induces the female to adopt the mating posture. A pheromone from male mice appears to stimulate early sexual development in young females, while a different pheromone from mature females suppresses such development. Pheromones are also used to mark trails, to indicate territorial boundaries, and to attract sexual partners (▷ pp. 188–9).

Acoustic signals

The main advantage of acoustic signals is that they can change very rapidly in pitch and intensity, and so can be used to convey a wide range of information. They also travel very fast and in all directions from the signaller, and are easy for the recipient to locate. The intensity of the signal – and hence the distance it travels – is enhanced by some ingenious devices. Mole crickets construct a horn-shaped burrow that increases the sound produced by the cricket some twenty times. Primates such as howler monkeys, and some birds, frogs and toads, have large vocal sacs that amplify the sound made by the animal. Some tree crickets cut holes in leaves and place themselves in the holes while singing; the wing vibrations that produce their 'songs' are transmitted to the larger surface area of the leaf, and are thus broadcast over a wider area.

In birds short, relatively simple acoustic signals are known as 'calls', while longer and more complex sounds are referred to as songs (▷ box, p. 152). The noises made by crickets, cicadas and some fishes are sometimes called songs, but they lack the complexity of birdsong. Humpback whales produce songs that generally last 10 minutes, although songs up to 30 minutes long have been recorded; they probably help to keep groups of whales together and to aid cooperation between them. Other whales and dolphins produce a variety of squeaks and whistles (▷ pp. 144–5). Because low sound frequencies travel further than higher ones, large animals such as whales and elephants use very low frequencies – below the human hearing range – to communicate over long distances. Sound travels especially well in water, and the songs of baleen whales can travel over hundreds of kilometres.

Visual and electrical signals

Visual signals are used by many different animals. They can be turned on and off very rapidly, but can generally only be used in daytime and are easily blocked by objects such as trees. Such signals are often bright or consist of jerky movements to make them more conspicuous. One of the claws of the male fiddler crab is enlarged and brightly coloured, and is waved in a characteristic way to attract females. The bright colours and patterns of butterfly wings and of many male birds attract mates over short distances, while bowerbirds build special mating areas which they adorn with brightly coloured objects to attract females (▷ pp. 94–5).

Some animals overcome the main limitation of visual communication by manufacturing their own light. As they fly at night, male fireflies produce flashes of light in characteristic patterns to which the females on the ground respond with their own flashes (▷ p. 62). Females of a different species mimic these responding flashes, and then eat the unsuspecting males as they fly down to mate with them. Many fishes and other marine animals living at depths to which sunlight cannot penetrate have evolved a bizarre array of luminous organs for a variety of purposes (▷ box, p. 221).

Electrical signals are used by some fishes that live in muddy rivers in South America and Africa. These signals can pass through solid objects and are used aggressively and in courtship, as well as

BODY LANGUAGE

The position of the body and the general bearing of birds and mammals can be used to convey information about their state and status. High-ranking or aggressive animals often carry the body erect, with ears up in dogs and wolves, and head held high in gulls. High-ranking rhesus monkeys may carry the tail in an S-shape over the back, and high-ranking wolves hold the tail away from the body. Low-ranking or submissive animals often adopt the opposite body posture. Submissive gulls hold the body low, with the wings retracted and the head withdrawn, while subordinate rats lie on their back exposing the chest and abdomen. A similar position is adopted by subordinate and juvenile wolves, or they may simply turn the head away from the dominant animal, keep the body, head and ears low, and hold the tail between the legs.

Mammals have the ability to move their eyebrows, lips, and cheeks, while some, such as dogs, cats, horses and other hoofed mammals, can move their ears. The differing positions of these various features allow a rich variety of expressions that can be used to convey aggression, appeasement, friendliness, fear and (in primates) anger and joy.

The ears are used in communication by elephants, horses and zebras, cats, and dogs. Aggression in horses and zebras is indicated by flattening the ears and opening the mouth, but dogs and cats signify this by raising their ears and opening the mouth with bared teeth. Flattened ears in cats and dogs indicates submission. Friendliness in all these animals is associated with erect ears, but more intense greeting is indicated by an open mouth in horses and a relaxed mouth in dogs and wolves. In elephants the ears are held flat against the body in relaxed animals, with the trunk curled up and back to indicate a defensive mood. Elephants indicate offensive moods with ears held away from the body and the trunk held out straight.

With their extraordinarily mobile faces, primates can produce a wide variety of expressions in different situations. Some are thought to be used in communication, although the function of others is less clear. Higher primates show the 'staring open-mouth face' when threatening another individual: the eyes are opened wide and the mouth is open, but the lips cover the teeth. A pouting expression is associated with begging in infant primates. A relaxed face with an open mouth indicates a playful mood in baboons and chimps, and may be related to laughing in humans.

Baring of the teeth occurs in several primate expressions associated with submission, distress or flight. In the 'silent bared-teeth face' or 'fear grin' seen in most primates, the corners of the mouth are drawn back to reveal the teeth. This expression is assumed by animals when submissive or when indicating appeasement, but it is also a sign of friendliness in chimpanzees and is reminiscent of the smile in humans. A similar expression is given in response to unpleasant tastes or smells in both primates and infant humans. It has been suggested that this expression originated as a means of ridding the mouth of unwanted food and became the fear grin of monkeys and apes, perhaps initially as a reaction to the 'fearful' smell of a dominant individual. It may then have evolved as a signal of submission and then of appeasement, and in baboons and chimpanzees it is used as a way of reassuring a subordinate and so promoting friendly contact. It is suggestive that in humans the smile can also be used for appeasement and for friendly greeting.

Dominant and submissive postures in wolves.

for purposes of navigation (▷ p. 171). The patterns of electrical discharge may be specific not only to a species but also to a particular sex, and they vary in different social situations.

Touch and vibration signals

Tactile signals can only be used close at hand. They are particularly important among primates as indications of friendship and appeasement (▷ box, p. 139). Grooming of one individual by another not only helps to remove skin parasites but serves to strengthen bonds between family members and partners. When rats huddle close together, they are probably helping to reinforce social bonds as well as keeping one another warm. Titi monkeys often sit close together with their tails entwined.

Vibration signals are effective over short distances. Male orb-web spiders indicate their presence to females by vibrating their webs in a characteristic way, thus dissuading females from treating them as prey (▷ pp. 68–9). Male water striders make ripples on the water surface that are detected by the antennae of recipients – rival males and potential mates. Mole rats bang their heads against the roofs of their underground tunnels to communicate with both rivals and mates – a method also employed by wood-boring insects, such as deathwatch beetles.

Do animals have language?

To qualify as language, a communication system must be symbolic and must be able to denote abstract ideas, as well as things and events that are distant in time and place. It used to be thought that language was an exclusively human attribute, but some animal communication systems show features in common with language. A well-known example is the 'dance language' of honeybees, which is symbolic and codes for a variety of information, allowing a successful forager to convey to her fellow workers the direction, distance and abundance of a food source (▷ box, p. 153).

Another example of possible language in animals is the controversial attempts to teach chimpanzees and gorillas to communicate with humans using speech or symbols. All attempts to teach chimps to talk have failed because they do not have the necessary vocal apparatus to produce sounds as we do, but attempts to teach them to use sign language have been more successful. A young female chimp called Washoe was taught well over a hundred symbols of the American Sign Language used by deaf people. She learned to use the symbols in the appropriate contexts and to string them together to indicate, for example, 'gimme sweet' or 'come open'. Another chimp, Sarah, was taught to use coloured plastic symbols to represent words, and could answer questions with the appropriate symbols. These experiments show that chimps can certainly learn to use abstract symbols to indicate objects and actions, but how far this – or the dance language of bees – can be equated with human language is an area of continuing debate.   GS

THE EVOLUTION OF COMMUNICATION

The rushing display of the western grebe. The various species of grebe show an astonishing array of ritualized mating displays. Often the non-communicatory origin of a display is obvious enough; in other cases – including the rushing display shown here – their origins are a mystery.

Many of the displays used in communication are similar to behaviour patterns that normally serve other purposes. During courtship male ducks of several species point the bill at the wing or draw it over with movements that are similar to those used in preening, while courting great crested grebes present nesting material to their mates. The process by which various forms of communicatory behaviour have evolved from other, often more mundane types of behaviour is known as *ritualization*. This often involves exaggeration of the movement or the development of bright colours or special anatomical features to enchance the effect of the signal. The peacock's tail is a prime example of such exaggeration and adornment.

Often the non-communicatory origins (if any) of ritualized communication signals are unknown, but in some cases we can guess at their affinities through comparative studies of similar behaviour in a number of related species. For example, in the pheasant family the male bobwhite quail picks up and offers food to his partner before mating; the domestic cock pecks at the ground without picking up food; and the impeyan pheasant bows before the female and pecks at the ground as he raises and spreads his enormous tail feathers. This impressive display may therefore have its origins in feeding behaviour.

Many different kinds of non-signal behaviour appear to have become ritualized. Various aspects of the autonomic (involuntary) nervous system, including urination, thermoregulation (temperature con-

trol) and respiration, seem to have provided a rich source of material for ritualization. Feather-raising in birds and hair-raising in mammals control heat loss, but they are also used – especially in association with exaggerated crests or hackles – to signal aggression. Exaggerated breathing movements are seen in the extending of gill covers during aggression and courtship in Siamese fighting fish. Incomplete movements, or *intention movements*, occur when an animal fails to complete a behaviour pattern. The baring of teeth in dogs is the first stage in biting, but it also signals aggression when there is exaggerated retraction of the lips. When an animal is motivated to perform two (or more) different actions simultaneously, it is said to be in 'motivational conflict' – it may alternate between the two, showing elements of both in an intermediate form of behaviour, or it may perform apparently irrelevant actions called *displacement behaviour*. In many cases such 'conflict behaviour' appears to have become ritualized. The zig-zag courtship dance of the male stickleback alternates between approaching and avoiding the female; the threat posture of cats shows elements of aggression (raised fur and body held erect) but also elements of fear (ears held low and back arched); during courtship in zebra finches the male appears to be torn between mating, aggression and fleeing, and often he makes apparently irrelevant movements similar to the bill-wiping seen during feeding. In all these cases the ritualized behaviour is performed in a stereotyped way, making its signal function clear to the recipient.

SEE ALSO

● UNDERSTANDING ANIMAL
   BEHAVIOUR p. 148
● TERRITORY, MATING, SOCIAL
   ORGANIZATION p. 156
● DEFENCE, DISGUISE AND
   DECEIT p. 166
● SENSES AND PERCEPTION
   p. 188

Territory, Mating, Social Organization

A shoal or school of fishes is essentially a defensive association that reduces the impact of predation. Given the limited range of visibility in water, the individuals in a closely clustered group are less likely to be detected by a predator than if the same number of individuals were evenly scattered over a larger area. A predator that does detect a shoal may be confused by the apparent size of the object it encounters, and may have difficulty in focusing its attention on a particular individual. If a predator attacks, the individuals within a shoal may split into two groups on either side of the assailant and rejoin once it has swum through, or they may dart suddenly in all directions from the centre of the shoal in a so-called 'flash expansion'.

Animals interact with other members of their own species in a wide variety of ways. Some animals, such as moles, hamsters and tigers, are largely solitary in their behaviour, only coming together to mate. Other animals form groups of varying size and structure: shoals of fish are essentially loose aggregations of individuals, while in primate groups, for instance, roles of social dominance and subordination can be recognized and there is often division of labour.

Whatever form of social organization they adopt, all animals require an adequate supply of various essential resources, such as shelter, food and nesting materials. In order to secure these in sufficient quantities, it may be necessary for one or more animals to keep a particular area for their sole use and to keep out other members of the same species. An area in which vital resources are defended in this way is known as a *territory*.

Territory

Territories may be held by any number of animals: hamsters and mice, for instance, hold their territories singly; gulls in pairs; hyenas, lions and prairie dogs in larger groups. While many animals, such as robins and badgers, hold large multi-purpose territories for feeding, mating and rearing of young, other territories are used for one purpose only. Male sage

---

TERMITES

Termites live in colonies (▷ illustration, p. 160), and like many bees, wasps and ants, they are 'eusocial' insects, with castes specialized for particular tasks within the colony. Termite colonies differ from those of bees in that the workers are of either sex and there is a permanent reproductive male (the king), as well as a queen.

A new termite colony is founded by a winged male and female (the king and the queen, or the *primary reproductives*). These meet during a 'nuptial' flight and then construct an initial nest cell in which they mate after shedding their wings. The royal couple care for the young of the first brood, and because there are no workers to help, these first offspring are rather emaciated. Nevertheless, they develop into workers and then take over the work of the colony: they construct the nest, gather the particles of wood that form the food of most species, nurse future broods, and feed the royal couple. The workers are sterile males or females; they have no wings, and their eyes are reduced or absent. The youngest workers are thought to nurse the brood, producing food for the larvae from large salivary glands. Older workers appear to forage and lay odour trails to suitable food sources.

Defence of the colony is the responsibility of members of the *soldier caste*. These appear to be sterile, wingless forms specialized solely for defence. They are males in some species, females in others. The particular form of the soldiers also differs from species to species. They may have enormous cutting mandibles for biting intruders, pointed mandibles for piercing, or thick, cylindrical heads that serve as plugs to stop up the galleries of the nest. In some soldiers the salivary glands produce chemicals that irritate, poison or ensnare intruders.

As the colony develops, the king and queen do little other than feed and reproduce. The abdomen of the queen becomes enormously swollen, and she and the king become restricted to a royal cell near the centre of the colony. After a period of time – perhaps many years – some larvae develop into *nymphs* with wing buds. These moult and become winged reproductive adults with functional eyes. Leaving the parent colony, they fly out to obtain mates and found new colonies.

In termites the basis of caste determination is not clear. In many species the removal of the king and queen from the colony results in the appearance of replacement reproductive forms, so sexual development must normally be inhibited in some way. This inhibition is probably due to pheromones (▷ p. 153), but this is not known for certain.

The royal couple: the queen, with enormously bloated abdomen, side by side with the king. Also in attendance are a number of workers and soldiers (the latter can be distinguished by their large heads).

grouse and Uganda kob, for example, gather in traditional mating grounds called *leks*, where each male defends a small mating territory. Females are attracted to the leks and prefer to mate with males holding particular territories, often those in the centre of the lekking area. Colonial nesting gulls, which have the abundant food resources of the sea at their disposal, defend an area around the nest solely for the purpose of bringing up their young.

Many animals patrol the boundaries of their territories regularly and attack any intruder, using various signals to indicate their possession of a territory. Many mammals, including dogs, badgers, hyenas, rhinos and deer, use dung, urine or scent to mark their territorial boundaries. Other methods include birdsong (▷ box, p. 152) and the boundary displays performed by male sticklebacks.

A territory is only worth defending if the benefits – in terms of food obtained or young successfully reared – outweigh the cost in time, energy and potential injury. The golden-winged sunbird, for instance, defends a territory containing its food source, nectar from *Leonotis* flowers. The energy it obtains from the flowers just exceeds that used in feeding and in defending its territory, so the bird makes a small net gain in energy.

Mating systems

The various activities associated with reproduction generally represent a greater investment in time and energy for females than for males, so females tend to be choosy about potential mates (▷ box, p. 159). As a result males usually have to compete among themselves for the attentions of females. *Courtship* is the means by which males advertise themselves and their intentions and by which females assess potential mates. Males are judged on their prowess in competing successfully against other males. The criterion of 'success' may sometimes appear somewhat whimsical – the magnificence of a bird's plumage, for instance, or the complexity of his song – but often it has a more practical foundation. If a female chooses a male who is capable of defending a large territory, she will probably have good resources of food with which to rear her young.

Whether a male mates with one or more females, or vice versa, often depends on the type of parental care the male provides (▷ p. 185). *Polygyny* is the mating system in which a male mates with more than one female. In polygynous species, males are often highly territorial (or become so during the breeding season), fighting and competing with one another to defend harems of several females (▷ below). After mating, males frequently show no parental care: in many mammals, such as red deer and elephant seals, and in lekking birds, the males mate with many females, which are then left to care for the young alone. Some males – red grouse and yellow-bellied marmots, for

— **A G G R E S S I O N** —

Male bighorn sheep fight hard to win high social ranking – the thundering clashes of their horns can be heard over several kilometres – but they also fight fair. They rush at each other head to head, so most of the jarring impact is absorbed by the specially thickened skull. In this show of 'limited aggression', potentially damaging blows to the flanks are generally avoided.

Animals compete for access to vital resources such as food, shelter and mates through *aggression* – behaviour that harms or threatens to harm an opponent. The form of aggression differs from species to species. Some have special weapons, such as the antlers of deer and the horns of cattle and beetles. Others, such as zebras and rats, use their feet to kick or wrestle. Certain corals grow over their opponents, while sea anemones sting each other with special stinging cells (▷ p. 53).

At first opponents generally threaten each other from a distance using ritualized signals: howler monkeys, stags and crickets call or bellow, fiddler crabs wave their claws, and some turtles and lizards bob their heads. If this fails to deter one or other of the opponents, they may theaten at closer range: deer walk parallel to each other, while Siamese fighting fish circle tail-to-head, lashing water at each other with their tails. Pushing contests are seen in many animals, including fishes, snakes, rhinos and deer. Usually one animal gives up before they come to blows and signals its acceptance of defeat by fleeing or showing submissive behaviour (▷ box, p. 154).

The winner may gain possession of a large territory with food and a nesting site, or a small mating area in which to attract females. In social groupings aggression may result in the formation of a *hierarchy* – an order of precedence for access to food and mates. Gregarious birds such as chickens form a linear hierarchy known as a *pecking order* in which animal A is dominant to animals B, C and D; B is dominant to C and D, and so on. Among primates hierarchies are not so rigid. In baboons several males may hold top rank, and lower-ranking males acting together may dominate one that would outrank each on its own (▷ box, p. 139).

While some fights end in injury or even death, most are settled by conventional displays. Such 'limited aggression' has long been a puzzle to scientists, since aggression would seem to be selected for in evolution. To explain how such limited aggression might evolve, games theory (▷ p. 148) has been used to formulate the 'hawk–dove' model. Suppose that all the animals in a population compete using conventional displays only – the 'dove strategy': at any sign of conflict rivals retreat and avoid injury. In such a population a genetic mutation that produced a 'hawk' – an individual that invariably fought dangerously – would result in hawks multiplying in subsequent generations, since such an individual would inevitably win all its aggressive encounters. The dove strategy is therefore not an evolutionarily stable strategy. However, as the proportion of hawks in the population rises, hawks will meet other hawks more often, and the benefits of winning will soon be outweighed by the cost in injury or death. In such circumstances doves, which flee and thus avoid injury, will fare better: they will then tend to increase in the population and hawks will again become rare. The hawk strategy is therefore not evolutionarily stable either. Both limited aggression and all-out aggression can therefore be seen to benefit individuals in certain circumstances, and so each will be selected for in evolution when those circumstances exist. In this situation the population will tend towards a point of equilibrium at which there is a more-or-less fixed proportion of hawks to doves, and this state of affairs will be evolutionarily stable. The balance between hawks and doves may be maintained by different animals consistently using different strategies or by animals changing strategy – for instance, playing dove when young and becoming hawkish when older.

instance – mate with a few females and defend territories in which the females raise the young. In some cases, where eggs are fertilized externally, for example in sticklebacks and midwife toads, the male may mate with several females and then care for the eggs himself.

Where the male is needed to help feed the young, as is the case with many birds (⊳ pp. 94–5), the common mating system is *monogamy*, where one male mates with a single female. The pairing may be for life as in swans, geese and eagles, or for a single season as in many garden birds. Monogamy also occurs in mammals such as wolves, since a lactating female cannot hunt and the male is required to bring back food for the female. Another example of monogamy in mammals is the marmoset. In this case the male helps to carry the twin young, so that the small female can preserve her energy for feeding.

A few birds have a mating system called *polyandry*, in which a female mates with more than one male, with each male looking after his own offspring. This may occur in birds such as the jaçana because the risk of predation is high. Alternatively, it may take place where abundance of food allows the female to lay several clutches of eggs in a season, as is the case with the spotted sandpiper, which can lay as many as five clutches in 40 days.

Competition between males for access to females may continue after mating. Male damselflies have a special appendage to remove the sperm of a previous mating from the female before he deposits his own. Some insects and rodents produce a copulatory plug that reduces the chance of other males mating effectively, while in fruit flies and some butterflies the male places an anti-aphrodisiac substance on the females that inhibits mating by other males.

NAKED MOLE RATS

The naked mole rats are unique among mammals in showing a social organization akin to that of social insects such as some bees and wasps, ants and termites. These rodents are found in the hot, arid regions of Kenya, Ethiopia and Somalia, and spend all of their lives underground. They are 3–9 cm (1⅕–3½ in) long and are hairless, apart from a few sparse hairs and whiskers. Their eyes and external ears are small, but they have very long incisor teeth that protrude in front of the lips and can be used for digging without earth getting into the mouth.

Up to 300 of these rodents live in a colony in which a caste system operates. The smallest and most numerous animals in a colony are the 'frequent workers', which do most of the work: they dig foraging tunnels at the level of the large plant tubers on which the colony feeds; they build the deeper nest area where the colony members rest together; they dig out the latrine site used by all members for urination and defecation; and they take food to colony members in the communal nest and help feed the young. Members of this caste may be of either sex, but they never breed.

A few of the small workers grow and join the next caste, the 'infrequent workers'. These assist in nest-building and foraging, but at half the rate of the frequent workers. However, they have an important role in de-fending the colony against attack. They too do not breed, but they are able to do so if the breeding animals are removed. The largest animals – the 'non-workers' – do little or no work apart from caring for the young. They spend most of their time asleep in the nest chamber near the queen, but they will rush out to guard the colony if it is invaded and are sometimes called 'soldiers'. The female non-workers are apparently unable to breed at all, but one or two of the males mate with the one breeding female, the queen. The latter does little work apart from suckling the young in the nest chamber, but every few hours she patrols the colony, sniffing or prodding members that she meets.

The queen reigns supreme for many years. She appears to suppress the reproductive capacity of the workers through her behaviour towards them and through odours in her urine; when other animals use the latrine, they smear the contents over their bodies. If the queen is removed, other females become reproductive. In these animals the benefits of living in a social group – communal food-gathering and nest-building, and the opportunity to help a relative to reproduce (⊳ box, p. 149) – appear to outweigh the costs of foregoing reproduction and living a life of servitude. A similar system of 'social contraception' is seen in social insects and some other animals, including marmosets.

Social organization

Living in social groups has many advantages. Important among these is that some degree of cooperation is possible – in finding and collecting food (⊳ p. 165), in looking out for danger, and in defence against predators (⊳ p. 167).

One of the simplest types of social group is the *family*, where both parents remain with the young for a given period of time. Examples of this kind of family are the groups formed by jackals, swans and many songbirds. Some birds – mallard ducks, for instance – and many mammals, such as bears and rodents, form *one-parent families*, where the females alone show parental care. In the case of some fishes and amphibians, it is the male that guards the young.

Extended families are formed when the offspring remain and help to rear succeeding generations. In Florida scrub jays, older siblings often act as 'helpers', assisting their parents in feeding and looking after the next brood. Social insects – some bees and wasps, ants, and termites (⊳ box) – live in large groups and show extreme division of labour, with only one individual, the queen, reproducing. Most members of a colony are sterile and do not reproduce; instead they devote themselves to foraging and to building, maintaining and defending the colony (⊳ p. 64). The naked mole rat – a species of rodent – is unique among mammals in exhibiting a type of social organization similar in many respects to that of social insects (⊳ box, p. 158).

Harems are composed of a single breeding male and a group of females, each with her own young. Permanent harems are formed by animals such as zebras and patas monkeys, whereas red deer, walruses and sea lions, for instance, form temporary harems that stay together only for the duration of the breeding season. For the rest of the year the females and their young generally form female groups. The males either live in *bachelor herds* – as is the case with red deer – or they may live alone.

Multi-male groups are formed by several animals, including common or savannah baboons, hyenas and lions. A number of males and females breed, although there may be a hierarchy in both sexes to determine which animals breed most often and when they will do so. At adolescence male baboons frequently leave their home groups and join another group (▷ box, p. 139). A male lion may abandon its own group to take over another group of females by force; often the male will then kill the young cubs in the pride, which causes the females to come on heat and so allows the new pride-holder to mate sooner than would otherwise have been possible (▷ box, p. 165).

The social organization of a species is not rigidly fixed. As with red deer and elephant seals, it often changes with the season. Many garden birds live in mono-

Walruses assembled at a traditional 'rookery' on Round Island, Alaska. While a male bird (for instance) is just as capable as a female of feeding the young, a male mammal does not have the necessary apparatus – mammary glands. It is probably for this reason that monogamy is much more common among birds, while many mammals (including the walrus) are polygynous – males mate with a number of females and frequently leave the females to look after the young alone. In polygynous species males have to compete fiercely for access to females, so marked sexual dimorphism often arises: female walruses, for instance, are considerably smaller than males and have shorter, thinner tusks.

gamous pairs only during the summer; in winter, species such as chaffinches and starlings live in large flocks, while robins, for instance, live alone. Female robins are indeed seldom seen during the winter – their whereabouts is a mystery.   GS

SEXUAL DIMORPHISM

In many animals males are larger, more aggressive and more striking in appearance than females: male elephant seals are bigger and heavier, peacocks have more colourful plumage, male stag beetles have enormous 'antlers' (▷ box, p. 63). This phenomenon, known as *sexual dimorphism*, is related to the different amounts of time and energy that each sex puts into reproducing.

Females tend to invest more time and energy in reproduction than males. In comparison to sperm, eggs (ova) take more time to produce and are more costly in terms of energy, and they are produced in smaller numbers (▷ p. 184). Females may provide yolk or milk to nourish their developing young, and may protect them by means of egg cases or shells, or by carrying them within their body. If they make a poor mating and lose their brood, they must invest more to replace it than a male. Females therefore tend to be choosy about their mates, and males have to take trouble to attract them and to demonstrate their suitability as mates through bright adornments, songs and displays. Females appear to choose males that are 'superior' in some way – those with the longest tails, perhaps, or the most complex songs, or the most acrobatic displays – and these characteristics then get passed on to their male offspring. In this way selection for sexual attractiveness or *sexual selection* leads to adaptations in males such as the large tail of peacocks, the intricate songs of reed warblers, and the aerial displays of courting eagles.

While the eggs of a female can all be fertilized by a single mating, males, with their capacity to produce vast amounts of sperm fairly rapidly, can mate with many females. The most successful males will be those mating with as many females as possible, but not all males can do this, and in some species males (but not females) vary enormously in their reproductive success. In elephant seals the top-ranking male mates with about 94% of the females in the colony; many other males may not mate at all. Males therefore compete among themselves to gain as many females as possible, and this competition has led to increasing size in males and to the possession of weapons such as horns, antlers and tusks.

Not all species show such dimorphism, however. In some the males are needed to care for the young: many male birds, for instance, are required to feed the nestlings, while small primates are often needed to carry the young. In these cases males also invest heavily in reproduction. Such species typically mate monogamously rather than polygynously. More males are able to mate, and competition and hence sexual dimorphism is less extreme. In a few species, such as the jaçanas (a family of birds), it is the males that care for the young and females mate with more than one male. In these cases females are larger and more aggressive than males and more active in courtship.

Animal Builders

Throughout the natural world we find examples of animals that create building works, from the protective coat of sand grains constructed by the single-celled amoeba, to the simple but elegant branch-sprung beds of chimpanzees and gorillas. While most animals build in order to gain shelter and protection, others do so with a more sinister motive – to win a meal by ensnaring their animal prey.

As a species, we doubtless consider ourselves unrivalled as builders, but there are many examples in the animal kingdom that should give us pause to think. For instance, if a worker termite were to grow to the size of a human and to scale up its buildings in the same proportion, the biggest mound would rise to a towering 1800 m (5900 ft) – nearly three times as tall as man's best effort to date.

Mineral structures

Minerals such as silica and calcium carbonate, which occur naturally in great abundance, are used to atonishing effect by invertebrate animals. Although not strictly buildings – what we see are in fact the external skeletons of the animals concerned – these mineral structures are spectacular nevertheless, ranging from the microscopic to the largest living structures on earth – the coral barrier reefs of Australia and Belize (▷ p. 215). Recent research suggests that in the late Jurassic period, some 140 million years ago, sponges (which have silica-based skeletons) formed vast masses at greater depths than corals, often 100–400 m (330–1300 ft) below the surface. Now fossilized, one of these giant reefs covers a great swathe from beyond the Black Sea north of Turkey westwards into Spain.

At the other end of the scale, the single-celled marine foraminifers absorb calcium to form their tiny limestone shells, adding on extra compartments as they grow. The warm-water radiolarians, also single-celled, use silica to similar effect, producing spiky skeletons in fantastic symmetry that give both buoyancy and protection.

Building with silk

Some animals use bodily secretions as building materials. The most versatile of these are silks, protein-based substances that dry on expulsion from the body to produce delicate threads that are in many respects stronger and more resilient than man-made fibres. Spiders use a range of silks with astonishing versatility (▷ p. 68). Most familiar are the webs built to trap insects, but there are other, lesser-known uses, one of the more exotic being the 'diving bell' constructed by a species of water spider. This spider weaves a mat of silk threads anchored to waterweed and fills it with air bubbles, which it then takes down, trapped between its abdomen and legs. It can stay under water for a day, darting out to catch small crustaceans and aquatic insects.

Insect silk is produced from the salivary glands of grubs, larvae and caterpillars. It is spat out as a continuous thread, so that if it is to be used to construct a nest or cocoon, hours of head movements are required to build up the structure. Each silkworm incorporates three or four kilometres of fine thread into its cocoon (▷ p. 64). Many other moth larvae use silk, often as a casing on which they put tiny stones, pieces of wood and other debris to provide camouflage. The aquatic larvae of caddis flies make similar casings. One of these builds a spiral sheath coated in sand, another decorates its tube with tiny snailshells, while yet another makes a trapping net in fast-flowing water, anchored to sticks, and lurks in the base to catch food as it is funnelled in.

The tailor ant of tropical Asia lives in trees, in nests of leaves held together by silk. In constructing the nest, some workers pull together the edges of adjacent leaves, while others use ant larvae as shuttles, squeezing them so that they secrete sticky silk and moving them crisscross from leaf edge to leaf edge.

Natural adhesives

The adhesive power of saliva and other sticky secretions called mucins are exploited by some animals. The male of the common stickleback of Europe gathers plant materials, anoints them with mucus produced from his kidneys until the material is stuck together, and then bores a hole through the clump just large enough for a female laden with eggs. A suitable female is tempted into the nest and prodded until she lays her eggs, at which point the male sheds his sperm to fertilize them.

Tree frogs have a problem in finding enough water in which to lay their eggs. An ingenious solution is seen in the Javanese flying frog, which whips up a froth from the mucus surrounding the eggs by rapid kicking movements of the hind legs. Nearby leaves are pressed onto the ball of foam created in this way, so forming a sling in which the fertilized eggs and then the developing tadpoles are protected. The central part of the foam mass dissolves away, leaving an artificial

---

TERMITE MOUNDS

Termites are the true masterbuilders of the insect world. The enormous mounds of some of the more advanced species form towering pinnacles in tropical and subtropical landscapes. Amongst the most imposing of all are the nests of the African *Macrotermes* species: in extreme cases, these may measure over 8 m (26 ft) in height, and even an average colony may number around two million individuals.

The walls of the mound of *Macrotermes bellicosus* are made of soil particles mined from below by the tiny worker termites and mixed with saliva to form a hard, brick-like substance, resistant to all but the most persistent assailant. The nest proper, lying in the central part of the mound, consists of a maze of passages giving access to innumerable chambers of various kinds, each with a special function in the running of the colony: brood chambers where the larvae hatch and are nourished; special 'gardens' where fungi are cultivated on combs of wood particles to provide food for the colony; and a royal cell, where the queen lives with the king, laying thousands of eggs to ensure the continuance of the colony.

As well as affording protection against intruders, the mound is so constructed as to provide a perfectly balanced 'microclimate', essential to the survival of its vulnerable inhabitants. The temperature, humidity and freshness of the atmosphere are all precisely regulated. The air within the nest, warmed by the metabolic activity of both termites and fungi to a steady temperature of around 30 °C (86 °F), moves by convection upwards into the attic area. It then passes down narrow channels within ridges in the mound walls, losing heat to the

Labels (clockwise from top): Attic; Ridge; Brood chamber; Royal cell; Brood chamber; Foundation; Pillar; Cellar; Fungus combs; Channels; Ridge; Duct to air channel

Temperature and carbon dioxide (CO_2) content of circulating air:

A	30 °C / 86 °F	CO_2 – 2·7%
B	25 °C / 77 °F	CO_2 – 2·7%
C	24 °C / 75 °F	CO_2 – 0·8%

exterior and exchanging carbon dioxide for oxygen as it does so. Finally the air – now cooler and fresher – enters the inhabited area from below, to be circulated anew. This sophisticated air-conditioning system has given these termites such complete independence from the world outside that they have become the most successful and widespread of all Africa's termite species.

pond in the middle. Eventually the whole ball is washed away by rainstorms, liberating the young frogs.

Wax, wood, clay and mud

Worker bees excrete wax flakes from glands on their bellies, which are chewed thoroughly with saliva and then nibbled into the six-sided cells of the honeycomb. By using a pair of position sensors – tufts of sensory hairs that register the angle of the head against the thorax – the workers are able to build against a true vertical and thus to determine the correct angles for the walls. The thickness of the cell walls is sensed by prodding each wall and testing how quickly it returns to flatness, while the width of the cells is determined by using the first pair of legs as dividers. Although the wax is extremely thin, each cell supports the cells above it, so that just 40 g (1½ oz) of wax can yield a comb over 35 cm by 20 cm (14 in by 8 in), capable of holding more than 1·5 kg (3·3 lb) of honey.

The paper wasps of the northern hemisphere create cellular combs using wood, chewed with saliva to pulp and then nibbled into a pellet or a thread of paper. Some of their nests may be up to 1 m (39 in) in width. The outsides are often highly ornate, with layers of different-coloured ribbons revealing how the nest grew as the wasps deposited paper from different sources along successive rims.

When clays are moistened with water or saliva, they can be moulded into strips or pellets. Potter wasps build the pellets they gather into tiny pots, in each of which they lay an egg, deposit a paralysed insect as food for the developing grub, and finally stop the mouth with another pellet of clay.

The most conspicuous users of clay, wood and soil are the termites (▷ box and p. 156). The sophisticated architecture of their mounds is unsurpassed in the insect world. Compass termites of Australia align the external parts of their nests so that the long axis always runs from north to south. In this way they are warmed by morning and evening sun but avoid being blasted by the sun at midday.

Nesting birds

Birds' nests are among the most familiar of all animal buildings, and many of them are among the most complex (▷ p. 94). Perhaps the most exuberant of all avian builders are the weaver finches, gregarious birds of the equatorial and southern regions of Africa. Males use grasses to make globular nests attached to the forks of trees, tying, knotting and interweaving the strands to construct a light but solid structure that hangs like a fruit from the tree. In the case of the village weaver, tens or even hundreds of nests may festoon the branches of a single tree, while other weavers build nests that are truly communal (▷ photo).

Another accomplished builder among the birds is the African hornbill. The females spend their early nesting period walled-up in holes in hollow tree trunks. A pair of birds build a door from a mixture of soil, saliva, excreta and food, leaving a gap just big enough for the male to bring food to the female and chicks, and for the female to defend the nest. As soon as the chicks are too big for single-parent feeding, the female breaks out and the chicks rebuild the door until they are fledged.

Animal excavators

Skilled burrowers are found throughout the animal world. The proficiency of moles is proverbial (▷ p. 121), but there are other mammals that rival them. Some rodents build underground cities of tunnels and chambers to accommodate their colonies. The plains viscacha of the pampas of Argentina, growing up to 90 cm (almost 3 ft) in length, riddles the plains with burrows and entrance holes that pose a serious threat to riders and their horses. A smaller rodent of the pampas, the tuco-tuco, spends most of its time underground, burrowing from one plant tuber or fleshy root to the next. Another prodigious burrower is the naked mole rat of Africa, which lives in underground colonies exhibiting one of the most bizarre of all mammalian social systems (▷ p. 158).

In the grasslands of the central USA the prairie dog (a rodent in spite of its name) is the most extensive colony-builder. These ground squirrels are highly vocal and social, with a distinctive warning bark and much chattering between individuals. A colony may consist of several thousand members, which live in a network of underground burrows, with several exits to confuse predators. The exit holes erupt from the tops of small mounds; these not only serve as sentry posts for lookouts while others are foraging, but are also shaped in such a way that they draw in the breezes that move over the ground in the prairies, so keeping the tunnels well ventilated. ML-E

Sociable weaver nests in Etosha National Park, Namibia. The massive nests of these weavers are truly communal, with a shared framework and roof but perhaps a hundred or more separate nests within. Resembling huge sponges when seen from beneath, these nests can sometimes weigh so much that they break the branches of their host tree.

SEE ALSO

- INSECT INTERACTIONS p. 64
- SPIDERS p. 68
- BIRDS: LIFE CYCLES p. 94
- TERRITORY, MATING, SOCIAL ORGANIZATION pp. 156–9
- CORAL REEFS p. 215

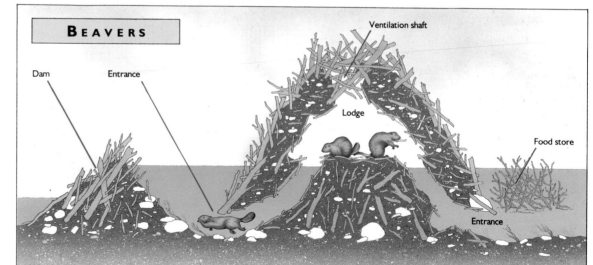

BEAVERS

Ventilation shaft
Dam
Entrance
Lodge
Food store
Entrance

With the exception of man, no mammal is capable of modifying its environment as dramatically as the beavers of North America and northern Eurasia. The aim of these accomplished engineers is to construct a family home, complete with moat, that is proof against predators and the rigours of winter.

To create an artificial lake, beavers obstruct the course of a stream by building a dam. This is made of felled logs, branches and twigs, shored up by layers of mud, gravel and larger stones. The upper surface of the dam is then waterproofed with a coating of mud, but the task of maintaining and raising the dam is continuous, and may spread over several generations.

Beavers may live in burrows dug into the banks of a stream, but frequently they construct special living quarters known as lodges within their artificial lakes. These are built along the same lines as their dams, but ventilation is provided by looser construction on the top of the mound. The floor of the lodge is above the level of the water, while access is provided by underwater tunnels. Nearby is a larder of branches and stems harvested during the summer and autumn. Beavers choose the site for their dam with care, ensuring that their lake is sufficiently deep that it does not freeze to the bottom even in the coldest days of winter. In this way they are certain to have uninterrupted access to their cache of food.

Finding Food

One of the most important distinctions between plants and animals is the different ways whereby they obtain the various materials they require in order to grow and function normally. While plants are able to meet all their nutritional requirements by taking up inorganic minerals from the soil and harnessing the sun's energy (▷ p. 31), every animal must obtain its nutrients and energy from its food, in the form of organic material garnered from the environment in which it lives. In ecological terms, all animals are consumers, directly or indirectly dependent for their livelihood on the organic matter produced by plants (▷ p. 194).

Feeding specialists are often superbly adapted for dealing with a particular kind of food. Feeding on grains and other firm vegetable matter, the weevils (a family of beetles) have gnawing mouthparts at the end of an extremely elongated snout, allowing them to bore through the protective coats of seeds and plants. So successful are they at plundering stored grain that as much as 80% of a crop may be lost through their activities.

Obtaining food of an appropriate kind and in adequate quantities is therefore an essential activity for all animals. Nevertheless, feeding is only one amongst many important activities in an animal's life, so the speed and efficiency with which it can gather food will determine how much time it can devote to other things – defending territories or nest sites, finding and courting mates, contesting its position in a social hierarchy, or warding off the unwelcome attentions of predators. Efficiency in acquiring food is thus central to an animal's survival and reproductive success, so individuals within a population better equipped for this activity have generally been favoured by the evolutionary process of natural selection

The world around us is teeming with an astonishing array of potentially nutritious substances, so for the brown rat – one of the most accomplished (and notorious) of all feeding generalists – the world is (almost literally) its oyster. But a rat's environment contains other things besides food: all sorts of toxic materials, for instance, and a good many pest-controllers. So a wise rat is a careful rat.

The secret of the rat's success is to combine an extreme readiness to investigate unfamiliar (and possibly nutritious) substances with a deep suspicion of them. In distinguishing between useful substances and harmful ones a rat is not thrown entirely upon its wits – like a specialist feeder, it has certain innate detector systems that guide it to some items (e.g. sweet ones) and cause it to reject others (e.g. bitter ones). But a rat will encounter numerous novel substances in its environment and requires a wide variety of different nutrients in order to avoid specific mineral deficiencies, so it cannot

rely entirely on such instinctive responses: its powers of discrimination must depend rather on a sophisticated general learning process.

A rat faced with an unfamiliar potential food item will typically sample it – take a small bit and then retire to await developments. If the rat feels unwell within the next hour or so, it will reject the new substance. It will avoid sampling more than one new item at a time, so that there is no confusion over which taste to associate with the adverse internal effects, and it is able to filter out other stimuli received during the relevant period (for instance, a familiar food taken at the same time or a strong visual stimulus). If no adverse reaction ensues, the rat may (tentatively) accept the food. Rat-exterminators often try to outmanoeuvre their intended victims by introducing a fresh bait without poison for a few days, in the hope that it will be sampled and accepted – but rats are remarkably adept at smelling a rat-catcher!

(▷ p. 13). As a consequence, many of the physical and behavioural adaptations that characterize particular species are directly related to the need for efficient food-gathering.

Feeding strategies

In considering an animal's role within an ecosystem, it is usual to categorize it on the basis of the kind of food that it eats – plant-eating cattle and aphids as herbivores, flesh-eating lions as carnivores, insect-eating ladybirds as insectivores, and so on (▷ pp. 194–5). However, when assessing the strategies that animals use to obtain food, it is useful to consider not only the type but also the range of foods selected. Many animals are *specialist feeders*, taking either a single kind of food or a narrow range of foods: koalas feed exclusively on the leaves of a few species of eucalyptus, pandas mainly on bamboo shoots, orb-web spiders on flying insects. At the other extreme, there are *generalist feeders* – equivalent, in ecological terms, to omnivores – such as rats, pigs, raccoons, badgers and some bears. These animals feed opportunistically on a wide range of foods and will typically assess the potential nutritive value of unfamiliar items that they encounter in their environment.

Generalist versus specialist

The obvious advantage of a generalist feeding strategy is that it gives great versatility. Not being tied to a particular kind of food, a generalist can usually turn to alternative food sources if certain types of food become scarce. However, the lack of such flexibility is only a problem for the specialist if its chosen food is liable to shortage and fluctuation. The moth *Mabra elephantophila*, which feeds by tickling the surface of an elephant's eye and sipping the tears, is in no danger of extinction provided that there is a good stock of elephants. In the same way, the survival of the many animals that exclus-

ively parasitize a single species is secure so long as their host species survives: dogs and dog fleas will stand and fall together. The capacity to exploit a wide range of food types also allows the generalist to invade and colonize new habitats. It is no accident that two of the best-known generalists – rats and humans – are among the most widely distributed species on earth.

Although limited in the foods that they can select, specialists are usually extremely well adapted to a particular feeding method. Such features as the stout claws and long tongues of anteaters (▷ p. 106), the long bills and tongues of hummingbirds (▷ pp. 92–3) and the filtering plates of baleen whales (▷ p. 145) have been carefully honed through evolutionary time to the point that they are now of near-perfect 'design' for extracting ants and termites from their nests, nectar from particular species of flower, and plankton from the sea. Generalists, on the other hand, must remain versatile and cannot afford the luxury of such specializations; as a result, they are frequently 'outgunned' with respect to a particular food type by specialists living in the same area.

For feeding specialists, recognition of appropriate food items is normally unproblematic. Focusing on one or a narrow range of food types, such animals generally rely on instinctive responses to suitable items – the required food-recognition mechanism is 'built-in' to the animal, i.e. encoded in the animal's genetic material (▷ p. 22). A koala does not need to learn to discriminate eucalyptus leaves from those of other trees, or to discriminate different species of eucalyptus. A similar system of innate responses would be unsuitable for a generalist, since it not only feeds on a wide variety of different foods but also needs the capacity to evaluate unfamiliar and potentially

The scant food and water resources of deserts mean that animals must be extremely efficient and well adapted if they are to survive. The gerenuk, a tall and graceful gazelle of the dry thorn-bush deserts of eastern Africa, uses its narrow muzzle and mobile lips to pluck leaves and shoots from trees and bushes, undeterred by the needle-sharp thorns. It never needs to drink, deriving all the water that it requires from its food.

nutritious substances. Rats and other generalists (▷ box) therefore rely on sophisticated recognition systems that combine innate responses and various forms of learning by experience.

Foraging patterns

Although some animals, such as spiders and aquatic filter-feeders (▷ p. 57), wait for their food to come to them, others must actively go in search of it. Often the food of such animals is unevenly distributed in the environment: a nectar-feeding insect finds collections of flowers on different plants at some distance from one another, while grazing animals – even in an apparently uniform grassland – detect patches of especially favoured sward. On the other

hand, prey animals may respond to the pressure of predation by adopting a scattered distribution.

Since survival may depend on the speed and efficiency with which food is found, it pays an animal to adopt a searching strategy that is suited to the distribution of its food. Whatever the distribution – whether patchy or scattered – a random search path is generally inefficient, as it usually involves an animal covering the same ground more than once. When food is evenly scattered, a searching animal may show a marked forward bias in its movement, making relatively few turns, so as to cover as much new ground as possible. A similar pattern may be profitable when food is patchily distributed, but as soon as a food item is detected, the way in which the animal moves may change: it usually turns more frequently and in the same direction, so that the search is concentrated in the vicinity of the find. Finding more food items will sustain or increase the rate of turning, but if no more food is found, the search path gradually straightens out, so taking the animal to an unexplored area. Such searching

ARMY ANTS

The army ants of Africa and Central and South America move through their forest habitats in cycles of activity. For 20 days the colony has a fixed nest site. The nest is often made of the bodies of workers holding onto each other, and at this time contains only pupae. About halfway through this stationary phase the queen lays a batch of 60 000–100 000 eggs. Workers forage for food both for the queen and for the workers forming the nest. This foraging is highly systematic to avoid going over ground that has been previously raided. Each day the raiding party sets off at an angle of 126° to the direction of the previous day's expedition. Over a period of about 15 days this pattern will cover all directions from the nest site with no overlap. The workers lay down trails of scent to enable them to return to the camp. At the end of 20 days all the eggs hatch at once and the new workers emerge from the pupal cases; this signals the start of the nomadic phase. For the next 15 days the colony moves through the forest, again avoiding areas previously visited. During the day the developing larvae are carried in the jaws of workers, and a temporary camp or 'bivouac' is constructed each evening. Foraging continues during this time to feed the growing larvae.

Despite the cinema images of landscapes blackened by marauding battalions of ants, the truth is rather more prosaic. Colonies may be large – up to 600 000 individuals – but when they forage, the workers advance slowly along a broad front at about 14 m (46 ft) per hour. Most larger active animals can easily avoid being trapped, but many invertebrates, including other ants, spiders, scorpions, large grasshoppers and caterpillars, may fall prey to the foragers. Any captured prey is dismembered and carried to the nest site. BDT

movements often take place within a feeding territory (▷ p. 156), which may be defended against intruders so long as it yields enough food to compensate for the energy used in its defence. A territory may be abandoned if food becomes scarce, and an animal may cease to defend it at times when there is an abundance of food. ML-E

Storing, hoarding or caching food may be a useful strategy, especially in situations where foraging conditions vary seasonally or from day to day. The great grey shrike or 'butcherbird' stores prey such as mice and lizards in a 'larder' by impaling them on the thorns of bushes – or even on the barbs of barbed-wire fences.

Predator and Prey

In order to obtain their food, many animals rely solely or partly on predation – killing and eating a member of another species. The supply of prey animals is never unlimited, so a predator will always find itself in some degree of competition not only with members of its own species but with other local predators dependent on the same resources. A predator that is better adapted than its rivals for dealing with its prey will stand a greater chance of surviving and reproducing successfully, with the result that over evolutionary time the skills and adaptations of predators have become ever more refined. At the same time, prey animals respond to increasing predatory pressure by evolving better defences – more effective camouflage, keener senses, greater speed, and so on (▷ p. 166). In effect, an 'evolutionary arms race' is being waged among and between predators and prey, with improvements in predatory capabilities being matched by more effective countermeasures.

The net result of these pressures is that the physical and behavioural adaptations of predators are typically more or less precisely 'tailored' to the demands of catching a certain kind of prey. Where a wide range of prey species is available, as in the tropics, an unspecialized and thus flexible predator might seem to be at an advantage, but in such circumstances there are usually a correspondingly large number of predators, so a 'generalist' might well lose out to specialists in regard to each prey type. Indeed, in diverse ecosystems where there are many different species of predator and prey, it is generally better for predators at least to reduce competition by some degree of specialization.

Sit-and-wait predators

Sit-and-wait or ambush predators conceal themselves in places that are likely to be visited by their prey and then seize the unsuspecting victim as it comes into range. Some use camouflage to escape detection. Praying mantises are coloured to blend in with the vegetation in which they lurk – green or brown to match living or dead leaves, or flower-coloured – and some even mimic the shape of leaves; as an item of prey (usually a fellow insect) moves into striking distance the mantis flicks out its powerful serrated fore legs to grab it. Crab spiders, perfectly matching the flowers in which they hide, seize visiting insects such as bees and butterflies, which they incapacitate with a lethal dose of venom.

Many ambush predators build or excavate structures, either as trapping devices or as a means of concealment. As well as web-building spiders (▷ p. 69), there are trapdoor spiders that hide in silken burrows, darting out to capture any insect that stumbles over their trip-lines. The tiny antlions (the larvae of lacewings) dig steep-sided pits in sandy terrain, lurking at the bottom in wait for their victims (usually ants) to walk by. Alerted by grains of sand falling into its lair, the antlion responds by catapulting sand with its head at the passing ant, which is swept down by the ensuing sand slide and then despatched by a venomous bite from the antlion's enormous pincer-like jaws.

Not content with relying on chance to bring their prey within range, some ambush predators have lures to entice them to their doom. The alligator snapping turtle has a worm-like appendage in its mouth that attracts inquisitive fishes, while anglerfishes have 'fishing rods' extending from their snouts that draw prey towards the enormous gaping mouth (▷ illustration, p. 221). The yellow-bellied sea snake floats on the surface of the sea like a piece of vegetation or a dead animal ready for scavenging; alerted by vibration and taste transmitted through the water, it strikes as small fishes come to investigate.

Opportunistic predator: a great skua hovers over a group of eiders, waiting for a chance to seize a chick. As well as taking the young and eggs of other seabirds (including other skuas), these swift and powerful birds prey on insects, rodents such as lemmings, and small birds. They are also pirates, harassing gulls, terns and other birds on the wing and forcing them to drop or disgorge their food.

Small and immobile prey

For predators that feed on relatively immobile prey, the main problem often lies not in capturing or subduing it but in detecting it in the first place. Such prey is often camouflaged or concealed in some other way, so predators have developed efficient searching strategies that enable them to maximize the number of prey items gathered in a given time (▷ p. 163). Wading birds such as curlews, plovers and oystercatchers that probe for marine worms, snails and shellfish in the soft sand and mud between the low- and high-tide marks may intensify their searches in areas where they have been successful; they also reduce competition among themselves by specializing in taking particular kinds of prey.

The large ant- and termite-feeding mammals, such as the American anteaters and the aardvark and pangolins of the Old World, do not generally have difficulty in locating their prey; their characteristic adaptations are such as to allow them to gain access to their prey (powerful claws on the front legs) and then to 'sweep' them up as rapidly as possible (long, sticky tongue). The necessity of bulk consumption for large animals feeding on small prey is even more starkly demonstrated by the giant marine filter-feeders, such as the basking shark and the baleen whales (▷ p. 145). These have undergone extreme adaptation to permit rapid harvesting of literally millions of tiny fishes and crustaceans.

Stealth and open pursuit

Many predators employ stealth to approach within striking range of their

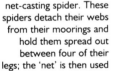

Sit-and-wait specialist: a female ogre-faced or net-casting spider. These spiders detach their webs from their moorings and hold them spread out between four of their legs; the 'net' is then used to intercept night-flying insects.

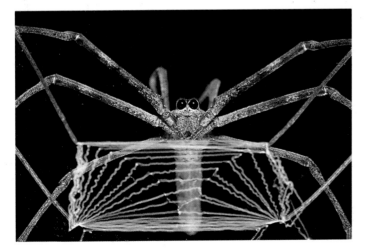

prey. Cuttlefishes and chameleons, for instance, are in many respects classic ambush predators. Both use highly sophisticated and versatile camouflaging techniques to escape detection, and they may indeed strike at suitable prey if it happens to come within range. However, if they locate an item of prey beyond their effective range, they may first stalk it, stealthily manoeuvring themselves into a position from which an attack can be launched. A cuttlefish approaches by inconspicuous movements of its lateral fins and then makes a sudden jet-propelled pounce on its prey (⊳ p. 57). Chameleons may make lengthy detours along the branches of a tree, sometimes losing sight of their prey in the process, before emerging in a place from which they can shoot their long, sticky tongue at their unsuspecting victim (⊳ p. 88).

Predators such as lions and cheetahs normally have to reveal themselves to their prey before launching a final attack. They can often outsprint their prey over relatively short distances, but their chances of success are greatly increased if they can minimize the extent of open pursuit by stalking. They make themselves as inconspicuous as possible by crouching low and approaching their prey head-on, and may move between clumps of vegetation and other natural cover as their prey looks away.

Hunting in conditions where visual perception is limited or impossible – at night, in dark or muddy waters, or underground – may be an effective strategy, provided of course that the predator itself has a sensory system well adapted to such conditions. The ability of owls to locate their prey in complete darkness is due to their extremely acute hearing (⊳ pp. 96–7), while oilbirds and many bats hunt by means of echolocation (similar in principle to radar), which involves emitting ultrasonic sounds and monitoring their echoes as they are reflected from obstacles (including items of prey) in their path (⊳ pp. 118–19). A similar system is used by dolphins, and is probably particularly important for river dolphins, which live and hunt in murky waters where visibility is very limited (⊳ p. 144). Several fishes are sensitive to electrical fields, and may be able to detect prey by the local distortions that they create in such fields. Pit vipers have a pair of heat-sensing organs on their snout that allow them to detect warm-blooded prey in total darkness; they are able to 'compute' the distance to the prey on the basis of temperature differences as small as 0·2 °C (0·36 °F) between each of the two organs.

Cooperative hunting

Cooperative hunting presupposes some degree of social organization, and so is only practised by social animals, such as wolves, African hunting dogs, hyenas, lions and killer whales. By hunting communally, such animals considerably increase the likelihood of a successful hunt and are able to subdue prey larger than they would be able to tackle on their own. However, these advantages are at least partially offset by the necessity of dividing the spoils among the members of the hunting party. Whether an animal hunts singly or in a group probably depends mainly on the nature of the available prey: in circumstances where small prey is relatively abundant, it may be more profitable for a predator to hunt on its own, while if larger prey predominates, joining a hunting group may be worthwhile.

Various social hunters, including porpoises, killer whales and wolves, herd or corral their prey by approaching in a scattered fan formation and then moving in to encircle it or drive it into a place from which escape is impossible. Predators such as lions and wolves may also drive their prey: some members of the hunting group manoeuvre themselves to the far side of the prey and then drive it in the direction in which their fellows are lying in wait. One of the most sophisticated hunting strategies, involving considerable division of labour, has been observed in chimpanzees hunting prey such as colobus monkeys (⊳ p. 141). Often the composition of a hunting group is altered to suit a particular kind of prey: spotted hyenas hunting wildebeest or Thomson's gazelle form groups of two or three, while packs intent on larger prey, such as zebra, may number 20 or more individuals.   ML-E

CANNIBALISM

Cannibalism – eating members of one's own species – is surprisingly widespread in the animal kingdom. Although more common among invertebrates and lower vertebrates such as fishes, frogs and toads, the phenomenon also occurs in some species of bird and mammal. Often cannibalistic behaviour is directed towards eggs and young, rather than adults. In many insects, for instance, and in frogs and toads, relatively few individuals survive to maturity in any case, and it may benefit a newly hatched larva or tadpole to feed on eggs or other tadpoles, thereby gaining a meal and eliminating a potential competitor (and a prospective cannibal). In breeding colonies of herring gulls there is a small proportion of specialist cannibals that gain virtually all their food by eating the eggs and chicks of their neighbours. Cannibalism may also occur as a response to population density: in the house mouse, for instance, and in certain butterfly species, the incidence of cannibalistic behaviour increases as their populations rise.

The reason for cannibalistic behaviour is generally the same as that for predation – simply to obtain food – but in some cases a less obvious motivation may be at work. For instance, when a male lion forcibly takes over a pride, he may kill and occasionally eat all the existing cubs. Having eliminated the offspring of the previous pride-holder, he then mates with the resident lionesses. The purpose behind such behaviour appears to be to create a situation in which the new pride-holder's offspring have a better chance of surviving – in other words, to increase his genetic representation in the next generation.

SEE ALSO

● FINDING FOOD p. 162
● DEFENCE, DISGUISE AND DECEIT p. 166
● PARASITISM AND SYMBIOSIS p. 168
● DIGESTION p. 178
● BIOSPHERE: ENERGY RELATIONS p. 194

Lionesses chasing a female Thomson's gazelle. Lions often stalk their prey prior to launching an open pursuit: the closer they can get before revealing themselves, the better their chances of success.

Defence, Disguise and Deceit

Few animals are without natural enemies. With the exception of animals at the top of food chains, such as killer whales, birds of prey, lions, bears and crocodiles, all predators are themselves the prey of other animals (▷ pp. 194–5). For most animals, therefore, it is critical to their survival that they have adequate defences. Individuals that have defensive adaptations that are in some way superior to those of others in the same locality are more likely to survive and prosper, so over the course of evolutionary time defensive attributes have become progressively more refined and effective.

SEE ALSO

- CHANGING COLOUR (BOX) p. 67
- ANIMAL COMMUNICATION p. 152
- TERRITORY, MATING, SOCIAL ORGANIZATION p. 156
- PREDATOR AND PREY p. 164

An animal typically shows a range of physical and behavioural adaptations that together make up its defensive system. Such systems are generally 'layered' – in other words, if one level of defence fails or is penetrated, another line of defence is activated.

Camouflage

Although animals that are dangerous or unpleasant to eat may advertise the fact (▷ box), for many animals the first line of defence is to remain inconspicuous and so avoid detection. Usually this is achieved by blending in with the environment by means of some form of camouflage (*crypsis*). In its simplest form this may amount to no more than an animal matching the colour of its background, but often more subtle techniques are employed. In normal light, because of the effect of shadow, a uniformly coloured animal would appear darker on its underside than on its back; to offset this effect, most camouflaged animals exhibit *countershading* – their coloration is darker on the upper surface, lighter beneath. The *disruptive coloration* of zebras and butterflyfishes – boldly striped in each case – appears conspicuous outside their natural setting, but in context (in the open savannah grasslands or against the kaleidoscopic backdrop of a coral reef) these patterns help to break up the animals' otherwise conspicuous outline.

Another way of escaping detection is to mimic some naturally occurring feature of the environment that a predator would normally regard as inedible. For instance, the caterpillar of the sphinx moth looks exactly like a small woody twig encrusted with spots of fungus, while another small moth closely resembles a bird's dropping as it rests on a leaf. Leaf and stick insects combine shape and colour to mimic every aspect of leaf and stem form (▷ photo).

Despite their obvious usefulness, both camouflage and camouflaging mimicry have their limitations. Animals employing these means of concealment are of course restricted to locations where the deception is effective. A number of animals have overcome this drawback by achieving some degree of versatility, allowing them to alter their coloration to match a variety of different backgrounds (▷ box, p. 67). A further problem for cryptic animals is that the effectiveness of their concealment depends on their remaining more or less motionless, and this will inevitably conflict with other essential activities, such as feeding and reproduction. Camouflaged animals tend to minimize this difficulty by restricting their periods of activity to times when predation is less intense: many moths, for instance, largely escape visually directed predators by resting motionless against tree trunks during the daytime, only becoming active under the cover of night. For many animals some degree of camouflage must be sacrificed in the struggle to gain a mate, as a result of the conflicting pressure of sexual selection (▷ p. 159).

Flight patterns

For more mobile animals, the initial response to detection by a predator is usually to flee. Rabbits and other animals that emerge from burrows to forage may make a bolt for the security of their burrow. Others may attempt to outdistance their assailant, the pattern of flight often depending on its proximity. Some moths can detect the ultrasonic pulses emitted by an echolocating bat (▷ pp. 118–19) before they are themselves detected, but they are unable to match the bat's speed; if the bat is some distance away, they may be able to get out of range by straight flight, but if the bat is close, they fly erratically or close their wings and drop to the ground. The natural evasion pattern of hares is to zig-zag and double back as they race away from foxes and other predators. Antelope also run erratically, with sporadic leaps and bounds, when pursuers such as lions or cheetahs get close enough for a pounce. Thomson's gazelle allow predators to approach within a certain distance before fleeing, apparently adjusting this 'flight distance' to match the ability of the predator concerned: thus they flee from African hunting dogs when they approach within about a kilometre, since these animals are noted for their capacity for sustained pursuit, but will let cheetahs, which are fast over short distances but have limited stamina, come very much closer before taking off.

Withdrawal, threat and bluff

Once detected by a predator, relatively slow-moving or immobile animals with body armour or some similar means of protection may sit tight, withdrawing or hiding vulnerable parts if possible. Turtles, snails and bivalve molluscs, such as clams and mussels, retract all their soft

Stick and leaf insects attempt to escape predation by mimicry. Some resemble heavily thorned plants, others fresh green leaves or drying yellow and brown plant material. Macleay's spectre of Australasia, shown here, lives on thornbushes and looks like dried, sun-shrivelled leaves and stems.

parts within their shell. Animals with protective spines, such as sticklebacks and hedgehogs, may erect their spines and (in the latter case) roll into a ball for extra protection.

Some animals attempt to startle their assailants by exposing areas of bright *flash coloration*, a ploy that may either cause a predator to withdraw or gain valuable seconds in which an animal may flee. Underwing moths, for instance, have brightly coloured hind wings, which are normally hidden by the camouflaged fore wings as they rest upon tree trunks; if their camouflage is penetrated, the moths suddenly reveal their patches of garish colour, often startling predatory birds into flying off. Hawkmoths achieve a similar effect by flashing large eye-spots on their hind wings, which may also be vividly coloured. The meadow grasshopper, as well as larger tropical species, has bright red hind wings, which are revealed as it leaps away from a predator; as the insect lands, the colour vanishes again, perhaps confusing the predator into believing that the insect itself has suddenly disappeared.

When an animal is cornered, it may attempt to intimidate a predator by some form of threat display. Often this is pure bluff. Toads and bullfrogs, for instance, swallow air and inflate their lungs and vocal sacs, so that they appear several times their normal size. Some harmless snakes are able to spread out their neck vertebrae, giving them the appearance of hooded cobras. In many cases, however, threat displays, like warning coloration, are genuine signals that a predator may be harmed if it does not back off. The rattle of a rattlesnake is a warning that a predator will do well to take seriously, as is the handstand display performed by a skunk, as a prelude to ejecting a spray of foul-smelling fluid.

Last-ditch defences

Many predators lose interest in dead prey or will not strike at it, so a number of animals, including various beetles, spiders, newts, snakes and opossums, may escape attack by pretending to be dead or *death-feigning* (▷ photo). When other defences fail, most animals respond to the close proximity of a predator by entering a paralysis-like state of extreme unresponsiveness known as *hypnosis* or *tonic immobility*; lasting anything from minutes to hours, this state appears to be induced by fear and presumably has a defensive function similar to that of death-feigning.

If an attack cannot be avoided, an animal may yet save itself by some system of damage limitation. Many butterflies have small eye-spots near the tips of their wings, which deflect bird pecks away from the head or body. Some animals are prepared to sacrifice some part of their body in order to save the whole: many lizards shed their tails (▷ p. 89) and echinoderms such as brittlestars cast off their arms (▷ p. 54). Ringed plovers and other ground-nesting birds may attempt to

distract a predator's attention from their eggs or chicks by fluttering away from them, pretending to have a broken wing.

An animal's final defence against attack is generally to fight back as best it can – with teeth, claws, feet, or whatever else it may have at its disposal. Often such retaliation is more or less futile, but not always. Porcupines thrash their quills and back onto an aggressor, while kangaroos and large flightless birds such as ostriches and cassowaries brace themselves and kick out. Many animals use chemicals as deterrents, but few do so with greater effect than the bombadier beetles. When under attack, these insects pump highly reactive chemicals, normally stored in separate sacs, into a discharge chamber, where they react explosively to form hot benzoquinones, which are forced out of the cloacal opening at up to 500 spurts a second.

Defence by associaton

Many animals derive protection from some form of social aggregation. Fishes reduce the impact of predation by shoaling (▷ p. 156), birds may succeed in driving off an attacker by mobbing (▷ box, p. 148), and warning signals such as alarm calls and tail flashes are used by various

birds and mammals (▷ box, p. 149). Among foraging mongooses and prairie dogs and in herds of grazing herbivores such as antelope, some individuals may act as sentinels, keeping a watch for predators and giving a warning signal if danger threatens. Animals may also benefit from associating with others that are better protected than themselves: for instance, aphids gain protection from ants, clownfishes from sea anemones (▷ p. 169). ML-E

Death-feigning may be effective as a defence since many predators lose interest in dead prey. The hog-nosed snake 'plays possum' by lolling on its back with its mouth open, simulating the slackness of death.

Parasitism and Symbiosis

There is no such thing as a truly independent organism. In all ecosystems, animals, plants and microorganisms interact and interfere with one another in a wide variety of ways. Some of these interactions are casual, short-term and unspecialized in nature. Others are the very special inter-relationships that occur between members of the same species in family or social groups. In addition to these types of association, however, almost all groups of living things involve themselves in highly intimate associations that form between *different* species and that typically result in clear-cut patterns of harm and benefit for the organisms concerned.

Associations of this kind are generally long-term interactions, often involving close physical contact between the species concerned and many special adaptations for life together. Two such associations – parasitism and symbiosis – require similar degrees of adaptation for the organisms concerned, but may be differentiated on the basis of their distinct ecological implications.

Types of association

In *parasitism*, one species, the *parasite*, lives in or on another species, the *host*. Internal parasites are called *endoparasites*, whereas those living on the outer surface of their hosts are known as *ecto-parasites*. The host is typically harmed by the presence of the parasite. The harm can take a number of unpleasant forms, but the net result of infection by a parasite is that the host survives less well and/or produces fewer offspring. The parasite, on the other hand, benefits from the relationship, usually gaining food or other resources from the host's body. The partnership is therefore an unequal one – the parasite gains, the host loses.

Symbiosis or *mutualism* is a more balanced partnership. In a typical symbiotic association, the species participating, called *symbionts* (or *symbiotes*), derive mutual benefit from it. It is usual for both species to survive and reproduce more successfully when living symbiotically than when living apart.

A third kind of association, known as *commensalism*, occurs when an organism of one species (the *commensal*) derives benefit from associating with an organism of a different species (the *host*), which itself receives neither benefit nor harm from the relationship. Barnacles, for instance, attach themselves to whales or sea turtles, thereby obtaining anchorage and transportation, while the whale or turtle apparently remains unaffected. A number of birds, including the cattle egret and the American cattle tyrant and cowbird, take advantage of the feast of insects flushed out by large grazing mammals (including domestic cattle); this feeding habit is clearly beneficial to the birds concerned and appears to have no adverse effect on the grazers themselves.

Types of parasite

All types of organism can lead parasitic ways of life. *Microparasites* – microscopic organisms such as parasitic viruses, bacteria, fungi and protists – typically multi-ply directly on a host. *Macroparasites* – larger organisms such as worms, insects, ticks, vertebrates and plants – usually produce offspring on a host that then leave it to establish new infections.

Parasites show adaptations for locating their preferred host species and for establishing themselves on or in them. Usually included among these adaptations are methods for feeding from host tissues and for maintaining transmission from host to host. The latter traits become tied together into often complex life cycles, sometimes – in the case of *multi-host* life cycles – involving several hosts of different species. In a typical multi-host cycle, a parasite infects different hosts at different stages of its development, undergoing specific adaptations as it passes from one host to another.

Human parasites

We think of ourselves, as a species, as being in control of our own ecological destiny, but we can in fact be infected by diseases caused by almost every kind of parasitic organism.

Among the microparasites, we can suffer viral infections ranging from the mild common cold to the fatal and incurable AIDS virus. We are prone to hundreds of different bacterial illnesses, from acne to bubonic plague. We also get skin diseases caused by fungi – athlete's foot, for example, and ringworm – as well as some very serious protist diseases, such as malaria and sleeping sickness.

Malaria and sleeping sickness are examples of parasites with multi-host life cycles. In each disease the infection is spread to new humans when an already infected mosquito (in the case of malaria) or tsetse fly (in the case of sleeping sickness) bites a person in order to feed on their blood. These insect hosts are called *secondary hosts*. Similarly, the insects themselves become infected when they remove blood from a person already suffering from the disease.

Human macroparasites cause several of the most serious tropical parasitic diseases. One such disease, filariasis, is caused by nematode roundworms (▷ p. 55), which are spread from person to person by mosquitoes. The presence of these worms causes inflammation and eventual blockage of the lymph vessels, which can result (in severe cases) in the dreadful disfigurement known as elephantiasis.

A typical parasitic life cycle

Another serious tropical disease, schistosomiasis (formerly known as bilharzia), is the result of infection by a flatworm blood fluke of the genus *Schistosoma*. The life cycle of this fluke, which infects more than 250 million people in the world today, well illustrates the types of adaptation that take place in the world of parasites.

The human disease is caused by pairs of worms, a male and a female locked together in permanent copulation, which

A common cuckoo in a reed warbler's nest. Many species of cuckoo are brood parasites, laying their eggs in the nests of other birds, a single egg per nest; the foster parents are then left to rear the cuckoo chick themselves. To reduce the likelihood of the egg being rejected, cuckoos specialize in parasitizing particular host species and their eggs closely resemble those of their hosts. Soon after hatching, the young cuckoo usually 'evicts' the host's own eggs or nestlings, thus giving the oversized imposter an unfettered supply of food from its diminutive providers.

live in the blood vessels of either the intestine or the bladder. The female worms in these pairs produce thousands of spined eggs, some of which escape from damaged blood vessels into the faeces or urine of an infected person. This is the parasite's subtle escape route for continuation of the life cycle.

Each egg contains a larva called a *miracidium*, which is stimulated to hatch when it reaches fresh water. It then swims to infect a freshwater snail. Within this secondary host, parasite numbers are dramatically increased by a form of cloning – thousands of new larvae called *cercariae*, genetically identical to the single invading miracidium, are produced in the snail's tissues.

The cercariae then emerge from the snail and swim in water for about 48 hours, until their finite food reserves are exhausted. If they come into contact with human skin during this period, the cercariae bore into the skin to establish a new human infection. Because of the organization of this life cycle, the disease is commonest in developing countries in the tropics, where sanitation is poor and where there is much human contact with infected fresh water – as in paddy fields and irrigation ditches.

Symbiosis

The mutually beneficial partnerships of symbiosis are used by a remarkably diverse range of organism pairings. Fungi and cells of algae collaborate in a delicately balanced metabolic interaction to produce a whole new life form – the lichens (▷ p. 34). These plant-like growths have a body based on fungal threads, but derive photosynthetic benefit from the algae embedded in the fungal base.

In the intestines of many animals (including humans) there are populations of symbiotic microorganisms – the so-called *gut flora* – some of which aid digestion. Termites and ruminants such as cows, for instance, are only able to digest their cellulose-rich diet because their guts are populated by a mixture of symbiotic bacteria and protists (▷ box, p. 112); these microorganism species are themselves only able to survive in the oxygen-free conditions of termite and ruminant guts. Termites even have nitrogen-fixing bacteria in their gut flora. With these they can obtain organic foods containing nitrogen derived from the gaseous nitrogen in the atmosphere (▷ p. 197).

The same metabolic trick is used by one family of flowering plants – the legumes (▷ p. 32). This very successful group of plants, with over 20 000 species, includes the peas, beans and acacias. These plants are able to grow well even in nitrate-poor soils because their roots possess nodules populated with nitrogen-fixing bacteria that initially invade the roots from the soil. The bacteria gain nutrients from the plant tissues, while the plant gets fixed nitrogen from the microorganisms – the type of two-way help typical of symbiosis.

Symbiosis occurs everywhere in the natural world. Many insects and even some bats and birds (such as hummingbirds) are linked to plants by the bond of pollination symbiosis (▷ p. 42). Certain ant colonies symbiotically protect the spiny acacias in which they build their nests. Sea anemones are locked in symbiotic pairings with hermit crabs (▷ p. 67) and clownfishes. The latter, immune to the anemone's stinging cells, live among its tentacles, so gaining protection from predators; in return, the clownfishes bring food to the anemone, and ward off various fishes that might otherwise attempt to nip a piece from the anemone's tentacles. A number of African birds known as honey-guides attract mammals – commonly ratels or honey badgers but sometimes humans – to a bees' nest and take their share of the spoils after the mammalian partner has plundered it.

There is even a suggestion that around 1500 million years ago the complex cells typical of higher organisms evolved from simpler bacteria-like cells by a symbiotic route (▷ pp. 6–7). If this theory is correct, we exist today because of a series of symbioses that first happened many millions of years ago. PW

Honeydew harvest. Many species of aphid are protected by associating with ants. In return the aphids provide the ants with food: when its abdomen is caressed by an ant in a characteristic pattern, an aphid exudes a droplet of a sugary substance called 'honeydew', which the ant then eats.

SEE ALSO

● SIMPLE LIFE FORMS p. 24–7
● WORMS p. 55
● INSECT INTERACTIONS p. 64
● PESTS AND PEST CONTROL p. 230

CLEANING SYMBIOSIS

A number of marine animals, including around 50 species of fish and a few species of shrimp, make their living as 'cleaners': they gain their food by removing parasites and food particles from the bodies of other animals, mainly fishes. The clients are generally very much larger than their cleaners and might normally regard them as suitable prey, so a degree of protocol is required to prevent misunderstandings. Recognizable by its distinctive uniform – often brightly coloured with bold stripes – a cleaner waits at a special cleaning station and performs a characteristic display; the client then indicates its readiness to cooperate, often by adopting an open-mouth posture. The cleaner then proceeds to go about its business, which may involve swimming right into the client's gaping jaws. This arrangement is clearly of considerable benefit to both parties, but it is also open to abuse. Some non-cleaners mimic the appearance and habits of a cleaner in order to dissuade larger fishes from eating them, and in some cases to provide an opportunity to dash in at close range and nip a piece from a trusting host.

A cleaner wrasse attending to a parrot fish.

Migration

Every animal needs to move from one place to another at some time in its life cycle. For a small animal such as an earthworm, such movements may be restricted to the few cubic metres of soil where it spends its entire life, but other animals may travel much greater distances, sometimes covering hundreds or even thousands of kilometres. Some movements, such as the mass population dispersals of locusts, are somewhat irregular, depending on local conditions that may vary from year to year (▷ pp. 172–3). Other animals, however, make highly predictable and usually repeated movements between different habitats known as *migrations*; these display a cyclical pattern, and are usually triggered by seasonal or other factors that recur at more or less fixed intervals.

Migration serves many different purposes. Some animals migrate to escape harsh winters or hot summers, others to find suitable breeding grounds. The need to find adequate food resources or to avoid predators may also be important factors. Whatever the reason, all migrations involve some active movement by the individuals concerned, often for many days. Smaller migrants, for example plankton and aphids, also make use of water or wind currents, while birds use trade winds and thermal upcurrents.

Some migrants such as birds move annually between summer breeding grounds and overwintering sites, often returning to the same areas each year, while salmon and eels take several years to complete their migrations. Most migrations involve a round trip between one habitat and another, but – especially in the case of insect migrants – it is not necessarily the same individuals that make the return journey.

Migration in invertebrates

Although many of the most spectacular migrations, such as those of birds, have a yearly cycle, other animals – particularly among the invertebrates – make regular migratory movements at much shorter intervals. Plankton gather in the surface waters at night to feed, but during the day move down to deeper, cooler waters, sometimes reaching depths of 1200 m (4000 ft). Here they save energy, as metabolism is reduced at lower temperatures, and they also avoid the fish that hunt by day at the surface. By moving between different currents they may also find new food sources.

Small crustaceans called sand hoppers live in wet sand just above the high-tide mark but feed on rotting seaweed higher up the shore. On moist nights they move many metres inland to feed, returning before day to their sandy refuges. Land crabs may travel up to 240 km (150 mi) to lay their eggs in salt water.

Among the most familiar insect migrants are various species of butterfly, including the painted lady and red admiral of Europe and the monarch, a large New World species easily recognized by its bright orange wings. During the winter monarchs cluster in large roosts on the trunks of trees in Mexico and the southern states of the USA. In spring they begin a leisurely migration northwards to breed on the milkweeds of North America. The outward journey takes many weeks and lasts through several generations. Some individuals reach as far north as Canada, while each year the odd specimen is blown off course and turns up in Europe. The return journey in the autumn is made by the latest generation at a much greater speed, averaging over 100 km (62 mi) a day.

Fishes, amphibians and reptiles

Fishes such as herring, plaice and cod undertake regular annual migrations. In the North Sea, for example, each population follows a seasonal anticlockwise migration cycle, feeding in one area and breeding in another. Knowledge of such movements is of great importance to fishermen, who may rely on it to make seasonal harvests.

Salmon make less frequent journeys. Adults spend most of their lives at sea, where they feed and grow to maturity, but they must return to fresh water to breed. They usually return to their native waters, being able to indentify a particular river by its distinctive odour, with which they became familiar as hatchlings. Most salmon die soon after spawning, but some Atlantic salmon make it back to the sea and survive to spawn again.

The adults of the North American and European eel live in rivers, but migrate to breed deep in the Sargasso Sea, an area of warm water in the Atlantic southeast of Bermuda. The very young eels, called leptocephali larvae, remain in the sea for 30 months, then metamorphose into small eels called elvers. These migrate in vast numbers up rivers, sometimes wriggling overland for short distances to reach their feeding grounds. Here they may remain for as long as 19 years before returning to the Sargasso Sea to breed and die.

Many newts, salamanders, frogs and toads migrate annually over distances of a few kilometres from their hibernation sites to the ponds or streams where they breed. Like salmon, such amphibians often show a preference for the particular body of water in which they themselves were spawned. Some reptiles, such as turtles, make much longer migrations, sometimes swimming thousands of kilometres in order to reach a particular beach on which to lay their eggs (▷ illustration).

Birds

Perhaps the most remarkable of all migrations are performed by birds. The arctic tern, probably the greatest migrant of all, journeys from one end of the earth to the other and back again each year (▷ illustration), while the wandering albatross circles the southern oceans for two years without making landfall, before returning to its island nesting site. Great shearwaters breed in the South Atlantic and then follow the prevailing winds in a clockwise direction, first to Newfoundland, then on to Greenland, and finally south again in the autumn.

European summer visitors, such as swallows, warblers and white storks, avoid the rigours of the northern winter by flying off in the autumn to central or southern Africa. Rather than fly directly across the Mediterranean, many birds cross at the Strait of Gibraltar or take a route around the eastern coast of the Mediterranean (▷ map of pied and collared flycatchers, p. 11).

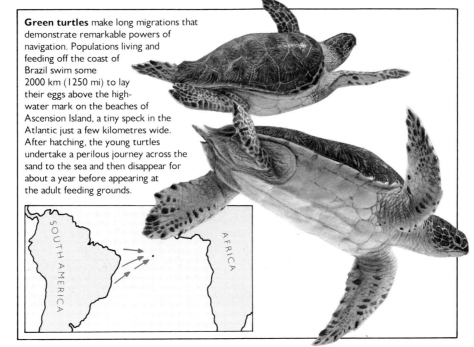

Green turtles make long migrations that demonstrate remarkable powers of navigation. Populations living and feeding off the coast of Brazil swim some 2000 km (1250 mi) to lay their eggs above the high-water mark on the beaches of Ascension Island, a tiny speck in the Atlantic just a few kilometres wide. After hatching, the young turtles undertake a perilous journey across the sand to the sea and then disappear for about a year before appearing at the adult feeding grounds.

SOUTH AMERICA

AFRICA

Blue wildebeest trekking across the plains of the Serengeti. Towards the end of the year, before the start of the wet season, vast herds of wildebeest, often mingled with zebra, topi and other migrants, set off south across the open plains of East Africa in search of water and fresh grass. They are able to sense rain up to 100 km (62 mi) away. In the spring, after the rains are over, they set off north and west to their summer feeding grounds.

Many migrants show great feats of speed and endurance. The tiny ruby-throated hummingbird migrates 1000 km (620 mi) across the Gulf of Mexico. There is no opportunity to stop and rest, and many birds complete the journey in 20 hours. One sandpiper flew 3680 km (2300 mi) from Massachusetts to the Panama Canal in 19 days – an average speed of 164 km (125 mi) per day.

Mammals

Large herbivores such as caribou, wildebeest and zebra are the most significant mammal migrants. On the plains of East Africa, wildebeest, zebra, antelope and elephants gather in huge numbers round water holes in the dry season, spreading out into smaller herds in the wet season. Caribou (reindeer) move from the arctic tundra in autumn to spend the winter in the great coniferous forest further south. In this way they not only gain shelter from the harsh winter but also reduce the impact of predation by wolves.

Other mammalian migrants include bats, whales and seals. Seals return each year to the same beaches to breed. Grey whales living in the eastern Pacific feed in the Bering Sea and move south to breed off California, where the water is warmer for the calves but less rich in food.

The timing of migrations

Migrations are often correlated with regular natural events such as the seasons or the phases of the moon. Before migrating, birds put on weight and show restless behaviour – even if caged they attempt to fly in the migratory direction. These changes are probably controlled by the pituitary gland, which affects weight gain and reproduction, and by the pineal organ, which appears to measure day length and to control annual rhythms of reproduction (⇨ p. 174).

The exact timing of departure tends to depend on environmental conditions such as changes in temperature and in the hours of daylight or decline in food supplies. Some animals appear to have an internal clock (probably associated with the pineal gland) that triggers migration at the appropriate time. In European garden warblers the internal clock apparently determines when the birds leave on their migration, when they change route, and when they stop flying.

Orientation and navigation

How animals find their way during migration is still something of a mystery. Close to the home site familiar landmarks undoubtedly play a part, and geographical features such as coastlines and mountain ranges may be followed for part of the way. Birds may also listen for infrasonic sounds – very low-frequency sounds produced by natural features such as waterfalls, waves and the wind over woodland and hills. Smell may also be important. Salmon can tell the particular river in which they hatched by its smell (⇨ above), and mammals may follow scent trails.

Many animals – insects and fishes as well as birds – use the sun as a compass for orientation, thus maintaining a particular direction. Such animals use their internal clock to compensate for the movements of the sun during the day. Experiments in which birds are placed in planetaria have shown that they can also use the stars for orientation. In addition, birds use the earth's magnetic field, which may be sensed by magnetic material in their bodies.

Most migrations can be accomplished by orientation alone, but many birds show evidence of *navigation* – the ability to reach a distant goal even if displaced from their original route. This means that the animals know the position of their home site, rather than just its general direction. They must therefore have some form of internal 'map' of their route, the nature of which is still unknown. It may involve the earth's magnetic field and, near to home, familiar landmarks, as well as guideposts offered by smell and sound. GS

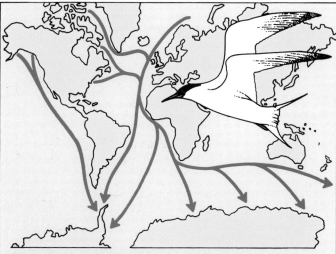

The arctic tern probably experiences more hours of daylight each year than any other creature on earth. Leaving its nesting sites along the northern coasts of America and Eurasia in the autumn, this bird embarks on an extraordinary journey of 16 000 km (10 000 mi) or more to its winter feeding grounds in the extreme southern Atlantic and Pacific. Having enjoyed the brief but productive southern summer, this tireless traveller sets off on the return journey in the following spring.

Population Dynamics

Individuals never live in total isolation. They share their environment with other species and with other individuals of the same species, so there is invariably some degree of competition for resources such as food and living space. When the resources of an area decline or become less attractive relative to the resources to be found elsewhere, many of the animals within the area tend to move on to places where conditions are more favourable.

THE CASE OF THE LYNX AND THE SNOWSHOE HARE

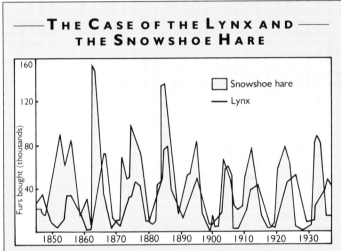

There are a number of examples in nature of populations undergoing regular fluctuations in density. Much research has been undertaken in attempts to understand such phenomena – artificial cycles have been created under controlled conditions and mathematical models have been produced to describe the patterns seen in cycling populations.

One of the best-known studies has focused on the lynx and the snowshoe hare of North America; the latter is the main herbivore in the area concerned and is preyed on by the lynx. In this case, evidence of population fluctuations is provided by statistics kept by the Hudson Bay Company of Canada, which record the number of lynx and hare furs bought over a period of many decades. These records clearly reveal a pattern of population peaks at roughly 10-yearly intervals, with rises and falls in lynx numbers generally lagging a year or two behind those of the hare.

For many years it was supposed that the predatory activity of the lynx was responsible for these population fluctuations, but more recently attention has focused on the hare and its food. As the hare populations increase, so food – especially high-quality food – becomes scarce. The situation is further aggravated by the fact that the new growth produced by severely grazed plants is rich in toxins, which dissuade continued grazing. The drop in the quality of the forage makes many hares sick, and these fall easy prey to the lynx. Thus in this case the driving force behind the population cycles seems to be the interaction of the herbivore and its food, while the lynx – rather than influencing the cycles – appears to be tracking the populations of hares.

When hare numbers are on the decline phase of the cycle, the lynx population must find alternative prey and so switches to grouse. In this way the 10-year cycle affecting the hare and its food induces similar-length cycles in other species living in the same area.

The pressures imposed by such fluctuating resources, in combination with other controlling factors, mean that populations – collections of individuals of a single species living in the same area (⇨ p. 194) – are never static. The density of such a population – the number of individuals in a given area – will therefore constantly be changing over time.

The resources available within an area will generally vary with the season in a more or less predictable fashion every year. Many animals respond to such seasonal fluctuations by migrating to areas where more suitable conditions prevail (⇨ p. 170). In addition to these regular changes, however, there are other influences, both in the environment and within populations of animals, that arise unexpectedly and less predictably, and these influences may also act as stimuli for animals to move out of an area.

If a particular area becomes overpopulated or if its resources unexpectedly decline, some of its inhabitants may be forced to move on in the hope of finding better prospects elsewhere. In other cases, usually in response to unexpected changes in the conditions outside an animal's normal range, there may be an opportunity for a species to extend its range beyond its former limits.

Factors controlling population

All species have the ability to multiply, and in unlimited conditions this leads to an ever-steepening curve of population density. Such growth is termed *exponential*, and is seen in human populations in a number of countries. In natural communities of coexisting plant and animal species, unlimited growth is not possible. When a new habitat is colonized by a species, its population gradually increases. This pattern of growth starts by being exponential, but as the population expands, so it begins to affect the environment about it in an adverse way. Principally, food becomes scarcer, and space more limited. This reduction in the quantity and quality of resources influences the population by slowing its rate of growth; this may come about by a reduction in the numbers of offspring or an increase in the death rate – through starvation or greater susceptibility to disease – or in the numbers leaving the area for new habitats.

Biological forces are also very important in controlling population density. Predators reduce numbers by eating some of the individuals in a population. This factor is particularly significant in invertebrate populations. In an invertebrate predator–prey situation, such as a ladybird feeding on aphids, the predator is able to feed on any prey individual, and therefore can be very important in controlling population density. In many vertebrate predator–prey relations, however, such as lions preying on wildebeest, the predators are

usually only capable of capturing the young, the old, or the sick – fit animals are unlikely to be caught. The old and the sick would die shortly anyway, and in the case of young animals killed by predators, in many instances more young are produced than the resources can sustain, so if some were not killed by predators, they would probably die of starvation instead. Predators therefore generally have little impact on vertebrate populations.

However, the population density of prey animals does have an effect on that of predators. For a predator, its prey is a resource to be exploited. If the predator population increases, its demands on the prey population increase also. This eventually leads to a decline in the number of prey, and in time brings about a reduction in the predator population as their resources diminish. The overall effect is for both predator and prey populations to control each other. A well-known illustration of this phenomenon is seen in the case of the lynx and the snowshoe hare of North America (▷ box).

Disease and parasites also have an effect on population density. Sometimes they may cause death directly, but more commonly they weaken the host and reduce its abilities to reproduce or survive. As in predator–prey relations, there is usually a natural balance in parasitic associations, since a decline in the host population will often reduce the resources available to the parasite.

Changes in climate can have a long-term influence on population dynamics (▷ p. 10), and in the shorter term abnormal weather conditions may also have an effect. For poikilothermic ('cold-blooded') animals, the temperature has to be sufficiently high that they can become metabolically active and so feed and reproduce, and even populations of homoiothermic ('warm-blooded') animals can be devastated by an unusually harsh winter. Drought has obvious effects, and even too much rain or wind can remove large numbers of insects from their habitats in trees and bushes. So population density is determined by a complex interaction of resource abundance, biological forces, climate and weather, and these can all vary over time.

Patterns of change

The effects of these various factors change through time, so that in any one place the density of a species will always be changing. These population cycles are sometimes easy to explain, for example where they correlate with the changing seasons, but in other cases, where the length of the cycle is more than one year, the reasons are less obvious. In Canada, snowshoe hares, lynxes, and a series of other species including willow grouse, foxes and snowy owls are linked in synchronous cycles of approximately 10 years (▷ box). In Europe foxes and various small mammals share a four-year synchronized cycle. Cycles of other lengths are seen in insects.

In other species population changes are more erratic, with long periods of low density occasionally interspersed by dramatic increases. African locusts show this pattern of random but prolonged population change. In this case a very specific combination of climatic conditions are necessary to congregate scattered locusts together in one place suitable for egg laying. This localized increase in density triggers a series of changes in the locusts so that they tend to stay together as a swarm. These swarms grow and create massive damage to crops.

A different pattern of population growth occurs when the area occupied by a species gradually increases to accommodate the growing population. This range expansion is well illustrated by the collared dove, which has colonized much of Europe during the past 50–60 years in a northwards expansion of range (▷ box).

'Boom-and-bust' dynamics

Some species, particularly those we call pests, operate in a radically different way from most animals. Instead of persisting in a particular area in harmony with the levels of resources and predatory pressure, they undergo what is called 'boom-and-bust' dynamics. When such a species finds a suitable habitat, such as a young growing crop, it multiplies very quickly, causing the population density to soar. This rapid growth swamps any potential control by predators and in time outstrips the food resources available. The population then crashes. For their survival, species of this kind rely on rapid growth and maturation by exploiting rich but short-term resources, so that as many individuals as possible reach adulthood. These then disperse to find other suitable habitats to exploit in the same way. BDT

SEE ALSO

- THE PLANET'S CHANGING FACE p. 8
- THE CLIMATIC FACTOR p. 10
- TERRITORY, MATING, SOCIAL ORGANIZATION p. 156
- MIGRATION p. 170
- THE BIOSPHERE: ENERGY RELATIONS p. 194

THE RISE AND RISE OF THE COLLARED DOVE

The collared dove is today an extremely common bird in Europe, with a range that extends throughout central and northern Europe, Britain and southern Scandinavia. It is also a familiar sight – and frequently a pest – in densely populated areas, including large cities such as London and Paris. However, the widespread occurrence of this species is an astonishingly recent phenomenon.

A relative of the pigeon, the collared dove feeds principally on seeds and grain, but will also take berries and fruit and scavenge for human leftovers. Prior to 1930, the species was almost exclusively Asiatic in distribution, penetrating into Europe no further than the Balkans. Over the past 60 years, however, it has made a spectacular extension of its range, spreading steadily in a northwesterly direction at a rate of tens of kilometres a year.

The reasons for this bird's relentless expansion are not fully understood. It is likely that the existence of an under-exploited ecological niche in areas adjacent to its former range has been an important factor in its success, but this still does not explain either the suddenness of the move or its remarkably 'unidirectional' nature. Progressive warming of northern zones may have played a part, or it may be that a genetic mutation in a population on the western periphery of the species' former range provided the initial stimulus to its inexorable conquest.

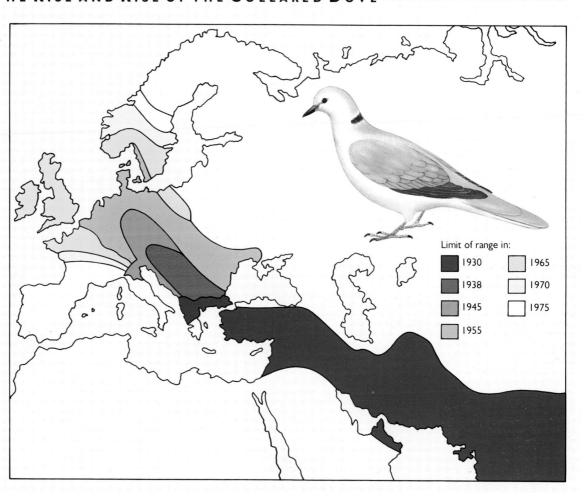

Limit of range in:
- 1930
- 1938
- 1945
- 1955
- 1965
- 1970
- 1975

The Rhythms of Life

Animals live in an ever-changing environment. Sometimes the changes are unexpected, but other important ones are very predictable. The daily rotation of the earth around its axis, the monthly orbit of the moon around the earth, and the annual orbit of the earth around the sun all impose regular changes in the physical environment. Such changes are most obvious in the daily cycles of light and darkness and the seasonal variations in climate.

The complete life histories of most animals – including activity patterns, reproduction and feeding habits – are adapted to fit in with these regular variations in the physical environment. Such rhythms in biological activity pervade the whole animal kingdom, from the simplest organisms to man.

Daily rhythms

Almost all living organisms, both plant and animal, show regular daily rhythms in virtually every aspect of their behaviour, physiology and biochemistry. Some animals are active at night (*nocturnal*), others during the day (*diurnal*). Activity patterns are often linked with feeding and other kinds of behaviour, and these in turn affect metabolism, growth and excretion. In humans – the object of the most extensive studies – daily rhythms are obvious in almost every system: heart rate, respiratory rate, mental activity, temperature, hormone secretion, kidney function, urine secretion, and cell division. Even single-celled organisms show daily changes in their behaviour and physiology.

A particularly interesting feature of most daily rhythms is that they will occur even when the environmental conditions are held constant. They may continue even in

The fiddler crabs of North America emerge at low tide to scavenge for food, and also undergo regular colour changes. Such rhythms are not just responses to the immediate state of the tide. Even under constant laboratory conditions, these animals pass through a cycle of activity every 12 hours or so, clearly showing that an in-built biological rhythm is at work.

cells and tissues cultured in test tubes. Rhythms of this kind that persist in the absence of any external environmental cue, continuing to show a periodicity of about 24 hours, are called *circadian rhythms*. Such rhythms are remarkably stable, and those associated with bodily function may continue without environmental cues for many days, weeks or even years. This internal, in-built sense of rhythm provides each animal with a biological clock, which can be used to predict the daily changes in the environment.

How are these 24-hour rhythms maintained? How can animals, tissues and even single cells tell the time? The answers to these questions are still largely unclear. Synchronization of the daily environmental changes with all the biochemical, physiological and behavioural rhythms within the body requires coordination. In most animals, this seems to depend initially on the ability to detect and follow the very predictable daily changes in light and darkness and sometimes in temperature. These external cues ('zeitgebers') allow the internal rhythms to be correctly synchronized with the outside world. Coordination within the body is then dependent on neural control and circulating hormones; in vertebrates, the pineal gland secretes melatonin at night and is believed to have an important role in coordinating circadian rhythms (▷ pp. 182–3).

Annual and seasonal cycles

Seasonal climatic changes impose many pressures on animals. Temperature, rainfall, humidity, and the availability of food and nesting material may all vary. How do animals cope with and adapt to such seasonal conditions?

The extent of seasonal effects depends on geographical location. Climatic changes in the tropics are much less extreme than those in temperate and polar regions. Where there are marked changes in the climate, animals may migrate to more favourable conditions (▷ p. 170). Animals that remain behind may adapt to the harsher environment. In mammals, thicker insulating layers of fur or fat may develop, which are lost again in the following spring. Sometimes colour changes occur: the plumage of birds such as the ptarmigan and the fur of various mammals including the arctic hare and the arctic fox become white in winter to camouflage the animals against their snowy background.

As well as surviving seasonal changes, the reproductive cycles of animals must be timed so that the young have the best chance of surviving to adulthood. Breeding in temperate regions is often restricted to the favourable conditions of the spring and summer. Sheep mate in the autumn so that the lambs are born in the spring and can grow when the pastures are rich. Nesting and breeding in birds normally start in the spring. In the nonbreeding seasons, the activity of the ovaries and testes of most animals stops (▷ p. 185).

SLEEP

Although we spend about a third of our lives asleep, the precise function of sleep is still unclear. It occurs in mammals other than man, although the periods of sleep may be quite different. Mammals such as sloths, armadillos and opossums may spend up to 80% of their lives sleeping, while small, active animals such as shrews virtually never sleep. In dolphins, one side of the brain sleeps at a time.

Most people go to sleep and wake up according to a fairly regular pattern. This partly reflects a response to the immediate environment but also involves an in-built rhythm of sleepiness and wakefulness. Even under entirely constant conditions and in the absence of any watch or clock, humans show a regular sleep–activity rhythm of about 25 hours. Almost all functions of the body show a regular variation in function and are coordinated with the sleep–activity rhythm: body temperature, urine flow and mental ability all decline at the expected time for sleep. The secretion of many hormones may also vary: growth-hormone secretion increases during the night but is low during the day.

In humans both mental and physical performance may be impaired when the rhythms of bodily functions are out of phase with the normal pattern of activity and the external cycle of changes in the environment. An example of this is the phenomenon known as jet lag, in which the body is suddenly transported to a new time zone. Similarly shift work may result in a change in the normal association between night and sleep. In both cases it may take several days for the internal body rhythms of alertness, temperature and metabolism to readjust to the new pattern of life.

It is vital that animals can predict the oncoming seasonal changes so that they are prepared for adverse or favourable conditions. So how do they know when to breed, migrate, become dormant or awake from dormancy? There is now good evidence that animals use the very predictable seasonal changes in daylight as a cue for timing their seasonal responses. The shortening days after the summer equinox stimulate reproductive activity in sheep and give the cue for hamsters to go into hibernation. In contrast, the lengthening days in spring stimulate birds to start building nests and laying eggs. There is often a critical length of day required to induce the change: quails, for instance, start breeding as soon as the daylength exceeds about 12 hours. In some poikilothermic ('cold-blooded') animals, such as reptiles and amphibians, changing environmental temperature may provide an additional clue to the season.

How does the daylength produce its effects? Animals that can detect light are able to use their in-built circadian rhy-

thms to measure its duration. Once the daylight hours have reached a critical length, the stimulation or inhibition of hormone secretion induces changes elsewhere in the body (▷ p. 183). In vertebrates, the pineal and pituitary glands are involved in these processes.

Lunar and tidal cycles

Although the writers of horror stories would have us believe otherwise, there is little evidence that the phases of the moon affect terrestrial animals. Although the human menstrual cycle is about 28 days, it can be very variable from cycle to cycle. It is not linked to any lunar rhythm and it is probably just coincidence that its average 28-day length approximates to that of the 29·5-day lunar cycle.

The moon does, however, affect many marine animals, especially those living on or near the shore. The rotation of the earth and the lunar cycle drive the daily and monthly cycle of tides. Animals living in the areas affected by tides have to adapt not only to the twice-daily tidal changes but also to the monthly cycle of high and low tides. The activity and feeding behaviour of many animals on the seashore are linked to the ebb and flow of the tides every 12.4 hours. Limpets are firmly attached to rocks when the tide is out, but actively graze the surfaces when the tide is in. The fiddler crabs of North America regularly emerge from their burrows at low tide in an activity cycle that persists in the absence of any environmental cue (▷ photo).

The breeding cycles of some marine animals are intimately linked to the phase of the moon. Some of the most dramatic illustrations of this phenomenon, known as 'moon-phase spawning', are seen in grunion (▷ photo) and the palolo worm. In the waters off Samoa, the palolo worm breeds in October or November at the beginning of the last lunar quarter. Towards dawn on the precise day, these animals, which normally live on the bottom, swim in vast swarms to the sur-

face, where eggs and sperm are released. This astonishing feat of timing is due to hormonal responses to the lunar cycle, which initially stimulates the development of mature eggs and sperm. A particular intensity of moonlight then attracts the animals to the surface and synchronizes swarming. Fishermen can also predict the timing of swarming and gather in their boats to take in a rich harvest of this local delicacy.

Dormancy and hibernation

When conditions become too hot, too cold or too dry, some animals respond by building up internal stores of energy, moving to a sheltered environment, and suspending their activity until favourable conditions return. This is particularly common in small terrestrial or semiterrestrial animals, such as insects, reptiles and amphibians. The environment of the sea and of other large water masses changes far less than on land and dormancy is rare. Although the process can variously be called hibernation, aestivation, dormancy or diapause in different animal groups, the fundamental characteristics are very similar: the rate of metabolism is reduced, heat or cold resistance is increased, and growth and developmental processes are slowed. The reduced metabolism of the animals means that they can survive for considerable periods on their stored energy reserves.

Some small mammals, such as dormice and bats, which are homoiothermic ('warm-blooded'), hibernate to survive cold winter conditions: heart rate and respiration slow dramatically and metabolism is reduced. In this torpid state, body temperature falls but is still controlled by the animal and is maintained at about 2–3 °C (about 36 °F). Hedgehogs may stop breathing for periods of up to an hour, separated by only a few minutes of respiration. Brown and black bears also spend most of the winter months hibernating, but do not undergo the extreme metabolic and respiratory changes found in smaller mammals. Diapause is a common feature of the life cycle of many insects and may occur at any larval stage.

Aestivation is a response to very hot, dry summer conditions. Earthworms hollow out chambers deep in the soil and surround themselves with a mucus coat. African lungfish dig cavities in the mud when their home rivers dry up and can survive in their 'cocoons' for many months until the rains return.   SM

Love on a moonlit beach . . . At spring tides, which coincide with full or new moons, millions of grunion (left) are carried up onto the beaches of the southwestern USA, where eggs and sperm are shed. The fertilized eggs are then buried in the sand, and will only hatch at the next spring tide two weeks later.

Saving up for a rainy day. For a few animals living in areas of unpredictable climate, the signal to breed or emerge from dormancy may come from a sudden change in the weather. In the very arid regions of Australia, zebra finches (pictured here) breed only immediately after a rainstorm and the young are fed on the fresh soft seeds of the grasses that develop following the rain.

How Animals Move

One of the most important characteristics of animals – and the one that most conspicuously distinguishes them from plants – is the ability to move. Many animals move actively, using muscles and skeletons or fine hair-like projections. Some creep along the ground or swim through water. Others fly or glide through the air, or burrow through earth, wood or even flesh.

Certain animals are carried passively by wind or water currents. Such animals often have projections such as fine hairs, or are flattened to increase their surface area. Jellyfish can move themselves through water by pulsations of the bell, but they can also be carried passively in the surface waters over great distances. Some very young spiders move by 'ballooning': climbing to the top of a twig or branch, they produce a fine silk thread that is caught by the wind, so carrying the tiny spiders away.

Movement in very small animals

Single-celled organisms (▷ p. 26) such as amoebas move by flowing. The outer jelly-like layer of the cell is converted at the rear end to a fluid, which flows forward and is converted back to jelly at the front. Other protists have either a whip-like structure called the *flagellum*, which undulates, or rows of hundreds of tiny hair-like projections called *cilia*, which beat one after another, so that waves

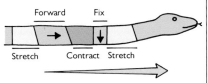

Pterosaurs were flying reptiles that arose in the Jurassic period (beginning 205 million years ago) and dominated the skies until their disappearance in the great extinction at the end of the Cretaceous, some 65 million years ago. Although many were bird-sized, others grew to gigantic proportions, with wing spans of up to 11 m (36 ft). Unlike the wings of birds and bats, the pterosaur's flight membrane was supported only along the leading edge, by a 'strut' formed by the arm and one extremely elongated finger. On the basis of their anatomy, it is thought that pterosaurs were somewhat slow flyers, gliding for the most part and with less control than modern birds or bats.

of movement sweep across the organism's body. Tiny flatworms crawl over surfaces by beating rows of cilia in a fine film of mucus that they secrete.

Muscles and skeletons

Most animals move by using muscles that contract against a rigid skeleton and so move parts of the body connected to them. As muscles are capable only of contraction, not extension, they are generally arranged in *antagonistic pairs*, where the contraction of one muscle extends another (▷ diagram).

In vertebrates, skeletons are internal and made of cartilage or bone. Invertebrates such as insects and crustaceans have hard external skeletons (▷ p. 58), while others – worms, coelenterates and molluscs, for instance – have internal fluid-filled (or *hydrostatic*) skeletons.

Crawling

Animals without tails, fins or legs often move by alternately shortening and lengthening either the whole or parts of their body. Leeches extend the body by contracting circular muscles running around the body while attached to a surface by a sucker at the rear end. The front sucker then attaches, the rear one releases, and the body is drawn forward by contraction of longitudinal muscles running the length of the body.

In earthworms the body is divided into segments, each with its own set of circular and longitudinal muscles and hydrostatic skeleton. When the longitudinal muscles are contracted, the segment is short and fat and pushes against the ground, being anchored by means of mucus and tiny bristles called *chaetae* (or *setae*). When the circular muscles contract, the segment is thin and elongated. During crawling, waves of contraction, first of one set of muscles then of another, pass down the worm's body. When burrowing, the chaetae at the rear end anchor the body in the burrow while the front part pushes forward between the soil particles.

Slugs and snails have a single large flat foot. Waves of muscular contraction pass over the foot, lifting small areas off the ground and moving them forwards. In snakes there are a number of different types of movement, depending on the species and on the kind of surface over which the animal is travelling (▷ diagram).

Movement in water

Long aquatic animals such as lampreys, eels and water snakes swim by throwing their body into lateral undulations that push back against the water – the method of movement probably used by early vertebrates. In most fishes, however, the tail is the main propulsive force. This pushes sideways and backwards against the water on each stroke, driving the fish forwards. The fins keep the fish stable and are used for fine adjustment of position. Marine mammals such as whales and dolphins use a similar method of propulsion (although the tail is set horizontally, and moves up and down), while walruses and seals flap their hind feet from side to

SNAKE MOVEMENT

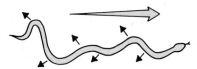

Serpentine movement is the most characteristic form of snake locomotion. The body is thrown into waves that travel from head to tail. At the crest of each wave the body pushes back (small arrows) against the ground, so moving the snake forward.

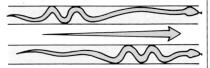

Concertina movement is used in confined spaces, such as a crevice or the burrow of a prey animal. The snake anchors its rear end by pressing tight loops against the confining wall and extends its front part. The front is then secured and the rear part drawn forward.

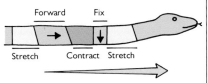

Straight-line movement is achieved by stretching and contracting alternate sections of the snake's underside, rather in the manner of an earthworm. Although slow, such movement is inconspicuous, and used – especially by heavy-bodied snakes – when stalking prey.

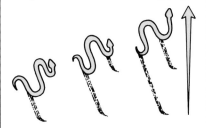

Sidewinding is used by certain snakes living in deserts, where the slippery sands do not give sufficient anchorage for other types of movement. The snake's weight is supported at two or three points only, from which successive loops of the body are swung forward.

side. Rays and flatfish move all or parts of their body up and down when swimming. In rays these undulations run down the side of the broad lateral fins, but in flatfish the whole body is flapped up and down.

Some animals row themselves through the water – sea lions use the front flippers like oars, while water boatmen (a kind of insect) use a pair of legs covered with fine hairs (▷ p. 59). Squids and their relatives move by jet propulsion: water is drawn into a large muscular-walled cavity, and then forced out in a particular direction through a narrow flexible funnel.

Walking, running and jumping

Movement on land often involves the use of legs. When walking, two-legged or *bipedal* animals – humans and birds –

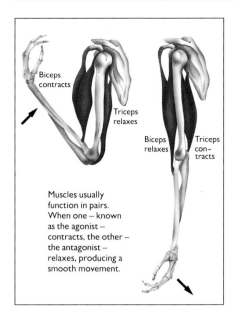

Biceps contracts

Triceps relaxes

Biceps relaxes

Triceps contracts

Muscles usually function in pairs. When one – known as the agonist – contracts, the other – the antagonist – relaxes, producing a smooth movement.

always keep at least one foot on the ground at any one time, but in running there are times when both feet leave the ground. Dinosaurs were also bipedal and presumably moved in a similar way.

Animals with four legs often show various gaits, which differ in the order in which the legs are moved and in their speed. Elephants, for instance, have two gaits – a walk and an amble (a fast walk), while horses have four gaits – a walk, a trot, a canter and a gallop. In most mammals walking involves moving one leg at a time – first a front leg, then the opposite hind leg, then the other front leg, and finally the other hind leg. In all other mammalian gaits, such as trotting, cantering, galloping and bounding, two, three and even all four legs can be off the ground at the same time. Long legs and large strides, separately or in combination, lead to increased speed in fast runners such as dogs, horses and cheetahs (⊳ illustration).

Many reptiles and amphibians have long bodies and hold their limbs out to the side. They bend their

body as they walk in a similar manner to the lateral undulations of fish, and move their legs in diagonally opposite pairs. Similar patterns of movement are seen in the fins of fish such as the coelacanth and the lungfishes, which are thought to be closely related to four-legged vertebrates.

Insects, with six legs, move the legs in sets of three, the first and third on one side and the second on the other. The other three legs form a triangle or rotating tripod on which the animal balances. Millipedes and centipedes, which may have a hundred legs or more, move the legs on each side of the body in order, one after another. Depending on the species, waves of leg movement travel backwards or forwards along the body, and opposite legs move either simultaneously or alternately.

Jumping is both a rapid means of moving from one place to another and an efficient one, in terms of energy consumption. Frogs, kangaroos, grasshoppers and fleas all hop or jump by pushing off the ground with long back legs. Kangaroos have elastic tendons in the legs that are stretched on landing and store energy to push the animal off as the tendons recoil. Fleas and grasshoppers store energy in an elastic protein in the legs by slow muscular contraction. A release mechanism triggers the recoil of the protein, catapulting the animal into the air. Fleas are the most accomplished leapers in the animal kingdom, being able to jump more than a hundred times their own height.

Gliding and flight

The use of extended flaps of skin for gliding or parachuting is seen in several groups of animals. Flying fish have enlarged lateral fins, flying amphibians use long webbed toes as parachutes, flying dragons (flying lizards) support a web of skin on elongated ribs, flying snakes flatten the body, and flying squirrels, possums and lemurs have a flap of skin between the fore and hind limbs. In these animals 'flight' is more or less uncontrolled and somewhat erratic.

True flight is seen today only in insects, birds and bats, but in prehistoric times there was a wide range of flying reptiles, some of them of gigantic proportions (⊳ illustration). The modifications necessary to achieve flight are very great, particularly in heavier animals. However, the advantages conferred by flight can readily be judged by the success of those animal groups that have achieved it.

All animals capable of controlled flight have wings that can move up and down and be held at different

JOINTS

A ball-and-socket joint, as in the mammalian hip and shoulder, allows movement in all directions, including rotation, but is susceptible to dislocation.

A saddle joint allows versatile movement in several directions, and – very significantly, in the case of the primate thumb joint – the 'opposition' of the thumb to the fingers that is characteristic of precise movements such as grasping.

A hinge joint, as in the knee and elbow, allows swinging movement, mostly in a single plane.

A pivotal joint is mainly restricted to rotational movement. Movement of the head from side to side is primarily due to a pivotal joint between the first and second neck vertebrae.

A condyloid joint, such as the wrist joint, allows both rotation and backward and forward movement.

A plane joint, such as that between the pelvis and the base of the spine, allows only very limited movement, except during pregnancy in mammals, when the pelvis expands to accommodate the growing foetus.

angles. When the wing is held at an angle to the air flowing over it, the air has to move further over the upper surface. This creates a lower pressure above the wing and higher pressure below it, thus producing a lifting force. This lift operates in gliding and flapping flight. When the wings are also moved downwards, both lift and a forward thrust are produced. On the upward stroke the wing is twisted to reduce reverse thrust.

Birds and bats have powerful chest muscles, attached to large projections on the breastbones, to produce strong downward wing movements on the power stroke. In birds the wings are formed from extended forearms covered with feathers (⊳ pp. 92–3). In bats, the hands are enlarged and the wing surface is a double flap of skin.

Insect wings project between the top and the sides of the thorax (⊳ p. 59). In fast-flying insects the flight muscles are not attached directly to the wing. One set of muscles, running from the roof to the floor of the thorax, contracts in such a way that the muscles pull the roof downwards and so raise the wing through a lever system at the base of the wing. The downstroke occurs when longitudinal muscles attached to each end of the thorax contract, so buckling the roof upwards. Once set in motion the muscles extend each other and then contract at a very high rate, giving flight speeds of up to 40 km/h (25 mph) in some moths.   GS

SEE ALSO

● INSECTS: FORM AND STRUCTURE p. 59
● FISHES: FORM AND STRUCTURE p. 72
● AMPHIBIANS: FORM AND STRUCTURE p. 79
● REPTILES: FORM AND STRUCTURE p. 85
● BIRDS: FORM AND STRUCTURE p. 92

Fast runners such as cheetahs (left) arch the spine upwards as they run to bring the hind feet as far forward as possible, thus increasing stride length and so speed. At the highest speeds, the hind legs land in front of the position previously occupied by the forelegs.

Digestion

Unlike plants, which are able to synthesize all the complex molecules that they require using simple inorganic molecules and energy from the sun (▷ p. 31), all animals are consumers and depend directly or indirectly on plants to provide their food (▷ p. 194). *Nutrients* are the substances present in food that are used as sources of energy and for all the processes involved in the growth, repair and maintenance of the body. Virtually all food, whether plant or animal in origin, consists of highly complex compounds that cannot be used by another animal unless they are first broken down into simpler compounds and absorbed into the body.

Both the body plan and the life style of every species are adapted to its particular way of obtaining food. All but the most simple animals have structural specializations in the form of a *digestive tract* (also called an *alimentary canal*), which is designed to deal with the problems of food intake and processing and absorption of nutrients.

Types of digestive system

The mechanisms of feeding can be very simple. Some parasites, such as tapeworms, simply absorb the nutrients they require from their hosts, already digested and in soluble form. However, the food of most animals is not in solution but solid. The mouth and digestive tract of such animals are therefore adapted to carrying out a series of separate functions:

1. Selecting and collecting food
2. Physically breaking up the food into small particles
3. Transporting and storing food
4. Chemically digesting the food into soluble nutrients
5. Absorbing the nutrients into the body
6. Eliminating unabsorbed and undigested food.

In simple organisms such as amoebas and sponges, microscopic algae and bacteria can be taken in directly and digested within the cell (▷ p. 26). Other animals, such as hydras, jellyfishes and flatworms, have very simple digestive tracts with only one opening: the mouth opens into a large cavity in which both digestion and absorption takes place; undigested material is then eliminated through the mouth.

In most animals, the digestive tract is a complete tube, with the mouth at one end and the anus at the other. The early parts of the tract are associated with storage and digestion, while later parts absorb the digested food. As it is processed, food is pushed through the tract by rhythmical contractions (*peristalsis*) of the muscles in the wall. The passage of food particles often requires lubrication, and this is provided by secretions such as saliva.

Selection and ingestion

Food selection is an extremely important part of feeding (▷ p. 162). Even in simple animals, the initial intake or *ingestion* of food particles is selective. In more complex organisms, the selection process may involve complex behavioural adaptations to locate suitable food sources, together with complex sensory mechanisms to identify particular types of food.

For many animals, food comes in the form of very small particles. Animals such as earthworms feed on small plant and animal food particles in the soil and have to ingest large amounts of material to obtain sufficient amounts of food. Many animals, including sponges and mussels, are *filter-feeders*, which have developed special mechanisms to trap tiny food particles suspended in water (▷ p. 57). It is not only small animals that depend exclusively or partially on particulate food. Herring and mackerel have structures (*gill rakers*) on their gills that act as a sieve to trap plankton, while the largest of all marine animals – baleen whales, basking sharks and whale sharks – are also filter-feeders

Animals that feed on bulkier food have to break it up first, to provide a larger surface area on which digestive juices can act. While only vertebrates have proper (bony) teeth, other animals have hard specialized mouthparts to perform a similar function. Amongst the methods used to break up food are chewing, rasping, scraping, boring and tearing. Some herbivores, such as aphids, have piercing-and-sucking mouthparts that enable them to feed on plant juices, while in various blood-feeders, including ticks, fleas, mosquitoes, leeches and vampire bats, a variety of functionally similar mouthparts have evolved by means of which they are able to feed on animal body fluids (▷ p. 59).

In some animals, the mechanical breakdown of food takes place in the *gizzard*. This is a special compartment early in the digestive tract with thick muscular walls. Earthworms have a gizzard that continuously grinds the ingested soil and plant material. Birds have no teeth but have a gizzard, and often swallow small stones to assist the grinding process. Birds also have a *crop* – a small storage compartment that precedes the gizzard.

Digestion

Digestion is the process by which the large, complex molecules in food are broken down into their simple constitu-

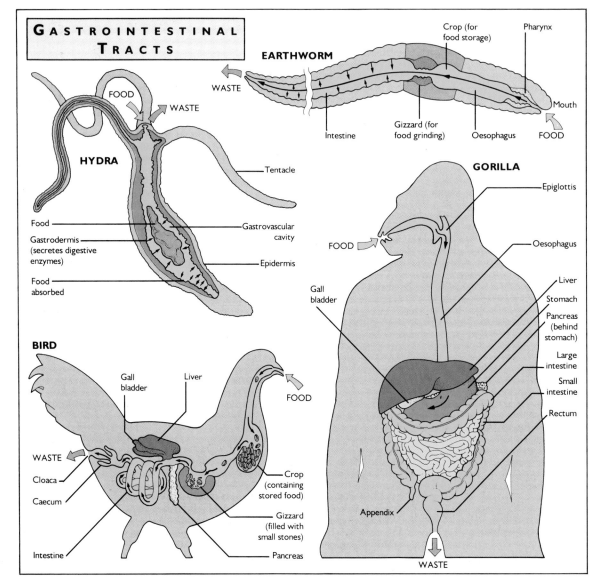

GASTROINTESTINAL TRACTS

EARTHWORM
Crop (for food storage)
Pharynx
WASTE
FOOD
WASTE
Mouth
Intestine
Gizzard (for food grinding)
Oesophagus
FOOD

HYDRA
FOOD
Tentacle
Food
Gastrovascular cavity
Gastrodermis (secretes digestive enzymes)
Epidermis
Food absorbed

BIRD
Gall bladder
Liver
FOOD
WASTE
Cloaca
Caecum
Crop (containing stored food)
Gizzard (filled with small stones)
Intestine
Pancreas

GORILLA
Epiglottis
FOOD
Oesophagus
Liver
Gall bladder
Stomach
Pancreas (behind stomach)
Large intestine
Small intestine
Rectum
Appendix
WASTE

ents. While this occurs within the cells of single-celled animals and sponges, animals with more complex digestive tracts secrete *enzymes* (proteins that act as catalysts in biochemical reactions) into the tract. The enzymes are specific, targeting and breaking down only certain types of carbohydrate, protein or fat. To prevent the enzymes from attacking the digestive tract itself, they are secreted in an inactive form and cannot function until they have been triggered by other chemicals in the digestive tract. In addition, mucus is secreted to line the walls of the tract. The various watery secretions mix with the food particles to form a semi-fluid *chyme*.

The *stomach* can act as a temporary storage place for food, where mixing and some initial digestion occurs. The stomach contents of vertebrates are usually acidic, so that most bacteria ingested with the food are killed. In ruminants, such as sheep and cows, parts of the stomach are specialized to allow fermentation of the plant material by symbiotic microorganisms (▷ p. 112). In vertebrates, the semi-digested food is released slowly into the first part of the small intestine, the *duodenum*. *Bile*, produced by the liver (▷ below) and stored in the *gall bladder*, is then released into the duodenum. This helps to break up (*emulsify*) large fat globules, again allowing a large surface area for enzyme action. Further enzymes are secreted by the *pancreas* and from glands in the intestinal wall.

Some foods are not easily digested. Most animals do not produce any enzymes for digesting cellulose, so animals that use cellulose as part of their diet are usually dependent on symbiotic microorganisms to produce the necessary digestive enzymes (▷ p. 169). The digestion of cellulose in mammals such as horses and rabbits occurs in the *caecum*, a blind-ending branch of the alimentary canal at the point where the lower part of the small intestine (*ileum*) meets the large intestine.

Absorption and elimination

The action of enzymes on the food ultimately produces small sugar, amino-acid and fatty-acid molecules (▷ below). These are absorbed across the wall of the digestive tract and carried to other areas of the body. In vertebrates, the inner surface of the ileum is covered by thousands of tiny, finger-like projections called *villi*, which provide a large surface area for absorption. Sugars and amino acids cross the villi and enter the blood, while fatty acids and glycerol enter the lymphatic system (▷ p. 181). In vertebrates, the blood draining the ileum is transported directly to the *liver*, where the amounts of free glucose and amino acids in the blood are regulated and excess glucose is stored as glycogen.

The lower regions of digestive tracts are specialized to eliminate unabsorbed and undigested material. In terrestrial animals, reabsorption of water from the chyme is essential. The large intestine of vertebrates consists of the *colon* and the *rectum*. These do not secrete enzymes, but

the colon is very important in the reabsorption of water. Normally the semi-solid waste (*faeces*) passes into the rectum and is expelled at intervals through the anus. If the intestinal contents pass through the colon too rapidly, only a small amount of water may be reabsorbed. This condition is known as *diarrhoea* and results in frequent defecation, watery faeces, and dehydration.

Dietary constituents

Although the precise source of nutrients may differ greatly between one animal and another, the fundamental requirements of most animals are very similar. This reflects the similar basic composition of all animal cells (▷ p. 20). The digestive systems of animals are designed to break down the complex macromolecules in food (carbohydrates, fats and proteins) into their basic components (sugars, fatty acids, glycerol and amino acids). These can then be absorbed and used in specific reactions within the body, reactions that either synthesize new chemical compounds necessary for making and maintaining cells, or that release the energy required by the animal.

Carbohydrates are based on sugar molecules and may be in the form of simple sugars (such as glucose, fructose, galactose and sucrose) or much more complex polysaccharides (such as starch, glycogen and cellulose). Dietary carbohydrates are a major source of energy, being oxidized to release carbon dioxide and water. Excess carbohydrate may be stored either as glycogen (for example, in the liver or in muscles) or converted into fat and stored in fat cells beneath the skin and elsewhere.

Fats are complex molecules consisting of fatty acids and glycerol. They are essential components in the structure of cell membranes (▷ p. 20), and are also very important as sources of energy, often being stored in preference to carbohydrates. Such internal stores of energy are particularly useful to animals whose food supply may be periodically limited. The migratory locust uses fat during its long flights (▷ p. 173), while various mammals lay down large fat stores before entering hibernation (▷ p. 175). Fat deposited beneath the skin of birds and mammals also has a very useful insulating function.

Proteins are made up of long chains of amino acids linked in a specific order that is genetically determined (▷ pp. 6–7). Proteins have several vital functions: they help to provide specific cell structure; as hormones, they act as signals to other cells (▷ p. 182); as enzymes, they regulate other cell activities; and they are the basis of motile activity in cells. Of the 20 or so different amino acids, about eight are essential in most animals, in the sense that they must be provided by dietary protein. The others are non-essential and can be synthesized from other amino acids.

Other vital dietary components are vitamins and minerals. *Vitamins* are complex chemical compounds that have no energy

EXCRETION

As well as food, water and minerals are essential to all animals. The percentage of water in different tissues varies widely – it accounts for 20% of bone tissue, 85% of vertebrate brain cells, and 95% of the body tissue of jellyfishes. Although water is often an integral part of food, it is lost in urine, faeces and through body surfaces. Many animals have to drink additional water to maintain their fluid balance. Mineral-salt concentrations are also very important: the total concentration of salts in the bodies of most marine invertebrates is equivalent to that of sea water (3.4%), while almost all vertebrates have less than 1% salt in their body fluids. The precise balance of water and salts is critical to the functioning of cells. In the face of variable food and water intake, regulation of the water and salt balance (*osmoregulation*) and excretion of other waste products is essential to maintain a constant internal environment.

Simple animals such as jellyfishes have no specialized excretory system and waste simply passes out by diffusion. More advanced animals, however, have specialized structures for this purpose. These all work in a similar way: fluid from the blood or the fluid surrounding cells (interstitial fluid) is collected, its composition adjusted by secretion or reabsorption of particular substances, and any waste expelled from the body. Invertebrates have tubular nephridia that perform this function, while in vertebrates it is carried out by the kidney, which consists of thousands of individual nephrons.

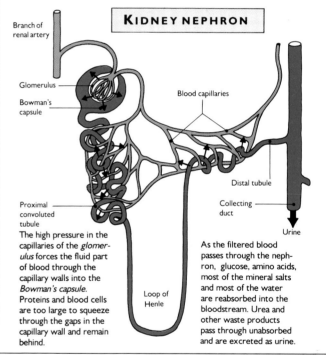

KIDNEY NEPHRON

Branch of renal artery

Glomerulus

Bowman's capsule

Blood capillaries

Distal tubule

Proximal convoluted tubule

Collecting duct

Urine

Loop of Henle

The high pressure in the capillaries of the *glomerulus* forces the fluid part of blood through the capillary walls into the *Bowman's capsule*. Proteins and blood cells are too large to squeeze through the gaps in the capillary wall and remain behind.

As the filtered blood passes through the nephron, glucose, amino acids, most of the mineral salts and most of the water are reabsorbed into the bloodstream. Urea and other waste products pass through unabsorbed and are excreted as urine.

value in themselves but are essential in small quantities for many chemical reactions. Different species vary both in the vitamins that they need and in their ability to synthesize them. For example, while dietary vitamin C (ascorbic acid) is necessary to man, most animals can synthesize it from glucose. Similarly, while vitamins D and K are essential to vertebrates, no role for them has yet been found in invertebrates.

A wide range of *mineral salts* is necessary for various chemical and physical processes in animals. Some are required in relatively large quantities, such as calcium in bones and shells, sodium in cells, and potassium in body fluids, while only trace amounts of others are necessary (zinc, for example, is an important component of many enzymes). Deficiencies in minerals or vitamins can cause serious health problems.   SM

SEE ALSO

● WHAT IS LIFE? p. 6
● CELLS p. 20
● HOW PLANTS FUNCTION p. 31
● INSECTS: FORM AND STRUCTURE p. 59
● NUTRITION AND FOOD SELECTION p. 162
● PREDATOR AND PREY p. 164
● PARASITISM AND SYMBIOSIS p. 168
● RESPIRATION AND CIRCULATION p. 180
● THE BIOSPHERE pp. 194–7

Respiration and Circulation

All animal cells require a constant supply of oxygen, so that food can be oxidized ('burnt') to release energy (▷ p. 21). At the same time, nutrients must be carried to each cell, and the various waste products of cellular activity – including carbon dioxide, the by-product of respiration – must be transported away.

In more advanced animals, special systems have developed to meet these needs. Respiratory systems allow sufficient oxygen to be obtained from the environment and carbon dioxide to be removed, while cirulatory systems are responsible for bringing oxygen and nutrients to cells and carrying away carbon dioxide and other waste products.

Respiration

The way an animal obtains oxygen from the environment is dictated by its basic body plan, its life style, and whether it lives in air or water. The amount of oxygen entering an animal and the amount of carbon dioxide lost depends on the surface area available for gas exchange. In small, simple animals with low energy requirements, such as sponges and hydras, gas exchange across the body surface is sufficient.

In larger animals, special respiratory systems have developed. Although these differ in structure, all are designed to provide large surface areas for the exchange of gases. A system of *ventilation* is also needed, whereby the oxygen-carrying air or water in contact with the respiratory surfaces is replaced at regular intervals. The more active an animal is, the greater its energy demand and the more oxygen it requires. Active, warm-blooded (homoiothermic) animals such as birds and mammals, which have high energy requirements, have respiratory systems with very large surface areas and efficient ventilation.

Oxygen does not diffuse very far in tissues, penetrating only about 1 mm ($^1/_{25}$ in) into tissues in sufficient amounts to maintain life. To ensure that sufficient oxygen reaches all cells, the respiratory surfaces are usually well supplied with blood. They are also very thin, to allow oxygen to diffuse in quickly.

Only small amounts of oxygen can be carried in simple solution, so many animals have special molecules called *respiratory pigments* in their blood. These molecules are able to combine with oxygen, picking it up at the respiratory surfaces and liberating it to the cells inside the body. The respiratory pigment in vertebrate blood – *haemoglobin* – contains iron atoms and is responsible for its red colour. It increases the capacity of the blood to carry oxygen by about 75 times. Although haemoglobin is also present in many invertebrates, some molluscs and arthropods have a blue, copper-containing pigment called *haemocyanin*. The metal atoms in both these molecules are important in binding oxygen. In vertebrates the respiratory pigment is contained within blood cells, but in many invertebrates the pigments are free in the blood fluid or *plasma*.

Respiration in air and water

Animals must obtain oxygen from their immediate environment – either from air or from water. Air is much lighter than water but contains far more oxygen – about 30 times as much for a given volume. Getting oxygen from air is therefore much easier than from water. A fish may use up to 20% of its total energy in just ventilating its gills, while a mammal may use only 1 or 2% of its energy in breathing air. Another advantage of air-breathing is that air has a low heat capacity and relatively little heat is lost despite the continuous ventilation of large respiratory surfaces. It is no coincidence that the only truly warm-blooded animals – birds and mammals – are air-breathers.

Breathing air does have some disadvantages, however. The large respiratory surfaces have to be kept moist, so that gases can dissolve in the surface layers before they diffuse across. Evaporation of water from these moist surfaces can be an important cause of water loss in terrestrial animals. Water lost in this way can easily be seen on cold days, when water vapour in the breath condenses in the cold air.

Respiratory structures

Gas exchange across the moist body surface occurs in many simple invertebrates, including most worms. These animals tend to be small, with a low metabolic rate and a low demand for oxygen. Movements of the animals or of the surrounding medium are sufficient to keep relatively fresh, oxygen-rich air or water in contact with the body surface. Most amphibians

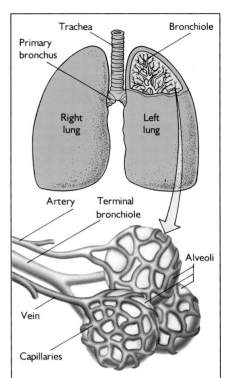

Respiration in mammals involves the use of two large, bellow-like lungs. Air is drawn down the *trachea* (windpipe) and passes into the lungs through a system of ever-narrowing, branching tubes (*bronchi* and *bronchioles*), which finally terminate in millions of tiny air sacs called *alveoli*. The wall of each alveolus is lined with a network of minute, thin-walled blood vessels called *capillaries*. As deoxygenated blood from the heart passes through the capillaries, it loses the carbon dioxide it is carrying, which diffuses into the alveoli to be breathed out from the lungs, and picks up oxygen; the oxygenated blood is then returned through veins to the heart, to be pumped around the body once again.

are also very efficient at 'skin-breathing', although it is not their only means of respiration (▷ p. 79).

Gills are more specialized respiratory structures found in aquatic animals. They are thin and often folded to provide a large surface area for gas exchange. While their outer surfaces are in contact with water, internally they usually have a network of blood vessels. Gills may be external, extending out into the water as in the tadpoles of amphibians, but in most vertebrates they are internal. In fishes, the gills are usually located inside a series of slits that perforate the wall of the pharynx, and are ventilated by water passing in through the mouth and out through the slits. In bony fishes, each gill consists of many thin filaments. Efficiency of gas exchange is maximized by passing blood through the gill filaments in the opposite direction to the flow of water (▷ pp. 72–3).

Insects are mostly small, active terrestrial animals. To avoid excessive water loss, they are covered by a hard, impermeable cuticle. Air enters the body through a unique system of narrow, branching tubes (*tracheae*) that carry air deep within the body (▷ illustration). The size of the

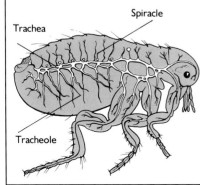

The tracheal respiratory system of an insect (flea). Faced with the constant danger of dehydration, insects have a unique respiratory system that requires only tiny openings to the air outside. Air enters the body through pairs of these openings (spiracles) situated along the abdomen, and is then carried by ever-narrowing tubes (tracheae and tracheoles) to the body cells deep within, where gases are exchanged.

openings to the exterior (*spiracles*) can be adjusted according to external conditions, and in active insects movements of the body help to move air in and out through the spiracles.

Lungs are larger respiratory structures adapted for exchange of gases with air. Some invertebrates have lungs. Spiders have *book lungs* – air-filled cavities in the abdomen into which air enters via a spiracle. A series of parallel, blood-filled plates hang in each cavity to exchange gases. In lower vertebrates, such as frogs and reptiles, the lungs are very simple sacs that are ventilated by movements of the body or mouth.

In birds and mammals, which have high metabolic rates, the lungs are complex and efficiently ventilated. The enormous surface area of mammalian lungs is due to millions of tiny, thin air sacs known as *alveoli* (⇨ diagram). Regular ventilation is achieved by contractions of the *diaphragm* – a muscular partition that separates the chest from the abdomen. The respiratory system of birds is arranged so that there is a one-way flow of air through the lungs. Air passes into a number of air sacs and through tiny, thin-walled tubes (*parabronchi*), where exchange of gases between air and blood takes place. The flow of air is maintained by movements of the chest and by the squeezing effect of muscles on the air sacs during flight.

Circulatory systems

Each animal cell requires an uninterrupted supply of oxygen and nutrients, and waste products must be continuously removed. Small, simple animals that live in a watery environment, such as sponges and flatworms, have no special circulatory systems. Their bodies are only a few cells thick and diffusion alone is sufficient. In larger animals, diffusion is not sufficient to meet the needs of all the cells in every part of the body, and circulatory systems have evolved to transport materials both between cells and between cells and the outside environment.

Circulatory systems typically consist of three components:

1. A fluid transport medium (*blood*) in which nutrients, gases and waste products are dissolved.
2. A mechanism by which the blood is pumped round the body (usually known as a *heart*).
3. A system of spaces through which the blood circulates.

Arthropods and many molluscs have *open circulatory systems*, in which the heart pumps blood into vessels that are open-ended. Blood flows out of the open vessels to bathe all the cells directly. The blood-filled spaces are called *haemocoels*. Blood returns to the heart via other open-ended vessels or through holes in the heart. Blood flow in such open systems is relatively slow.

Vertebrates, echinoderms and molluscs such as octopuses and squids have *closed circulatory systems*. The blood passes through a continuous network of flexible blood vessels, and can be pumped more quickly. Large vessels (*arteries*) branch

into smaller and smaller vessels, carrying blood away from the heart to all parts of the body. The smallest vessels (*capillaries*) are quite permeable and allow fluid to leak out to form the tissue fluid (*interstitial fluid*) that surrounds all cells. The exchange of gases, nutrients and waste materials between cells and the blood takes place by diffusion through the interstitial fluid. Vertebrates have a separate system of small vessels, the *lymphatic system*, by which excess interstitial fluid is collected and returned to the blood. After passing through the capillaries, blood returns to the heart through a network of *veins*.

Pressure is necessary to force blood through vessels and around circulatory systems. This pressure may simply be generated by movements of the surrounding muscles in the body. Return of blood through the veins in the human leg is assisted by the contractions of the leg muscles during walking. In more advanced animals, part or parts of the circulatory system are specialized as one or more hearts to pump the blood. In the simplest cases, rhythmical contractions (*peristalsis*) of the muscles in the vessel walls force the blood along. Annelid worms have a number of vessels that are specialized to act as hearts. *Valves* are used in hearts and in veins to ensure that the flow of blood proceeds in the right direction.

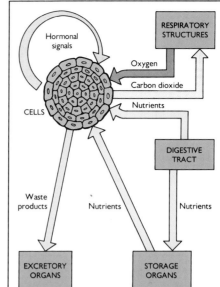

Functions of blood and circulatory systems. As well as serving various functions in relation to the transport of gases, nutrients, waste products and hormones, blood and circulatory systems play a number of other vital roles:

1. Helping to maintain salt and water balance between all parts of the body.
2. Helping to transfer heat and regulate body temperature in warm-blooded (homoiothermic) animals.
3. Defending against invading organisms (by means of white blood cells and antibodies).
4. Repairing tissue damage and preventing blood loss (through coagulation).

More specialized hearts consist of strong, muscular chambers. Although these have evolved from contractile blood vessels, they are able to pump much larger volumes of blood with greater force. Such structures range from the relatively simple heart of a fish to the high-pressure systems that are found in higher vertebrates such as birds and mammals (⇨ illustration).   SM

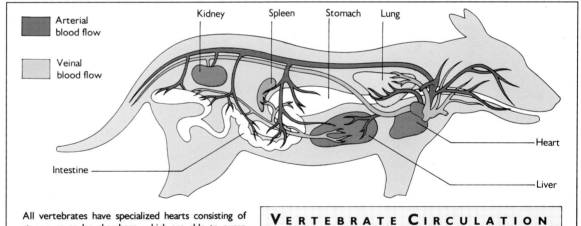

All vertebrates have specialized hearts consisting of strong, muscular chambers, which are able to pump large volumes of blood with considerable force. Lower vertebrates, such as fishes, have a *single circulatory system*. In a fish such as a shark, deoxygenated blood from the veins is collected in a thin-walled chamber (*atrium*) and then passed to a muscular, thick-walled chamber (*ventricle*), which contracts forcibly to send it to the gills, where it gains oxygen, and then straight on to the rest of the body. Because the passage of blood from the heart to the organs and tissues takes place via the gills, the pressure in such a system is rather low.

As vertebrate life changes from aquatic to terrestrial and air-breathing, the circulation becomes more complex. In birds and mammals, the heart has two atria and two ventricles, which serve a *double circulatory system* (above and right). Deoxygenated blood from the organs and tissues returns to the right side of the heart, which pumps it to the lungs. The oxygenated blood then returns to the left side of the heart, which pumps it out with great force to the rest of the body. This high-pressure flow of oxygenated blood to active cells allows them to maintain the high metabolic rate characteristic of birds and mammals.

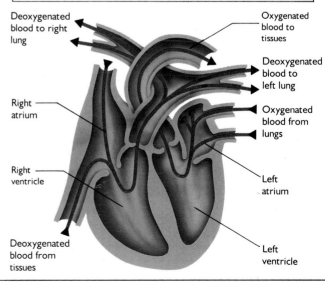

VERTEBRATE CIRCULATION

Nervous and Endocrine Systems

In a multicellular animal, the activity of all the cells must be coordinated if the body is to work as a single functional unit.

An animal must be able to respond as a whole to specific sensory stimuli such as heat and touch that are received at a single point on the body. At the same time, the development and functioning of the various tissues and organs involved in physiological processes such as digestion, reproduction and respiration must be precisely coordinated.

The coordination of animal cells – which may be widely separated in the body – depends on two basic mechanisms. Nervous systems based on vast networks of intercellular connections provide a means of extremely rapid communication between cells, while endocrine systems allow communication between cells in different tissues by means of chemicals (hormones) carried in the blood.

Communication occurs even within the single-celled protozoans. Chemical signals pass backwards and forwards between the cytoplasm and nucleus to coordinate the cell's internal activities (▷ pp. 20–1). Such organisms can respond to food and other stimuli in the environment, and their movement is controlled. In the amoeba, the cytoplasm flows in one direction at a time, while in other protozoans the beating of cilia is coordinated. In multicellular animals, however, the problem of communication is increased. Each cell must continue to regulate its own internal functions, but cannot do so in isolation from the other cells in the body.

Nervous systems

A *nervous system* is a network of cells adapted for very rapid communication over long distances. The cells involved – nerve cells or *neurones* – have extremely long extensions from their cell bodies along which tiny electrical signals can pass at great speed. Nerve cells receive and integrate information received from the various sense organs and then pass on signals to coordinate movement and behavioural responses.

The simplest nervous system is the *nerve net* found in coelenterate animals such as hydras. The nerve cells are scattered throughout the body and there is no central coordinating point. From each nerve cell there are a number of long projections, and an irregular, interconnecting network of these nerve fibres surround the body. Signals are transmitted in all directions through this diffuse network, becoming less intense as they spread from the source. The stronger the initial stimulus, the more nerves in the net will be affected. In animals that have more complex motile systems (i.e. systems involved with movement), some of the nerves are linked together in distinct tracts. For instance, echinoderms such as starfishes have a ring of nerves around the mouth and a large nerve extending down each arm.

As animals become more complex, nerve cells increase in number and greater organization develops. Not only do collections of nerve fibres occur together in nerve tracts but the cell bodies of neurones are grouped together in *ganglia*. With this greater organization, the nerves develop specific functions. *Afferent nerves* carry information from sensory receptors in the periphery towards a *central nervous system* (CNS). In vertebrates, the CNS comprises the brain and the spinal cord. *Efferent nerves* carry signals from the CNS to outlying systems: stimulation of muscle systems (by means of *motor nerves*) causes contraction (▷ p. 176), while stimulation of glandular systems causes secretion (▷ below).

Within the central nervous system, linking the afferent and efferent nerves, are large numbers of special nerves known as *interneurones*. These form a complex but highly organized network of interconnecting nerve cells and are responsible for integrating the information coming from all sources. All the patterns of movement and behaviour in animals are the result of the way the interneurones process and pass on the incoming signals.

The bodies of most animals exhibit bilateral symmetry – in other words, the body can be divided into similar right and left halves. Coelenterates (e.g. jellyfish) and echinoderms (e.g. starfish) are exceptions, being radially symmetrical. Bilateral symmetry influences the structure of the nervous system. In planarian flatworms, two main nerve cords run from the ganglia in the head towards the tail, with other nerves running transversely. In segmented animals, such as annelid worms and arthropods (crustaceans, spiders, insects, etc.), a pair of ventral nerve cords run along the length of the body. These two nerve tracts are often fused together so closely that they appear as one. The cell bodies of interneurones and motor nerves are concentrated in ganglia in each body segment. These ganglia receive afferent fibres from outlying receptors and control the muscles in that segment.

Because most animals move forwards, the mouth and sense organs are concentrated at the front end. This leads to the formation of a head (*cephalization*). To integrate all the incoming information, large concentrations of nerve cells in the head region develop into a distinct *brain*. Different regions of the brain may become specialized to perform different functions. This specialization is most pronounced in vertebrates.

In simple vertebrates, the brain is divided into three main regions. The *forebrain* receives information from the nose. In dogfishes, for instance, this area is very well developed since they mainly hunt

---

THE NERVE CELL

The functional unit of the nervous system is the nerve cell or *neurone*. Many millions of these work together to form an intricate network that far outstrips in complexity the circuitry of the most advanced electronic computer. Neurones vary greatly in shape, but typically consist of a cell body from which there are a variable number of very thin, branching projections (*dendrites*) and a single, much longer extension (the *axon*).

Communication along nerves depends on tiny electrical signals. These are sent as discrete electrical pulses (*impulses*) that can be propagated along the axons. Each impulse is continuously regenerated as it passes along the axon and can therefore be transmitted over great distances. Information about the strength of the original signal is encoded in the frequency of impulses in each nerve axon and in the number of nerves that are carrying such impulses.

Axons normally terminate at specialized junctions called *synapses*, where they link up with the cell bodies or dendrites of other neurones, or with secretory cells or muscles. At the synapse, the electrical impulses passing along the axon cause the release of tiny amounts of chemicals known as *neurotransmitters*, which diffuse across to the next cell. Many different chemicals are known to act as neurotransmitters: acetylcholine, noradrenaline, dopamine, glycine and serotonin are common examples. The neurotransmitters bind to specific receptors and cause new electrical or chemical signals to be initiated in the neighbouring cell. The final electrical output signal of each nerve cell depends on the nature and amount of stimulation it receives from other nerves. In muscles, the electrical signals cause contraction (▷ p. 176).

Some chemical substances produce much more diffuse effects, acting beyond the immediate vicinity of the cells from which they were released and changing the levels of excitability in whole groups of neurones. These are not neurotransmitters in the strict sense, since their action is not confined to the synapse where they were released; they are therefore described as *neuromodulators*.

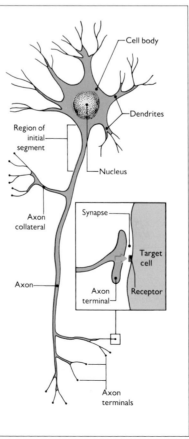

Cell body
Dendrites
Region of initial segment
Nucleus
Axon collateral
Axon
Synapse
Target cell
Axon terminal
Receptor
Axon terminals

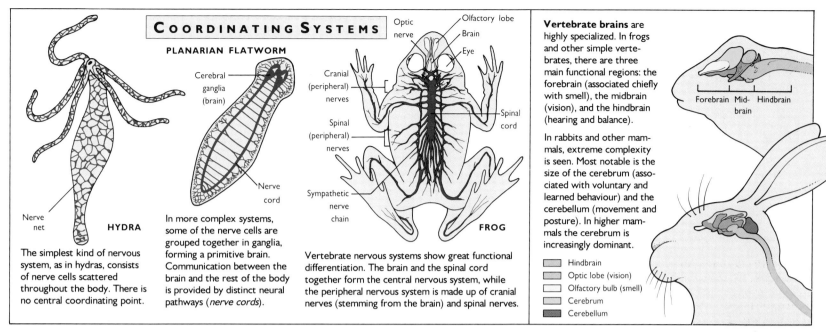

COORDINATING SYSTEMS

PLANARIAN FLATWORM

Cerebral ganglia (brain)

Nerve cord

Nerve net

HYDRA

Nerve net

The simplest kind of nervous system, as in hydras, consists of nerve cells scattered throughout the body. There is no central coordinating point.

In more complex systems, some of the nerve cells are grouped together in ganglia, forming a primitive brain. Communication between the brain and the rest of the body is provided by distinct neural pathways (*nerve cords*).

Optic nerve
Olfactory lobe
Brain
Eye
Cranial (peripheral) nerves
Spinal (peripheral) nerves
Spinal cord
Sympathetic nerve chain

FROG

Vertebrate nervous systems show great functional differentiation. The brain and the spinal cord together form the central nervous system, while the peripheral nervous system is made up of cranial nerves (stemming from the brain) and spinal nerves.

Vertebrate brains are highly specialized. In frogs and other simple vertebrates, there are three main functional regions: the forebrain (associated chiefly with smell), the midbrain (vision), and the hindbrain (hearing and balance).

In rabbits and other mammals, extreme complexity is seen. Most notable is the size of the cerebrum (associated with voluntary and learned behaviour) and the cerebellum (movement and posture). In higher mammals the cerebrum is increasingly dominant.

Forebrain Mid-brain Hindbrain

Hindbrain
Optic lobe (vision)
Olfactory bulb (smell)
Cerebrum
Cerebellum

their prey by scent. The *midbrain* receives information from the eyes, while the *hindbrain* receives information from the ears, the organs associated with balance, and the skin. During embryonic development, the roof of the hindbrain enlarges to become the *cerebellum*, which coordinates balance and movement. The floor of the hindbrain thickens into the *medulla oblongata*, controlling 'involuntary' actions such as breathing and heart beat.

As the life style and behaviour of vertebrates become more complex, so also does their brain. In mammals, two cerebral hemispheres (together forming the *cerebrum*) develop from the forebrain, and in more advanced species, the surface area of the hemispheres is increased by numerous folds or *convolutions*. In man, this area of the brain is associated with intelligent behaviour, language, memory and consciousness.

Endocrine systems

All cells produce various chemicals that are secreted through the cell membrane to the exterior. Sometimes these are waste products of the cell's activities, such as carbon dioxide, but often they are specific chemical secretions that act as signals to affect other cells. There is a complete continuum in the range over which such chemical signals can act. Local (*paracrine*) signals diffuse between neighbouring cells to regulate how the cells within a tissue develop, differentiate, and function as a unit. A number of cells (*endocrine cells*) have become adapted to produce secretions that act as chemical signals over much greater distances. A *hormone* is a chemical signal produced by a particular type of cell and carried by the blood. The structure of each hormone is highly specific, and the 'target' cells that respond to it have special receptor molecules that recognize and bind with the hormone concerned. The binding of the hormone to the receptor causes changes in the metabolic or genetic machinery of the target cell in such a way as to alter its

activity. Endocrine cells producing particular hormones are often grouped together in *endocrine glands*. Some chemical signals are used over even wider distances: pheromones are chemicals produced by animals and carried in the air to communicate with other animals of the same species (⊳ pp. 152–5).

Chemical signals are used by all multicellular animals and are particularly important in the control of growth, development and reproduction, and in coordinating metabolism and water and salt balance. An example of the highly specific targeting of hormones is seen in insects, where juvenile hormone is secreted to keep the animal in an immature state, while ecdysone stimulates the moulting that accompanies growth (⊳ p. 62).

In vertebrates, many different hormones are known, affecting almost every bodily function. Examples include insulin from the pancreas, which controls the levels of blood sugar; growth hormone from the pituitary gland and thyroxine from the thyroid gland, both of which control general body growth and metabolism; and oestradiol and testosterone from the ovaries and testes, which influence secondary sexual characteristics, such as breast growth and hair distribution. The same hormone may have different functions in different species: in fishes, for example, prolactin from the pituitary is important in salt and water balance, but in mammals it controls milk secretion.

Integration of nervous and endocrine systems

For the body to work as an integrated unit, the nervous and endocrine systems must be exactly coordinated. This allows processes such as reproduction to be linked with changes in the external environment in such a way that breeding, for example, occurs at the best time for survival of the young. In vertebrates, stressful stimuli detected by the sense organs initiate release of hormones from

the adrenal gland to prepare the animal for any sudden activity, such as fighting or fleeing.

Some nerves produce hormones, so providing the link between the neural and endocrine systems. These neurones (known as *neurosecretory cells*) are similar to normal nerves in that they receive inputs from receptors and other nerves, but instead of releasing locally acting neurotransmitters at the terminals of the nerve fibres (⊳ box), they secrete hormones into the blood. Produced in the cell bodies of the neurones, these hormones pass down the axons to be stored at the nerve terminals and are released during neural activity. Many invertebrate hormones, including those involved in the control of growth and reproduction, are produced by neurosecretory cells. The ability of crustaceans such as crabs to undergo rapid colour changes in response to the colour of their background is due to the release of neurosecretory hormones that affect pigment cells in the skin.

In vertebrates, the main link between the nervous and endocrine systems is provided by the *pituitary gland*. This is under the control of the *hypothalamus*, an area of the forebrain that is important as an integrating centre, receiving inputs from many other areas of the brain. Various neurones in the hypothalamus are specialized for neurosecretion. Some of these produce oxytocin or anti-diuretic hormone (involved in reproductive and excretory activities respectively), which pass down the axons to be released in the posterior pituitary. Others produce hormones that pass via a local blood supply to the anterior pituitary, where they control the release of each of the anterior pituitary hormones. Since many of the anterior pituitary hormones control other endocrine glands (for instance, thyroid-stimulating hormone controls the thyroid, and gonadotrophic hormones control the ovaries and testes), the hypothalamus indirectly controls much of the whole endocrine system. SM

SEE ALSO

● WHAT IS LIFE? p. 6
● CELLS p. 20
● LEARNING AND INTELLIGENCE p. 150
● THE RHYTHMS OF LIFE p. 174
● GROWTH AND DEVELOPMENT p. 186
● SENSES AND PERCEPTION p. 188

Sex and Reproduction

One of the most fundamental characteristics of living systems is the ability to reproduce. This has ensured the survival of many different and diverse patterns of life over thousands of millions of years, even though particular individuals live only for a short time. Through the processes of reproduction, animals (and plants; ▷ pp. 33–43) pass on their genes to their offspring, and it is on the basis of genetic changes over time that evolution proceeds (▷ pp. 12–17). The importance of reproduction to animals is reflected in the amount of time and energy devoted to preparing for and supporting reproductive processes.

The end result of reproduction is a generation of new individuals, which are in turn capable of reproducing. To accomplish this, reproduction involves many different processes: not only must new individuals be produced, but they must also survive and grow, developing the special structural and behavioural characteristics of their parents as they do so. Reproduction is therefore integrated into the whole life cycle of every animal.

SPERM AND EGGS

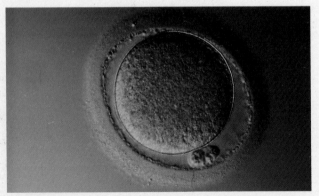

A human ovum (egg) in the process of being penetrated by a sperm (top left).

The eggs produced by the female are typically large because they contain a considerable amount of stored material to support the early development of the embryo (▷ p. 186). This represents a considerable drain on the resources of the mother and sets a limit on the number of eggs that can be produced.

Sperm are much smaller, motile cells which are adapted to find, penetrate and fuse with the eggs. The small head of the sperm contains the nucleus, and a bag of enzymes (the *acrosome*) at the very front of the sperm is important in helping the sperm to penetrate through the various protective layers that often surround the egg. A long flagellum forms the tail of the sperm and provides the mechanism for swimming. The small size of sperm means that only limited resources are required to produce many millions of them. Once released, the motility of sperm increases the chances that eggs and sperm will meet. Complex molecular mechanisms allow the sperm to fuse with an egg of the appropriate species and prevent other sperm from fertilizing the same egg.

Asexual reproduction

Cell division is a normal part of growth in all multicellular animals. Exact copies of the chromosomes containing the genetic material are produced (by mitosis) and the cell divides to form two identical (diploid) daughter cells (▷ p. 22). The daughter cells in turn grow and divide. This replication process is the basis of the asexual reproduction that occurs in many lower organisms (▷ pp. 24–7).

Asexual reproduction involves just a single parent, which produces offspring by splitting or dividing the parent body. All the offspring produced in this way are identical both to the parent and to each other. This process is related to the ability of some simple animals to regenerate lost parts. Starfishes, for example, can regenerate lost arms, and the lost arm may itself regenerate into a new starfish. In the same way, if the body of a flatworm is broken into several pieces, each piece may develop into a new individual. A form of asexual reproduction called *budding*, in which a small part of the body grows and separates from the parent to develop into a new individual, is common in sea anemones, sponges and many marine worms. The ability to reproduce asexually allows animals to reproduce very rapidly under favourable environmental conditions.

Sexual reproduction

The majority of animals reproduce by the more complex process of sexual reproduction. New individuals are not formed by simple division from a single parent but as the result of the fusion of two sex cells (*gametes*), each of which is derived from a different parent. The parents are usually sexually differentiated, with two different forms (male and female) each producing a particular type of gamete. The internal ducts that store and deliver the gametes differ between the sexes, and the areas surrounding the external openings of the tracts (the *external genitalia*) are often specialized to meet the needs of mating and fertilization. The two sexes may also differ in their behaviour patterns, particularly in relation to territory selection, mating, and parental behaviour (▷ below and ▷ pp. 156–9).

The two cells that fuse to form the new individual are highly specialized. Eggs (*ova*) are produced by the female parent and sperm (*spermatozoa*) by the male. In multicellular animals, the gametes are produced in the *gonads* – the *ovary* of the female and the *testis* of the male. Within these organs, a special group of 'germ cells' divide (by meiosis) to produce the gametes; both eggs and sperm carry equal amounts of genetic information and are haploid – each has half the genetic complement of the parent (▷ p. 22). Fusion between the two gametes at fertilization produces a new, single-celled individual (the *zygote*), which is diploid – i.e. it has a full genetic constitution.

Although the processes involved in sexual reproduction are much more complicated than those of asexual reproduction, the majority of animals reproduce sexually. This suggests that in evolutionary

terms sexual reproduction offers selective advantages to those species using it. The most obvious advantage is that sexual reproduction allows genetic traits from each parent to be combined in the offspring. This constant mixing of genetic profiles promotes variety in the population and provides a broad base on which the evolutionary pressures of natural selection can operate.

Reproductive variations

While two sexes is the most common situation, there are exceptions. The individuals of some invertebrate species are *hermaphrodite* – that is, they can produce both eggs and sperm. Although each individual can produce both sets of gametes, self-fertilization is not common; instead, fertilization usually involves the eggs of one individual and the sperm of another.

Parthenogenesis is a specialized form of reproduction in which eggs develop into new individuals without the need for fertilization. Parthenogenesis occurs sporadically throughout the animal kingdom, for example in rotifers (minute aquatic invertebrates) and in some insects and lizards. In some species it is the only form of reproduction, but often sexual reproduction may also occur. In honeybees unfertilized eggs develop into females (workers or queens), while males (drones) result from fertilized eggs (▷ p. 63). In some animals, such as jellyfishes, corals and aphids, the combination of different forms of reproduction within the life history is even more complicated, with one or more asexually produced generations alternating with a sexual one.

Mating and fertilization

There is very great variety in the ways in which individuals of the two sexes interact to achieve fertilization. In many animals, including sea urchins and the majority of worms, fishes and amphibians, eggs and sperm are simply liberated into the surrounding water and fertilization occurs in the water (*external fertilization*). However, neither the production nor the release of eggs and sperm occurs at random. Individuals respond both to cues in the environment and to the behaviour of other individuals (▷ pp. 174–5). These responses aim to ensure that the individuals concerned are of the same species and that the release of the eggs and sperm is synchronized in such a way as to minimize the dilution effect of the external medium and to maximize the chances of fertilization. Such interactions between individuals often develop into complex patterns of mating behaviour.

In many species, the gametes are not simply released to the exterior; instead *internal fertilization* occurs. Eggs are retained within the body of the female and the male's sperm are transferred to her. In some cases, including many crustaceans and newts, the sperm are packaged in a *spermatophore*, which is deposited by the male and picked up by the female. More direct transfer of sperm requires intimate physical contact between individuals (*copulation*). The attainment and maintenance of such close physical contact

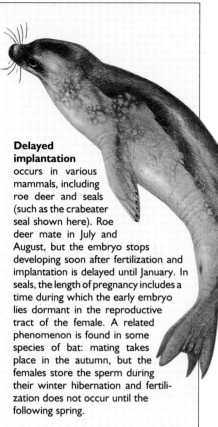

Delayed implantation occurs in various mammals, including roe deer and seals (such as the crabeater seal shown here). Roe deer mate in July and August, but the embryo stops developing soon after fertilization and implantation is delayed until January. In seals, the length of pregnancy includes a time during which the early embryo lies dormant in the reproductive tract of the female. A related phenomenon is found in some species of bat: mating takes place in the autumn, but the females store the sperm during their winter hibernation and fertilization does not occur until the following spring.

involve complex behavioural interactions between the mating partners, and the actual transfer of sperm in copulation may be dependent on various structural adaptations. The delivery of sperm is facilitated by *intromittent organs* (such as a penis) that can be inserted into the female genital tract. The delivery of sperm relatively close to the eggs within the female tract reduces sperm loss while increasing the chances of fertilization.

Egg protection and parental care

Although fertilization results in a single cell potentially capable of developing into a new individual, reproduction is not yet complete. Growth, development and maturation of the young are necessary to complete the reproductive life cycle.

Animals that lay eggs are termed *oviparous*. The fertilized eggs and the developing young are particularly susceptible to loss through predation and adverse conditions in the environment. In many marine invertebrates and fishes, which simply liberate their eggs into the surrounding water, very large numbers – sometimes millions – of eggs are produced by a single individual to offset the large losses. Other animals have developed mechanisms to help protect and support their growing young. This may take the form of eggs with large food reserves, often encased within protective coverings. Such protection is often most obvious in terrestrial animals such as reptiles and birds (amniotes; ⊳ p. 86); in these cases the eggs must also provide protection against excessive water loss. The area in which eggs are laid (a nest) is often specially selected and prepared to

help protect the eggs and developing young.

Viviparous animals do not lay eggs but produce live young. The fertilized eggs develop within the very protected environment of the mother's body. Nourishment may be supplied by secretions from the adult, or there may be special structures to allow the exchange of nutrients between the parent and the developing young. In placental mammals, the embryo becomes implanted in the wall of the uterus, and the placenta develops to allow the exchange of oxygen, nutrients and waste products (⊳ pp. 186–7). The development of the fetus during pregnancy (*gestation*) means that the young are generally born in a relatively well-developed state (in contrast to marsupial mammals; ⊳ pp. 103–5). The length of pregnancy in placental mammals varies from species to species, from 16 days in the hamster to around 20 months in elephants.

The chances of the offspring surviving are often increased by behavioural adaptations in which one or both parents help to look after the young (⊳ pp. 156–9). The degree of such parental care varies considerably. In some animals it simply involves guarding the eggs and nest site against predators, and may extend to guarding the newly hatched young. Birds and mammals feed and care for their young until the offspring are comparatively large and well developed. In mammals, this is associated with the production of milk to feed the young in the early stages (⊳ p. 98). Although birds and mammals produce only a few young at a time, the high degree of parental care means that a relatively large proportion of them survive to adulthood.

Timing of reproduction

It is vital that reproduction is coordinated both within the life cycle of each animal and with other animals of the same species. Coordination with the environment is also particularly important, since most animals live in situations in which the climatic and nutritional conditions are not stable throughout the year.

Animals usually identify the appropriate time to breed by detecting and responding to changing environmental conditions (⊳ pp. 174–5). Changes in temperature, rainfall and food supply sometimes provide the cues, but many animals use the more predictable seasonal changes in the number of hours of daylight. In many birds and mammals, the lengthening days in spring initiate the hormonal changes that stimulate the ovaries and testes to produce gametes. The ability of humans to breed at any time of the year reflects their very high degree of parental care and their capacity to control their own environment.

Coordination of reproductive processes within the body is largely achieved by hormones (⊳ p. 183). As well as producing gametes, the gonads of both sexes produce hormones that affect the reproductive tract, other secondary sexual characteristics such as plumage, and behaviour. Changes in sexual behaviour and interactions between individuals allow mating to be synchronized with the production of gametes. Most female mammals show regular cycles of willingness to mate (*oestrous cycles*), which are associated with the cycles of egg production. The fact that sexual behaviour in humans is not restricted to the time of egg release probably reflects the role of sexual behaviour in maintaining the bond between partners. The *menstrual cycle* of human females is linked to the underlying cycle of hormonal changes associated with egg production.   SM

SEE ALSO

● GENETICS p. 22
● SIMPLE LIFE FORMS pp. 24–7
● PLANTS pp. 33–43
● INSECTS: LIFE CYCLES p. 62
● FISHES: LIFE CYCLES p. 74
● THE WORLD OF AMPHIBIANS p. 80
● REPTILES: LIFE CYCLES p. 86
● BIRDS: LIFE CYCLES p. 94
● TERRITORY, MATING, SOCIAL ORGANIZATION p. 156
● THE RHYTHMS OF LIFE p. 174
● GROWTH AND DEVELOPMENT p. 186

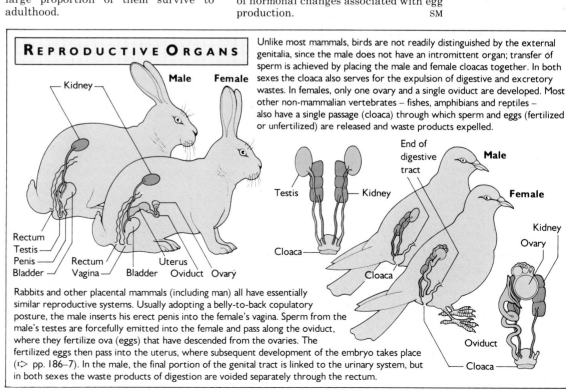

REPRODUCTIVE ORGANS

Unlike most mammals, birds are not readily distinguished by the external genitalia, since the male does not have an intromittent organ; transfer of sperm is achieved by placing the male and female cloacas together. In both sexes the cloaca also serves for the expulsion of digestive and excretory wastes. In females, only one ovary and a single oviduct are developed. Most other non-mammalian vertebrates – fishes, amphibians and reptiles – also have a single passage (cloaca) through which sperm and eggs (fertilized or unfertilized) are released and waste products expelled.

Rabbits and other placental mammals (including man) all have essentially similar reproductive systems. Usually adopting a belly-to-back copulatory posture, the male inserts his erect penis into the female's vagina. Sperm from the male's testes are forcefully emitted into the female and pass along the oviduct, where they fertilize ova (eggs) that have descended from the ovaries. The fertilized eggs then pass into the uterus, where subsequent development of the embryo takes place (⊳ pp. 186–7). In the male, the final portion of the genital tract is linked to the urinary system, but in both sexes the waste products of digestion are voided separately through the rectum.

Growth and Development

The processes of growth and development pose some of the most fundamental questions in biology, as well as some of the most perplexing. How does a single cell – the fertilized egg – develop into an adult? How does this single cell produce a complex multicellular organism with many different tissues, each with its own structure and function? Despite the complexity of such matters, the processes that underlie them are now beginning to be better understood.

Cell division is a fundamental part of the process by which a fertilized egg cell (*zygote*) divides and becomes organized into a complex organism (▷ p. 22), but it is not sufficient on its own. Individual cells must also undergo *differentiation*, in which they become specialized in structure to perform particular tasks, and then be arranged in specific patterns in each organ. Furthermore, attainment of the final adult form may require changes in body proportions by differential growth in different organs.

Early embryonic development

The large size of eggs compared to other cells reflects the need for them to have sufficient stored nutrients (*yolk*) to allow development to proceed until another source of food is available. The eggs of mammals are relatively small because the embryo receives nourishment directly from the mother at an early stage. The eggs of many marine mammals are also small, but in these cases it is because the embryos rapidly develop into free-living larvae that can feed themselves. In contrast, birds and reptiles have comparatively huge eggs because of the large amount of yolk required for growth before the young hatch.

The fusion of sperm and egg cells at fertilization provides the correct complement of genetic material (DNA) that will determine the precise characteristics of the individual (▷ p. 22). Before fertilization the egg cell is relatively dormant. However, entry of the sperm stimulates the start of intense cell activity and very rapid development, involving a number of swift, simultaneous cell divisions (*cleavages*; ▷ diagram). At each division, the cells (*blastomeres*) are halved in size, the rapid pace of cell division allowing little time for any real increase in cell size.

The pattern of cleavage differs in different animal groups. The blastomeres of eggs with little yolk, as in mammals, are all of a similar size. The eggs of fishes and amphibians contain much larger amounts of yolk, concentrated in the bottom half: this slows the rate of cleavage and so produces larger blastomeres in this area. In birds and reptiles, the yolk is so large that cleavage is restricted to a thin layer at the top of the egg.

During development, complexity develops gradually. The rapid succession of cleavages eventually transforms the fertilized egg cell into a multicellular embryo (the *blastula*, or *blastocyst* in mammals), consisting of a hollow ball of cells surrounding a fluid-filled cavity (the *blastocoel*). The formation of the blastocoel is an important step in development because cells on the outside of the blastocoel come to be arranged in sheets around the cavity. This structural arrangement provides the basis for the next major step: *gastrulation*. Sheets of cells start to move relative to each other, so setting up new configurations and relationships. Eventually three distinct cell layers become established: an outer sheath or *ectoderm*, the inner *endoderm*, and the *mesoderm* lying between them.

Once gastrulation is complete, a process known as *organogenesis* begins to take place, in which specific organs develop to serve particular functions (respiration, excretion, circulation, digestion, and so on). Different organs and tissues each develop from a particular cell layer. In vertebrates, the ectoderm gives rise to the outer layers of skin and hair. Thickening folds of the ectoderm come together to form the neural tube, which differentiates into the brain and the spinal cord. Cells from the mesoderm give rise to the inner layer of skin (*dermis*), muscles, the skeleton, the heart and circulatory system, kidneys, and the reproductive system. The lining of the digestive tract and the liver develop from the endoderm. The respiratory system also develops as an outpocketing from the endoderm. In mammals, after specific organ systems have developed, the embryo becomes known as a *fetus*.

Not all the cells of the blastula of amniotes (mammals, birds and reptiles) develop into the embryo itself. Some form membranes that lie outside the body of the true embryo (▷ diagram). In mammals, these form the placenta and the membranes that surround and protect the embryo within the uterus, while in reptiles and birds they are involved in respiration, in utilizing the yolk, and in the disposal of waste products within the egg.

Control of development

Every cell in the body is genetically identical, containing the same sets of chromosomes inherited from each parent (▷ p. 22). Gradually, however, particular groups of cells in different locations in the embryo become committed to particular paths of development. This process (*determination*) is irreversible, and commits the cell and all its progeny to the same developmental pathway. In the course of differentiation, specific groups of genes are activated, while others become inactive. The different gene activity in each cell type dictates the final size, shape and functional characteristics of the cell.

Some cells do not undergo any final differentiation but are set aside as *stem cells*.

DEVELOPMENT OF THE MOUSE EMBRYO

Successive divisions or *cleavages* of the fertilized egg while still within the mother's oviduct result in the formation of a ball of 16 or more cells known as a *morula*, which enters the uterus after about 72 hours. A little later, a fluid-filled cavity (*blastocoel*) appears within the morula, at which stage the egg becomes a *blastocyst*. The greater part of the blastocoel is bounded by a single layer of cells known as the *trophectoderm*, which contributes to the extra-embryonic parts (structures outside the embryo itself); on one side, however, there is a concentration of cells called the *inner cell mass*, some of which will develop into the embryo proper. By the fifth day the blastocyst has become implanted in the inner wall of the uterus.

The embryo reaches the *egg cylinder stage* as the inner cell mass moves downwards into the blastocoel, thus forming the *yolk sac cavity*. (The yolk sac of mammals, though containing virtually no yolk, is clear evidence of their reptilian origins; in mice, it has a significant role in providing accessory nutrition to the embryo.) A subdivision of the inner cell mass produces a lower part (the *embryonic ectoderm*), from which the embryo proper develops, and an upper part (the *extra-embryonic ectoderm*). From the latter a mass of cells proliferates to form the *ectoplacental cone*, which establishes close contact with the mother's blood vessels and later forms part of the placenta.

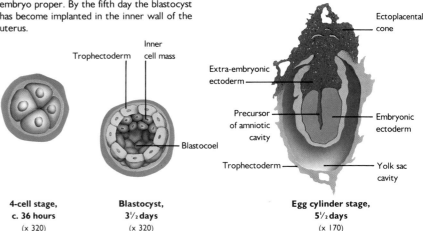

4-cell stage, c. 36 hours (× 320)

Trophectoderm
Inner cell mass
Blastocoel

Blastocyst, 3½ days (× 320)

Ectoplacental cone
Extra-embryonic ectoderm
Precursor of amniotic cavity
Trophectoderm
Embryonic ectoderm
Yolk sac cavity

Egg cylinder stage, 5½ days (× 170)

These cells divide to provide a supply of cells to replace or increase the population of differentiated cells. In vertebrates, stem cells in the bone marrow continuously replace lost red and white blood cells.

The fertilized egg is *totipotent* – it can form every cell type in the body. In most animals, each of the blastomeres at the two-cell stage can also form complete embryos if separated. The same may be true for the blastomeres at the four- and eight-cell stages. However, gradually the fate of blastomeres becomes restricted. In mammals, determination has begun by the blastula stage, with only the inner cell mass contributing to the embryo (▷ diagram).

What determines the developmental pathway a cell will take? In some cases, factors within the egg cytoplasm (▷ p. 20) may be important, as different parts of the dividing egg become distributed to different blastomeres. More generally, the position of a cell within the embryo and its relationship to its neighbours determine its fate: the particular environment of a cell and the various chemical interactions between cells induce specific developmental pathways. During the gastrulation of the vertebrate embryo, sheets of cells interact with each other: for instance, the mesoderm induces the differentiation of the overlying ectoderm into neural tissue. As development proceeds, with the formation of new cells in new positions, the greater structural complexity allows more and more opportunities for new interactions and the determination of other cell types.

Growth

Growth itself is not easy to define and may describe different processes in different situations. In the developing embryo, there is no initial increase in overall size because the first few cell divisions result in smaller and smaller cells. When an increase in the size of a tissue or organ does occur, it may be due to an increase in either cell number or cell size.

In the case of bone and cartilage, growth results from the deposition of material outside the cells. Bone undergoes continual remodelling, being reabsorbed and laid down again throughout life. Some tissues do not grow in size even though there is continued cell division: cell death may balance the equation. In both the skin and the lining of the gut, continued cell division replenishes lost cells. Some tissues, such as the liver and skin, can regenerate lost tissue if damaged. At the other extreme, nerve tissue exhibits very few cell divisions after the initial period of growth and (at least in mammals) shows almost no ability to regenerate.

Life histories

The transition between the single-celled zygote and the multicellular adult takes many different forms in different animals. In some cases, the changes involved occur gradually as part of a continuing process of growth and differentiation. Thus in mammals the mature fetus in the uterus resembles the adult in general body shape, although changes in body proportions occur during subsequent growth.

The situation is not as simple as this in many other animals, where the adult stage is not reached by progressive growth and gradual differentiation. Instead, intermediate larval stages occur, which differ radically from the adult in both structure and life style. In such cases, the final adult form is attained by means of a major transformation of structure (*metamorphosis*).

Larval forms are to be found in both vertebrates and invertebrates. A tadpole is an example of a vertebrate larva that is transformed into a mature adult (such as a frog) with a very different structure and life style (▷ p. 80). In invertebrates, larval forms are a usual feature of development and a number of larval forms may follow each other before the adult stage is reached. A caterpillar is a larva that metamorphoses during the pupal stage into an adult butterfly (▷ p. 62). Arthropods, such as insects and crabs, have a hard external skeleton that restricts growth, so necessitating a step-by-step increase in size at times of moult (▷ p. 58). The development of many marine invertebrates, including molluscs, echinoderms and crustaceans, proceeds through a number of free-living larval stages, each of which differs considerably from the next. These larval forms are very important for ensuring adequate dispersal of the young.

Adulthood and sexual maturity represents the climax of the process of growth and development. As reproduction takes place in the new generation, the cycle of life has turned full circle.   SM

SEE ALSO

- THE EVIDENCE FOR EVOLUTION p. 14
- THE BUILDING BLOCKS OF LIFE: CELLS p. 20
- THE BLUEPRINT OF LIFE: GENETICS p. 22
- INSECTS: LIFE CYCLES p. 62
- FISHES: LIFE CYCLES p. 75
- THE WORLD OF AMPHIBIANS p. 80
- REPTILES: LIFE CYCLES p. 86
- BIRDS: LIFE CYCLES p. 94
- TERRITORY, MATING, SOCIAL ORGANIZATION p. 156–9
- BIOLOGICAL CYCLES AND RHYTHMS p. 174
- SEX AND REPRODUCTION p. 184

The formation of a membrane called the *amnion* creates the *amniotic cavity*, a fluid-filled enclosure that isolates the embryo from the extra-embryonic structures above it and provides a protective cushion against shock. A structure known as the *allantois* grows from the rear end of the embryo into the extra-embryonic coelem or *exocoelem*, a cavity bounded by the amnion below and another membrane (the *chorion*) above. The allantois eventually fuses with the chorion and the ectoplacental cone to form the *placenta*, a structure that links the embryo with the maternal blood supply, thereby providing it with oxygen and nutrients, and allowing the disposal of waste products. Unlike most mammalian embryos at a similar stage, which float free within the amniotic cavity attached to the placenta only by the *umbilical cord*, the mouse embryo has its ventral (front) surface surrounded by the yolk sac cavity and its dorsal (back) side facing inwards.

The *primitive streak*, first appearing at about 6½ days, soon becomes a centre of rapid growth. The front end of this gives rise to the *notochord*, a rod-like structure running down the central axis of the embryo that marks the foundation of the vertebral column. Along the length of the notochord a series of paired segments (*somites*) develop, which subsequently give rise to bone and muscle. At 7 days a trough begins to deepen along the back of the embryo, parallel to the notochord; this is the *neural groove*, which later closes over to form a tube, the basis of the central nervous system. At this stage the embryo is S-shaped, facing to the left, and the fore- and hindguts have appeared as pockets at either end of the embryo. The intrusion of the foregut produces the *head fold*; the rapid growth of the neural folds in this area indicates early development of the brain. The heart, just in front of the head fold, is also growing rapidly at this time.

Between 8 and 9 days the embryo rotates through 180°, thereby assuming the concave position characteristic of other mammalian embryos. The walls of the midgut, previously attached to the yolk sac, are brought parallel and fuse to form a tube; this then links up with the fore- and hindguts, the blind ends of which subsequently break through to form the mouth and anus respectively.

About 20 days after fertilization the young mouse is born, with eyes closed and no hair. By day 10 it has a full covering of hair, and a day or two after that it opens its eyes. Within 6 to 8 weeks of birth the mouse becomes a fertile adult.

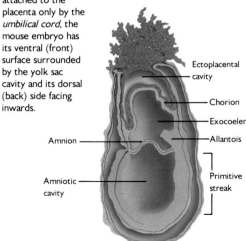

Primitive streak stage, 7¼ days
(x 125)

Labels: Ectoplacental cavity; Chorion; Exocoelem; Allantois; Amnion; Amniotic cavity; Primitive streak

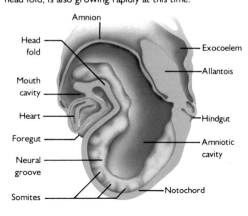

Embryo at 8 days
(x 60)

Labels: Amnion; Head fold; Mouth cavity; Heart; Foregut; Neural groove; Somites; Exocoelem; Allantois; Hindgut; Amniotic cavity; Notochord

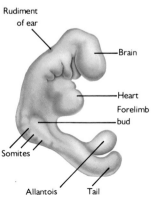

Embryo at 9 days
(x 45)

Labels: Rudiment of ear; Brain; Heart; Forelimb bud; Somites; Allantois; Tail

Embryo at 16½ days
(x 9)

Senses and Perception

Animals gain information about what is going on in the world around them and within their own bodies through their senses. This information is received by various receptor cells as sensory stimuli and is transformed into nerve impulses that are transmitted to the central nervous system (▷ p. 182). Sensory information passed to the brain may result in an awareness of the stimulus (*sensation*), which can be studied in animals by observing changes in their behaviour. Alternatively, it may result in the formation of an internal image of the stimulus (*sensory perception*), which is difficult to study in non-human animals.

The various senses are based on the sensitivity of different types of sensory cell to different forms of energy. Sensations of taste and smell, for example, are due to receptors responsive to chemical energy,

Some male moths, such as the atlas moth shown here, are extraordinarily sensitive to chemical signals known as pheromones. Sensory cells in the antennae respond over distances of several kilometres to just a few molecules of the sex pheromone released by females.

while vision depends on receptors sensitive to the electromagnetic energy of light. Often sensory cells of a single type are grouped together as *sense organs*. Greater refinement within a particular sense depends on different groups of the receptor cells concerned being sensitive to different qualities, such as different colours of light or different sound frequencies. Fixing the position of a sensory stimulus in space is often achieved by comparing the sensory data from two or more groups of receptors in different locations.

Sensory signals

When a stimulus reaches a sensory cell, it causes a change in the electrical potential across the cell, the size of the potential varying with the strength of the stimulus. The potential change is then transmitted across the cell or passed onto a neighbouring cell, where nerve impulses are produced. The intensity of the stimulus is encoded in the rate at which nerve impulses are produced.

Impulses may be produced for the full duration of the stimulus (a *tonic response*) or only at the beginning (a *phasic response*). Not all sense organs, therefore, faithfully report how long the stimulation lasts. The sensation of our clothes fades soon after we put them on as a result of this process (known as *adaptation*), so preventing the nervous system from becoming overloaded with unnecessary information. Some sensory cells are *phasitonic* – temperature receptors, for example, signal a change in temperature with a rapid burst of impulses and then signal the level of temperature at a lower but consistent rate of impulses related to the actual temperature.

The lowest intensity of stimulation that any cell responds to is its *threshold level*. This can be very low: the receptors of the vertebrate eye respond to just a few photons of light, while the mammalian ear can detect sounds that cause movement within the ear as small as the diameter of a hydrogen atom.

Impulses from the peripheral sensory cells may excite or inhibit activity in subsequent sensory neurones (nerve cells), and these may interact with each other so that the information is processed before or after reaching the brain. More central sensory neurones may respond to complex aspects of the stimulus, such as curved edges, the direction of movement, or a particular type of sound. Cells responding to particular aspects of the stimulus are called *feature detectors*.

Taste

Sensitivity to chemical stimuli (*chemo-*

The fleshy barbels around the mouth of many catfishes and some other bottom-feeding fishes are not only highly sensitive to touch but also bear numerous taste buds. They are used to probe the bottom silt and sediment in search of prey.

reception) is found in all organisms, including bacteria and protozoans. The sense of taste or *gustation*, which is generally concerned with detecting food, requires contact with the source of the chemical. If the source is distant from the animal, the sense of smell or *olfaction* is involved.

The gustatory (taste) cells of invertebrates may be found on any part of the body – near the mouth of nematode worms, on the tentacles of octopuses, on the horns of snails, and on the legs of insects. They are often grouped together in organs called *sensilla*, and have long terminal processes (*dendrites*; ▷ p. 182) that extend to a common pore in the skin or cuticle. In insects the opening may be in a pit in the cuticle or at the end of a long sensory hair. True (dipteran) flies have sensory-hair organs containing four types of chemosensory cell. Each cell is most sensitive to a particular chemical – water, sugar, the positive ion of a salt, or the negative ion of a salt – but will also respond to a lesser extent to other compounds. The combined responses of all four types of cell enable the insect to recognize a wide range of different substances.

In vertebrates gustatory receptors are grouped together in *taste buds*. These are commonly found on the tongue, but may be on the roof of the mouth and in the throat, and in some bottom-dwelling fishes they occur either on projections from the pectoral fins that touch the bottom or on sensory 'barbels', such as those of catfishes (▷ illustration). Humans can distinguish four basic taste qualities – sweet, salt, sour and bitter – each of which is most strongly detected on a specific area of the tongue. As in invertebrates, each type of cell is most sensitive to one type of chemical and responds to a lesser degree to others, so that together they code for a wide range of substances.

Smell

The detection of distant objects, including food, predators, mates, migratory land marks and the home base, often involves olfaction (smell). In insects most of the olfactory receptors are on the antennae, which may bear between 40 000 and 200 000 receptors. The sensory cells may occur singly or in groups. The branched terminal dendrites of the cells extend into a fluid-filled projection in the cuticle, which is perforated by thousands of tiny pores. Some cells (called *odour specialists*) respond to a single chemical only, often a pheromone (▷ pp. 152–5), and may be extremely sensitive (▷ photo). A pheromone-specialist cell in the male silk-moth responds to a single molecule of a pheromone released by the female, and the animal responds behaviourally at levels of 200 molecules. Other cells respond to a narrow range of substances, and still others (*broad generalists*) to a wide range of substances. Together these can code for a great variety of chemicals.

The olfactory receptors of vertebrates lie in cavities. In many fishes, these are blind-ending nasal sacs on the head. In lungfishes and all higher vertebrates, the nasal sacs open internally to the pharynx, so water or air can be drawn in a continuous stream over the sensory cells, which in air-breathing animals are bathed in mucus. Chemicals in solution in the water or dissolved in the mucus come into contact with the tips of the cells, which end in a number of tiny projections (cilia). Single cells (equivalent to the insect broad generalists) appear to respond to a wide variety of chemicals, but in different ways to different chemicals. Odour specialists have not yet been found in vertebrates.

In mammals the olfactory membrane lies at the top of the nose. In keen-scented mammals such as rabbits and dogs, the membrane is greatly folded to increase its surface area and hence the number of receptor cells. In man the olfactory membrane covers 5 cm² (¾ sq in) and contains about 5 million olfactory cells, but in German shepherd dogs it is 150 cm² (23 sq in) and contains nearer 220 million sensory cells.

Many terrestrial vertebrates have a second olfactory area, the *vomeronasal organ* or *Jacobsen's organ*, which lies in the ventral region of the nose or in the roof of the mouth. In mammals such as mice, guinea pigs and cats, it appears to be associated with the detection of sex pheromones. In snakes odours are picked up on the tongue, which is then inserted into the openings of the vomeronasal organ in the roof of the mouth.

Temperature reception

The mechanism of temperature reception or *thermoreception* is not well understood. Insects, especially those such as ticks that are parasitic on mammals, can detect small changes in temperature. Some fishes have temperature receptors on the head, at the bases of jelly-filled ducts called the *ampullae of Lorenzini*. Mammals and birds have thermoreceptors in the skin, tongue, central nervous system and deep organs. There are two kinds of skin receptor in mammals – *warm receptors* that respond to an increase in temperature over the range 22–45 °C (72–113 °F) and *cold receptors* that respond to a decrease in temperature over the range 12–35 °C (54–95 °F). The latter also respond to temperatures of 45–50 °C (113–122 °F), and this may give rise to the sensation of 'paradoxical cold' at these high temperatures.

Photoreception

The ability to detect light depends on the

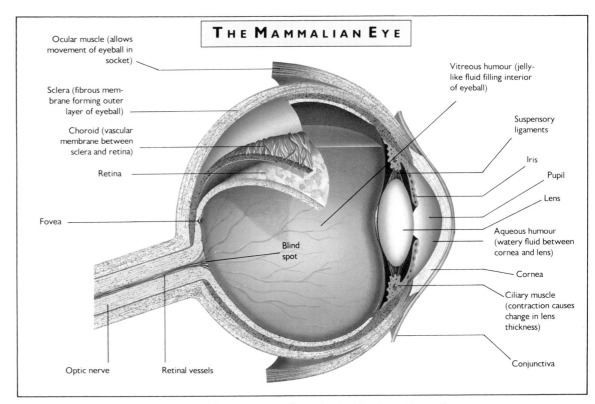

THE MAMMALIAN EYE

Ocular muscle (allows movement of eyeball in socket)

Sclera (fibrous membrane forming outer layer of eyeball)

Choroid (vascular membrane between sclera and retina)

Retina

Fovea

Blind spot

Optic nerve

Retinal vessels

Vitreous humour (jelly-like fluid filling interior of eyeball)

Suspensory ligaments

Iris

Pupil

Lens

Aqueous humour (watery fluid between cornea and lens)

Cornea

Ciliary muscle (contraction causes change in lens thickness)

Conjunctiva

presence in light-sensitive receptor cells of special molecules called *photopigments*, which absorb light energy and change their shape, ultimately causing a change in electrical potential across the cell membrane. Different pigments absorb a different range of light wavelengths across a band of wavelengths from 400 nanometres (nm) (violet) to 700 nm (red), which is therefore the spectrum of visible light. Insects, however, are able to detect shorter wavelengths in the ultraviolet region. The most familiar pigment is *rhodopsin*, which occurs in the rod cells (▷ below) of vertebrates and absorbs light across a broad spectrum.

Light sensitivity or *photoreception* is widespread in the animal kingdom. It occurs in protozoans such as euglena, which have a photopigment spot near the base of the flagellum and swim towards light. In a few animals, light-sensitive cells occur singly over the body surface, but more often they are grouped together as eyes. The simplest type of light-sensitive organ is the *ocellus*, a group of cells in a shallow pit. Such organs detect light intensity and sometimes direction, and occur in a wide range of invertebrates from coelenterates (jellyfishes, etc.) to insects.

True vision is the ability to form an image of the light source. There are various ways of doing this. In one system, similar in principle to a pinhole camera, a narrow beam of light is allowed to pass through a tiny aperture (pinhole) onto a photoreceptive sheet, which is made up of an array of photoreceptors arranged on a curved surface behind the pinhole.

The vertebrate eye

In vertebrates, the sheet of photoreceptors and associated cells – the *retina* – is arranged on the inside of a sphere, the *eyeball* (▷ diagram). Light from the visual field enters through a curved outer *cornea* and passes through a *lens*, both of which bend light rays to focus them on the retina, where an inverted image is formed (which is re-inverted by the brain). Each photoreceptor detects light from a minute part of this image. The lens can be moved or changed in shape by muscles in such a way that the eye can adjust to both near and far objects. Between the cornea and the lens lies the *iris*, an opaque area with a central aperture – the *pupil* – which regulates the amount of light entering the eye. In bright light, muscles in the iris contract, reducing the size of the pupil. In dim light the pupil expands to allow more light into the eye.

COLOUR VISION

In 1802 the English physicist Thomas Young suggested that, just as the full spectrum of colours can be made from mixing paints of the three primary colours (red, yellow and blue), so the eye must contain receptors selectively sensitive to each of these colours and that the sensation of different colours results from the stimulation of these three receptors in different combinations. The basic principle underlying this hypothesis (known as the *trichromacy theory*) was confirmed in 1964, when three classes of pigments were found in the cone cells of the goldfish. Each of these is maximally sensitive to blue, green or red.

In humans the equivalent colours are in fact blue, green and yellow. The electrical output of each type of cone depends on the amount of light energy absorbed, and the sensations of colour appear to depend on the proportions of the electrical response from the three classes of cones. If one type of cone cell is defective or absent, vision for that colour is affected. The most common type of colour blindness is the inability to distinguish red from green.

A similar basis for colour vision occurs in insects, which can also distinguish ultraviolet light. Different cells in each ommatidium are sensitive to different wavelengths of light, and some can respond to the plane of polarization of light. This allows them to determine the position of the sun from a small patch of blue sky, even if the sun itself is covered by cloud.

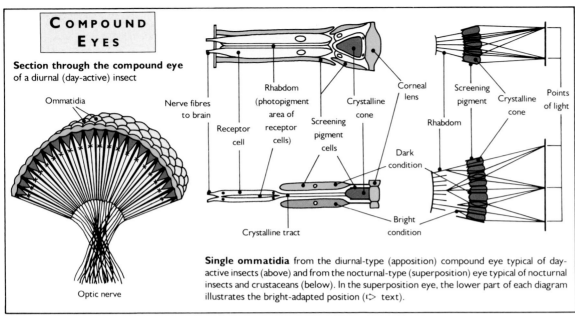

Section through the compound eye
of a diurnal (day-active) insect

Ommatidia

Nerve fibres to brain

Receptor cell

Rhabdom (photopigment area of receptor cells)

Screening pigment cells

Crystalline tract

Crystalline cone

Corneal lens

Dark condition

Bright condition

Screening pigment

Crystalline cone

Rhabdom

Points of light

Optic nerve

Single ommatidia from the diurnal-type (apposition) compound eye typical of day-active insects (above) and from the nocturnal-type (superposition) eye typical of nocturnal insects and crustaceans (below). In the superposition eye, the lower part of each diagram illustrates the bright-adapted position (▷ text).

The retina of many vertebrates contains two different types of photosensitive cell, rods and cones. The smaller *cones* are responsible for colour vision (▷ box, p. 189) and are mostly found in a small area opposite the pupil (the *fovea*). In humans they are closely packed together in this region and allow the resolution of fine detail such as print, but operate only in bright light. *Rods* are found mainly towards the outer regions of the retina. These can operate in dim light but do not give the fine discrimination of the cones. Some animals, such as chickens, have only cones in the retina, while others, including non-primate mammals, have only rods and so lack full colour vision.

Compound eyes

A different mechanism for forming an image is found in crustaceans, insects and some other arthropods. These animals have *compound eyes*, which consist of thousands of separate optic units called *ommatidia* arranged in a hemisphere, each of which receives light from a slightly different part of the visual field (▷ diagram). An ommatidium consists of a lens behind which is a group of six to eight photoreceptive cells, arranged in a circle around the central axis of the ommatidium. The inner edges of these cells contain photopigment.

In insects active by day, each ommatidium is separated from the others by a sleeve of light-absorbing (but non-photosensitive) pigment; this acts as an opaque shield around the ommatidium, which therefore acts as an independent unit and contributes a small spot to the total image, which is a mosaic of tiny dots like the photographs printed in this book. Some crustaceans and nocturnal insects have compound eyes in which the lens is separated from the receptor cells by a clear crystalline tract. The screening pigment surrounding each ommatidium can be expanded in bright light to form a complete sleeve or contracted in dim light to allow the receptor cells of one omma-

tidium to collect light from adjacent ommatidia through the clear tract. This increases the light-gathering power of the eye, which can therefore perform well in dim light, but decreases the sharpness of the image.

Touch and pressure

Receptor cells sensitive to mechanical energy (*mechanoreceptors*) respond to stretching or distortion of a flexible membrane caused by pressure, movement, vibration, or sound waves. Some mechanoreceptors are simply bare nerve endings in the skin, which respond to touch or pressure, but there are other types that have complex associated structures that determine the type of stimulus to which the cell responds. In vertebrates, elaborate accessory structures of this kind are found in the ear and its associated organs, and it is by means of these that sound, gravity and movement can be detected (▷ below).

In vertebrates and invertebrates such as leeches, external pressure or touch stimulates bare nerve endings in the soft skin. In mammals there are various ways in which mechanoreceptor endings may be encapsulated. The *paccinian corpuscle* (found, for example, in the fingertips) is composed of concentric layers of membranes like an onion, and responds to pressure and vibration. In arthropods, mechanoreceptor cells are connected to stiff hairs that articulate with the cuticle in such a way that any force displacing the hair stimulates the sensory cell. The hairs of mammals and the feathers of birds respond in a similar way to displacement caused by touch or pressure.

Balance and movement

A variety of sensory information can be used to determine position with respect to the environment, including input from the eyes, from the skin, and from sensory receptors in the muscles. Many animals also have special organs sensitive to gravity. The simplest of these is the *stato-

cyst*, a fluid-filled cavity in the wall of which are cells with projecting hairs or cilia. A mass of dense crystals (often sand grains or calcareous secretions) are glued together to form a heavy *statolith*, which weighs down on the hairs, making them bend and so stimulating the sensory cells in a particular region. Statocysts are found in most invertebrates (but not in insects). Lobsters have a statocyst organ at the base of each antenna. This organ also has sensory hairs that are embedded in two fluid-filled canals lying at right angles to one another. When the animal moves, the fluid in the canals lags behind through inertia, so bending the hairs and signalling angular acceleration in different planes.

In vertebrates, gravity and movement are usually detected by two *otolith organs*, which together with the semicircular canals (which detect angular acceleration) and the cochlea (which detects sound) form the *labyrinth* or *inner ear*. All of these structures have similar sensory-hair cells, the hairs of which are in contact with an overlying gelatinous membrane. In the otolith organs, the membrane contains calcium carbonate crystals. The hair cells are arranged in two or three patches in different planes on the walls of the two fluid-filled chambers, and are sensitive to movement in particular directions and so signal position with respect to gravity.

Attached to the top of the two otolith chambers are three *semicircular canals*, one in each plane. Where each canal joins the upper chamber, there is a patch of sensory-hair cells, the hairs of which are embedded in one of three gelatinous

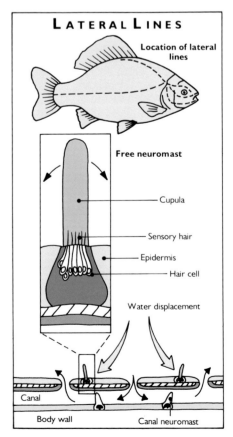

LATERAL LINES

Location of lateral lines

Free neuromast

Cupula

Sensory hair

Epidermis

Hair cell

Water displacement

Canal

Body wall

Canal neuromast

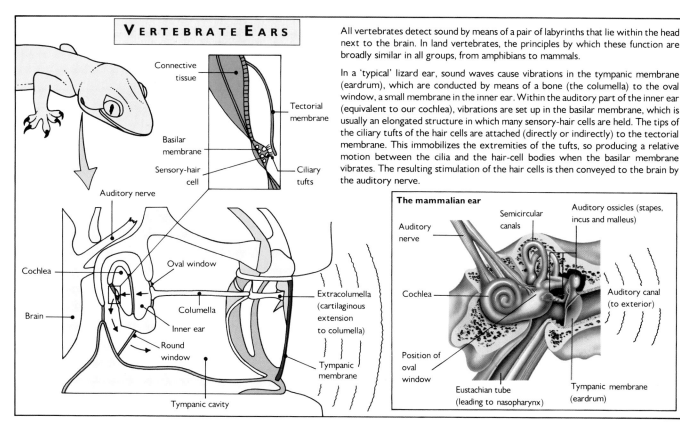

VERTEBRATE EARS

Connective tissue

Tectorial membrane

Basilar membrane

Sensory-hair cell

Ciliary tufts

Auditory nerve

Cochlea

Oval window

Brain

Columella

Inner ear

Round window

Tympanic cavity

Extracolumella (cartilaginous extension to columella)

Tympanic membrane

All vertebrates detect sound by means of a pair of labyrinths that lie within the head next to the brain. In land vertebrates, the principles by which these function are broadly similar in all groups, from amphibians to mammals.

In a 'typical' lizard ear, sound waves cause vibrations in the tympanic membrane (eardrum), which are conducted by means of a bone (the columella) to the oval window, a small membrane in the inner ear. Within the auditory part of the inner ear (equivalent to our cochlea), vibrations are set up in the basilar membrane, which is usually an elongated structure in which many sensory-hair cells are held. The tips of the ciliary tufts of the hair cells are attached (directly or indirectly) to the tectorial membrane. This immobilizes the extremities of the tufts, so producing a relative motion between the cilia and the hair-cell bodies when the basilar membrane vibrates. The resulting stimulation of the hair cells is then conveyed to the brain by the auditory nerve.

The mammalian ear

Auditory nerve

Semicircular canals

Auditory ossicles (stapes, incus and malleus)

Cochlea

Auditory canal (to exterior)

Position of oval window

Eustachian tube (leading to nasopharynx)

Tympanic membrane (eardrum)

SEE ALSO
● ANIMAL COMMUNICATION p. 152
● NERVOUS SYSTEMS p. 182

cupulae (cup-shaped structures) that lie across the lumen (passage) of each canal. As the head accelerates in a particular plane, the fluid in the canals lags behind through inertia, so pressing against the cupulae and bending the attached hairs. The combined responses of all three canal organs give the direction of angular acceleration. When the fluid in the canals moves at the same velocity as the head, the cupula positions are restored and any sensation of acceleration is lost, so constant velocity is not signalled.

Fishes have an additional sensory system to detect movement or vibrations in the water – the *lateral-line system* (⊳ diagram). This consists of groups of hair cells that project into the water or into a fluid-filled canal open to the water. The tips of each group of cells are embedded in a cupula, which is moved by displacement of the external fluid.

Sound

The ability to detect sound is limited largely to insects and vertebrates. Insects can detect low-frequency sound by displacement of sensory hairs, such as those on the abdominal cerci (sensory appendages) of cockroaches. More complex sensory cells called *chordotonal sensilla* detect vibration or movement between two parts of the body. Such cells occur in the base of the antennae of male mosquitoes in a special organ that detects the wing beats of females. In other insects, such as locusts and moths, they are attached to membranes, which are vibrated by sound waves. In some moths, just two sensory cells are used to detect the ultrasonic cries of the bats that prey on them, while in locusts different groups

of cells respond to different sound frequencies.

Lower vertebrates such as fishes detect sound through the otolith organs (⊳ above). The walls of the chambers are vibrated by sound, which may be transmitted directly from the water via the body tissues or may cause vibration in the swimbladder. Some fishes, including minnows and catfishes, have a series of bones extending from the swimbladder to the inner ear.

In terrestrial vertebrates (⊳ diagram), sound waves enter the ear by first vibrating the round *tympanic membrane* or *eardrum* on or near the surface of the head. In mammals the vibrations are passed to the inner ear by a chain of three small bones or ossicles (the malleus, incus and stapes); in other land vertebrates this is achieved by a single bone (the columella, equivalent to our stapes), which sometimes has a cartilaginous extension. The inner end of the columella or the bony chain abuts against a small membrane, the *oval window*, which transmits the vibrations to the inner ear, where they stimulate sensory-hair cells. The vibrations reaching the tympanic membrane are amplified by being transmitted to the much smaller oval window, and by lever action of the bony chain in mammals. In marine mammals sound is believed to reach the middle or inner ear via the lower jaw and an adjacent fat body.

Amphibians have two groups of hair cells sensitive to sound, which lie within the chambers containing the otolith membrane. In reptiles, birds and mammals, the sensory hairs are in a duct leading off from these chambers. The hair cells are arranged on a basilar membrane (⊳ dia-

gram), which in birds and mammals is ribbon-shaped and runs the length of the elongated duct (the *cochlea*). In mammals the cochlea is coiled.

In mammals, the whole basilar membrane vibrates at low frequencies of 100 to 1000 hertz (Hz) and so all the hair cells are stimulated. At frequencies above about 1000 Hz the membrane is thrown into a series of waves, which reach their maximum amplitude at the base of the cochlea for higher frequencies and at the top for lower frequencies. Mammals therefore distinguish the pitch of sounds by the extent to which the whole or various parts of the membrane vibrate in response to sound.   GS

SIXTH SENSES

Some fishes produce electric currents for purposes of communication or navigation, and are able to detect changes in the electric field with specialized receptor cells sensitive to electrical energy (*electroreceptors*). Distributed over the head and body, these are located in pits that are open to the exterior through tiny pores in the skin. The most sensitive receptors are in the *ampullae of Lorenzini*, jelly-filled ducts on the head and snout, which respond to changes in the field of just a few millivolts. The duck-billed platypus can also detect electrical fields, probably produced by the muscle contractions of its invertebrate prey. The detectors, also in jelly-filled pits, are in the skin covering the bill.

Some pit vipers, including rattlesnakes, can detect infrared radiation emanating from the bodies of their mammalian prey. Small pits just below the eyes contain cells that can detect temperature increases of 0.002 °C. The sensory fields of the pits overlap, allowing these animals to judge the position and distance of their prey and to strike accurately at a mouse 40 cm (16 in) away in the dark.

Various animals, including bees and migratory birds (⊳ p. 171), are able to sense magnetic fields, which they use as a guide to navigation. This ability is presumably due to specialized sensory cells within the body, but these have yet to be discovered.

ECOSYSTEMS
AND
HABITATS

The Biosphere Energy Relations

No animal or plant lives independently of other animals and plants, but constantly interacts with other organisms around it. Each individual reacts to the other members of its own species that it encounters. It may cooperate in feeding, or it may court and mate with them. Alternatively, there may be competition for food or for potential mates, which may give rise to aggressive behaviour. Such groups of individuals belonging to the same species are called *populations*.

Species also interact with each other. To an animal, one species of plant may represent a source of food, while another provides shelter or a breeding site. Another animal species may be a predator or an item of prey, it may be a competitor for space or food, or it may be neutral as far as

ENERGY FLOW IN THE BIOSPHERE

SOLAR ENERGY

Green plants — PRODUCERS

Herbivores

Carnivores

Top carnivores — CONSUMERS

Death and defecation

Scavengers
(Woodlice, millipedes, etc.) — CONSUMERS

Death and defecation

ENERGY LOST THROUGH NATURAL PROCESSES

Decomposers
(Fungi and bacteria) — CONSUMERS

Energy enters the biosphere in the form of solar energy, which is used by the primary producers – predominantly green plants – to fuel the photosynthetic reactions by which atmospheric carbon dioxide is converted into organic sugars. Virtually all other organisms – animals, fungi and most bacteria – are consumers, ultimately relying on the organic matter of plants for their support.

such direct effects are concerned. To a plant, grazing animals represent a threat to be deterred, while others serve as agents of pollination or fruit dispersal, and so must be encouraged (▷ p. 42). Collections of different animal and plant species living in the same area are called *communities*.

The ecosystem model

No organisms are unaffected by the non-living environment. The climate of an area affects the survival and the breeding success of both plants and animals (▷ pp. 10, 198). The geology of the area determines its landscape and the soils that develop, and these in turn control the range and abundance of various chemicals that both plants and animals need to build and sustain their bodies (▷ pp. 6, 31–3, 196). The ecologist must therefore consider living communities in their non-living setting; each such complex is termed an *ecosystem*. The whole range of ecosystems that characterize our planet together form the *biosphere* – the 'layer' around the earth where all life forms exist; and the biosphere itself can be seen as a single vast ecosystem.

Ecosystems are never in fact entirely independent of one another, but they do provide a useful unit for the ecologist to study. Many useful observations of the ways in which the natural world operates have been made by the study of ecosystems. For example, one can trace the flow of energy through an ecosystem, and this helps to explain the way it functions and often provides clues to its best management. Similarly, one can follow the movement of different chemicals around an ecosystem (nutrient cycling; ▷ p. 196), and this may assist in understanding the consequences of cropping certain components from the system (for instance, mowing a meadow or extracting timber from a forest). An understanding of the behaviour of energy and materials in an ecosystem is thus valuable both for its management and for its conservation.

The energy for life

The existence and continuation of life on earth depends on a constant supply of energy, and this is provided by the radiant energy received across space from the sun. However, not all living organisms are able to take full and direct advantage of this energy source. An insect or a lizard may be able to bask in the warmth of the sun and so raise its temperature and make it more active (▷ p. 87), but this does not provide the energy needed for the maintenance and running of the animal's cells and organs. This is provided by *cell respiration* (▷ p. 21) – the chemical breakdown of energy-rich substances in the animal's body that have been derived ultimately from its feeding activities. Animals feed in a variety of ways (▷ below), but directly or indirectly they depend for their food on green plants, which gain their energy directly from the sun.

The process by which solar energy is trapped by green plants is *photosynthesis*

(▷ pp. 31–3). In this process atmospheric carbon dioxide is converted into energy-rich sugar molecules. Although some of these sugars are needed straightaway by the plant as an energy source for its growth or for taking up nutrients from the soil, they may also be used in the development of the structure of the plant (for instance in the construction of cell walls) or stored for future use, either as starch (chemically a string of glucose molecules linked together) or as an oil (as in oil-seed rape). The range of possibilities is very considerable, but the most obvious outcome of this process is that the plant gains in weight.

Measuring productivity

The best way of measuring the input of energy into an ecosystem is to record the gain in weight by vegetation in a particular area over a measured period of time. This is called the *primary productivity* of that vegetation. It is often measured in kilograms per square metre (kg/m^2) per annum, or some other convenient set of units. To avoid the confusion that would be caused by the varying water content of plant tissues (depending on whether or not they had recently received water), samples are usually dried before weighing to give results in terms of *dry weight*. There is also the problem that some materials are richer in energy than others (oils, for example, are richer in energy than cellulose, for a given weight); this can be overcome by expressing results in energy terms, for example, kilojoules per square metre (kJ/m^2) per annum.

While primary productivity is the rate at which energy is fixed by plants, the quantity of material present in an ecosystem at any given time (per unit ground area) is called the *biomass*. This term can be used of either plant or animal matter. Different types of ecosystem vary greatly in their productivity (▷ box).

Food chains and webs

Plants are thus the basis of all other energy reactions in the biosphere. They manufacture their own food and are called *producers* or *autotrophs* ('self-feeders'). Animals, fungi and most bacteria are dependent on the organic matter of plants for their support and are known as *consumers* or *heterotrophs* ('fed by others'). Some animals (*herbivores* or *primary consumers*) feed upon living plant tissues; others (*carnivores*, or *secondary* and *tertiary consumers*) prey upon other animals (herbivores or other carnivores). Others still (*scavengers*) obtain their energy from the dead remains of plants or other animals, while in the final process of decomposition certain fungi and bacteria – the *decomposers* – make use of every last piece of dead material that still contains usable energy.

Energy is passed from one organism to another in a sequence, such as plant → herbivore → carnivore → top predator. This is called a *food chain* and each feeding level is referred to as a *trophic level*. The number of trophic levels that can exist in an ecosystem is limited, and it

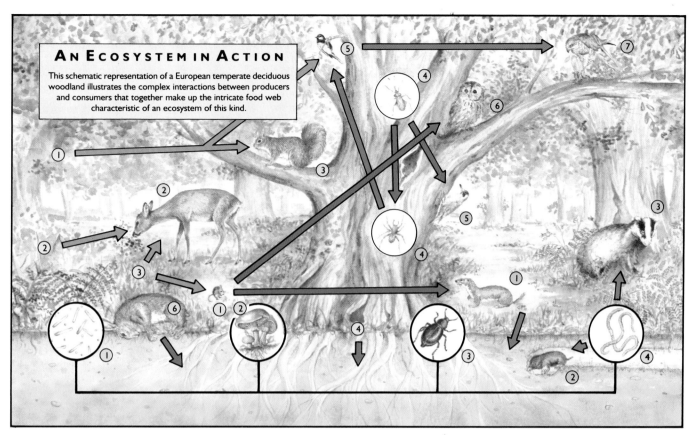

AN ECOSYSTEM IN ACTION

This schematic representation of a European temperate deciduous woodland illustrates the complex interactions between producers and consumers that together make up the intricate food web characteristic of an ecosystem of this kind.

Producers
① Trees ② Shrubs
③ Grasses ④ Leaf litter

Herbivores
(primary consumers)
① Mice ② Deer
③ Squirrel
④ Plant-feeding insect
⑤ Great tit ⑥ Rabbit

Carnivores
(secondary and higher consumers)
① Weasel ② Mole
③ Badger ④ Spider
⑤ Woodpecker ⑥ Owl
⑦ Sparrowhawk

Decomposers and Scavengers
① Bacteria ② Fungi
③ Beetle ④ Earthworms

⇨ Producer to primary consumer

⇨ Primary to secondary consumer

⇨ Other consumer-to-consumer transfers

⇨ Defecation and dead material to decomposers and scavengers

is rare to find more than five, even in a complicated system. This limit is a consequence of the loss of energy at each transfer stage (▷ below), and is also related to the instability that results in an ecosystem if these chains become elongated. Even so, the feeding relationships in an ecosystem are usually much more complicated than can be represented by the straightforward linear movement of energy suggested by a simple food chain. A herbivore often feeds on several different types of plant, and carnivores may eat many different kinds of animal. Many organisms are *omnivorous*, taking energy at a variety of trophic levels. So the movement of energy in an ecosystem is better represented as an intricate *food web* rather than a linear food chain.

The flow of energy in an ecosystem is governed by the physical laws of energy exchange (the laws of thermodynamics). The first law states that energy cannot be created or destroyed, so it is always possible in principle to construct a balance sheet for the energy flow through an ecosystem. The second law describes the way in which energy transfers are always inefficient and involve energy wastage: when an animal consumes a plant, only part of the available energy is transferred – often only about 10% is added to the body of the consumer, the rest being lost in undigested material or dissipated as heat in respiration. The implication of this law in ecology is that energy is lost at each transfer and less becomes available at each successive trophic level in an ecosystem. As a result, there are usually fewer individuals and a lower biomass from one trophic level to the next. An

exception to this is the aquatic ecosystem, where fish may attain a higher biomass than the microscopic plants (phytoplankton) that support them. The reason for this is that there is a rapid turnover of the plant material, so that the energy supply is maintained even though the quantity available at any one moment may be low.

Diversity and richness

The number of species found in an ecosystem is called its *richness*. This is sometimes referred to as *diversity*, but this term should really be used only when some account is taken of the abundance of the different species as well as their number. The factors determining the species richness and diversity of an ecosystem are not well understood. The total energy entering the system by primary production is often related to diversity, as in the case of tropical rainforests (▷ p. 200). But some ecosystems, such as reed swamps, are not very diverse, yet are highly productive. Another factor that influences diversity is the structural complexity of the vegetation. If there is a complicated spatial arrangement of plants in layers (as in a tropical forest), there are usually more opportunities for animals to make their living in a wide variety of ways; in other words, there are more ecological niches present in such an ecosystem. The *ecological niche* is the role that an organism has adopted within an ecosystem – where it lives, what it feeds on, when it feeds, and so on. Some species (most notably, mankind) have very wide niches and have adopted generalist strategies when it comes to making a living. Other species are highly specialized and are said to occupy narrow niches. Diverse eco-

systems usually contain a large number of species, each with a narrow ecological niche.

The stability of an ecosystem was once thought to be related to its diversity. A diverse ecosystem, it was argued, could survive damage (such as the loss of one species) because its complexity allowed its predators to turn to other sources of food. But in fact diverse ecosystems may be unstable and highly sensitive to disturbance, as in the case of tropical forests, which are easily damaged but difficult to repair (▷ p. 201). Simple ecosystems, on the other hand, may be able to recover more rapidly from disturbance.   PDM

SEE ALSO

● WHAT IS LIFE? p. 6
● THE CLIMATIC FACTOR p. 10
● CELLS p. 20
● BACTERIA p. 24
● HOW PLANTS FUNCTION p. 31
● FUNGI p. 35
● POPULATION DYNAMICS p. 172
● BIOSPHERE: BIOGEOCHEMICAL CYCLES p. 196
● CLIMATE AND VEGETATION p. 198
● ECOSYSTEMS pp. 200–21

ECOSYSTEM PRODUCTIVITY

Different ecosystems vary greatly in their net primary productivity – the total amount of material assimilated by autotrophs, minus the material lost in respiration. Forests, swamps and estuaries are the most productive systems, and there is a strong productivity gradient from the tropics (high) to the poles (low).

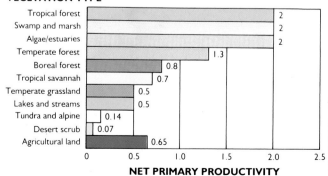

VEGETATION TYPE

Vegetation Type	Value
Tropical forest	2
Swamp and marsh	2
Algae/estuaries	2
Temperate forest	1.3
Boreal forest	0.8
Tropical savannah	0.7
Temperate grassland	0.5
Lakes and streams	0.5
Tundra and alpine	0.14
Desert scrub	0.07
Agricultural land	0.65

NET PRIMARY PRODUCTIVITY
in kilograms per m² per year (average values)

The Biosphere Biogeochemical Cycles

Energy arrives in an ecosystem as sunlight, flows through the ecosystem from one trophic level to another, and is finally dissipated as heat and lost back into the environment (▷ pp. 194–5). But the materials from which animals and plants are constructed behave in a very different way. Although there is some input and output of chemical elements to and from ecosystems, most of them generally remain within the system and are recycled after use. So while we speak of energy flow, the movement of nutrients within an ecosystem is described as nutrient cycling.

Living organisms need a wide range of different elements for the construction of their bodies (▷ pp. 6, 20), and they obtain these either from their non-living environment or from the food they eat. Some of these elements, such as carbon, hydrogen, oxygen, nitrogen, calcium and phosphorus, are needed in relatively large quantities, while others are needed in smaller amounts.

Obtaining nutrients

Plants obtain their mineral nutrients largely from the soil through their roots, while carbon and oxygen are easily available to them from the atmosphere (▷ pp. 31–3). Some of the elements needed are in short supply in certain types of soil, so deficiencies in elements such as calcium, phosphorus, nitrogen, potassium and iron may be experienced under some conditions. Plants that live in nutrient-poor habitats, such as bogs and heathlands, are often able to accumulate nutrients that exist in low concentrations and to conserve them by extracting them from leaves before they die and fall off.

Animals depend largely on their food for their basic supply of elements (▷ pp.

Plant matter is the direct or indirect source from which animals draw the bulk of the mineral nutrients they require. Sometimes, however, there are specific mineral deficiencies that must be made up. In certain areas of central Canada, for example, moose obtain most of the sodium they need by grazing on aquatic plants in the summer months, since these are richer in sodium than most terrestrial plants.

178–9), so they draw directly or indirectly on vegetation to satisfy their needs, just as they do for the energy they require. However, they may find that some of the elements that they need in relatively large quantities, such as calcium for bones or (in the case of some molluscs) shells, are not available in adequate quantities. This has been observed in the case of the element sodium, which is not required by plants but is needed by animals for nerve function. Herbivores usually make up any deficit in this element quite easily by taking in salt (sodium chloride), either in their drinking water or by eating soil or licking rocks, but there are some situations in which this does not provide an adequate supply and the shortfall must be made up in other ways (▷ photo). In human beings, nutrient deficiencies may result in unpleasant conditions, such as goitre, which develops as a result of iodine deficiency in certain areas of the world.

Nutrient cycling

Fortunately, the elements that go to make up living organisms are released back into the environment when they die as a result of the processes of decomposition. The living biomass of an ecosystem can be regarded as a reservoir of elements. Growth in the biomass represents a transfer of elements from the non-living to the living reservoir, while death and decay provide a return route. The release of elements from plant and animal tissues generally takes place in the soil and is controlled by the activity of decomposer organisms – certain bacteria and fungi that obtain their energy from the breakdown of these tissues (▷ p. 195). The elements are usually released in the form of relatively simple compounds with little energy left in them – carbon as carbon dioxide, nitrogen as ammonia, and so on. Some of these may be lost from the ecosystem either in gaseous form or as dissolved material in the soil water.

The loss of elements from the soil as water drains through it – a process known as *leaching* – is an important aspect of nutrient cycling. Soils with a high content of organic matter (humus) or clay are generally better at retaining nutrients than those that are sandy and have little humus. This is because the clay and humus particles in a soil have the ability to form bonds with many of the mineral elements dissolved in the soil water. These bonds are relatively loose, however, and the roots of plants (together with their associated communities of microbes; ▷ below) are able to break the bonds as they forage in the soil for nutrient supplies and to absorb the elements for re-use.

Some loss inevitably occurs in every ecosystem as water drains through the soil and enters deep underground aquifers or flows into streams and rivers. If this loss were not made good, ecosystems would gradually become impoverished in their nutrient capital, and plants and animals would find it increasingly difficult to satisfy their needs. However, there are sources of new nutrient supply to eco-

systems, which are normally adequate to make up any deficit – namely rainfall and the rocks that underlie the soil.

Rainfall contains a range of elements, some derived from the sea and others from dust and soil that have been eroded from other ecosystems. Rocks are an important source of chemical elements, which become available to plants as they are broken down by the physical and chemical processes of degradation. These include frost shattering, splitting by plant roots, and the chemical solution process that accompanies the movement of water through the soil. The gradual release of elements from the rocks is termed *weathering*.

In some ecosystems there is a significant input of nutrients from the droppings of animals that use the site for breeding, roosting or other temporary accommodation. Thus communal birds such as herons and egrets may feed in a wetland and then transport large quantities of many elements, including phosphorus, into the wooded areas that they use for roosting and nesting. Similarly, hippopotamuses often feed on the vegetation along riverbanks during the night and defecate into the waters where they reside during the daytime.

Two nutrient cycles – involving the elements carbon and nitrogen – are of particular importance because they are concerned with elements needed in large quantities. They also have some peculiarities that make them of especial interest.

The carbon cycle

Most ecosystems have a large reservoir of carbon (C) in their biomass – the bodies of the plants (and to a lesser extent the animals) that they contain. This reservoir is built up by the process of photosynthesis, which takes carbon dioxide (CO_2) out of the atmosphere (▷ pp. 31–3). However, the atmospheric reservoir of carbon dioxide is not very large: only about 0.035% by volume of the atmosphere is made up of carbon dioxide. This means that plants have to be very effective scavengers of this gas in order to concentrate it in their tissues and convert it into organic material. As they respire, the gas is released back into the atmosphere, and the plant tissues that are consumed by animals or pass to the decomposer microbes also eventually become degraded in the same way. So much of the carbon cycle takes place between the living biomass and the atmosphere.

Under waterlogged conditions – in bogs and swamps – some of the organic matter that passes to the soil may not be decomposed but remains for long periods in organic form. This is because fungi and bacteria generally do not operate as effectively in an environment where oxygen is in short supply, as is the case when soils become waterlogged. Some of these organic 'fossil' deposits (hydrocarbons; compounds containing only hydrogen and carbon) have survived for hundreds of millions of years, and have been compacted into coal or remain buried as pockets of gas and oil. This

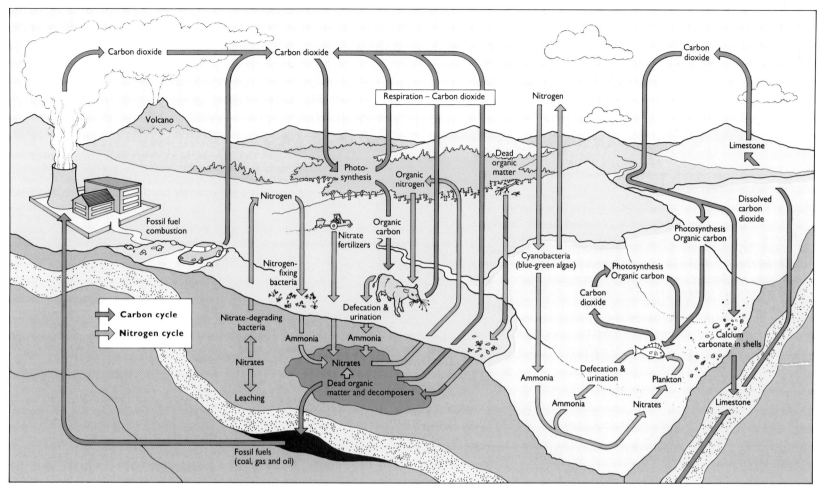

source of energy has been tapped by humans to fuel the industrial development of the world.

The oceans also represent a large reservoir of carbon, since carbon dioxide is soluble in sea water. Both plants and animals make use of this source of carbon: microscopic marine plants (phytoplankton) use it for photosynthesis, while many plants and animals – including microscopic planktonic forms – build shells and protective coats from calcium carbonate (lime; $CaCO_3$). When these planktonic organisms die, their shells fall as sediment to the sea floor, where the calcium carbonate from their coats eventually – over millions of years – becomes consolidated as limestone. Over the course of geological time, these deep-sea sediments eventually become elevated back into the atmosphere and are weathered, so releasing once more their carbon as carbon dioxide.

In fact, therefore, there are two carbon cycles in operation. In the short-term cycle between atmosphere and biosphere it takes a carbon atom about 100 000 years (on average) to make a complete circuit. In the long-term cycle between the atmosphere and the rocks via the oceans, it may take a carbon atom some 250 million years to complete its journey.

The nitrogen cycle

Nitrogen, unlike carbon, is an extremely abundant element of the atmosphere, making up almost 80% of the atmosphere

by volume. However – again unlike carbon – it cannot be obtained or 'fixed' directly from the atmosphere by green plants. Although molecular nitrogen (N_2) is a rather inert gas, some microbes are nevertheless able to reduce it to ammonia (NH_3). In this form it can become incorporated into proteins, and it is principally for the synthesis of these that nitrogen is required in the bodies of living organisms ($\Rightarrow$ p. 6). Some of the microbes that are able to conduct this reaction, such as the bacterium *Azotobacter* and cyanobacteria (blue-green algae), live independent lives in soil or water. Others have developed special (symbiotic) relationships with other organisms and conduct their nitrogen fixation in close association with a host, protected and sometimes even fed by the host tissues in exchange for donating some of the ammonia gained for the host's own protein production ($\Rightarrow$ p. 169). Perhaps the most familiar example is the association between *Rhizobium* bacteria and members of the legume (pea) family of flowering plants, where the bacterium occupies nodules in the roots ($\Rightarrow$ pp. 31–3). But other plants have also developed such associations, incuding alders, cycads, sea buckthorn, and even some aquatic ferns such as *Azolla*. Similar associations are also known to occur with other microbes, particularly between certain algae and fungi, in which case the species mixture is called a lichen ($\Rightarrow$ pp. 34–5).

These various symbiotic associations are

extremely important in maintaining a global supply of nitrogen to living organisms, so that they can build their proteins. The nitrogen is passed to the soil when the plant or consumer animal dies and decays, or when the animal urinates or defecates. The ammonia released is then oxidized by soil bacteria to nitrates (NO_3^- compounds) – a process termed *nitrification* – and in this form it can be reabsorbed by plant roots and continue the cycle. Nitrates are soluble, however, and are not easily retained by soils, so they are easily lost in drainage water. Since they contain oxygen, nitrates can be used as an oxygen source for respiration by some of the bacteria inhabiting waterlogged soils. As a result of this reaction nitrogen gas is released back into the atmosphere.

Nitrogen lost from soils as a result of these processes or as a result of harvesting plants and animals from agricultural ecosystems may need to be artificially replaced if high productivity is desired, as is the case with arable crops and in the production of hay or dairy produce ($\Rightarrow$ p. 226). Atmospheric nitrogen can be fixed industrially, but a great deal of energy has to be expended to achieve the high temperatures and pressures needed to carry out the reaction involved. Fertilizers produced in this way are widely used in agriculture. At the moment it is still more economic to use artificial fertilizers than to recover nitrogen from human sewage, which remains a major source of nitrogen loss. PDM

SEE ALSO

● WHAT IS LIFE? p. 6
● CELLS p. 20
● HOW PLANTS FUNCTION p. 31
● SYMBIOSIS p. 169
● DIGESTION p. 178
● BIOSPHERE: ENERGY RELATIONS p. 194
● ECOSYSTEMS pp. 200–21
● ATMOSPHERIC CHANGES p. 224
● POLLUTION p. 226

Climate and Vegetation

Climate can be defined as the average weather conditions experienced in a given region, and is based on records compiled over a number of years – conventionally 30 years. Climate thus describes the conditions that can normally be expected in a particular location. The weather is expected in general to conform to these conditions, but exceptions and extremes may occur periodically. The vegetation found in a particular region is more likely to correspond with the climate than with any given set of weather conditions.

It will be obvious from a glance at the accompanying map that climate is far from being uniform over the surface of the globe. There are many variables that result in the rather complex pattern of climate that exists.

Factors determining climate

Probably the most important factor determining climate is the amount of energy arriving from the sun. In the equatorial regions the sun is always high in the sky at noon and sometimes directly overhead, whereas at the poles the sun can never be observed at a high angle even in midsummer. As a result, the intensity of sunlight is greater in the low latitudes, for the light penetrating to the earth's surface is spread over a smaller area and passes through less of the atmosphere in order to reach the surface. This accounts for the generally warmer conditions near the equator and the cooler ones towards the poles.

This unequal distribution of energy over the earth's surface results in an instability in the atmosphere and sets up convectional movements in air masses. Cold air over the poles is dense and sinks, while warm air at the equator becomes less dense and is forced upwards. The general effect of this circulation of air masses is that rising equatorial air cools with altitude and its rich content of water vapour begins to condense and fall as rain; hence the equatorial regions are both hot and wet. At the poles the falling air produces high-pressure systems in the atmosphere, which are accompanied by very little precipitation; for this reason the poles can in effect be seen as cold deserts.

The atmospheric circulation of air is actually much more complex than this simple model suggests. The air masses rising over the equator descend again long before they reach the poles. They fall mainly over the regions of the Tropics of Capricorn and Cancer, forming high-pressure systems; these again result in very dry conditions, but this time accompanied by high temperatures, since the energy input from the sun in these regions is high. Many of the world's hot deserts are situated on the lines of the Tropics.

While some of the air descending over the Tropics is deflected back towards the equator, much of it moves outwards towards the poles. These warm air masses collide with descending and spreading polar air in the mid-latitudes; this results in unstable climates, often dominated by the development of low-pressure areas (depressions) over the oceans, which are then carried over the continental landmasses. The general course of movement of these mid-latitude air masses is from west to east, as a consequence of the spin of the earth on its axis.

A further complication is the uneven distribution of the oceans and landmasses over the surface of the earth. This affects climate because land areas heat up and cool down much faster than the oceans. For this reason the interiors of continents become hot in summer and cold in winter, whereas parts of the land surface lying close to the oceans are protected from such conditions by the sea, which has the effect of damping down the extreme temperature variations. These milder areas are said to experience oceanic conditions, in which frost and summer heat are less frequent.

The oceans are also the source of moisture for rainfall, so oceanic areas generally receive more precipitation. As one moves into the interiors of continents, rainfall becomes scarcer. Ocean currents can also have an influence on this process, since a cold ocean current, especially when combined with offshore winds, can produce dry climates even in regions close to the sea. This can be seen in southwestern Africa, where the Benguela Current is associated with a coastal desert.

Warm ocean currents can also have a strong influence on climates, as in the case of the North Atlantic Drift, which brings warm water from the Caribbean into the Arctic Ocean. This keeps the Scandinavian coast ice-free and warm, in contrast to the equivalent latitudes on the west coast of North America, where landmasses block the northward movement of

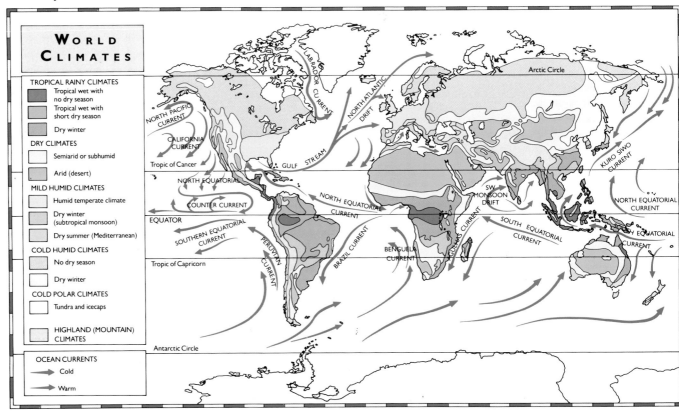

WORLD CLIMATES

TROPICAL RAINY CLIMATES
- Tropical wet with no dry season
- Tropical wet with short dry season
- Dry winter

DRY CLIMATES
- Semiarid or subhumid
- Arid (desert)

MILD HUMID CLIMATES
- Humid temperate climate
- Dry winter (subtropical monsoon)
- Dry summer (Mediterranean)

COLD HUMID CLIMATES
- No dry season
- Dry winter

COLD POLAR CLIMATES
- Tundra and icecaps

HIGHLAND (MOUNTAIN) CLIMATES

OCEAN CURRENTS
→ Cold
→ Warm

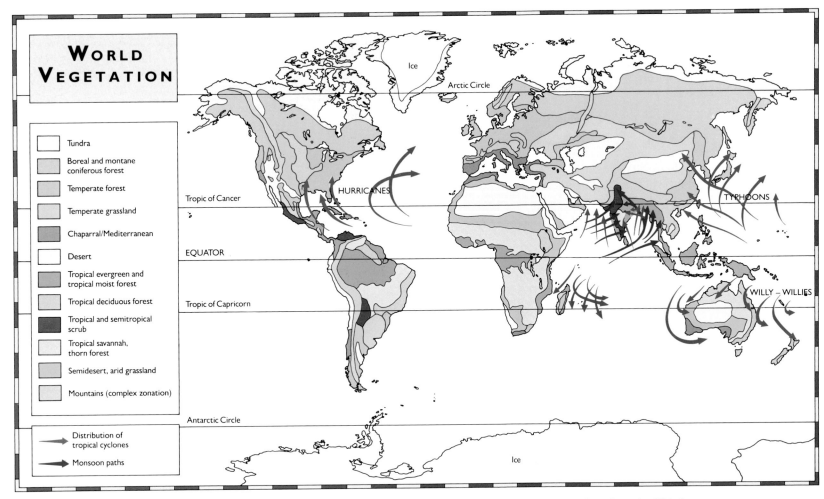

WORLD VEGETATION

Tundra

Boreal and montane coniferous forest

Temperate forest

Temperate grassland

Chaparral/Mediterranean

Desert

Tropical evergreen and tropical moist forest

Tropical deciduous forest

Tropical and semitropical scrub

Tropical savannah, thorn forest

Semidesert, arid grassland

Mountains (complex zonation)

Distribution of tropical cyclones

Monsoon paths

the North Pacific Current and the seas are very cold.

Mountain ranges can modify climate, since temperature falls with increasing altitude. They also modify the general movement of air masses. For example, the Himalaya and the Tibetan plateau block the monsoon rains from the south and create desert conditions in their rain shadow (⇨ p. 204).

The seasons

All parts of the earth experience some change in climate during the passage of the year, as the earth completes its cycle of the sun. These changes are small in the region of the equator, but as one moves from the equator to progressively higher latitudes, the seasonal variation in climate becomes more pronounced. The main cause of seasonal variation is the tilt of the earth upon its axis. When the tilt leans towards the sun, the northern hemisphere experiences its summer and the sun is overhead at the Tropic of Cancer. When the tilt is away from the sun, summer occurs in the southern hemisphere and the sun is overhead at the Tropic of Capricorn.

Classification of climates

There have been many attempts to classify climates, but all are rather artificial, since climate does not occur as a range of discrete types but as a series of continuous variables. The accompanying map shows a convenient classification system. The tropical climates are divided into rainy and dry types, and between these two extremes lie dry-season climates in which dry and wet conditions occur in succession at different times of year. The temperate areas (between the Tropics and the Arctic and Antarctic Circles) experience a complex series of climates, including summer-dry (mediterranean) and mild-humid types. There are also the colder temperate climates of the northern continental temperate areas (there are no landmasses over what would be the southern equivalents of these regions). Finally there are the polar climates of the very high latitudes, which are cold and dry.

Vegetation and climate

From the accompanying maps it is evident that the global distribution of vegetation types broadly follows that of the major climate types. However, the pattern of vegetation over the face of the earth is even more complicated than the climate, so it is no less difficult to formulate a satisfactory classificatory scheme. The most effective system of vegetation classification for use on a global scale depends not upon the taxonomic relationships of plants (i.e. what families they belong to) but upon the general structure and form that a particular plant community assumes. The terms 'forest', 'scrub', and 'grassland' are examples of this type of system in its broadest form; in each case the terms describe the dominant life form of the vegetation in a given region.

The tree as a life form is sensitive to

climatic stresses such as drought. This is because the buds of the tree, which are the centres of new growth, are held high above the ground and are exposed to dry air and wind. Areas of drought or extreme cold cannot support such sensitive forms of plant life, so in such areas trees are replaced by dwarf shrubs or – in the case of the tundra – by cushion-forming plants that escape cold winds by moulding themselves to the contours of the ground (⇨ p. 210). Regions that are cold or dry for only part of the year may support broadleaved trees, but only those that lose their leaf canopy in the unfavourable period. Evergreen needle-leaved trees – the conifers (⇨ p. 38) – are more efficient than most others at coping with more prolonged cold and snowfall, so they are dominant in the cold-temperate areas.

Herbaceous plants may also be important in the structure of vegetation. Most forests have a herbaceous understorey, and grasslands are dominated by herbaceous plants that die back to ground level during an unfavourable, prolonged dry period. This capacity also renders them less sensitive to other pressures, such as fire and grazing, which are frequently encountered in both tropical and temperate grassland zones.

Vegetation can thus be classified as an assemblage of plant life forms, and each of these vegetation types has its characteristic fauna. These units are called *biomes*, and the main biomes of the earth will be described in detail on the following pages (⇨ pp. 200–11).   PDM

SEE ALSO

● THE CLIMATIC FACTOR p. 10
● HOW PLANTS FUNCTION p. 31
● PLANT TYPES pp. 34–47
● THE BIOSPHERE pp. 194–7
● ECOSYSTEMS pp. 200–21
● ATMOSPHERIC CHANGES p. 224

Tropical Forests

Although the tropical forests occupy only about one fifth of the earth's land surface, they contain about 1·5 million species of plants and animals – over half of the known total. Since these forests are among the least known of the world's biomes, they may well prove to hold an even greater proportion of the species diversity of our planet.

Rain forest (right) on Monte Verde, Costa Rica. This small Central American country has been highly innovative in its approach to forest reserves, linking them with protected 'corridors' that allow the movement of species, which in turn helps to maintain the diversity of plants and animals (▷ p. 233).

True tropical rain forest is confined to a narrow belt around the equator, from 4° N to 4° S. Here there is little seasonal change in climate, and temperature and precipitation are always high: temperatures average around 20–25 °C (68–77 °F) all year round, and the annual rainfall is between 1500 and 4000 mm (60 in and 160 in). The leaves of most trees are evergreen, that is they tend not to fall in unison leaving the trees bare. This is partly a response to the lack of any unfavourable period for growth, but, as the leaves last longer, it also provides a mechanism for conserving nutrients. Although the overall rainfall is high, the upper part of the high canopy can become hot and dry, so leathery, evergreen leaves can be useful because of their ability to resist drought.

As one moves further from the equator, the climate begins to show seasonal vari-ation. Certain weeks in the year may have little or no rain, and this places a new stress upon the forest. The trees are less tall and may lose their leaves during the dry spell as a means of water conservation. Seasonal changes can also provide environmental cues for the initiation of flowering, fruit production and breeding among animals.

By about 15° on either side of the equator there is a very marked seasonal pattern of rainfall, with wet and dry seasons, and the vegetation is severely limited by the drought. Here the forests border on to the savannah grasslands (▷ p. 202), the precise boundary often being determined by grassland burning on the part of man.

Biomass, structure and productivity

The tropical forests contain the largest bulk of living material per unit area of ground found in any biome, often amounting to over 45 kg/m². The productivity is similarly high – up to 3.5 kg/m² per year. This huge quantity of biomass is arranged in a complex spatial structure (▷ diagram).

In a mature forest very little light penetrates to the ground and hence there is little ground vegetation. Most reports of 'impenetrable jungle' refer either to areas recovering from disturbance or the fringes of the rivers, where low growth may well be luxuriant. Occasional sunflecks may pass though small canopy gaps and reach the forest floor, and these are very important for the photosynthesis of any plants growing at ground level. Some plants overcome this problem by using others for support, for example lianas or climbers can develop to a length of 200 m (660 ft) or more, extending up from the ground to form winding trailers in the upper canopy layers. Epiphytic plants (▷ below) also ensure they receive sufficient light by using trees as platforms. Wherever a new clearing is formed, for example by an old tree falling, there is intense competition among saplings to reach the canopy – and the light – before their neighbours. Tropical tree species have astonishing growth rates, and do not put out extensive branched canopies until they reach the light.

Soils and nutrients

Because of high temperatures, microbial activity in the soil is very high, so any organic litter falling from the canopy is rapidly decomposed. This releases mineral elements into the soil, and under conditions of high rainfall these can easily be leached (washed) out of the soils and into rivers, for the soils of the tropical forests have very little capacity to retain the nutrients released into them. This is mainly because the clays that serve this function in temperate soils are degraded and lost in the high prevailing temperatures, with the result that the soils become hardened and enriched with a residue of iron and aluminium, forming the red-coloured *lateritic* soils.

Plants overcome this problem of nutrient scarcity and loss by rooting in the upper soil layers where the decomposition is taking place and absorbing the liberated nutrients before they can be lost. The precious nutrients are then stored within the bodies of the trees, and the long-lasting leaves help to reduce the rate of their wastage.

Falling organic litter may become lodged in the junctions of tree branches and form a peaty soil in which other plants can become established. Such plants include orchids and bromeliads (▷ pp. 44–7), and are described as *epiphytic*, i.e. they use other plants as a means of support and have no roots in the ground. These organic pockets form their own microecosystems with their own nutrient turnover, and may become quite rich in nutrients. Some tree roots are sensitive to the enriched trickles of water that descend from these pockets and actually grow up the trunks of other trees so that they can tap into this unexpected resource. Such upwardly growing roots are very rare in any other type of ecosystem.

TROPICAL DIVERSITY

There are many possible explanations for the extremely high diversity of the tropical forest. The high productivity and massive biomass supported is undoubtedly an important factor. A large biomass in an ecosystem allows a high degree of structural complexity, and this encourages the evolution of many specialized insect species, each occupying its own preferred microhabitat more efficiently than any of its possible competitors. In addition, separate communities can develop within each of the canopy layers (▷ main text). However, not all of the highly productive ecosystems of the world are diverse (reed swamps, for example, are productive but not diverse). So other factors must be involved.

The stability of the climate, both seasonally and long term, may permit higher diversity. At one time it was believed that the tropics had retained a stable climate throughout the last ice age (reaching its coldest about 20 000 years ago), and that these stable conditions allowed a greater period for species to diversify. But it is now known that the rain forests became drier at that time and were probably much fragmented. The argument can be turned on its head and the fragmentation could itself be regarded as a source of evolutionary diversification, since evolution could have proceeded separately in each forest fragment – much as it does on isolated islands (▷ p. 214) – before the fragments joined together again.

A final source of novelty is the mosaic of patches within the forest. Wet and dry regions and sites in different stages of recovery following the death of a tree, or in an area where trees have been blown down, means that species can occupy different stages in time as well as locations in space. Whatever the reason for the diversity of the forest, this wealth of species must be regarded as a vital resource for the future of mankind.

Biological diversity

The proverbial richness of the tropical forest extends into every kingdom of living organism – plant, animal, fungal, bacterial and protist. The numbers of tree species alone can be remarkably high, as for instance in the Brazilian rain forests, where 300 tree species have been found within an area of 2 km² (¾ sq mi). Even the epiphyte communities high among the trees support their own great range of algae, fungi and invertebrate animals.

On the dark forest floor most animals feed upon the rich rain of organic matter from above. These are the *detritivores*, like millipedes, that begin the process of decomposition finished off by microbes. The fungi, as well as some plants, also derive energy from the degradation of the organic detritus. Other plants live as parasites.

Many animals make a living on the forest floor as omnivores, taking whatever food comes their way. Important among these are the ants that feed on anything available from seeds and fungi to fallen caterpillars. Some large herbivores, such as the capybara, tapirs, deer and antelopes, live on the forest floor, providing food for predators such as the jaguar.

In the canopy are found many fruit-eating species, including monkeys, bats, and birds such as toucans and parrots. These may be very specific in their habitats, confining their activities to particular layers of the forest where they find their preferred food. Thus the canopy layers have their own communities of insects, birds and mammals as well as the epiphytes and climbers that find a home there.

Management problems

The greatest problem associated with the tropical forests is the rate at which they are being destroyed. Selective logging for timber, clear felling for pulp production and destruction by burning for subsistence arable farming or commercial cattle ranching are all taking their toll on the world's dwindling tropical forests. It is difficult to be sure how much forest is lost each year, but it probably lies between 75 000 and 140 000 km² (29 000 and 58 000 sq mi) – up to twice the area of Ireland. This is over 1% of the total area of the world's tropical forests, and the rate is accelerating each year.

Apart from the concurrent loss of species, the destruction of the considerable biomass of the forest by burning puts large quantities of carbon dioxide into the atmosphere, adding to the greenhouse effect and global warming (⮑ p. 224). Where crops or grass for cattle are planted the nutrient-poor tropical soils soon become exhausted, and taking away the canopy from above the soil allows penetration of solar radiation, baking the soils and destroying their structure. The result is a virtual desert, and the ranchers, farmers and foresters thus move on to clear further areas. Rain falls directly on to the soil surface and causes extensive erosion, silting up rivers. The run-off may occur in flash floods, resulting in the destruction of agricultural land and often

the loss of human life downstream. Finally, the water cycle of the equatorial areas may be upset, for less water is returned to the atmosphere by plant transpiration (⮑ p. 31), ultimately leading to reduced rainfall.

There is much concern to develop systems of rational exploitation of the tropical forests that would combine economic gain for local people with ecological stability. Experiments in which the forest is exploited in a patchwork and then allowed to recover are being conducted in Papua New Guinea, but so far the results have not been promising. Areas clear-felled for pulpwood do recover if allowed to regenerate naturally, but many species are probably lost and the economic return to the foresters is not as great as when fast-growing plantations of exotic (i.e. non-native) trees are planted instead.

Greed for short-term profits by ranchers and logging companies, together with the population pressure on poor peasant farmers that forces them to encroach on the forests, currently seem to be the major driving forces in encouraging forest stripping. Until adequate controls can be enforced on commercial interests and until the much harder problems of structural poverty in the Third World can be addressed, the future of the tropical forests looks extremely grim. It is as a renewable resource, carefully managed, that the tropical forests can make their greatest long-term yields. Instead of clear felling, continuous harvesting of products – such as essential oils for the pharmaceutical industry – can generate a constant income for local people, without damaging the stability of the ecosystem.   PDM

SEE ALSO

- THE BIOSPHERE pp. 194–7
- CLIMATE AND VEGETATION p. 198
- TROPICAL GRASSLAND p. 202
- TEMPERATE FORESTS p. 207
- ATMOSPHERIC CHANGES p. 224
- VANISHING HABITATS p. 228
- NATURE CONSERVATION p. 232

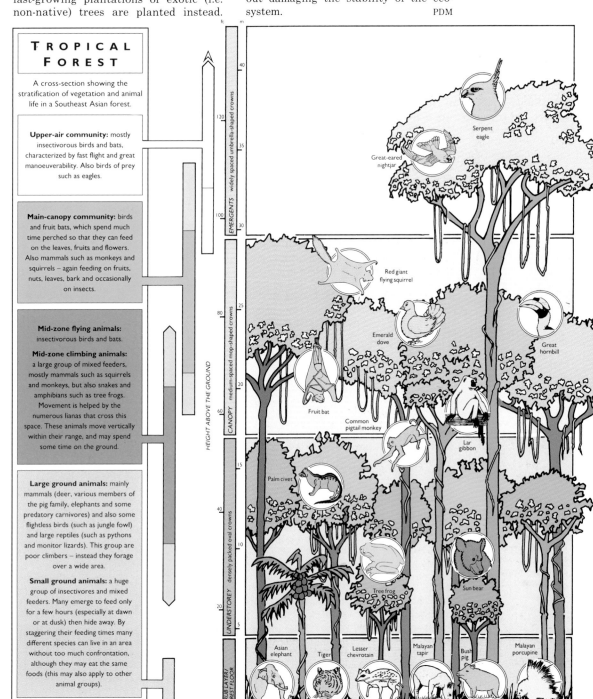

TROPICAL FOREST

A cross-section showing the stratification of vegetation and animal life in a Southeast Asian forest.

Upper-air community: mostly insectivorous birds and bats, characterized by fast flight and great manoeuverability. Also birds of prey such as eagles.

Main-canopy community: birds and fruit bats, which spend much time perched so that they can feed on the leaves, fruits and flowers. Also mammals such as monkeys and squirrels – again feeding on fruits, nuts, leaves, bark and occasionally on insects.

Mid-zone flying animals: insectivorous birds and bats.

Mid-zone climbing animals: a large group of mixed feeders, mostly mammals such as squirrels and monkeys, but also snakes and amphibians such as tree frogs. Movement is helped by the numerous lianas that cross this space. These animals move vertically within their range, and may spend some time on the ground.

Large ground animals: mainly mammals (deer, various members of the pig family, elephants and some predatory carnivores) and also some flightless birds (such as jungle fowl) and large reptiles (such as pythons and monitor lizards). This group are poor climbers – instead they forage over a wide area.

Small ground animals: a huge group of insectivores and mixed feeders. Many emerge to feed only for a few hours (especially at dawn or at dusk) then hide away. By staggering their feeding times many different species can live in an area without too much confrontation, although they may eat the same foods (this may also apply to other animal groups).

Tropical Grasslands

Many anthropologists believe that it was within the tall tropical grasslands – the savannah – that our distant ancestors adopted an upright stance to assist them in hunting and to keep a lookout for their predators. The grasslands of the tropics remain rich in herbivores and their predators, which make them important habitats for large mammals. The high productivity of these grasslands and their capacity to support big herbivores has also made them particularly valuable for pastoral farming, so careful human management has become very important for their conservation and survival. This type of ecosystem is more easily cropped than a forest, so the conversion of tropical forests to grassland by felling and grazing is an increasingly popular option, but is rarely successful in the long term (▷ p. 201).

High temperatures persist throughout the year in the tropical grasslands, averages varying seasonally between 15 and 25 °C (59 and 77 °F). The rainfall is not so even: the annual average of 600–700 mm (24–28 in) is mainly confined to the strongly marked, five-month wet season. The wet season alternates with a dry season, and the latter limits the growth of trees. The wet season marks the beginning of the main growth period for the grasses, which then become dead and dry during the subsequent dry period. At this time the whole ecosystem becomes prone to fire, often started deliberately by human populations. Fire, in fact, may play an equally important role to climate in determining vegetation.

The trees that survive fire and drought are usually deciduous, losing their leaves during the dry season, or evergreen with hard leaves, well adapted to drought.

Tropical grasslands are most characteristic of central and eastern Africa and South America south of the Amazon basin. But much of India and some northern areas of Australia also bear savannah vegetation.

Biomass, structure and productivity

The biomass of savannah grasslands varies with season, being highest in the early stages of the dry season, when soil moisture is still sufficient to promote growth. Much of this biomass lies under the surface of the ground in the root tissues. The above-ground portion consists of tall grasses, often exceeding 2 or even 3 m (6½–10 ft) in height, and trees, which vary in density from one area to another depending mainly on the frequency of fires. The total biomass often attains levels of 5 kg/m^2 and the produc-

tivity may be as high as 2 kg/m^2 per year, which is about the same as that of the boreal forests (▷ p. 208). It is remarkable that such a high productivity can be achieved by a relatively low biomass, and this illustrates the efficient energy conversion that is possible in the grassland ecosystem. Much of this energy is then passed on to the large grazing animals – the next step in the food web. This direct movement of large quantities of energy into meat production is precisely why human agriculturalists have found grassland systems so much more useful than forests (▷ p. 212).

Because the period of high productivity is relatively brief, herds of large grazing animals often have to migrate from one area to another to take advantage of this productive time. Grazers and browsers also have very specific feeding preferences, so the total resources of the savannah are often divided up among many different species (▷ below).

Soils and nutrients

High productivity and rapid production of leaf litter from the grasses and trees provides a rich source of organic matter for the soils, and many decomposer and detritivore (detritus-eating) animals are active in using this rich source of energy. Termites are particularly important consumers of dead vegetation, though they may process the material first by growing fungi on the detritus to digest it for them – an unusual example of animals practising agriculture.

The cycle of nutrients in the soil is, therefore, normally rapid, but some of the grasses secrete chemicals that supress the activity of certain bacteria in the soil and so slow down the release of nitrogen. This may be a mechanism for preventing other species tapping in on this important resource.

Biological diversity

The grasses dominate the savannahs. This family of highly successful plants has the remarkable property of growing from the leaf bases, so that if the tips are grazed off they can continue growing from below. This is an extremely efficient adaptation

The sheer scale (right) of the African grasslands can be seen in this picture, with Kilimanjaro (5895 m / 19 340 ft) rising some 4500 m (15 000 ft) above the surrounding plains. The large mammals of the savannah require vast areas to support their populations, and maintaining such large areas is proving difficult in the face of human pressure.

to cope with incessant grazing for, as long as green tissue remains above the ground surface, growth can continue. The deep rooting systems also protect that part of the plant from fire, so ensuring survival into the next wet season.

The trees of the savannah are equally well adapted to the sequence of wet and dry seasons. In Australia, fast-growing trees like eucalyptus take advantage of the wet period and then become dormant. The most typical savannah trees of the Australian, Asian, South American and African grasslands are the acacias, their flat-topped form lending a characteristic aspect to the landscape. Many acacia species have formed a close association with certain ants. The leaves of these trees have nectar-secreting glands on the leaf stalks, thus encouraging the presence of ants. In return, the ants attack any animal that attempts to browse on the branches. Ants also play an important part in the dispersal of seeds in the savannah. Another characteristic tree of the African savannah is the baobab, with its crown of spindly branches and huge

swollen trunk that acts as a massive water-storage organ.

The most remarkable feature of the animal life is the abundance and diversity of grazing animals. This is made possible by the high degree of specialization of the various species, both in their food requirements and the sequence in which they feed. In the African savannah some species, such as the giraffe and elephant, browse high up on the trees, whereas others, such as antelopes and the black rhinoceros, browse lower down. Other species, such as wildebeest, zebra and the white rhinoceros, graze the grasses. Often there is a distinct sequence in which the grazers take their turn to operate on an area of grass: for example, plains zebra is often replaced by brindled gnu and then by Thompson's gazelle. This division of the resources both in space and in time enables more species to exist within a given area.

Insect herbivores abound, especially the locusts. These periodically form immense swarms that strip vegetation even more

effectively than the herds of large herbivores (⊳ p. 173). Other insects, like dung beetles, feed upon the large amounts of faecal material deposited by the mammalian grazers.

The abundance of herbivorous life provides the basis for complex food webs, at the summit of which sit the large cats, such as lions and cheetahs. But less conspicuous predators also make a living out of smaller prey. Anteaters consume the termites, while birds like the secretary bird prey upon snakes. There is also a living to be made by scavenging on the dead remains of other animals, and here vultures and marabou storks – and also hunter-scavengers such as jackals and hyenas – come into their own.

Management problems

The use of savannah for cattle grazing is one of the most serious threats to its wildlife. Some of the people of the savannah, such as the Masai of Kenya, are almost totally dependent on cattle and cow-products for their livelihood. Naturally, stock-breeding peoples are unwilling to share the resources of these grasslands with wild competitors. Not only can this cause the local extermination of natural grazers, it can also result in the interruption of the traditional routes of the migrating herds. The use of savannah for agricultural purposes has also disrupted the tropical grasslands of South America, and populations of the flightless rheas have declined as a consequence.

Grazing by a single domestic species may result in changes to the composition of the savannah flora, for cows do not graze on as wide a range of species as the assemblage of native fauna. Cattle concentrate their attentions on the more palatable grasses and neglect coarse, toxic and spiny species. Ultimately this can lead to the invasion of the grasslands by scrub vegetation and the loss of its potential productivity.

In many ways the cow (originally a temperate forest animal) is not a very suitable animal for grazing these tropical grasslands efficiently. Quite apart from its inability to utilize all types of vegetation, it is also subject to many pests and diseases, such as tsetse fly. Higher yields of meat can be achieved by ranching and harvesting the natural grazers, and much effort is now being expended on developing this approach to pastoralism in Africa.

The use of fire to manage tropical grasslands can also create problems. Some nutrient elements, such as nitrogen, are lost in smoke, leaving the soils poorer. As yet there are few studies on the amounts of carbon dioxide produced by tropical grassland burning, but it is entirely possible that this annual event has a marked effect on the levels of this gas in the atmosphere, so adding to the greenhouse effect (⊳ p. 224). Fuel for cooking is usually obtained from the savannah trees, and this can also lead to over-exploitation, although plantations of eucalyptus and acacia can provide the fuel needed for this purpose.   PDM

SEE ALSO

● MIGRATION p. 170
● THE BIOSPHERE pp. 194–7
● CLIMATE AND VEGETATION p. 198
● DESERTS p. 204
● MAN-MADE HABITATS p. 212
● VANISHING HABITATS p. 228
● NATURE CONSERVATION p. 232

Deserts

Deserts cover more than one third of the earth's land surface. Such areas are characterized by a deficiency of water, chiefly as a result of low rainfall, and consequently are able to support little vegetation. On the margins of many of the world's deserts are areas of slightly less extreme climate, where some woody scrub vegetation is able to survive, depending largely on the grazing pressures imposed by local populations and their domestic animals.

The hot deserts of northern and southern Africa, the Middle East, central Australia, and the western coastal areas of North and South America fall within the subtropical high-pressure belts north and south of the equator; these areas experience high temperatures throughout the year, but especially in summer, and very low rainfall. In many parts of these deserts no rain at all may fall for several years at a time.

The dry climate of other deserts is due to their geographical isolation from the sea, which prevents moisture-bearing winds from reaching them. This condition may be aggravated by the shape of the landscape: for example, because moist air coming in from the sea precipitates on mountains as rain or snow, the air that has passed over the mountains is dry and may cause the formation of so-called 'rain-shadow' deserts. The deserts north of the Himalaya are examples of this effect. Deserts of this kind may be quite cold in winter. In the deserts of Afghanistan and Iran, for example, frosts are common, and much of the precipitation comes as winter snow. Winter conditions are even colder in the Gobi Desert of central Asia, where temperatures may fall below –20 °C (–4 °F).

Biomass, structure and productivity

In extreme cases, deserts may be bereft of all living matter – their biomass is literally zero. Even in less harsh conditions, vegetation may be confined to the drainage channels, or 'wadis', that carry the water from any occasional rain showers. This is called 'contracted vegetation'. In regions of desert scrub, the biomass often falls between 5 and 25 kg/m², and consists mainly of branched, deciduous shrubs, such as the creosote bush of North America, that put out their leaves in the early spring following winter showers.

The density of vegetation in arid lands is sparse, and often there is a very regular spacing of plants. The struggle for water, and therefore space, is so great that new colonists are excluded by intense root competition and sometimes by chemical exudates from other plants. The structure of the vegetation is thus simple, but it still determines to a large extent the patterns of small mammals and invertebrate animals.

Soils and nutrients

The water present in desert soils tends to move upwards rather than draining through them, as is the case in wetter areas. This is caused by the high rate of evaporation at the surface. As a result, soluble salts within the soil are carried up to the surface and may crystallize as the water evaporates, forming incrustations of gypsum, salt or lime.

Organic matter is scarce in desert soils, and this means that nutrient turnover is poor. Inconspicuous life forms may play an important role, such as lichens that are able to capture or 'fix' gaseous nitrogen from the atmosphere. During the night these are scraped off the rocks on which they grow by snails. As the snails retreat beneath stones to escape the daytime heat, they carry the nitrogen with them and discharge it into the soil in their faeces, thus enriching the soil for plant growth.

Biological diversity

Diversity is low in the desert environment. However, the plants and animals that survive there are of great interest because of their high degree of adaptation to such extreme conditions.

Many plants have developed elaborate methods of protecting themselves from excessive heat and drought. Some root systems are deep and extensive, while others form a fibrous mat in the surface soils to catch the dew. The latter system is typical of most of the New World cacti, which also enhance their tolerance to drought by reducing their leaves to spines and by storing water in their fleshy, evergreen stems. Some desert plants are able to grow very rapidly at the onset of rain, and the seeds of many are able to lie dormant for several years, awaiting conditions sufficiently damp for germination. The euphorbias – an important plant group in the deserts of the Old World – have stems filled with sticky latex, which is highly resistant to evaporation.

Most desert animals are nocturnal or active during the cooler periods of dawn and dusk; to avoid the daytime heat, they seek shade or hide away in burrows, which are often located around the base of shrubs, where there is greater protection from the sun. Certain large animals, such as the Arabian oryx, can survive on a very small water budget, while small mammals may derive their entire water input from the foliage they eat. Some desert reptiles excrete crystalline uric acid to conserve water.

Management problems

An ecosystem with a very low level of productivity can support only a limited grazing pressure, but pastoral agriculture is common among the peoples of the desert and surrounding areas. There is also a great demand among these people for firewood, which adds to the strain on the system. The outcome of these pressures is a further reduction in biomass and productivity, causing desert conditions to spread into surrounding areas – a process known as *desertification*. Attempts to increase the productivity of desert areas by irrigation have often failed because adding water to the soils only serves to enhance the movement of salts to the surface, resulting in salinization.   PDM

Cacti in the Organ Pipe Cactus National Monument, Arizona, USA. The column-like saguaro (left and background) can grow to around 15 m (50 ft) and reach an age of up to 200 years. The jumping cholla (right foreground) is a close relative of the prickly pear, or Indian fig, an important food plant in tropical and subtropical countries.

Mediterranean Climates

A number of warm-temperate areas of the world are characterized by a mediterranean climate, in which hot, dry summers alternate with mild, wet winters. This type of climate is found not only in the Mediterranean basin, after which it was named, but also in California, Chile, South Africa and parts of southern Australia. Such areas typically have sufficient water resources to support tall scrub or even forest vegetation, but a high frequency of fire, often accompanied by overgrazing, has generally reduced them to low scrub or open heath, variously called chaparral, maquis, garrigue or fynbos in its different geographic locations.

Mediterranean-type vegetation in the Massif de l'Esterel, Côte d'Azur, France (right). Fire is a major threat in regions of mediterranean climate; the area of France shown here has been ravaged by fires in recent years, signs of which can be seen in the background.

In the lands that border on the Mediterranean, the pattern of lower-biomass vegetation developing in the wake of deforestation dates back at least to Bronze Age times. More remote areas of similar climate, such as the Cape region of South Africa, have suffered a similar plight since their discovery and exploitation by Europeans in the 16th century.

Biomass, structure and productivity

Like the desert scrub vegetation into which it may merge, the plant life of mediterranean areas is dominated by a mixture of deciduous and evergreen shrubs. Depending upon winter rainfall, this scrub may be quite dense and even include patches of woodland such as evergreen and deciduous oaks, pistachios, pines and cypresses. This vegetation has a higher biomass than desert scrub, often exceeding 25 kg/m², and has a relatively complex structure that supports a wealth of wildlife. Where grazing is intense, however, as in much of the Mediterranean basin, the vegetation has become degraded and simplified into a low cover of open grass and heathland that dries to an almost desert appearance in the hot summer. This is termed 'garrigue'.

Soils and nutrients

Unremitting heat leaves soils dry in summer, and the deciduous leaves of shrubs are often rich in aromatic compounds that are extremely flammable. As a result, fire is a regular feature of this habitat and has a major impact on nutrient cycling, releasing nutrients into the soil and atmosphere. Where soils are sandy, they do not retain nutrients well and can become depleted, leading to an acid heathland vegetation.

Biological diversity

All areas of mediterranean climate have in common a very rich flora, consisting of drought-tolerant aromatic shrubs, a rich variety of bulb plants, and many species resistant to fire and grazing. Shrubs include the rockroses in Europe, sagebrush in California, and the proteas in South Africa and Australia. Fire often causes a constantly recurring pattern of replacement among these plants. Amongst the bulb plants that thrive in this habitat are fritillaries, grape hyacinths and cyclamens; these grow and flower in the spring and then become dormant through the hot, dry summer.

Regions of mediterranean climate are rich in bird and other animal life, including insectivorous warblers, reptile-eating birds of prey, seed-eating rodents and predatory snakes. As with many of the plants, some animals are active mainly through the winter, becoming dormant during the summer heat. Among the large mammals, roe deer are native to the area in Europe, as are predators such as the lynx, wild cat and wolf, but these are now close to extinction in many areas. Wild boar and tortoise are also very characteristic of the shrub-dominated heathlands. In North America jack rabbits and pumas occupy this habitat. In Australia the kangaroo is the major mammalian grazer, while in South Africa hyraxes and duikers are important.

Management problems

Historically the greatest problem in mediterranean habitats has been forest clearance, but now intensive grazing and repeated fires probably have the greatest impact on the environment. Grazing initially suppresses shrubs and replaces them with grasses, but continued grazing can result in the total loss of vegetation cover. Often goats are involved in this process, but they are associated with extremely degraded habitats simply because no other grazer can survive at this advanced stage. Fire in the scrub becomes ever more likely as the biomass grows. Increasing quantities of volatile compounds in the leaves eventually make fire inevitable. Future climatic changes will probably lead to hotter, drier summers, and this habitat will become even more prone to fire.

Some trees are retained in parts of the Mediterranean because of their economic importance. Examples include the cork oak, from which the bark is stripped and used for cork making, the olive, and the stone pine, the seeds of which are edible. However, where forestry is practised for pulp-wood production, the Australian eucalypts have proved most successful in this climate and now dominate large areas throughout the world. PDM

SEE ALSO

- THE BIOSPHERE: ENERGY RELATIONS p. 194
- VEGETATION–CLIMATE RELATIONS p. 198

Temperate Grassland

Some temperate areas receive too little rainfall to carry forest cover, and these areas usually have a vegetation dominated by grasses and herbaceous plants, and a fauna in which grazing animals form a conspicuous part. The success of grazers in these habitats, together with the rich soils, have made such areas attractive to human agriculturalists, who have substituted domestic grazers and cultivated grasses (the cereals) for their wild equivalents. In North America, for example, the bison and native grasses of the prairie were replaced by cattle and wheat during the 19th century.

SEE ALSO

- THE BIOSPHERE pp. 194–7
- CLIMATE AND VEGETATION p. 198
- TROPICAL GRASSLAND p. 202
- DESERTS p. 204
- MAN-MADE HABITATS p. 212
- VANISHING HABITATS p. 228
- NATURE CONSERVATION p. 232

Usually the temperate grasslands lie in the central parts of the major continents, far from the sea, where rainfall is low (around 400 mm / 24 in) and where there is a wide seasonal range of temperature – from a monthly mean in summer of around 20 °C (68 °F) down to −15 °C (5 °F) in winter. Such conditions mean that there is a relatively short growing season in the spring before the hot drought sets in.

Such grasslands have been given different names in various parts of the world: steppes in Asia, pampas in South America, prairie in North America, and veld in South Africa. In the South Island of New Zealand grassland has developed in the rain shadow east of the Southern Alps. Sometimes the grasslands are so dry that they merge into deserts, as in the case of the area to the north of the Gobi Desert in Central Asia, where winter cold is intense, or the prairies in the American Southwest, where the hot summer is the most stressful time.

Biomass, structure and productivity

The plant biomass of the temperate grasslands varies greatly through the year. Springtime is the period of greatest productivity, though grass species often replace one another through the growing season as the more heat- and drought-tolerant species take over the growth. As the summer drought sets in, however, the grasses dry and die back to the roots. This annual death of the above-ground vegetation keeps biomass low (about 3 kg/m²), but the spring productivity and the general palatability of the vegetation allows it to support a rich grazing fauna.

Where rivers pass through the grasslands there is often a ribbon of forest development along its sides, where the water supply is reliable through the summer.

Soils and nutrients

The hot summers lead to high evaporation, and this can lead to the water in the soil moving upwards and depositing salts at or near the surface as it evaporates. But soils are generally fertile and very rich in organic matter as a result of the grass litter that is constantly being incorporated by the activities of the soil fauna. In appearance these soils are dark brown or even black in the upper layers.

Biological diversity

Besides the grasses, there are many wild plants in the prairie grasslands, most of which cope with summer drought by becoming dormant as bulbs, corms, tubers or simply as seeds. The high grazing pressures have resulted in the evolution of some spiny and distasteful plants, and the disturbance of the soil by burrowing and trampling has provided opportunities for other plants to exploit temporary habitats. Many of these have now become very successful weeds when spread around the world by humans.

In their natural state these grasslands were rich in large herbivores: the saiga antelope and European bison of the steppes, the North American bison and pronghorn, the rheas (large flightless birds) and the guanaco of the pampas. Many of these large herbivores have been severely reduced in numbers by human farming activity. There are also many smaller mammals, such as ground squirrel, prairie dog, mole rat and many small rodents. These are preyed on by coyotes and foxes and by large predatory birds such as buzzards.

Management problems

The great potential of the temperate grasslands for human exploitation has been evident ever since the development of agriculture (▷ p. 212). Grassland ecosystems are ideal for providing a high yield of animal products from domestic grazers, so the pastoral activities of these regions often have a long history. The replacement of wild herbivores by domestic ones has not always proved efficient, however. Domestic animals are often more fastidious in their diets, so fewer of the plant species are consumed and distasteful ones may multiply. Overgrazing can reduce the grasslands to a desert condition and a similar result has occasionally been achieved by the conversion of prairies to arable agriculture, as in the case of the 'Dust Bowl' experience of the southern Midwest states of North America in the 1930s, when wind erosion denuded vast areas of topsoil. The rich soils make these regions very attractive for the growing of cereal crops, but the soils are easily degraded. If the climate continues to become warmer as a result of the 'greenhouse effect' (▷ p. 224), then the temperate grassland regions of the world are likely to be converted into desert, with very serious consequences for the world's food supply.   PDM

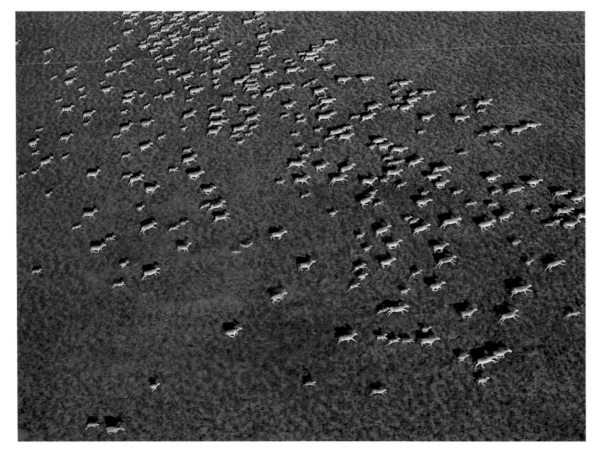

Saiga antelope on migration on the steppes of Kazakhstan. The saiga's adaptations to cold winters include the growth of a thicker, woollier coat, and curious downward-pointing nostrils that may help it to warm and moisturize air prior to inhalation. At one time overhunted for its horns – which the Chinese regard as having a medicinal value – the saiga is now protected, and has a total population in excess of 1 million.

Temperate Forests

The temperate areas of the world, especially in the northern hemisphere, have suffered very considerably at the hands of humanity. Many of our great civilizations have developed in the temperate zone, and the consequence has been a very heavy impact on the forests that once occupied much of these latitudes. In Europe only a small fragment remains of what was once a continuous forest cover. The advent of agriculture some 10 000 years ago in the Near East and its spread across Europe and western Asia about 4000 years later has devastated the forests of the temperate Old World (▷ p. 212). In eastern North America the destruction began later – from the start of European colonization in the 17th century – so more of the forest has survived to the present day.

The temperate zone is wide and contains many different types of forest. In very wet areas with rainfall in excess of 1500 mm (60 in) per year – such as the west coast of the South Island of New Zealand, Tasmania, southern Chile and the Pacific northwest of America – there exist temperate rain forests, consisting of either broadleaved evergreen or coniferous trees. In regions of mediterranean climate (▷ p. 205), mixed forests of conifers and broad-leaved evergreens were once widespread but, unlike the temperate rain forests, these were hard-leaved woodlands, adapted to prolonged summer drought. In much of Europe and eastern America the characteristic forest is dominated by deciduous broad-leaved trees, while further north and in more continental regions, such as the Great Lakes area of North America, there is an admixture of conifers.

Biomass, structure and productivity

The biomass of the temperate forest is only about half that of the tropical rain forests, ranging from 25 to 30 kg/m². The productivity is also less at around 1 kg/m² per year, which reflects the fact that there is usually a period unfavourable for growth during the temperate year. This may be due to summer drought in the warm temperate zone, or winter cold and frost in the higher latitudes. In the case of winter cold, when temperatures fall below about 5 °C (41 °F) water becomes too cold for plants to absorb easily, and they effectively suffer drought. The characteristic response is the dropping of leaves. In the areas that suffer summer drought the leaves are usually retained, but the plant becomes dormant until autumn rains relieve the stress.

The temperate forests have a simpler structure than the tropical forests (▷ p. 200); below the canopy layer there is usually just one subsidiary layer of shrubs and saplings. As a result, light penetration is reasonably good, allowing a ground layer of herbaceous plants and mosses to thrive.

Soils and nutrients

Temperate forest soils are mainly well supplied with water, and the overall direction of water movement is downwards through the soil. The effects of leaching may be to remove nutrients into lower layers, especially where coniferous species grow and earthworms are rare. However, in the deciduous forests there is an input of nutrient-rich leaf litter that feeds a healthy earthworm population. The earthworms ensure a good mixing of the soils, with the result that few nutrients are lost from the soil by leaching. The abundance of clay and humus in these soils also helps to retain nutrients. All these factors mean that the soil is a more efficient nutrient reservoir in temperate forests than is the case in the tropics, which have often lost their clay and humus content (▷ p. 200).

Biological diversity

Although the structure of temperate forest (▷ above) is simpler than that of tropical forest, it is still sufficiently complex to permit the development of diverse assemblages of animals. The seasonal variations in climate operate as a brake on the development of diversity, however, and the higher latitudes lose many of their bird species during the winter when they migrate to lower latitudes. Other animals, such as hedgehogs and chipmunks, hibernate in the cold winters of the north, while snails aestivate through the hot summers of the south (▷ p. 175).

Temperate rain forest in southern Chile. Such forests occur in particularly wet areas of the mid-latitudes, from Tasmania to British Columbia. They may be dominated by coniferous or broad-leaved trees, but all are characterized by an abundance of ferns and mosses.

All of the temperate forests contain a variety of large mammals, from herbivorous squirrels, deer and pigs, to omnivorous bears and badgers, and carnivorous foxes and wolves. The destruction of habitats has reduced the populations of some of these, or forced others (such as the red deer in Britain) to adapt to different habitats.

Management problems

The greatest problem facing the temperate forests is their fragmentation following thousands of years of human pressures. In Europe there remains only a very small proportion of the original forest cover, and the remaining fragments have usually suffered considerable disturbance. As a consequence some animals, such as the brown bear in Spain and the wolf in Scandinavia, are now on the brink of extinction. Small areas of forests cannot support viable populations of these animals and the journey between one forest fragment and another involves crossing alien grasslands and agricultural lands where there is much danger from man. Large areas of forest need to be conserved to protect such species.

Forests are also suffering from air pollution in many of the more industrial regions, such as Europe and North America. Exhaust gases from motor cars and from industry have acidified the atmosphere and acid rain (▷ p. 224) can severely damage leaf surfaces, so permitting the invasion of fungal infections.   PDM

Boreal Forest

The colder regions of the temperate climatic zone, those parts north of about latitude 55°N, are not able to support deciduous forest, but are clothed instead with evergreen needle-leaved coniferous trees. This biome is called the *boreal forest*, or *taiga* (a Russian word). Only a few deciduous trees, such as birch, aspen and larch, can cope with the extremely cold winters (which may fall to $-60\,°C$ / $-76\,°F$) and the abundance of snowfall, conditions to which most conifers are well adapted ($\triangleright$ p. 39). The warm summers are extremely short, often only three months long, which allows very little time for growth.

The northern fringes (right) of the boreal forest in the interior of Alaska. The abandoned beaver ponds in the foreground have created a surrounding wetland unsuitable for conifers, although shrub willows thrive in their place. Eventually peatlands will establish themselves by the process of hydroseral succession ($\triangleright$ pp. 218–19), and the conifers will re-invade. In the background, the forest gives way to tundra on the lower levels of the mountains.

The boreal forests form a complete girdle around the northern hemisphere, but are not represented in the southern hemisphere because there are no land masses at the appropriate latitudes. The boreal forests extend south along the major mountain chains in North America, and are found as isolated patches on the European and Asian mountains. Various species of pine and spruce dominate the boreal forests of Europe and western Asia, but species of larch and tamarack (North American larches) are more important in the peatland regions of Siberia and North America, where larch trees can grow as far north as 70°N. In the interiors of continents, the taiga borders directly onto steppe and prairie grassland, sometimes with a belt of aspen parkland separating them. In Europe and eastern North America the southern boundary of the coniferous forest contacts the deciduous forest and may form an extensive mixed forest, as in the Great Lakes area of Canada and the United States. To the north of the taiga is the tundra ($\triangleright$ p. 210).

The timber from the boreal forests has been extensively exploited for construction and for paper pulp. As a consequence, there remain very few locations in Europe where untouched spruce forest remains, but parts of Siberia and North America are still in a virgin condition.

Biomass, structure and productivity

The productivity of the boreal forest – about $0.5\,kg/m^2$ per year – is only about half that of the deciduous temperate forest, but much of this productivity is diverted into the formation of the woody trunks of the trees, so the timber-producing potential of the region is considerable. The biomass of 20–$50\,kg/m^2$ is concentrated in the trees, and there is relatively little layering in the forest structure. Often a ground cover of dwarf shrubs, such as blueberry or bilberry, is found, together with an abundance of mosses, club mosses and lichens.

The short growing season is one important reason why evergreen foliage is more efficient than a deciduous canopy. The evergreens are able to begin their photosynthesis as soon as conditions become warm enough in spring; there is no need for the delay entailed in the development of new foliage. Very little photosynthesis is conducted during the winter, even on sunny days, despite the fact that green leaves are in place throughout the winter.

The conical structure of the trees assists in the shedding of the large loads of snow that would otherwise accumulate on the branches and eventually cause structural damage. High winds can result in the collapse of whole areas of conifer woodland, and sometimes these areas of wind damage advance in a wave pattern, to be followed by waves of recovering forest.

Fire is also an important and regular natural feature of the coniferous forests. Most conifers are extremely flammable because of the resins in their foliage and wood. But they are also fire resistant in the sense that they can survive fires, either resisting the fire by means of their thick, fibrous bark, or shedding new seeds to germinate when the cones open following burning. These adaptations give conifers a distinct advantage over most deciduous trees under conditions of regular burning. Burning cycles of about 70–100 years prevailed in the North American forests prior to the arrival of European settlers and were probably caused by lightning strikes.

Soils and nutrients

Although the precipitation (mainly snowfall) is not excessive, the short summer and generally low temperatures keep evaporation low. The result is that some soils become waterlogged, and this in turn means that the rate of decomposition of leaf litter is slow, which leads eventually to the formation of peatlands. Some regions of the boreal forest consist entirely of vast areas of peats with a cover of flood-tolerant evergreen trees, such as the black spruce of Canada.

In regions with better drainage, the

excess water drains through the soils, leaching out their nutrient content. Even iron and aluminium compounds are carried down through the soil by the water that has been made acid by the leaf litter of needle foliage on the soil surface. The breakdown products of the resins and tannins contained in conifer needles is unpalatable to earthworms and these rarely survive beneath this type of canopy. In their absence there is no means by which the soils can become mixed, hence distinct layers develop within the soil profile. Leached chemicals are deposited further down the profile, where they may become evident as a band of red material, the so-called 'iron pan'. This type of soil profile is termed a *podzol*.

The low nutrient content of the soils in much of the boreal region provides an additional advantage to the evergreen rather than the deciduous leaf. Evergreen leaves usually last for two or more years, so the nutrients used in their construction are conserved more efficiently than in leaves that must be renewed each growing season.

Some grazing animals may experience difficulties in obtaining sufficient minerals from their plant diet to maintain their growth. This is particularly true of large mammals such as the moose. Its diet of coniferous foliage (together with willow, birch and aspen) provides very little of the sodium that is needed for nerve function in animals (but is not needed by plants). Some of the deficit can be made up by licking soil and mud, but in the case of the moose, much of its sodium requirement seems to be satisfied by the consumption of aquatic vegetation, which has a higher content of sodium than coniferous foliage. This fact may well account for the feeding behaviour of moose in the summer, when they spend considerable amounts of time grazing in shallow water (▷ photo, p. 196).

Biological diversity

Generally, the further one proceeds towards the poles the lower the number of plant and animal species one finds. This is certainly true of the boreal forest, for the number of tree species is very limited and the hard winters restrict the number of animal species that can survive in this habitat. Hibernation is the answer for animals such as bears (▷ p. 175), but many birds, for example, migrate into the boreal forest for the productive summer, then leave to avoid the winter stress. This is particularly true of insectivorous species, such as some flycatchers and warblers. Conversely, caribou and the wolves that prey on them move south in winter from the tundra into the boreal forest.

Herbivorous animals include browsers, such as red deer and moose, and feeders on the cones of the trees, such as crossbills and squirrels. The cones of pine and spruce take several years to mature (▷ p. 39), so they are a reliable source of food, and birds like crossbills and grosbeaks have evolved highly specialized bills and feeding habits based upon the exploitation of cones. The mandibles of the crossbill's beak actually cross over and are used to prise open the woody scales of cones, while their tongues are used to extricate the edible seeds. Squirrels are not confined to cones in their diet, but are perfectly prepared to take the eggs and young of birds. The ground-dwelling animals are preyed upon by wolves, lynx and wolverine, while squirrels in the canopy are hunted by the pine marten. The canopy birds, as well as the mammals, are hunted by a large and agile hawk, the goshawk, which is found throughout the boreal forest zone of the Old and New Worlds. In mid-winter, birds are less frequent in the forest and goshawks then focus their attention on the squirrels. One other resident of the boreal zone, the beaver, can have a considerable impact on the entire ecosystem. By building dams (▷ illustration, p. 161) it raises local water tables and can cause the extension of wetland areas (▷ photo).

Some herbivorous grazers, notably the caterpillars of certain moths, sometimes undergo population explosions and completely defoliate whole areas of forest, leaving scenes of extensive devastation. Some of the birds of the boreal forest, including waxwings, crossbills and nutcrackers, may also experience periodic population explosions or eruptions, spreading thousands of kilometres into surrounding areas.

Management problems

Over-exploitation by man for forestry is a major threat to the survival of natural boreal forest. Boreal peatlands are also suffering from human exploitation, both by drainage for further forestry and as a source of energy (▷ p. 217). Russia is particularly dependent on peat as an energy as well as a horticultural resource. The burning of the peat, however, results in the release of carbon dioxide into the atmosphere, adding to the greenhouse effect (▷ p. 224). Even in their natural condition the boreal peatlands may be contributing to the greenhouse gases of the atmosphere by releasing methane. There have been suggestions that the beaver, by extending the area of wetland within the forest, may also be making its small contribution to the greenhouse effect.

Management of the taiga for conservation may well entail periodic burning. If natural fires are prevented, then the biomass becomes so large and flammable that the eventual and inevitable accidental fire can be disastrous. Well-intentioned conservationists in North America have often protected these forests from fires with the result that periodic catastrophes occur, as in the Yellowstone fires of 1989.

The low nutrient status of the boreal forest and its acid soils have made the area particularly sensitive to the acidification of the atmosphere and of the rainfall by pollution (▷ p. 224). The release of acid pollutants into the air in the industrial regions of Europe and North America have had serious repercussions in the acidification of soils and water bodies in Scandinavia and Canada. In some cases this may even result in direct damage to the foliage of the trees (▷ photo, pp. 224–5).   PDM

SEE ALSO

- CONIFERS p. 38
- THE BIOSPHERE pp. 194–7
- CLIMATE AND VEGETATION p. 198
- TEMPERATE GRASSLANDS p. 206
- TEMPERATE FORESTS p. 207
- TUNDRA AND ALPINE ECO-SYSTEMS p. 210
- WETLAND ECOSYSTEMS p. 216

THE OLD CALEDONIAN FOREST

Much of the Highlands of Scotland was once covered in a forest dominated by Scots pines, with juniper, bilberry and heather also being prominent, together with birches. By the late Middle Ages much of the forest had been felled, and since then there has been a steady shrinkage of the remaining areas. Now only a few fragments remain, notably around the Cairngorm Mountains.

The largest herbivore of the forest was the red deer, but with the loss of its natural habitat it was forced to adapt to moorlands and mountains. Nevertheless, the extinction of the wolf, the red deer's main predator, by the 18th century means that red deer now require regular culling. Unfortunately, management of red deer populations is not always adequate, and the deer are a major barrier to the natural regeneration of the Old Caledonian Forest, feeding as they do on the shoots of young Scots pine. The ecological problem is one of imbalance between the large deer populations and the small areas of forest; in larger areas of forest the pressure from the deer would be dissipated and young trees given a chance to grow.

Tundra and Alpine Ecosystems

The polar regions lie beneath high-pressure air masses, which cool and descend over these very high latitudes. The cooling air is dry and very little precipitation falls, leading to the development of what is called *polar desert*. The winters are long, dark and very cold. High winds bear abrasive ice crystals that can strip the surface off any plant projecting above the general level of vegetation. Hence trees cannot grow except in a stunted dwarf form, clinging to the surface of the ground or to rock surfaces. The summer has long days and may be quite warm and dry, leading to water shortages in some locations.

The northern polar area differs from the southern in two important ways. The North Pole lies in the middle of an ocean, the Arctic Ocean, and there is no extensive land mass in the Arctic, but both North America and Eurasia have northern fringes extending into the high latitudes. The Antarctic, on the other hand, is occupied by a large continental land mass, covered by an ice cap. The other difference lies in the ocean currents of the area. Ocean currents in the North Atlantic drift northeast and the warm waters find their way into the Arctic Ocean, keeping it relatively warm in summer and restricting the development of sea ice. This keeps the European and western Asian tundra regions warmer than their American counterparts. It also permits the development of extensive areas of ice-free land, which is rare in the Antarctic, even at similar latitudes.

In many respects high mountains in lower latitudes – the so-called alpine areas – have a very similar climate to the tundra. Low temperature and ice-blasting are found here too, but some features are quite different. In the alpine habitat days are shorter in summer and longer in winter. Because the sun is higher in the sky, daytime temperatures can rise much higher, so there is a greater daily fluctuation in temperature. Precipitation is also usually higher, so a substantial snow cover may accumulate in winter, protecting plants and animals from the worst of the low temperatures. But the generally similar climate has led to the evolution and survival of a very comparable flora and fauna in the arctic and alpine tundra regions.

About 20 000 years ago the earth was in the middle of the last ice age, and at that time many of the cold-tolerant plants and animals were widely dispersed over North America, Europe and Asia. Since then – particularly in the last 10 000 years – the earth has become warmer and forests have spread across much of these continents, while the arctic/alpine species have retreated. Many of these species have now become confined to the polar tundra regions, with outlying populations still surviving on the mountain peaks of more southerly regions (▷ box, p. 10).

Biomass, structure and productivity

The adverse climatic conditions, particularly during the long winter night, severely limit the potential productivity of the tundra. The area can be regarded as a cold desert not only because of its low precipitation, but also because of its low levels of plant production. Vegetation is normally present in the tundra in all but the most exposed and unstable sites, and a biomass of about 3 kg/m^2 – similar to that of desert scrub – is often attained. The productivity of this plant cover is about $0.1–0.4 \text{ kg/m}^2$ per year, which is also comparable to that of the semi-arid zone.

In structure, tundra vegetation consists of low shrubs and perennial herbs, often having a hemispherical, cushion-shaped form. Buds projecting beyond this low canopy are subjected to the abrasion of ice particles carried by the high winds, and also to the drying effects of such wind on the living tissues. These wind effects have led to the evolution of species with low, ground-hugging growth forms in which buds are kept close to the soil. One might suppose that plants would keep their delicate buds underground under such conditions and that the climate would favour bulbs and corms, but the soil is frozen throughout the winter and only the surface layers thaw in summer, so underground organs are at risk of permanent freezing. Annual plants are also rare, having difficulty in completing their life cycles within the limited period suitable for growth. Only one annual plant, the Iceland purslane, has been successful in the arctic tundra, where it grows on disturbed screes.

The same limitations on growth form apply in alpine tundra, but since the sun is usually higher in the sky during the summer, higher temperatures are generated and alpine plants are normally more tolerant of hot summer temperatures than truly arctic ones. The structure of the vegetation is, however, similarly low and simple.

In the high latitudes the tundra abuts onto boreal forest (▷ p. 208), the two usually being linked by a zone of birch forest. In alpine areas there is often a sharp boundary between boreal forest and tundra, forming a distinct tree line. The altitude of this transition is determined primarily by climate (▷ diagram), although where forest has been modified by domestic grazing animals the sharpness of the boundary can be blurred.

During the summer, large mammals such as caribou (reindeer) in the arctic and red deer, ibex, bighorn sheep and chamois in the lower-latitude mountains migrate into the tundra areas to take advantage of the new growth of plant material. The intense grazing pressures add to the other factors that combine to keep vegetation low.

Soils and nutrients

Like the boreal forest, the arctic tundra is generally wet despite very low rain and snow fall. This is because low temperatures keep down the rate of evaporation. Low-lying areas become swampy and begin to develop peat as the organic matter fails to decompose. The tundra peats and soils are subjected to freezing during the winter. This not only results in the lower layers (often only 1 m / 3 ft below

Bilberries in their autumn colouring, just beyond the boundary between forest and tundra vegetation.

the surface) being permanently frozen throughout the year – the *permafrost* – but also in the development of various strange features. Ice wedges penetrate down into the soil in winter and then thaw to leave waterlogged cracks during the summer, and often these cracks are arranged in a characteristic manner, forming polyhedral shapes that look like a honeycomb when viewed from the air. On slopes the stones are forced to the surface of the peaty soils, mainly because the first frosts of winter begin beneath the stones in the ground and the expansion of the water as it freezes beneath the stones pushes them upwards to the surface. Here they often become arranged in stripes following the contours of the slope.

In the peaty hollows of the tundra the ice within the peat may heave the surface into a massive system of organic mounds with frozen cores called *palsas*. Palsas pass through a series of growth and degradation phases in an unceasing cycle. They grow as the ice in their centres accumulates each winter and fails to thaw completely each summer. But as their surface is raised it becomes eroded and cracked, leading to the exposure of the black peat beneath the crust of lichens and mosses. The dark peat then absorbs heat during the summer days and causes the meltdown of the ice core, resulting in the collapse of the palsa and the formation of a water-filled crater where the core was situated. These cycles lead to the development of complex patterns in the vegetation of the tundra.

In alpine tundra the picture is very different, with soils building up in isolated hollows and crevices among the rocks and screes. Because of the sloping topography, water run-off is usually rapid, although marshes may form in hollows where peat builds up. The soils are not deep enough to have a permanently frozen layer. Rock-type has a crucial influence on flora, very different plants growing in, for example, limestone and granite areas.

Biological diversity

Extreme climatic conditions, low productivity and simple architecture in the vegetation structure combine to give the tundra a low biological diversity. The tundra habitat is also a relatively young one. Only for the past 2 million years or so has the earth experienced sufficiently cold climates to generate ice caps – so there has been relatively little time for the evolution of forms that are suited to these conditions. Nevertheless, the habitat is remarkable for the adaptations found among plants and animals to cope with such a stressful environment.

Dwarf shrubs are among the most conspicuous members of the flora, including dwarf birch and many species of dwarf willow, and, in the alpine regions, smaller forms of rhododendron. Crowberry and various bilberries and blueberries are also found, many of which lose their leaves in winter in response to the perpetual night and display a remarkable array of autumnal colouring in the fall.

Herbaceous plants include cushion- and mat-forming saxifrages, stonecrops, campions, lady's mantles and cinquefoils. Many of these are capable of withstanding considerable drought during the warm summer when rainfall is rare. Some plants, especially those of the alpine tundra, have spectacularly coloured flowers, the best known of which are the gentians. Such large, conspicuous flowers are associated with insect pollination and there is evidently considerable competition among plants to advertise their presence and attract pollinators.

Insects are abundant in the arctic summer, including pollen-feeding flies and bees and blood-sucking midges, mosquitoes and blackflies. Butterflies are much more abundant in the alpine sites than in the arctic because of the warmer summers and also because of the higher productivity and diversity of the managed grasslands and meadows of many alpine regions.

Insectivorous and seed-eating birds migrate into the tundra regions to breed during the summer. They take advantage both of the supply of food and the long days, which allow greater time for collecting food for their young. The seed-eaters include snow bunting and Lapland bunting, and the insectivores include the swallow, bluethroat and some species of warbler, together with some of the smaller wading birds that feed their young on insects. Many migratory species of ducks and geese also breed in the tundra, where insects form an important part of the early diet of the young.

Small herbivorous mammals are frequent in number if not in diversity of species. The lemming is best known in arctic regions (⊳ p. 172), together with the snowshoe hare (⊳ p. 172), while alpine areas have populations of marmots. The abundance of these small mammals provides a living for the arctic fox, hawk owl and snowy owl. Other predatory birds, such as the gyrfalcon, feed on avian prey, while wolves follow migrating herds of caribou from the boreal forests and prey upon them in their tundra breeding grounds.

Management problems

The discovery of oil in the North American tundra created a series of problems for its continued survival and conservation. The sheer pressure of human activities, including the use of tracked vehicles over frozen ground, has resulted in erosion and damage to vegetation which heals only slowly in such a region of low productivity. The construction of extensive pipelines has also interfered with the migration routes of caribou, which are nervous in approaching these strange new structures.

In alpine tundra sites a similar problem is caused by the development of ski slopes, involving damage to vegetation in their construction and the compaction of vegetation and soil during their use.

A new problem threatening the tundra is the prospect of global warming (⊳ p. 224). It is possible that a warmer, greenhouse climate will permit the extension of boreal forest north into the arctic and will raise the tree lines on mountains, thus restricting the area occupied by tundra. This may well prove to be one of the most threatened biomes in a warmer world.   PDM

Lac Gris (left) at 4500 m (15 000 ft) in the Ruwenzori Mountains of Central Africa. In the high equatorial mountains of Africa – Mount Kenya, Kilimanjaro and the Ruwenzori – the lower parts of the alpine zone are characterized by giant forms of normally small plants, with lobelias and groundsels growing up to 2 m (6½ ft) high. Similarly, in the tropical Andes giant members of the daisy family occur.

SEE ALSO

● THE BIOSPHERE pp. 194–7
● CLIMATE AND VEGETATION p. 198
● BOREAL FOREST p. 208
● ATMOSPHERIC CHANGES p. 224

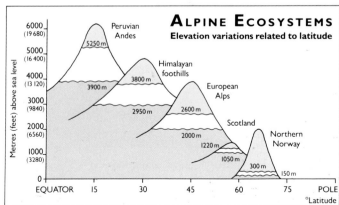

ALPINE ECOSYSTEMS
Elevation variations related to latitude

Man-made Habitats

There is no area of the land surface of the earth that has been unaffected by the human species. Pollutants and pesticides have been found wherever a search has been conducted (▷ p. 226), and recent changes in the atmosphere ensure that the human impact is felt globally (▷ p. 224). Even more remarkable is the fact that large portions of the earth's surface are now covered by habitats that are essentially man-made. These range from modified and intensively grazed grasslands to urban habitats that are almost devoid of any photosynthetic input.

The omnivorous hunting and gathering activities of early humans proved one of the great reasons for the success of this species. A lack of specialization or, in ecological terms, a broad niche, gave considerable advantages to *Homo sapiens* in the struggle for existence, especially during the extensive changes in global climate that have been experienced over the last 2 million years. In hunter-gatherer societies, populations were limited by food supply, but the development of agriculture after the end of the last ice age 10 000 years ago provided reliable sources of nutrition and so encouraged population growth.

Agricultural habitats

Domestication of plants and animals demanded in its wake the manipulation of the environment to suit the needs of the domesticated species. Plants such as wild cereals are annual grasses that thrive best in disturbed soils that are unshaded by trees. Such habitats were common in the Middle East following the end of the ice age, but as the climate grew warmer and woodland spread into such areas, it became necessary to remove the tree canopy and plough the ground to maintain or to increase productivity. As the cereal crops were carried to new areas, such as northern Europe, cultivation involved the clearance of virgin forest.

A similar requirement attended the domestication of grazing animals (▷ Domesticated Bovids, p. 114). Sheep and goats are natives of semi-arid scrub and steppe and they thrive best in tree-free habitats. The ancestor of the cow, the wild aurochs, was actually a forest animal, but it also could achieve higher levels of productivity when herded in open grasslands than it could free-grazing in forests or kept in stalls and fed with forest produce. Forest destruction in the temperate areas thus became inevitable in Europe with the spread of agriculture.

The North American story is similar. The domestication of maize in Central America provided a basis for arable agriculture, and its spread northwards involved the clearance of small areas of woodland for cultivation. Grazing animals were not, however, domesticated and the impact on forest cover remained small until the start of significant European settlement in the 17th century.

Industrialization in Europe from the later 18th century had a major impact on agriculture, both in the provision of new machinery that could improve the efficiency of agriculture, and in offering new opportunities for the manufacture of synthetic fertilizers to improve the fertility of soils. The increases in productivity resulting from such developments have contributed considerably to the growth of the human population over the past 150 years. Human beings now represent about 4% of the total animal biomass of the planet, and they consume about 0·7% of the earth's primary productivity.

The arable ecosystem is characterized by extreme levels of disturbance. Vegetation is not normally allowed to develop for more than one growing season before it is harvested, the soil turned over and a replacement crop sown. Plant diversity is thus maintained at a low level, ideally at one species (a *monoculture*), but in practice there are usually 'contaminant' weed species present. These are often themselves annual plants with high levels of seed production and excellent seed longevity in the soil, which provides them with the best chance of survival in the long term in such a 'catastrophic' habitat.

Energy flow in the arable ecosystem is basically very simple, since a short, linear food chain has been created, leading directly from the primary producer – the crop plant – to man. But modern farming has introduced a new complication to the energy equation with the use of fossil fuels to power the machinery used in agriculture and to manufacture the fertilizers that are applied to the land. Taking

Green but not 'green', arable monocultures are deserts as far as species diversity is concerned. In contrast, another man-made habitat, the motorway verge, has a great diversity of both plants and animals, as they are usually not directly subjected to herbicides and insecticides. The frequency of small birds of prey such as kestrels hovering on the edges of motorways bears witness to the richness of this habitat.

Heather moors such as these in Yorkshire, England, are also a man-made habitat. Cleared of trees in prehistoric times and used for common pasture, in the 18th and 19th centuries they were enclosed by landowners for grouse shooting. The heather is maintained by regular burning, grouse preferring to feed on the shoots of young plants.

this component of the equation into account, every unit of energy we consume in the form of potatoes has required the application of a similar amount of energy in the form of fossil fuel to grow the potatoes: in effect we have converted the energy of fossil fuels into the energy content of potatoes. These energy subsidies in modern agriculture may seem ridiculous when presented in these terms, but they make economic sense while potatoes are worth more than oil or coal. Such energy subsidies also permit higher yields of food, which have contributed to increasing human populations.

Animal life in the agricultural ecosystem is not encouraged by management practices, which may include the use of pesticides. Invertebrate herbivores would find in an unprotected arable ecosystem an ideal opportunity to invade and expand their populations, given a monoculture of an appropriate food resource. If the cultivated area is small and is surrounded by other semi-natural ecosystems, such as verges and hedges, or even woodland patches, then there will usually be sufficient populations of predatory species in the neighbourhood to keep the invertebrate pests under control. But the loss of such diversity as field systems become bigger and hedges less frequent is increasingly common, and this reduces the potential for natural pest control. Vertebrates are uncommon in such ecosystems, as most of the birds and mammals that may be encountered in arable fields also require the shelter of surrounding ecosystems. The exceptions are a few steppe-dwelling birds such as the stone curlew and the little bustard that have adopted arable fields as an effective steppe substitute. However, modern mechanized techniques of land management have brought new threats to the breeding of these ground-nesting species.

In the pastoral ecosystem man is harvesting products (meat, milk or wool) at the second trophic level (▷ p. 194). This is less efficient than the direct use of primary production, but can be carried out in habitats and on soils where arable agriculture would not be profitable. If one considers the total grazing pressure on the earth's primary production, about 14.5% is due to the activity of domestic animals.

The spread of pastoral habitats has provided an opportunity for many plants and animals – especially those from continental grassland biomes – to extend their ranges. But the use of high levels of fertilizers and herbicides has had the effect of encouraging the growth of robust and fast-growing grasses that dominate the vegetation and exclude smaller and slower-growing plants. The overall effect of these management techniques is to reduce the grassland diversity in the course of raising its productivity.

Hay meadows are relatively rich in species when they are not intensively managed, but again the use of fertilizers and also the practice of drainage of wetter areas has reduced their biological diversity. One bird species that has suffered badly as a result of meadow 'improvement' in Europe is the corncrake. This was once an abundant species of hay meadows but is now restricted to the few areas where traditional methods of farming are retained.

Urban habitats

The development of cities followed the growth of agriculture in the course of human history and was in part dependent on it. An assured source of food provided an opportunity for the development of specialized activities on the part of individuals, and certain societies thus developed more complex systems of interrelationships and dependence between their members. With this mutual interdependence, communal living became essential.

Urbanization meant some fundamental changes in ecosystems whereby the primary production of an area was sacrificed and the resident organisms (people) imported all their required energy from outside. The physical structure of the environment in some ways came to resemble rocky cliffs, and it is not surprising that some of the animals that joined humanity in its new environment were species like the starling, pigeon and jackdaw – all cliff-nesting birds in their original state. These species have adopted the ledges and crannies of the concrete jungle as roosting and nesting sites. Human waste, together with that derived from domestic animals such as dogs, became a reliable source of energy for such species and other scavengers, and omnivores like kites, rats, mice and, more recently, foxes and raccoons have taken advantage of the opportunities available in the cities. Some of these species have

been carried around the world and are widespread in their urban distributions. Meanwhile new organisms are constantly adapting their behaviour patterns and becoming urban in their habits. In Europe, the black-headed gull has increasingly become a city species in the last few decades, as has the red fox. Unlike its country counterparts, the urban fox lives not by hunting but by scavenging waste food and eating invertebrates like earthworms from garden compost heaps.

Modern cities are deserts in many respects. Not only are they deficient in primary production (with the exception of gardens and parks), they also have climates that are low in relative humidity and that may experience higher temperatures than surrounding areas. Low temperatures in winter may still be up to 3 °C (5·4 °F) higher than in the surrounding countryside. This may well be the reason why subtropical birds such as the ring-necked parakeet have been successful in maintaining populations in cities as far north as New York and London following accidental escapes from captivity.

If the city is generally a desert, parks and gardens are oases. Many species find refuge in these habitats and the diversity of wildlife, particularly birds, may exceed that of many 'natural' habitats as a result of the careful management of artificially inflated diversities of plant species. The construction of aquatic habitats in cities may prove particularly attractive to bird species, from waterfowl to herons, as can be seen in cities as widely separated as New York, Cairo and New Delhi.   PDM

Cities provide a happy hunting ground for a wide variety of species, even mammals as large as these raccoons in the USA.

Islands

Islands may occur in almost any latitude and climate, but all have in common certain features that make their study rewarding. First of all, islands are often difficult to reach and colonize, especially for animals and plants with poor methods of dispersal. Survival and maintenance of a substantial population may also be difficult, especially on a small island where resources are limited. Furthermore, some islands are subject to natural catastrophes, following which recovery is slow because of the difficulties involved in further immigration. In all these ways islands present a considerable challenge to colonizing organisms.

SEE ALSO

- EVOLUTION pp. 12–17
- POPULATION DYNAMICS p. 172
- THE BIOSPHERE pp. 194–7
- CORAL REEFS p. 215
- NATURE CONSERVATION p. 232

The Iriomote cat is unique to the small island of Iriomote, east of Taiwan. Species such as this that evolve in isolation on small islands are particularly vulnerable. Only discovered in 1967, the Iriomote cat has suffered habitat loss from agricultural and tourist development, and its population has fallen to about 80 individuals.

But islands also provide opportunities. The same problems that make immigration difficult may also exclude predators, parasites and diseases, reducing pressures on some animals and plants. They may also limit the number of competitors for ecological niches, which permits those species that do survive to diversify into a range of unexpected ecological roles in the community.

Thus islands, especially remote islands, provide particularly useful sites in which to study biogeography and evolution. In addition, the principles of island biogeography have been successfully applied to the selection and design of nature reserves (▷ p. 233).

Immigration and extinction rates

Islands usually contain fewer species than an equivalent area of the mainland at the same latitude. This can be accounted for by a consideration of the immigration rate of species and their extinction rate. A large island has a higher immigration rate than a small one because it is more likely to be encountered by a wanderer, and this serves to enhance the final equilibrium level of species. It is in fact quite difficult to prove this, but experiments with islands in the Florida Keys – in which all animals were removed from islands of different sizes and the subsequent recolonization care-

fully monitored – have shown it to be the case. It has also been demonstrated for islands in Canadian lakes during winter frosts, when the tracks of animals can be followed over the ice and immigration rates measured.

Not only is immigration dependent on island size, but also on distance from the mainland or from other islands. The more remote an island the less likely it is that a lost or wandering creature will make a landfall there.

Extinction rate is also related to island size. A very small island may not even be able to support much land vegetation, so only marine or maritime species can live there. The bigger the island, the greater its resources and opportunities for survival and the building up of viable populations. But other factors are also important in survival, such as the diversity of different habitats. An island with a wide range of topography and geology and different soil types, or even different microclimates, is more likely to supply the needs of a greater variety of species and thus reduce the level of extinction. The final number of species on an island is determined by an equilibrium being attained between the processes of immigration and extinction.

Some islands have been formed by catastrophic geological upheaval, as in the case of volcanic eruptions, such as formed the island of Surtsey off the coast of Iceland. Others develop more gently with the build-up of corals (▷ opposite). But in both of these cases the islands initially bear no land vegetation and they are dependent on immigration for the process of development to begin. Some other islands are not in this position. They may have once been joined to a mainland and have been isolated by rising sea levels. These start their existence with a relatively rich flora and fauna, and may find themselves *supersaturated* with species – having more than they can support in the long term – in which case the course of extinction proceeds faster than that of immigration.

Isolation and evolution

Once isolated from other populations, species may evolve along quite unexpected lines, leading to the peculiarities that have proved so interesting to biologists like Darwin in the past (▷ p. 12). The powers of dispersal that bring an animal to an island in the first place may now prove an unnecessary feature, perhaps even a dangerous one. In the absence of predators many birds become flightless, such as the kiwi and notornis of New Zealand, or the unfortunate dodo of Mauritius. Flightless insects are even more abundant.

The opportunities provided under conditions of low competition may lead some species to evolve in a radiating manner, developing along a number of different lines at the same time to assume a variety of different roles. Thus an ancestral species can lead to the development of a

A silversword – one of the Hawaiian tarweeds – in flower high on a dormant volcano on Maui Island. Only plants with seeds that can be borne by wind, water or birds can colonize islands as remote as the Hawaii group. Once there, an individual species can often diversify rapidly to fill a range of vacant ecological niches. The 41 species of Hawaiian tarweed include shrubs, trees, lianas and cushion plants, all thought to have descended from a single Pacific Coast tarweed. Individual species have adapted to a variety of habitats from very dry to very wet, and from sea level to high altitude.

whole suite of species with different structures and ecological niches, such as are found in the finches of the Galápagos Islands and the honeycreepers of Hawaii (▷ pp. 16–17). Similar examples are found in the plant kingdom, where some herbaceous species, like the tarweeds of Hawaii, have even evolved into tree forms, taking advantage of the lack of robust competitors that have already assumed that ecological role (▷ photo).

In this way unexpectedly large members of certain groups can evolve, like the tree groundsels of St Helena in the South Atlantic, or the giant Komodo dragon, an outsize lizard from some Indonesian islands, or the now extinct moa, a giant flightless bird from New Zealand. These have adopted a role normally taken by other types of plant and animal on the mainland areas, and are restricted to the confined regions in which they have evolved. Species restricted in their distribution in this way are said to be endemic to those areas.

However, species that have evolved in conditions of low competition often find themselves particularly vulnerable to more vigorous introduced species – for example, the virtually flightless kakapo of New Zealand has been brought to the verge of extinction by the predations of introduced rodents and cats. The loss of one species may put others at risk of extinction, as in the case of the dodo on the island of Mauritius. The seeds of one tree species, *Calvaria*, no longer germinate effectively on the island, and it is possible that they need to pass through the gut of the dodo before they can do so. Extinction of one species can lead to a cascade of other extinctions.   PDM

Coral Reefs

Coral reefs constitute one of the most diverse and spectacular types of natural habitat. They are formed by symbiotic colonies of corals and algae (▷ p. 53), which have limestone (calcium carbonate) skeletons. As individual organisms die, their skeletal components remain; new generations grow on these, so building up sometimes massive structures.

SEE ALSO

- MARINE INVERTEBRATES pp. 52–7, 66
- FISH pp. 70–77
- ISLAND ECOSYSTEMS p. 214
- MARINE ECOSYSTEMS p. 220

Coral reefs are restricted to warm water – more than 21 °C (70 °F) – and are thus sub-tropical and tropical in distribution. Both algae and reef-forming corals also require light, and so reefs are further restricted to shallow, clear water. Corals do not thrive in water containing sediment, and are intolerant of fresh water. They are thus absent near estuaries, and passages through reefs are often found at river mouths. Some coral reefs are in decline due to greater amounts of sediments entering the sea because of changes in land use.

Types of reef

There are three main types of reef: fringing reefs, barrier reefs, and atolls. *Fringing reefs* are the most common. They extend directly out from the shores of most tropical countries and are distinguished from *barrier reefs* by the fact that the latter have a wide lagoon separating them from the land. The Great Barrier Reef off the eastern coast of Australia is the largest structure made by any living organism, with a total length of 2027 km (1257 mi), and a lagoon 150 km (90 mi) wide. *Atolls* are round or horseshoe-shaped structures with a central sheltered lagoon. It is thought that most atolls began as fringing reefs around small islands that sank slowly enough for the coral's upward growth to keep pace with the rate of sinking. The largest coral atoll

is Kwajalein in the Marshall Islands: the 283 km (175 mi) long arc of coral encloses a lagoon of 2850 km² (1100 sq mi).

Reef zones

A typical fringing reef has several zones. Nearest the land is the shallow, sheltered *back reef* area with coral rubble, coral heads, and sea-grass meadows. Next is the *reef crest*, where the water is very shallow and waves break on the reef. Conditions are too rough for corals, and this part is dominated by encrusting coralline algae. The effects of waves are less strong in deeper water, and here corals dominate the area called the *reef front*. Those corals nearer the surface are strongly built to withstand surge conditions, while those growing in deeper water or in more sheltered conditions are more delicately made. Often the reef front is broken up into spurs and grooves, complicating the pattern of water movement and providing conditions for a variety of corals. Coral diversity is here at its greatest, but diversity declines with depth as few corals are adapted to live in the dim conditions of deep water, and gradually corals become less important in the community. The deep reef often falls away vertically as a *reef cliff*, which, though made of ancient coral, bears a community without living corals, being dominated instead by sponges (▷ p. 52) and sea whips (▷ p. 53).

Barrier reefs have a similar structure on their outer edges, while the inner, sheltered side may have a whole variety of different types of community associated with it, from diverse coral communities to sea-grass meadows and mangrove swamps (▷ pp. 216–17). Atolls also have a similar reef on the side facing the prevailing winds, but on the lee side and in the lagoon itself fragile corals come closer to the surface. The sandy bottom of the lagoon forms an entirely different kind of habitat.

Biological diversity

As many as 150 species of coral may be found on a single atoll or reef, each with a slightly different ecological niche and structure. Some are delicately branched, others flat and plate-like, while some form massive boulders. Their polyps may be large or small with long or short tentacles adapted for capturing different kinds of plankton. Corals derive their nutrition from two main sources: planktonic organisms captured at night, and material derived from a symbiotic relationship with small algal cells living in their tissues. These algal cells receive the benefit of living in a protected environment, and also use the nitrogenous wastes of the coral's cells for their own growth, while in return the coral receives excess carbohydrates and lipids produced by the plant cells. In some as yet imperfectly understood way the presence of algal cells also enhances skeleton-forming ability among the corals.

The reef by day is alive with a spectacular diversity of fish species of all shapes and sizes, a diversity made possible by the

great range of environmental and feeding opportunities provided by the complex structure of the reef. The abundance of fish is also an indication of the productivity of the environment. Many reef-dwelling fish are brightly coloured and obvious – in some cases this is a warning coloration, but for most it is a means of distinguishing a mate among other species.

Most fish are visual feeders and thus only active by day. This has a major influence on the organization of the activities of other reef animals. By day the reef is dominated by hungry fish, so few mobile invertebrates appear, no planktonic animals are detectable, and most corals keep their tentacles retracted for safety. However, the symbiotic algal cells will be photosynthesizing actively at this time, and sea fans, sea whips, sea anemones, sponges and a host of other sedentary invertebrates can be seen, many exhibiting warning coloration to show that they are unpleasant to eat.

At night the picture changes dramatically, the threat of predation by fish being largely removed by darkness. Planktonic and other swimming invertebrates such as squid emerge from cavities in the reef. Bottom-living crawling animals such as molluscs, sea-urchins, brittlestars and a variety of crabs emerge to feed, worms extend their tentacle fans, and the corals too extend their tentacles to feed on the plankton.

The intense predator pressure and competition for space amongst sedentary organisms has led to a wide variety of defensive adaptations in reef animals. Many forms are poisonous and have spines or other venom-injecting organs. It is known that corals fight for space using stinging cells, poisons and other means.

Many animals avoid predation by actually living within the reef. Some sponges, clams, worms and sea-urchins actually bore holes in the limestone, while others take advantage of unoccupied holes.

Build-up and breakdown

Actively feeding healthy corals lay down calcareous skeleton as a part of their growth. The rate of growth depends upon the type of coral, but can be several centimetres a year in some species. This – together with the contribution from a range of calcareous algae – is how reefs are built. At the same time, organisms of various kinds may be boring into the coral, and reefs may be seriously undermined by such activities. The effects of such forces are most obvious after heavy wave action following storms, when great destruction may occur. Although delicate corals suffer most from these destructive forces, detached pieces can survive and regenerate to restore the reef. Another threat to coral reefs has come in the form of the crown-of-thorns starfish (▷ p. 54), which feeds on coral. Between 1959 and 1971 a population explosion of these starfish destroyed the living coral in a large section of the Great Barrier Reef. **RHE**

Coral reefs are some of the most beautiful natural habitats, and as such have suffered from tourist pressure in many parts of the world. Divers and snorkellers can easily damage delicate corals inadvertently, but deliberate removal of specimens is a more severe problem on many reefs.

Wetland Ecosystems

Wetland ecosystems develop wherever water accumulates and yet emergent vegetation can maintain itself. Such ecosystems are found both around the margins of freshwater bodies (lakes and ponds), and in saline areas along the edges of the oceans. Although it is convenient to divide these habitats into fresh and saline, there may be a gradual transition from one type to the other in estuarine regions, where the influence of salt water diminishes as one proceeds inland.

Wetland ecosystems are found in all parts of the world, wherever water accumulates, even in the more arid regions. In very wet areas they may even be found on plateaux and sloping ground, but in drier areas they are restricted to valley sites receiving drainage water.

The one feature that all wetlands have in common is a waterlogged soil, although this may not be the case throughout the year. Some wetlands are essentially seasonal. All wetlands bear a cover of vegetation that is tolerant of long periods of flooding, and some families of plants, such as the sedges (Cyperaceae), are found in almost all the wetland systems of the world. Others are climatically limited and may be found only in certain areas, like the tropical mangroves.

Wetlands are most important for the high diversity and peculiarity of their animal life. This feature, combined with their vulnerability to human impact, has brought this type of ecosystem to the forefront among the concerns of conservationists.

Biomass, structure and productivity

Wetland ecosystems include both highly productive and very unproductive types. Tropical swamps are among the most productive systems in the world with values regularly around 3 kg/m^2 per year, and with some measurements in excess of 10 kg/m^2 per year being recorded. They may thus exceed even tropical forest and coral reef ecosystems in their productivity. Yet they achieve this with a biomass far smaller than that of the forests, often amounting to only 5 kg/m^2. Even the temperate region swamps are highly productive, sometimes reaching 2.5 kg/m^2 per year.

Swamps are ecosystems in which water is in luxuriant supply throughout the year. Even in dry seasons the water level is above the surface of the soil. They may be dominated by tall herbs, reeds and sedges, or they may become invaded by trees, in which case they have a more complex structure and microclimate. Forested swamps have a higher biomass than herbaceous ones and can be found both in freshwater and saline ecosystems. In North America the term *swamp* is sometimes reserved for forested wetlands, while herb-dominated wetlands are called *marshes*.

In areas of high precipitation the rate of build-up of dead matter in the soil is so great that the whole region becomes elevated in the form of massive domes raised above the influence of the drainage water from the catchment area. This type of wetland is termed *bog*. In some areas, such as the tropical regions of Southeast Asia, these elevated bogs are forested, but in oceanic regions of the temperate zone bogs are dominated by the sphagnum mosses. These treeless bogs are low in biomass and also in productivity, often attaining only 0.1 kg/m^2 per year.

The structure of treeless bogs is relatively simple: cushions of dense moss with occasional dwarf woody shrubs alternate with pools of open water. These pools may be arranged in a linear fashion, forming chains along the contours of the slopes of the bog surface.

Soils and nutrients

The soils of wetlands are unusual in that material is constantly being added to them from outside the ecosystem, either in the form of inorganic mineral materials eroded from surrounding catchment areas, or as organic matter derived from the litter of local plants or carried into the ecosystem from elsewhere. As these materials are added to the soil, the bulk of the soil itself increases and its surface becomes raised with respect to the water table, so that flooding may become less frequent with the gradual development of the ecosystem. This is the case in both freshwater and saline wetlands. Thus a salt marsh, developed in calm coastal areas, gradually accumulates silt and mud carried to the site by water. But as the soil surface rises the region is flooded less frequently, so the source of suspended material that adds to the soil mass gradually becomes less reliable until an equilibrium is reached. In this equilibrium state, occasional sea flooding prevents the invasion of non-saline species, but the rate of soil build-up has effectively ceased because of the rarity of that flooding. Only a change in sea level can upset this balance.

In freshwater systems, a similar process takes place in the early stages, for rivers and streams bring silt into the swamps and the very presence of the plants causes an increase in sedimentation. To this they add the organic matter of their own litter, and the surface of the sediment rises until it exceeds the water table. At that stage the decomposition rate increases because of the penetration of air into the upper layers of soil. But the invasion of bog mosses acts like a sponge and water is held up in the soil once more. At this stage the dead mosses are the major source of new soil, and the surface becomes totally dependent on rainfall for its nutrient supply because it has been elevated above the influence of nutrient-rich drainage water. Rain water is usually acid and nutrient-poor, so the soil assumes the same character, assisted by the nutrient-scavenging and acidifying properties of the sphagnum moss itself. So the succession of vegetation in freshwater wetlands leads from nutrient-rich to nutrient-poor soils.

Biological diversity

The early stages in wetland succession are relatively poor in terms of plant diversity. Both the freshwater swamps and the pioneer communities of salt marshes are poor in plant species. Sometimes only one or two species dominate the ecosystem and are responsible for the bulk of their high productivity. But their soils are often rich in invertebrate life, which attracts a diverse range of birds, and this is often the most conspicuous and important aspect of wetlands when considered from the point of view of wildlife conservation.

The temperate salt marshes are particularly important because of their abundant wading birds and waterfowl – especially ducks and geese. Both of these groups use mudflats and salt marshes as a source of winter food, when freshwater wetlands may be frozen. Many of the waterfowl and waders that breed in the tundra regions of the high latitudes spend their winters in low-latitude wetlands. Different species of waders vary in their leg and beak length and this determines the water depth in which they can forage and the depth to which they can probe, so each species has its characteristic feeding location and method. Waterfowl also have different techniques of obtaining their food. Some can dive and are able to exploit food resources even when covered by sea water, while others are only able to up-end and can feed only in shallow water. The latter are limited by their neck length when it comes to feeding underwater. Many find it more convenient to feed on the surface of mudflats when the water has receded.

Tropical saline wetlands may be forested, bearing mangrove trees. These, in

Mangrove swamp is a typical wetland habitat found in coastal areas in the tropics.

common with all wetland plants, have to face the problem of obtaining enough oxygen to survive in an environment which, being waterlogged, has little air penetrating down to the roots. The mechanism that has evolved in the mangrove trees is the development of upwardly directed roots that rise above the mud and water surface and can obtain oxygen directly from the air above.

Mangrove swamps are rich in bird life and also in other animals. Birds of the mangrove include the spectacular scarlet ibis and also fish-eating species like the darter or anhinga. One primate species is characteristic of the mangroves of Borneo, namely the proboscis monkey. A reptile that thrives in such habitats is the saltwater crocodile, the largest of all the crocodilians, found mainly in Southeast Asia and northern Australia. Fish exploit the mangroves when the tide brings them into the complex habitat of upturned roots, and one species, the mudskipper, remains when the tide recedes, being able to propel itself over mud surfaces using flips of its tail.

The freshwater swamps also have a rich bird life, including many species of herons. Herons are able to move among tall reedbeds on long legs, and some, like the bittern, are well camouflaged from predators by streaked plumage and by their habit of pointing their bills upwards when danger threatens. Herons feed on fish and frogs, but other swamp birds may feed mainly on the insects that abound in this habitat. Nesting may prove a problem for such species, and birds like the reed warbler and the red-winged blackbird hang their nests among the reeds. Predatory birds hunt over reed beds, including the widespread marsh harrier and the very restricted Everglade kite of the Florida Everglades.

The replacement of swamps and wet woodlands by open bog vegetation as the wetland succession proceeds results in a decrease both in plant and animal diversity. In many respects these bogs resemble the tundra, and several tundra species, including wading birds, may be found breeding well to the south of their main range.

Management problems

The problems faced by wetlands are twofold. Those that receive drainage water are in danger of pollution either from industrial or urban contaminants or from excessive nutrient flushing from agricultural land. These water-borne problems can severely damage wetlands either directly (as in the case of oil spills on salt marsh sites) or by increasing the productivity of certain fast-growing species at the expense of other, slower-growing species (as happens when nitrogen and phosphate fertilizers are inadvertently added to swamps and marshes).

The second problem is that the commercial exploitation of wetlands demands a lowering of their water table. Hence, drainage has damaged or destroyed many different types of wetland ecosystem in many parts of the world. The swamp

THE OKAVANGO DELTA

The Okavango Delta in northwestern Botswana is one of the largest wetland areas in the world, varying from 16 000 km² (6175 sq mi) in the dry season to 22 000 km² (8500 sq mi) in the wet. Unlike other rivers that fail to reach the sea the Okavango does not end in a salt pan, as the delta has a few outlets that help to wash away salts. The delta contains a constantly changing pattern of habitats, including open water, seasonal and perennial swamps dominated by papyrus reed beds, floodplain grasslands, and countless islands where trees can become established. This habitat richness provides homes to a wealth of fauna, including hippos, crocodiles, and even elephants, together with numerous species of antelope, and a huge variety of fish and water birds. However, only a small part of the delta is protected, and in this largely desert country there is an increasing demand for the water of the Okavango River for irrigation and industrial uses.

forests of Sarawak, for example, are being drained for agricultural reclamation; most of the fenlands of England have been drained for a similar purpose. Many of the extensive bogs of central Ireland, Canada and Russia have been drained for peat exploitation. This resource has proved valuable as an energy source and also as a horticultural commodity – but peat is a non-renewable resource, and peat bogs take hundreds if not thousands of years to regenerate. Many of the bogs that are too shallow or inaccessible to exploit for peat extraction have been drained for forestry plantations. Even if exploitation of a wetland is restricted to a certain area, the drainage of this area may also affect the surrounding wetland areas, eventually drying them out.

Wetlands are thus a threatened habitat on a global scale, and conservationists have placed this type of ecosystem high on the list of priorities for protection against further damage.   PDM

SEE ALSO

● THE BIOSPHERE pp. 194–7
● CLIMATE AND VEGETATION p. 198
● BOREAL FOREST p. 208
● FRESHWATER ECOSYSTEMS p. 218
● MARINE ECOSYSTEMS p. 220
● VANISHING HABITATS p. 228

Freshwater Ecosystems

The diversity of freshwater habitats in terms of size, age and water quality is immense. There is a basic subdivision into flowing and still water, and the range can be judged from the fact that a temporary puddle left after seasonal rains, a small garden pond and the enormous Great Lakes of North America are all habitats for aquatic life. Equally a rivulet and the Amazon are both flowing-water habitats. This diversity provides different challenges and opportunities.

The smallest pools capable of forming a habitat are probably those forming in tree forks in a tropical rain forest. These might contain 500 ml (30 cubic in), while in contrast the largest lake on earth in terms of volume is Lake Baikal in Siberia, which contains 23 000 km³ (5518 cubic miles) of water. In terms of area, however, Lake Superior is larger, with an area of 82 350 km² (31 787 sq mi). The longest rivers in the world are the Amazon and the Nile, which are both about 6500 km (4000 mi) long. The Amazon, however, has the largest flow, discharging between 120 000 m³ and 200 000 m³ (160 000 and 260 000 cubic yds) of water per second into the Atlantic. Its flow is so great that 80 km (50 mi) off shore the water is still fresh.

Still water

Lakes, ponds and pools are formed by water filling a cavity. Such cavities may be created by man, by animals such as the beaver, by local land changes such as earth slips, or by movements of the earth's tectonic plates ($\triangleright$ p. 8). Thus they may be very new or very old and accordingly contain little life or complex communities. The type and quantity of the life dwelling in the aquatic habitat is affected by the chemical content of the water, and this in turn depends on the nature of the local soils and rocks. The terms eutrophic, dystrophic and oligotrophic are used to describe lakes with different chemical characteristics.

Eutrophic lakes are commonest in lowland areas. They are found on base-rich soils derived from soft rocks (such as limestone) from which minerals may leach. The water – often called hard water – has a high content of inorganic salt ions such as calcium, potassium and magnesium, and the nitrogen and phosphorus necessary for plant growth are also plentiful. Sediment washing into such lakes from the surrounding area is also nutrient rich. All these factors contribute to a high productivity of plant and animal life. The water is often greenish and opaque because of the planktonic plant growth, and the margins are densely populated with emergent plants. While the surface waters have high oxygen levels due to photosynthesis, deep down oxygen may be absent owing to large quantities of decaying plant material. The presence of calcium aids plant decomposition, so contributing to the cycling of nutrients and hence productivity, and is also beneficial to worms and snails, which need calcium.

Dystrophic lakes are typical of acid heathlands and peat moors. They have a characteristic brown colour caused by the presence of humic acid as well as plant material. Such pools have a high organic content, but calcium and other salt ions are lacking, and the few animals that are found are mostly insects.

Oligotrophic lakes are filled with water containing little in the way of nutrients. They are situated upon hard, base-poor, rock (such as granite) from which few ions are released. Water from such lakes – which are usually in mountainous areas – is called soft water and is relatively unproductive. There are few plankton, and partly for this reason the water is clear. The bottom consists of mineral sediments rather than a thick layer of decomposing plant material. Animals that require calcium are rare or absent, decomposition is slow, and fringing plants are less luxuriant. Oligotrophic lakes are vulnerable to acid rain ($\triangleright$ pp. 224–5), which has destroyed all fish life in many such lakes in Scandinavia.

Animals begin to invade standing water almost as soon as it is formed. Garden ponds may be invaded by flying insects minutes after first being filled, and the wind-borne eggs and cysts of small invertebrates and protozoans soon follow to begin the colonization process. As more habitats such as weeds and mud become available so the diversity of creatures increases until the system is ecologically saturated. The species present change with time: the early colonists are replaced as a result of competition, predation or changing conditions in the pond. Conditions in all bodies of standing water change over time as the pond or lake itself begins to fill in by the process of *hydroseral succession* ($\triangleright$ illustration). The rate at which this occurs varies considerably according to the type and size of the pond or lake.

Lakes and ponds contain free-floating and bottom-dwelling plants and animals. The phytoplankton consists of unicellular algae, mostly diatoms, which are abundant in spring and summer and are fed upon by zooplankton, small protozoans, rotifers (tiny multicellular invertebrates), crustaceans, and other small animals. These in turn are fed upon by *nektonic* fishes (i.e. fishes living in the middle depths), which complete the short food chain. In water shallow enough to allow growth and photosynthesis, large rooted plants are found, some of which live completely submerged except for their flowering stalks, while others are emergent, living in water that only covers their bases. On and around these plants and in and on the bottom itself are found a great variety of *benthic* (i.e. bottom-dwelling) animals, including protozoans, sponges, flatworms, segmented worms and crustaceans, but molluscs and particularly insects predominate. Many of these animals show elaborate adaptations for their particular habitat or way of life.

The animals of the bottom have a variety of food types available to them. Some such as snails are grazers – grazing not on the large plants on which they are found, but on the microscopic algae on their surfaces. Other animals feed on the large quantities of detritus on the bottom and on plankton, while yet others prey on the first two types. The vegetation, although not used for food, has great importance in providing a refuge from predators such as fish.

Flowing water

Flowing water is a very different habitat from still water, and poses different problems. Plankton is virtually absent, and thus the only plants are bottom-rooted forms. Animals must either be powerful swimmers or bottom dwellers.

Flowing-water habitats are usually divided into two basic categories: streams and rivers. *Streams* are generally cool and shallow, and often have a bottom of gravel, stones or boulders. *Rivers* are deeper, warmer and have a silty bottom. Rivers also change their characteristics along their courses. Those that arise in mountains have three sections. Where slopes are steep there is an upper course

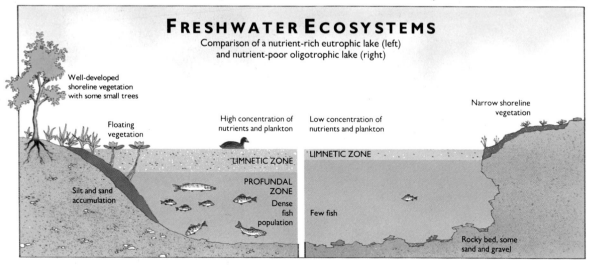

FRESHWATER ECOSYSTEMS

Comparison of a nutrient-rich eutrophic lake (left) and nutrient-poor oligotrophic lake (right)

Well-developed shoreline vegetation with some small trees

Floating vegetation

High concentration of nutrients and plankton

Low concentration of nutrients and plankton

Narrow shoreline vegetation

LIMNETIC ZONE

LIMNETIC ZONE

Silt and sand accumulation

PROFUNDAL ZONE

Dense fish population

Few fish

Rocky bed, some sand and gravel

FROM LAKE TO BOG

All bodies of standing water are continuously evolving and many, particularly eutrophic lakes, have a short life span. This is because they are continuously being filled with sediments and progressing towards the creation of new land surface as part of the process called *hydroseral succession*. As plants root and grow in the sediments of shallow water they both add to the production of detritus and act as traps for sediment, so enhancing the rate of sediment build-up and altering the environment available for animals. The rate of this succession is variable: some ponds in areas of high productivity will fill and disappear in 50 years, while in deep oligotrophic lakes succession is imperceptible.

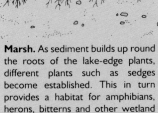

Lake edge. Water lilies, duckweed and pondweed float on the surface, while reeds and bulrushes fringe the water. The bottom provides a home for molluscs, worms, insects and other invertebrates. The water itself is dominated by plankton, small crustaceans and fish, while the surface animals include water voles, moorhens and surface-dwelling insects. Various birds, including the reed warbler, nest in the reeds.

Marsh. As sediment builds up round the roots of the lake-edge plants, different plants such as sedges become established. This in turn provides a habitat for amphibians, herons, bitterns and other wetland species.

Wooded bank. As the sediment builds up further, sedges form tussocks and water-tolerant trees (principally alder and willow) provide a habitat for various birds such as the siskin.

Raised bog. Bog mosses (sphagnum) invade the alder forest and rapidly build up peat on a base of logs and old roots. The result is raised bog, characterized by sphagnum tussocks and small open pools. Other plant species include heather, sundew and bog asphodel. Moorland birds predominate.

(the head waters) that is swift-flowing and stream-like. There follows a transitional zone in the lower slopes of the hills, and finally a slow meandering section in the plains beyond the mountains. Rivers arising in non-mountainous areas only have the latter two zones.

Conditions in rivers vary according to flow, which is itself dependent upon rainfall, and likely to vary from day to day and from season to season. Strength of current, sediment load, oxygen supply and temperature are all dependent upon flow. In most parts of the temperate latitudes flows vary considerably with the seasons. The winter is wetter and so flow is usually greater, but in high latitudes the greatest flows follow the thawing of snow in spring. In areas with very distinct wet and dry seasons rivers may dry up completely for long periods, only to become turbulent torrents during the wet-season rains.

The flora and fauna in a river reflects the reliability of water supply. Rivers that dry out completely are devoid of aquatic vegetation, and have fauna capable of resisting drought conditions as cysts (⊳ p. 27) or resting stages of various kinds (⊳ p. 175). Even where flow is reliable, life cycles may be adjusted so that delicate stages only occur when flow is likely to be suitable. Rivers with reliable flow have a more diverse fauna and flora.

The chemical quality of the water of rivers depends upon the nature of the rocks of the area drained by the river. Streams and rivers originating on hard impermeable rocks contain little in the way of inorganic ions. For this reason the effect of acid rain (⊳ pp. 224–5) is most marked in such rivers, and has had a profound effect on the fauna of many northern rivers. Rivers whose waters mostly derive from softer permeable rocks usually contain an abundance of minerals, often basic in

nature, which buffer the effect of acid rain. In general such rivers also have higher plant productivity.

The nature of the adaptations of river animals depends upon where they live. The animals from the head waters are adapted for strong flows, most being flattened to reduce resistance and having strong attachment mechanisms to allow them to grip rocks and move about to feed. Classic examples are freshwater limpets, leeches, water shrimps, mayflies, and stone flies. Even the fish of such waters may have suckers formed from their ventral fins to enable them to grip rocks. Many animals have preferred current speeds, and their distribution across and along a river reflects their preferences. Some of the animals in such habitats are *suspension feeders*, which sieve small particles from the water, and these have an abundant food supply brought to them by the flow of the river; suspension feeders such as black fly and caddis fly larvae are often present in large numbers. Snails and other animals graze the surfaces of boulders or feed on particles trapped beneath boulders and stones, while various insect larvae and worms (such as leeches) have a predatory or parasitic way of life. In lesser current flows the diversity of animals becomes greater, and obvious adaptations for flow are no longer seen. Instead, in animals that are found in mud in intermediate flow, and in almost all animals in slow-flowing rivers, adaptations for obtaining oxygen in a low-oxygen environment have evolved. Examination of many mud-dwelling animals reveals the presence of red oxygen-carrying haemoglobin pigments, while some animals, for example mayflies that have adapted to these conditions, have large mobile gills.

The invertebrate animals of streams and rivers mostly feed on plant detritus

brought to them in the water currents, and on the bacteria on the surfaces of this decaying plant material. Many rely almost entirely on the bacteria for their sustenance. Others process the actual sediments, from which they extract organic material, or graze on the surface of rocks and fringing plants. Although some fish are herbivorous, most are predatory, specializing in invertebrates or other fish. RHE

A mountain stream in Cape Province, South Africa. Generally cool and shallow, with rocky or stony beds and a fast flow, streams provide a very different kind of habitat from the lower reaches of a river, where the water is warmer and slower-flowing, and the bottom often muddy.

Marine Ecosystems

The oceans and seas together form a major ecological unit, the marine *biome*. Covering 71% of the earth's surface and with an average depth of 3700 m (12 140 ft), the marine biome is easily the world's largest. It has immense chasms – deep-sea trenches – up to 11 000 m (36 000 ft) deep, greater than the height of the highest land mountains, and there are also massive underwater mountain chains extending through all the oceans, occasionally breaking the surface as islands.

The sea is the habitat in which life first arose (▷ p. 6), and all parts of the sea, even the greatest depths, are inhabited by animals today. Sea water is an excellent medium in which to live. It is buoyant, and contains plenty of oxygen, even at great depths. It also contains an abundance of most of the chemicals necessary for life – 35% of full-strength sea water consists of salts, mostly sodium and chlorine. The sea is slow to warm or cool – on average the annual variation in surface-water temperature is about 10 °C (18 °F), and at a depth of 20 m (66 ft) the annual variation may be as little as 1 or 2 °C (2 or 4 °F). Thus the sea is a stable environment, requiring little or no need for temperature-control mechanisms in the creatures that live in it. Sea water is also transparent, thus allowing sufficient light for plant photosynthesis to penetrate down to a depth of 100–200 m (330–660 ft).

Ocean currents

The seas and the air above them are in continuous interaction, generating the world's wind and ocean-current circulation patterns. All the earth's oceans have a distinct circulation pattern (▷ map, p. 198). At the surface of the North Atlantic, for example, there is a clockwise circulation driven by sub-tropical trade winds and the prevailing westerlies of temperate latitudes. Such currents require balancing counter-currents at greater depths. Water that finds its way to high latitudes becomes cold and dense and sinks, often to the bottom. It becomes part of a deep circulation flowing towards equatorial regions and mingling with other waters. Thus all the waters of the oceans, from the surface to the greatest depths, are in continual movement. Some flows like those of the Gulf Stream are rapid, reaching rates of 2 m (6½ ft) per second, but even deep ocean currents may travel at 0·2 m (8 in) per second. Such currents in addition to carrying oxygen and nutrients also carry larvae and spores, and may be a navigation aid to migratory species.

The pelagic environment

The marine biome divides naturally into

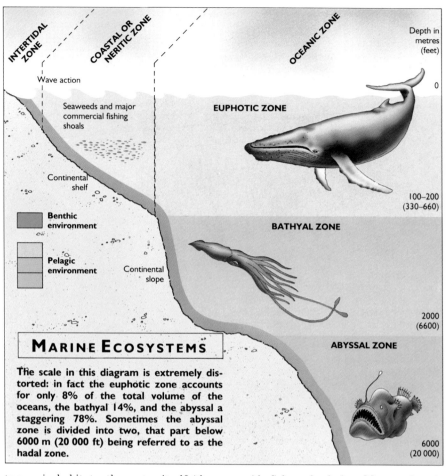

INTERTIDAL ZONE · COASTAL OR NERITIC ZONE · OCEANIC ZONE

Depth in metres (feet)

Wave action

Seaweeds and major commercial fishing shoals

Continental shelf

EUPHOTIC ZONE

0

100–200 (330–660)

BATHYAL ZONE

Benthic environment

Pelagic environment

Continental slope

2000 (6600)

ABYSSAL ZONE

MARINE ECOSYSTEMS

The scale in this diagram is extremely distorted: in fact the euphotic zone accounts for only 8% of the total volume of the oceans, the bathyal 14%, and the abyssal a staggering 78%. Sometimes the abyssal zone is divided into two, that part below 6000 m (20 000 ft) being referred to as the hadal zone.

6000 (20 000)

two main habitats, the water itself (the *pelagic* environment) and the bottom (the *benthic* environment), each populated by specially adapted organisms.

The pelagic habitat can be divided into three zones. The uppermost is the part in which there is enough light for plants to photosynthesize and grow. This is the *euphotic zone*, which extends down to about 100–200 m (330–660 ft). Beneath it lies the *bathyal zone*, extending from 200 m (660 ft) to about 2000 m (6600 ft), in which there is some light in the upper reaches (down to about 800 m / 2600 ft) but not enough for plants to grow. Beneath this again is the largest zone, the *abyssal zone*, to which no light ever penetrates.

Some 90% of the primary production of the sea – on which the sea fisheries and all other sea creatures depend – comes from single-celled microscopic algae (particularly diatoms; ▷ p. 34) floating in the surface waters. Collectively known as phytoplankton, these algae account for an astonishing 30% of the world's total primary production. Phytoplankton are most abundant in shallow water and where currents upwell against continents, conditions in which there is the best supply of essential nutrients.

The animals of the pelagic zone are of two types: *plankton* and *nekton*. Planktonic animals drift with the currents; nektonic animals swim actively and are usually much larger. Typical planktonic animals include copepods and krill (tiny crustaceans), jellyfish, sea gooseberries and salps; typical nektonic animals include squid, fish and whales. Many nektonic species, including baleen whales, feed on the plankton, and a high proportion of the world's commercial fishing is for such plankton-feeding species. Deep-sea nektonic animals are all predatory, feeding on other co-existing animals, and have elaborate adaptations for this way of life (▷ box). Sinking and being visible to predators are problems common to all pelagic animals. Swimming – which is costly in energy terms – is minimized by weight reduction using gelatinous tissues, fat droplets and even gas bladders, while disguise is achieved by disruptive patterning, countershading or becoming virtually transparent (▷ p. 166).

The benthic environment

The benthic environment extends from the furthest upward extent of the sea's influence to the bottom of the abyssal trenches. Animals are found throughout, but plants are limited to shallow water by their requirement for light. The bottom may be of solid rock or of particulate material ranging from boulders through the sand of open ocean beaches to the fine ooze associated with estuaries, mangrove swamps and the deep sea. The type of bottom is dictated by water movement: areas subject to wave action or brisk flows have bare rock or course substrates, while areas with gentle flow are muddy. In general deeper water is more likely to be dominated by fine sands and muds, but there are rocky areas even in the deep seas. This diversity of habitats provides a variety of feeding possibilities and thus supports different communities of animals.

The uppermost environment, the *intertidal zone*, is the area regularly exposed to the air by the movement of the tides. Many habitats exist from rocky shores and coral reefs (▷ p. 215) to mud flats and mangrove swamps (▷ pp. 216–17). Their common feature is that organisms living there are subjected to the rigours of aerial existence, including desiccation, damaging ultraviolet radiation from the sun, rainfall, and less stable temperatures. In addition most intertidal areas are subject to some wave action. The intertidal zone is thus one of the most challenging of marine environments, and only adaptable species can thrive there.

Rocky shores, particularly in temperate latitudes, are dominated by multicellular species of algae – the seaweeds (▷ p. 34). The zone inhabited by a seaweed is dependent upon its tolerance of desiccation. Kelps or oarweeds are found at the low water mark but do not extend higher, and with them will be found many delicate red and green algae. Above them – often in very clear zones – will be found a number of the brown algae known as wracks, as well as some tolerant red and green species and some marine lichens. Seaweeds found high on the shore – which may be exposed to the air for days at neap tides – can put up with major water loss, and may even be killed by continuous immersion. Which species of seaweed is found at any level is also dependent upon wave action. Some seaweeds with strong holdfasts and with fronds that are either long and streamlined or short and sturdy are adapted to withstand a high degree of buffeting and are found in exposed conditions. Others with longer, branched fronds are less suited for such conditions, and are found instead in sheltered areas. The animals of rocky shores are principally sedentary *suspension feeders* (i.e. animals such as worms, barnacles and mussels that sieve small particles from the water) and mobile grazers (e.g. limpets and winkles), and may also be zoned on the shore in accordance with their tolerance of desiccation. Some can live out of water for more than a week. The number of species of both plants and animals decreases from low water upwards.

Sandy and muddy shores show little sign of plant life unless seagrasses are present, as otherwise most of the plants are microscopic. Animal life also appears to be absent, but large numbers of many species of burrowing clams, crustaceans, worms and sea urchins may lie hidden in the sand or mud. These animals also often show zoning patterns on the beach, but the reasons for these are less clear. Desiccation is no longer a problem, but oxygen supply may be, particularly in organically enriched sand or muddy beaches. Fewer species are adapted for life in mud, but those that are present are often found in high densities.

The majority of marine animals are found in the shallow productive waters of the continental shelves. This is the *coastal* or *neritic zone*, defined by a 100–200 m

(330–660 ft) depth limit. The bottom in these areas can be of many types, each with a characteristic group of species associated with it. Areas with coarser sediments are dominated by suspension feeders (▷ above), while areas with fine sediments are dominated by *deposit feeders*, which process the actual sediment in some way. The worms, crustaceans, clams, brittlestars, and other animals that make up this fauna provide food for bottom-feeding fish and other commercially important predators such as crabs and lobsters. Because fish have preferences for different food species it is common to find particular fish associated with particular bottom types. Fish will gather in areas where the productivity of the bottom fauna is high. In general the productivity of bottom-dwelling animals depends on local primary productivity, and also declines with depth and distance from land. Thus most fisheries for bottom-dwelling fish are in shallow waters near the shore.

Beyond the continental shelf – which on average extends about 100 km (60 mi) from the shore – the benthic environment, like the pelagic environment, is divisible into *bathyal* and *abyssal zones* (▷ above). In the bathyal zone temperature and salinity vary over the year, while in the abyss both remain constant. For the most part the bottom consists of fine

sediment, and thus the animals of these depths are mostly deposit feeders and predators. However, specialized suspension feeders do occur on rocky outcrops. Both the quality and quantity of food available to the animals processing the sediments is low, and much of it is the undigested remains of food eaten by surface dwellers. This can be broken down by bacteria, and many animals depend for their nutrition on these bacteria. The food supply diminishes with depth and so does the biomass and the average size of the animals. Their growth is slow and their ability to reproduce reduced – but they may be very long lived.

Recent studies have revealed that at points deep in the ocean volcanic activity results in the release of sulphurous hot water from vents in the bottom. Certain species of bacteria are able to use sulphur as their source of energy (▷ p. 25), and these bacteria provide an abundant source of nutrition for strange communities of animals. These animals may be large, fast growing and have a great reproductive capacity – in stark contrast to surrounding areas. Similar highly productive areas are being discovered associated with other sulphur- and methane-rich seeps elsewhere in the oceans. Some scientists think it possible that life first evolved round these deep-sea thermal vents (▷ p. 6).   RHE

SEE ALSO

● WHAT IS LIFE? p. 6
● BACTERIA p. 24
● PROTISTS p. 26
● ALGAE p. 34
● MARINE INVERTEBRATES pp. 52–7, 66
● FISHES pp. 70–7
● MARINE MAMMALS pp. 142–5
● CORAL REEFS p. 215

DEEP-SEA FISHES

Deep-sea fishes – fishes living below 200 m (660 ft) – show the most elaborate and spectacular adaptations to their environment of any marine creatures. They live in a habitat where almost every other living thing is an enemy, where meals are likely to be infrequent, and where there is little if any light. In these conditions they must be highly efficient predators, be able to conceal themselves effectively, and to minimize energy expenditure. The mouth often has widely opening jaws and sharp recurved teeth to ensure large prey can be captured and never escape, and in many species there is a highly extensible gut to **allow digestion of large meals. Many are bioluminescent, i.e. they can produce their own light: some have luminous lures in front of the mouth to attract prey, and most (except those that live at the greatest depths) have ventral photophores arranged in a pattern so that the light they produce matches the light penetrating from above, making them invisible from below. In contrast to their large jaws, the fish themselves are usually small, with all unnecessary organs reduced to save energy. The results of these various adaptations are some of the strangest-looking of all creatures.**

The gulper eel (centre) is 61 cm (2 ft) long and lives at depths greater than 1400 m (4500 ft). Sloane's viperfish (top) is half the length, and has a luminous extension of its dorsal fin that acts as a lure. *Linophryne arborifera* (which does not have an English name) is one of the smaller angler fishes (bottom), with a luminous 'fishing rod' on its snout. The elaborate chin barbel, resembling seaweed, has a sensory function.

THREATS
TO THE
PLANET

Atmospheric Changes

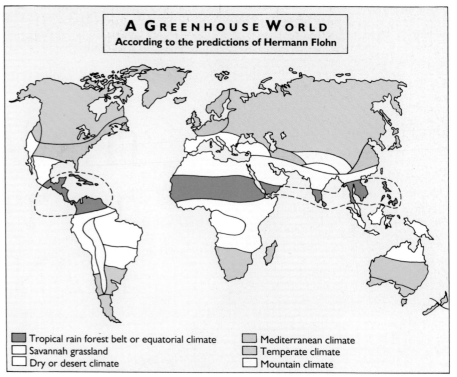

A GREENHOUSE WORLD
According to the predictions of Hermann Flohn

■ Tropical rain forest belt or equatorial climate   ■ Mediterranean climate
□ Savannah grassland   ■ Temperate climate
□ Dry or desert climate   □ Mountain climate

One might suppose that the sheer volume of the atmosphere and its turbulent nature make it an ideal sink for all the gaseous waste material that humanity can cast into it. It seems almost unbelievable that one species can actually modify the composition of the atmosphere over the course of just two centuries. Yet this is precisely what human beings have achieved. Despite its great mass and mobility, the atmosphere now contains a significant proportion of human waste – sufficient to influence its behaviour and to create problems for all those species, including our own, that rely upon the atmosphere for their continued existence.

Human beings have added two types of material to the atmosphere. One type, including the gases carbon dioxide and methane, are natural components of the atmosphere, and we have simply accelerated one stage in the great carbon cycle of the earth (▷ p. 196) resulting in a relatively rapid increase in the atmospheric load of these gases. The other type of material is normally very scarce or absent from the atmosphere and has been generated by mankind as a result of industrial and domestic activities. These substances include oxides of nitrogen and sulphur, and the chlorofluorocarbons (CFCs). All of these compounds have become much more abundant in the atmosphere, either locally or globally, and have given rise to a variety of problems both for human beings and the other occupants of the planet.

Carbon dioxide and the greenhouse effect

Carbon dioxide (CO_2) is a small but significant component of the lower layers of the atmosphere (the troposphere), constituting about 0.035% by volume (350 parts per million). We add to it every time we breathe (▷ p. 180), and plants remove it when they photosynthesize (▷ p. 33), so its level is in fact very variable according to where and when we take our measurements. In summer, when plants are growing most actively, the level may fall by about 16 parts per million (ppm), but locally there is an even greater variation from day to night, perhaps by as much as 40 ppm. The height above the ground at which measurements are taken is also important, since the soil 'respires' as the invertebrate animals and microbes decompose the organic litter it contains and pump out CO_2 into the atmosphere. Against this background of variation it is not easy to assess the rate of change to the overall carbon dioxide level in the atmosphere, but measurements at a remote testing site in Hawaii over the past 30 years have shown a steady and accelerat-

Conifers damaged by acid rain (right) in the Great Smoky Mountains National Park, eastern USA. The acidity strips the leaves of their protective covering of wax and, besides the direct damage that the acid causes to their cells, the leaves are left unprotected from desiccation and from fungal and bacterial infections.

ing tendency for the concentration of the gas to rise – from 315 ppm in 1960 to 350 ppm in 1990.

The study of air trapped in bubbles in the ice caps of Greenland and the Antarctic has also enabled scientists to reconstruct the long-term history of CO_2 in the atmosphere. A hundred years ago the level was only 290 ppm – a level that had persisted since the end of the last ice age, some 10 000 years ago. So the beginning of the rise of this gas can be correlated fairly well with the increasing impact of human beings. When burnt, organic material releases CO_2, so the use of fossil fuels for industry, domestic heating and automobiles has increased the output of the gas, as has the extensive burning of tropical forests to make way for agriculture (▷ p. 200). It is difficult to assess which of

these processes has contributed the greater proportion, but current estimates suggest that the industrial output of CO_2 is about three times that derived from the destruction of forest.

The increase in atmospheric CO_2 over the last century may seem quite trivial (0.006% of the total atmosphere), but the gas has certain properties that make such a small change very important in its effect. Carbon dioxide absorbs energy very strongly in the long wavelength range (infrared or heat radiation). When energy arrives at the earth from the sun, much is in a short wavelength form (visible light) and passes straight through the atmosphere to the earth's surface. There it is absorbed and re-radiated in a long wavelength form (heat), which is more efficiently trapped by the CO_2 in the atmosphere. So light energy is effectively converted to heat and held in by a thermal blanket in precisely the same way that is found in greenhouses, since glass operates in a similar way to CO_2. The term *greenhouse effect* is thus used to describe the global rise in temperature that might result from increased CO_2 levels.

As yet the temperature effect is very small (of the order of 0.5 °C / 0.9 °F), and it is difficult to be sure that the observed rise is not simply a minor fluctuation caused by other varying factors. But if the CO_2 levels continue to rise at their current rate (0.4% per annum) we could see a rise of 2 °C (3.6 °F) by the year 2030, which would mean a significantly warmer world with very different patterns of vegetation and agriculture. Among other problems, this could raise the sea level by more than 1 m (3¼ ft) as a result of the melting of part of the ice caps.

The ultimate outcome of these changes, however, is difficult to predict with certainty, for raised atmospheric CO_2 could lead to faster uptake by the oceans, in which the gas is soluble, and also by the world's vegetation, which is constantly demanding the gas for photosynthesis. One problem with the latter process, however, is the deforestation that is taking place globally, for this will itself release more CO_2 into the atmosphere and also reduce the potential of vegetation to act as a carbon sink. Cloud cover may increase and atmospheric and ocean circulation patterns change in a warmer world, and all of these considerations serve to complicate the picture when predictions are attempted.

Other greenhouse gases

Although carbon dioxide is the most important contributor, it is not the only gas responsible for the greenhouse effect. Perhaps about 71% of the heating is due to CO_2, the remainder being caused by methane (9%), carbon monoxide (7%), CFCs (10%) and oxides of nitrogen (3%).

Methane is only a tiny constituent of the atmosphere (about 1.7 ppm), but it is a much more efficient absorber of infrared than CO_2, so it plays a significant part in the greenhouse effect. It is also increasing as a result of human activity, being released from the wet paddy fields of Asia and also from our herds of domesticated cattle in the form of intestinal gas (▷ box, p. 112). The CFCs are entirely synthetic in origin, being used in refrigeration, aerosols and polystyrene packing. CFCs are only present in the atmosphere in minute traces (about 0.0003 ppm), but are influential because of their extremely efficient infrared absorption capacity. They also have a destructive effect on the ozone layer (▷ below).

Acid rain

The greenhouse effect is not the only problem caused by atmospheric pollutants. Many of the chemicals released into the air during the combustion of fossil fuels are acidic in reaction and can lead to the phenomenon known as *acid rain*. The most important gases in this process are the oxides of sulphur and nitrogen. Both of these elements are important components of proteins in living plants and animals, and hence of their fossil remains, so when organic matter (such as oil or coal) is combusted they are released as their oxides. Sulphur dioxide (SO_2) is one of the most abundant of these pollutants, and when it dissolves in rain water it forms sulphurous acid, which in turn can be oxidized in air to give sulphuric acid. Similarly the oxides of nitrogen can lead to the formation of nitric acid. Fossil-fuel power stations are probably responsible for almost two thirds of the SO_2 in the atmosphere, and car exhausts give rise to almost half of the nitrogen oxides.

The very specific sources of these pollutants mean that they are generated in the greatest quantity in industrialized areas with high densities of population – but the turbulence of the atmosphere soon carries the problems to other parts of the globe. Perhaps the main areas to suffer the effects of this type of pollution are Scandinavia and Canada, for these are the regions that lie downwind of the world's great concentrations of industry, the United States and northern Europe. Canada and Scandinavia also suffer from the problem of having acid rocks and poor soils, so acid rain is not neutralized when it reaches the ground, but goes on to acidify rivers and lakes, leading to fish death. Perhaps the worst effects are felt in spring when acid snow melts and there is a sudden flush of acidity into the waterways. But even before reaching the ground the acid rain can damage buildings and, even more significantly, the vegetation. Trees often show the first signs of damage, losing their leaves in mid-summer and then dying (▷ photo).

Other air pollutants

The results of air pollution are not restricted to the greenhouse effect and acid rain. Motor exhausts inject many other toxins into the atmosphere at precisely the height where they cause most damage. Heavy metals, especially lead, are released by the combustion of some petrol additives. These become concentrated in the atmosphere along roadsides and there are fears that they may be of sufficient concentration to cause damage to humans, especially children, living permanently in urban sites. Carbon monoxide, the product of incomplete burning of gasoline, is extremely toxic and very irritating to the lungs. Even more irritation can be caused by ozone (O_3), a gas generated in our streets when car fumes with their content of nitrogen oxides are acted upon by high light intensities. Ozone is highly reactive and is a particular problem to people with respiratory problems such as asthma. Since it requires strong light for its formation, it is commonest in the cities of lower latitudes, from Los Angeles and New Delhi to Sydney, but even more northerly cities like London have an ozone problem in summer.

INTERNATIONAL ACTION

To combat the threat of global warming and the destruction of the ozone layer, an agreement, known as the Montreal Protocol, was reached by representatives of the main industrialized nations meeting in Montreal, Canada, in 1987. Under the terms of the Protocol, the use of CFCs was to be halved by 1998 and the use of halon – halogenated aliphatic hydrocarbons used in fire fighting – was to be set at 1986 levels by 1992. However, by July 1990, scientific evidence had suggested that these targets could be insufficient to stem global warming, and an agreement was reached between the developing nations and the developed world to achieve a 50% reduction in the use of CFCs by 1995, an 85% cut by 1997 and a total ban by 2000. Halons will also be banned by 2000. Many nations wanted an even faster phasing out of these substances, but the USA, Japan and what was then the USSR claimed that it would be impossible to meet earlier targets. A fund of $240 million has been set up to help the economies of developing countries adapt to the ban on these substances.

CFCs and the ozone layer

Ironically, the same gas, ozone, that causes problems near the ground is vital in the higher layers of the atmosphere, the stratosphere, for the maintenance of life. At a height of between 20 and 50 km (12 and 30 mi) ozone forms a layer that shields the earth from the effects of harmful, very short wavelength radiation from the sun, the ultraviolet rays. These rays in contact with the skin can penetrate and damage cells and may result in cancer development. The ozone shield is itself in danger of damage from air pollution, this time from CFCs (▷ above), which degrade to release chlorine atoms, which in turn react with ozone, breaking it down to oxygen. The ozone destruction seems currently to be concentrated over the South Pole region, but there are signs that an Arctic 'ozone hole' is also developing.

The replacement of the CFCs, the cutting down of the use of automobiles, and a switch to a cleaner system of energy generation than is provided by oil- and coal-fired power stations, are the essential steps towards the improvement of atmospheric pollution problems.   PDM

SEE ALSO

● CLIMATE AND VEGETATION p. 198
● TROPICAL FORESTS p. 200
● POLLUTION (LAND AND WATER) p. 226
● PESTS AND PEST CONTROL p. 230

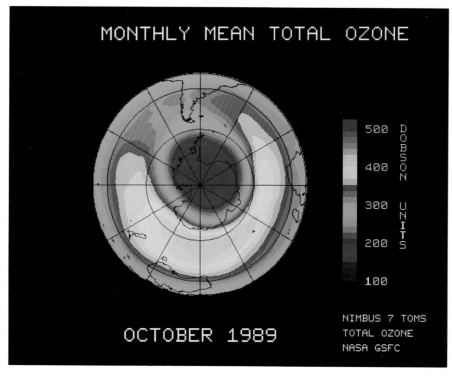

MONTHLY MEAN TOTAL OZONE

OCTOBER 1989

NIMBUS 7 TOMS
TOTAL OZONE
NASA GSFC

A satellite map using computer-imaging to show the severe depletion or 'hole' in the ozone layer above Antarctica in October 1989. First detected in 1979, the hole has been bigger in each subsequent year, reaching its maximum in October, the Antarctic spring. Since this image was made, signs of an ozone hole developing above the Arctic have also been detected.

Pollution

Human beings are damaging the surface of the earth and its water bodies in two main ways. Firstly, we are manipulating natural cycles of elements (▷ p. 196), sometimes increasing the rate at which they take place, sometimes creating concentrations of elements that are beyond the planet's natural capacity to cope.

Secondly, we are creating and distributing new combinations of elements, many of them toxic to life. This latter process may result from the attempt to destroy selectively those organisms that harm our domesticated animals, our crops or ourselves, or it may simply be the outcome of our rapid industrial development and the creation of waste materials.

Misuse of fertilizers

An example of our modification of the natural cycles of elements is the application of fertilizers to agricultural land to increase crop productivity. When we repeatedly harvest crops such as wheat, maize or sugar cane, we remove many elements from the soil, including nitrogen, phosphorus, potassium and calcium. The soil replenishes these by the natural breakdown of rock and from the elements contained in rainfall, but if harvesting proceeds at a faster rate than replenishment, fertilization becomes necessary. Some of these elements may be difficult and costly to obtain, often requiring industrial processes that are expensive in energy (▷ p. 212), as is the case in the fixation of nitrogen from the atmosphere to supplement the natural fixation in soils (▷ pp. 32 and 196). Mining may also be necessary, bringing a host of problems of its own.

Burning oil wells in Kuwait are reflected in a lake of oil. The retreating Iraqi forces set fire to over 500 oil wells in February 1991, and it took nearly a year for the last one to be extinguished. The effects on the climate of the region are not yet clear, although the dumping of vast amounts of crude oil directly into the Persian Gulf had an immediate effect on marine life.

EUTROPHICATION

When the nitrates from artificial fertilizers are washed through the soil and into bodies of water, they are not immediately damaging to plant and animal life. This type of pollution initially results in an enrichment (*eutrophication*) of water bodies that actually stimulates wild plant growth, especially that of aquatic algae, just as it stimulates the growth of the crops for which it was intended. But as the surface mat of algae increases in thickness over the water, it cuts out light to lower layers and so isolates the deeper waters from the atmosphere above. Ageing and dying algae sink down and decay on the lake bottom, and in doing so take up oxygen from the surrounding water. Eventually decay becomes incomplete through lack of oxygen, and hydrogen sulphide and methane are generated. This fouls the waters to such an extent that oxygen-demanding invertebrate animals and fish can no longer survive. The enrichment of a lake with fertilizers can thus cause its death.

Green algae growing on a pond that has undergone eutrophication.

Unfortunately, a large amount of the chemicals sprayed onto our fields to promote crop growth finds its way into the soil. The soil retains some chemicals quite efficiently, including potassium and calcium, but is less efficient at holding nitrates in storage. Unless these are taken up by plants soon after application, they may be washed ('leached') through the soil and into streams, rivers and lakes, resulting in eutrophication of the water (▷ box), and ultimately the death of living things.

The destruction of lake life is not the only consequence of excessive nitrates in water bodies. They can also cause direct harm to human beings, especially young children, by generating carcinogenic compounds in the gut and by interfering with blood transport of oxygen.

The solution to this problem is clearly a less profligate use of fertilizers, which makes economic as well as environmental sense. Fertilizers should only be applied when the rate of crop growth is greatest, the period during which demand for nutrients is at its height; they should never be sprayed onto bare soils. Care should also be taken when land use is changed. If grassland is ploughed, for example, the decay of dead plant material is greatly stimulated and large quantities of nutrients, including nitrates, are released into the soil. Some soil scientists believe that the ploughing of old grassland is a greater source of eutrophication pollution than the misuse of fertilizers.

Human sewage and heavy metals

Human sewage is a further source of eutrophication in many waterways, despite the fact that such material could in principle be used to advantage. If it were used directly as a fertilizer, for example, it could save energy and money currently spent on fertilizer manufacture. This solution is not without its problems, however. One problem is that the sewage contains large populations of bacteria, some of which can cause disease. Another relates to the chemical contaminants found in human sewage. One of the most important groups of such chemicals consists of the heavy metals, including copper, lead, zinc and cadmium; mercury can also be an important component. The content of these elements in sewage may not be high, but if the sludge is applied to soils for a number of years, they may build up and become concentrated in the crops grown upon them. Even the construction of domestic housing on the sites of old sewage farms may become impracticable because of the risk that vegetables will be grown on garden soils and subsequently consumed by humans.

Many of these metals are used in industry and have to be extracted at great expense from the rocks in which they naturally occur. Such mining activities can themselves create surface concentrations of elements that are inimical to life. The spoil heaps of such mines may contain quantities of metals that are not economic to extract but are sufficiently large to render life difficult. Very few species of plant, for example, can cope with high levels of lead in the environment, so the natural establishment of vegetation on such sites can be very slow, often taking decades or even centuries. Meanwhile the metal residues may be washed into waterways and contaminate human water supplies.

Energy production

Even more persistent and sinister is the menace of radioactive pollution. Radioactive materials occur naturally in the earth, but their concentration for the controlled production of energy creates new hazards. The danger lies both in the waste materials produced in nuclear power stations and in the inherent risk of accidents that attends such establishments.

Another material that is extracted from deep within the earth and is used abundantly in industry is oil. This is a perfectly 'natural' commodity in the sense that it has been formed geologically without human intervention, but in bringing it to the surface in large quantities it can become a hazard. Oil was produced by the accumulation and partial decomposition of the bodies of microscopic organisms many millions of years ago. When exposed to air and the natural activity of microbes, it will decay slowly, but the immediate effects of its contamination of land, or more especially water, can be distressing. Most obvious among these is the damage done to the plumage of birds, to the respiratory systems of fish, and to coral reefs. Unfortunately, the emulsification and dispersal of oil using detergents (compounds created by man and not found in the natural environment) can result in damage to the environment that has even more severe and longer-lasting consequences.

These examples of naturally occurring elements and compounds that are being concentrated locally by human activity or whose cycles are being artificially accelerated may thus be accompanied by problems in which other materials that have been created by man for a wide variety of purposes are the major offenders. Detergents, used in industry as well as for domestic purposes and in the treatment of oil spillages, are an example of these. They are designed to reduce surface tension around oily droplets and to make it easier to mix oils and water, but their effect in nature is to disturb the chemistry of oxygen uptake from water by invertebrates and fish – with the result that they can kill a water body very rapidly.

Pesticides

Detergents are not intended to kill living things, but many of the compounds we

Fields of pyrethrum in Rwanda, Central Africa. An extract from this daisy-like flower has proved to be an effective natural insecticide, with less damaging effects on the environment than its synthetic predecessors.

release into the environment are specifically designed to do so. Pesticides are an important component of modern farming and strong economic arguments are put forward for their use. Records of improved crop yields following pesticide introduction are impressive. In Japan, for example, rice production in 1949, before the use of pesticides, involved losses of about 600 kg per hectare (535 lb per acre) to pests. By 1955, following the introduction of pesticides this had been reduced to 150 kg/ha (135 lb/acre). Similar figures for potato crops in the United States show a jump in production from less than 10 tonnes/ha (4 tons/acre) in 1941 to over 22 tonnes/ha (9 tons/acre) in 1961. It is not surprising, therefore, that farmers welcomed the advent of pesticides.

Experience has shown, however, that the side effects of pesticide use in the natural world can be considerable. Mercury dressings on seeds, used to reduce fungal infection, had dramatic and devastating effects on seed-eating birds. A famous case in Japan in the 1950s led to the realization that humanity is not divorced from the natural world and that a polluted environment can lead to human suffering and death. Mercury from an industrial plant was ejected into coastal waters, where it became concentrated at the top of the marine food chain, with the result that many fishermen and their families were poisoned by eating contaminated fish. Persistent insecticides, such as DDT – which was extensively used for delousing in World War II and which may well have saved many lives – have been found to be insidious in their gradual accumulation in the body, becoming serious contaminants of the milk of breast-fed babies. DDT has also had destructive effects at the top of natural food chains, where it accumulates, for example, in birds of prey, which then lay very fragile eggs that often do not survive to hatch.

Pesticides are unlikely to disappear from modern agriculture, but they are being replaced by compounds that are more specific in their target organisms and less persistent in the environment. Many are now based upon naturally occurring compounds, such as the insect deterrents found in some plants (e.g. pyrethrum, a daisy-like flower), since these seem to have less catastrophic effects beyond their intended purpose. There are other approaches, such as 'integrated pest management' (▷ p. 231), that can also reduce the reliance on pesticides.

A similar threat to that of pesticides is posed by the use of antibiotics in medicine, where naturally occurring compounds, or synthesized materials with a similar structure, are increasingly being used in the prevention and cure of disease. The value of such microbial 'pesticides' is immense and few would wish to see the neglect of such a powerful weapon against harmful organisms, but their use can pose new problems. If they are spread too widely through excessive use, as for example in the feedstuffs of domestic animals, they give the bacteria that cause diseases the opportunity to adapt genetically in order to survive their impact. In other words, new strains of bacteria evolve that are immune to the antibiotics, and these can pose a serious health risk. Cautious and restrained use is evidently required.   PDM

SEE ALSO

● MAN-MADE HABITATS p. 212
● ATMOSPHERIC CHANGES p. 224
● PESTS AND PEST CONTROL p. 230

A THREAT FOR THE FUTURE

In the field of biotechnology, a new type of pollution is now emerging as a potential risk, as new strains of microbe are being manufactured for a wide range of purposes (▷ pp. 23 and 25). The complex biochemistry of structurally simple organisms is being harnessed to produce a whole range of chemicals in a cheap and efficient fashion. In effect, microbes are now being domesticated (though yeast, of course, has long been used in just such a way), and their genetic constitution is being manipulated by a system of sub-cellular surgery. The release of genetically manipulated microbes and even higher organisms into the natural environment poses the threat of 'genetic pollution', and should only be undertaken following careful experimentation, for the balance of natural ecosystems could become upset in unpredictable ways. We can only hope that the lessons of the past have been learned to some extent and that caution will be applied in the use of this new tool to improve the quality of human life.

Vanishing Habitats

Any attempt to save a species from extinction must begin by looking at the habitat of that species. Every plant and animal has its own special requirements from its environment and its own part to play in the organization of the ecosystem to which it belongs – its own niche. The niches of different organisms interlock to produce a complex entity, the community – which may be stable or fragile. The destruction of any part of this interactive system can result in the collapse of the whole structure. The removal of a top predator from an African savannah, for example, can lead to overgrazing and habitat degradation. For this reason, all the components of a habitat need to be taken into consideration when formulating plans for its management.

The greatest threat to the biological diversity of our planet is the constant erosion of habitats that is taking place as ever more of the land surface is subjected to human exploitation. Indeed, the problem is not a new one; it began when human beings first found it possible to manipulate their environment to their own ends (▷ p. 212).

Forest clearance

Even before the advent of agriculture, the use of fire to encourage the growth of new vegetation and thus to attract large grazing animals became a major force in the process of habitat degradation. Fire reduced forests in western Europe to heathland, moorland and blanket bog. The establishment of pastoral systems of agriculture ensured that many such areas of reduced biomass were maintained in that condition, by preventing the regeneration of trees and the recovery of the forest.

In North America the same process of major habitat modification dates from the early 17th century, when European colonization began in earnest. Before that time, the native population created local clearances for cultivation and may have encouraged more widespread fire, but it was the influx of Europeans that heralded the extensive forest clearance that still continues to this day. The redwood and Douglas fir forests of the northwest Pacific coast are among the last remaining extensive areas of virtually untouched forest on the continent, and even these are being rapidly eroded as timber exploitation takes its inevitable and destructive course.

A consequence of these depredations is that virgin temperate forest has been reduced to a collection of fragments, and a similar process is now under way – at a much greater pace – in the tropical regions. Tropical forests once occupied about 6% of the earth's land surface, but now only about half of this remains. Probably around 50% of the world's species of plant and animal are native to this type of habitat, amounting to perhaps 15 million species (the figure is very uncertain as only a very small fraction is yet known to science). The Atlantic forests of the eastern seaboard of South America have already all but vanished. Only 2% of these rich tropical forests now remain, as scattered and vulnerable reserves.

In Madagascar the position is currently critical. Being an isolated island off the African coast, its flora and fauna is largely endemic, having evolved within the island setting. About 84% of the original tropical forest of Madagascar has now been cleared, and it is estimated that 25% of the surviving animal species in the remaining portion, still rich in animal life, will become extinct within the next 50 years or even sooner.

The great rain forests of West Africa have already suffered badly, although much remains in the basin of the Zaïre River. Some national parks exist further west along the coastal regions, as in the Ivory Coast, but the practical problems involved in the protection of these areas are difficult to solve and erosion of the forest by illegal felling is rapid.

The problems of the Amazon rain forest are well known. Much land is being cleared for beef-cattle production with disastrous environmental consequences, especially on the hydrology (water movements) of the region: the change from forest to grassland reduces the rate of transpiration (i.e. evaporation of water from leaves; ▷ p. 31), but increases run-off water, thus eroding soils and reducing the potential productivity of the land (▷ p. 201). A similar process in the Himalaya over the past 35 years has affected the lives of the 400 million people who inhabit the plains to the south, subjecting many of these areas to periodic flooding as the run-off waters pour down from the hills. Some efforts have been made to halt and even reverse this process by planting fuel-wood forests on the Himalayan foothills and protecting them from grazing and destruction by cattle. Sustainable forests can ease the pressures for fuel wood and reduce the rate of deforestation.

The economic problems of many of the countries that still have forest resources must be recognized, but in some cases the retention of forest can prove an efficient economic investment. In Rwanda, for example, the fragments of montane forest that provide a home for the much-reduced populations of gorillas have attracted so many tourists that they have become the country's greatest earner of foreign currency. Generally, however, it is difficult

Deforestation in the Amazon basin. Although most clearance of tropical forest is for agricultural purposes, this site was cleared to make way for the mining of manganese ore. Even though the surrounding forest may be left largely untouched, the construction of access roads can disrupt territories and split populations of animals, reducing possibilities for breeding variety.

to use economic arguments for conservation to counter those of the potential exploiter, since it is easier to set a price on a volume of timber than to project the earnings from tourism likely to be generated by a given area of forest.

Grasslands and wetlands

Forests are not the only habitat to suffer as a result of human activity. The world's grasslands have also been extensively abused both as a consequence of overgrazing by domestic stock and by ill-advised conversion to arable agriculture. The 'Dust Bowl' experience of the American Midwest in the 1930s is well documented (▷ p. 206), and a similar process has taken place in Central Asia. In the 1950s and 1960s the Soviet Union attempted to increase agricultural production in the steppes (grasslands) of Kazakhstan, but wind erosion soon stripped away the topsoil. The dry soils were irrigated with scarce supplies of local water, with the result that salts were drawn up through the soil by evaporation and the soil consequently became highly saline. The upshot has been that virtual desert conditions cover some 130 000 km² (over 50 000 sq mi) of land.

THE GOLDEN LION TAMARIN

The story of the golden lion tamarin is an object lesson in the dangers of habitat loss. Once widespread in Brazil's coastal rainforest, the tamarin has declined with its habitat. Only 2% of the coastal rainforest remains, the rest having been cleared for agriculture and the urban sprawl north and south of Rio de Janiero. The result of this devastation is that only about 500 tamarins survive in the wild, mostly in a reserve northeast of Rio de Janeiro – but even this reserve is under pressure from squatters and cattle from adjoining ranches. However, there has been a successful programme of captive breeding, leading to reintroductions into the wild, and as part of the same project a programme of public education has been undertaken, including the encouragement of landowners to protect their private woodlands. There are still sufficient numbers of tamarins to make a viable population, so the golden lion tamarin may yet survive – unlike many other species whose habitat and numbers have fallen below the critical level.

Desertification – the reduction in the productive potential of land to a level where it is effectively worthless – has proceeded rapidly as a consequence of such mismanagement. The proportion of the earth's land surface that is desert has risen over the past 70 years from about 9% to over 23%. This can only partially be blamed on changing climate: much of the area now classified as desert could have remained productive if it had been spared the additional stress of overgrazing.

At the other end of the hydrological spectrum, wetlands are rapidly disappearing around the world, both in tropical and in temperate areas. The tropical mangroves, for example, are particularly vulnerable (▷ pp. 216–17). Although economically vital as a habitat for breeding fish and for coastal protection, their loss receives little attention, chiefly because of their inaccessibility and hence the lack of ecological information on their wildlife. Tropical wetlands are not alone in their vulnerability: in the United States, for example, approximately 54% of the original wetlands have been destroyed, mainly by drainage for agriculture.

Semi-natural habitats

The developed nations have already inflicted severe damage on their native habitats and have often destroyed them completely by constructing cities. Now these same nations are actively engaged in the degradation of 'semi-natural' habitats that are still valuable for their wildlife potential, despite man's role in their development (▷ p. 212). Originally brought about by forest clearance, many of these habitats, including heathland, grassland and some types of managed woodland, may nevertheless be rich in species and have an inherent interest of their own. But even these are becoming increasingly fragmented by the development of intensive agriculture, urban sprawl, and the expansion of road systems that often cut habitats in two and isolate the resulting patches from one another.

One particular danger is that habitats that have been managed in a particular fashion for several centuries are now being brought into more intensive production. Ancient grasslands that may never have been fertilized artificially sometimes become so nutrient-rich as a result of such fertilization that robust species assume dominance and many of the diverse assemblage of plants derived from years of stable management lose out in competition and disappear. Similarly, old woodlands that were harvested for coppice wood, or which provided rough grazing for pigs, have often developed a characteristic and rich flora and fauna. However, these are irretrievably lost if the woods are cleared, or even if their management is neglected and they are no longer cropped.

Heathland and moorland need periodic burning to maintain a healthy growth of heather, and this is a traditional practice in the management of such areas for grouse production (▷ photo, p. 213). The fire cycle needs to be kept up in a

regular fashion about every 10 to 15 years, as too frequent firing can lead to bracken invasion, while if no burning is practised, the heather becomes moribund and over-mature.

The bogs of temperate areas are rapidly being lost, mainly as a result of drainage for forestry or for peat extraction (▷ p. 217). They generally require little management, except perhaps the control of invasive trees such as birch after they have suffered a degree of drainage and surface drying. But in cases where bogs receive drainage water, careful control of the surrounding catchment area is necessary. Eutrophication (nutrient enrichment; ▷ p. 226) of the catchment area causes the bog receiving the run-off water to become overrich, and this can severely affect the vegetation of the peatland itself. The claim that many of the peatlands of Europe are at least partly man-made does not detract from their value as refuges for wildlife.

The fact that a habitat is semi-natural rather than truly virgin and pristine can no longer be used as an excuse for further degradation. Otherwise we shall end up with a monotonous landscape of urban sprawl and rural monoculture (▷ photo, p. 212) with a limited range of species and little potential either for research or for recreation.   PDM

The grizzly bear, like many other large animals (especially predators), needs a large territory in order to survive, and is hence more vulnerable to habitat loss than many smaller species.

SEE ALSO

- TERRESTRIAL ECOSYSTEMS pp. 200–11
- MAN-MADE HABITATS p. 212
- ISLAND ECOSYSTEMS p. 214
- AQUATIC ECOSYSTEMS pp. 216–21
- NATURE CONSERVATION p. 232

Pests and Pest Control

All pests are man-made. In the natural world communities of plants and animals coexist together in a complex web of interactions that regulates their individual population sizes. Too large a number of a particular herbivore may lead to a corresponding increase in the species that eat them, or may lead to the greater ease with which a disease can spread (▷ p. 172). In these and other ways natural populations of species are controlled by the features of their habitat and the other species that share it with them.

Species become pests when they move into the artificial environments we create and damage our crops, stores or other resources. In other words, 'pests' is the name we give our animal and plant competitors, the latter often being called 'weeds'. Pests also include those species that threaten our health.

The red-billed quelea is not only the commonest wild bird in the world – with an estimated population in excess of 10 billion – but also the most destructive, ravaging cereal crops in many parts of Africa. Part of its success is attributable to its rapid reproductive rate: it breeds three or four times a year, and its eggs hatch in 12 days.

The nature of pests

In one sense pests are just species living in the wrong place, but they are able to do this because they are highly adaptable and can take advantage of the new opportunities that we create when we grow crops, store food, use timber, indeed do all the things we do. Two of the most impor-tant features of a pest are that it can spread easily from place to place, finding new man-made opportunities, and that it can then rapidly multiply to exploit these opportunities. These are characteristics of species that ecologists call 'r' strat-egists. In natural environments their life-styles are adapted to finding and col-onizing new habitats, in other words they are pioneer species in the natural succes-sion of flora and fauna. Characteristically 'r' species are small, putting most of their energy into producing offspring rather than building a large body size.

Among animals, the majority of pests are insects. Important mammal pests include rats and mice (▷ box, p. 134), which eat, damage and soil stored produce, and rab-bits (▷ box, p. 132), which feed on food crops. Birds also damage crops: for example, bullfinches eat buds in the spring, and pigeons feed on cabbages and similar crops. Pigeons and starlings are a nuisance because they foul buildings, and sparrows can get into food stores.

Some pests impinge directly on us rather than attacking our crops. Mosquitoes are pests because they bite us. In some parts of the Arctic they become very abundant during the summer months and make life miserable for anything living there. But in the tropics mosquitoes are more than just a nuisance, as they may transmit a variety of diseases, including malaria, yellow fever, dengue fever and viral ence-phalitis. Other blood-feeding insects – tsetse flies, black flies and triatomine bugs – can also transmit diseases when they bite. Control of these insects can either take the form of insecticidal spray-ing of their resting or breeding sites, or the use of physical or chemical barriers such as nets and insect repellants.

Monocultures and pesticides

We simplify the environment around us, taking away the complexity that is neces-sary to maintain a balanced community. When we grow crops we plant vast fields of just a single species – *monocultures* (▷ p. 212). On a global scale we have simplified much of the land surface by planting monocultures of just a few species we use as food, such as rice, wheat and maize. These crops are harvested at least once per year, and this harvesting involves the removal of everything from the fields, which are then ploughed, ferti-lized and replanted. There is no succes-sion towards a more complex community, only the maintenance of a simple single-species plant community. These huge tar-gets are easily found by those species that can feed on the crop (for example aphids). Predators and parasites will eventually find the colonizing herbivores, but by then the herbivores will have substan-tially increased, by which time the crop is damaged and its value reduced. Harvest-ing and clearing the land removes the build-up of predators and parasites of the herbivores.

We control the growth of pests in our crops by using pesticides (▷ p. 227) – herbicides against weeds that compete with the crop for water and nutrients, fungicides against parasitic fungi, and insecticides against the insects feeding on the crop plants. Whilst this can provide some relief to the problems of pests on the crop, it tends to lead to greater difficulties in the longer term. The insect pest is reduced by the application of insecticide, but it is not 100% effective and the pest is able to increase again. Insecticides may also be more harmful to the predators and

parasitic insects than the herbivores (⊳ p. 227). One result of the rapid rate of multiplication of the pest is that resistance can develop quickly, particularly when the pest is being frequently challenged by a particular insecticide; this has happened in the case of mosquitoes and DDT, for example. The insecticide treatments become less effective and more insecticide is used, thus speeding the process of resistance. Insecticides are also potentially harmful to us, either directly if living near areas being sprayed, or indirectly through residues in the food.

The farmer also has a second problem apart from resistance, and that is to do with costs and profits. A field of wheat, for example, will have a market value. Set against this value is the price of producing it: fuel, labour, equipment depreciation, seed and fertilizer costs, and the costs of harvesting and transport. The

Duckweed covering a waterway in the Somerset Levels of Southwest England. Weed species are usually capable of rapid reproduction. Some, including most annual species, produce large numbers of seeds, while others use vegetative means – for example, in duckweed each disc can give rise to daughter plants, and given the right conditions duckweed can double the area it covers every 2½ days. When such a 'carpet' forms, the light is blocked from other aquatic plants, and the amount of oxygen available in the water is reduced.

BIOLOGICAL CONTROL

In addition to the natural predators and parasites found in a pest's own habitat, various other biological controls have proved useful against pest species – particularly insects. The introduction of non-native predators has sometimes proved effective, but can lead to unforeseen consequences – as has happened with the cane toad. This large amphibian was brought to Australia to keep down insects in the sugar cane plantations, but has now multiplied out of control, even threatening other biological-control programmes involving the use of beetles to process cattle dung. Several bacteria and other microorganisms that cause disease in insects have been successfully introduced into insect populations, for example, *Bacillus thurungiensis* (often shortened to BT) has been successful against caterpillars on food crops. The polyhedrosis virus is used against defoliating caterpillars on trees, and certain fungi have been used with limited success against aphids and against some weeds. Among mammal pests, the disease myxomatosis initially proved very effective against rabbits (⊳ box, p. 132).

Several other biological methods are now used against pests, including the use of hormones that inhibit development in particular species, and the introduction of large numbers of males sterilized by gamma rays into a population. Experiments are also being conducted with genetically engineered strains that transmit lethal genes. The use of pheromones – volatile chemicals produced by one insect that affect the behaviour of another of the same species – have proved particularly effective. Usually pheromones are used to attract the opposite sex. A number of these pheromones can now be synthesized and are used in conjunction with a trap to catch or monitor pests such as flour moths, flour beetles and cockroaches. They are particularly effective when the pest is rare, and are commonly used in food-storage areas and public hygiene programmes.

difference between these costs and the market value of the crop represents the profit to the farmer. Every time the pest numbers increase they weaken the crop and affect its value, reducing the profit margin. Use of insecticides can limit pest damage but at the cost of more labour, fuel, spraying equipment and the chemicals. Again the profits are reduced.

Ecological solutions

The answer to this catalogue of problems has been obvious to ecologists for many years, but is only now beginning to be appreciated by a wider audience – particularly by farmers. Instead of simplifying the habitat it must be made more complex. This can be accomplished in several ways. In the interests of efficiency there has in the past been a tendency to think big. Large fields allow the use of big machines that speed up the processes of ploughing, sowing and harvesting. In the process hedgerows were removed. By replacing hedgerows and making smaller fields there is an immediate increase in the complexity of the habitat and, more importantly, a permanent network of havens for a wide variety of plant and animal species that can move in and out of the crop (⊳ p. 212). An experimental variation on this theme is the construction within fields of long, low earth banks, which are left to be colonized by natural vegetation and also by animals. If limited spraying is still necessary then these banks and hedgerows must be avoided. Complexity can also be achieved by planting several crops together. Strip farming, as this is called, can give good yields and provides a far greater abundance and variety of predatory or parasitic insects, which in turn keep pests under better control. The downside of these approaches is a reduction in the efficiency of planting and harvesting the crops.

These two approaches – use of insecticides and the use of natural enemies to control the pest – have been combined in what is called *integrated pest manage-*

ment. Natural methods are augmented by a limited and careful use of insecticides. The insecticides used are short lived, and are only applied when most predators of the insect pests are protected from the chemicals – for example, at a stage in the predator's life cycle when it is in the soil rather than on the plant.

A matter of attitude

Despite our attitude to pests we should realize that it is not their fault that we find them so obnoxious. Flies are considered as pests because they foul our food and transmit a range of enteric diseases. Yet as consumers of animal corpses they play a major role in speeding up the process of decomposition, essential in the recycling of nutrients (⊳ p. 196). We have to realize that pests are pests because we made them so. This is nowhere better illustrated than in the marine fisheries, where seals and dolphins – animals with great popular appeal – are sometimes considered as pests by fishermen because they compete with man for the fish. Perhaps we should begin to recognize that the only real pest is ourselves.   BDT

SEE ALSO

● INSECT INTERACTIONS p. 64
● RABBITS AND MAN p. 132
● OF MICE AND MEN (BOX) p. 134
● PARASITISM AND SYMBIOSIS p. 168
● POPULATION DYNAMICS p. 172
● MAN-MADE HABITATS p. 212
● POLLUTION p. 226

Oriental cockroaches are among several species of cockroach that have become a pest in many parts of the world, moving as they do between sewage and waste systems and kitchens, carrying with them organisms that cause enteric (intestinal) diseases. However, their penchant for food fragments is simply an extension of their natural role as scavengers, a vital link in food chains in the wild. It is humans that have given them the opportunity to become pests.

Nature Conservation

The extent to which the natural world is under threat makes it imperative that measures are taken to control the damage done to the environment as a whole. In particular, it is essential that some areas of natural habitat are set aside for special protection, to ensure the survival of as many species of animal and plant as possible. There is a limit, of course, to how great an area of our planet we can afford to use in this way, since the human population itself demands both living space and an adequate provision for food production and industrial activity. It is the task of the conservationist to attempt to balance these sometimes conflicting demands.

SEE ALSO

- TERRESTRIAL ECOSYSTEMS (MANAGEMENT PROBLEMS) pp. 200–11
- ISLAND ECOSYSTEMS p. 214
- WETLANDS (MANAGEMENT PROBLEMS) p. 216
- POLLUTION pp. 224–7
- VANISHING HABITATS p. 228

Conservation is not simply preservation; rather it involves the wise use of resources, a kind of informed stewardship that aims to pass on the planet to our descendants in at least as good a condition as we inherited it. Conservation, therefore, is rarely a passive process in which we sit back and allow nature to take its course; generally it involves the active management of resources to maximize the benefits to the widest range of species, including our own.

Why bother with conservation?

There are several reasons why we should be concerned to promote conservation:

(1) To promote biodiversity. The loss of a species is irreversible – any mistake made in this area can never be put right. From a selfish point of view, we can never tell how valuable a lost species might one day have been to mankind, as a source of food, drugs or scientific information.

(2) To permit renewable exploitation. Human beings will always need to draw upon the resources of the natural world for raw materials, energy and food. The repeated harvesting of nature, whether forestry or fishery, requires careful management to ensure that resources do not become depleted or lost.

(3) To maintain and regulate the global environment. Conservation demands a clean and healthy environment that will ensure the survival of the natural world and with it our own survival.

(4) To conduct research and education. The advance of scientific knowledge is vital to the three points made above, and the maintenance of nature reserves and parks provides an opportunity to study the ecosystems of the world and to understand their function and balance. Such lessons can then be applied to our own species.

(5) To provide leisure and recreational activities for mankind. The world would be a dull place without the variety that nature provides, so the quality of human life is closely dependent on the diversity of the natural world.

All of these are reasons why we should be concerned to conserve our environment, even though the process may be economically costly and require a check on development. One important aspect of conservation, especially in these days of rapid human population expansion and worldwide industrial development, is the establishment of nature reserves to ensure that all of the natural world is not lost or irreparably damaged by these processes. But the selection, design and management of such reserves presents the conservationist with a number of problems.

Choosing sites and species

The selection of areas for particular attention demands a process of evaluation, but it is far from easy to establish objective criteria upon which such value judgements can be based. One important aspect of a prospective nature reserve is that it should be rich in species, and this often means that it needs to be large and diverse in the habitats it contains. But size alone does not guarantee diversity, and some small reserves may nevertheless be valuable.

A second important question in the evaluation of sites is how easily they can be replaced. Some habitats, such as birch woodland invading old quarries, are fairly young and can be replaced within a matter of years if they are destroyed or damaged. Other habitats, such as virgin forest or open peatlands, may have taken several thousand years to reach their current state of development, and must be regarded as essentially irreplaceable. Clearly, the latter type of ecosystem should be rated more highly than the former.

Some habitats are particularly fragile, being easily damaged even by relatively minor interference. Among these are many wetland systems that are easily disturbed and polluted. Others, such as short-turf grasslands and heathlands, can withstand a certain amount of pressure from human activity, including recreational use. Particular attention needs to be paid to the protection of fragile and vulnerable environments, and their management needs to be adjusted accordingly.

Considerable public support and attention is given to rarity as a factor in comparing different species, especially if a particular species is threatened on a global scale. But even among rare species, it is usually the large and glamorous examples that are the objects of greatest concern, often to the detriment of the remainder. Pandas and Californian condors make the headlines, while many obscure plants and insects become extinct

American bison, or plains buffalo, grazing in Yellowstone National Park, Wyoming, USA. Until the advent of European settlers, probably in excess of 50 million bison roamed the prairies of North America, their vast migrating herds providing the economic mainstay of the Plains Indians. By the end of the 19th century, however, following decades of unbridled slaughter by Europeans for the animal's meat and hide, the species had been brought to the very verge of extinction, with no more than a few hundred surviving in total. Careful management in protected reserves has since allowed the few remaining animals to rebuild their numbers, which now run into the tens of thousands.

each day as the South American rain forest is destroyed.

There is also a danger of parochialism when rarity is given too great a prominence in selecting habitats and species that warrant attention. In the British Isles, for example, a small heathland bird, the Dartford warbler, is the object of considerable attention because it is restricted as a breeding bird to a few southern sites. During hard winters it may draw close to extinction and become the object of deep concern. But this bird is in fact quite widely distributed in the south and west of Europe, and in Spain it can be regarded as fairly common. In Britain it is at the northern edge of its geographical distribution and evidently finds conditions only marginally tolerable. Although there is no harm done by attempts to retain such local rarities, they should not be given priority over species that may seem locally abundant and yet are scarce or restricted on an international scale. The bluebell, for example, is a very common plant of British oakwoods, but globally it is very distinctly confined to a corner of northwest Europe; the bluebell wood should therefore be a prime habitat for the attention of British conservationists.

Habitat requirements

It has now become clear that rare species are best conserved by giving attention to the maintenance of their habitats, and sometimes their habitat requirements are considerable. Vultures in Spain, for example, and condors in California can only be sustained by very large tracts of country over which to forage. They also need to be assured of a supply of carrion food, which may in turn demand some untidy farming practices in which dead animals are left out on the hills to rot.

At the other end of the spatial scale, the maintenance of frog diversity in a tropical rain forest may entail the assurance of small, wet, muddy hollows for breeding. Although the frogs avail themselves of such locations for only a few days each year, the survival of many species depends on them. In studies of frog requirements in Central America, for example, it was found that the peccary (a kind of forest pig; ▷ p. 113) neatly answers the needs of many frogs, since these animals enjoy wallowing in wet areas and thereby maintain the open, muddy ponds the frogs appreciate. In this case, conservation of frogs depends on conservation of peccaries. The web of conservation ecology can be very complex.

The island model

There have been several attempts to put the selection and design of nature reserves on a more scientific and objective basis by applying the ideas of island biogeography (▷ p. 214). In many ways nature reserves are similar to islands. Just as an island is a unit of habitat isolated by water, so, for instance, a fragment of woodland in an open, agricultural setting represents a habitat cut off from other such areas by intervening regions of alien vegetation.

Among other things, the theory of island biogeography suggests that large reserves will be preferable to small ones and that similar habitats should not be situated too far from one another, so that exchange of populations can take place. 'Island chains' may facilitate this movement between reserves, and the establishment of corridors from one reserve to another has been attempted, as in Costa Rica, where a 3–6 km (2–4 mi) wide strip of forest joins an upland with a lowland forest reserve. This strip provides a migration route for the animals and birds of the forest. The splitting of a reserve in two, on the other hand, as when a major road is constructed through a forest, can isolate the populations of animals and plants on either side and thus effectively halve the size of the reserve.

Just as it is possible for islands recently cut off from the mainland by rising seas to be supersaturated with species, so it may be that some of our modern fragmented habitats in temperate and tropical areas lose species as they assume a new equilibrium level. Experimental forest islands in Brazil have been losing species at a rate that corresponds quite closely to their area. A plot of 1400 hectares (14 km²/5·4 sq mi) was found to have lost 14% of its breeding bird species in the first year of isolation, while a plot of 250 hectares lost 41% and one of 21 hectares 62%.

The influence of reserve size on extinction rate is therefore clear enough. What is not yet clear is whether some of our important older reserves in the temperate regions, such as the American National Parks, are still losing species by the same process. Some of our larger and more demanding species may well be at risk. The grizzly bear, for example, has a home range of about 3000 km² (1158 sq mi), so the maximum population even of large areas of reserve can never be very big. This raises the question of just how big a population of animals needs to be to retain its viability. Some have suggested a minimum of 500 individuals, otherwise genetic weakness from inbreeding becomes a problem. But some species have fallen well below this level and yet managed to recover, often with human assistance. We should not be complacent, however, as certain species are allowed to drop to critically low levels.

The human factor

Once reserves are selected, they must be managed to maintain their vitality and diversity. The 'hands off' approach to nature conservation has often proved to be counterproductive, as in the grasslands and heathlands of Europe and the forests of America. The latter have suffered a number of devastating fires in recent years as a consequence of neglect over several decades. What is needed is a programme of prescribed burning, which can prevent the excessive build-up of flammable material and at the same time serves to diversify the habitat and maintain the conditions needed for many successional species of plant and animal.

CONSERVATION AT SEA

Whale hunting: the final flurry

The challenges facing the conservationist at sea are in some ways different from those on land. Man is not in direct competition with marine species for the use of habitats as such, but the effects of his activities can be highly damaging nevertheless. The principal dangers to marine life stem from pollution and from the exploitation of the seas as a source of food.

The most serious pollution of the seas is limited to shallow coastal waters, which account for less than a tenth of the total area of the oceans but are home to well over 90% of all marine life forms. The dumping at sea of toxic chemicals, nuclear waste and human sewage may pose a direct threat to the life of marine organisms, or – more insidiously – poisonous waste products may become concentrated as they pass along food chains, eventually reaching top consumers such as large fishes and mammals, including man (▷ p. 227). Accidental pollution, such as oil spillage, can also spell disaster for marine life, interfering with the respiratory systems of fishes and killing off coral reefs. So too can the frequent responses to oil pollution, such as the use of detergents as dispersal agents, which may disturb the chemistry of oxygen uptake from water by both invertebrates and fishes.

The oceans have long been harvested as a source of food for mankind, but technological advances over the past few decades have led to overexploitation, with profound ecological implications. The advent in the 1950s of factory fishing fleets, equipped with sonar to detect shoals of fishes, enormous hydraulically operated nets and refrigeration facilities, brought about a massive but unsustainable increase in annual catches, leading to a dramatic slump in the early 1970s. Even more emotive has been the hunting of whales for their meat and valuable oils (▷ p. 145). One by one the species of great whale have been brought to the brink of extinction, their numbers declining as methods of hunting them have improved.

The measures needed to remedy these problems are obvious enough, but difficult to implement effectively. A comprehensive and internationally agreed policy of ocean management is required, and bodies such as the United Nations and the International Whaling Commission have made progress towards this goal. The oceans represent a vast and sustainable resource for mankind, but only if the marine environment is kept healthy and is not overtaxed for short-term gain.

The human species, whether we like it or not, is now a major factor of influence in the ecology of this planet. This influence can be exceedingly harmful to other forms of life, as we release pesticides and the waste products of our society into the environment. But – provided that we apply ourselves to the careful management of what remains of the planet's biosphere – we can become a constructive factor in maintaining the great diversity of life on earth.   PDM

aardvark (*Orycteropus afer*), an ant- and termite-feeding mammal of sub-Saharan Africa, so unusual and distinctive that it is placed in an order of its own (Tubulidentata); ▷ pp. 101, **109** (box), 164.

aardwolf, an atypical member of the HYENA family; ▷ p. 125 (box).

Aaron's rod or **great MULLEIN** (*Verbascum thapsus*), a tall hairy biennial with yellow flowers found on dry soils throughout Europe, Asia and the Himalaya. See also MULLEIN.

abdomen, in insects (and other arthropods) the hindmost part of the body, behind the thorax or cephalothorax (▷ p. 59). The belly region of a vertebrate's body (the area bounded by the diaphragm in mammals) may also be termed the abdomen.

Abies, ▷ FIR.

abiotic, non-living; usually refers to the non-living parts of an ecosystem, e.g. minerals soil, atmosphere, etc., as opposed to the living (BIOTIC) components.

abomasum, the fourth chamber of a ruminant's stomach; ▷ p. 112.

abscisic acid, a plant hormone with a variety of important effects. It is partly responsible for ABSCISSION; induces dormancy in seeds and buds; causes the closure of leaf stomata in dry conditions (▷ p. 31, box); and plays a role in various growth and development processes (▷ p. 33).

abscission, the shedding of an organ (leaf, fruit, flower) from the plant caused by the disintegration of cell walls in an area called the abscission layer at the base of the organ. ABSCISIC ACID plays a role in this process.

absorption, the process whereby the products of digestion are absorbed and carried around the body; ▷ p. 179.

abyssal zone, a marine zone; ▷ pp. 220–1.

acacia, any of around 800 species of the genus *Acacia* in the family Mimosaceae (formerly part of the Leguminosae). They are evergreen, often spiny, trees and shrubs of tropical and subtropical dry scrub. They are used to stabilize sand dunes and for timber production in arid regions. Early Australian settlers used to build huts of wattlework plastered with mud, hence the common name wattle. The flowers of *A. dealbata* are sold by florists under the name of MIMOSA. The gum arabic tree (*A. nilotica*) is a source of gum arabic; ▷ p. 203.

acanthodian or **spiny shark,** ▷ pp. 70–1.

accentor, any of 13 PASSERINE birds of the subfamily Prunellinae, in the same family as the SPARROWS. They are sparrow-sized, brown/grey birds with strong bills living in low vegetation of Eurasia. The European hedge sparrow, or dunnock (*Prunella modularis*), feeds on insects, seeds and berries.

accidental, an organism that appears occasionally as a casual, non-persistent visitor beyond the normal limits of its distribution. Compare EXOTIC, INDIGENOUS.

accipiter, any of 48 BIRDS OF PREY of the genus *Accipiter* (family Accipitridae, order Ciconiiformes), with round wings, a long tail and long legs. They include the SPARROWHAWKS and GOSHAWKS.

-aceae, a standard ending indicating family status in plants (▷ p. 45).

Aceraceae, a dicotyledonous plant family, including maples and sycamore; ▷ p. 45.

achene, a dry, INDEHISCENT (i.e. not opening) one-seeded fruit developing from one CARPEL (e.g. a strawberry 'pip', lettuce 'seed').

achillea, any of around 200 species of herbs in the temperate Old World genus *Achillea* in the daisy family, Asteraceae (formerly Compositae). Yarrow (*A. millefolium*) is a medicinal herb containing the alkaloids achillein and moschatin and has been used to prevent bleeding.

acidic, rich in hydrogen ions, so reacting with basic compounds. Measured by pH, all values below 7 (neutrality) are acidic. Often refers to waters and soils from base-poor areas. Compare BASIC.

acid rain, rain containing various acids as a result of industrial processes; ▷ pp. 207, 209, 218, 219, **225**.

acid soil, soil with an acidic reaction due to its excess of hydrogen ions. It is usually poor in calcium and often in other elements, and may suffer from aluminium toxicity.

Aconitum, a north temperate genus of around 300 species of perennial herbs in the buttercup family, Ranunculaceae. The flowers are characteristically hooded, e.g. monkshood (*A. napellus*) and wolfbane (*A. vulparia*). All species are very poisonous, containing the alkaloid aconitine.

acorn worm, any invertebrate of the subphylum Hemichordata (phylum Chordata), worm-like, mud-dwelling animals with a body in three parts – proboscis, collar and trunk. Their supposed relationship with the CHORDATES is based on a series of gill slits in the trunk and a supporting rod (notochord) in the collar, but this link is no longer universally accepted and they are often put in a separate phylum; ▷ p. 51 (box).

acouchi, either of two large, short-tailed, paca-like caviomorph RODENTS of the genus *Myoprocta* (family Dasyproctidae). Inhabiting tropical forests in South America, they feed on fruits and seeds. See also AGOUTI.

acquired characteristics, characteristics acquired by an organism during its lifetime. In his theory of evolution Lamarck proposed that such characteristics can be inherited; ▷ p. 12, and WEISSMAN.

actinopterygian, ▷ RAY-FINNED FISH.

active transport, the movement of molecules from a region of lower concentration to a region of higher concentration. Cell membranes (▷ p. 20) are able to do this. Compare DIFFUSION.

adaptation, 1. any characteristic of a living organism that has evolved to perform a particular function. Adaptations improve an organism's chances of surviving and/or breeding successfully (or those of its close genetic relatives) in the environment it inhabits, and are therefore maintained in a population as a result of natural selection; ▷ pp. 12–17. **2.** a physiological process by which various sensory stimuli fade over time; ▷ p. 188.

adaptive radiation, the diversification of a group of related organisms into a variety of habitats, resulting in the evolution of a number of new groups with distinct modes of life; ▷ pp. 16, **17** (box), *214*.

addax (*Addax nasomaculatus*), a greyish-white ANTELOPE with long, spirally twisted horns, inhabiting desert areas of northern Africa.

adder, ▷ VIPER.

adipose tissue, ▷ FAT.

admiral, any of several NYMPHALID butterflies of the genus *Vanessa* (subfamily Nymphalinae), occurring worldwide. Larvae are nettle- or thistle-feeders.

ADP, adenosine diphosphate, a complex molecule that can be converted into ATP.

adrenal gland, in vertebrates, an endocrine GLAND situated near the kidneys, responsible for the secretion of various hormones, including ADRENALINE.

adrenaline, in vertebrates, a HORMONE from the adrenal gland, secreted in response to stressful stimuli and preparing an animal for sudden activity, such as fighting or fleeing; ▷ p. 183.

advanced, (of an organism or structure) specialized or considerably modified from the ancestral form; ▷ p. 60 (box).

adventitious, (of a structure of a plant) produced at an uncharacteristic position, such as adventitious roots, or buds on leaf margins.

advertisement, a form of conspicuous DISPLAY usually shown by territorial male animals to attract females and repel rivals (▷ p. 17).

aerobic, functioning only in the presence of oxygen. Compare ANAEROBIC.

aerobic bacteria, bacteria dependent on oxygen; ▷ p. 25.

aestivation, a state of suspended bodily activity, generally a response to hot, dry summer conditions; ▷ pp. 73 (lungfish), 83 (dinosaurs), **175**.

afferent nerve, a nerve cell carrying information from a sensory receptor towards the central nervous system; ▷ p. 182.

African hunting dog, ▷ HUNTING DOG.

agamid, any of 300 species of LIZARD of the family Agamidae, widely distributed in warmer parts of the Old World. Most species are insectivorous, and many can change colour rapidly for camouflage and display. See also FLYING DRAGON, FRILLED LIZARD, THORNY DEVIL and WATER DRAGON.

agaric, any species of BASIDIOMYCETE fungi (▷ p. 35) in the orders Agaricales or Russulales. They are mushroom-shaped FUNGI with gills. Several are grown commercially for food (e.g. common mushroom, oyster mushroom, SHIITAKE).

Agave, an American genus of around 300 species of plants in the family Agavaceae with rosettes of large, tough, fleshy leaves suited to their arid habitats. The century plant or American aloe (*A. Americana*) produces a few leaves annually, eventually flowering and dying after as much as 50 years. See also SISAL.

aggregation, a group of animals independently attracted to a particular environmental feature such as food or water, and showing no social organization; ▷ p. 156.

aggression, behaviour that harms or threatens to harm another member of the same species, typically in defence of a TERRITORY, in competition for access to females (▷ COURTSHIP), or in establishing a position within a social HIERARCHY; ▷ pp. 154, **157** (box).

agnathan, ▷ JAWLESS FISH.

agouti, any of 10 large, short-tailed, paca-like caviomorph RODENTS of the family Dasyproctidae, inhabiting forests and grasslands in South America. They feed on fruits, seeds and succulent vegetation. See also ACOUCHI.

agriculture, the rearing of domesticated animals and the cultivation of crops by humans; ▷ pp. **212–13**, **230–1**. For the impact of agriculture on natural ecosystems ▷ pp. 200–11, 217, 228–9.

air bladder, ▷ SWIM BLADDER.

air sac, 1. another name for ALVEOLUS. **2.** any air space serving to lighten a heavy structure, as in the bones of birds (▷ p. 92).

alarm call, ▷ ALARM RESPONSE.

alarm response, a response to signs of danger that serves as a warning signal to other animals; often a specific acoustic, olfactory or visual signal, such as a cry, a PHEROMONE or a TAIL FLASH.

albatross, any of 14 species of the family Procellariidae (order Ciconiiformes), large gliding birds of southern oceans, characterized by their long, thin wings. They habitually follow ships, feeding on fish and crustaceans. The wandering albatross (*Diomedea exulans*) is the largest, with a wing span exceeding 3 m (10 ft); ▷ pp. 90, *91*, 92, 93, 170.

alder, any of around 35 species of the genus *Alnus* in the birch family, Betulaceae. They are trees or shrubs principally of temperate regions and mainly in damp habitats forming a dense scrub (▷ p. *219*). Tannin for treating hides and dyeing linen may be extracted from the bark of some species, e.g. black alder (*A. glutinosa*); ▷ pp. 197, *219*.

alder fly, any NEUROPTERAN insect of the widely distributed family Sialidae. The darkly coloured adults are sluggish fliers. Larvae are aquatic and carnivorous, with a series of long-fringed gills along the abdomen.

-ales, a standard ending indicating order status in plants (▷ p. 45).

alfalfa or **lucerne** (*Medicago sativa*), a very important leguminous fodder crop (family Fabaceae). The subspecies *sativa* with blue to violet flowers is most widely grown, but there is a yellow flowered subspecies (*falcata*) and a number of other species of *Medicago* may also be grown as alfalfa.

algae, a term generally applied to lower plants that do not possess any special modifications for life outside water. Unicellular forms are usually classified as protists, while multicellular forms are usually called seaweeds; ▷ pp. **26–7**, 34, 215, 226 (box).

algin, any of several alginic acids, complex polysaccharide-type polymers of acids of certain sugars found in brown algae (seaweeds); commercially important in food as thickeners.

alimentary canal, ▷ DIGESTIVE TRACT.

Alismatidae, a monocotyledonous plant subclass; ▷ p. 45.

alkaline soil, soil with an alkaline reaction due to low concentration of hydrogen ions. It is usually rich in calcium or magnesium (serpentine soils), but may suffer from iron or phosphorus deficiency.

allantois, a membrane that acts as a respiratory and excretory organ in reptile and bird eggs. In placental mammals it eventually forms part of the PLACENTA; ▷ pp. *86*, *187*.

allele, a sequence of nucleotide bases that makes up a gene; ▷ p. 22.

alligator, either of two species of CROCODILIAN reptile of the genus *Alligator* (family Alligatoridae, which also includes the CAYMANS). The larger American alligator (*A. mississipiensis*) is farmed for its skin and meat, and has become a pest in some urban areas of the USA; the smaller Chinese alligator (*A. sinensis*) is critically endangered (▷ p. *89*).

Allium, a genus of the family Alliaceae, including CHIVE, GARLIC and ONION.

allspice, pimento or **Jamaica pepper** (*Pimenta dioica*), a tropical evergreen New World tree. The dried berries are used as a spice, the flavour of which resembles a mixture of CLOVE, CINNAMON and NUTMEG.

almond (*Prunus amygdalus*), a tree of the rose family, native to western Asia and cultivated in southern Europe, North Africa, South Africa and California for its nuts. Only sweet almonds from *P. amygdalus* var. *dulcis* are edible, although oil of sweet almond is extracted from both these and bitter almonds (var. *amara*).

aloe, 1. any of around 100 species of shrubby or xerophytic (drought-resistant) plants in the tropical Old World genus *Aloe* in the lily family, Liliaceae. Bitter aloe is extracted from the leaves of the Curaçao aloe (*A. vera*). **2.** ▷ AGAVE.

alpaca (*Lama pacos*), a domesticated South American ARTIODACTYL mammal (family Camelidae), descended from the wild GUANACO. Wool and cloth are made from its dark, shaggy hair; ▷ p. 115 (box).

alpine, occurring at high altitude, between the tree line and the permanent snow line; ▷ pp. 10–11, 210–11.

alternation of generations, a reproductive process in various organisms by which a sexually produced generation is followed by one or more asexual generations. Alternation of generations between GAMETOPHYTE and SPOROPHYTE is found in all plants (▷ pp. 33, 34, 36–7), but is much less obvious in the SEED-BEARING PLANTS (▷ pp. 38–41). Alternation of generations also occurs in many invertebrate animals, such as corals and aphids (▷ pp. 53, 63, **184**).

altruism, any behaviour, such as a TAIL FLASH, that is performed by an animal for the benefit of others at some cost to itself, 'cost' and 'benefit' generally being defined in terms of the ability of the animals concerned to survive and reproduce successfully. The evolution of such behaviour is usually explained by the notion of KIN SELECTION; ▷ pp. **149** (box), 153.

alveolus or **air sac,** any of numerous tiny swellings at the termini of the bronchioles in the lungs that increase the surface area for gas exchange with the blood; ▷ p. 180 (box).

alyssum, any of around 100 species of often matforming herbs in the mainly Mediterranean genus *Alyssum* in the cabbage family, Brassicaceae (formerly Cruciferae). Some species, such as golden alyssum (*A. saxatile*), are cultivated as rock plants.

Amanita, a widely distributed genus of AGARIC fungi, including poisonous (death cap and destroying angel), hallucinogenic (fly agaric) and edible (blusher) species; ▷ p. 35.

amaranth, any of around 60 species of herbs in the tropical and temperate genus *Amaranthus* in the family Amaranthaceae. Love-lies-bleeding (*A. caudatus*) has pendulous red inflorescences. Several amaranths of the western USA are known as TUMBLEWEEDS.

amaryllis, any of around 75 species of bulbous plants in the genus *Hippeastrum* in the family Amaryllidaceae. They have large funnel-shaped flowers and are popular house plants.

Amastigomycota, one of the two divisions of true fungi; ▷ p. 35.

Ambiortus, ▷ p. 90.

ambrosia or **ragweed,** any of around 40 species of annual to perennial aromatic medicinal herbs in the New World genus *Ambrosia* in the daisy family Asteraceae (formerly Compositae).

ambrosia beetle, any BARK BEETLE of the family Platypodidae. They are so named because they cultivate a fungus on wood chewings and excrement on which the larvae feed.

ambush, a method of PREDATION that involves waiting at a place likely to be visited by an item of prey and seizing it as it comes into range. Such predation often requires CAMOUFLAGE to avoid detection by prey, and various devices such as WEBS and LURES; ▷ p. 164.

American blackbird, any of about 100 New World PASSERINE birds of the tribe Icterini, related to the BUNTINGS. Many have dark glossy plumage, often offset by patches of very bright colour. The family includes the New World ORIOLES, MEADOWLARKS, GRACKLES and COWBIRDS, as well as 'blackbirds', of which the redwing blackbird (*Agelaius phoeniceus*) is one of the most common species of North America (▷ p. 217).

amino acid, a building block of protein, consisting of one or more carboxyl groups (-COOH) and one or more amino groups (-NH$_2$) attached to a carbon atom. There are over 80 naturally occurring amino acids, 20 of which occur in proteins; each amino acid is distinguished by a different side chain (an 'R' group). Organisms can synthesize many amino acids, but there are some that have to be obtained from the diet; ▷ pp. 6 (box), 7, 22, 179.

ammonia, a colourless, pungent gas (NH$_3$) whose presence in the early atmosphere probably contributed to the beginning of life (▷ p. 7). It also plays an important role in the nitrogen cycle (▷ p. 197).

ammonite, any extinct CEPHALOPOD mollusc of the order Ammonoidea, related to the cuttlefishes and squids. Often large, they were formidable predators; the spiral-coiled chambered shell was a buoyancy device (▷ NAUTILUS). They dominated the seas during the Mesozoic period (before fishes came to dominate), and are familiar fossils in chalk and limestone areas; ▷ pp. 5, 57.

ammonium, (used as a modifier) containing NH$_4^-$ or NH$_4^+$; ▷ p. 31.

amnion, an inner protective membrane surrounding the embryos of reptiles, birds and mammals and creating a fluid-filled amniotic cavity. On account of this structure these three animal groups are referred to as 'amniotes'; ▷ pp. *86*, *187*.

amniote, a reptile, bird or mammal (▷ AMNION).

amniotic cavity/fluid, ▷ AMNION.

amoeba, any member of the protozoan genus *Amoeba* (etc.); ▷ pp. 26–7, 176.

amoeboid movement, the type of movement used by amoeba; ▷ p. 27, 176.

amphibian, any member of the vertebrate class Amphibia, whose living representatives are classified in three orders: ANURA (frogs and toads); Urodela (SALAMANDERS, including NEWTS and SIRENS), and Gymnophiona (CAECILIANS). Although the smallest of the major vertebrate groups, with around 4000 species, the amphibians are of considerable interest, particularly since they played a central role in the evolution of life on earth, being the first vertebrates to leave water to live life on land; ▷ pp. **78–81**.

amphioxus, any LANCELET of the genus *Amphioxus*.

amphisbaenian or **worm lizard,** any of about 100 species of virtually blind, limbless, burrowing REPTILES of the suborder Amphisbaenia, related to the snakes and (true) lizards. They feed principally on worms and other small invertebrates; ▷ p. 85.

ampullae of Lorenzini, jelly-filled sensory ducts found on the heads of fishes; ▷ pp. 189, 191 (box).

anabolism, ▷ METABOLIC PATHWAY.

anaconda (*Eunectes murinus*), the great water BOA of South America, one of the largest and certainly the most massive of all snakes. An average adult grows to around 6 m (20 ft), and much longer specimens (nearly 9 m/30 ft) have been recorded; ▷ p. 89.

anaerobic, functioning in the absence of oxygen. Compare AEROBIC.

anaerobic bacteria, bacteria that can only function in the absence of oxygen; ▷ p. 25.

anal, of or near the anus.

analogous, (of an organ or other structure in two different organisms) not deriving from a common evolutionary ancestor, even though similar in appearance; ▷ p. 14.

anapsid, any REPTILE of a group defined by the absence of openings in the temporal region of the skull (▷ pp. 82, 84), represented today only by the CHELONIANS (turtles and tortoises).

anaspid, ▷ p. 70.

anatomy, the study of the structure of bodies.

anchovy, any of about 200 fishes of the family Engraulidae, belonging to the same order as the HERRINGS. Like small herrings in appearance, anchovies occur in huge shoals in the surface waters of temperate and tropical seas around the world; they are important prey for larger fishes such as tuna, and many species are valuable commercially; ▷ p. 66.

anemone or **windflower,** any of around 80 species of perennial herb in the temperate genus *Anemone* in the buttercup family, Ranunculaceae. Many are cultivated ornamentals, but they are toxic due to the presence of the lactone, anemonin.

anemone fish, any of 26 DAMSELFISHES of the subfamily Amphiprioninae, found in coastal tropical waters of the Indo-Pacific. They live amongst the stinging tentacles of sea anemones, thereby gaining some protection from predators. They are themselves protected from the anemone's stings by the chemical nature of the mucus covering their bodies.

angelfish, any of 74 PERCIFORM fishes of the family Pomacanthidae, inhabiting tropical reefs and often brightly coloured; they are distinguished by a strong, sharp, backward-facing spine at the angle of the gill cover. The name 'angelfish' is sometimes applied to various other fish groups, both marine and freshwater; ⟹ p. 74.

angel shark, any of about 10 marine SHARKS of the family Squatinidae, occurring in the Atlantic and Pacific Oceans. They have large flattened pectoral fins, giving them an appearance intermediate between that of a shark and a ray. They are sometimes known as 'monkfishes'.

angiosperm, any seed-bearing plant in which the seed is enclosed in an ovary, i.e. any flowering plant; ⟹ pp. 38, **40–1**, 42–7. Compare GYMNO-SPERM.

angler fish, common name given to various fishes within the order Lophiiformes, which are found from deep to shallow water in tropical and temperate seas. These fishes typically have large heads with wide mouths, armed with many rows of sharp teeth, and a 'lure' formed from an elongated dorsal fin ray tipped with a fleshy bait, with which they attract small fishes towards their huge jaws. In deep-sea anglers the lure may be luminous. The shallow-water anglers of the family Lophiidae (also known as monkfishes or goosefishes) have become popular as food fishes in recent years; ⟹ pp. 75, 76 (box), *221*. See also FROGFISH, BATFISH, FOOTBALL FISH.

angwantibo or **golden potto** (*Arctocebus calabarensis*), a slow-moving, nocturnal lower PRIMATE (family Lorisidae) inhabiting rainforest in western Africa; ⟹ p. 137.

anhinga, ⟹ DARTER.

Animalia, the animal kingdom.

animals, ⟹ pp. 50–1.

anise (*Pimpinella anisum*), an eastern Mediterranean herb belonging to the carrot family, Apiaceae (formerly Umbelliferae). Oil of anise is extracted from the fruits (aniseed) and used medicinally and as a flavouring. A similar flavouring is obtained from FENNEL and from the unrelated star anise (*Illicium verum*).

aniseed, ⟹ ANISE.

ankylosaur, any of a group of heavily armoured quadrupedal ORNITHISCHIAN dinosaurs; ⟹ p. 83.

annelid (or 'segmented worm'), any invertebrate animal of the phylum Annelida, comprising over 9000 species in three classes and including polychaete worms, earthworms and leeches; ⟹ pp. **55**, 218.

anoa (*Bubalus depressicornis*), a small, wild species of CATTLE resembling a miniature water buffalo, with a black coat and white stockings. It is found only in mountain forests on Sulawesi.

ant, any HYMENOPTERAN insect of the family Formicidae, which comprises in excess of 5000 species worldwide. They are all social with a well-developed caste system of one or more worker castes. Each nest contains a single reproductive queen, who is tended by workers. The larvae are loose in the nest, not held in cells as in bees and wasps, and they are frequently moved around by workers; ⟹ pp. 63, 66 (tailor and honey ants);

⟹ pp. 153, 160, 163 (box). See also ARMY ANT, DRIVER ANT, FIRE ANT, LEAFCUTTER ANT.

Antarctic, 1. areas lying south of the Antarctic Circle (66° 33′ S). The area receives 24 hours daylight in midsummer and 24 hours night in midwinter. It contains the continent of Antarctica. **2.** the Antarctic floral region; ⟹ p. 30 (map).

ant bear, another name for AARDVARK.

antbird, any of 244 PASSERINE birds inhabiting forests and scrubland in Central and South America, belonging to the families Thamnophilidae and Formicariidae (the ground antbirds). Most species are drab-coloured, but the males of some have bright patches on their heads. The upper bill is hooked over the front of the lower. Their name comes from the habit seen in many species of feeding on insects stirred up by marauding army ants.

anteater, any edentate mammal of the family Myrmecophagidae, comprising four species that range from southern Mexico to northern Argentina. All species are specialized ant- and termite-feeders; the smaller species – the lesser anteaters or tamanduas – are arboreal forest-dwellers; ⟹ pp. **106–7**, 164, 203. See also PANGOLIN and ECHIDNA.

antechinus, any of about 10 small DASYURID marsupials of the genus *Antechinus*, widespread in Australia and New Guinea. They feed on insects and other invertebrates.

antelope, common name given to many BOVID mammals, typically graceful, long-legged grazers and browsers in the subfamilies Hippotraginae, which includes oryxes, kobs, waterbuck and wildebeest, and Antilopinae, which includes smaller antelopes such as impala and gazelles. The spiral-horned antelopes, such as eland and bongo, belong in the subfamily Bovinae, which also includes CATTLE; ⟹ pp. 114, **115**, 203, *206*. See also PRONGHORN.

antenna, one of a pair of appendages projecting from the head of insects, crustaceans, etc. They are generally highly mobile and have a sensory function, but they are sometimes modified for other uses, such as swimming and defence; ⟹ pp. 59, 66, 67.

anterior, at the front end (of a body or structure).

anther, the pollen-producing part at the top of the stamen in a flower; ⟹ p. 41.

antheridium (pl. -dia), the organ producing male gametes (sex cells) in spore-bearing plants (⟹ p. 36) and fungi; commonly a spherical structure on a short stalk. Compare ARCHEGONIUM.

Anthocerotae, the bryophyte class comprising the hornworts; ⟹ p. 37.

anthozoan, any marine COELENTERATE of the class Anthozoa; they may be solitary (e.g. sea anemones) or colonial (e.g. corals). In all anthozoans the polyp phase is dominant throughout the life cycle, and many secrete a calcareous skeleton; ⟹ p. 53.

anthracosaur, any member of a group of extinct tetrapods (four-limbed vertebrates) that flourished 350–270 million years ago and are believed to be closely related to early reptiles and hence to all higher AMNIOTES; ⟹ p. 78.

anthropoid, a higher primate (monkeys, apes, man); ⟹ p. 136.

anthropomorphism, the tendency to attribute human emotions or other characteristics to animals, usually on the basis of non-human behaviour that superficially resembles some human behaviour pattern. See also EMOTION.

antibiotic, any drug, such as penicillin, that is effective against bacterial infection; ⟹ p. 227.

antibody, any defensive protein (immunoglobulin) that is produced in an organism in response to the presence of a foreign chemical (antigen) and that counteracts its effects.

antigen, ⟹ ANTIBODY.

antirrhinum, ⟹ SNAPDRAGON.

antler, one of two bony structures that grow from the head of male deer (and female reindeer) and are shed annually. The name is sometimes applied to similar-looking structures in other animals, e.g. stag beetles; ⟹ pp. 116–17, *152*.

antlion, any NEUROPTERAN insect of the family Myrmeleontidae. Adults are similar to slow-flying DRAGONFLIES but have antennae. The larvae, also known as doodlebugs, dig conical pits in which to trap their prey (usually ants); ⟹ p. 164.

anuran, any AMPHIBIAN of the order Anura, comprising about 24 families and approximately 3500 species of FROG and TOAD (the 'tailless' amphibians). Although anurans are structurally rather uniform, they exhibit a great diversity of habit, and terrestrial, arboreal, burrowing and totally aquatic forms are known. Most anurans have a free-swimming tadpole stage, but many, including the MIDWIFE TOADS and DARWIN'S FROG, show some form of parental care (⟹ p. 81); ⟹ pp. **78–81**. See also GLASS FROG, MARSUPIAL FROG, POISON-DART FROG, SPADEFOOT TOAD and TREE FROG.

anus, the opening at the end of the digestive tract through which waste products are discharged; ⟹ p. 179.

aorta, the great artery that carries oxygenated blood from the left ventricle of the heart to be distributed throughout the body via branching arteries.

ape, any higher PRIMATE of the superfamily Hominoidea, an exclusively Old World group (except for man) that is conventionally divided into the lesser apes (9 species of gibbon), the 4 species of great ape (orang-utan, gorilla and chimpanzees), and man; ⟹ pp. 14, 136, **140–1**.

apex, the tip of a structure such as a root or shoot.

aphid, any HOMOPTERAN insect (bug) of the family Aphididae. This worldwide group of over 3600 species is of major economic importance both for the direct damage they cause to crops and for the viral diseases they transmit. The life cycle is often complex, with periods of sexual reproduction alternating with asexual (parthenogenic) development, so that populations can increase dramatically in a short period of time. Several different plant species may be used during a year. Winged adults are dispersed in wind currents; ⟹ pp. 62, 63, 65.

Apiaceae, a dicotyledonous plant family (formerly Umbelliferae) in which the flowers are arranged in UMBELS. Members include carrot, parsley, hogweeds, celery, hemlock, etc.; ⟹ pp. 45, 46.

appeasement, behaviour that inhibits or reduces AGGRESSION between members of the same species. Appeasement postures are often the opposite or antithesis of aggressive ones (⟹ box, p. 154).

appendix, ⟹ CAECUM.

apple, any of around 35 species of shrubs and trees and their derivatives in the north temperate genus *Malus* (rose family, Rosaceae). The many varieties of cultivated apple are derived principally from *M. sylvestris* (⟹ CRAB APPLE) and *M. pumila*. Varieties have been bred as apples for dessert, cooking or cider making or as attractive ornamentals.

apricot (*Prunus armeniaca*), a tree of the rose family, Rosaceae, cultivated for its peach-like fruit. It is grown principally in the USA, France, Spain, Hungary and Turkey. It is also grown for its spring blossom.

apterygote, any INSECT of the group Apterygota, the 'wingless insects', which comprises four orders: thysanurans (bristletails), diplurans, collembolans (springtails) and proturans; ⟹ pp. 60–1.

aqueous humour, in the mammalian eye, ⟹ p. 189.

aquilegia, ⟹ COLUMBINE.

arachnid, any ARTHROPOD of the class Arachnida, which comprises over 60 000 species and includes spiders, scorpions, and ticks and mites; ⇨ pp. 68–9.

arapaima, a BONY TONGUE; ⇨ p. 77.

Araucaria, a southern hemisphere genus of around 14 species of conifers with hard, sharp, overlapping scale-like leaves. The Chile pine or monkey puzzle (*A. araucana*) is hardy and is planted as an ornamental in northern temperate regions (⇨ p. 39). Several species, including the monkey puzzle and Paraná pine (*A. angustifolia*) are used as a source of timber.

Araucariaceae, a family of conifers including KAURIS (genus *Agathis*) and monkey puzzles (genus *ARAUCARIA*); ⇨ p. 39.

arbor-vitae or **western red cedar** (*Thuja plicata*), a coniferous tree native to North America. It may grow to 60 m (200 ft) in height and is planted for timber and as a hedge plant.

arbutus, any of around 14 species of the genus *Arbutus*, belonging to the heather family, Ericaceae. They are shrubs or small trees found in North and Central America and Southern Europe, and have edible red fruits from which the alternative name 'strawberry tree' derives.

archaebacteria, a group of bacteria of very ancient origins that some scientists distinguish from other bacteria on the basis of certain distinct properties; ⇨ pp. 24, 25.

Archaeopteryx, the earliest known bird; ⇨ pp. 5, 90 (box), *91*.

archegonium (pl. -onia), the female sex organ of spore-bearing plants (⇨ p. 36) and some gymnosperms (⇨ p. 38), commonly composed of a neck-like region leading to a swollen base containing the female gamete (sex cell).

archerfish, any of five PERCIFORM fishes of the family Toxotidae, occurring in marine and fresh waters in the tropical Indo-Pacific region. Their name derives from their ability to knock insects off overhanging vegetation by expelling a jet of water from the mouth; ⇨ pp. 76, 151.

archosaur, any of a group of DIAPSID reptiles (Archosauria or 'ruling reptiles') that includes crocodilians, dinosaurs and pterosaurs; ⇨ p. *84*.

Arctic, areas lying north of the Arctic Circle (66° 33´ N). The area receives 24 hours of sunlight in midsummer and 24 hours darkness in midwinter. The Arctic Ocean is frozen in winter.

arctic/alpine distribution, the occurrence of the same species at high latitudes and high altitudes; ⇨ pp. 10, 210.

arctic fox, white fox or **blue fox** (*Alopex lagopus*), a CANID inhabiting tundra areas of northern Eurasia, North America, Greenland and Iceland; ⇨ p. 210. It has thick fur, which is white or bluish in winter (⇨ p. 174).

arctic heath, tundra vegetation dominated by dwarf evergreen shrubs of the Ericaceae (heather family).

arctic scrub, tundra vegetation dominated by dwarf deciduous shrubs, such as dwarf willows and birches.

Arecaceae, a monocotyledonous plant family (formerly Palmae), including palms; ⇨ p. 45.

Arecidae, a monocotyledonous plant subclass; ⇨ p. 45.

argentine or **herring smelt,** any of about 18 species of the family Argentinidae, related to the SALMONIDS. They are midwater fishes of continental slopes, feeding on fishes and crustaceans.

argonaut or **paper nautilus,** any CEPHALOPOD mollusc of the genus *Argonauta*, of tropical and subtropical seas. It has a papery external shell.

aril, a fleshy outgrowth on a seed derived from a region of the ovary, the funicle. Nutmeg is an aril.

Aristotle (384–322 BC), Greek philosopher. In addition to his work on logic, philosophy, ethics, literature and politics, he also shaped biological thinking in his *History of Animals*, *Parts of Animals*, and *Generation of Animals*. A great classifier, he dissected numerous animals to see how they were related; ⇨ pp. 6, 18.

armadillo, any edentate mammal of the family Dasypodidae, comprising 20 species that occur throughout much of South and Central America and into southern USA. Distinguished by their body armour of bony plates, armadillos are omnivorous and versatile, adapting to a wide range of habitats. Some of the smaller species are known as pichiciegos; ⇨ pp. *9*, 106, **107**.

army ant, legionary ant or **driver ant,** any predatory tropical ANT of the subfamily Dorylinae. They move through the forest without establishing a permanent nest site, although temporary nests or bivouacs are made by workers interlocking their bodies; ⇨ p. 163 (box).

army worm (*Leucania unipuncta*), a NOCTUID moth whose larvae appear in large numbers and cause considerable damage to a variety of crops.

arnica, any of around 30 species of perennial herbs in the north temperate and arctic genus *Arnica* in the daisy family, Asteraceae (formerly Compositae). Arnicine is obtained from this genus for the treatment of bruises.

arrow-poison frog, alternative name for POISON-DART FROG.

arrowroot (*Maranta arundinacea*), a tall perennial reed-like plant of tropical America. Edible starch is extracted from the rhizomes, and crushed rhizomes were formerly used to treat arrow wounds.

Artemisia, a genus of around 400 species of mostly herbaceous perennials in the daisy family, Asteraceae (formerly Compositae), native to Europe, Asia and America. Common sagebrush (*A. tridentata*) is the dominant plant in some parts of the southwestern USA (⇨ p. 205). Many are aromatic and several are used for culinary and medicinal purposes. Tarragon (*A. dracunculus*) is used in salads. Mugwort (*A. vulgaris*), wormwood (*A. absinthium*) and tarragon are used to flavour vermouth.

artery, a blood vessel by which oxygenated blood is carried from the heart; ⇨ 181.

arthropod, any member of the phylum Arthropoda, the largest and most important invertebrate group. The main subgroups are the crustaceans, insects, centipedes and millipedes, and arachnids (spiders, scorpions, ticks and mites), but it also includes a number of minor groups, such as velvet worms, horseshoe crabs, SEA SPIDERS and WATER BEARS; ⇨ pp. 58–69.

artichoke, 1. globe artichoke (*Cynara scolymus*), a tall perennial herb with large grey leaves and flowers resembling a thistle. The flower buds are edible. **2.** Jerusalem artichoke (*Helianthus tuberosus*), a tall perennial herb with daisy-like flowers and edible tubers.

artificial selection, the basis of plant and animal breeding where humans act as the selective agent and choose what they consider to be the fittest members of a species to interbreed. In agriculture, artificial selection became widespread during the 18th century, while in the later 20th century genetic engineering revolutionized what is possible.

artiodactyl, any hoofed mammal (UNGULATE) of the order Artiodactyla (the even-toed ungulates), comprising some 200 species in 10 families: pigs (Suidae, 8 species); peccaries (Tayassuidae, 3 species); hippopotamuses (Hippopotamidae, 2 species); camels, llamas, etc. (Camelidae, 6 species); chevrotains (Tragulidae, 4 species); musk deer (Moschidae, 3 species); deer (Cervidae, about 38 species);

the giraffe and okapi (Giraffidae, 2 species); cattle, antelope, sheep, goats, etc. (Bovidae, about 128 species); and the pronghorn (Antilocapridae, 1 species). All artiodactyls are primarily or exclusively herbivorous, and with the exception of pigs, peccaries and hippopotamuses, all are RUMINANTS; ⇨ pp. 110 (box), **112** (box), 112–17.

arum, any of around 15 species of perennial herbs in the European and North African genus *Arum* in the family Araceae. The inflorescence is surrounded by a sheath or spathe. Cuckoopint or lords-and-ladies (*A. maculatum*) is common in woods and on hedgebanks. Members of other genera in the Araceae – such as *water arum* and *arum lily* – may also be referred to as arums.

arum lily (*Zantedeschia aethiopica*), a cultivated ornamental plant in the arum family, Araceae. It has a showy white spathe or sheath surrounding the inflorescence.

aschelminth, any invertebrate of the phylum Aschelminthes, a large group of mainly rather obscure animals, including ROTIFERS and SPINY-HEADED WORMS. Aschelminths have unsegmented, bilaterally symmetrical bodies and a complete digestive tract, but no circulatory or respiratory system; most have adhesive glands.

ascomycete, any member of the class Ascomycetes, which comprises fungi with sexual spores produced in an ascus; (⇨ p. 35).

Ascomycotina, a subdivision of the fungal division Amastigomycota; ⇨ p. 35.

ascus (pl. asci), the structure in the ascomycete fungi producing the sexual spores (ascospores; ⇨ p. 35). In higher forms they are located on fruiting bodies, the ascocarps.

aseptate, ⇨ HYPHA.

asexual reproduction, a means of REPRODUCTION that does not involve the fusion of male and female sex cells (gametes) (⇨ p. 184). Plants may reproduce asexually by VEGETATIVE means, unicellular organisms by FISSION, and many fungi and simple animals, such as sea anemones and corals, by BUDDING. See also PARTHENOGENESIS.

ash, 1. any of around 70 species of trees usually with pinnate leaves in the mainly north temperate genus *Fraxinus* in the olive family, Oleaceae. A hard, durable timber is obtained from several species, e.g. the European ash (*F. Excelsior*) and white ash (*F. americana*). **2.** any of a number of unrelated trees superficially resembling ash, such as mountain ash – a name applied to both rowan and the giant gum (*SORBUS aucuparia*), dogberry (*S. americana*), and *EUCALYPTUS regnans*).

Asian wild dog, ⇨ DHOLE.

asparagus, any of around 300 species of perennial herbs, shrubs and climbers in the genus *Asparagus*. Garden asparagus (*A. officinalis*) is grown for its protein-rich shoots, and asparagus fern (*A. plumosus*) as an ornamental.

aspen, the name applied to five of the 35 species of POPLAR (e.g. European Aspen, *Populus tremula*). They are characterized by smooth pale bark and flattened leaf stalks, which enable the leaves to tremble in light breezes; ⇨ p. 208.

asphodel, 1. any of around 12 species of herbs in the genus *Asphodelus* (family Liliaceae) found from the Mediterranean to the Himalaya. **2.** bog asphodel, any of around 6 species of herbs in the north temperate genus *Narthecium* (family Liliaceae); ⇨ p. 219.

aspidistra (*Aspidistra lurida*), a herb in the lily family, Liliaceae. It has dark green oblong leaves and was popular as a house plant in the 19th century.

ass, either of two PERISSODACTYL mammals of the family Equidae – the African wild ass (*Equus africanus*), the ancestor of the domesticated donkey, and the Asiatic wild ass (*E. hemionus*). There are a number of distinct varieties or subspecies of the latter, including the onager, which is reddish-brown with a prominent dorsal stripe, and inhabits desert and dry steppe from the Middle East to China and western India; and the kiang, the largest subspecies, which is grey-brown with a dark dorsal stripe and inhabits arid, mountainous areas of Tibet and the surrounding region; ▷ p. 111.

assassin bug, any HETEROPTERAN insect (bug) of the large family Reduviidae, all of which are predatory or blood-feeding. Most prey on other arthropods, but some tropical species (e.g. those of the genera *Rhodnius* and *Triatoma*) feed on human blood and may transmit the organism responsible for Chaga's disease.

association, ▷ PLANT ASSOCIATION.

associative or **association learning** ▷ CONDITIONING.

aster, any of around 500 species of the genus *Aster* in the daisy family, Asteraceae (formerly Compositae). They are perennial herbaceous plants of temperate regions found in a range of habitats from salt marsh to alpine meadow. About 12 species are cultivated, including the Michaelmas daisies.

Asteraceae, a dicotyledonous plant family (formerly Compositae) in which the flowers are arranged in multiple flower heads. Members include daisies, chrysanthemum, thistles, lettuce, etc.; ▷ pp. 45, 46.

Asteridae, a dicotyledonous plant subclass; ▷ p. 45.

asthenosphere, ▷ MANTLE and p. 8.

atlas moth, any SATURNIID moth of the genus *Attacus*, mainly found in the tropics. They are among the largest of moths, with wing spans of up to 25 cm (10 in); ▷ p. *188*.

atmosphere, the envelope of gases that surrounds the earth, consisting of 78% nitrogen, 21% oxygen, less than 1% argon, 0.03% carbon dioxide, and comparatively minute quantities of neon, helium, methane, krypton and hydrogen. The amount of water vapour varies with climatic and weather conditions. The early atmosphere of the earth was very different, initially containing little or no oxygen (▷ pp. 6–7). Industrial processes are currently contributing to changes in the atmosphere (▷ pp. 224–5).

atoll, a round or horseshoe-shaped coral reef, surrounding a lagoon; ▷ p. 215.

ATP, adenosine triphosphate; ▷ p. 21.

atrium (pl. atria), a chamber in a heart in which blood is collected from the veins and passed to a ventricle; ▷ pp. *181*.

aubergine or **eggplant** (*Solanum melongena*), a perennial herb in the potato family, Solanaceae cultivated in the tropics and subtropics for its fruits, which are used as a vegetable.

aubretia, aubrietia or **aubrieta,** any of around 15 species of perennial herbs in the Mediterranean genus *Aubrietia* in the cabbage family, Brassicaceae (formerly Cruciferae). *A. deltoides* is a popular ornamental.

auditory nerve, a group of nerve fibres by which sensory nerve impulses are conveyed from the inner ear to the brain; ▷ p. *191*.

auk, any of 22 stout sea birds with small stubby wings and powerful webbed feet, of the family Laridae (order Ciconiiformes). They live in the seas of the northern hemisphere. Auks are very powerful underwater swimmers; their flight is fast, low and straight. Most are colonial nesters on cliff ledges. The flightless great or giant auk (*Pinguinus impennis*) recently became extinct; ▷ pp. 92, 96. See also AUKLET, GUILLEMOT, PUFFIN, RAZORBILL.

auklet, any of six species of AUK with small bills. They occur around the shores of the north Pacific and tend to breed in burrows or rock crevices.

aurochs (*Bos primigenius*), an extinct species of CATTLE of northern Africa and Eurasia, believed to be the ancestor of most breeds of domestic cattle; ▷ pp. **114,** 212.

Australasia, a term that covers Australia and New Zealand, together with New Guinea and adjacent islands.

Australian, 1. a faunal region; ▷ p. 50 (map). **2.** a floral region; ▷ p. 30 (map).

australopithecine, any pre-human HOMINID primate of the genus *Australopithecus* ('southern ape'), which lived in eastern and southern Africa and died out about 1 million years ago; ▷ p. 5.

autonomic, relating to the 'involuntary' part of the nervous system, i.e. that part that is not under conscious control (e.g. heart and gland function); ▷ p. 155.

autotomy, a defensive ploy that involves shedding an expendable body part, as when a lizard casts off its tail; ▷ pp. **54, 89,** 167.

autotroph, producer or **primary producer,** any organism capable of synthesizing the complex compounds necessary for life out of simple inorganic molecules, in most cases by photosynthesis (▷ box, p. 33). Autotrophs include some bacteria (▷ p. 24–5) and protists (▷ pp. 26–7, 34), and most plants (▷ pp. 30–3). All other organisms are HETEROTROPHS (i.e. consumers); ▷ p. **194.**

auxin, a hormone involved in the growth of plant cell walls; ▷ p. 32.

avens, 1. mountain aven (*Dryas octopetala*), a creeping dwarf shrub of mountains in north temperate regions. Popular as a rock plant. **2.** wood aven (*Geum urbanum*) and water aven (*G. rivale*), perennial herbs of north temperate regions.

Avery, Oswald T. (1877–1955), American microbiologist, who first established the role of DNA in inheritance; ▷ p. 22.

Aves, the class comprising the birds.

avocado, the green to purple pear-shaped fruit of the Central American evergreen tree, *Persea americana*. The fruit has a single large seed surrounded by highly nutritious flesh.

avocet, any of four species of WADER, with long legs and long, very thin upturned bills, found worldwide on marshes, lakes and sea coasts. They are closely related to STILTS; ▷ pp. 96, 97 (box).

avoidance, behaviour that allows animals to escape actual or potential danger. It may be INNATE or learned by CONDITIONING.

axis, the main central part of a plant, i.e. the main stem and root, from which branches etc. develop outwards.

axolotl (*Ambystoma mexicanum*), a MOLE SALAMANDER from Lakes Xochimilco and Chalco near Mexico City, a well-known neotenic species, i.e. one that is capable of breeding while still in the larval state; ▷ pp. 78, 79, **81.**

axon, a single long extension from a nerve cell by which impulses are conducted from the cell body; ▷ p. 182 (box).

aye-aye (*Daubentonia madagascariensis*), a rare and highly distinctive nocturnal LEMUR of Madagascan rainforest; ▷ pp. 136, **137.**

ayu (*Plecoglossus altivelis*), a northern Pacific fish related to the SALMONIDS. Like the salmon, it moves into fresh water to breed. It is the most highly valued food fish in Japan.

azalea, a name, previously of generic rank, used by gardeners to describe some small deciduous and evergreen species of the genus RHODODENDRON.

Azolla, a genus of water ferns; ▷ pp. 37 (box), 197.

babirusa (*Babyrousa babyrussa*), a wild PIG with large tusks and nearly naked skin, inhabiting moist forest and riversides in Sulawesi and nearby islands in Southeast Asia.

baboon, any large, often brightly coloured Old World cercopithecine monkey of the genus *Papio* and related genera, such as the HAMADRYAS and CHACMA; ▷ **139,** 149, 151, 154 (box), 157 (box), 159. See also DRILL and MANDRILL.

bachelor herd, a group of males living together outside the breeding season, as in species that form temporary HAREMS; ▷ p. 158.

bacillus (pl. -li), any rod-shaped bacterium; also a generic name for spore-producing forms of bacteria, e.g. *Bacillus subtilis*; ▷ p. 24.

backswimmer, any HETEROPTERAN insect of the family Notonectidae. These insects are similar to WATERBOATMEN but swim on their backs. They are carnivorous, feeding on insects and tadpoles.

bacteria (sing. bacterium), a diverse group of prokaryote single-celled organisms; ▷ pp. 24–5.

bacteriophage or **phage,** a virus that only attacks bacteria; ▷ p. 24.

bacteroids, the modified forms of nitrogen-fixing bacteria inside a root nodule; ▷ p. 32 (box).

badger, any of several large, stockily built, burrowing MUSTELIDS, typically with black and white markings on the head. They are generally nocturnal and solitary, but the Eurasian badger (*Meles meles*) is social. The Southeast Asian ferret badgers (*Melogale* species) are small and ferret-like, while the hog badger (*Arctonyx collaris*), also of Southeast Asia, has a mobile pig-like snout; ▷ pp. **131,** *195,* 207. See also RATEL.

bagworm moth, any LEPIDOPTERAN insect of the family Psychidae, the larvae of which are plant-feeders and protected within a tubular silken bag coated with plant fragments.

baked bean, ▷ HARICOT BEAN.

balance, the sense whereby an animal is able to maintain a correct bodily orientation with respect to gravity; ▷ pp. 190–1.

baleen or **whalebone,** a horny, flexible material arranged in a series of thin sheets and hanging from the upper jaw of baleen whales, used in filtering plankton from the water; ▷ p. 145.

baleen whale, ▷ CETACEAN.

balm or **balsam,** any of various sweet-smelling oily resinous substances produced by a number of unrelated plants, e.g. lemon balm (*Melissa officinalis*) and bergamot (*Monarda didyma*).

balsa (*Ochroma pyramidale*), a rapidly growing lowland tropical American tree. The cells of the heartwood have very thin walls resulting in an extremely light timber.

balsam, ▷ BALM

Baltica, an ancient continent; ▷ pp. 8–9.

bamboo, any of around 1000 species of mainly tropical perennial woody grasses (family Peaceae) in about 70 genera. They grow in thick clumps and spread by branched underground stems. The stems are used for many purposes such as building, for making musical instruments, fishing rods and masts, and as garden canes; ▷ p. 46.

banana, any species of giant treelike herbs in the tropical genus *Musa*. The fruits are rich in starch and form a staple crop as well as being important for export; ▷ p. 47. See also PLANTAIN.

bandicoot, any of 15 MARSUPIALS of the family Peramelidae, widespread in Australia, Tasmania and New Guinea. They feed on fruits, seeds and invertebrates. See also RABBIT-EARED BANDICOOT.

banksia, any of around 50 species of evergreen trees and shrubs in the Australasian genus *Banksia* in the family Proteaceae. The flowers, in dense heads, are rich in nectar, hence the Australian name 'honeysuckle'.

banteng (*Bos javanicus*), a reddish-brown species of wild CATTLE with white stockings. It inhabits thickets and forests in parts of Indochina, Borneo,

Java and Bali, and has been domesticated as Bali cattle.

banyan (*Ficus benghalensis*), a large tree of the FIG genus with aerial roots that give the appearance of pillars; native to the Himalayan foothills but much planted in India.

baobab either of two tropical trees with swollen trunks and thick root-like branches: *Adansonia digitata*, native to Africa, and *A. gregorii* of Australia; ⊳ pp. **44**, 203. The woody fruit shells are used as vessels and glue is made from the pollen.

barb, any of the numerous projections on either side of the quill of a bird's feather that together make up the vane; ⊳ p. **92**, 93.

barbary ape, a tailless MACAQUE (not an ape) of Gibraltar and northern Africa; ⊳ p. 139.

barbel, 1. any of several slender sensory organs around the mouth of certain fishes (e.g. catfishes) used for locating food; ⊳ p. **188**. **2.** any of several fishes of the genus *Barbus* in the CARP family, so named because of the small tentacle-like organs (barbels) at the corners of the mouth.

barberry, any of around 450 species of the genus *Berberis* in the family Berberidaceae. They are spiny deciduous or evergreen shrubs mainly of the northern hemisphere. Many are grown as ornamentals.

barbet, any of 82 species of small, usually colourful tropical birds belonging to several families related to the WOODPECKERS. They have long sensory bristles surrounding the strong bill, which in most species is used to eat fruit; ⊳ p. 91.

barbicel, ⊳ p. 92.

barbule, ⊳ p. 92.

bark, tissues in woody stems and roots external to the vascular CAMBIUM (including secondary PHLOEM), or its corky part only. It provides a protective layer for trees and shrubs.

bark beetle, any BEETLE of the families Scolytidae and Platypodidae, closely related to the WEEVILS. They are cylindrical in shape and bore into the bark of trees where it joins the wood. They may be highly damaging to trees, either directly or by transmitting diseases (e.g. Dutch elm disease). See also AMBROSIA BEETLE.

barley (*Hordeum vulgare*), a grass of temperate regions cultivated as a cereal, chiefly for animal feed. Much of the balance is used for malt for beer and spirits. Like wheat it is one of the oldest cultivated plants and is derived from wild races of *H. spontaneum* in southwest Asia.

barnacle, any marine CRUSTACEAN of the class Cirripedia. Most live as sedentary filter-feeders, but some are parasitic on crabs, starfishes and jellyfishes; ⊳ pp. 14 (box), **66**, 221.

barracuda, any of 18 marine PERCIFORM fishes of the family Sphyraenidae, found in all tropical and warm seas. They have elongated bodies and jaws, and the latter are studded with sharp teeth. They are large, active predators (up to 1.8 m/6 ft), and have been known to attack humans; ⊳ p. 75 (box).

barrier reef, a type of coral reef; ⊳ p. 215.

basic, the opposite of ACIDIC. Low in hydrogen ions, therefore reacting with acids. Measured on a scale of pH, where values above 7 (neutrality) are increasingly basic. Often refers to rocks and soils rich in calcium or magnesium.

basidiomycete, any member of the class Basidiomycetes, which comprises fungi (including mushrooms and toadstools) with sexual spores borne on a basidium; ⊳ p. **35**.

Basidiomycotina, a subdivision of the fungal division Amastigomycota; ⊳ p. 35.

basil, 1. sweet basil (*Ocimum basilicum*), an aromatic annual herb used to flavour food. It is in the family Lamiaceae (formerly Labiatae) and is native to western tropical Asia. **2.** some other species in the genus *Ocimum*.

basilar membrane, a membrane in the vertebrate inner ear; ⊳ p. *191*.

basilisk, any small American IGUANID lizard of the genus *Basiliscus*. Long-limbed with a whip-like tail, these lightly built lizards are mainly tree-dwelling, but they can run rapidly on their hind legs when startled, even over water.

basket star, an alternative name for BRITTLESTAR; ⊳ p. 54.

basking shark (*Cetorhinus maximus*), an extremely large marine SHARK, growing up to 14 m (46 ft) in length. It lives at or near the surface in all seas outside the tropics, sieving plankton by means of its gill rakers; ⊳ p. 76.

bass, any of 50 or so PERCIFORM fishes of the family Percichthyidae, marine, brackish and freshwater fishes from tropical and temperate waters worldwide. They are closely related to the SEA BASSES.

bat, any mammal of the order Chiroptera, one of the largest groups of mammals and the only one to have achieved true flight. The order is divided into two unequal suborders: the Megachiroptera, containing a single family with about 170 species of FRUIT BAT or flying fox; and the Microchiroptera, containing all other bats – around 800 species in 18 families – most of which are capable of ECHOLOCATION; ⊳ pp. 14, 43, **118–19**. For individual bat families, see under DISC-WINGED, FALSE VAMPIRE, FREE-TAILED, FUNNEL-EARED, GHOST-FACED, HORSESHOE, LEAF-NOSED, MOUSTACHED, SHEATH-TAILED, SHORT-TAILED, SLIT-FACED, SUCKER-FOOTED, VAMPIRE and VESPERTILIONID.

Batesian mimicry, ⊳ MIMICRY.

batfish, any member of the family Ogcocephalidae, related to the ANGLER FISHES. Like the FROGFISHES, they rest on the bottom and use a lure to attract prey. They have pointed snouts and flattened bodies, and they move over the sea bed on pectoral fins on limb-like extensions.

bathyal zone, a marine zone; ⊳ pp. 220–1.

bay laurel or **sweet bay** (*Laurus nobilis*), an evergreen shrub or small tree native of the Mediterranean region. It was used in classical times for making ceremonial wreaths. Its leaves are used for flavouring.

beach flea, a SAND HOPPER.

beachgrass, ⊳ MARRAM GRASS.

beaked whale, any marine CETACEAN mammal of the family Ziphiidae (18 species, including the bottlenose whales), medium-sized toothed whales found in all oceans; ⊳ pp. 142, 143 (box), 144, **145**.

beak or **bill,** the projecting jaws of a bird, covered in hardened keratin, or a similar-looking structure in other animals, such as fishes and turtles; ⊳ pp. **93**, 96.

bean, any of various leguminous plants of the family Fabaceae (formerly part of the Leguminosae); ⊳ box, p. **32**. The seeds are generally kidney-shaped, are rich in protein, and are used for human and animal consumption. See also BROAD BEAN, FRENCH BEAN, HARICOT BEAN, KIDNEY BEAN, LIMA BEAN, MUNG BEAN and SOYA.

bean sprout, ⊳ MUNG BEAN.

bear, any mammal of the family Ursidae, which contains eight species, including the giant panda and the largest members of the order Carnivora (the grizzly and polar bears). Bears are widely distributed, mainly in the northern hemisphere, but are absent from Africa; ⊳ pp. *9*, 122, **128–9**, 207, 209, *229*, 233.

beaver, either of two large, social, semi-aquatic sciuromorph RODENTS of the family Castoridae – the Eurasian beaver (*Castor fiber*) and the American beaver (*C. canadensis*); ⊳ pp. **134**, *161*. See also MOUNTAIN BEAVER.

bedbug, any HETEROPTERAN insect (bug) of the blood-feeding family Cimicidae, especially *Cimex lectularis*, which is widespread and feeds on man. Bedbugs emerge at night and feed on people while they sleep; ⊳ pp. 62, 64.

bedstraw, any of a number of annual or perennial herbs with whorled simple leaves in the widely distributed genus *Galium* in the family Rubiaceae, e.g. hedge bedstraw (*G. mollugo*).

bee, any HYMENOPTERAN insect of the superfamily Apoidea, characterized by having many branched (plumose) hairs on the body. The group contains both solitary and social forms, but all feed on nectar and pollen. Nectar is chemically altered and concentrated to form honey, which, together with pollen, is fed to the larvae. The best known of the social forms are the HONEYBEE and BUMBLEBEE. Both solitary and social bees make cells, which are used to house the larvae and store food. In some solitary species the cells are made of plant material (e.g. leafcutter bee), but the social species use wax for this purpose; ⊳ pp. *61*, 63, 64, 149 (box), 153, 161. See also CARPENTER BEE and CUCKOO BEE.

beech, 1. any of around 10 species of deciduous trees in the north temperate genus *Fagus* in the family Fagaceae. The bark is smooth and grey and the spreading canopy gives a dense shade. The hard timber is used for furniture, but soon decays when exposed to the weather. **2.** southern beech, any species of the related southern hemisphere genus *Nothofagus*, which includes both deciduous and evergreen trees.

bee dance, ⊳ DANCE LANGUAGE and p. 153.

bee-eater, any of 26 species of slender birds with long tails and pointed, downwardly curved bills, belonging to the family Meropidae (order Coraciiformes) and occurring in the Old World and Australia. The plumage is usually brilliant blues and greens contrasted with reds and yellows. They catch insects on the wing – often bees – and squeeze out the sting by rubbing the bee against a branch; ⊳ pp. 91, *152*.

bee fly, any fast-flying DIPTERAN insect of the family Bombyliidae. They are very hairy, resembling bumblebees. The larvae are parasitic on bees and wasps.

beeswax, ⊳ WAX.

beet, any of around 6 species of herbs in the European and Asian genus *Beta* in the family Chenopodiaceae. Several cultivars of *Beta vulgaris* are of economic importance. The root of sugar beet provides nearly half the world production of sugar. The mangelwurzel or mangold is used for animal feed, and beetroot for human consumption. See also SPINACH and CHARD.

beetle, any ENDOPTERYGOTE insect belonging to the order Coleoptera, in terms of numbers of species, the largest order in the animal and plant kingdoms. The order contains some of the largest insects (GOLIATH and HERCULES BEETLES, over 15 cm/6 in long), as well as some of the smallest (ptiliid beetles less than 0.5 mm/1–50 in long). Despite their very varied habits, adult beetles are quite similar in structure. The tough, compact and obviously successful body plan features mandibulate (biting) mouthparts and hardened fore wings that protect the functional hind wings and the abdomen. The larvae are more varied in body form, and usually live a different life style to that of the adults. Terrestrial or aquatic, some are fast-moving and predatory; others caterpillar-like, feeding on plant material; others still virtually legless, feeding inside timber or seeds; ⊳ pp. 60–1, *63*, 64, 65, *162*, *195*. For individual named beetle groups, see under BARK, BOMBARDIER, CARDINAL, CARRION, CHICK, DEATHWATCH, DERMESTID, DIVING, FIREFLY, GROUND,

LADYBIRD, LEAF, LONG-HORNED, OIL, ROVE, SCARA-
BAEOID, TENEBRIONID, WEEVIL, WHIRLIGIG, WOOD-
WORM.

beetroot, ▷ BEET.

begonia, any of around 350 species of tropical or
subtropical herbs in the genus *Begonia* in the
family Begoniaceae. They have asymmetrical
leaves and are often succulent in habit. Many are
popular indoor and summer bedding plants.

behaviour, in the context of animal studies, 'behaviour' embraces everything that animals do, both
actively and passively (e.g. sleeping). The study of
animal behaviour so defined is the field of the
science of ethology; ▷ pp. 148–9.

belemnite, any extinct CEPHALOPOD mollusc of the
order Belemnoidea. Common in the Mesozoic era,
they had an internal chambered shell, which forms
the bullet-shaped fossils seen in Cretaceous rocks
today. Belemnites were probably rather like
modern squid in appearance and way of life; ▷ p.
57.

belladonna, ▷ DEADLY NIGHTSHADE.

bellbird, 1. any of three species of COTINGA of the
genus *Procnias*. **2.** the Australian bellbird (*Oreoica
guttaralis*), a PASSERINE of the subfamily Pachy-
cephalinae, in the CROW family.

bellflower, any of around 300 species of usually
perennial herbs in the genus *Campanula* in the
family Campanulaceae. They are native to north
temperate regions and mountains in the tropics.
The harebell (the 'bluebell' in Scotland; *C. rotundi-
folia*) is common in dry grassy places.

beluga, 1. (*Delphinapterus leucas*), also called white
whale, a medium-sized (up to 6 m/20 ft) toothed
whale (CETACEAN) of northern waters (family
Monodontidae); ▷ p. 145. **2.** a large STURGEON of
the Black and Caspian Seas.

bent grass, any of around 200 species of the mostly
temperate genus *Agrostis*. Many of these grasses
are important in agriculture, and varieties of *A.
canina* are used for lawns.

benthic environment, the zone at the bottom of
aquatic ecosystems. Benthic animals are bottom-
dwelling; ▷ pp. 218–21. Compare NEKTON,
PLANKTON.

berberis, ▷ BARBERRY.

bergamot, 1. *Monarda didyma*, a North American
perennial aromatic herb, the flowers and leaves of
which are used in flavouring. **2.** Oil of bergamot, an
essential oil extracted from the peel of a subspecies
of the Seville orange (*Citrus aurantium bergamia*).

Bermuda grass (*Cynodon dactylon*), a grass with a
wide distribution, used in lawns, and for pasture
and binding sand dunes.

berry, any fruit that contains many small seeds; ▷
p. 43.

betel nut palm (*Areca catechu*), a palm widely
cultivated in the tropics for its nut, which is
chewed in the leaf of the betel PEPPER as a mild
stimulant.

bettong, a short-nosed Australian RAT KANGAROO.

Betula, ▷ BIRCH.

bichir, any primitive African freshwater RAY-FINNED
FISH of the genus *Polypterus*, closely related to the
REEDFISH. It has armour-like scales and a row of
small 'finlets' along the back, and can survive in
oxygen-poor water by breathing through a lung;
▷ pp. 70–1.

bighorn sheep (*Ovis canadensis*), a stocky, pale to
dark-brown wild SHEEP, with massive, curved
horns. It mostly inhabits alpine areas in North
America; ▷ pp. *157*, 210.

bilberry, blaeberry, BLUEBERRY or **whortleberry**
(*Vaccinium myrtillis*), a low-growing shrub of the
heather family, Ericaceae. It is common on poor
soils, particularly in upland areas, in Europe and

Asia. The berries are collected to eat fresh or for
cooking.; ▷ pp. 208, 209 (box), *210*, 211.

bilby, another name for RABBIT-EARED BANDICOOT.

bile, a fluid secreted by the liver, stored in the gall
bladder, and released into the duodenum to help
break up (emulsify) fats; ▷ p. 179.

bilipid layer, a double layer of lipids, as found in cell
membranes (▷ p. *20*) and surrounding viruses
(▷ p. 24).

bill, ▷ BEAK.

billfish, any large marine PERCIFORM fish of the
family Istiophoridae, including the sailfishes
(genus *Istiophorus*, 2 species), which have very
high, sail-shaped dorsal fins (▷ p. *73*); the spear-
fishes (*Tetrapturus*, 6 species); and the marlins
(*Makaira*, 2 species). In all the snout is extended
into a long, rounded spike. They are powerful
swimmers, and very popular as game fishes. The
black marlin ranges throughout the warmer waters
of the Indian and Pacific Oceans, while the blue
marlin is worldwide in tropical and temperate seas;
they feed on schooling fishes and squid.

binary fission, ▷ FISSION.

bindweed, the name of several twining weeds, e.g.
field bindweed (*Convolvulus arvensis*) and hedge
bindweed (*Calystegia sepium*), both in the family
Convolvulaceae, and black bindweed (*Fallopia
convolvulus = Polygonum convolvulus*) in the
family Polygonaceae.

binocular vision, vision that involves the use of
both eyes in the same visual field, thus allowing
accurate judgement of distance and good perspective.

binturong (*Arctictis binturong*), a large, civet-like
VIVERRID carnivore with a shaggy black coat, found
in the forests of Southeast Asia; ▷ p. 123.

bio-, prefix indicating life or living organisms.

biochemistry, the study of the chemical processes
within living organisms.

biogeochemical cycle, any of various cycles by
which elements essential to life circulate between
living things and the non-living environment; ▷
pp. 196–7.

biogeographic regions, regions of the world that
are reasonably discrete and self-contained in terms
of the taxonomic groups of plants and animals they
contain; ▷ maps, pp. 30 (plants), 50 (animals).

biogeography, the study of animal and plant distributions and the causes that underlie these patterns (▷ pp. 8–11). Usually applied to studies on
a large scale, while ECOLOGY covers smaller-scale
patterns, but the two areas overlap.

biological clock, an in-built periodicity, or sense of
rhythm, in the physiological processes of a living
organism that is not dependent on external cues;
▷ p. 174.

biological control, the use of biological agents to
control the population levels of a pest organism;
▷ p. 231.

biology, the study of living organisms; there are
many branches, including anatomy, biochemistry,
botany, cytology, ecology, embryology, ethnology,
genetics, histology, microbiology, molecular biology, palaeontology, physiology, taxonomy, zoology.

bioluminescence, the production of light by living
organisms. It is common in deep-sea fishes (▷
box, p. 221) and also occurs in some insects (e.g.
fireflies; ▷ p. 62).

biomass, the quantity of organic material (plants
and animals) present in an ecosystem at any given
time, measured in kilograms per square metre; ▷
p. 194. For the biomass of individual terrestrial
ecosystems ▷ pp. 200–11. See also PRIMARY PRO-
DUCTIVITY.

biome, a vegetation type with its associated animal

life that is defined by life forms, e.g. tall trees, dwarf
shrubs, grasses, etc. Major biomes include oceans,
tropical rainforest, savannah, prairie, tundra,
desert, etc.

biosphere, the 'layer' around the earth where all life
forms exist; ▷ pp. 194–7.

biosynthesis, the synthesis of complex molecules
out of simpler materials by living organisms.

biota, the total living organism component of a
community, including plants, animals, fungi and
microbes.

biotechnology, the technical application of the
capacities and products of living cells to provide
goods and services. Long-established biotechnology techniques include the use of yeast in brewing
and bread making. More recently, scientists have
begun to use genetic-engineering techniques; ▷
pp. 23, 25, 227 (box), 231 (box).

biotic, related to living organisms, as in biotic
factor, ecological factors caused by living organisms (e.g. competition), and biotic component, the
living material within an ecosystem.

bipedal, moving on two limbs.

birch, any of around 60 species of the north temperate and arctic genus *Betula* in the family Betulaceae. Many species such as the silver birch (*B.
pendula*) flourish on acid heaths, which they
rapidly colonize by virtue of their prodigious production of wind-borne seeds. Their silver bark and
slender branches make them attractive ornamental
trees, and their branches are also useful to the
gardener to make besum brushes; ▷ pp. 43, 210,
211.

bird, any vertebrate of the class Aves, comprising
over 9500 species; ▷ pp. *15* (embryology), **90–7**,
178 (digestion).

bird-eating spider, ▷ TARANTULA.

bird louse, ▷ BITING LOUSE.

bird of paradise, any of 42 PASSERINE birds of the
tribe Paradisaeini, in the same family as the CROWS.
They are medium to large (25–110 cm / 9–43 in),
arboreal, fruit-eating forest birds, famous for the
spectacular plumage in the males of most species,
which often includes elaborate plumes and extended tail feathers. They live in Australia, New
Guinea and neighbouring islands. In some species
the males display alone; in others the males congregate to dance in communal LEKS; ▷ p. 94.

bird of prey, common name given to birds that hunt
and kill other animals (especially vertebrates) for
food. Such birds typically have keen senses and
sharp, powerful beaks and talons. They are conventionally divided into the diurnal birds of prey (▷
EAGLE, BUZZARD, FALCON, HAWK, KITE, OSPREY, SECRE-
TARY BIRD) and the nocturnal birds of prey (▷
OWLS); ▷ pp. *91*, 94, 95, **96**, 97, *151*.

bird's-nest fern (*Asplenium nidus*), an epiphytic
fern; ▷ p. 37 (box).

bird's-nest fungus, any one of around 5 genera of
basidiomycete fungi (▷ p. 35) in the order
Nidulariales. The spores are contained in packets
or 'eggs' within a nest-like structure.

birdsong, a more or less complex, species-specific
pattern of sounds produced by many birds, usually
males, to defend a territory or attract a mate; ▷
pp. 94, **152** (box), 154.

bisexual, ▷ HERMAPHRODITE.

bison, either of two massively built cattle-like BOVID
mammals of the genus *Bison*: the American bison or
plains buffalo (*B. bison*), formerly widespread over
the North American prairies, and the European
bison, or wisent (*B. bonasus*), now found only in
isolated reserves; ▷ pp. 114, 206, *232*.

biting louse or **feather louse,** any EXOPTERYGOTE
insect of the order Mallophaga, comprising some
2800 species. Wingless external parasites, mainly of

birds, they have biting mouthparts and eat pieces of feather and dead skin, and possibly drink blood. They are small, elongate and flattened, fitting between the feather barbules and thus avoiding removal during preening; ⇨ pp. *61*, 64.

bitter, a basic taste quality; ⇨ p. 188.

bittern, any of 14 species in the family Ardeidae (order Ciconiiformes) closely related to HERONS. Bitterns are camouflaged and usually hide amongst water reeds; ⇨ p. 97 (box).

bitterweed (*Helenium amarum*), a North American weed in the daisy family, Asteraceae (formerly Compositae). It is toxic to grazing animals.

bivalve, any of some 14 000 MOLLUSCS of the class Bivalvia, characterized by a shell made of two parts (or 'valves') connected by an elastic hinge; ⇨ pp. **56,** *57*.

black bear, 1. (*Selenarctos thibetanus*), Asiatic or Himalayan black bear, a small herbivorous BEAR with black fur, inhabiting forests in central and eastern Asia. **2.** (*Ursus americanus*), American black bear, a large black BEAR inhabiting forest and woodland, and widespread in North America (⇨ p. *128*).

blackberry, ⇨ BRAMBLE.

blackbird (*Turdus merula*), a distinctive THRUSH in which the males are black with striking orange bills and rings around the eyes. It lives chiefly in the western Palaearctic, and is much rarer in the east and Australasia. There are a wide variety of other black birds with the same common name – many are ORIOLES and GRACKLES of the New World. See also AMERICAN BLACKBIRD.

blackbuck (*Antilope cervicapra*), a small fawn or dark-brown/black ANTELOPE, with long, spiral horns in the males. It is widespread in India.

blackcurrant (*Ribes nigrum*), a small shrub in the family Saxifragaceae with strongly flavoured black fruits rich in vitamin C. Some other species of *Ribes* are also called blackcurrant.

black-eyed Susan, 1. gloriosa daisy (*Rudbeckia hirta*), a plant popular for the herbaceous border. **2.** *Thunbergia alata*, an annual herb that has yellow or white flowers with a velvet-black centre. It is an easily raised pot plant.

blackfly, 1. any small robust DIPTERAN insect of the family Simuliidae, sometimes called 'buffalo gnats'. Both sexes have piercing mouthparts, and many are vicious blood-feeders. In Africa they are carriers of the filarial worm *Onchocerca*, which causes river blindness (⇨ p. 64). The larvae live in fast-flowing streams, anchored by a pad of silk at the posterior end, and feed by filtering particles from the water; ⇨ pp. 211, 219. **2.** common name for black-coloured APHIDS.

blackthorn or **sloe** (*Prunus spinosa*), a thorny shrub or small tree belonging to the rose family, Rosaceae, common in hedgerows. The blue-black fruits are used to flavour gin.

black widow, any SPIDER of the genus *Latrodectus* (family Theridiidae), highly venomous web-spinning spiders of Europe, North America, Africa and Australasia. Their bite causes great pain and may occasionaly be fatal.

bladder, 1. (in animals) a membranous distensible sac that acts as the receptacle of a liquid (e.g. urine) or a gas (⇨ SWIM BLADDER). **2.** a sac-like part or organ of various plants (e.g. bladderwrack).

bladderwort, any of around 120 species of aquatic, epiphytic or twining herbs in the widely distributed genus *Utricularia* in the family Lentibulariaceae. The much divided leaves bear bladders that trap small animals; ⇨ box, p. 47.

bladderwrack (*Fucus vesiculosus*), a seaweed belonging to the brown algae (⇨ p. 34) with fronds up to 90 cm (3 ft) long containing many bladders. It is common on rocks between tide marks in temperate and cold regions.

blaeberry, ⇨ BILBERRY.

blanket bog, a peat-forming ecosystem, fed entirely by rainfall and therefore acidic in its water. Such bogs develop under conditions of very high precipitation so that even plateaux and slopes are waterlogged.

blastocoel, a fluid-filled cavity in an early animal embryo (blastula); ⇨ p. 186.

blastocyst, the BLASTULA of a mammal; ⇨ p. 186.

blastomere, in embryology, any cell resulting from the cleavage (division) of the fertilized egg (zygote); ⇨ p. 186.

blastula, an early animal embryo, at the stage at which a cavity (blastocoel) has appeared in the dividing ball of cells; ⇨ pp. 51 (box), 186.

blenny, any small PERCIFORM fish of the family Blenniidae (combtooth blennies, 300 species) and several related families. They are mostly marine fishes found worldwide, especially in warm shallow seas; they are among the most numerous of intertidal rockpool inhabitants. The body is scaleless, while the rather filamentous pectoral fins are composed of a few long rays united at their bases under the throat. Many species bear branched tentacles over the eyes. One group – the sabre-toothed blennies – have enlarged fangs in the mouth.

blewit (*Lepista saeva*) and **wood blewit** (*L. nuda*), two edible species of AGARIC fungi.

blind mole rat, any burrow-living myomorph RODENT of the genus *Spalax* (family Muridae) of Asia and North Africa; ⇨ p. 135. See also MOLE RAT and ZOKOR.

blister beetle, or **Spanish fly** (*Lytta vesicatoria*), a southern European OIL BEETLE. Cantharidin, a substance that causes an increased blood flow when applied to skin, is extracted from the dried wing covers of this beetle.

blood, the fluid in which nutrients, oxygen and waste products (including carbon dioxide) are dissolved and carried around the body. In vertebrates, the blood consists of a fluid medium (plasma) in which cells are suspended. The white blood cells (or leucocytes) are important in combatting invading organisms, while the red blood cells (or erythrocytes) carry the RESPIRATORY PIGMENT; ⇨ pp. 180–1. See also OSMOREGULATION.

blood cell, ⇨ BLOOD.

blood-feeding, ⇨ pp. *59*, *64*, 68, 119, **178**.

blood vessel, any tube by which blood is carried to or from the heart, i.e. an artery, vein or capillary; ⇨ 181.

blowfly, any of various metallic blue or green DIPTERAN insects of the family Calliphoridae, which also includes the FLESH FLIES. The species known as 'bluebottles' and 'greenbottles' have a worldwide distribution and are important in the decomposition of carcasses. Their larvae are early colonizers of corpses and dramatically reduce their bulk in the first few weeks after death.

blue, name given to a number of LYCAENID butterflies, often with a bluish metallic upper wing surface and orange or white spots on the underwing.

bluebell, 1. wild HYACINTH (*Hyacinthoides non-scripta*), a common European woodland herb of the lily family, Liliaceae; ⇨ p. 233. **2.** the bluebell of Scotland is the harebell (⇨ BELLFLOWER).

blueberry, 1. any of several American species of the genus *Vaccinium* in the heather family, Ericaceae. They are commercially grown in the USA for their fruit; ⇨ pp. 208, 211. **2.** another name for BILBERRY.

bluebird, any of three smaller THRUSHES with bright blue backs and tails, living in North America. The FAIRY BLUEBIRD of the Orient is unrelated.

bluebottle, ⇨ BLOWFLY.

blue bull, another name for NILGAI.

bluefish (*Pomatomus saltatrix*), a PERCIFORM fish, the sole member of the family Pomatomidae, a voracious predator occurring in tropical and warm-temperate seas; ⇨ p. 76.

blue fox, a variety of ARCTIC FOX.

bluegrass, an American name for species of the genus POA.

blue-green algae, ⇨ CYANOBACTERIA.

bluethroat (*Erithacus svecica*), a slim robin-like THRUSH of the Palaearctic, with a bright blue and red throat; ⇨ p. 211.

blue whale (*Balaenoptera musculus*), a baleen whale, the largest of the RORQUALS and the largest animal ever to inhabit the earth; ⇨ pp. 142, *144*, **145**.

boa, any non-venomous constricting snake (CONSTRICTOR) of the subfamily Boinae (family Boidae), mainly confined to the New World tropics. Boas are medium-sized to large snakes, and one of them – the ANACONDA – is amongst the largest of all snakes; ⇨ p. 89. See also PYTHON.

boar, a male PIG. See also WILD BOAR.

bobcat (*Felis rufus*), a stocky, reddish-brown CAT with dark markings and a short tail. It inhabits thickets, swamps and rocky areas over much of North America.

bobwhite, either of two New World QUAILS. Their common name is derived from their call – 'bob-bob-wheet'; ⇨ p. 155 (box).

body cell, ⇨ SOMATIC CELL.

body language, a form of COMMUNICATION involving body posture, facial expression, etc.; ⇨ pp. 110, 141, **154** (box).

bog, a type of wetland ecosystem; ⇨ pp. **216–17,** *219*.

bog moss or **sphagnum moss,** any species of the widely distributed genus *Sphagnum*. It has high water absorbing properties attributable to dead cells in the leaves and is important in the formation of bogs, the peat from which is cut for fuel and for use in horticulture; ⇨ pp. 216, 217, *219*.

bog rush, any of around 300 species of the widely distributed genus *Juncus*; ⇨ RUSH.

bole, the trunk of a tree.

bolete, any species of the order Boletales of the basidiomycete fungi (⇨ p. 35). They are mushroom-shaped fungi with pores, and a number are edible including the highly prized cep (*Boletus edulis*).

bombardier beetle, common name given to several tribes of BEETLES in the family Carabidae. The adults use enzyme-catalysed reactions to generate oxygen as a propulsive agent to eject noxious compounds forcibly from the anus; ⇨ p. 167.

bone, a hard, calcium-containing connective tissue formed by special cells called osteocytes. Bone forms the majority of the adult skeleton in most vertebrates; ⇨ p. 70.

bongo (*Tragelaphus euryceros*), a chestnut-brown ANTELOPE with white stripes, a long tail and short horns. It inhabits forest in Africa.

bonobo, another name for the pygmy CHIMPANZEE; ⇨ p. 141.

bontebok (*Damaliscus dorcas*), a reddish-brown ANTELOPE with white legs, belly, rump and face, and small horns. It inhabits grassland in southern Africa.

bony fish or **osteichthyan,** any FISH of the class Osteichthyes, one of the major groups of living fishes. Bony fishes are distinguished from JAWLESS FISHES and CARTILAGINOUS FISHES by the possession of an internal skeleton of endochondrial bone, i.e. bone that replaces cartilage during the course of development. The bony fishes are subdivided into two principal subclasses: the fleshy-finned fishes (LUNGFISHES and COELACANTHS), and the RAY-FINNED FISHES, the latter group comprising the great majority of living species; ⇨ pp. 5, **70–7**.

bony tongue, any primitive tropical freshwater TELEOST fish of the family Osteoglossidae (order Osteoglossiformes), some of which are able to breath air. The arapaima (*Arapaima gigas*) is one of the largest freshwater fishes; ⇨ p. 77.

booby, any of six species of large tropical and southern-hemisphere sea birds, in the same family as the GANNETS. The body is usually dull brown and white, but the webbed feet are bright red or blue; ⇨ p. *91*. See also GANNET.

booklouse, any small EXOPTERYGOTE insect of the order Psocoptera, containing some 2000 species of worldwide distribution. Several species are commonly found among stored cereal products and among books, etc. – hence the common name – but more often they occur in vegetation or on tree bark. They may be winged or wingless; ⇨ p. *61*.

book lung, a respiratory structure in spiders; ⇨ p. 181.

boomslang (*Dispholidus typus*), a large southern African COLUBRID snake, one of relatively few venomous colubrids; ⇨ p. 89.

borage (*Borageo officinalis*), a medicinal and culinary herb in the family Boraginaceae. Its blue flowers are used to decorate cold drinks, and are candied for cake decoration.

boreal, a floral region; ⇨ p. 30 (map).

boreal forest or **taiga,** the coniferous forest found north of about latitude 55° N, and also further south at high altitudes; ⇨ pp. 199 (map), **208–9**.

botany, the study of plants.

botfly, any of various DIPTERAN insects of the family Oestridae, which also includes the WARBLE FLIES. The larval stages of these rather hairy flies are internal parasites of mammals, inhabiting the nasal cavities; ⇨ p. 64.

Botrytis, grey mould, a genus of fungi imperfecti; ⇨ p. 35.

bottlebrush, 1. any of around 25 species of shrubs and trees in the Australasian genus *Callistemon* in the eucalyptus family, Myrtaceae. They are characterized by their colourful long protruding stamens. **2.** bottlebrush grass, any of around 9 species of grass in the genus *Hystrix*. Some are cultivated for use in bouquets.

bottlenose dolphin (*Tursiops truncatus*), the familiar performing DOLPHIN.

bottlenose whale, ⇨ BEAKED WHALE.

bottle tree, any of around 11 species of tropical Australian trees in the genus *Brachychiton* with bottle-shaped trunks, e.g. Queensland bottle tree (*B. rupestris*).

bouganvillea, any of around 18 species of climbing shrubs and small trees in the South American genus *Bouganvillea* in the family Nyctaginaceae. The colourful bracts round the flowers make them popular ornamentals.

bovid, any of around 128 ruminating ARTIODACTYL mammals of the family Bovidae, the largest group of hoofed mammals (UNGULATES), including cattle, sheep, goats and antelope. Bovids are widely distributed in North America and the Old World, especially Africa, and have adapted to a very wide range of habitats, from desert to arctic tundra. All species are strictly herbivorous and most are social; ⇨ pp. 110 (box), 112 (box), **114–15**.

bowerbird, any of 20 birds of the family Ptilonorhynchidae, songbirds living in Australia and New Guinea. Male bowerbirds build special 'bowers' or display grounds to attract females in the breeding season; ⇨ p. *94*. See also CATBIRD.

bowfin (*Amia calva*), a primitive North American

freshwater RAY-FINNED FISH, often called a living fossil, since it is the sole survivor of a once-widespread group. It can live in stagnant water, breathing air by means of a lung; ⇨ pp. 70–1.

box (*Buxus sempervirens*), a European and North African small-leaved evergreen shrub or small tree noted for its use in topiary work, wood engraving and making musical instruments. The related Cape box (*B. macowanii*) of South Africa produces wood of similar quality.

boxfish, any of about 30 tropical marine fishes of the family Ostraciidae, related to the PUFFERS. Their bodies are encased in a hard shell of closely fitting polygonal plates, with openings for the eyes, mouth, fins and tail. They are sometimes known as 'cowfishes' and 'trunkfishes'; ⇨ pp. 72, 77.

BP, abbreviation for 'before the present', often used by geologists and palaeontologists instead of AD or BC.

brachiopod, a LAMP SHELL.

brachiosaur, a gigantic SAUROPOD dinosaur of the genus *Brachiosaurus*, the largest land animal ever known; ⇨ p. 83.

bracken (*Pteridium aquilinum*), a widely distributed species of fern (⇨ pp. *36*, 37) growing mainly on light acid soils. Its vegetative spread by creeping rhizomes is a problem to farmers, and because it is toxic it is avoided by grazing animals thus encouraging its spread (⇨ p. 229).

bracket fungus, any of various basidiomycete fungi (⇨ p. 35) with a fruit body forming a shelf on the timber on which it is growing, e.g. forester's foe.

bract, 1. a leafy organ at the base of a flower or inflorescence; usually green, but bright red in *Poinsettia*. **2.** a leafy organ protecting the sex organs in bryophytes (mosses, liverworts, hornworts).

brain, a concentration of nerve cells acting as the central coordinating point of a nervous system, progressively complex and functionally differentiated in more advanced animals; ⇨ pp. 182–3.

bramble, any of several species of the genus *Rubus* (family Rosaceae), particularly *R. fruticosus* and its many subspecies. This species and others with black berries are also referred to as blackberries. They are trailing thorny shrubs, though some thornless cultivars have been developed.

branchiopod, a FAIRY SHRIMP.

branchiuran, or 'fish louse', a group of CRUSTACEANS that live as external parasites of fish. They are oval in shape and have large suckers for attachment.

brassica, any of around 40 species of the principally Mediterranean genus *Brassica* in the family Brassicaceae (formerly Cruciferae); ⇨ p. 46. The genus includes many important crop plants, though from comparatively few species, e.g. BROCCOLI, BRUSSELS SPROUT, CABBAGE, CALABRESE, CAULIFLOWER, KALE, KOHLRABI, MUSTARD, RAPE, SWEDE and TURNIP. The related RADISH may also be referred to as a brassica.

Brassicaceae, a dicotyledonous plant family (formerly Cruciferae), including cabbage, turnip, mustard, wallflower, etc.; ⇨ pp. 45, **46**.

brazil nut, the seed of *Bertholletia excelsa*, a large tree that grows in the Amazon basin. The nuts are gathered from the wild trees and exported.

breadfruit (*Artocarpus altilis*), a tropical tree in the family Moraceae native to Polynesia. The globe-shaped fruit is formed from the whole inflorescence and is eaten as a vegetable.

bream (*Abramis brama*), a deep-bodied European fish of the CARP family, with a protrusible tubular mouth for bottom-feeding. It can live for some time out of water, which facilitates its transport to market.

briar or **brier, 1.** a prickly bush, especially a wild

ROSE. **2.** *Erica arborea*, a small Mediterranean tree that has swellings on the roots ideal for making pipe bowls.

brier, ⇨ BRIAR.

brine shrimp, ⇨ FAIRY SHRIMP.

bristletail, common name given to some or all of the THYSANURAN insects.

bristleworm, ⇨ POLYCHAETE.

brittlestar or **serpent star,** ECHINODERM of the class Ophiuroidea, comprising some 2000 species; ⇨ pp. 54, 215, 221.

broad bean, field bean or **tick bean** (*Vicia faba*), a species of VETCH in the leguminous family Fabaceae, grown in temperate regions principally for its protein-rich seeds, which are used as a vegetable or as animal feed.

broad generalist, a specialized type of olfactory cell; ⇨ p. 188.

broad-leaved forest, usually applied to temperate deciduous forest, the trees of which largely have broad deciduous leaves; ⇨ pp. 45, **207**.

broccoli, a cultivar of *BRASSICA oleracea* var. *botrytis*, the immature inflorescence of which is harvested in spring and eaten as a vegetable.

brolga or **native companion** (*Grus rubicundus*), a grey species of CRANE standing 1.5 m (5 ft), which lives a nomadic life on the plains of Australia.

brome, any of around 50 species of the mainly temperate grass genus *Bromus*. Hungarian brome (*B. inermis*) is cultivated as a fodder crop.

bromeliad, any of around 40 species of tropical herbs in the American genus *Bromelia* (family Bromeliaceae). Most have rosettes of stiff spiny leaves and many are epiphytic; ⇨ p. 44.

bronchiole, any of numerous tiny branching tubes in lungs; ⇨ 180.

bronchus (pl. -chi), in air-breathing vertebrates, either of the two large branches by which air is conveyed from the TRACHEA (1) into the lungs; ⇨ p. *180*.

brontosaur, any giant SAUROPOD dinosaur of the genus *Apatosaurus*, with a long neck and tail; ⇨ p. 83.

bronzewing, any of four species of Australian PIGEON, with iridescent markings on the wings that 'flash' when the bird is disturbed.

brood parasite, a bird that lays eggs in the nest of another species; ⇨ pp. 95, *168*.

broom, any species of shrub in the three leguminous genera *Cytisus, GENISTA* and *Spartium* (family Fabaceae). In common they have obvious green photosynthetic stems and small inconspicuous leaves.

brown, any of various NYMPHALID butterflies of the subfamily Satyrinae, occurring worldwide. The larvae feed on grasses.

brown algae, any seaweed belonging to the algal division Phaeophyta; ⇨ pp. 34, 221.

brown bear (*Ursus arctos*), a large, omnivorous BEAR, widely distributed in Eurasia and North America. A number of varieties or subspecies are recognized, including the grizzly bear (*U. a. horribilis*) and the Kodiak bear (*U. a. middendorffi*); ⇨ pp. **128**, *129*, 229.

brown earth, a soil profile in which there is full mixing of organic humus with the lower soil layers, usually as a result of the presence of earthworms. It often develops under temperate deciduous forest vegetation, such as oak.

brown-tail (*Euproctis phaeorrhoea*), a LEPIDOPTERAN insect (moth) of the family Lymantriidae, whose larvae are serious pests of a range of tree species in temperate areas of the world.

browsing, a FORAGING strategy involving cropping leaves, shoots, and other vegetation raised above the ground; ⊳ pp. 99 (early mammals), 108–17 (hoofed mammals), *163*, 202–3 (tropical grasslands), 209 (boreal forest). Compare GRAZING.

brush turkey, any of six MEGAPODES of the genus *Alectura*, esp. *A. lathomi*, a solitary, fruit- and insect-eating bird with black plumage; ⊳ p. 95.

brussels sprout, a cultivar of *BRASSICA oleracea* (*gemmifera* group), the swollen axillary buds of which are harvested in winter and eaten as a vegetable.

bryophyte, any member of the plant division Bryophyta, comprising mosses, liverworts and hornworts; ⊳ p. 37.

bryozoan, polyzoan, ectoproct or **sea mat,** any aquatic colonial invertebrate of the phylum Bryozoa (or Polyzoa). Individual bryozoans are tiny, but their colonies can be massive. Characteristically each individual constructs a calcified protective chamber and by asexual division creates either large mat-like colonies on seaweeds, or sprawling or upright colonies on rocks or the sea bottom. Most individuals have a large contractile crown of tentacles (a lophophore) for feeding, but some are modified for a protective function. Their ability to attach themselves to any hard substrate and grow rapidly makes them important fouling organisms.

buckeye, any American species of the genus *Aesculus* such as the sweet buckeye (*A. octandra*). See also HORSE CHESTNUT.

buckthorn (*Rhamnus catharticus*), a thorny shrub or small tree of fen, scrub and hedges, and of woods on chalky or limestone soils; native to Europe and North Africa.

buckwheat (*Fagopyrum esculentum*), an annual herb in the dock family Polygonaceae; grown as a grain crop like a cereal, and also for fodder and as green manure.

bud, a very compact stem with densely packed young leaves or flower parts. Buds can develop into shoots or flowers. Sometimes the term is applied to any protuberance.

budding, a form of ASEXUAL REPRODUCTION found in fungi (⊳ p. 35) and in many simple animals (⊳ p. 184).

buddleia or **butterfly bush,** any of around 70 species of evergreen or deciduous trees and shrubs in the tropical and temperate genus *Buddleia*. The popular garden shrub, *B. davidii*, is very attractive to butterflies.

budgerigar (*Melopsittacus undulatus*), a familiar small, long-tailed PARROT of Australian grasslands. The wild form is green, and flocks of many thousands may be seen feeding at dawn and dusk.

buffalo, either of two massively built species of CATTLE – the dark-brown, herd-dwelling, wild African buffalo (*Synceros caffer*), and the larger, black Asiatic water buffalo (*Bubalus arnee*), which is now widely domesticated. The American BISON is sometimes incorrectly referred to as a buffalo.

Buffon, Georges (1707–88), French naturalist. His 44-volume *Natural History, General and Particular* (1749–1804) is one of the first systematized works of comparative anatomy.

bug, general term applied colloquially to any insect, but entomologically to any HEMIPTERAN.

bugle, any of several species of herbs in the temperate Old World genus *Ajuga* in the family Lamiaceae (formerly Labiatae). Bugle (*A. reptans*) and pyramidal bugle (*A. pyramidalis*) are grown in rockeries.

bugloss, any of a number of species of herb in the genera *Anchusa* and *Echium* (family Boraginaceae), e.g. bugloss (*A. arvensis*), viper's bugloss (*E. vulgare*) and purple viper's bugloss (*E. plantagineaum*) which has become a serious weed in Australia.

bulb, a compact underground stem bearing fleshy leaves; ⊳ pp. **44**, 205, 206.

bulbul, any of 137 PASSERINE birds of the family Pycnonotidae, mainly fruit-eating forest birds of the Old World tropics. Most have a patch of hair-like bristles on the back of the head.

bulldog bat or **fisherman bat,** either of two large microchiropteran BATS of the family Noctilionidae, restricted to the tropics of Central and northern South America. These bats have long legs and strong claws with which they seize fish from water, but they also prey on insects. The unrelated fish-eating bat (*Pizonyx vivesi*) is one of the VESPERTILIONID BATS.

bullfinch, any of six stout Palaearctic FINCHES. Some are partial to buds, which makes them pests in some parts of their range; ⊳ pp. 97 (box), 230.

bullfrog, name given to a number of large frogs (ANURANS) with a loud, deep croak, such as the American bullfrog (*Rana catesbeiana*); ⊳ pp. 79, 167.

bull, 1. any sexually mature male BOVID mammal. **2.** the male of certain other mammalian species, e.g. elephants, walruses, whales. See also NILGAI.

bulrush, 1. *Schoenoplectus lacustris* in the family Cyperaceae, is the true bulrush, a semi-aquatic herb with round stems, which are used for basketwork, mats and chair seats; ⊳ p. *219*. **2.** ⊳ REEDMACE.

bumblebee, any BEE of the genus *Bombus* (etc.) in the family Apidae, occurring in temperate climates. The nests are built in holes in the ground and consist of a collection of rounded wax cells containing honey or developing larvae. The colony does not survive the winter – overwintering queens start new nests each year. See also CUCKOO BEE.

bunting, any of over 250 PASSERINE birds in the subfamily Emberizinae, in the same family as the FINCHES, small, finch-like, predominantly seed-eating birds with short conical bills. The great majority of these live in the New World, where they are often called 'sparrows' and 'finches', and they include the famous Galápagos finches, studied by Charles Darwin as an example of local adaptive radiation (⊳ p. 15). The remaining species live in the Old World; ⊳ p. 211.

burbot (*Lota lota*), the only freshwater fish in the COD family, found in lakes and rivers in Eurasia and North America.

burnet, any LEPIDOPTERAN insect of the genus *Zygaena* (family Zygaenidae), essentially Palaearctic in distribution. They are often coloured black and red, and are day-flying.

burning bush, 1. dittany or gas plant (*Dictamnus albus*), a woody perennial the flowers of which in warm weather produce a vapour that may ignite – a possible explanation for Moses' vision. **2.** *Kochia trichophylla*, an annual with leaves that turn red in autumn. **3.** wahoo tree (*Euonymus atropurpureus*), a North American shrub with crimson fruit.

bur or **burr,** a roughened hairy or spiny seed on the fruit of many plants (⊳ p. 43). Burs are usually globular, e.g. the collection of fruits of a thistle.

burrfish, ⊳ PORCUPINE FISH.

burrowing, living underground in excavated nests or burrows provides both shelter and protection from predators, and has proved a successful way of life for many animals.
Text references: p. **161**; earthworms 55, 176; caecilians 78–9; amphisbaenians 85; armadillos 107; moles 121; badgers 131; lagomorphs 132; mole rats 134, 158 (box).

burying beetle, ⊳ CARRION BEETLE.

bush, ⊳ SHRUB.

bushbaby, another name for GALAGO; ⊳ p. 137.

bushbuck (*Tragelaphus scriptus*), a brown ANTELOPE with white markings and spiral horns, widespread in Africa in areas of dense cover.

bush cricket or **katydid,** any ORTHOPTERAN insect of the family Tettigoniidae, a large, predominantly tropical group of over 5000 species. Bush crickets live among vegetation, and many are well camouflaged, looking like leaves. The females have a sword-like ovipositor, and eggs are usually laid inside plant tissues. As in crickets, sounds are produced by rubbing one wing against another, and there are auditory organs on the fore legs. Many species are winged. See also GRASSHOPPER.

bush dog (*Speothos venaticus*), a stocky, short-legged, dark brown CANID with small ears, broad face and short tail, inhabiting forests in South America.

bush wolf, another name for the COYOTE.

bustard, any of 25 species of large, heavy, ground-dwelling birds with long necks, short tails and broad bills, belonging to the family Otididae (order Gruiformes) and living in scrub and grasslands of the Old World and Australia. The African kori bustard (*Ardeotis kori*) may weigh up to 18 kg (40 lb) and is the heaviest bird capable of flying; ⊳ pp. 92, 213.

busy Lizzie, ⊳ *IMPATIENS*.

butcherbird, ⊳ SHRIKE and CURRAWONG.

butter bean, ⊳ LIMA BEAN.

buttercup, ⊳ *RANUNCULUS*.

butterfly, name given to LEPIDOPTERAN insects of the superfamilies Papilionoidea and Hesperioidea. They typically have clubbed antennae and are active by day, resting with the upper surfaces of the wings closed together over the back, but the distinction between butterflies and MOTHS has no taxonomical significance; ⊳ pp. *59* (box), 62, *63*, 211.

butterfly bush, ⊳ BUDDLEIA.

butterfly fish, 1. any PERCIFORM fish of the family Chaetodontidae, brightly coloured, laterally flattened, deep-bodied fishes associated with coral reefs. **2.** (*Pantodon buchholzi*), an African freshwater fish related to the BONY TONGUES. **3.** any venomous-spined marine fish of the genus *Pterois*, related to the SCORPION FISHES.

butterwort, any of around 46 species of perennial herbs in the northern hemisphere and South American genus *Pinguicula* in the family Lentibulariaceae. The leaves catch insects with sticky glands; ⊳ p.47 (box).

buttonquail, any of 17 species of small quail-like birds belonging to the family Turnicidae (order Turniciformes) living in the grasslands of the Old World. Unlike true quails, these birds lack a hind toe; ⊳ p. *91*.

buttonwood, ⊳ PLANE.

buzzard, any of 27 medium-sized BIRDS OF PREY with rounded wings and long tails, belonging to the genera *Buteo* and *Parabuteo* (family Accipitridae, order Ciconiiformes). They are found everywhere except Australia; ⊳ pp. 92, 206.

cabbage, a cultivar of *BRASSICA oleracea* with a condensed main stem giving rise to a head of tightly packed leaves; ⊳ p. 46.

cacao, ⊳ p. 47.

cachalot, another name for the large SPERM WHALE.

cacomistle, another name for RINGTAIL.

cactus, any of around 800 species of plants in the family Cactaceae, within which the number of genera is controversial. The cacti are mainly from tropical and subtropical America, and they are strongly adapted to arid conditions (⊳ p. 204), having succulent photosynthetic stems, which are usually spiny. The flowers are often large and brightly coloured. The spineless Christmas cactus (*Schlumbergera* x *buckleyi*) is a popular house plant. *Nopalea cochenillifera* is used as the food plant of the cochineal insect. See also CHOLLA, MESCAL, PRICKLY PEAR and SAGUARO.

caddisfly, any INSECT of the order Trichoptera; about 5000 species are known worldwide. The adults resemble moths with long hairs on the wings; they are weak fliers, usually found close to water. The larvae are aquatic, and many build protective cases from sand grains, pieces of twig, etc.; ▷ pp. *61*, 160, 219.

caecilian, any AMPHIBIAN of the order Gymnophiona, comprising some 150 species in 5 families. Caecilians are limbless, worm-like amphibians occurring in tropical South America, Africa, India and Southeast Asia; most are burrowing, but one group is aquatic; ▷ pp. **78–81**.

caecum, a blind-ending sac at the junction of the small and large intestines, terminating in the appendix (▷ p. 179); the site of cellulose digestion in non-ruminant mammals such as rabbits and horses (▷ pp. 105, 108, 110, 132).

caffeine, an alkaloid with stimulant properties; ▷ p. 47.

caiman, variant spelling of CAYMAN.

calabash, either of two unrelated plants, the dried fruits or GOURDS of which are used as containers. The bottle gourd or calabash (*Lagenaria siceraria*) is an Old World trailing vine in the cucumber family, Cucurbitaceae, and the calabash tree (*Crescentia cujete*) is a New World species in the family Bignoniaceae.

calabrese, a cultivar of *BRASSICA oleracea* var *italica*, the immature inflorescence of which is harvested in autumn and eaten as a vegetable.

calcareous, containing CALCIUM CARBONATE; chalky.

calcium, a metallic element (symbol Ca), abundant especially in the form of calcium carbonate ($CaCO_3$). It is required by animals in their diet (▷ p. 179), and is an essential constituent of bone, shells, etc.

calcium carbonate, lime ($CaCO_3$), a calcareous material that provides the principal component of the shells of many protists (▷ p. 27), of corals (▷ p. 215) and of LIMESTONE and chalk; ▷ p. 197.

calimite, an extinct tree-like plant related to the HORSETAILS.

call, a short, relatively simple COMMUNICATION signal; ▷ p. 154.

calliopsis, ▷ COREOPSIS.

calyx, 1. the collective term for the SEPALS, the outer whorl of leaf-like organs of flowers (▷ p. 41); usually green, but in some species coloured. **2.** a crown of numerous feeding 'arms' in marine animals such as sea lilies (▷ p. 54).

cambium, a secondary MERISTEM extending down the length of a stem or root. It is responsible for producing new cells, which differentiate into XYLEM (making up the wood in trees) and PHLOEM (the secondary cambium of woody plants) or the periderm of the bark; ▷ p. 32.

Cambrian period, a geological period lasting from 570 to 510 million years ago; ▷ p. 4.

camel, either of two large ruminating ARTIODACTYL mammals of the family Camelidae, both adapted to life in arid desert conditions: the domesticated one-humped Arabian camel or dromedary (*Camelus dromedarius*) and the two-humped Bactrian camel (*C. bactrianus*) of Central Asia; ▷ p. **115** (box).

camellia, any of around 84 species of trees and shrubs of the Far Eastern genus *Camellia* in the family Theaceae. They are important as ornamentals (▷ *JAPONICA*) and as a source of tea (▷ p. 47) and edible oil.

camomile or **chamomile** (*Chamaemelum nobile*), sometimes called lawn or sweet camomile, a scented European herb in the daisy family, Asteraceae (formerly Compositae). It is used medicinally as a tonic, and is sometimes used in lawns instead of grass.

camouflage or **crypsis** body markings, coloration or patterning that allow an animal to blend in with its background and so escape detection. Camouflage is usually defensive, reducing the probability of detection by predators, but it may also be used to deceive prey, especially by AMBUSH predators. Text references: p. **166**; colour change 57, 67 (box); crustaceans *66*; fishes 77; amphibians 81; reptiles *88*, 89; birds 95, 96 (box); tapirs *111*; pigs *112*; predatory camouflage 164.

campanula, ▷ BELLFLOWER.

campion, any of a number of species of annual to perennial herbs in the genus *SILENE*, e.g. red campion (*S. dioica*); ▷ p. 211.

canary, any of 11 African and western Palaearctic FINCHES. The cagebird canary was domesticated in the 16th century from *Serinus canaria*.

candirú (*Vandellia cirrhosa*), a tiny South American freshwater CATFISH, notorious for its habit of entering the urinary tracts of human bathers; ▷ p. 76.

candytuft, ▷ IBERIS.

cane, the stem of some plants, particularly large GRASSES, e.g. SUGAR CANE and BAMBOO. It may also refer to stems of RASPBERRY and slender PALMS, e.g. RATTAN.

cane rat, either of two large, stocky caviomorph RODENTS of the genus *Thryonomys* (family Thryonomidae), with blunt noses and long tails. They are found in marshy and savannah habitats in Africa.

canid, any mammal of the family Canidae (order Carnivora), comprising 35 species and including dogs, wolves, jackals, foxes and others. Canids typically favour open habitats, and are adapted for fast, long-distance running in pursuit of prey. They are found on every continent except Antarctica. Most are social and largely carnivorous, but a few are more omnivorous; ▷ pp. 122, **126–7**.

canine, (literally 'dog-like') in many mammals, especially carnivores, any of the four sharp, pointed teeth (two in the upper jaw, two in the lower) lying between the incisors and premolars; ▷ pp. 98 (box), 122–31.

Canna, a genus of around 55 species of tropical perennial herbs in the family Cannaceae. Some species are cultivated for their starchy rhizomes or for colouring and spice.

cannabis or **hemp** (*Cannabis sativa*), an erect annual herb in the family Cannabidaceae. The subspecies *sativa* is cultivated for its fibre, and the subspecies *indica* for psychotropic drugs (marijuana, hashish). The seeds are used in birdseed and for the production of hemp seed oil.

cannibalism, eating members of one's own species, usually as a source of food; ▷ p. 165 (box).

canopy, the upper layers of a plant community, often applied to the leaf cover of a forest, but equally applicable to other communities, such as grassland; ▷ pp. 200–1, 207, 208.

cantaloupe, ▷ MELON.

caper (*Capparis spinosa*), a trailing Mediterranean shrub, the flower buds of which are pickled for use in flavouring.

capercaillie (*Tetrao urogallus*), a typically forest-dwelling GROUSE of the western Palaearctic.

capillary, a minute, thin-walled blood vessel; ▷ pp. 180, 181.

capillary action, the process by which water is drawn along narrow tubes or other spaces as a result of the forces between the water and the sides of the tube. This can lead to water in narrow confines being drawn up against the force of gravity.

capsicum, any of around 50 species of pepper in the tropical American genus *Capsicum* in the potato family, Solanaceae. Several species are now widely cultivated, but most cultivars are derived from *C. annuum*. Hot peppers include cayenne, chilli, tabasco and paprika. Sweet peppers may be red (pimentos or pimientos) or green.

capsule, a containing structure. The term is applied to different structures in different groups: the outer layer of some bacteria (▷ p. 24), the spore-producing organ in bryophytes (▷ p. 36), a DEHISCENT fruit of some flowering plants (e.g. poppies), and various anatomical structures in animals.

capuchin, any lively, inquisitive New World monkey of the genus *Cebus* (family Cebidae), so named because of the cowl or 'hood' of thick hair on top of the head; ▷ p. 138.

capybara (*Hydrochoeris hydrochaeris*), the largest living RODENT. The stocky, semi-aquatic capybara has a blunt muzzle, and short ears and tail. It inhabits dense vegetation near water in South America (▷ pp. 134, *135, 201*).

caracal (*Felis caracal*), a reddish-brown to grey CAT with long legs and ear tufts. It is widespread in Africa, Arabia and central Asia.

carapace, the hard protective shell plate of crustaceans such as crabs (▷ p. 66) or the upper part of a turtle's shell (▷ p. 85).

caraway (*Carum carvi*), a biennial, chiefly Mediterranean, umbelliferous herb cultivated for its fruits, which are used for flavouring.

carbohydrate, any of a large number of organic compounds, including SUGARS and POLYSACCHARIDES; ▷ pp. 6 (box), 20, 31, **179**.

carbon, a non-metallic element (symbol C) that is the basis of all organic compounds (▷ p. 6). The carbon cycle is one of the crucial nutrient cycles that maintains life on earth (▷ pp. 196–7).

carbon cycle, the biogeochemical cycle involving carbon; ▷ pp. 196–7.

carbon dating, ▷ RADIOCARBON DATING.

carbon dioxide, a colourless, odourless gas (CO_2) that plays an important role in the biosphere. Its presence in the atmosphere contributed to the beginning of life (▷ p. 6). It is a waste product of respiration in plants (▷ pp. 32–3) and animals (pp. 180–1), but is used by plants in photosynthesis (▷ p. 33), on which virtually all other life on earth depends via the carbon cycle (▷ pp. 196–7). The accumulation of carbon dioxide in the atmosphere through the burning of fossil fuels contributes to the greenhouse effect (▷ p. 224).

Carboniferous period, a geological period lasting from 355 to 300 million years ago; ▷ p. 4.

carbon monoxide, a poisonous gas (CO) released in exhaust fumes. It is also a greenhouse gas; ▷ p. 225.

Cardamine, a cosmopolitan genus of around 120 species in the cabbage family, Brassicaceae (formerly Cruciferae). It includes lady's smock or cuckoo flower (*C. pratensis*), which is cultivated in rockeries, and the bittercresses.

cardamon (*Elettaria cardamomum*), an Indian herbaceous perennial in the ginger family, Zingiberaceae. Its dried fruits are used as a spice.

cardinal or **redbird** (*Richmondena* or *Pyrrhuloxia cardinalis*), a North American BUNTING with bright red plumage.

cardinal beetle, any BEETLE or the family Pyrochroidae. This small, mainly northern-temperate family comprises about 100 species, mainly red and black, medium-sized beetles.

caribou or **reindeer** (*Rangifer tarandus*), a large DEER, with branched antlers in both sexes. It inhabits forest and arctic tundra in North America and Greenland (where it is known as caribou) and Eurasia (reindeer), and has also been domesticated; ▷ pp. *152*, 171, 209, 210, 211.

carina, or **keel** in birds, the ridge-like extension of the breastbone to which the flight muscles are attached; ▷ pp. 90 (box), *92*.

carnassial, a large, long cutting tooth in the mammalian carnivores; ▷ p. *122*.

carnation, cultivars of the genus *DIANTHUS* developed for the cut-flower industry. Border carnations are almost directly descended from wild species, e.g. clove PINK (*D. caryophyllus*), whilst perpetual carnations are the result of hybridization.

carnivore, 1. a flesh-eating organism; ▷ p. 194. **2.** any mammal of the order Carnivora, comprising some 230 species in 7 families: CATS; VIVERRIDS (civets and relatives); HYENAS; CANIDS (dogs and relatives); BEARS; PROCYONIDS (raccoons and relatives); and MUSTELIDS (weasels and relatives). Typically flesh-eating, members of the Carnivora have exploited a wide range of habitats, and their natural distribution extends to all continents except Australasia (where they have been introduced) and Antarctica. Some authorities place the PINNIPEDS (seals, etc.) in this order; ▷ pp. **122–31.**

carnosaur, any of a group of heavily built, large-headed THEROPOD dinosaurs; ▷ p. 83.

carob or **locust** (*Ceratonia siliqua*), a leguminous evergreen tree native to the eastern Mediterranean. Its black pods are used in confectionery and as animal feed.

carp (*Cyprinus carpio*), an important freshwater fish, valuable as a food fish in some parts of the world, although now eaten less widely than in the past. Ornamental varieties, such as the Japanese koi carps, are prized pond fish, while other varieties are valued as sport fish. Its original range is thought to have been central European and Asian, but it has been introduced widely elsewhere and is now one of the most widespread of all fishes. The 'carp family', Cyprinidae (order Cypriniformes), is an extremely successful, predominantly freshwater family, with over 2000 species, and includes MINNOWS, GOLDFISH, GUDGEON, ROACH, RUDD and TENCH; ▷ p. 75.

carpel, the female reproductive organ of a flower, consisting of ovary, style and stigma (▷ p. 41).

carpenter bee, any BEE of the family Xylocopidae, mainly tropical and subtropical in distribution. These large dark-coloured solitary species lay their eggs in tunnels bored into wood or plant stems.

carpet, name given to various GEOMETRID moths.

carragheen (*Chondrus crispus*), a seaweed belonging to the red algae (▷ p. 34) with a flat branched thallus. It is common on northern Atlantic rocky shores between three-quarters tide-level to below low water.

carrion beetle, any BEETLE of the family Silphidae. Most feed on carrion although some are vegetarian. Those of the genus *Nicrophorus* bury the corpses of small animals in which they have laid their eggs. They are sometimes called 'burying beetles' or 'sexton beetles'.

carrot, a biennial herb belonging to the family Apiaceae (formerly Umbelliferae). The wild carrot (*Daucus carota*) is known in the USA as Queen Anne's lace. The cultivated variety (*D. carota* var. *sativus*) is grown as a root vegetable, and is rich in carotene.

cartilage, a firm, translucent, elastic tissue formed by special cells called chondrocytes, which makes up the bulk of the skeleton of young vertebrates and most of which is converted to bone in adults. Fishes such as sharks retain a cartilaginous skeleton throughout life (▷ p. 70).

cartilaginous fish or **chondrichthyan,** any fish belonging to the class Chondrichthyes, which includes SHARKS, RAYS and CHIMAERAS. They are so named because their entire skeleton is made up of cartilage, in contrast to BONY FISHES. The great

majority of species are marine predators, often favouring the relatively shallow waters of the continental margins. Fertilization is internal; some species lay eggs, but many give birth to live young. The cartilaginous fishes are of limited economic importance, and a few species are dangerous to man (▷ pp. 5, **70–7**).

Caryophyllaceae, a dicotyledonous plant family, including pinks; ▷ p. 45.

Caryophyllidae, a dicotyledonous plant subclass; ▷ p. 45.

cashew (*Anacardium occidentale*), a tropical tree native to Mexico and Brazil but introduced to the Old World. The fruit ('cashew apple') may be eaten raw or in preserves or it may be roasted until the outer parts open to reveal the cashew nut kernel.

cassava or **manioc** (*Manihot esculenta*), a tropical shrub in the family Euphorbiaceae. The starch-rich roots form a staple diet, particularly in dry areas, though if undercooked the cyanogenic glycoside they contain may cause poisoning. Tapioca is made from the extracted starch.

cassowary, any of three species of the family Casuariidae (order Struthioniformes), found in the dense rainforests of Queensland and New Guinea. The double-wattled cassowary (*Casuarius casuarius*) is the largest of these flightless birds, standing at 1.5 m (5 ft) and feeding mostly on fruits; ▷ p. 97 (box).

caste, in social insects, a functionally and structurally differentiated form specialized for a particular task within the community, such as a queen, worker or soldier; ▷ pp. 64–5, 153 (box), **156** (box), *160*, 163 (box). A similar system exists in naked mole rats (▷ box, p. 158).

castor oil plant (*Ricinus communis*), a tree in the family Euphorbiaceae grown in temperate, subtropical and tropical areas for oil extracted from the seed and as an ornamental.

cat, any mammal of the family Felidae (order Carnivora), which comprises 34 species of typically solitary, flesh-eating predator. Most cats fall into one of two categories – the big cats of the genus *Panthera* (lion, tiger, leopard, etc.) and the small cats of the genus *Felis* (lynx, ocelot, bobcat, serval, etc.); ▷ pp. 15, 122, **124–5**, 154 (box), *214*. For the extinct sabre-toothed cat ▷ p. *100*.

catabolism, ▷ METABOLIC PATHWAY.

catalpa, any of around 11 species of the genus *Catalpa* native to Asia, North America and the West Indies. The Indian bean tree (*C. bignonioides*) is a popular ornamental.

catalyst, any substance that helps to drive a chemical reaction without itself being changed. Proteins that act as catalysts are known as ENZYMES.

catastrophism, a geological theory popular in the early 19th century but now discarded in favour of UNIFORMITARIANISM; ▷ p. 12.

cat bear, another name for the RED PANDA; ▷ p. 130.

catbird, 1. either of two MOCKINGBIRDS with calls like the mew of a cat. **2.** any BOWERBIRD of the genus *Ailuroedus*, such as the green catbird (*A. crassirostris*), with a cat-like call.

catchfly, any of a number of species of annual to perennial herbs in the genera SILENE and *Lychnis* (family Caryophyllaceae), e.g. alpine catchfly (*S. quadrifida*) and red alpine catchfly (*L. alpina*). See also CAMPION and PINK.

caterpillar, the LARVA of a butterfly or moth; ▷ p. *63*

catfish, any of around 2400 species of the order Siluriformes. They are mostly tropical freshwater fishes, particularly numerous in South America, but some are found in temperate regions and a few are marine. Most (but not all) are recognizable by

long, whisker-like sensory barbels around the mouth. Some are armoured with bony scales, but the majority are scaleless. Many possess strong, sharp fin spines, and in some cases these are venomous. Others are able to produce electric currents capable of stunning. In many parts of the world catfishes are a vital source of protein for humans, and they are also popular aquarium fishes; ▷ pp. 73, 77, *188*. See also SHEATFISH, CANDIRU and (unrelated) WOLFFISH.

catkin, an INFLORESCENCE (usually hanging) in which small, reduced, unisexual flowers are borne on a central stem, as in willow or birch (▷ p. *42*).

catmint or **catnip** (*Nepeta cataria*), a strongly scented perennial herb in the mint family, Lamiaceae (formerly Labiatae). It is native to Europe, Asia and North Africa and is popular in gardens.

cattle, common name given to large BOVID mammals of the genus *Bos* and related genera in the subfamily Bovinae, such as gaur, banteng, yak, kouprey, bison and buffalo. Domestic cattle are mostly descended from the wild aurochs; ▷ pp. **114–15**, 203, 212.

caudal, of or towards the tail or the posterior part of the body.

cauliflower, a cultivar of *BRASSICA oleracea* var botrytis, the dense immature white inflorescence of which is eaten as a vegetable.

cavefish, common name given to about 40 species in at least 13 families, all of which are adapted to environments where there is little or no light; they are typically colourless and blind or virtually blind, and often have highly sensitive body surfaces. In North America the name is often applied to members of the family Amblyopsidae, related to the TROUT-PERCHES.

caviar, ▷ STURGEON.

caviomorph, any RODENT of the suborder Caviomorpha – the 'cavy-like' rodents; ▷ p. 133.

cavy, any of several small, stocky, short-legged caviomorph RODENTS of the genus *Cavia* (family Caviidae), inhabiting grassland and rocky areas in South America. The guinea pig is a domesticated cavy; ▷ pp. 133, 134–5. See also MARA.

cayenne, ▷ CAPSICUM.

cayman, any of 5 species of South American CROCODILIAN reptile of the genera *Caiman* and *Paleosuchus* (family Alligatoridae), closely related to the true ALLIGATORS, though generally smaller.

Ceanothus, a North American genus of around 50 species of predominantly blue-flowered shrubs and small trees. A tea-like beverage may be extracted from the leaves of *C. americanus*.

cedar, any of four species of conifers in the genus *Cedrus* from the Mediterranean and western Himalaya, e.g. cedar of Lebanon (*C. libani*).

celandine, 1. lesser celandine or pilewort (*RANUNCULUS ficaria*), a perennial herb with club-shaped root-tubers. **2.** greater celandine (*Chelidonium majus*), a perennial herb in the poppy family, Papaveraceae.

celeriac (*Apium graveolens* var. *rapaceum*), a variety of CELERY which has a turnip-like root with a similar taste and is similarly used.

celery (*Apium graveolens* var. *dulce*), an umbelliferous temperate herb grown for its leaf stalks, which are used in salads or for flavouring. See also CELERIAC.

cell, the basic biological unit of all living things; ▷ pp. 6–7, **20–1.**

cellulose, a polysaccharide carbohydrate made up of unbranched chains of many glucose molecules (i.e. a POLYMER). Its fibrils form the framework of plant cell walls (▷ p. 21), and it is also the principal constituent in the shells of certain protists (▷ p. 27). Cellulose is generally important in the diets of herbivorous animals, many of which cannot produce the enzymes necessary for its digestion and therefore enlist the help of cellulose-digesting bacteria and protists (▷ pp. 110, **112**, 132, 169).

cell wall, an additional layer outside the cell membrane in bacteria, plants (made of cellulose) and fungi (made of chitin); ▷ pp. 21, 24 (box), **32**, 50.

Cenozoic era, a geological era lasting from 65 million years ago to the present; ▷ p. 4.

centipede, any of some 2800 predatory ARTHROPODS of the class Chilopoda, characterized by a body comprising 15 or so segments, each bearing a pair of legs; ▷ pp. 68 (box), 177.

central dogma, ▷ WEISSMAN.

centriole, a type of organelle in animal and some plant cells; ▷ pp. 7, 20 (box), 21.

cepe, ▷ BOLETE.

cephalization, in evolution, the formation of a head through the concentration of feeding and sensory organs and nervous cells at the front end of the body; ▷ p. 182.

cephalochordate, a LANCELET.

cephalopod, any MOLLUSC of the class Cephalopoda, which includes squid, cuttlefish, octopuses and the NAUTILUS; ▷ pp. 56–7.

cephalothorax, the combined head and thorax of an arachnid or higher crustacean; ▷ pp. 66, 69.

cercopithecine, any Old World monkey of the subfamily Cercopithecinae (family Cercopithecidae), such as macaques, baboons and vervets; ▷ p. 139.

cercus (pl. cerci), one of a pair of sensory structures projecting from the end of the abdomen of many insects and some other arthropods; ▷ p. 191.

cereal, any species of grass cultivated for its edible grain; ▷ pp. 45–6, 206, 212.

cerebellum, in vertebrates, a part of the brain associated with voluntary movement and balance; ▷ p. 183.

cerebral hemisphere, one of the two halves of the CEREBRUM; ▷ p. 183.

cerebrum, in mammals, a part of the brain associated with voluntary and learned behaviour, increasingly dominant in higher mammals; ▷ p. 183.

cestode, a TAPEWORM; ▷ p. 55.

cetacean, any aquatic (mainly marine) mammal of the order Cetacea, which is divided into two major groups: the toothed whales (suborder Odontoceti, 72 species), which include dolphins, porpoises and most of the smaller whales; and the baleen whales (suborder Mysticeti, 10 species), which include the large to very large filter-feeding whales; ▷ pp. 142, 143 (box), **144–5.**

chacma (*Papio ursinus*), a savannah BABOON of southern and eastern Africa.

chaeta or **seta** (pl. -ae), any of numerous tiny bristles on the bodies of annelids such as earthworms, used in locomotion; ▷ pp. 55, 176.

chafer, common name given to many SCARABAEOID BEETLES of the family Scarabaeidae that feed on plant material. They are often pests, especially as larvae. The tip of the abdomen typically protrudes from under the wing covers. The rose chafers are often a brilliant metallic colour, while cockchafers are more dull.

chaffinch, any of three FINCHES of the genus *Fringilla*, which, unlike most true finches, feed insects to their young. The common chaffinch (*F. coelebs*) is a bird of gardens and parks of the western Palaearctic; ▷ p. 152 (box).

Chambers, Robert (1802–71), Scottish writer and publisher. He was the first person to apply the geological theory of uniformitarianism (▷ p. 12) to biology, proposing in 1844 that life first evolved

out of inorganic dust and that it is progressing in a linear fashion towards a higher state.

chameleon, any of around 85 species of mainly tree-dwelling LIZARD of the family Chamaeleontidae. These highly specialized insectivores, famous for their capacity to change colour rapidly, are mainly restricted to Africa and Madagascar; ▷ pp. 88, 89, 165.

chamois (*Rupicapra rupicapra*), a nimble, GOAT-like artiodactyl with pale to dark-brown fur and dark, slender, curved horns. It inhabits steep, alpine areas of Europe, Asia Minor and the Caucasus; ▷ pp. 115, 210.

chaparral, a type of mediterranean vegetation found in California; ▷ pp. 199 (map), **205.**

characin, any freshwater fish of the family Characidae (order Characiformes), a large and diverse group of some 840 species, related to the catfishes and the cyprinids (the carp family). They occur mainly in South and Central America, and in Africa; most are carnivorous, including the notorious PIRANHAS.

chard, the edible stem or leaf midrib of a number of plant species, e.g. Swiss chard (▷ SPINACH), globe ARTICHOKE and SALSIFY.

charlock or **wild mustard** (*Sinapis arvensis*), an annual herb in the cabbage family, Brassicaceae (formerly Cruciferae). It is a serious weed in most temperate regions.

charr, any of seven SALMONIDS of the genus *Salvelinus*, which are either freshwater or migrate to fresh water to spawn.

chat, any of about 150 PASSERINE birds in the subfamily Muscicapinae, closely related to the Old World FLYCATCHERS and the THRUSHES. Chats are often colourful with combinations of black, russet and grey. For species referred to as chats but living in America, ▷ WOOD WARBLER.

cheetah (*Acionyx jubatus*), a slim, lithe CAT with a spotted coat and long tail, reputedly the fastest land animal. It is now largely restricted to African savannah; ▷ pp. 125, 165, *177*, 203.

chelicera (pl. -erae), either of a pair of appendages in spiders and other arachnids, situated immediately in front of the mouth and used for manipulating food.

chelicerates, a group of arthropods including arachnids and horseshoe crabs, distinguished by the possession of CHELICERAE; ▷ p. *58*.

chelonian, any REPTILE of the order Chelonia, an ancient group that includes over 200 species of turtle, tortoise and terrapin. The characteristic feature of the group is the hard, bony shell that encases the body; ▷ pp. 15, **82–9.**

chemoreception, sensitivity to chemical stimuli, i.e. taste and smell; ▷ pp. 188–9.

chemotrophic bacteria, a group of AUTOTROPHIC bacteria that perform chemical transformations other than photosynthesis to obtain energy; ▷ p. 25.

chernozem, the soil type developed in temperate grassland biomes (steppe, prairie, etc.), usually black in colour because of the abundance of organic matter.

cherry, any of a number of species of the genus *Prunus* (family Rosaceae) with flowers in clusters. Flowering cherries are popular ornamentals, particularly in Japan. Sweet cherries are derived from the gean or wild cherry (*P. avium*) and sour cherry varieties from sour cherry (*P. cerasus*); ▷ pp. 43, 46.

chervil, any of a number of annual to perennial herbs in the genera *Anthriscus* and *Chaerophyllum* (family Apiaceae, formerly Umbelliferae), e.g. COW PARSLEY or wild chervil, salad chervil (*A. cerefolium*) and rough chervil (*C. temulentum*).

chestnut, 1. sweet chestnut (*Castanea sativa*), a

large tree native to southern Europe, North Africa and southwest Asia. It was introduced to Britain and was coppiced to produce timber for fencing. Its nuts are eaten roasted or in stuffing. **2.** ▷ HORSE CHESTNUT.

chevrotain or **mouse deer,** any of four small, deer-like, ruminating ARTIODACTYL mammals of the family Tragulidae, found in forests and swamps in tropical Africa and Southeast Asia; ▷ pp. **116–17,** *201.*

chickpea (*Cicer arietinum*), a legume grown for its protein-rich seeds, which may be eaten raw or cooked.

chickweed, any of a number of herbs in the genera *Cerastium*, *Stellaria*, *Holosteum* and *Myosoton* (family Caryophyllaceae), e.g. field mouse-ear chickweed (*C. arvense*), chickweed (*S. media*), jagged chickweed (*H. umbellatum*) and water chickweed (*M. aquaticum*).

chicory (*Cichorium intybus*), a herbaceous perennial, related to ENDIVE, in the daisy family, Asteraceae (formerly Compositae). Its root may be eaten as a vegetable or roasted and ground for use in coffee or as a coffee substitute.

Chile pine, an alternative name for monkey puzzle; ▷ ARAUCARIA.

chili pepper, ▷ CAPSICUM.

chimaera, any of about 30 marine CARTILAGINOUS FISHES of the subclass Holocephali, found in cooler seas worldwide. They have grinding toothplates, smooth skin, and a single gill slit. They practise internal fertilization and lay eggs in horny capsules; ▷ pp. 73, 75.

chimpanzee, either of two species of the genus *Pan* (family Hominidae, which includes the gorilla and man) – the common chimpanzee (*P. troglodytes*) and the pygmy chimpanzee or bonobo (*P. paniscus*); ▷ pp. *101*, 136, 140, **141**, 151 (box), 154 (box), 155.

chinch bug (*Blissus leucopteris*), a HETEROPTERAN insect (bug) of the family Lygaeidae, a pest of cereal crops in the USA.

chinchilla, either of two soft-furred, agile caviomorph RODENTS of the genus *Chinchilla* (family Chinchillidae) with large ears and long, furry tail. They are native to arid areas of the Andes, but are farmed worldwide for their valuable fur.

chinchilla rat, either of two small, softly furred, rat-like caviomorph RODENTS of the genus *Abrocoma* (family Abrocomyidae) from eastern South America.

chipmunk, any ground-dwelling, squirrel-like RODENT of the genera *Tamias* and *Eutamias* (family Sciuridae). They have long tails and cheek pouches, and inhabit forests and scrub in North America and northern Asia.

chitin, a complex sugar molecule (polysaccharide), the principal component of arthropod EXOSKELETONS (▷ pp. 58, 66) and of the cell walls of most fungi (▷ p. 35). Chitin-like substances are also found in the shells of some protists (▷ p. 27).

chiton, any MOLLUSC of the class Amphineura, found worldwide grazing on rocky surfaces. They are easily recognized by their eight shell plates, and are sometimes known as 'coat-of-mail shells'.

chive (*Allium schoenoprasum*), a north temperate herb (family Alliaceae) grown for its onion-flavoured leaves, which are used in salads.

chlorofluorocarbon (CFC), any of several synthetic compounds implicated in the greenhouse effect and the destruction of the ozone layer; ▷ p. 225.

chlorophyll, a light-absorbing molecule (pigment) that helps to drive photosynthesis; ▷ p. 33 (box).

Chlorophyta, a plant division comprising the green algae; ▷ pp. 30, 34.

chloroplast, a type of organelle in plant cells, containing the chlorophyll necessary for photosynthesis; ▷ pp. 20 (box), *21*, 33 (box).

chloroxybacteria, green-coloured prokaryote organisms structurally similar to cyanobacteria (▷ p. 25). It has been suggested that they could be the precursors of the chloroplasts in eukaryote cells (▷ box, p. 7); ▷ p. 24.

cholesterol, a white crystalline substance (sterol), an essential and major component of cell membranes. Animal fat and dairy produce are rich in cholesterol, which may build up on the walls of arteries and restrict blood flow.

cholla, any of a number of species of CACTUS with cylindrical stems in the genus *Opuntia*. Their reticulate woody skeletons are used for making ornaments; ▷ p. 204. See also PRICKLY PEAR.

chondrichthyan, ▷ CARTILAGINOUS FISH.

chordate, any member of the phylum Chordata, which includes the protochordates (ACORN WORMS, TUNICATES and LANCELETS) and the VERTEBRATES. All chordates are characterized at some time of their development by a notochord (a stiffening skeletal rod running down the back of the animal), a double dorsal nerve cord, and gill clefts in the pharynx, but these features do not remain in adult vertebrates; ▷ pp. 18 (box), 51.

chorion, a protective membrane in reptile and bird eggs. It forms part of the PLACENTA in placental mammals; ▷ pp. 86, 187.

choroid, in the mammalian eye, ▷ p. 189.

chough, either of two small rock-dwelling CROWS which (unusually) have bright red or yellow bills and legs. They live in the Palaearctic.

Christmas rose (*Helleborus niger*), a HELLEBORE that produces white or cream flowers in winter and early spring.

chromatophore, a specialized skin cell containing pigment, which is dispersed or concentrated to change the colour of an animal (e.g. chameleon, squid); ▷ pp. 57, 67 (box), 89.

chromoplast, a type of organelle; ▷ p. 20 (box).

chromosome, a coiled structure in the nucleus of cells, containing a large number of genes; ▷ pp. 20, **22**, *24*.

chrysalis, the PUPA of a butterfly or moth; ▷ p. 63.

chrysanthemum, the horticultural name for some 200 species of herbaceous plants of the daisy family, Asteraceae (formerly Compositae), including genera such as *Tanacetum* (which includes pyrethrum; ▷ p. 227) and *Leucanthemum* as well as *Chrysanthemum* itself. Autumn-flowering chrysanthemums are derived from the Asian species *Dendranthema morifolium*.

Chrysophyta, a division of unicellular algae, including the DIATOMS; ▷ p. 34.

chuckwalla (*Sauromalus obesus*), a desert-dwelling IGUANID lizard of southern USA and Mexico; it retreats into a crevice when threatened, and inflates its body to wedge itself tight.

chyme, a semi-fluid mixture of partially digested food particles and watery secretions that forms during digestion; ▷ p. 179.

cicada, any HOMOPTERAN insect (bug) of the family Cicadidae, mainly tropical or subtropical and containing about 4000 species. These moderately large insects are well known for the noisy, monotonous calling of males (▷ p. 62). Juveniles live underground, feeding on sap from plant roots.

cichlid, any of about 700 PERCIFORM fishes of the family Cichlidae; they are generally small, perch-like fishes, inhabiting fresh and brackish waters in Central and South America, Africa and coastal India. They have adopted many different modes of feeding and are especially diverse in the African Rift Valley Lakes, where there are species filling virtually every available niche. Cichlids are also of interest for their nest-building habits and parental care; some species (the so-called mouthbrooders; ▷ p. 74) keep their eggs or hatched young in their mouths for protection. They are among the most popular of all aquarium fishes, and in their native ranges they are a vital source of food; ▷ p. 75.

ciliary muscle, in the mammalian eye, ▷ p. 189.

ciliate, (of a cell) possessing CILIA.

cilium (pl. cilia), a hair-like structure found in organisms such as protists and sponges that is used in locomotion, filter-feeding, etc.; ▷ pp. *7*, **26–7**, **176**.

cineraria, 1. any of various garden plants derived from *Pericallis hybrida*, itself derived from *P. cruenta*, *P. lanata* and other species. *Pericallis* is closely related to SENECIO. **2.** *Cineraria*, a genus closely related to SENECIO, with around 20 species from Africa and Madagascar.

cinnamon (*Cinnamomum zeylanicum*), a shrub native to East and Southeast Asia and Indonesia. The bark from young shoots is fermented, dried and rolled as cinnamon sticks, which are used as a spice.

cinquefoil, ▷ *POTENTILLA*.

circadian rhythm, a biological rhythm that shows a periodicity of about 24 hours, i.e. a daily rhythm; ▷ p. 174.

circulation, the process whereby oxygen- and nutrient-carrying BLOOD is transported to the body cells and deoxygenated blood is returned to the heart, to repeat the cycle; ▷ p. 181.

cirripede, a BARNACLE.

citric acid, an acid found in citrus fruits; it is also the first molecule formed in the TCA cycle (▷ p. 21).

citrus fruit, the juicy fruit of several genera (e.g. *Citrus* and *Fortunella*) in the family Rutaceae, native to Southeast Asia but now widely cultivated in tropical and subtropical areas of the world. They are rich in vitamin C. ▷ GRAPEFRUIT, KUMQUAT, LEMON, LIME, MANDARIN, ORANGE and UGLI.

civet, any cat-like VIVERRID carnivore of the genus *Viverra* and related genera, typically omnivorous, solitary and nocturnal, found in forested areas of Africa and southern Asia; ▷ p. 123. See also PALM CIVET, WATER CIVET, OTTER CIVET.

civet cat, another name for RINGTAIL.

clade, a classificatory group used in cladistics; ▷ p. 18.

cladistics, a method of classification; ▷ pp. 18–19.

cladogram, a branching diagram used in cladistic classification; ▷ p. 19.

clam, any burrowing marine or freshwater BIVALVE mollusc of the genus *Venus* (etc.); ▷ pp. **56**, *57*, 215, 221.

clamworm, another name for RAGWORM (especially US).

clasper, a male copulatory structure in some insects and fishes.

class, a classificatory group between phylum (or division) and order. For example, all mammals belong to the class Mammalia. The names of all plant classes end in -opsida; ▷ pp. 18, 45.

classical conditioning, ▷ CONDITIONING and p. 150.

classification, any of various means of ordering organisms into groups. The science of biological classification is known as taxonomy; ▷ pp. 18–19.

claw, a solid hooked nail, or a pincer-like organ on the end of a limb; ▷ pp. *59* (insects), 67 (crabs), 68 (scorpions), 84 (dinosaurs), **93**, 96 (birds). .

clawed dinosaur, another name for DEINONYCHOSAUR.

clay, the finest particles of soil, consisting of latticed crystalline structures composed of silicon and aluminium oxides. They have properties of retaining both mineral nutrients and water in the soil.

cleaner fish, name given to any small fish, including many WRASSES, that lives by removing parasites and other unwanted material from larger fishes; ▷ pp. 76, **169** (box).

cleaner shrimp, ▷ pp. 66, 67.

clearwing moth, any LEPIDOPTERAN insect of the family Sesiidae; scales are largely absent from the wings, making them transparent. They are often mimics of bees or wasps.

cleavage, in embryology, the successive division of the fertilized egg (zygote); ▷ pp. 186–7.

cleavers, goosegrass or **sticky willie** (*Galium aparine*), a scrambling annual herb with fine prickles that will cling to clothing. See also BEDSTRAW.

cleg, ▷ HORSEFLY.

clementine, ▷ MANDARIN.

clematis, any of around 250 species of mainly woody climbers in the principally north temperate genus *clematis* in the buttercup family, Ranunculaceae. There are many garden cultivars. See also OLD MAN'S BEARD.

click beetle, any BEETLE of the family Elateridae, comprising some 7000 species, distributed worldwide. These elongated beetles are characterized by their habit of feigning death and then suddenly jumping in the air with a snapping noise.

climate, the average weather conditions experienced in a given region; ▷ pp. **198–9**. For the impact of climate on evolution ▷ pp. 10–11, and for the climates of various terrestrial ecosystems ▷ pp. 200–11.

climatic zone, any of several divisions of the earth based upon overall precipitation and temperature; ▷ p. 198 (map).

climax vegetation, vegetation that is in equilibrium with its environment and no longer undergoing SUCCESSION. Determined mainly by climate, but also modified by soil, grazing pressures and land use.

climber, any plant that uses other plants or other structures for support. Woody climbers are often called lianas; ▷ pp. **44–5**, 201.

clinid, any of 75 PERCIFORM fishes of the family Clinidae, mainly from shallow waters in temperate regions, with relatively few species in the tropics. They resemble BLENNIES, but the body is scaled. Some bear live young, although most are egg-laying.

cloaca, the common exit chamber into which the intestinal, urinary, and reproductive canals discharge, especially in birds, reptiles and amphibians; ▷ pp. 93, 94, 98 (box), 102, 120, *185*.

clone, any member of a population of genetically identical cells derived from a single cell by mitosis (▷ p. 22), or of genetically identical individual organisms resulting from asexual reproduction, such as may occur in bacteria (▷ p. 25), protists (▷ p. 27), plants (vegetative reproduction; ▷ p. 33), fungi (▷ p. 35), and various invertebrates. See also p. 21.

clothes moth, either of two LEPIDOPTERAN insects of the family Tineidae, the larvae of which are household pests, feeding on clothing, carpets, furs, etc; ▷ p. 64.

cloudberry (*Rubus chamaemorus*), a temperate herb related to BRAMBLE with edible orange-coloured fruits.

clouded leopard (*Neofelis nebulosa*), a yellow, brown or grey CAT with black markings, found in dense forest in eastern and Southeast Asia.

cloud forest, a forest type developed at high altitude

on tropical mountains, such as parts of the Andes, where condensation provides a constant supply of water.

clover, any of a number of annual or perennial herbs in the mainly north temperate genus *Trifolium* in the pea family, Fabaceae (formerly Leguminosae). Some species, e.g. white clover (*T. repens*) and alsike clover (*T. hybridum*) are cultivated as fodder crops or as green manure.

clove tree (*Syzygium aromaticum*), a tropical evergreen tree. The flower buds, when dried, are the cloves used as spice and the name derives from the French 'clou', a nail, which a clove resembles.

club moss, any species of the families Lycopodiaceae and Selaginellaceae in the division Lycophyta; ⊳ pp. *37*, 208.

coagulation, (of blood) clotting; ⊳ p. *181*.

coal, a carbon-based mineral formed over many millions of years as a result of the gradual compacting of partially decomposed plant matter. Most of today's coal measures were laid down during the Carboniferous period (355–300 million years ago); ⊳ pp. *5, 36, 196*.

coati, any social, omnivorous PROCYONID mammal of Central and South American forest and scrub, especially the ringtail coati (*Nasua nasua*); ⊳ p. 130.

cobnut, ⊳ HAZEL.

cobra, any highly venomous ELAPID snake of the genus *Naja* or *Ophiophagus*, occurring in tropical parts of Africa and Asia. When alarmed, they give a threat display by expanding the ribs in the neck region (the 'hood'); ⊳ pp. *85, 89*.

coca (*Erythroxylum coca*), a tropical tree or shrub native to the Andes. The narcotic, cocaine, is extracted from the leaves.

coccid, any HOMOPTERAN insect (bug) of the superfamily Coccoidea. Females are flattened, degenerate and scale-like, and are more or less permanently anchored to plants by their mouthparts. Males are winged and non-feeding, while juveniles resemble females; ⊳ p. 63 (box). See also SCALE INSECT, MEALY BUG.

coccolithophorid, any species of alga in the family Coccolithophoridaceae. Their walls are covered with plates of calcium carbonate, called coccoliths, which are a major component of chalk.

coccus (pl. cocci), any spherical bacterium. They may grow in pairs (e.g. diplococcus), chains (e.g. streptococcus), or clusters (e.g. staphylococcus); ⊳ p. 25.

coccyx, the bone at the base of the human spine, presumed to be the vestigial remains of a tail; ⊳ p. 14 (box).

cochlea, in vertebrates, a tube in the inner ear (coiled in mammals) by which sound vibrations are encoded as nerve impulses; ⊳ p. *191*.

cockatiel (*Nymphicus hollandicus*), a small, longtailed COCKATOO of Australasian scrubland; a grey bird with a red spot on its yellow cheeks.

cockatoo, any of 15 Australasian PARROTS with erectile head crests and short tails. See also COCKATIEL.

cockchafer, ⊳ CHAFER.

cockle, name given to some 250 species of BIVALVE mollusc; they are found in estuarine conditions, where they are often very abundant. The members of several genera are used for food.

cock-of-the-rock, either of two species of stocky, ground-dwelling COTINGA of the genus *Rupicola*. The colourful males gather to show off their scarlet and orange head crests in a communal mating ground (LEK).

cockroach, any EXOPTERYGOTE insect of the suborder Blattaria (order Dictyoptera), comprising about 4000 species known mainly from the tropics.

They are omnivorous and nocturnal, and can be very abundant. The head is protected by a large plate, and they may or may not be winged. Some are important pests in the transmission of intestinal diseases in domestic situations; ⊳ pp. *61, 231*.

cocoa, a powder made from cocoa beans; ⊳ p. *47*.

coco de mer or **Seychelles coconut** or **double coconut** (*Lodoicea maldivica*), a palm tree of the Seychelles; ⊳ p. *43*.

coconut palm (*Cocos nucifera*), a tall palm tree of the tropics, belonging to the family Arecaceae; ⊳ pp. *43, 45*. See also COCO DE MER.

cocoon, a protective covering surrounding eggs or larvae, such as the silk envelope which an insect larva forms about itself and in which it passes the pupal stage; ⊳ p. *65*.

cod, common name given to many distantly related fish groups, but especially to members of the family Gadidae, and in particular to *Gadus morhua*, an extremely important food fish (length up to 1.2 m/4 ft) ranging throughout the North Atlantic and into the Arctic, with a separate race recognized in the Baltic. Cod migrate in huge numbers between the north and south of their range, and between deep and shallow waters; commercial interest has led to extensive knowledge of their most prolific feeding grounds. The cod order (Gadiformes) is almost exclusively marine and largely northern-hemisphere; many species favour shallow coastal waters. Codfishes typically have small scales and soft-rayed fins (often three dorsal and two anal), and many have a sensory barbel on the chin; ⊳ p. 74. See also HADDOCK, HAKE, BURBOT, WHITING, POLLACK, SAITHE, LING and GRENADIER.

codon, any triplet of nucleotide bases in a DNA molecule that codes for a particular amino acid; ⊳ p. 22.

coelacanth, any BONY FISH of the order Crossopterygii, a formerly successful group in both freshwater and marine habitats, but today represented by just a single species, *Latimeria chalumnae*. This important group is believed to lie close to the ancestry of land vertebrates; ⊳ pp. 70 (box), *71, 74 78*.

coelenterate, any animal of the phylum Coelenterata, which includes sea anemones, corals, jellyfishes and hydrozoans. Coelenterates are the simplest animals with differentiated cell layers, and lack most of the organ systems that characterize higher animals; ⊳ p. 53.

coelom, the cavity that forms in, and is lined by, the mesoderm during development; it separates the body wall and digestive tract in animals more advanced than lower worms; ⊳ pp. 186–7.

coelurosaur, any of a group of lightly built THEROPOD dinosaurs, the reptilian group from which birds are thought to have evolved; ⊳ pp. *82, 83*, 90.

coevolution, the process by which two or more (unrelated) species evolve in adaptation to one another; ⊳ p. 17 (box).

coffee tree, any of several species of the genus *Coffea*, particularly *C. arabica*; ⊳ p. 47.

cohort, a classificatory group sometimes used between class and order (⊳ p. 18). For example, the cohort Glires includes the orders Lagomorpha (rabbits, hares, pikas) and Rodentia (rodents).

cola or **kola,** any of around 50 species of trees and shrubs in the tropical African genus *Cola*. *C. nitida* is cultivated for its cola nuts, which contain stimulative compounds. Similar compounds are synthesized for use in drinks.

cold-blooded, ⊳ POIKILOTHERMIC.

cold desert, ⊳ POLAR DESERT.

cold receptor, a specialized type of temperature-sensitive cell; ⊳ p. 189.

coleopteran, any ENDOPTERYGOTE insect belonging to the order Coleoptera, the BEETLES; ⊳ p. *61*.

collagen, an insoluble, inextensible protein that occurs in fibres in connective tissue (e.g. tendons) and in bones.

collard, an American cultivar of ⊳ BRASSICA *oleracea* var *acephala* intermediate between CABBAGE and KALE.

collembolan or **springtail,** any APTERYGOTE insect of the order Collembola, a widespread group commonly occurring in soil, leaf litter, decaying vegetation, grassland, on the surface of freshwater and marine pools, and on tree bark. They feed on a variety of microscopic particles such as pollen grains and algal cells, as well as decaying plant and animal material. Most are small, and sometimes colourful. Except for soil-dwellers, they are equipped with a forked structure at the end of the abdomen that enables them to jump, hence their common name; ⊳ pp. *10, 61*.

colobine, any Old World monkey of the subfamily Colobinae (family Cercopithecidae), comprising langurs and colobus monkeys; ⊳ p. 139.

colobus, any Old World colobine monkey of the genera *Colobus* and *Procolobus*, highly arboreal, leaf-eating monkeys of equatorial Africa. They are typically slender with long, silky hair and long tails; ⊳ pp. **139**, 141. See also LANGUR.

colon, in vertebrates, the large intestine excluding the rectum; ⊳ p. 179.

colorado beetle (*Leptinotarsa decemlineata*), a highly destructive LEAF BEETLE that feeds on potatoes. Originally a rarity living in North America on native vegetation in the Rockies, it rapidly spread across to the East Coast as the Midwest was settled, and thence by ship to Europe.

colour change, ⊳ pp. 57, 67 (box) and CAMOUFLAGE.

coltsfoot (*Tussilago farfara*), a perennial Old World herb in the daisy family, Asteraceae (formerly Compositae).

colubrid, any SNAKE of the family Colubridae, the largest group of snakes with over 1000 species distributed worldwide. Most colubrids, such as EGG-EATING SNAKES the GRASS SNAKE, GARTER SNAKES and WHIP SNAKES, are harmless, but there are a few venomous species, including the BOOMSLANG.

colugo or **flying lemur,** either of two mammalian species of the family Cynocephalidae (order Dermoptera). Feeding on leaves and other plant matter in the forests of Indonesia and the Philippines, colugos are able to glide for considerable distances from tree to tree; ⊳ pp. **101**, 137 (box).

columbine, any of around 100 species of poisonous perennial herbs in the north temperate genus *Aquilegia* in the buttercup family, Ranunculaceae. The flowers have long spurs containing nectaries. Several species and their derivatives are popular garden plants.

columella, the bone in non-mammalian vertebrates by which vibrations are passed from the tympanic membrane (eardrum) to the inner ear; ⊳ p. *191*.

coly or **mousebird,** any of six small, drab-coloured birds with long tails and fluffy plumage belonging to the order Coliiformes. They live in African scrubland south of the Sahara, where they creep and crawl along branches feeding on fruit, buds and nectar; ⊳ p. 91.

comb jelly, sea walnut or **sea gooseberry,** any marine invertebrate of the phylum Ctenophora, a small group of beautiful, delicate creatures closely related to sea anemones and jellyfishes. Commonly gooseberry- or walnut-shaped, their name comes from the eight ciliated bands or combs that run down the body; these have a locomotory function, and are often iridescent. Comb jellies capture food by trailing two long, highly contractile tentacles with sticky elastic side branches through the water; ⊳ p. 220.

Commelinidae, a monocotyledonous plant subclass; ⇨ p. 45.

commensalism, an association between species from which one species (the commensal) derives benefit, while the other (the host) remains unaffected; ⇨ p. 168.

communication, (in animal behaviour) the use of various kinds of signal (chemical, auditory, visual, etc.) by an animal to convey information about its current state to another animal, in such a way that the latter's behaviour is affected. The functions of communication include bringing the sexes together for reproduction (e.g. COURTSHIP), establishing positions in social HIERARCHIES, and anti-predator defence (e.g. WARNING COLORATION, ALARM RESPONSE); ⇨ pp. 152–5.

community, a collection of different plant and animal species living in the same area; ⇨ p. 194.

comparative anatomy, the study of the anatomical differences and similarities between organisms; ⇨ pp. 14, 18–19.

competition, a type of behaviour occurring between two or more animals that require the same scarce resources, such as food (⇨ FORAGING) or space (⇨ TERRITORY and pp. 172–3). Competition between members of the same species often leads to AGGRESSION.

Compositae, ⇨ ASTERACEAE.

composite, any member of the plant family ASTERACEAE.

conch, name given to various large herbivorous marine snails that are found commonly in sea-grass beds in tropical waters; ⇨ p. 56.

conditioning, a form of LEARNING by association whereby an animal comes either to respond to a neutral stimulus that reliably precedes a significant one (classical or Pavlovian conditioning) or to behave in a certain way because of the consequences that usually ensue (instrumental conditioning or trial-and-error learning); ⇨ pp. **150–1**, 162 (box).

condor, either of two very large, crested New World VULTURES. The Andean condor (*Vultur gryphus*) is huge, weighing up to 14 kg (31 lb) with a wing span of 3 m (10 ft); ⇨ pp. 232, 233.

condylarth, any extinct mammal of the order Condylarthra, a group of primitive UNGULATES (hoofed mammals) that appeared some 65 million years ago and are thought to have been the ancestors of the living ungulates (horses, cattle, antelope, etc.) and of the so-called primitive ungulates such as elephants and hyraxes; ⇨ p. 108.

cone, 1. (in plants) a compact collection of bracts or scale-like leaves bearing the spore-producing structures. In the gymnosperms (e.g. the conifers) these develop in situ into the male and female organs; ⇨ pp. 38–9. **2.** any of numerous cone-shaped cells in the retina of many vertebrates, sensitive to bright light and colour; ⇨ pp. 97, **190**. Compare ROD.

cone shell, any tropical marine GASTROPOD of the genus *Conus* (etc.), characterized by a smooth cone-shaped shell; ⇨ p. 56.

coney, a name sometimes applied to a HYRAX or a PIKA.

conflict behaviour, a type of behaviour occurring when two or more opposing tendencies (e.g. COURTSHIP and AGGRESSION) are aroused simultaneously. It is often expressed as INTENTION MOVEMENTS or DISPLACEMENT BEHAVIOUR; ⇨ p. 155 (box).

conger eel, any EEL of the family Congridae (about 110 species); most species have pectoral fins but lack pelvic fins, and the body is scaleless.

congo eel or **amphiuma,** any of three species of eel-like SALAMANDER of the family Amphiumidae, occurring in southeastern USA. They are permanently aquatic or semi-aquatic, with gill slits and tiny fore and hind limbs.

conidium (pl. -idia), an asexual spore produced by certain fungi; ⇨ p. 35.

conifer, any member of the plant division Coniferophyta; ⇨ pp. 38–9, 208–9.

conjugation, a primitive form of sexual reproduction found in some bacteria, algae, protozoans, and fungi; ⇨ pp. 25, 27.

conjunctiva, in the mammalian eye, ⇨ p. *189*.

conservation, the active management of the natural resources of the planet without depleting them, and the protection of threatened species and habitats; ⇨ pp. 232–3.

constrictor, any of around 80 species of non-venomous SNAKE in the family Boidae that kill their prey by constriction, including PYTHONS and BOAS; ⇨ pp. *88*, **89**.

consumer, ⇨ HETEROTROPH.

continental drift, the process by which the continents have moved to their present positions, and by which they continue to move; ⇨ p. 8.

continental shelf, the margin of a continental plate that extends out beneath the sea. Generally shallow, it gives way to the continental slope where the sea deepens rapidly. This slope marks the boundary of the plate; ⇨ pp. 220, 221.

convergent evolution, the process by which unrelated organisms have evolved similar characteristics in adapting to similar circumstances; ⇨ pp. 14, **17** (box).

convolvulus, ⇨ BINDWEED.

cooperative behaviour, (in social animals) combined and coordinated activity between members of a species that promotes efficiency in foraging and hunting (⇨ p. 165), in defence against predators (⇨ p. 167), etc.; ⇨ pp. 158–9. Such behaviour is most developed in social insects such as termites, bees, wasps and ants.

coot, any of nine black, stocky water birds, with a white patch on the head at the base of the bill, belonging to the family Rallidae (order Gruiformes) and occurring worldwide. The legs and feet are long, the latter with lobes between the toes to assist swimming.

copepod, any minute marine or freshwater CRUSTACEAN of the class Copedoda, a major constituent of the zooplankton; ⇨ pp. 66–7, 220.

copper, name given to a number of LYCAENID butterflies, usually with a coppery metallic upper wing surface.

coprophagy, ⇨ REFECTION.

copulation, close physical contact between male and female in order to transfer sperm, occurring in many animals practising internal FERTILIZATION, including birds, mammals and most insects; ⇨ pp. 184–5.

coral, any of numerous marine ANTHOZOANS (coelenterates), the majority of which are colonial and secrete skeletons that are horny (⇨ OCTOCORAL) or calcareous (stony or reef-forming corals); ⇨ pp. 53, 215.

coral reef, a structure consisting of the skeletons of huge numbers of corals; ⇨ pp. 53, **215**.

coral snake, any venomous New World ELAPID snake of the genera *Micrurus* and *Micruroides*. They are small, brightly coloured snakes, each species having a different pattern of red, yellow and black rings; ⇨ p. 89.

cord grass, any of around 17 species of the mainly temperate genus Spartina. These grasses grow on tidal mudflats and the hybrid Townsend's cord grass (*S.* x *townsendii*) is planted to bind mud; ⇨ p. 46.

corella, any of four small white COCKATOOS with rather short bills. They form large flocks.

Coreopsis, an African and New World genus of around 120 species in the daisy family, Asteraceae

(formerly Compositae), including useful summer- and autumn-flowering annuals and perennials for the garden, e.g. calliopsis (*C. tinctoria*).

coriander, the fruit of *Coriandrum sativum*, a Mediterranean herb in the carrot family, Apiaceae (formerly Umbelliferae). It has a spicy flavour and is therefore used in cooking and in liqueurs and gin.

cork, the outer layers of waterproof, spongy, air-filled dead cells on woody stems and roots. Cork provides a protective layer, being impervious to air and water. The large amounts of cork produced by the cork oak (*Quercus suber*) are exploited commercially; ⇨ pp. *32* and 45.

corm, a swollen stem base; ⇨ pp. 44, 206.

cormorant, any of about 35 species of large, dark-coloured water birds (family Phalacrocoracidae, order Ciconiiformes) with long hook-tipped bills. Their wings are short but powerful, giving a fast straight flight. Habitually the wings are held outstretched to dry as the birds sit. The feet are set far back for strong swimming; ⇨ pp. 90, 91, 95. See also SHAG.

corn, 1. (in traditional European usage) any of the main cereal crops grown in Europe: wheat, barley, oat and rye. **2.** (in American usage) maize, which in Britain is more often referred to as sweet corn.

corncrake, ⇨ CRAKE.

cornea, a transparent membrane at the front of the vertebrate eye; ⇨ pp. 97, **189**.

cornel, ⇨ DOGWOOD.

cornflower (*Centaurea cyanus*), a cornfield weed in the daisy family, Asteraceae (formerly Compositae). It is native to Europe and is grown as an ornamental. See also KNAPWEED.

corolla, the collective term for the whorl or whorls of petals of a flower (⇨ p. 41).

cortex, the outer layer of an organ (e.g. brain, kidney, adrenal gland). In the vertebrate brain, the cortex is concerned with sensory and motor functions; ⇨ pp. 32, 80, 140.

cosmopolitan, having a very wide geographical range throughout the world.

cosmos, any of around 25 species of annual and perennial herbs in the New World genus *Cosmos* in the daisy family, Asteraceae (formerly Compositae). Some are grown in gardens.

cotinga, any of 61 PASSERINE birds of the subfamily Cotinginae, in the same family as the TYRANT FLYCATCHERS. They are medium-sized, usually arboreal birds, feeding on fruit and insects in the upper canopies of Central and South American forests. Female cotingas are usually drab-coloured, but the males of many species are very colourful and some have crests and wattles. See also COCK-OF-THE-ROCK, UMBRELLA BIRD, BELLBIRD.

Cotoneaster, a north temperate Old World genus of around 50 species of shrubs and small trees in the family Rosaceae. Many are grown in gardens for their attractive red berries.

cotton, any of a number of species of shrub in the subtropical genus *Gossypium* in the mallow family, Malvaceae, grown chiefly for the production of cotton, which is a fibre attached to the seeds. The seeds themselves are a source of oil and cattle feed. Species grown commercially include *G. barbadense* and *G. hirsutum*; ⇨ p. 43.

cotton grass, any of around 20 species of perennial herbs in the mainly arctic and north temperate genus *Eriophorum* in the family Cyperaceae.

cottonmouth or **water moccasin** (*Agkistrodon piscivorus*), a highly venomous aquatic or semi-aquatic PIT VIPER of southern USA.

cotton stainer (*Oxycareus hyalipennis*) a HETEROPTERAN insect (bug) of the family Lygaeidae, which feeds on developing cotton-flower buds and produces a red staining in the resultant cotton.

cottontail, any of several RABBITS of the genus *Sylvilagus* found in North and Central America. They are mostly solitary and nocturnal.

cotyledon, the 'leaf' on seed embryos, sometimes modified for food storage (e.g. in peas). After germination the cotyledon can remain underground (as in peas) or be carried above ground (e.g. French beans). The number of seed leaves provides one of the principal distinctions between the two classes of flowering plants: the Liliopsida (the monocotyledons) have one, and the Magnoliopsida (the dicotyledons) have two; ⟾ pp. **38, 41,** 45.

coucal, any of 25 ground-dwelling cuckoo-like birds of the family Centropodidae (order Cuculiformes), living in Australia and building domed nests.

couch grass, any of a number of species of the temperate genus *Elymus (Agropyron)*. The creeping rhizomes of the sand couch (*E. farctus*) make it important in the stabilization of dunes, but those of twitch (*E. repens*) make that species a pernicious weed.

cougar, another name for PUMA.

countershading, a type of CAMOUFLAGE, usually to offset the effect of shadow; ⟾ pp. 77 (fishes), **166**.

courgette or **zucchini** (*Cucurbita pepo*), a dwarf bush variety of MARROW, the immature ribbed fruit of which is eaten as a vegetable and may be pickled.

courser, any of eight small, long-legged, delicate WADERS with a large head and small bill, belonging to the family Glareolidae (order Ciconiiformes). They live in arid lands of Africa, Asia and Australia.

courtship, the various behaviour patterns that precede and accompany mating. Courtship often involves males competing among themselves to attract mates, and advertising to females their readiness to mate and inducing in them a similar state of readiness; ⟾ pp. **157, 159** (box). Other references: insects 62; crustaceans 66; arachnids 68–9; fishes 74, 149; amphibians 80; reptiles 86; birds 94; deer 116, *152*.

cow, ⟾ CATTLE.

cowbird, any of six AMERICAN BLACKBIRDS, some of which are associated with cattle (⟾ p. 168); most lay their eggs in the nests of other birds.

cowfish, ⟾ BOXFISH.

cow parsley, wild chervil or **keck** (*Anthriscus sylvestris*), a biennial herb in the carrot family, Apiaceae (formerly Umbelliferae). It is a common hedgerow plant in temperate regions.

cowrie, any (mainly) tropical GASTROPOD of the genus *Cypraea*. The attractive shells of the smaller species of these marine snails were once used as currency on certain Pacific Islands.

cow shark, any of about five marine SHARKS of the family Hexanchidae, of warm and temperate seas. They have six or seven gill slits; *Hexanchus griseus* grows up to 5 m (16½ ft).

cowslip, 1. a species of PRIMULA. **2.** (in the USA) any of a number of species of the related genus *Dodecatheon*, which is like a primrose, but with reflexed petals.

coyote, prairie wolf or **bush wolf** (*Canis latrans*), a CANID resembling a small wolf with a grizzled grey coat, widely distributed in open country in North America and northern Central America; ⟾ p. 206.

coypu or **nutria** (*Myocastor coypus*), a stocky, rat-like caviomorph RODENT with webbed hind feet and a long cylindrical tail. It is a semi-aquatic herbivore indigenous only to South America, but it has been introduced elsewhere.

crab, any of more than 4500 DECAPOD crustaceans of the genus *Cancer* (etc.). They have a reduced abdomen that is concealed beneath a short broad cephalothorax, and the first pair of limbs is modified as pincers; ⟾ p. **67,** 215, 220. See also FIDDLER CRAB, HERMIT CRAB, HORSESHOE CRAB, SPIDER CRAB.

crab apple, any of a number of sour APPLE trees. *Malus sylvestris* is the crab apple found wild in Europe and the American crab apple is *M. coronaria*.

crab-eating fox (*Dusicyon thous*), a small CANID with grey-brown fur and short ears. Inhabiting woodlands and grasslands in South America, it feeds on arthropods and small vertebrates.

crake, any of 50 slender, marsh-dwelling birds, belonging to the family Rallidae (order Gruiformes) and occurring worldwide. Crakes have short bills, but are otherwise similar to the related RAILS; ⟾ p. 213.

cranberry, any of several species of low-growing shrubs in the genus *Vaccinium*, in the heather family Ericaceae. The American cranberry (*V. macrocarpon*) is grown commercially for its fruit.

crane, any of 15 species of the family Gruidae (order Gruiformes), long-legged birds with long, slender necks and bills, occurring in North America, the Old World and Australia. These elegant birds are often brightly coloured in both sexes and perform elaborate courtship rituals; ⟾ p. *91*. See also BROLGA.

crane fly, any DIPTERAN of the family Tipulidae, generally large, long-legged insects of worldwide distribution. The larvae are subterranean and feed on roots (in which case they are called 'leatherjackets') or aquatic. They are commonly known (in Britain) as 'daddy-longlegs'.

cranesbill, ⟾ GERANIUM.

cranium, (in vertebrates) the skull, or the part of it that encloses the brain.

crawfish, US name for SPINY LOBSTER.

crawling, as a means of locomotion, ⟾ p. 176.

crayfish, any freshwater lobster-like DECAPOD crustacean of the genera *Astacus* and *Cambarus*. Crayfish are omnivorous, and found worldwide in streams and lakes. They are important as food and widely farmed; ⟾ p. 67.

creationism, the belief held by religious fundamentalists that each kind of organism was individually created by a supernatural being, rather than through the process of evolution (⟾ pp. 12–13).

creeper, any plant that has long stems that trail along the ground, e.g. CREEPING JENNY, or that climb by means of adhesive pads or adventitious roots, e.g. VIRGINIA CREEPER.

creeping jenny or **moneywort** (*Lysimachia nummularia*), a creeping perennial herb in the primrose family, Primulaceae. It is native to Europe.

creosote bush, any of around 4 species of the South American genus *Larrea* in the family Zygophyllaceae; ⟾ pp. 30, 204. Extracts of these strongly scented shrubs are used in the treatment of rheumatism.

cress, any of a number of species of the cabbage family, Brassicaceae (formerly Cruciferae), the young shoots of which are used in salads. They include the true cress (*Lepidum sativum*) and WATERCRESS.

Cretaceous period, a geological period lasting from 135 to 65 million years ago; ⟾ p. 4.

cricket, any ORTHOPTERAN insect of the family Gryllidae. Crickets are primarily ground-dwelling and do not fly; the wings are modified to produce sounds by rubbing them against one another (⟾ STRIDULATION), and there are auditory organs on the fore legs (⟾ p. 62). They are omnivorous, and some are frequent inhabitants of human dwellings (e.g. the house cricket, *Acheta domesticus*); ⟾ p. 154. See also BUSH CRICKET, MOLE CRICKET.

Crick, Francis (1916–), English geneticist, who with James WATSON established the structure of the DNA molecule in 1953, one of the greatest scientific breakthroughs ever achieved (⟾ p. 22). In 1962

he shared the Nobel Prize for Physiology and Medicine with Watson and Maurice Wilkins, who with his assistant Rosalind Franklin had done much important work on the chemical composition of the nucleic acids. Crick went on to establish the role of transfer RNA, and the basis of the genetic code in triplets of nucleotide bases (⟾ p. 22). He also enunciated the 'central dogma' (⟾ WEISSMAN).

crinoid, a SEA LILY; ⟾ p. 54.

croaker, ⟾ DRUM.

crocodile, any of 13 species of large CROCODILIAN reptile of the family Crocodylidae, of mainly tropical distribution in the Old and New Worlds (⟾ pp. **84–9,** 217). Both crocodiles and ALLIGATORS have a large pair of teeth near the front of the lower jaw, which in crocodiles are visible when the mouth is closed. The name 'crocodile' is sometimes used in a broad sense to refer to any crocodilian, including alligators and the GAVIAL.

crocodile bird or **Egyptian plover** (*Pluvianus aegypticus*), a short-legged wading bird belonging to the family Glareolidae (order Ciconiiformes), living around rivers and lakes of North and central Africa. It is said to walk in and out of the mouths of resting crocodiles.

crocodilian, any REPTILE of the order Crocodilia, an ancient group containing just 21 living species. The order comprises CROCODILES, ALLIGATORS, CAYMANS and the GAVIAL; ⟾ pp. **82–9**.

crocus, any of around 75 species of herb in the temperate Old World genus *Crocus* in the iris family, Iridaceae. Crocuses are popular garden ornamentals. See also CORM, SAFFRON.

crop, in the digestive tract of birds and many invertebrates, a food-storage compartment preceding the gizzard; ⟾ pp. 59, 93, **178**.

crossbill, any of four Palaearctic and American FINCHES, which have odd-looking bills crossed over at the tips and used to twist seeds from pine cones; ⟾ pp. 96, 209.

cross-breeding, breeding between parents of different races, maintaining genetic variation and reducing the expression of harmful recessive genes (e.g. those that lead to genetic diseases) through inbreeding (⟾ p. 23, box).

crossing-over, a process that increases genetic diversity; ⟾ p. 23.

crossopterygian, ⟾ COELACANTH.

crow, collective name given to about 115 PASSERINE birds of the tribe Corvini (family Corvidae, the 'crow' family). They are medium to large, heavy perching birds, which have relatively long legs, strong feet and (in most) powerful black bills. These omnivorous birds tend to build simple nests consisting of little more than a platform of twigs. True crows include members of the genus *Corvus*, which are usually black and found worldwide. See also JACKDAW, ROOK, RAVEN, JAY, MAGPIE and CHOUGH.

crowberry (*Empetrum nigrum*), a low-growing evergreen shrub in the family Empetraceae. It is a plant of moors and mountains in cool temperate regions. The black or red fruits are collected for preserves; ⟾ p. 211.

crowfoot, ⟾ RANUNCULUS.

crucifer, any member of the plant family BRASSICACEAE (formerly Cruciferae).

Cruciferae, ⟾ BRASSICACEAE.

crust, the upper layer of the earth, above the MANTLE. Continental crust has an average depth of 30–40 km (18½–25 miles), but oceanic crust is thinner at 5–9 km (3–5½ miles) deep. The crust together with the upper, solid part of the mantle forms the lithosphere, which has a depth of 250 km (155 miles); ⟾ p. 8.

crustacean, any ARTHROPOD of the subphylum

Crustacea, which comprises some 42 000 in 10 classes; among its major members are copepods, barnacles, decapods (crabs, shrimps and prawns), sandhoppers, and woodlice; ▷ pp. **66–7**, 218, 220, 221.

crypsis, ▷ CAMOUFLAGE.

cryptodire, any member of the larger of the two suborders of CHELONIAN reptiles (turtles, etc.), which withdraw the head into the shell by making a vertical bend in the neck. The other suborder is the PLEURODIRES.

ctenidium (pl. -idia), the gill structure of gastropod molluscs, consisting of a row of filaments providing a large surface area for gas exchange; ▷ p. 56.

ctenophore, a COMB JELLY.

cuckoo, any of about 100 species of the order Cuculiformes, which includes the families Coccyzidae (American cuckoos) and Cuculidae (Old World cuckoos). They are powerful birds, with slightly curved bills, long tails, and pointed wings and feet, in which two toes point forwards and two backwards. The cuckoos of the Old World are tree-dwellers, and include at least 50 species that lay their eggs in the nests of other species, sometimes mimicking the host eggs; ▷ pp. *91, 95, 149, 168*.

cuckoo bee, any BUMBLEBEE of the genus *Psithyrus*. They have no pollen baskets on their legs and cannot secrete wax; they live in the nests of other bumblebees, destroying the host brood and replacing it with their own.

cuckooflower. 1. lady's smock (▷ CARDAMINE). **2.** another name for RAGGED ROBIN.

cuckoopint, ▷ ARUM.

cuckoo shrike, any of over 70 PASSERINE birds in the CROW family, closely related to the Old World ORIOLES. They are cuckoo-like birds with long tails, pointed wings and powerful bills, living secretively in the tropical forests of the Old World and Australia. Characteristically, they have a patch of dense but loosely attached feathers on the rump.

cuckoo spit, ▷ FROGHOPPER.

cucumber (*Cucumis sativus*), an annual vine in the family Cucurbitaceae grown from the tropics to temperate regions for its fruit, which is used in salads. Gherkins for pickling come from a cultivar.

Cucurbitaceae, a dicotyledonous plant family, including courgette, cucumbers, marrows and pumpkins; ▷ p. 46.

cud, ▷ RUMINATION.

cultivar, a VARIETY of a cultivated plant species that has been bred for agricultural or horticultural purposes.

cultural, or **social transmission** a LEARNING process whereby a pattern of behaviour (or 'tradition') is passed from generation to generation within a population; ▷ pp. *150*, **151**.

cumin (*Cuminum cyminum*), an annual Mediterranean herb of the carrot family, Apiaceae (formerly Umbelliferae), cultivated for its fruits, which are used in flavouring.

cup fungus, any of various ascomycete fungi (▷ p. 35) with cup-shaped fruit bodies.

curassow, any of 13 pheasant-like birds of the family Cracidae (order Craciformes), living in the rainforests of Central and South America. They have distinctive crests, and the males are usually black with bright bills; ▷ pp. *91*, 97 (box). See also GUAN.

curlew, any of seven medium-sized WADERS, with a long, downwardly curved bill used for probing mud for worms. Occurring worldwide, they are well known for their haunting whistles. See also STONE CURLEW.

currant, 1. the dried fruit of the grape (*Vitis vinifera* var. *carinthiaca* or var. *apyrena*). **2.** see BLACKCURRANT and REDCURRANT. **3.** flowering currant (*Ribes sanguineum)*, an ornamental shrub.

currawong, any of three PASSERINE birds, in the same family as the CROWS and closely related to the WOODSWALLOWS. They are large stocky birds, usually black, white and grey, with powerful bills. They are found mainly in Australian forests, but some have moved to city parks. The name is derived from their song.

currents, ocean, ▷ pp. 198, 220.

cusimanse (*Crossarchus obscurus*), a stocky, dark brown, mongoose-like VIVERRID with a long, mobile snout, inhabiting forests and swampy areas in West Africa.

custard apple or **sweetsop** (*Annona squamosa*), a small tree native to tropical America cultivated for its edible fruit, which has a custard-like pulp. The related Bullock's heart (*A. reticulata*) is also called custard apple. See also SOURSOP.

cuticle, 1. (in plants) an outer covering secreted by the EPIDERMIS of leaves, stems and young roots. Its main constituent is usually cutin, a fatty, waterproof substance, and its function is to restrict water loss. **2.** (in animals) a protective layer, such as the hard layer covering the epidermis of arthropods (the EXOSKELETON) and of other invertebrates such as roundworms (▷ p. 55).

cuttlefish, any marine CEPHALOPOD mollusc of the genus *Sepia* (etc.), predatory animals usually found in shallow inshore waters; ▷ pp. 57, 165.

cutworm, any NOCTUID moth of the genus *Argrotis*, whose larvae are important pests of cotton crops.

Cuvier, Baron Georges (1769–1832), French naturalist. He realized that the discovery of fossils – such as the giant dinosaur bones found in Ohio – meant that many now extinct creatures had once roamed the earth. However, he rejected evolutionary ideas. His work of comparative anatomy, *The Animal Kingdom Arranged in Accordance with Structure* (1817–30), was highly influential; ▷ p. 12.

cyanobacteria or **blue-green algae,** a group of bacteria capable of photosynthesis in the same way as green plants; ▷ pp. 6, 7, **25**, 197.

cycad, any member of the plant division Cycadophyta; ▷ pp. **38–9**, 40, 197.

cyclamen, any of around 15 species of perennial herbs in the Mediterranean to western Asian genus *Cyclamen* in the primrose family, Primulaceae. Cultivars of *C. persicum* are grown as house plants; ▷ p. 205.

cynodont, any member of an extinct group of advanced MAMMAL-LIKE REPTILES, belonging to the lineage that gave rise to the mammals; ▷ p. 98.

cypress, any species of conifer in the genera *Chamaecyparis* (*6 species*) and *Cupressus* (around 20 species) and their hybrids. The popular Leyland cypress (x *Cupressocyparis leylandii*) is a hybrid between the two genera; ▷ p. 205. See also SWAMP CYPRESS.

cyprinid, any fish of the family Cyprinidae, the CARP family; ▷ p. 77.

cyst, 1. a protective structure used for reproductive or resting purposes in various organisms such as some protists and worms; ▷ p. 27. **2.** a kind of diseased growth occurring on or within the body.

cyto-, prefix indicating a cell.

cytology, the study of cells.

cytoplasm, the material making up all of a eukaryote cell apart from the nucleus; ▷ p. 20.

cytoskeleton, an important structure within eukaryote cells; ▷ pp. **20**, *21*, 27.

cytosol, the solution between the organelles in eukaryote cells; ▷ pp. 20 (box), *21*.

dab (*Limanda limanda*), a European FLATFISH (family Pleuronectidae), relatively small in length (up to 35 cm/14 in) but very abundant. Although difficult to prepare (it must be skinned), it is considered a delicacy. The name is also given to

various other small flatfishes, such as the sand dabs of American Pacific coastal waters; ▷ p. 76. See also FLOUNDER.

dabchick (*Tachybaptus ruficolis*), the smallest of the GREBES, unusual in that the neck is short.

dace (*Leuciscus leuciscus*), a European fish of clearwater streams, belonging to the CARP family. In North America the name is applied to several other members of the same family.

daddy-longlegs, colloquial name for CRANEFLY (British) and HARVESTMAN (US).

daffodil, ▷ NARCISSUS.

dahlia, any of around 15 species of perennial herbs in the New World genus *Dahlia* in the daisy family, Asteraceae (formerly Compositae). Many cultivars have been developed and are attributed to *D. pinnata*.

daisy, any of around 15 species of the genus *Bellis* and species of other genera of the family Asteraceae (formerly Compositae) with flowers like daisies, such as michaelmas daisy (*Aster*), daisy bush (*Olearea*), and oxeye daisy (*Leucanthemum*). The common small daisy found on lawns is *B. perennis*.

damselfish, any of about 235 PERCIFORM fishes of the family Pomacentridae, found on reefs and in shallow waters in tropical and temperate seas worldwide. They are small, deep-bodied and often highly colourful. See also ANEMONE FISH.

damselfly, any EXOPTERYGOTE insect of the suborder Zygoptera, order Odonata (DRAGONFLIES). Adults have widely spaced eyes, and (unlike typical dragonflies) can fold their wings over their abdomen. Nymphs have three flattened tail filaments, which are used as gills.

damson (*Prunus domestica insititia*), a tree of the family Rosaceae with a sour blue-black fruit like a small plum, which is used in pies and preserves.

dance language, a complex, symbolic COMMUNICATION system used by honeybees; ▷ pp. **153** (box), 155.

dandelion, any of an indefinite number of perennial weeds in the genus *Taraxacum* in the daisy family, Asteraceae (formerly Compositae).

Daphne, a temperate genus of around 70 species in the family Thymelaeaceae of mostly poisonous shrubs with scented flowers.

darter, snakebird or **anhinga,** any of four species of the family Anhingidae (order Ciconiiformes), occurring worldwide, superficially similar to CORMORANTS, but with long, sinuous necks and straight bills used to spear fishes. The feet are set far back, allowing them to swim strongly low in the water; ▷ p. 217.

Darwin, Charles (1809–82), English naturalist who originated the theory of evolution by natural selection; ▷ pp. *12–13*, *14–17*. He extended his theory to human beings in *The Descent of Man* (1871), tracing their ancestry among anthropoid ancestors. He also worked in many other fields, producing important work, for example, on barnacles.

Darwin, Erasmus (1731–1802), British physician, grandfather of Charles Darwin, and originator of an early theory of evolution; ▷ p. 12.

Darwinism, the theory of evolution by natural selection; ▷ p. 13. See also NEO-DARWINISM, SOCIAL DARWINISM.

Darwin's finches, ▷ p. 15 and BUNTING.

Darwin's frog (*Rhinoderma darwinii*), an ANURAN amphibian of western Argentina and southern Chile, males of which brood the young in the vocal sac and eject them as froglets. The name is sometimes applied to the other species in the family (*R. rufum*), which broods in a similar manner but ejects late-stage tadpoles; ▷ p. 81.

dassie, another name for HYRAX, especially a rock hyrax.

dassie rat, another name for ROCK RAT (2).

dasyurid, any MARSUPIAL mammal of the family Dasyuridae, which comprises nearly 50 species and most of the carnivorous marsupials of Australasia, including the Tasmanian devil and the quoll; ▷ p. 105.

date palm (*Phoenix dactylifera*), a subtropical palm grown in North Africa and the Middle East for its edible fruit; ▷ p. 45.

dating methods, ▷ p. 4.

daughter cell, any cell produced by mitosis; ▷ p. 22.

dayfly, another name for MAYFLY.

DDT, dichlorodiphenyltrichloroethane, an insecticide now no longer used in the developed world owing to its toxic effects on wildlife; ▷ pp. 227, 231.

deadly nightshade or **belladonna** (*Atropa belladonna*), a perennial herb in the potato family, Solanaceae. It is native of Europe, western Asia and North Africa, and is cultivated as the source of the drug atropine.

dead-nettle, any of around 40 species of annual or perennial herbs, superficially resembling stinging NETTLES, in the European, temperate Asian and North African genus *Lamium* in the family Lamiaceae (formerly Labiatae). Hemp-nettles are in the related genus *Geleopsis*.

deal, 1. white deal, timber obtained from species of *Abies* (FIR). **2.** red or yellow deal, timber obtained from Scots PINE (*Pinus sylvestris*).

death cap. ▷ AMANITA.

death-feigning, or **thanatosis** pretending to be dead in order to escape attack by a predator; ▷ p. 167.

death's-head moth, any HAWK MOTH of the genus *Acherontia*, such as *A. atropos* of Europe, with markings reminiscent of a human skull on the body. They are able to produce sound and have been seen stealing honey from hives.

deathwatch beetle (*Xestobium rufovillosum*), a BEETLE (family Anobiidae) that tunnels into hardwoods (e.g. oak) used in furniture and building construction. Its name derives from the ticking courtship noises it produces by hitting its head on the walls of its tunnel; ▷ pp. 62, 155.

decapod, any member of the class Decapoda, the largest and most important group of CRUSTACEANS, with some 10 000 species. The group includes crabs, spider crabs, shrimps, lobsters, prawns and crayfish; ▷ p. 67.

deciduous, (of plants) characterized by the seasonal shedding of leaves by abscission. Many broadleaved trees in higher temperate latitudes shed their leaves in winter, entering a dormant phase (▷ p. 207). In lower-latitude temperate zones related species may be evergreen (i.e. the leaves tend not to fall in unison), while in tropical forests the leaves of most trees are evergreen (▷ p. 200). Most conifers are evergreen, their leaves being specially adapted to winter conditions (▷ p. 39); a notable exception is the larch.

decomposer, any organism (including scavenging animals, fungi and bacteria) that depends upon dead plant and animal materials for its supply of energy. Decomposers break down organic matter to carbon dioxide by their respiration and they release elements such as calcium and nitrogen to be recycled in ecosystems; ▷ pp. 194–5.

deer, any of some 38 finely built, ruminating ARTIODACTYL mammals of the family Cervidae, widely distributed in the New and Old Worlds. The most distinctive feature is the annually shed antlers, carried by males of most species; ▷ pp. *9*, 17, **116**, *152*, 157, *195*, 205, 207, 209 (box), 210.

defence, the means by which animals and plants protect themselves from attack by predators. Defence encompasses an enormous range of physical and behavioural adaptations in animals, including camouflage, mimicry, flash and warning coloration, threat, death-feigning, autotomy, retaliation, mobbing, and schooling; ▷ pp. 164–5, **166–7**, 169. Other references: insects 65; fishes 76–7; amphibians 81; reptiles 89; monotremes 102; armadillos and pangolins 107; hedgehogs 120.

deforestation, the felling of areas of forest, a major ecological problem in many parts of the world; ▷ pp. 201, 207, 209, 212, 225, **228**.

degu or **octodont,** any of several small caviomorph RODENTS of the family Octodontidae. They have long tails and soft fur, and are chinchilla-like in appearance; they live colonially in the Andes.

dehiscent, (of plant structures) spontaneously bursting open to release their contents. Dehiscent structures include anthers (releasing pollen) and certain fruits, e.g. peas (releasing seeds). Compare INDEHISCENT.

deinonychosaur or **clawed dinosaur,** any carnivorous bipedal THEROPOD dinosaur of the genus *Deinonychus*; ▷ pp. 83, **84**.

deinothere, any member of an extinct group of trunked mammals (proboscideans) related to the living elephants; ▷ p. 108.

deletion, a form of mutation in which material – anything from a single base to a piece of DNA carrying a number of genes – is removed from a chromosome (▷ pp. 22–3).

delphinium, any of around 250 species of the north temperate genus *Delphinium* in the buttercup family, Ranunculaceae. Perennial garden cultivars are derived from *D. elatum* and *D. grandiflorum*. LARKSPUR belongs to a closely related genus.

dendrite, one of a number of branching projections from a nerve cell by which impulses are conducted to the cell body; ▷ p. 182 (box).

dendrochronology, a method of dating relatively recent material by examining tree rings. Depending on climatic conditions, the distance between annual rings varies, and these sequences can be correlated to date wood found at archaeological sites. The very old bristlecone pines (▷ p. 30) have provided sequences going back thousands of years. Dendrochronology provides a useful double check on dates provided by radiocarbon methods (▷ p. 4).

denticle, a small tooth-like structure, such as the tiny scales of sharks.

dentine, a calcium-containing material similar to but harder and denser than bone, which forms the bulk of teeth.

deoxyribonucleic acid, ▷ DNA.

deposit feeder, any aquatic animal that feeds by processing sediments; ▷ p. 221.

dermestid beetle, any BEETLE of the family Dermestidae. These insects are normally scavengers, feeding on dried-out corpses, but they are also important pests in museums, storage areas and domestic dwellings, where they feed on hides, leather, hair, wool, and foods such as bacon. Larvae have long, loose bristle-like appendages (setae) and are called 'woolly bears'.

dermis, in vertebrates, the inner layer of skin, containing connective tissue, blood vessels, etc.

desert, an extremely arid area, with little vegetation; ▷ pp. 199 (map), **204**. See also POLAR DESERT.

desertification, the process of turning productive land into desert; ▷ pp. 201, 203, **204**, 206, 224 (map), **229**.

desman, the common name given to 2 of the 29 species in the mammalian family Talpidae (order Insectivora), all the others being MOLES. Desmans resemble moles in general appearance but have various adaptations for a semi-aquatic life style. They are restricted to small areas of southern Europe and Russia; ▷ pp. 120, **121**.

determination, in embryology, an irreversible process by which a cell becomes committed to a particular developmental pathway; ▷ pp. 21, **186–7**.

detritivore, any organism that feeds on small particles of dead organic matter; ▷ pp. 201, 202.

Deuteromycotina, the 'fungi imperfecti', a subdivision of the fungal division Amastigomycota; ▷ p. 35.

deuterostome, ▷ p. 51 (box).

development (embryological), ▷ pp. 186–7. See also GROWTH.

devil ray, ▷ MANTA.

Devonian period, a geological period lasting from 410 to 355 million years ago; ▷ p. 4.

dewlap, the pendulous skin under the throat of oxen, dogs, etc., or the fleshy wattle of a turkey.

dextrose, another name for GLUCOSE.

dhole, Asian wild dog or **red dog** (*Cuon alpinus*), a brown, dog-like CANID with a bushy, black tail. It lives socially in forests in eastern and Southeast Asia, including India.

dialect, a local variation in the songs or other vocalizations of animals such as birds (▷ p. 152) and frogs (▷ p. 80).

diamond bird, ▷ PARDALOTE.

diana monkey (*Cercopithecus diana*), a strikingly coloured GUENON; ▷ p. 139.

Dianthus, a genus of around 300 species in the family Caryophyllaceae native to Europe, Asia and South Africa. Some, e.g. sweet william (*D. barbatus*), CARNATIONS and PINKS, are cultivated for their attractive flowers.

diapause, a period of suspended growth and development that occurs in some animals, especially insects; ▷ p. 175.

diaphragm, a partition in an organ or organism, especially the thin muscular sheet that separates the chest and abdominal cavities in mammals, which assists in respiration; ▷ p. 181.

diapsid, any REPTILE of a group defined by the presence of two openings in the temporal region of the skull (▷ pp. 82, *84*); the group includes most living reptiles (crocodiles, lizards and snakes) as well as the extinct dinosaurs. See also PARAPSID.

diarrhoea, ▷ p. 179.

diatom, any of a group of unicellular algae possessing silica walls and forming filaments or colonies. Some species possess a single FLAGELLUM to aid movement, while others move by gliding; ▷ pp. 26, 34, 218, 220.

Diatryma, ▷ p. 90.

Dicentra, a north temperate genus of around 19 species of perennial herbs in the family Fumariaceae. A number, such as bleeding heart (*D. spectabilis*), are grown as ornamentals.

dicotyledon (often abbreviated to **dicot**), any member of the flowering-plant class Magnoliopsida. The main distinguishing feature is a double seed leaf (cotyledon); ▷ p. 45. Compare MONOCOTYLEDON.

dictyopteran, any EXOPTERYGOTE insect of the order Orthoptera, which includes COCKROACHES and MANTIDS; ▷ p. 61.

dicynodont, any member of an extinct group of MAMMAL-LIKE REPTILES, belonging to the lineage that gave rise to the mammals; ▷ p. 98.

Dieffenbachia, a genus of around 30 species in the

arum family, Araceae, from tropical America. The acrid juice of mother-in-law's tongue (*D. seguine*) was used in the West Indies to torture slaves, who were made to bite the plant.

differentiation, within a developing embryo, the process whereby cells become specialized to perform particular functions; ⊳ pp. 21, 186.

diffusion, the movement of molecules from a region of higher concentration to a region of lower concentration. GAS EXCHANGE across respiratory surfaces takes place by diffusion, and in small animals diffusion alone is sufficient to meet their oxygen needs; ⊳ pp. 180–1. Compare ACTIVE TRANSPORT. See also OSMOSIS.

digestion, the process in animals whereby complex molecules in food are broken down into simpler compounds that can be absorbed into the body and used as a source of energy, for growth, etc.
  Text references: **178–9**; protists 26; worms 55, *178*; insects 59; arachnids 68–9; birds 93, *178*; marsupials 104–5; sloths 107; elephants 108; ungulates 110 (box); ruminants 112 (box); rabbits, etc. 132; rodents 133.

digestive tract or **alimentary canal,** in animals, the series of structures that are specialized to deal with the intake, processing and absorption of nutrients; ⊳ pp. 178–9. See also DIGESTION.

digger wasp, any solitary WASP belonging to various genera in the family Sphecidae. They burrow into the soil to make nests, which they stock with insects to nourish the larvae.

digit, a finger or toe.

Digitalis, a genus of around 20 species of biennial or perennial herbs native to Europe, the Mediterranean region and Central Asia. Foxglove (*D. purpurea*) is the source of the drug digitalin.

dik-dik, any of three small, grey-brown, browsing ANTELOPES of the genus *Madoqua*, inhabiting arid, evergreen scrub in Africa. The males possess very short horns.

dill (*Anthemum graveolens*), an aromatic herb of the carrot family, Apiaceae (formerly Umbelliferae), native to India. The seeds are used for their aniseed flavour, and the stems and leaves yield a medicinal oil.

Dilleniidae, a dicotyledonous plant subclass; ⊳ pp. 40, **45**.

dimorphism, ⊳ SEXUAL DIMORPHISM.

dingo (*Canis dingo*), or 'wild dog', a short-haired, dog-like CANID with pale brown fur. Living in social groups, it is widespread in Australasia and parts of Southeast Asia; ⊳ p. 126.

dinoflagellate, any member of the protistan group Dinophyceae (or Dinophyta). They are predominantly flagellate marine algae, often brown in colour, and form part of the phytoplankton; ⊳ pp. 26–7.

dinosaur, any REPTILE of the orders Saurischia and Ornithischia that dominated life on earth for over 160 million years until their extinction 65 million years ago; ⊳ pp. **82–4**, 87 (box),

dioecious, (of a plant species) having male and female reproductive organs on separate plants, e.g. cycads, gingko (⊳ p. 39), willows. Compare MONOECIOUS.

diplodocus, any giant SAUROPOD dinosaur of the genus *Diplodocus*; ⊳ p. 83.

diploid, having two copies of each different chromosome in normal body cells (apart from sex cells). All the cells in the SPOROPHYTE stages of plants are diploid; ⊳ p. 22, 149 (box), 184. Compare HAPLOID.

dipluran, forktail or **twintail,** any APTERYGOTE insect of the order Diplura, containing over 600 species of worldwide distribution. The majority are very small soil- and litter-dwelling forms, but the largest reach 5 cm (2 in) in length. Their name

derives from their two tail filaments (cerci), which may be either long and sensory, or short, strong and pincer-like, and used in catching prey; ⊳ pp. 60–1.

dipnoan, ⊳ LUNGFISH.

dipper, any of five PASSERINE birds of the family Cinclidae, round-bodied birds with short, uptilted tails, living in fast-flowing streams of the northern hemisphere and the Andes. They are the only passerines to use short wings and powerful feet to catch insects underwater; ⊳ p. 97.

dipteran, any INSECT of the order Diptera, the so-called 'true flies', comprising over 85 000 described species (⊳ p. 61). The fore wings are well developed, but the hind wings are tiny, forming gyroscopic structures (halteres) that act to stabilize flight. The majority of species are day-flying, feeding on nectar and pollen or on rotting vegetation or carrion. Some are specialized blood-feeders on birds and mammals, while others are predatory on other invertebrates, especially other insects. The larvae usually live in moist conditions, and some are aquatic.

disc-winged bat, either of two microchiropteran BATS of the family Thyropteridae, occurring in tropical Central and South America. These insectivorous bats derive their name from suction pads attached to the wrists and ankles, which assist in climbing smooth banana leaves. By means of these pads, they are able to roost head upwards. See also SUCKER-FOOTED BAT.

displacement behaviour, behaviour that is apparently irrelevant to a particular situation, as when a courting bird breaks off to preen itself or feed. Such behaviour often results from an animal being in motivational CONFLICT; ⊳ p. 155 (box).

display, a largely stereotyped, species-specific behaviour pattern that functions as a visual COMMUNICATION signal. Displays are often RITUALIZED, and may serve many purposes, such as COURTSHIP and DEFENCE; ⊳ pp. 152–5.

disruptive coloration, a type of CAMOUFLAGE, as in zebras, that helps to break up an animal's outline; ⊳ p. 166.

distribution, the geographical range of an animal or plant. It can be affected by many factors, such as the physical limitations of the organism, the presence of barriers such as seas and mountains, competition from other species, and evolutionary history.

diurnal, associated with the daylight hours, as opposed to the night. Compare NOCTURNAL.

diver or **loon,** any of five species of the family Gaviidae (order Ciconiiformes), aquatic birds, with long bills and powerful webbed feet placed far back on the body. They dive for fish, and live on northern lakes in summer and at sea in winter; ⊳ pp. 90, *91*, 93, 95, 96, 97.

divergence, change in a presumed ancestral species resulting in two or more descendant species; the splitting of a taxonomic group into two or more parts; ⊳ pp. 12–17.

diversity, the number of species found within an ecosystem; ⊳ pp. **195**, 200–11.

diving beetle, any BEETLE of the family Dytiscidae, medium to large freshwater beetles predominantly from the Palaearctic region. About 4000 species are known. Both adults and larvae are carnivorous.

division, 1. any of several high-ranking classificatory groups into which the plant kingdom is divided. The names of plant divisions all end in -phyta; ⊳ pp. 18, **30** (table), 45. **2.** a classificatory group between phylum and class in the animal kingdom; ⊳ p. 18.

DNA, deoxyribonucleic acid, the basic genetic material of most living organisms; ⊳ pp. 6–7, 15 (box), 16, 19, 20, **22–3**, 24 (viruses), 25 (bacteria).

dobsonfly, any large NEUROPTERAN insect of the family Corydalidae (⊳ p. *61*). Their aquatic larvae are carnivorous, with well-developed mandibles; they are used by anglers as bait.

dock, any of a number of species of annual to perennial herbs in the temperate genus *Rumex* in the family Polygonaceae. Curled dock (*R. crispus*) is a serious weed. ⊳ SORREL.

dodo (*Raphus cucullatus*), a flightless turkey-like bird of the family Raphidae, which inhabited the island of Mauritius until its extinction near the end of the 17th century. The dodo and the solitaires of nearby Rodriguez and Réunion (also extinct) were related to the PIGEONS; ⊳ p. 97 (box), 214.

dog, the common name given to many CANIDS, especially the larger, more social species (in contrast to the more solitary FOXES) and often used generically for any canid. The many breeds of domesticated dog (*Canis familiaris*) are all believed to be descended from the GREY WOLF; ⊳ pp. *14*, **126–7**, 150, 154 (box), 155.

dogberry, 1. a species of DOGWOOD. **2.** American mountain ash, *SORBUS americana*.

dogfish, any small SHARK of various families, including the smooth dogfishes (family Triakidae) and the spiny dogfishes (family Squalidae). The sandy dogfish (*Scyliorhinus canicula*) is a common fish in European shallow waters; ⊳ p. 74.

dog's mercury, ⊳ p. 42.

dogwood or **cornel,** any of around 45 species of trees and shrubs of the genus *Cornus* in the family Cornaceae. Found in temperate regions, they are cultivated for the winter colour of their young growth, e.g. *C. alba*, and for their showy bracts, e.g. *C. florida*. Some species, such as dogberry (*C. sanguinea*) provide valuable timber.

dolphin, any small marine toothed whale (CETACEAN) of the family Delphinidae (32 species), found in all oceans. The familiar performing animal is the bottlenose (*Tursiops truncatus*); ⊳ pp. 17, 142, 144, **145**. See also RIVER DOLPHIN.

domestication, the process by which wild species have been harnessed by human beings for their own uses, principally agriculture. Some wild species have been radically modified by ARTIFICIAL SELECTION; ⊳ pp. 114 (box), **212**.

dominance, a state of superiority or high status (in relation to a lower-ranking or subordinate individual) within a social grouping or HIERARCHY.

dominant, (of an allele) being expressed in preference to a RECESSIVE allele where a pair of alleles are not identical; ⊳ p. 23.

dopamine, ⊳ p. 182 (box).

dormancy, a state of suspended activity and reduced metabolism in an organism, usually as a response to unfavourable external conditions. The SPORES and CYSTS of various simple organisms are able to remain dormant for long periods, as are the seeds of seed-bearing plants (⊳ pp. 38–41). For dormancy in animals ⊳ p. 175. See also AESTIVATION, HIBERNATION.

dormouse, any of several small, mouse-like myomorph RODENTS of the family Gliridae, with long, furry tails. Found in Eurasia and Africa, they hibernate in winter.

dorsal, relating to or situated near the back.

dory, any of 10 marine fishes of the family Zeidae (order Zeiformes), from the Atlantic, Indian and Pacific Oceans. They have deep, laterally compressed bodies, obliquely upturned mouths, and large eyes, giving them a very doleful expression.

dotterel, either of two WADERS of the family Charadriidae, small, slender-billed birds feeding mainly on insects. They are found in the Palaearctic and Neotropical regions.

double coconut, ⊳ COC DE MER.

double helix, a double spiral structure, as in the DNA molecule; ▷ p. 22 (box).

Douglas fir, any of around seven species of conifer in the genus *Pseudotsuga*; ▷ p. 39. *P. menziesii* (= *P. douglasii*) is grown both for timber and as an ornamental tree.

douroucouli, another name for NIGHT MONKEY.

dove, any of approximately 155 of the smaller, more slender and graceful species belonging to the order Columbiformes, which also includes the PIGEONS. They are found worldwide. Many have elaborate courtship displays and some, such as the peaceful dove (*Geophilia placida*), have beautiful and subtle markings; ▷ pp. **173** (box), *201*.

dowitcher, any of three WADERS of the genus *Limnodromus* in the family Scolopacidae. They are long-legged, short-tailed birds resembling SNIPE but are more vividly coloured. They are found in North America and Asia.

down, the soft fine feathers that form the principal insulative layer of birds; ▷ pp. 92–3.

dragonet, any of about 130 marine PERCIFORM fishes of the family Callionymidae, found in warm inshore waters worldwide. They lie on the sea bed; the males are often brilliantly coloured and perform elaborate courtship displays.

dragonfish, common name given to at least three families of small, elongated deep-sea fishes. They all have black, usually scaleless bodies, sharp teeth, and rows of luminous organs. Most also have a barbel on the chin.

dragonfly, any EXOPTERYGOTE insect of the order Odonata, comprising over 5000 species and of worldwide (especially tropical) distribution. Generally large and brightly coloured, adults are predatory, catching flying insects in their legs; they are often territorial, patrolling their patch and seeing off other individuals. Nymphs are aquatic and also predatory, catching prey with a 'mask' (labium), which is shot out like a tongue; ▷ pp. 60, *61*, *62*.

drey, the arboreal nest of a squirrel, made from twigs, moss and leaves.

drill (*Mandrillus leucophaeus*), a baboon-like Old World cercopithecine monkey, a West African forest-living species similar to the MANDRILL but with a less brightly coloured muzzle; ▷ p. 139.

driver ant, ▷ ARMY ANT.

dromedary, another name for the Arabian CAMEL; ▷ p. 115.

drone, in social BEES, a male whose sole function is to mate with the queen.

drongo, any of 24 PASSERINE birds in the subfamily Dicrurinae (family Corvidae), closely related to the MONARCH FLYCATCHERS. They are medium-sized, aggressive, arboreal birds, with black iridescent plumage, stout slightly hooked bills, and very long tail feathers. They live in wooded country of Africa, the Orient and Australia.

drum or **croaker,** any of 210 PERCIFORM fishes of the family Sciaenidae, marine, brackish and freshwater fishes from tropical and temperate seas. They are mostly inhabitants of shallow coastal waters, and are well known for their sound production, using their swim bladder as a resonating chamber.

drupe, an INDEHISCENT fruit consisting of an outer layer (skin), a fleshy middle layer, and a stony inner layer, within which there is a single seed. Examples include plum, peach, cherry and pistachio; ▷ p. 43.

dry rot, timber decay in buildings due to the basidiomycete fungus (▷ p. 35) *Serpula lacrymans*. The fungus needs damp wood to become established, but because the decay results in more moisture the fungus can spread to otherwise dry wood. Compare WET ROT.

duck, any of 130 species of the family Anatidae (order Anseriformes), which includes GEESE and SWANS. Ducks are adapted to living on or near water, and occur worldwide. They have insulating down under their plumage, and webbed feet. The sexes are usually differently coloured; ▷ pp. *91*, *93*, *96*, *97* (box), *155* (box), *211*. See also SHELDUCK, WIGEON, TEAL, MALLARD, POCHARD, EIDER, SCOTER and MERGANSER.

duck-billed dinosaur, another name for HADROSAUR.

duck-billed platypus, ▷ PLATYPUS.

duckweed, any of around 16 species of tiny aquatic herbs in the widely distributed genus *Lemna* in the family Lemnaceae; ▷ pp. *219*, *213*. Australian duckweed belongs to the genus *Wolffia*; ▷ p. 30.

dugong (*Dugong dugon*), a herbivorous marine mammal, the only species in the family Dugongidae (order Sirenia, which also includes the MANATEES). Dugongs are found in tropical and subtropical coastal waters in the Indian Ocean, the Red Sea and the western Pacific; ▷ p. **142**.

duiker or **duikerbok,** any small, antelope-like BOVID mammal of the subfamily Cephalophinae, typically occurring in forest and thicket in sub-Saharan Africa; ▷ pp. 115, 205.

dung beetle, common name given to several SCARABAEOID BEETLES that develop and live in dung, including the SCARABS.

dunnart, any of 13 small, mouse-like carnivorous/insectivorous DASYURID marsupials of the genus *Sminthopsis*. They are widespread in Australia.

dunnock, ▷ ACCENTOR.

duodenum, the first part of the small intestine extending from the stomach to the jejunum, into which the bile and pancreatic ducts empty and in which most of the enzymatic digestion of food occurs; ▷ pp. 178–9.

Dutch elm disease, ▷ ELM.

duyker, variant spelling of DUIKER.

dwarf bean, ▷ FRENCH BEAN.

dystrophic, (of lakes) having a high organic content, but low concentrations of mineral salts; ▷ p. 218.

eagle, any of about 30 large, powerful BIRDS OF PREY in the family Accipitridae (order Ciconiiformes) found worldwide and characterized by feathered legs, and by brows above the eyes to cut down glare. The Palaearctic golden eagle (*Aquila chrysaetos*) is a magnificent soarer, searching for hares and game birds. The American bald eagle (*Haliaeetus leucocephalus*) feeds largely on fish; ▷ pp. *92*, *96*, *97*, *151*, *201*.

eardrum, a non-technical name for the tympanic membrane; ▷ p. 191.

earthworm, any terrestrial ANNELID of the class Oligochaeta. They are very important in circulating nutrients in the soil; ▷ pp. **55**, 176 (movement), *178* (digestion), 195, 207, 209 (role in ecosystems).

earwig, any EXOPTERYGOTE insect of the order Dermaptera, comprising over 1200 species worldwide, with little variation between species. Their hind wings are fan-like, complexly folded, and protected by short hardened fore wings. On the tip of the abdomen there is a pair of forceps, used for defence, holding prey, and in courtship. Earwigs are omnivorous, and can be a minor horticultural pest by damaging flower blossoms; others may be of value by eating aphids. Some species display parental care of the developing young; ▷ pp. 60–1.

ebony, the hard, heavy, dark heartwood obtained mainly from species of small trees in the genus *Diospyros*, e.g. Macassar ebony (*D. ebenum*), the heartwood of which is black. See also PERSIMMON.

ecdysone, in insects, a HORMONE that initiates MOULTING; ▷ pp. 62, 183.

echidna or **spiny anteater,** either of two species of monotreme mammal of the family Tachyglossidae, occurring exclusively in Australasia. Both species – the long-beaked and the short-beaked – are covered in long protective spines, and are principally insectivorous; ▷ pp. 98 (box), 100, **102**.

echinoderm, any marine invertebrate of the phylum Echinodermata, comprising about 6000 species and including sea stars (starfishes), sea lilies, sea urchins, brittlestars, and sea cucumbers; ▷ pp. 51 (box), **54**.

echolocation, a system of non-visual orientation and detection involving the emission of high-frequency sounds and monitoring their echoes from intervening objects (including items of prey). Echolocation is used by most bats (▷ pp. 118, 165), many toothed whales (▷ p. 144), and a few birds (▷ p. 97).

ecological gradient, the continuous variation in space of an ecological factor, such as light, temperature, humidity, etc. Different organisms display different limits and optimums for growth along such a gradient.

ecology, the study of living organisms in relation to their physical and biological environment.

ecosystem, a complex system involving a living community and its non-living setting; ▷ pp. 194–5, 200–21.

ectoderm, the outer cell layer in a developing animal embryo, giving rise to nervous tissue, etc.; ▷ p. 186.

ectoplacental cone, ▷ p. *186*.

ectoplasm, the gel-like outer layer of the CYTOPLASM of many amoeboid cells and some plant (especially meristematic) cells. Compare ENDOPLASM.

ectoproct, ▷ BRYOZOAN.

ectothermic, ▷ POIKILOTHERMIC.

edelweiss (*Leontopodium alpinum*), a European alpine herb in the daisy family, Asteraceae (formerly Compositae), with petal-like bracts round the flower heads.

edentate, any mammal of the order Edentata, a predominantly South American group that includes anteaters, sloths and armadillos; ▷ pp. *101*, **106–7**.

eel, any fish of the order Anguilliformes, comprising some 600 species in 19 families. They are characterized by their snake-like body form and the absence of pelvic (and sometimes pectoral) fins. This body form allows them to insinuate themselves into small spaces under stones, between vegetation, etc., and some species are able to wriggle over land in a serpentine fashion. Eels are marine fishes found in all oceans, but the members of the best-known family (Anguillidae) move between fresh and salt water, and are often called 'freshwater eels'. The larvae and young of the European freshwater eel undertake a long migration across the Atlantic, between the Sargasso Sea and European rivers (▷ p. 170). The larvae (known as leptocephali) are transparent with ribbon-shaped bodies (▷ p. 74). The name 'eel' is also given to various elongated fishes from other orders, such as the SPINY EELS and the electric eel (▷ KNIFEFISH). See also MORAY, CONGER, GULPER.

eelgrass, any of around 12 species of perennial grass-like herbs of shallow salt water in the temperate genus *Zostera* in the family Zosteraceae.

eelworms, a group of ROUNDWORMS, including both free-living and parasitic forms; e.g. the wheatworm forms galls in wheat seeds.

efferent nerve, a nerve cell carrying information from the central nervous system to outlying systems, such as a motor nerve stimulating a muscle system; ▷ p. 182.

efficiency, (in ecology) the degree to which the energy available to an animal is actually used and assimilated in its diet. It is measured by the ratio of energy assimilated to energy consumed; ⇨ p. 195.

egg, a round or oval reproductive body produced by many animals, containing the developing embryo, a food supply (yolk) and various associated structures, and surrounded by some form of protective shell or coat; ⇨ pp. 86 (reptiles), 95 (birds), **184–7.** Confusingly, the term 'egg' may also be used to refer to an unfertilized egg cell or OVUM.

eggar, ⇨ LAPPET MOTH.

egg cell, ⇨ OVUM.

egg cylinder, ⇨ p. *186.*

egg-eating snake, any African or Indian COLUBRID snake of the genera *Dasypeltis* and *Elachistodon* that feeds almost exclusively on eggs, breaking the shells by muscular action in the neck.

eggplant, ⇨ AUBERGINE.

eglantine, ⇨ SWEET BRIAR.

egret, any of eight species in the family Ardeidae (order Ciconiiformes), similar in appearance to the closely related HERONS. Their breeding plumage includes long delicate wing plumes. A widespread species is the common egret (*Casmerodius albus*); ⇨ pp. 97 (box), 168.

eider, any of four heavily built diving DUCKS, circumpolar in distribution. The king eider (*Somateria spectabilis*) is one of the most spectacularly coloured ducks; ⇨ pp. 93, 97 (box).

eland (*Taurotragus oryx*), a large, stocky, grey or brown ANTELOPE with black and white leg markings, a short mane and spiral horns, inhabiting grassland in Africa. The giant eland (*T. derbianus*) is largely confined to game parks in Namibia and is endangered (⇨ p. *114*).

elapid, any venomous SNAKE of the family Elapidae, containing some 200 species of widespread distribution outside Europe. The group includes COBRAS, CORAL SNAKES, KRAITS and MAMBAS. Like the VIPERS, elapids have fangs at the front of the mouth, but they are short and fixed in position (⇨ p. *87*).

elder (*Sambucus nigra*), a European and North African shrub in the family Caprifoliaceae. It is common in wasteland and roadsides. Both the flowers and fruits are used in home wine making.

electromagnetic sense, the ability to use electrical or magnetic fields for purposes of communication, orientation, aggression, etc.; ⇨ pp. 77, 80, 102, **154–5,** 171, 191 (box).

electroreceptor, a sensory receptor sensitive to electrical energy; ⇨ pp. 77, 102, 191 (box).

elephant, either of two trunked mammals of the family Elephantidae, the largest living land animals: the African elephant (*Loxodonta africana*) and the Indian elephant (*Elephas maximus*); ⇨ pp. *9, 101,* **108–9,** *154,* 162, *201, 202–3.*

elephant grass (*Pennisetum purpureum*), a grass originating from Uganda but now cultivated throughout the tropics as a fodder crop.

elephantiasis, ⇨ p. 168.

elephant seal, either of two large earless SEALS of the genus *Mirounga*. The southern elephant seal (*M. leonina*) is the largest seal; ⇨ p. *143,* 157, 159 (box).

elephant shrew, any mammal of the family Macroscelididae (order Macroscelidea), comprising some 15 species of widespread distribution in Africa. Elephant shrews are shrew-like in appearance, with extremely long, mobile snouts, and feed mainly on invertebrates; ⇨ pp. *101,* 120, **121** (box).

elimination, the expulsion of waste products from the body by excretion and defecation; ⇨ p. 179.

elk, the name commonly used for the MOOSE in the Eurasian part of its range. In American usage, the WAPITI is usually referred to as an 'elk'; ⇨ p. 116.

elm, any of around 30 species of large tree in the genus *Ulmus* native to north temperate regions and mountains of tropical Asia. *U. minor* was once common in southern England but was decimated by Dutch elm disease caused by a virulent strain of the ASCOMYCETE fungus, *Ceratocystis ulmi*, the spores of which are spread by elm bark beetles (*Scolytus* spp.).

embryo, a developing plant or animal from first cleavage of the fertilized egg (zygote) to seed germination (in plants), birth or hatching; ⇨ pp. 14, *15,* **186–7** (animals), 36, 38, 41 (plants).

embryology, the development of an embryo, or the branch of science concerned with the study of embryos; ⇨ pp. 14, *15,* **186–7.**

embryo sac, the structure in the ovule of the flowering-plant ovary in which the ovum (egg) and other specialized cells are formed; ⇨ p. 41.

emergent, 1. a tree that rises above the forest canopy; ⇨ p. 201. **2.** any vegetation that rises above water; ⇨ pp. 216–17.

emigration, a movement of animals out of an area. In contrast to MIGRATIONS, such movements are usually non-cyclical (not involving a return journey) and occur as a consequence of more or less unpredictable factors, such as food shortage or population density; ⇨ pp. 135, **172–3.**

emotion, a number of emotions or 'inner feelings', such as friendliness, anger, frustration, joy and FEAR, are often ascribed to animals, especially domestic animals and higher primates. However, although some or all of these emotions (notably fear and frustration) may be experienced by many animals, their inner feelings are in fact unknowable, and a degree of ANTHROPOMORPHISM is generally involved in our perception or interpretation of them.

emperor moth (*Saturnia pavonia*), a large European SATURNIID moth, with eye-spots on the wings.

emu (*Dromaius novaehollandiae*), a large nomadic flightless bird of the Australian scrubland, growing to 2 m (6½ ft). It has three toes and a partially feathered neck, and belongs to the same family as the CASSOWARIES; ⇨ p. 97 (box).

Enaliornis, ⇨ p. 90.

enamel, a glossy substance composed of calcium phosphate that forms a thin, very hard layer covering the teeth of vertebrates.

endemic, (of a species) native to a particular, usually restricted, geographical region, either through having evolved there or as a result of having become extinct elsewhere; ⇨ p. 214.

endive (*Cichorum endivia*), an annual or biennial herb, related to CHICORY, in the daisy family, Asteraceae (formerly Compositae). Its young leaves are used in salads.

endocrine system, a system of GLANDS that produce chemicals (hormones) that are transported in the blood and carried to distant tissues, whose activity they modify. The system is principally involved in the control and regulation of growth, reproduction, the internal environment, and energy production; ⇨ p. 183.

endocytosis, the importation of materials into a cell; ⇨ p. 21. Compare EXOCYTOSIS.

endoderm, the inner cell layer in a developing animal embryo, giving rise to the respiratory- and digestive-tract linings, etc.; ⇨ p. 186.

endomembrane system, the principal linked membrane within cells apart from the outer cell membrane; ⇨ p. 21.

endoparasite, an internal parasite; ⇨ p. 168.

endoplasm, the inner CYTOPLASM of cells. Compare ECTOPLASM.

endoplasmic reticulum, a complex network of membranes within eukaryote cells; ⇨ pp. **20–1** (box), 21.

endopterygote, any PTERYGOTE insect of the group Endopterygota, comprising nine orders and over 85% of all known insect species. Familiar endopterygotes include beetles, butterflies and moths, flies, and bees, ants and wasps. The group is characterized by complete METAMORPHOSIS, i.e. juveniles bear no similarity to adults, and the adult form is attained via larval and pupal stages; ⇨ pp. 60–1, 62–3.

endosperm nucleus, ⇨ p. 41.

endothermic, ⇨ HOMOIOTHERMIC.

energetics, the study of how energy flows within an ecosystem: the routes along which it moves, how efficiently it is used, and where it is stored (⇨ pp. 194–5).

energy flow, the flow of energy through an ecosystem; ⇨ pp. **194–5,** 212–13 (agricultural ecosystems).

enzyme, any protein that acts as a catalyst in a biochemical reaction; ⇨ pp. 6 (box), 179.

Eocene epoch, a geological epoch lasting from 53 to 34 million years ago; ⇨ p. 4.

eosuchian, any member of an early group of DIAPSID.

Ephedra, a genus of gnetophyte plants; ⇨ p. 39.

ephemeral, a species that completes several life cycles in a season, and can spend considerable time in a dormant state (such as a seed) awaiting appropriate conditions.

epidermis, 1. (in plants) the thin protective tissue surrounding young roots, stems and leaves. Stem and leaf epidermis secretes a CUTICLE. In older stems and roots the epidermis is often replaced by CORK; ⇨ p. *32.* **2.** (in animals) the outermost layer or layers of cells. In vertebrates, the epidermis covers the inner layer of skin (DERMIS), and is often formed from dead, hardened cells.

epiglottis, a thin plate of flexible cartilage in front of the opening to the upper windpipe (larynx); it folds back and protects the opening during swallowing.

epiphyte, any plant that grows on other plants without damaging them; ⇨ pp. 44, *46,* 200.

epoch, the smallest unit of geological time, into which the two most recent geological periods are divided; ⇨ p. 4.

equid, any PERISSODACTYL mammal of the family Equidae (7 species), comprising horses, asses and zebras; ⇨ pp. 110–11.

Equisetum, the genus comprising the horsetails; ⇨ p. 37.

era, the largest unit of geological time, divided into periods; ⇨ p. 4.

ergot, a disease of cereals and other grasses caused by ascomycete fungi (⇨ p. 35) of the genus *Claviceps.* Infected grain contains alkaloids with both poisonous and therapeutic properties.

Erica, a genus of around 500 species of evergreen shrubs generally referred to as heaths (but also including bell HEATHER) in the family Ericaceae. Most are native to South Africa but some are Mediterranean or European.

Ericaceae, a dicotyledonous plant family, including heathers, rhododendrons, cranberry, etc.; ⇨ p. 45.

ermine moth, any LEPIDOPTERAN insect (moth) of the genus *Spilosoma,* in the same family as the TIGER MOTHS. They have dark spots on light-coloured wings.

erythrocyte, a red BLOOD cell.

esparto, a fibre derived from the leaves of two Mediterranean grasses: alfa (*Stipa tenacissima*) and *Lygeum spartum.*

essential oil, any of various volatile oily liquids obtained from plants by distillation or volatilization. Possessing the flavour and smell of the particular plant from which they are extracted, essential oils are used in flavouring and perfumery.

estuary, the lower, tidal section of a river where it enters the sea; ⇨ pp. 216–17.

Ethiopian, a faunal region; ⇨ p. 50 (map).

ethology, the scientific study of animal behaviour; ⇨ pp. 148–9.

eubacteria, rod-shaped or spherical members of the Eubacteriales. Some motile forms have many flagella; ⇨ p. 24.

eucalyptus or **gumtree,** any of around 175 species of evergreen trees and shrubs in the mainly Australasian genus *Eucalyptus*. These trees predominate in the forests of Australia and many, such as the ironbarks (e.g. *E. fergusoni* and *E. crebra*) are important for hardwood timber production. The flowers are a major nectar source for honeybees; ⇨ pp. 30, 203, 205.

euglenoid, any member of the protistan group Euglenophyta, classifiable either as algae or as flagellate protozoans. Euglenoids are large flagellates with a flexible PELLICLE instead of a cell wall. They are green or colourless; ⇨ pp. 26–7, 34.

eukaryote, any organism whose cell or cells are eukaryotic, i.e. in which the genetic material is contained within a distinct nucleus. Protists and all higher organisms are eukaryotes; ⇨ pp. 7, **20**. Compare PROKARYOTE.

Euphorbia, a tropical and temperate genus of around 2000 species of trees, shrubs or herbs with a poisonous milky juice. Many herbaceous forms are referred to as spurges, and many species are adapted to arid conditions (⇨ p. 204). Poinsettia (*E. pulcherrima*) is a popular Christmas pot plant.

Euphorbiaceae, a dicotyledonous plant family, including euphorbias and spurges; ⇨ p. 45.

euphotic zone, a marine zone; ⇨ p. 220.

eurypterid, a SEA SCORPION.

Eustachian tube, in the mammalian ear, ⇨ p. *191*.

eutherians, the PLACENTAL MAMMALS; ⇨ pp. 100–1.

eutrophic, (of a lake) having nutrient-rich water; ⇨ p. 218.

eutrophication, a process caused by excess nitrates that leads to the death of all life in bodies of water; ⇨ p. 226 (box).

evening primrose, any plant of the American genus *Oenathera*, family Onagraceae, particularly *O. biennis*, which has large yellow flowers that open in the evening.

even-toed ungulate, ⇨ ARTIODACTYL.

evergreen, ⇨ DECIDUOUS.

evolution, any of various theories that explain the origin of species through gradual changes in ancestral groups. The generally accepted theory today is Darwin's theory of evolution by natural selection; ⇨ pp. 12–17.

excretion, the process whereby waste products are removed from body fluids and expelled from the body; ⇨ pp. 26 (protists), 59 (insects), 88 (reptiles), 93 (birds), **179** (box). See also OSMOREGULATION.

exocytosis, the export of materials out of a cell; ⇨ p. 21. Compare ENDOCYTOSIS.

exopterygote, any PTERYGOTE insect of the group Exopterygota, comprising 16 orders and roughly 12% of all known insect species. Familiar exopterygotes include grasshoppers, dragonflies, earwigs and aphids. The group is characterized by incomplete METAMORPHOSIS, i.e. juveniles resemble adults and there is no pupal stage; ⇨ pp. 60–1, 62–3.

exoskeleton, the hard external SKELETON of insects, crustaceans and other arthropods, formed from a protective layer (the CUTICLE) that is based on the polysaccharide CHITIN; ⇨ pp. **58**, 176.

exotic, a species derived from another area that has been introduced into a site. If it persists without constant management it is said to be naturalized. Compare INDIGENOUS.

extinction, 1. any of various processes – such as changes in climate, habitat, predators, etc. – by which a species dies out; ⇨ pp. 4–5, 84 (box), 214, 228–9, 232–3. **2.** the process by which a learned behavioural response wanes through lack of REINFORCEMENT; ⇨ p. 150.

extracolumella, in the vertebrate ear, ⇨ p. *191*.

eye, a more or less complex organ of SIGHT, based on a group of associated photosensory cells; ⇨ pp. 189–90.

eyebright, any of around 200 species of semiparasitic annual herbs in the temperate genus *Euphrasia*. Extracts of *E. officinalis* used to be used as an eyewash.

Fabaceae or **Papilionaceae,** a dicotyledonous plant family (formerly part of the Leguminosae, the legume family), including peas, beans, peanut, clovers, brooms, lupins, etc.; ⇨ pp. *32* (box), 45.

faeces, the semi-solid waste products of digestion discharged through the anus; ⇨ p. 179.

Fagus, ⇨ BEECH (1).

fairy or **brine shrimp,** any small, primitive CRUSTACEAN of the class Branchiopoda, found in abundance in temporary pools; ⇨ p. 67.

fairy bluebird, either of two PASSERINE birds of the genus *Irena*. They are starling-sized birds, the male of which has a beautiful ultramarine back. See also BLUEBIRD.

fairy fly, any minute parasitic WASP of the family Mymaridae, of worldwide distribution. Among the smallest of all insects, fairy flies lay their eggs in the eggs of other insects; one host egg is sufficient for the complete development of the fairy fly.

fairy ring, a circle or ring of fungal fruiting bodies (e.g. toadstools) – commonly of *Marasmius oreades* – formed on the periphery of the underground mycelium (network of fungal filaments).

fairy wren, any of 15 PASSERINE birds in the family Maluridae; tiny, delicate, ground-dwelling foragers with long, cocked tails, living in Australia and New Guinea. The males show beautiful plumage of turquoise, reds, blacks and whites.

falcon, any of about 40 powerful BIRDS OF PREY in the family Falconidae (order Ciconiiformes), which also includes the CARACARAS. They have pointed wings and a square tail, and are distributed worldwide. The peregrine falcon (*Falco peregrinus*) 'stoops' by circling high up and then folds its wings back to dive at prey. During the stoop the bird can reach speeds of 290 km/h (180 mph); ⇨ p. 96. See also KESTREL, HOBBY.

falconet, any of four species of small FALCONS, living in the Orient and feeding chiefly on insects.

Fallopian tube, the upper part of the female reproductive tract in mammals, receiving the ova (eggs) after ovulation and where fertilization occurs. The paired Fallopian tubes lead into the uterus.

false vampire bat, any of five microchiropteran BATS of the family Megadermatidae, widely distributed in tropical regions of the Old World. The group is one of the few predominantly carnivorous bat families (⇨ p. 119). The American false vampire is one of the New World LEAF-NOSED BATS.

family, 1. a classificatory group between order and genus. For example, the gorilla, chimpanzees and humans make up the family Hominidae within the order Primates. In zoology, family names end with -idae, and in botany with -aceae; ⇨ pp. 18, 45. **2.** a form of social organization in which both parents remain with the young for a given period. Often the female alone shows parental care (one-parent families), and occasionally the offspring may remain to help rear succeeding generations (extended families); ⇨ p. 158.

fanalouc (*Eupleres goudotii*), a brown, fox-like VIVERRID with a narrow, pointed muzzle and a bushy tail. It is found in rainforest in Madagascar.

fan worm, any tube-dwelling marine POLYCHAETE of the family Sabellidae; they have a fan of filterfeeding tentacles, which they withdraw rapidly into their tubes if disturbed; ⇨ p. *55*.

farming, ⇨ AGRICULTURE.

fat, any of a number of simple lipids. Fats are important energy-storage molecules found in virtually all organisms. They consist of one glycerol and three fatty-acid molecules. They are also important for their insulating and protecting qualities. In animals fats are stored in specialized tissue called adipose tissue, and in plants (in the form of oils, which are liquid fats) are particularly abundant in seeds, but are also found in other parts of the plant; ⇨ pp. **31**, 32, 45, **179**.

fatty acid, any of a number of organic molecules that are components of lipids, and, with glycerol, form basic components of FATS.

fauna, animals collectively, especially all the animals in any one location.

faunal region or **zoogeographic region,** any of several regions into which the world is divided on the basis of the similarities of their animal inhabitants; ⇨ p. 50 (map).

fear, a state of MOTIVATION arising (innately or through learning) as a response to certain stimuli, and usually resulting in some form of DEFENCE behaviour. See also EMOTION.

feather, any of the keratin-based epidermal structures of a bird that together make up its plumage; ⇨ pp. 90, **92–3**, 96 (box).

featherback or **knifefish,** any African or Southeast Asian freshwater fish of the family Notopteridae, in the same order as the BONY TONGUES. They are distinguished by a long, undulating anal fin (supposedly resembling a blade, hence 'knifefish') and a tiny feather-like dorsal fin.

feather louse, ⇨ BITING LOUSE.

feather star, an alternative name for SEA LILY.

feathertail possum (*Distoechurus pennatus*), a small, omnivorous, arboreal MARSUPIAL of the family Burramyidae.

feature detector, a specialized type of sensory cell; ⇨ p. 188.

feeding, ⇨ FORAGING.

felid, any mammal of the family Felidae (order Carnivora), the CAT family; ⇨ pp. 122, **124–5**.

femur, the long bone nearest the body in the hind or lower limb in vertebrates, or the third segment from the base of an insect's leg (⇨ box, p. 59).

fen, a wetland in which the summer water table is at or below the ground surface, but is waterlogged in winter. Nutrient supply is by groundwater flow.

fennec fox (*Vulpes zerda*), a small, fox-like CANID with large ears, bushy tail and pale fur. It is nocturnal, and found in the deserts of North Africa, Sinai and Arabia; ⇨ p. *126*.

fennel (*Foeniculum vulgare*), a herb belonging to the carrot family, Apiaceae (formerly Umbelliferae), naturalized in most temperate countries. The leaves and seeds are used for flavouring fish. See also ANISE.

fenugreek (*Trigonella foenum-graecum*), an annual Old World leguminous herb. It is used as a fodder crop and the seeds are ground for use in curries.

feral, (of a previously domesticated animal) having reverted to the wild. Feral animals have created problems for certain wild species, especially on islands (⇨ p. 214).

fer-de-lance (*Trimeresurus* or *Bothrops atrox*), a large (up to 2.5 m/8¼ ft) PIT VIPER that ranges widely from southern Mexico through South America.

fermentation, the stepwise breakdown of glucose

or other organic molecules under anaerobic (oxygen-free) conditions to yield energy and various metabolic intermediates – such as alcohol and other industrial products; ▷ pp. **25**, 112 (box).

fern, any member of the plant division Pteridophyta; ▷ pp. 36–7.

ferret, name given to several small, brownish MUSTELIDS, with elongated body, short legs and a long bushy tail. The black-footed ferret (*Mustela nigriceps*) is found in North America; *M. putorius furo* is the domesticated form of the polecat.

fertilization, in SEXUAL REPRODUCTION, the fusion of sex cells (GAMETES) to form a ZYGOTE. In animals, fertilization may take place externally, as in most fishes and amphibians such as frogs and toads, or internally, as in insects, birds and mammals (▷ pp. 184–5). Fertilization in plants takes a variety of forms; ▷ pp. 36, 38, 41.

fertilizer, any material, natural or artificial, added to soil to increase its fertility; ▷ pp. 212, 213, **226**.

fescue, any of around 300 species of the temperate grass genus *Festuca*. Some species such as red fescue (*F. rubra*) are used in lawns.

fetus or **foetus,** a mammalian embryo following the development of various major organ systems, when recognizably similar to the adult form; ▷ pp. 185, **186**.

feverfew (*Tanacetum parthenium*), an aromatic perennial herb used in herbal medicine, the Latin origin of the common name meaning 'that which puts fever to flight'. See also CHRYSANTHEMUM.

fibre, 1. (in higher plants) an elongated, tapered SCLERENCHYMA cell of high tensile strength found in the vascular tissues (XYLEM, PHLOEM) and sometimes other tissues. Fibres of commercial use include flax and hemp. 2. ▷ MUSCLE.

Ficus, ▷ FIG.

fiddler crab, any DECAPOD crustacean of the genus *Uca*, the males of which have one of their pincers greatly enlarged; ▷ pp. 154, 174–5.

fieldfare (*Turdus pilaris*), a northern Palaearctic THRUSH of open pastureland. Unlike many of the larger thrushes, it is colourful with a grey head and chestnut back.

fieldmouse or **wood mouse,** any of several long-tailed, seed-eating myomorph RODENTS of the genus *Apodemus* (family Muridae), including *A. sylvaticus*, widespread in grasslands and forests in Eurasia. There are also numerous South American fieldmice (*Akodon* species, etc.).

fig, any of around 2 000 species of trees, shrubs and climbers in the pantropical genus *Ficus*; ▷ box, p. 43, for the details of the flowers and their pollination. The fig fruit develops from the whole inflorescence. The edible fig (*F. carica*) has probably been cultivated for more than six millenia.

fighting fish, any of several small tropical freshwater PERCIFORM fishes of the family Belontiidae, especially the Siamese fighting fish (*Betta splendens*). In Thailand males of this species have been selectively bred for aggressiveness and are placed together in staged combats, on which bets are laid; ▷ pp. 155 (box), 157 (box).

fig wasp, any small WASP of the family Agaonidae, a group intimately involved in the pollination of various species of fig; ▷ p. 43 (box).

filariasis, ▷ p. 168.

filbert, ▷ HAZEL.

filefish, ▷ TRIGGERFISH.

filter- or **suspension-feeding,** sieving, straining or capturing floating particles from water, an important feeding method for many marine animals; ▷ pp. **57** (box), 178, 219, 221. Other references: sponges 52; coelenterates 53; echinoderms 54; worms *55*; molluscs 56, *57*; crustaceans 66–7; fishes 76.

fin, any of the projecting structures by which fishes swim, steer, maintain balance, etc. Fins are supported by rays (cartilaginous, horny or bony), which may be soft or hard. The dorsal, caudal (tail) and anal fins are single, while the pectoral and pelvic fins are paired; ▷ pp. **72**.

finback, another name for RORQUAL.

finch, common name given to a variety of small primarily seed-eating PASSERINE birds with pointed conical bills. Many New World BUNTINGS are called finches. True finches belong to the subfamily Fringillinae (family Fringillidae), and although they occur worldwide, they are found mainly in the temperate northern hemisphere, where they are migratory. True finches have nine primary wing feathers and all but three of the 125 or so species feed seeds to their young (most other seed-eaters switch to an insect diet during rearing). See also CHAFFINCH, CANARY, SISKIN, GOLDFINCH, GREENFINCH, LINNET, REDPOLL, CROSSBILL and BULLFINCH. Other birds of the Old World called finches have 10 primary wing feathers and include the estrildine finches (▷ WAXBILL), ploceine finches (▷ WEAVER, QUELEA, WHYDAH) and the passerine finches (▷ SPARROW); ▷ pp. 155 (box), *175*.

finfoot or **sungrebe,** any of three species of rail-like waterbirds with long-lobed toes, long neck and powerful bill, belonging to the family Heliornithidae (order Gruiformes). They are found in the Neotropics, Africa and Asia; ▷ p. 93.

fin whale (*Balaenoptera physalus*), a baleen whale, one of the RORQUALS; ▷ p. 145.

fir, any species of conifer in the genera *Abies* (around 50 species) and *Pseudotsuga* (around 7 species; ▷ DOUGLAS FIR), both in the family Pinaceae; ▷ p. 39. *Abies* species are north temperate trees and many provide commercially important timber, which is referred to as DEAL. Turpentine and resin is obtained from the bark.

fire. For the impact of fire on ecosystems, and adaptations of the vegetation ▷ pp. 39, 201, 202, 205, 208–9, 228.

fire ant, any ANT of the genus *Solenopsis*, a widespread group of predatory mound-building ants. They can be pests, readily entering homes and capable of giving a painful sting.

fire-bellied toad, any of several discoglossid ANURANS related to the MIDWIFE TOADS.

firefly, any BEETLE of the mainly tropical family Lampyridae (1700 species). These nocturnal beetles produce spectacular light shows as part of their courtship display. The light is produced by a special organ on the abdomen, by the breakdown of a luminescent substance, luciferin, by the enzyme luciferase. They are sometimes referred to as 'glowworms', especially the European species *Lampyris noctiluca*; ▷ pp. 62, 154.

fireweed, ▷ WILLOWHERB.

fish, common name given to the members of a very diverse assemblage of cold-blooded (poikilothermic) aquatic vertebrates. The 'characteristic' features of fishes include gills for respiration, fins for steering and locomotion, and a covering of scales, but these features are far from being common to all fishes. The name 'fish' describes a life form rather than a particular taxonomical grouping, and the many thousands of fish species belong to several distinct evolutionary lines; the dominant groups of living fishes are the BONY FISHES and the CARTILAGINOUS FISHES; ▷ pp. *15*, **70**–7, *197*, *215*, 217, **218–21**.

fisherman bat, ▷ BULLDOG BAT.

fishing cat (*Felis viverrina*), a pale brown CAT with dark spots and partially webbed feet, inhabiting forests and marshes in Southeast Asia and India. It feeds on fish and other prey.

fish louse, ▷ BRANCHIURAN.

fission, an asexual method of cell division that does not involve mitosis (▷ p. 22). It occurs in prokaryotes such as bacteria and in some primitive eukaryotes such as protists; ▷ pp. 25, 27.

fixation, the process whereby an element in an inorganic form (such as carbon in carbon dioxide, or nitrogen gas) is converted to an organic form by a living organism; ▷ pp. **32** (box), 169, 196–7.

flag, any of a number of IRIS species and cultivars, e.g. yellow flag (*Iris pseudacorus*) and blue flag (*I. versicolor*).

flagellate, (of a cell) possessing FLAGELLA.

flagellum (pl. -ella), a whip-like structure projecting from a cell that acts as a locomotive organ in single-celled organisms; ▷ pp. 7, **26–7**, *176*.

flamingo, any of five species of the family Phoenicopteridae (order Ciconiiformes), slender water birds with very long necks and legs. The bill is large and angled, and is held upside down and used to filter minute aquatic animals and plants from the water. The greater flamingo (*Phoenicopterus ruber*) of Central and North America, Africa and the Mediterranean is the most common species; ▷ pp. *91*, *93*, 96 (box).

flash coloration, a defensive ploy that aims to startle a predator by the sudden exposure of brightly coloured body parts that are normally covered; ▷ p. 167.

flatfish, any of about 538 (almost exclusively) marine fishes of the order Pleuronectiformes, mainly occurring in relatively shallow waters (under 200 m/656 ft) of tropical and temperate seas. Young flatfishes are symmetrical, but during development one eye migrates over the head to lie next to the one on the other side; the flattened adult then swims with its eyeless side facing downwards. Usually this is the left side, but sometimes it is the right, and some groups contain individuals that are either right or left. Most flatfishes lie on the bottom, and many are well known for their ability to change colour to merge with their background. Many species are important food fishes; ▷ pp. 72, 74, *76*. See also FLOUNDER and SOLE.

flatworm, any simple invertebrate of the phylum Platyhelminthes, with a flattened body and a single opening to the gut. They are both free-living (e.g. turbellarians) and parasitic (e.g. tapeworms, flukes); ▷ pp. 55, 168, *183*, 218.

flax or **linseed** (*Linum usitatissimum*), an annual herb in the family Linaceae cultivated for its stem fibre (used for linen), its seed oil, and as a source of animal feed. Many other species of *Linum* are also referred to as flax.

flea, any INSECT of the order Siphonaptera, containing 1400 known species. They are totally parasitic as adults, feeding on their host's blood; they are laterally flattened, allowing them to move through fur easily, and they can jump by means of specialized, highly elongated hind legs. The larvae feed on skin flakes and other protein-rich food in the host's nest. Fleas transmit diseases such as plague in humans and myxomatosis in rabbits; ▷ pp. 60 (box), 64, 177, *180*.

flehmen, curling of the upper lip of mammals (generally males) when exposed to urine from e.g. oestrous females. It probably enhances perception by the VOMERONASAL ORGAN.

flesh fly, any moderately sized, chequered grey DIPTERAN insect of the genus *Sarcophaga* in the family Calliphoridae, which also includes the BLOWFLIES. The larvae feed on decomposing flesh.

flight, 1. as a means of locomotion, ▷ pp. 59, 84, 90–7, 118–19, *176*, **177**. 2. running from a predator, an initial defensive ploy used by most mobile animals; ▷ p. 166.

flocking, behaviour that results in birds coming, remaining and acting together through social attraction, probably for defence against predators (▷ pp. 97, 167).

flora, plants collectively, especially all the plants in any one location.

floral region or **phytogeographic region,** any of several regions into which the world is divided on the basis of the similarities of their plant inhabitants; ▷ p. 30 (map).

flounder (*Platichthys flesus*), an abundant European FLATFISH (family Pleuronectidae), important as a food fish. It penetrates fresh water and is found in the lower reaches of rivers, as well as in the shallow waters of most European seas. Especially in US and Canadian usage, the name 'flounder' may be applied to any flatfish of the families Bothidae (the 'left-eyed flounders', including TURBOT) and Pleuronectidae (the 'right-eyed flounders', including HALIBUT, PLAICE and DAB).

flour beetle, any of various small reddish-brown TENEBRIONID BEETLES of the genus *Tribolium*, which can be pests in flour mills.

flower, the part of a plant containing the sexual organs. True flowers are only found in the division Magnoliophyta (the flowering plants or angiosperms); ▷ pp. **40–1**, 42–7.

flowering plant, any member of the plant division Magnoliophyta, i.e. any ANGIOSPERM; ▷ pp. 38, **40–1**, 42–7.

fluke, any of nearly 6000 FLATWORMS of the class Trematoda, an extremely important group of internal and external parasites; ▷ pp. **55**, 168.

fly, common term applied (with no taxonomical significance) to numerous flying insects. More specifically, however, the name is given to the 'true' flies of the order Diptera (▷ DIPTERAN).

fly agaric, ▷ AMANITA.

flycatcher, any of 155 PASSERINE birds in the subfamily Muscicapini (family Muscicapidae), closely related to the CHATS, insectivorous birds of the Old World and Australasia. Most are small with long tails and large eyes. Those that catch insects on the wing have broad bills, while those that pick insects from the ground have thin bills; ▷ pp. 11, 209. See also MONARCH FLYCATCHER, TYRANT FLYCATCHER.

flying dragon, (*Draco volans*), a Southeast Asian AGAMID lizard that has a membrane supported by folding skin-covered ribs, allowing it to glide considerable distances from tree to tree; ▷ p. 84.

flying fish, any of 48 species of the family Exocoetidae (order Cyprinodontiformes), which are able to make gliding 'flights' over the surface of the sea. The family includes both four-winged and two-winged forms, depending on whether the pelvic fins are enlarged in addition to the pectoral fins; ▷ p. 72. Members of certain other groups have also devised ways of leaping clear of the surface and gliding.

flying fox, ▷ FRUIT BAT.

flying frog, any TREE FROG of the family Rhacophoridae, found in South and Central America and Southeast Asia. These frogs are capable of controlled gliding using outstretched webbing between the fingers and toes; ▷ p. 160.

flying gurnard, any of four tropical marine fishes of the family Dactylopteridae (order Dactylopteriformes). They have large blunt heads protected by hard bony plates, and the body is covered with hard scales. The pectoral fins are greatly expanded, giving rise to the (unsubstantiated) notion that these fish are capable of flying or gliding.

flying lemur, another name for COLUGO.

fodder crop, any crop grown to feed livestock either directly by grazing or by feeding as hay or silage. The most important are grasses such as ryegrass, and legumes such as lucerne.

foetus, variant spelling of FETUS.

follicle, 1. a dry, DEHISCENT fruit found in certain plants (e.g. delphiniums). It is formed from a single CARPEL, which splits laterally at maturity releasing seeds. **2.** (in animals) any small sac or cavity in the body (e.g. a hair follicle), usually with a secretory or excretory function.

food chain or **food web,** a pattern of feeding relationships (i.e. energy transfers) within an ecosystem; ▷ pp. 194–5.

football fish, any of four deep-sea ANGLER FISHES of the family Himantolophidae. They have spherical bodies studded with bony plates, and their lure is large and luminous.

foraging, behaviour associated with gathering food, from the passive FILTER-FEEDING of many marine animals, to active PREDATION; ▷ pp. 50, 156–7, **162–5**, 168–9, 170, 172–3, 178–9. Other references: insects 64–5; arachnids 68–9; fishes 75–6; amphibians 81; reptiles 89; birds 96; mammals 102–45.

foraminifera, a group of amoeboid protozoans, many of which have a calcareous shell. The remains of these shells have sedimented in vast numbers to form chalk; ▷ pp. 5, 27, 160.

forest, an area, usually extensive, in which trees are the dominant plants; ▷ pp. 200–1, 207, 208–9.

forester, any LEPIDOPTERAN insect of the genus *Ino* (family Zygaenidae), essentially Palaearctic in distribution. They are metallic-green, day-flying moths.

forget-me-not or **scorpion grass,** any of around 50 species of annual or perennial herbs in the temperate genus *Myosotis* in the family Boraginaceae.

forktail, common name for a DIPLURAN insect.

forsythia, any of around 6 species of shrubs, some popular in gardens, in the Asian genus *Forsythia* in the olive family, Oleaceae.

fossa (*Cryptoprocta ferox*), a brown, cat-like, semiarboreal VIVERRID with retractile claws. It inhabits forests in Madagascar; ▷ p. 123.

fossil, the impression of a long-dead organism in a deposit of sedimentary rock; ▷ pp. 4, 12–15.

four-eyed fish, any of three small fishes of the family Anablepidae, occurring in fresh and brackish waters in South America. Each of the large eyes (including both retina and cornea) are divided across the middle, with the upper half adapted for vision in air, the lower half for vision in water; they are set on top of the head, so that the fish can swim at the surface using both visual faculties at the same time. They are live-bearing fishes; ▷ p. 73 (box).

fovea, a small area in the retina of the vertebrate eye where the cone cells are concentrated; ▷ pp. 97, *189*, **190**.

fox, the common name given to many CANIDS, especially the smaller, more solitary species (in contrast to the more gregarious DOGS); ▷ pp. 126–7, 206, 207, 213. See also FENNEC FOX, GREY FOX, RED FOX.

foxglove, ▷ DIGITALIS.

francolin, any of 40 plump partridge-like GAMEBIRDS of open country in Africa, belonging to the family Phasianidae. Unlike partridges, there is a spur on the foot.

frangipani (*Plumeria rubra*), a small central American tree in the family Apocynaceae. It has fragrant flowers and is cultivated as an ornamental.

frankincense tree (*Boswellia carteri*), a tropical tree that is tapped to obtain a milky resin, which is used as an ingredient of incense and perfumes.

free-living, (of an animal) able to move freely; or non-parasitic. Compare SESSILE.

freesia, any of around 20 species of perennial herbs growing from corms in the South African genus *Freesia* in the iris family, Iridaceae.

free-tailed bat, any of about 80 microchiropteran BATS of the family Molossidae, found in tropical and subtemperate parts of the world. Free-tailed bats are so called because of their long tail, which projects from a loose sheath formed by the wing membrane. They form huge colonies in caves, where the great accumulations of dung, or guano, is mined as fertilizer (▷ p. 119).

french bean, string bean or **dwarf bean,** a usually low-growing cultivar of the legume, *Phaseolus vulgaris* (family Fabaceae), the pods of which are eaten before maturity. See also KIDNEY BEAN, HARICOT BEAN and RUNNER BEAN; p. 205.

fresh water, water in which the salinity is low; ▷ pp. 218–19.

friarbird, any of 17 species of larger Australasian HONEY-EATER belonging to the genus *Philema*, slender birds with large bills, often with a knob at the base.

frigate bird or **man-of-war bird,** any of five sea birds of the family Fregatidae (order Ciconiiformes), living in all oceans. They are excellent fliers on slender wings spanning 2 m (6½ ft), and have a deeply forked tail. Males have a bright-red, extensible throat pouch beneath a long hooked bill; ▷ pp. *91*, 96.

frilled lizard, (*Chlamydosaurus kingii*), an Australian AGAMID lizard notable for its large neck ruff which can be expanded as a form of threat display (▷ p. 167).

Frisch, Karl von (1886–1983), Austrian scientist, one of the pioneers of modern ethology (the study of animal behaviour). His best-known research resulted in the discovery of the dance language of honeybees (▷ box, p. 153). In 1973 he shared the Nobel Prize with LORENZ and TINBERGEN for his contribution to ethology; ▷ p. 148.

frit fly (*Oscinella frit*), a DIPTERAN insect of the family Chloropidae; the larvae of this small fly cause considerable losses to cereal crops.

fritillary, 1. any NYMPHALID butterfly of various genera in the subfamily Nymphalinae, occurring worldwide. **2.** any of a number of herbs growing from bulbs in the north temperate genus *Fritillaria* in the lily family, Liliaceae. Snake's-head fritillary (*F. meleagris*) is popular in gardens; ▷ p. 205.

frog, the common name given to many of the ANURAN amphibians (the group that also includes TOADS). The term is often used to distinguish smooth-skinned, long-legged jumping species from warty-skinned, short-legged forms, which are generally called 'toads', but the distinction is imprecise and has no taxonomical significance; frequently closely related species belonging to the same family are indifferently termed frogs or toads. In a restricted sense, the name 'frog' may be used for the 'true' frogs of the family Ranidae, such as the common European frog *Rana temporaria*; in a broader sense it is sometimes used to refer to any anuran; ▷ pp. **78–81**, *183*, 233.

frogfish, any of 60 species of the family Antennariidae, related to the ANGLER FISHES. Well disguised by their short lumpy bodies and loose warty skin, they rest on the sea bed, manoeuvring themselves by means of robust limb-like pectoral fins and using a fleshy flap to lure prey within range; ▷ p. 75.

froghopper or **spittlebug,** any HOMOPTERAN insect (bug) of the family Cercopidae, comprising over 1500 species of plant-feeding insect, of worldwide distribution. The juveniles (nymphs) surround themselves with 'cuckoo spit', a stable droplet of foam that protects them from predation and desiccation.

frogmouth, any of 14 insect-eating forest birds of the families Podargidae (Australasian species) and Batrachostomidae (Asian species), in the same order as the OWLS. They roost during the day by camouflaging themselves against tree bark. At night they forage for insects with a bill that is extremely wide and surrounded with sensory hairs; ⇨ p. 97.

frond, 1. a large leaf, usually divided (e.g. in ferns, palms). **2.** the leaf-like body of some seaweeds (e.g. bladderwrack) and lichens.

fructose, a sugar found in honey and many fruits; ⇨ p. 179.

frugivore, a fruit-eating animal.

fruit, the seed-containing structure that develops from the ovary of a flower, usually after fertilization. Some parts of the fruit may derive from other parts of the flower. A great variety of fruits have developed to appeal to different animal dispersers (⇨ pp. 41, 43).

fruit bat or **flying fox,** any of around 170 BATS of the family Pteropodidae, making up the smaller of the two suborders of bats (Megachiroptera). Fruit bats are found in tropical and subtropical parts of the Old World and Australia. Most are active at dawn or dusk, feeding principally on fruit, and only a few are capable of ECHOLOCATION; ⇨ pp. *14,* 118–19, *201.*

fruit fly, any small DIPTERAN insect of the family Drosophilidae, of worldwide distribution. Adults are attracted to fermenting fruits, drinks, sap flows and fungi. They are used extensively in laboratories and have been especially important in genetic research.

fruiting body, an organ in fungi and lower plants given over to reproductive functions. The visible parts of mushrooms and toadstools are the fruiting bodies; ⇨ p. 35.

fuchsia, any of around 100 species of shrubs and small trees in the mainly tropical and subtropical genus *Fuchsia* in the family Onagraceae. Fuchsias are popular ornamentals.

fugu, ⇨ PUFFER.

fulmar, either of two species of the genus *Fulmarus* (family Procellariidae, order Ciconiiformes), small gull-like birds that glide on stiffly held wings, rarely flapping; ⇨ p. 96.

fumitory, any of around 55 species of annual herbs in the mainly European and Mediterranean genus *Fumaria* in the family Fumariaceae.

Fungi, ⇨ MYCETEAE, FUNGUS.

fungicide, ⇨ PESTICIDE.

fungi imperfecti, the subdivision Deuteromycotina of the fungal division Amastigomycota; ⇨ p. 35.

fungus (pl. fungi), any member of the kingdom Myceteae or Fungi; ⇨ p. **35.** Other text references: 30 (distinction from plants), 34 (in lichens), 39 (symbiosis in conifers), 43 (association with seeds), 194–5 (role in food webs), 201 (tropical forests), 202 (grown by termites), 230 (as pests).

fungus gnat, any small DIPTERAN insect of the family Mycetophilidae; they have a wide geographical range and over 2000 species have been described. They are usually found in damp places, and the larvae feed on fungal material on decomposing vegetation or timber.

funnel-eared bat or **long-legged bat,** any of about eight microchiropteran BATS of the family Natalidae, found in tropical parts of the New World. The common name of these insectivorous bats comes from their large ears, the outer surfaces of which bear glandular projections.

fur beetle, ⇨ DERMESTID BEETLE.

furcula ('wishbone'), in birds, a part of the skeleton formed from the fused clavicles (collarbones), important in the functioning of the wing mechanism; ⇨ pp. 90 (box), *92.*

furniture beetle, ⇨ WOODWORM.

fur seal, any of various eared SEALS that have fine, dense fur and have often been hunted for their skins; ⇨ p. 143.

furze, ⇨ GORSE.

fynbos, dwarf-shrub vegetation, dry in summer, of the mediterranean-climate area in South Africa.

gadfly, ⇨ HORSEFLY.

Gaia, the name given by James A. Lovelock to the complex interaction of organic and inorganic components of our planet and its biosphere that provide it with great resilience and ability to cope with a great deal of change while still retaining living organisms.

gait, a particular pattern of leg movement in walking or running, such as an amble or a gallop; ⇨ p. 177.

galactose, a simple SUGAR; ⇨ p. 179.

galago or **bushbaby,** any of several agile, tree-dwelling lower PRIMATES of the family Lorisidae, found in sub-Saharan Africa; ⇨ p. 137.

Galápagos Islands, a group of volcanic islands in the eastern Pacific. Darwin's studies of the fauna there led him towards his theory of natural selection; ⇨ pp. 15, *86* (giant turtle), 96 (frigate bird).

Galen (AD 130–201), Greek scientist and physician. His interest in comparative anatomy led him to dissect many animals. However, he applied his findings to human anatomy and physiology, leading to many errors that became medical orthodoxy until the 16th century.

gall, an abnormal outgrowth of a plant, caused by cell multiplication and enlargement. Galls may be induced by bacterial infection (e.g. crown gall) or insect attack (e.g. oak gall).

gall bladder, a membranous, muscular sac associated with the liver and in which bile made in the liver is stored. The bile ducts conduct bile into the duodenum, where it assists digestion.

gallinule, any of four water birds with a bright bill and head, belonging to the family Rallidae (order Gruiformes), found in the Neotropics, North America and Africa.

gall midge, any DIPTERAN insect of the family Cecidomyidae, small delicate flies with whorls of hairs on the antennal segments. Larval life styles are very variable. Many plant-feeders produce deformities and galls in their food plants.

gall wasp, any solitary WASP of the family Cynipidae, of worldwide distribution; they are usually quite small and feed on plants. The processes of egg-laying and larval development cause excess growth in plant tissues, producing a gall that protects the young wasp.

gamebird, a general term applied to any bird that is hunted as game. The 'typical' gamebirds of the order Galliformes include GROUSE, PARTRIDGES, PHEASANTS and TURKEYS. They are generally plump, chicken-like birds, and most nest and feed on the ground; ⇨ pp. 90, 95, 97 (box).

games theory, method of investigating animal behaviour by treating the various available behavioural options as 'strategies', as if in a game, and assessing which is the most successful or beneficial; ⇨ pp. 148, 157 (box).

gamete or **sex cell,** a HAPLOID cell, either male (SPERM) or female (OVUM), one of each type fusing at fertilization to produce a DIPLOID cell (zygote), which develops as an embryo to produce a new individual. For references ⇨ REPRODUCTION.

gametophyte, the stage in a plant life cycle that produces GAMETES (sex cells). It alternates with an asexual SPOROPHYTE generation; ⇨ ALTERNATION OF GENERATIONS.

ganglion (pl. -lia), an encapsulated group of nerve-cell bodies; ⇨ p. 182.

gannet, any of three species of the family Sulidae (order Ciconiiformes), closely related to the BOOBIES. Large sea birds with long pointed wings and a long strong bill, they dive from great heights, feeding on fishes and squid; ⇨ pp. *91,* 93, 95, 96.

gar, any primitive RAY-FINNED FISH of the family Lepisosteidae, large North American freshwater fishes with heavily armoured bodies and long snouts armed with sharp teeth. These powerful fish-eating predators are unpopular among fishermen, whose nets they damage; ⇨ pp. 70–1.

Gardenia, a tropical and subtropical genus of evergreen shrubs and trees in the family Rubiaceae. Cape jasmine (*G. jasminoides*) – not a true JASMINE – is grown for its showy fragrant flowers.

garfish, ⇨ NEEDLEFISH.

garlic (*Allium sativum*), a herb of the family Alliaceae, grown for its bulbs or cloves, which are strong flavoured and used in cooking. See also RAMSONS.

garrigue, a type of mediterranean vegetation; ⇨ p. 205.

garter snake, any harmless COLUBRID snake of the genus *Thamnophis,* especially *T. sirtalis,* one of the commonest snakes in North America.

gas exchange, the reciprocal movement or DIFFUSION of oxygen and carbon dioxide across a respiratory structure; ⇨ pp. 180–1.

gas plant, ⇨ BURNING BUSH.

gastropod, any MOLLUSC of the class Gastropoda, comprising about 35 000 species and including SLUGS, SEA SLUGS, SNAILS, PERIWINKLES and LIMPETS; ⇨ pp. 5, **56.**

gastrulation, process of cell migration in an early animal embryo, resulting in the formation of three distinct cell layers (ectoderm, mesoderm, endoderm); ⇨ p. 186.

gaur (*Bos gaurus*), a massively built, brown to black species of wild CATTLE with white stockings. It inhabits forested hills in southern India and Indochina.

gavial (*Gavialis gangeticus*), a CROCODILIAN occurring in various large Indian rivers. It has an extremely long and slender snout with many small teeth, ideal for catching fish, its main food.

gazelle, common name given to about 18 small, elegant, agile ANTELOPES, including those of the genus *Gazella.* Gazelles are found in open habitats in Africa, Arabia, India and eastern Asia; ⇨ pp. 115, *165,* 166, 203.

gean, the wild CHERRY.

gecko, any of over 600 species of mainly small, insectivorous LIZARD of the family Gekkonidae, common in tropical regions. They are typically nocturnal, with large eyes and slit-like pupils. Many species have ridged pads on their broad, rounded toes that act as suckers, allowing them to walk upside-down on smooth surfaces.

gemsbok (*Oryx gazella*), an ORYX native to southern Africa. It has a broad black band on its flanks.

gene, the basic unit of inheritance, consisting of a linear section of a DNA molecule; ⇨ p. 22.

gene pool, any interbreeding population of organisms; ⇨ p. 16.

generalist, an animal that is relatively unspecialized, e.g. in its feeding; ⇨ pp. **162–3,** 164.

genet, any small cat-like VIVERRID carnivore of the genus *Genetta,* usually boldly striped or spotted, found in wooded areas of Africa and southern Europe; ⇨ p. 123. See also WATER CIVET.

genetic drift, the process of gradual change in a population by neutral mutations; ⇨ p. 16.

genetic engineering, a branch of biotechnology in which genetic material is manipulated; ⇨ pp. 23, 25, 227 (box), 231 (box).

genetics, the study of the mechanisms of inheritance; ⇨ pp. 22–3.

Genista, a mainly Mediterranean genus of around 75 shrubs in the pea family, Fabaceae (formerly part of the Leguminosae); one of the three genera of

BROOMS. A number, such as Dalmatian broom (*G. dalmatica*) are grown as ornamentals.

genitals or **genitalia,** ⊳ REPRODUCTIVE SYSTEM.

genome, the sum of all the genetic information contained within an organism; ⊳ pp. **22,** 23 (human genome project).

genotype, the total genetic constitution of an individual organism, comprising the entire set of alleles in its genome, i.e. including recessive characteristics that are not expressed; ⊳ p. 23. Compare PHENOTYPE.

gentian, any of around 400 species of mainly perennial herbs in the mainly alpine genus *Gentiana* in the family Gentanaceae. They are popular with alpine gardeners for their usually blue flowers; ⊳ p. 211.

genus, a classificatory group between family and species. For example, *Homo habilis, H. erectus* (both extinct) and *H. sapiens* make up the genus *Homo* within the family Hominidae (gorilla, chimpanzees and humans); ⊳ p. 18.

geological timescale, the chronology used by geologists and palaeontologists; ⊳ pp. 4–5.

geology, the study of the origin, history, structure and composition of the earth. Geological studies of sedimentary rocks are important in dating the fossils contained in them (⊳ p. 4), and the study of continental drift has done much to explain the past and present distribution of groups of organisms on the planet (⊳ pp. 8–9).

geometrid, any LEPIDOPTERAN insect (moth) of the large family Geometridae, comprising over 12 000 species. Both larvae and adults are camouflaged. Larvae are elongated with only a final pair of abdominal legs present; they often adopt a twig-like posture to escape detection. They move in a looping motion, giving rise to such common names as 'looper', 'measuring worm' and 'inchworm'.

geranium, 1. any of around 400 species of herbs in the mainly temperate genus *Geranium* in the family Geraniaceae. The shape of the fruit gives rise to the common name of cranesbill for a number of species, e.g. meadow cranesbill (*G. pratense*). Herb Robert (*G. robertianum*) is a common wayside plant. **2.** a commonly used term for the related PELARGONIUM.

gerbil, any of some 81 mouse-like, burrowing myomorph RODENTS in the family Muridae, with long hind limbs and long, furry tails. They are found in arid habitats in Africa and Central Asia.

gerenuk (*Litocranius walleri*), a reddish-brown browsing ANTELOPE with long neck and limbs, and curved horns in the males. It inhabits arid bush in Africa; ⊳ p. *163.*

germ cell, any of a group of cells that divide by meiosis (⊳ p. 22) to form sex cells or GAMETES; ⊳ pp. 21, 184.

germination, 1. the commencement of growth of a seed into a seedling, generally following water uptake and the release from dormancy. The term is also applied to similar processes in fungi and spore-bearing plants involving the commencement of growth from the spore into e.g. fungal hypha or fern prothallus. **2.** the process in seed-bearing plants by which pollen, absorbed into the ovule, forms the pollen tube; ⊳ pp. **38, 41.**

gerygone or **flyeater,** any of 19 species of Australasian warbler (family Acanthizidae, order Passeriformes), which – unusually for members of this family – are migratory.

gestation, in live-bearing mammals, the period of pregnancy, i.e. from conception to birth; ⊳ pp. 103, 104, 108, 135, 143, **185.**

gharial, alternative name for the GAVIAL.

gherkin, ⊳ CUCUMBER.

ghost-faced bat or **moustached bat,** any of about eight microchiropteran BATS of the family Mormoopidae, insectivorous species that range through tropical Central and South America. Their alternative name is derived from the fringe of hair surrounding the mouth.

giant salamander, either of two species of SALAMANDER of the genus *Andrias* (family Cryptobranchidae). Giant salamanders have broad, flattened heads and bodies, and are the largest living salamanders, reaching up to 2 m (6½ ft) in length. They are restricted to the cold waters of deep mountain rivers, one species in China, the other in Japan. See also HELLBENDER.

gibbon, any (lesser) APE of the family Hylobatidae, which includes nine species all occurring in the forests of Southeast Asia and Indonesia. The largest gibbon is known as the siamang (*Hylobates syndactylus*); ⊳ pp. 136, **140.**

Gila monster (*Heloderma suspectum*), a large (up to 60 cm/24 in), heavily built, brightly coloured lizard of arid parts of southwestern USA and Mexico. It is unusual for its venomous bite, passing venom down grooves in its teeth.

gill, an internal or external respiratory structure used for GAS EXCHANGE in water; ⊳ pp. 56 (molluscs), 72–3 (fishes), 79 (amphibians), **180–1.**

gill raker, a structure on the gills of fishes such as mackerel and herring that sieves plankton from water; ⊳ p. 178.

gill slit, one or more openings in the pharynx region of fishes, containing the gills.

ginger (*Zingiber officinale*), a tropical perennial herb in the family Zingiberaceae. The fleshy rhizomes are used either whole or powdered as flavouring.

gingko or **maidenhair tree** (*Gingko biloba*), the only surviving species of the division Ginkgophyta; ⊳ pp. 30, 38, **39.**

ginseng, the powdered rhizomes and roots of ginseng (*Panax pseudoginseng*) and American ginseng (*P. quinquefolius*), in the ivy family, Araliaceae, used as a stimulant.

giraffe (*Giraffa camelopardalis*), a social ruminating ARTIODACTYL mammal of sub-Saharan Africa, the tallest land animal; ⊳ pp. **117, 203.**

gizzard, in birds and many invertebrates, a compartment in the digestive tract in which food is broken down by a grinding mechanical action; ⊳ pp. 93, **178.**

glacial period or **ice age,** a period of cold climate, characterized by the growth of the ice caps. There have been many glacial periods over the last million years, separated by warmer interglacial periods; ⊳ pp. 5, **10.**

gladiolus, any of around 300 species of perennial herbs growing from corms in the Old World genus *Gladiolus* in the iris family, Iridaceae. There are many garden cultivars.

gland, any organ producing a secretion that is either carried by the blood (an ENDOCRINE gland) or passed via a duct to where it is needed (e.g. salivary and mammary glands); ⊳ p. 183.

glass frog, any of about 65 species of TREE FROG of the family Centrolenidae, of Central and South America. The heart and other internal organs can be seen through the transparent/translucent skin, giving them excellent camouflage in water; ⊳ p. 81.

glass snake (*Ophisaurus apodus*), European limbless lizard, related to and similar in appearance to the SLOW WORM. Like other lizards, it can shed its tail when seized (⊳ p. 89), but in spite of its name, it does not break into many pieces and it is not a snake.

glider, any of seven small, arboreal, forest-dwelling MARSUPIALS of the family Petauridae. They have a thin, furry membrane between fore and hind limbs that helps them to glide between trees.

gliding, as a means of locomotion, ⊳ pp. *72,* 79, 84, 104, *134,* 137 (box), **177.**

global warming, ⊳ GREENHOUSE EFFECT.

globeflower (*Trollius europeaus*), a perennial herb with globe-shaped yellow flowers in the buttercup family, Ranunculaceae; native to Europe, the Caucasus and the American Arctic.

glow-worm, ⊳ FIREFLY.

glucose, a simple sugar, an essential source of energy produced by plants via photosynthesis and obtained by animals from plants; ⊳ pp. 6 (box), 21 (in cells), 31–2 (in plants), 33 (photosynthesis box), 179 (dietary).

glutton, another name for WOLVERINE.

glycerol or **glycerin,** a simple energy-rich lipid, a basic component of phospholipids (⊳ box, p. 20), and, with fatty acids, of FATS.

glycogen, the form in which carbohydrates are stored in the liver and muscles of animals. It is a polysaccharide carbohydrate consisting of glucose units and is readily converted to glucose; ⊳ pp. 21, 179.

glycolysis, a metabolic pathway involving the transformation of glucose into a utilizable form of energy; ⊳ p. 21.

glyoxisome, a type of organelle; ⊳ p. 20 (box).

glyptodont, any extinct South American edentate mammal of the genus *Glyptodon* (and related genera), some of which survived into historical times. They resembled and were related to the living ARMADILLOS, but some attained giant proportions; ⊳ pp. *99, 106.*

gnat, a term with a similar connotation to MIDGE and often interchangeable with it. See also FUNGUS GNAT.

gnetophyte, any member of the division Gnetophyta; ⊳ pp. 30, **39,** 40.

Gnetum, a genus of gnetophyte plants; ⊳ pp. **39,** 34.

gnu, another name for WILDEBEEST.

goat, common name given to various sure-footed, agile BOVID mammals of the genus *Capra* and related genera in the subfamily Caprinae (which also includes SHEEP), such as the ibex (*C. ibex*); ⊳ pp. 114, **115,** 205, 212.

goatfish or **red mullet,** any of 55 marine PERCIFORM fishes of the family Mullidae, found in tropical and warm-temperate seas worldwide. They are often brightly coloured (predominantly red), and on the chin there are two long barbels, which are used to probe the sand or mud for invertebrate prey. They are important food fishes.

goby, any of about 800 PERCIFORM fishes of the family Gobiidae, found in all seas, especially in coastal and intertidal waters; some live in brackish water, and a few are freshwater. They are small fishes, mainly 2–5 cm (about 1–2 in) in length, with rounded heads and tapering tails. The pelvic fins are united to form an adhesive or sucking disc on the belly just behind the head.

godwit, any of four long-legged WADERS of the genus *Limosa,* occurring worldwide. Unlike the related SNIPE, the long bill is used to search underwater; ⊳ p. 97 (box).

golden bat, ⊳ SUCKER-FOOTED BAT.

golden cat (*Felis aurata*), a brown to grey CAT with dark markings and long legs. It inhabits forest and scrub in Africa.

golden eye, an adult green LACEWING.

golden mole, any mammal of the family Chrysochloridae (order Insectivora), comprising 18 species that occur in southern and central Africa. All species are specialized as burrowers, feeding on earthworms and other invertebrates; ⊳ pp. 120, 121.

golden monkey, either of two Old World colobine monkey of the genus *Rhinopithecus* (⊳ p. 139). They inhabit mountainous forest in central and western China, and have a golden mane and orange hands and feet.

golden rod, any of around 120 species of perennial herbs in the mainly American genus *Solidago* in the daisy family, Asteraceae (formerly Compositae).

goldfinch, any of four strikingly coloured black and gold FINCHES of North America and the Palaearctic.

goldfish (*Carassius auratus*), a small member of the CARP family, originally from eastern Europe and Russia, now widely kept in aquaria. Many ornamental varieties have been developed.

Golgi apparatus, a type of organelle in eukaryote cells; ⇨ pp. **20** (box), *21*.

goliath beetle, any SCARABAEOID BEETLE of the genus *Goliathus* (family Scarabaeidae) of tropical Africa, amongst the heaviest and largest insects.

gomphothere, any member of an extinct group of trunked mammals (proboscideans) related to the living elephants; ⇨ p. *108*.

gonad, in animals, an organ in which GAMETES (sex cells) are produced, i.e. a TESTIS or an OVARY.

Gondwanaland, an ancient supercontinent; ⇨ pp. *4*, **8–9**.

goose, name given to various birds in the family Anatidae (order Anseriformes), but the distinction between geese and DUCKS (also included in this family), is not clear-cut. There are about 14 species of true geese, such as the Canada goose (*Branta canadensis*), confined to the northern hemisphere. They are typically large birds and graze grass; ⇨ pp. *91*, 95, 96, 97 (box). See also SWAN.

gooseberry (*Ribes uva-crispa* or *R. hirtellum*), small spiny shrubs in the family Saxifragaceae. They are cultivated in temperate regions for their fruit, which is usually picked before ripening; ⇨ p. *43*.

goosegrass, ⇨ CLEAVERS.

gopher, an alternative name for GROUND SQUIRREL or an abbreviation of POCKET GOPHER.

goral (*Nemorhaedus goral*), a nimble, GOAT-like artiodactyl with long, buff to red-brown fur and small, black horns. It inhabits steep, wooded mountains in Central and southern Asia.

gorilla (*Gorilla gorilla*), a herbivorous great APE of central Africa, the largest of the primates and man's closest relative after the chimpanzees; ⇨ pp. *136*, *140*, **141**, 155, *178*, 228.

gorse, any of around 15 species of spiny shrubs in the European and north west African genus *Ulex* in the pea family, Fabaceae (formerly part of the Leguminosae). *U. europeaus*, also called furze or whin, is common on rough acid grassland.

goshawk, any of 18 relatively large ACCIPITERS found worldwide. The elegant grey and brown goshawk (*Accipiter gentilis*) of the western Palaearctic hunts gamebirds and rabbits; ⇨ pp. *93*, 209.

Gould, Stephen Jay, ⇨ PUNCTUATED EQUILIBRIUM.

gourami (*Osphronemus goramy*), a heavy-bodied freshwater PERCIFORM fish, the sole member of the family Osphronemidae, occurring in Southeast Asia; one ray of the pelvic fin is drawn out into a long filament. A number of other, related fishes are also referred to as 'gouramis'.

gourd, the fruit of any of a number of species, mostly in the cucumber family, the Cucurbitaceae. In the tropics these gourds may be used for food (e.g. West Indian gourd, *Cucumis anguria*) or as utensils. See also CALABASH and LOOFA.

grackle, any of six relatively large AMERICAN BLACKBIRDS. They are gregarious birds, and some have become urbanized.

gradualism, the evolutionary process as envisaged by Darwin (⇨ p. *13*), i.e. descent with gradual modification as a consequence of natural selection. Compare PUNCTUATED EQUILIBRIUM.

grafting, (in plants) the insertion of part of one plant (the graft) into another plant (the stock), from which the graft subsequently grows. The technique is commonly used in horticulture to achieve a combination of the best characteristics (e.g. disease resistance and fruit form) of two different varieties.

grain, ⇨ CEREAL.

Graminae, ⇨ POACEAE.

graminivore, a grass- or grain-eating animal.

grape, the fruit of the GRAPEVINE.

grapefruit (*Citrus* x *paradisi*), an evergreen CITRUS FRUIT tree thought to have arisen from a seedling in Barbados; now an important commercial crop, particularly in the USA.

grape hyacinth, any of various members of the genus *Muscari*, family Liliaceae. *M. botryoides* is frequently cultivated for its clusters of small, blue, grape-like flowers; ⇨ p. *205*.

grapevine, any of several vines of the genus *Vitis* (especially cultivars of *V. vinifera*), whose fruit is grown mainly for wine; ⇨ pp. *43*, 45, 46.

graptolites, an extinct group of marine animals, variously thought to be related to HYDROZOANS or to ACORN WORMS. Their skeletal remains are of importance in identifying the age of rocks; ⇨ p. *5*.

grass, any of around 9000 species in the monocotyledonous family Poaceae (formerly Graminae). Many areas of both tropical and temperate regions are dominated by GRASSLANDS; ⇨ pp. **45–6** (biology), 202 (tropical grasslands), 206 (temperate grasslands), 212 (man-made grasslands), 229 (threats).

grasshopper, general term for many ORTHOPTERAN insects. Long-horned grasshoppers, distinguished by their long antennae, belong to the family Tettigoniidae (BUSH CRICKETS), while the short-horn grasshoppers belong to the large family Acrididae (9000 species), which includes the LOCUSTS. Acridids produce sound by rubbing their hind femora against the edge of the fore wing and lay their eggs in pods in the ground; ⇨ pp. 59, 62, 167, 177.

grassland, an area, usually extensive, in which grasses are the dominant plants; ⇨ pp. **202–3**, **206**, 212, 229.

grass snake, any harmless COLUBRID snake of the genus *Natrix*, especially *N. natrix*, one of the commonest and most widespread snakes in Europe. These snakes spend much of their lives in water.

gravitropism, the ability of plants to react to gravity; ⇨ p. 33.

gray (fox, etc.), ⇨ GREY.

grayling, 1. any of four SALMONIDS of the subfamily Thymallinae, northern-hemisphere freshwater fishes recognizable by their high dorsal fins. **2.** (*Hipparchia semele*), a European NYMPHALID butterfly (subfamily Satyrinae). The larvae feed on grasses.

grazing, a FORAGING strategy involving cropping grass and other ground vegetation; ⇨ pp. 99 (early mammals), 104 (kangaroos), 108–17 (hoofed mammals), 132 (rabbits, etc.), 142 (sirenians), **163**, *171*, 202–3 (tropical grasslands), 206 (temperate grasslands), 209 (boreal forest), 210 (tundra), 212–13 (domesticated animals), 218 (freshwater ecosystems). Compare BROWSING.

grebe, any of about 20 swimming and diving birds of the family Podicipedidae (order Ciconiiformes). Most have long necks and short tails, and they live on rivers, lakes and at sea. The great crested grebe (*Podiceps cristatus*) has ear tufts and performs elaborate courtship displays; ⇨ pp. 90, 91, 93, 94, 95, 155 (box).

green algae, any unicellular alga or seaweed of the division Chlorophyta; ⇨ pp. **34**, 221.

green bean, any bean eaten while still green, such as the FRENCH BEAN and the RUNNER BEAN.

greenbottle, ⇨ BLOWFLY.

greenfinch, any of four gregarious FINCHES, the most common of which is the western Palaearctic greenfinch (*Carduelis chloris*), often seen in gardens and parks.

greenfly, common name for green-coloured APHIDS.

greengage (*Prunus* x *italica*), a green plum believed to be the result of hybridization between a DAMSON and the European PLUM.

greenhouse effect, the process by which the accumulation of carbon dioxide and various other gases in the atmosphere may lead to global warming; ⇨ pp. 201, 203, 205, 209, 211, **224–5**.

greenlet, ⇨ VIREO.

greenling, any of nine North Pacific marine fishes of the family Hexagrammidae, related to the SCORPION FISHES. Some have five rows of lateral-line sense organs on the flanks, and several are commercially important in the Far East.

green monkey, ⇨ GUENON.

green pepper, ⇨ CAPSICUM.

greenshank (*Tringa nebularia*), a large WADER of the family Scolopacidae, with long, green legs and a slender body, which is usually held horizontally. It lives in marshes in the Old World and Australia.

gregarious, (of animals) tending to form groups with distinct social relations between members.

grenadier or **rat-tail,** any of 260 deep-sea fishes of the family Macrouridae, belonging to the COD order. They have short, robust heads and bodies, and long, thin tails.

grey fox (*Urocyon cinereoargenteus*), a grey-coloured CANID, found in wooded and brush habitats in North, Central and northern South America.

grey matter, ⇨ SPINAL CORD.

grey whale (*Eschrichtius glaucus*), a baleen whale (CETACEAN), the sole member of its family, found in the coastal waters of the northern Pacific; ⇨ pp. **145**, 171.

grey wolf, ⇨ WOLF.

griffon, any of five Old World VULTURES of the genus *Gyps*. The griffon vulture (*Gyps fulvus*) stands over 1 m (39 in) tall.

grivet, ⇨ GUENON.

grizzly bear, a subspecies of BROWN BEAR; ⇨ pp. **128**, *229*, 233.

grooming, the care and maintenance of the body surface, such as a bird's feathers or a mammal's fur, by cleaning, waterproofing (⇨ PREENING), removal of parasites, etc. In social species, including many primates, mutual grooming is important in strengthening group and family bonds; ⇨ pp. 93, 108, 139 (box), 141, **155**.

grosbeak or **cardinal,** any of 42 PASSERINE birds of the tribe Cardinalini, closely related to the BUNTINGS. They are small birds of the Americas, resembling buntings but with stouter bills used to crush seeds; ⇨ p. 209.

ground beetle, any BEETLE of the family Carabidae, comprising about 12 000 species, distributed worldwide. Usually dark-coloured and variable in size, they are frequently found under stones, logs, etc., and are capable of running fast; they are generally predatory or carrion-feeding. See also TIGER BEETLE.

ground elder, goutweed or **bishop's weed** (*Aegopodium podagraria*), a perennial herb in the carrot family, Apiaceae (formerly Umbelliferae), with elder-like leaves and creeping rhizomes. It is distributed in Europe and temperate Asia and is a persistent garden weed.

groundhog, another name for WOODCHUCK.

ground ivy (*Glechoma hederacea*), a creeping perennial herb native to Europe, Asia and Japan in the family Lamiaceae (formerly Labiatae).

groundnut, ⇨ PEANUT.

groundsel (*SENECIO vulgaris*), an annual or overwintering weed of cultivated ground and waste places. A number of tree-like species of *Senecio* are referred to as tree groundsels (⇨ p. *211*, *214*), as is also *Baccharis halimifolia*.

ground squirrel, or 'gopher', common name given to several species of ground-dwelling, long-tailed, squirrel-like RODENTS of the genera *Spermophilus*, *Spermophilopsis* and *Xerus* (family Sciuridae). Ground squirrels are widespread in North America, Africa and northern Asia; ⇨ pp. **133–4**, 206.

grouse, any of 11 typical GAMEBIRDS in the family Phasianidae, plump birds living in open ground in the northern hemisphere. In winter the bill and feet are feathered. The males display in communal leks, strutting and inflating throat air-sacs to attract females; ⇨ pp. *91*, 94, 97 (box), *213*. See also CAPERCAILLIE and PRAIRIE CHICKEN.

growth: plants 32–3; animals 51, **186–7**; arthropods 58; insects 60, 62–3; crustaceans 66; fishes 74; amphibians *80*, 81; turtles 85; birds 95; mammals 100; monotremes 102; marsupials 103; elephants 108; bats 119; insectivores *120*; bears 128; rodents 134–5, **186–7**; primates 136; whales 144.

grub, an insect larva, especially one that is thick and soft-bodied.

grunion, ⇨ SILVERSIDE.

grunt, any of about 175 PERCIFORM fishes of the family Haemulidae, mainly marine fishes from the Atlantic, Indian and Pacific Oceans. They have the general appearance of SNAPPERS, and some (known as 'sweetlips') have thick fleshy lips.

grylloblattid or **ice bug,** any EXOPTERYGOTE insect of the order Grylloblattodea, comprising only 16 species known from mountainous areas of North America, Siberia and Japan. They are omnivorous, and prefer temperatures close to 0 °C (32 °F). They have features in common with crickets and cockroaches, and may be a relic of the stock from which these groups arose; ⇨ p. *61*.

grysbok, either of two small, reddish-brown, browsing ANTELOPES with small horns, belonging to the genus *Raphicerus*. They inhabit woodland and thicket in eastern and southern Africa.

guan, any of 22 tree-dwelling birds of the family Cracidae, which also includes the CURASSOWS. They live in the forests of Central and South America; ⇨ p. *91*.

guanaco (*Lama guanicoe*), a ruminating South American ARTIODACTYL mammal (family Camelidae), related to the Old World camels and the wild ancestor of the domesticated llama and alpaca; ⇨ p. **115** (box), 206.

guanine, one of the four nucleotide bases in DNA; ⇨ p. 22.

guano, the dried excrement of bats and fish-eating sea birds, often used as a fertilizer; ⇨ p. 119.

guard cell, either of two cells bordering a stomatum; ⇨ p. 31 (box).

guava (*Psidium guajava*), a tropical and subtropical shrub or tree in the family Myrtaceae cultivated for its edible aromatic fruits.

gudgeon (*Gobio gobio*), a small, widely distributed European freshwater fish of the CARP family. It has sensory barbels on its down-turned mouth and feeds on the bottom; it is eaten in some parts of Europe.

guelder rose (*VIBURNUM opulus*), a shrub with clusters of tubular white flowers common in woods and hedgerows in northern Europe.

guenon, any of 20 species of Old World cercopithecine monkey of the genus *Cercopithecus*, omnivorous, often brightly coloured monkeys widespread

in sub-Saharan Africa. The common guenon *C. aethiops* varies considerably in its coloration, and is known as the vervet, grivet and green monkey in different geographical regions; ⇨ p. 139, *148*, 153.

guillemot or **murre,** any of five large AUKS, up to 45 cm (18 in) long with straight, pointed bills, of Holarctic distribution; ⇨ p. 97 (box).

guinea fowl, any of six small GAMEBIRDS of the family Numididae, living in the open woodland of Africa. *Numida meleagris*, the tufted guinea fowl, is the ancestor of the domestic guinea fowl.

guinea pig, a domesticated form of CAVY.

gull, any of about 50 species of sea bird, solidly built with strong webbed feet, belonging to the family Laridae (order Ciconiiformes); they live and nest colonially along shorelines worldwide and also by large inland waters (⇨ p. 157). Gulls are usually white, with variable markings; the bill is powerful and used for scavenging; ⇨ pp. 17, *91*, 96, 148 (box), 154 (box), 213.

gulper, any of six species of deep-sea EEL of the families Saccopharyngidae and Eurypharyngidae. These eels are highly modified, with reduced skeletons, no pelvic fins, and enormously expanded jaws that enable them to swallow prey larger than themselves; ⇨ p. 221.

gum, a sticky, water-soluble substance (usually made of polysaccharides) exuded by plants, often from specialized secretory cells (e.g. on the trunks of plum trees).

gum arabic tree, ⇨ ACACIA.

gumtree, a term applied to EUCALYPTUSES and various other trees that produce gum.

gundi, any of five small caviomorph RODENTS with long, furry tails (family Ctenodactylidae). They are found in rocky, arid habitats in North and northeast Africa.

gunnel, any of about 13 small marine PERCIFORM fishes of the family Pholididae, found in the North Pacific and North Atlantic.

guppy (*Poecilia reticulata*), a small LIVEBEARER from Central and South America, widely introduced elsewhere, partly to control malarial mosquitoes, whose larvae it eats. It is a prolific breeder and a very popular aquarium fish; ⇨ p. 74.

gurnard, ⇨ SEA ROBIN and FLYING GURNARD.

gustation (adj. gustatory), the sense of TASTE.

gut flora, symbiotic microorganisms that assist the digestive processes of many animals; ⇨ p. 169, RUMINATION and CAECUM.

gutta-percha (*Palaquium gutta*), tree from Southeast Asia and Indonesia containing a rubber-like latex used for insulating electric cables and in golf balls.

gymnosperm, any seed-bearing plant in which the seed is not enclosed in an ovary; ⇨ pp. **38–9**. Compare ANGIOSPERM.

gymnure, another name for MOONRAT.

gypsophila, any of around 125 species of herbs and dwarf shrubs in the Eurasian, Egyptian and Australasian genus *Gypsophila* in the pink family, Caryophyllaceae.

gypsy moth (*Lymantria dispar*), a LEPIDOPTERAN insect of the family Lymantriidae, whose larvae are serious pests of a range of tree species in temperate areas of the world.

gyrfalcon (*Falco rusticolus*), a very large FALCON found in northern tundra regions; ⇨ p. 211.

habitat, the surroundings in which an organism is found, such as woodland, water, grassland, etc. Habitats can be subdivided into microhabitats, such as rotting logs, pond mud, grass stems, etc.

habituation, a LEARNING process whereby an animal's responsiveness to a stimulus decreases if the stimulus is given repeatedly without REINFORCEMENT, such as food or a painful stimulus; ⇨ p. 150.

hadal zone, a marine zone; ⇨ p. *220*.

haddock (*Melanogrammus aeglefinus*), a North Atlantic fish in the COD family, a smaller version of the cod, distinguished by a black spot on each flank.

hadrosaur or **duck-billed dinosaur,** any herbivorous bipedal ORNITHOPOD dinosaur of the genus *Anatosaurus*. Hadrosaurs were semi-aquatic, with webbed feet; ⇨ p. 84.

Haeckel, Ernst (1834–1919), German biologist, whose work in comparative embryology provided important evidence for evolution; ⇨ p. 14.

haemocoel, in arthropods and many molluscs, a cavity through which the blood circulates, nourishing the organs and tissues; ⇨ pp. 59, 181.

haemocyanin, a RESPIRATORY PIGMENT present in the blood of molluscs and crustaceans; ⇨ p. 180.

haemoglobin, a RESPIRATORY PIGMENT present in the blood of vertebrates and many invertebrates; ⇨ 73, 96 (box), **180**.

hagfish, any JAWLESS FISH of the family Myxinidae, a small group (about 20 species) of exclusively marine, eel-like fishes. They are mainly bottom-dwelling, either predatory/parasitic or feeding on the flesh of dead or dying fishes; ⇨ pp. 70, 75.

hairstreak, name given to a number of LYCAENID butterflies; larvae feed on trees or shrubs.

hake, any of seven temperate, mainly deep-water fishes of the family Merlucciidae, related to the COD but with more elongated bodies and large jaws armed with strong teeth.

halibut, any of 18 FLATFISHES of the tribe Hippoglossini (family Pleuronectidae), including the Atlantic halibut (*Hippoglossus hippoglossus*), the largest of the flatfishes, reaching 2.5 m (over 8 ft) in length and 300 kg (661 lb) in weight. See also FLOUNDER.

halosaur, any deep-sea fish of the family Halosauridae (15 species), in the same order as the SPINY EELS (1) and found in all oceans. They are related to the true EELS, and like them have a leptocephalus-type larva (⇨ p. 74).

haltere, a club-shaped organ, modified from the original hind wing, which helps maintain equilibrium during flight in DIPTERAN flies.

hamadryas (*Papio hamadryas*), a BABOON of northeastern Africa and Arabia, with long silvery hair covering most of the front of the body; ⇨ p. 139.

Hamamelidae, a dicotyledonous plant subclass; ⇨ pp. *40*, 45.

hammerhead, 1. any of 10 predatory marine SHARKS of the family Sphyrinidae, occurring in all temperate and tropical seas. The head is laterally expanded into a hammer shape, with the eyes set at the outer edges. The largest species can grow to around 6 m (20 ft). **2.** (*Scopus umbretta*), the only species of the family Scopidae (order Ciconiiformes). This short brown water bird of Africa resembles a stork with a heavy squat bill.

hamster, the common name for some 24 short-tailed, mouse-like, hibernating myomorph RODENTS of the family Muridae, widely distributed in Eurasia in steppe habitats.

haploid, having only one copy of each chromosome, as occurs in sex cells and in the GAMETOPHYTE stages of plants; ⇨ p. 22. Compare DIPLOID.

haptotropism, ⇨ TENDRIL, TROPISM.

hard water, water with a high concentration of inorganic salt ions; ⇨ p. 218.

hardwood, wood from broadleaved trees (oak, walnut, teak, mahogany, etc.), generally of better quality than the SOFTWOODS of conifers; used for furniture-making, outdoor construction, etc.; ⇨ p. 45.

hare, any lagomorph mammal of the genus *Lepus* (family Leporidae), such as the snowshoe hare (*L. americanus*) and the arctic hare (*L. arcticus*). In contrast to RABBITS, hares are typically solitary, produce young that are fully furred and active at birth, and live in shallow surface depressions (not burrows). The American jack rabbits belong to the genus *Lepus* and are hares in all but name; ⊳ pp. **132**, 172 (box).

harebell, ⊳ BELLFLOWER.

harem, a permanent or temporary form of social organization consisting of a single breeding male with several females, each with her own young (e.g. zebras and many primates); ⊳ p. 158.

haricot bean, any cultivar of the legume, *Phaseolus vulgaris*, family Fabaceae, the seeds of which are harvested when mature and stored dry or processed as baked beans. See also FRENCH BEAN, KIDNEY BEAN and RUNNER BEAN.

harrier, any of 12 long-tailed, long-winged HAWKS of the genus *Circus*, occurring worldwide. The northern-hemisphere marsh harrier (*C. aeruginosus*) flies low and slow to drop onto small reptiles and mammals; ⊳ p. 217.

harrier-eagle (*Circaetus cinereus*), a medium-sized EAGLE of the African scrubland, conspicuous because of its barred tail.

harrier-hawk, either of two large blue-grey BIRDS OF PREY belonging to the family Accipitridae (order Ciconiiformes), living south of the Sahara and in Madagascar.

hartebeest, either of two fawn to reddish-brown ANTELOPES of the genus *Alcelaphus*, with long, curved horns. They inhabit grassland and woodland in Africa.

hart's-tongue fern (*Phyllitis scolopendrium*), an evergreen fern with narrow undivided fronds, native to Europe and Asia; family Polypodiaceae.

harvestman, any long-legged, spider-like ARACHNID of the order Opiliones, found in grassland and other vegetation in tropical and temperate areas. They are often abundant in late summer – hence their name.

harvest mouse, common name given to several small myomorph RODENTS, all of the family Muridae, that are typically found in grassland and cornfields: the tiny Eurasian species *Micromys minutus* (⊳ p. 135) and several New World species of the genus *Reithrodontomys*.

hashish, ⊳ CANNABIS.

hatchet fish, 1. any deep-bodied, laterally flattened freshwater fish of the family Gasteropelicidae (9 species) of Central and South America, related to the CHARACINS. They can beat their pectoral fins so fast that they are capable of powered flight over short distances. **2.** any deep-water, oceanic fish of the family Sternoptychidae (27 species). They have deep, laterally flattened bodies, silver in colour, and have rows of spectacular light organs on the belly.

haw, ⊳ HAWTHORN.

hawk, a term applied rather loosely to many small to medium-sized BIRDS OF PREY (family Accipitridae, order Ciconiiformes) Such species usually have pointed wings, long tails and bare legs, and they fly fast and straight; ⊳ pp. 96, 97. See also GOSHAWK, HARRIER-HAWK, MERLIN, SPARROWHAWK.

hawk moth or **sphinx moth,** any large LEPIDOPTERAN insect of the family Sphingidae, comprising over 1000 species of worldwide distribution. They are powerful fliers and able to hover at flowers; ⊳ pp. 166, 167.

hawkweed, any of around 15 000 species of perennial herbs in the mainly northern hemisphere genus *Hieracium* in the daisy family, Asteraceae (formerly Compositae).

hawthorn or **may,** any of around 200 species of tree or shrub in the north temperate genus *Craetagus* in the family Rosaceae. Most are thorny. When in bloom the shrubs are covered with flowers, making

them attractive ornamentals. Pink may is a variety of the midland hawthorn (*C. laevigata*). The red, blue-black or yellow fruits are known as haws.

hazel, 1. any of around 15 species of shrubs and trees in the north temperate genus *Corylus* in the family Corylaceae. The hazel or cobnut native to Britain is *C. avellana* but the Kentish cob or filbert is the introduced *C. maxima*; ⊳ pp. 42, 43. **2.** buttercup winter hazel (*Corylopsis pauciflora*), an Asian shrub cultivated for its fragrant catkin-like flowers produced in spring. **3.** ⊳ WITCH HAZEL.

head fold, ⊳ p. 187.

hearing, one of the basic SENSES. The capacity to detect sound appears to be largely restricted to insects and vertebrates; ⊳ pp. 191, 73 (fishes), 80 (amphibians), 88 (reptiles), 97 (birds), 98 (mammals), 118 (bats), 124 (cats), 144 (whales).

heart, any specialized structure by which BLOOD is pumped around the body; ⊳ pp. 59, 79, 93, **181**.

heartsease, an alternative name for wild PANSY.

heartwood, the central region of wood in trees; it is older wood, no longer involved in conducting sap, and often impregnated with tannins and other chemicals. Compare SAPWOOD.

heath, 1. ⊳ ERICA, HEATHER. **2.** any of various NYMPHALID butterflies of the subfamily Satyrinae, occurring worldwide. The larvae feed on grasses.

heather, 1. ling or heath (*Calluna vulgaris*), a mainly European low evergreen shrub, family Ericaceae. It has purple flowers and grows on acid soils, especially on heaths and moors, where honeybees are taken for the production of heather honey. There are many cultivated varieties with different colours of flowers and foliage; ⊳ pp. 43, 209 (box), *219*. For the maintenance of heather moors, ⊳ pp. *213*, 229. **2.** bell heather (*ERICA cinerea*), a similar plant, also in the family Ericaceae.

heathland, an area of land dominated by dwarf shrubs usually belonging to the Ericaceae (heather family). It is more specifically applied to the lowland sites of northwestern Europe in which heather (*Calluna vulgaris*) dominates; ⊳ pp. 205, *213*.

heat (sensitivity to), ⊳ THERMORECEPTION. See also INFRARED RADIATION.

Hebe, a mainly New Zealand genus of evergreen shrubs and small trees related to VERONICA. A number are grown as ornamentals.

hedgehog, the common name given to up to 12 of the 17 species in the mammalian family Erinaceidae (order Insectivora), the other 5 being MOONRATS. Distinguished by their covering of spines, hedgehogs feed on a range of animal and plant material. They occur in Europe, Asia and Africa, and in colder parts of their range they may hibernate; ⊳ pp. *101*, **120**.

heliotrope, 1. any of around 250 herbs and shrubs in the tropical and temperate genus, *Heliotropium* in the family Boraginaceae. **2.** winter heliotrope (*Petasites fragrans*), a Mediterranean perennial herb in the daisy family, Asteraceae (formerly Compositae).

heliozoan, a freshwater protozoan. It lacks a shell but sometimes has a skeleton containing silica. It moves by rolling and captures prey with pseudopodia (⊳ p. 27).

helix, a spiral structure. The structure of DNA is a double helix; ⊳ p. 22 (box).

hellbender (*Cryptobranchus alleghaniensis*), a North American salamander, placed in the same family as the closely related GIANT SALAMANDERS.

hellebore, any of around 20 species of perennial herbs in the European and west Asian genus *Helleborus* in the buttercup family, Ranunculaceae. The CHRISTMAS ROSE and lenten rose (*H. orientalis*) are popular garden plants.

helleborine, any of a number of European orchids

with flowers resembling those of a HELLEBORE, particularly species of *Epipactis* and *Cephalanthera*.

hemichordate, ⊳ ACORN WORM.

hemipenis, one of a pair of pockets, often containing thorn-like spines, which lie in the skin adjacent to the cloaca in snakes and lizards; during copulation, one or other is extruded into the cloaca of the female to guide sperm.

hemipteran, any EXOPTERYGOTE insect of the large order Hemiptera, comprising some 60 000 known species, commonly known as 'bugs'. Readily identified by the piercing-sucking mouthparts, many are major pests of plants, others are blood-feeders; they are vectors of many plant and animal diseases. There are some aquatic species, but most are terrestrial. The order has a worldwide distribution and is divided into two major categories, HOMOPTERANS and HETEROPTERANS; ⊳ p. **61**.

hemlock *Conium maculatum*, a Eurasian biennial herb in the carrot family, Apiaceae (formerly Umbelliferae). It grows in damp places and is highly toxic. See also HEMLOCK SPRUCE.

hemlock spruce, any of around 10 species of the conifer genus *Tsuga* in the family Pinaceae. They are native to North America (e.g. western hemlock, *T. heterophylla*) and southern and eastern Asia. The crushed leaves of some species smell like HEMLOCK; ⊳ p. 39.

hemp, ⊳ CANNABIS.

henbane (*Hyoscyamus niger*), a Eurasian and North African strong-smelling annual in the potato family, Solanaceae.

Hennig, Willi (1913–76), German entomologist and founder of the classificatory method known as cladistics (⊳ p. 18).

Hepaticae, the bryophyte class comprising the liverworts; ⊳ p. 37.

herbicide, ⊳ PESTICIDE.

herbivore, any animal that feeds on plants; ⊳ p. 194.

herb Robert, ⊳ GERANIUM.

hercules beetle (*Dynastes hercules*), a SCARABAEOID BEETLE (family Scarabaeidae) from Central and South America, one of the largest insects (about 15.5 cm/6 in long). Much of this length is accounted for by a pair of long opposing horns, one on the head, the other on the prothorax.

herding, behaviour in mammals that results in large social groups, allowing shared vigilance and defence against predators; ⊳ pp. 108, 110, 112 (box), **167**, 171.

hermaphrodite or **bisexual** (adj. or noun), an individual that has both male and female reproductive organs, a common condition in plants (⊳ p. 42) and invertebrate animals (⊳ p. 184).

hermit crab, any DECAPOD crustacean of the families Coenobitidae and Paguridae; they have soft abdomens, which they must protect with a borrowed mollusc shell; ⊳ pp. **67**, 169.

heron, any of 38 species in the family Ardeidae (order Ciconiiformes), which also includes the EGRETS and BITTERNS; they occur worldwide. Long-legged wading birds, they fly with the neck bent in an S-shape. They stalk fishes and amphibians, using their long bill as a spear. The grey heron (*Ardea cinerea*) is a common species in the northern hemisphere; ⊳ pp. *91*, 93, 97 (box), 217.

herring, either of two species of the genus *Clupea* of the family Clupeidae (order Clupeiformes) – the Atlantic herring (*C. harengus*) and the Pacific herring (*C. pallasi*). In both the back is brilliant blue, the flanks silver (sometimes golden). They are plankton-feeders, spawning in shallow waters during the summer months, when vast numbers of their eggs may cover the sea bed. Herrings are preyed on by most larger marine vertebrates and by sea birds. They are shoaling fishes, supporting major fisheries; ▷ p. 66.

herring smelt, ▷ ARGENTINE.

Hesperornis, ▷ p. 90.

Hess, Harry, US geologist who originated the theory of plate tectonics; ▷ p. 8.

hessian, ▷ JUTE.

heteropteran, any HEMIPTERAN insect (bug) of the suborder Heteroptera. These bugs typically have thickened fore wings with a membranous peak or apex, which sit flat on the abdomen and overlap. Most are plant-feeding and many are important pests (▷ CHINCH BUG, COTTON STAINER), while others are blood-feeding or predatory (▷ BEDBUG, ASSASSIN BUG). A number of families are aquatic and generally predatory or scavenging (▷ POND SKATER, WATER BUG, WATER BOATMAN).

heterostracan, ▷ p. 70.

heterotroph or **consumer,** any organism – including many bacteria and protists (▷ pp. 24–7, 34), fungi (▷ p. 35) and all animals – that is incapable of synthesizing the complex organic compounds necessary for life out of simple inorganic molecules, and therefore must obtain these compounds by consuming other organisms. All heterotrophs are thus ultimately dependent on the AUTOTROPHS or producers, principally plants; ▷ p. **194**.

heterozygous, having different alleles in each of the two copies of a gene; ▷ p. 23.

hexapod, a 'six-legged' arthropod, a term used in a proposed reclassification of insects that separates some of the apterygote or wingless groups from (other) insects; ▷ p. 60.

hibernation, a state of dormancy (suspended bodily activity and reduced metabolism), particularly common in small terrestrial animals, usually as a response to harsh winter conditions; ▷ pp. **175**, 119 (bats), 120 (hedgehogs), 128–9 (bears), 134 (squirrels).

hibiscus, any of around 250 species of herbs, shrubs and small trees in the tropical, subtropical and warm temperate genus *Hibiscus* in the mallow family, Malvaceae. Some species are of economic value and many are ornamentals.

hickory, the name given to some North American species of large trees in the genus *Carya*, e.g. shagbark hickory (*C. ovata*) and smooth-bark hickory (*C. glabra*). Tool handles are made from the tough elastic timber. See also PECAN.

hide beetle, ▷ DERMESTID BEETLE.

hierarchy, among social animals, an order of precedence for access to vital resources, usually food or mates. The structure and rigidity of hierarchies vary greatly from species to species. Rankings (positions of dominance and subordination) within a hierarchy are often established by AGGRESSION, and may be reinforced by various forms of DISPLAY, indicating submission, appeasement, etc.; ▷ pp. 141, 153, 154 (box), **157** (box).

hippopotamus, either of two non-ruminating ARTIODACTYL mammals of the family Hippopotamidae: the massively built common hippopotamus (*Hippopotamus amphibius*), a gregarious, amphibious grazer of sub-Saharan Africa, and the pygmy hippopotamus (*Choeropsis liberiensis*) of the forests of West Africa; ▷ pp. 112, **113**.

histology, the microscopic study of plant and animal tissues.

hoarding, a FORAGING strategy involving the storage of food, either in a central cache (or 'larder') or scattered throughout a territory; ▷ pp. *88*, 132, *163*.

hoatzin (*Opisthocomus hoazin*), a large, ungainly bird placed in its own family within the CUCKOO order. It lives in the forests of northern South America, where it feeds on leaves. Superficially cuckoo-like, it has a tall wispy crest. The young have claws on their wings to help them clamber through trees.

hobby, any of four FALCONS of the genus *Falco*, occurring in the Old World. They are fast-flying acrobatic birds and superb hunters; ▷ p. 96.

hog, another name (especially US and Canadian) for PIG; ▷ pp. 112–13.

hog-nosed bat or **Kitti's hog-nosed bat** (*Craseonycteris thonglongyai*), a tiny microchiropteran BAT, the only member of its family. It was first discovered in Thailand in 1973; an adult is about the size of a bumblebee and weighs around 2 g (1/14 oz) – the smallest known mammal.

hogweed, cow parsnip or **keck** (*Heracleum sphondylium*), an Old World biennial herb of the carrot family, Apiaceae (formerly Umbelliferae). Its edible leaves enjoyed by herbivores. Giant hogweed (*H. mantegazzianum*) has a sap that sensitizes skin to sunlight, causing blisters.

Holarctic, a combination of faunal regions including the Palaearctic and the Nearctic regions; covers North America, Europe and much of Asia (▷ map, p. 50).

holly, 1. any of around 400 species of mostly evergreen trees and shrubs of the widely distributed genus *Ilex*. The European holly (*I. aquifolia*) has dark spiny leaves and bright red berries and is used for Christmas decoration. See also MATE. **2.** any of various unrelated plants such as sea holly (*Eryngium maritimum*), which has holly-like leaves.

hollyhock, any of a number of tall herbs in the genus *Alcea* in the mallow family, Malvaceae. *A. rosea* has many garden cultivars.

Holocene epoch, a geological epoch lasting from 10 000 years ago to the present; ▷ p. 4.

homeostasis, the tendency of an organism to actively maintain a relatively constant internal environment in the face of changing external conditions.

hominid, any APE of the family Hominidae, which is now taken to include not only MAN and his immediate extinct precursors but also the gorilla and chimpanzees; ▷ pp. 5, 141.

homoiothermic or **endothermic** ('warm-blooded'), relating to an animal that uses energy generated by metabolism to maintain its body temperature within narrow limits, usually above and more or less independently of the environmental temperature. Birds and mammals are the only homoiothermic animals or 'homoiotherms'; ▷ pp. **87**, 93, 98 (box), 102 (box), 106, 109, 142.

homologous, (of an organ or other structure) deriving from a common evolutionary ancestor; ▷ p. 14.

homologous chromosomes, chromosomes that pair during meiosis (▷ p. 22), each pair possessing a similar sequence of genes. In every cell (other than sex cells) in most higher organisms the chromosomes are in homologous pairs, one of each derived from the mother and the other derived from the father.

homopteran, any HEMIPTERAN insect (bug) belonging to the suborder Homoptera. They are very varied, but typically have fore wings of similar composition along their length. The group includes FROGHOPPERS, TREE HOPPERS, LEAFHOPPERS, CICADAS, WHITEFLIES, APHIDS and COCCIDS.

homozygous, having identical alleles in each of the two copies of a gene; ▷ p. 23.

honesty (*Lunaria annua*), a herb of the cabbage family, Brassicaceae (formerly Cruciferae), with large white paper-thin disc-like fruits and therefore popular for flower arranging.

honey, a sweet, thick golden-brown fluid made by bees from the nectar of flowers (▷ p. 64). It is stored in honeycombs and is used to feed young and adults.

honey badger, another name for RATEL.

honey bear, another name for SUN BEAR and KINKAJOU (1).

honeybee (*Apis mellifera*), a social BEE, introduced worldwide. Selective breeding from specific strains has attempted to maximize honey production (greatest in more aggressive races), while retaining docility; ▷ pp. 63, 64, 153 (box), 161.

honeycreeper, 1. any of five species of Neotropical TANAGER, with sharp downwardly curved bills. **2.** any of 30 PASSERINE birds of the tribe Drepanidini, closely related to the FINCHES, small arboreal birds of Hawaii. These have a remarkable variety of bill shapes and sizes, each specialized for a different food, from seed-cracking to nectar-feeding (▷ box, p. 16). Several species are under threat of extinction.

honeydew, a sugary fluid secreted by aphids in a symbiotic association with ants; ▷ pp. 65, *169*.

honey-eater, any of about 180 PASSERINE birds of the family Meliphagidae, arboreal birds characterized by deeply cleft, brush-like tongues, living in Africa, Australasia and the southwest Pacific. They feed on nectar and insects.

honey fungus (*Armillaria mellea*), an AGARIC FUNGUS that parasitizes trees, shrubs and herbaceous plants. Bootlace-like rhizomorphs enable it to spread through the soil; ▷ p. 35.

honey guide, any of 17 small birds of the family Indicatoridae (order Piciformes), living in tropical forests of Africa and Asia. Honey guides have a symbiotic relationship with the RATEL (▷ p. 169) and are brood parasites (▷ p. 168).

honey possum, long-snouted phalanger or **noolbenger** (*Tarsipes rostratus*), a tiny, mouse-like Australian MARSUPIAL (family Tarsipedidae). It has grasping hands and feet, a prehensile tail, and a long snout with a brush-tipped tongue, adapted for feeding on nectar and pollen; ▷ pp. 43, *99*.

honeysuckle, 1. any of around 200 species of shrubs and woody climbers in the north temperate genus *Lonicera* in the family Caprifoliaceae. Many are grown as ornamentals and for their fragrance. **2.** an Australian name for BANKSIA.

hoof, a hard structure made of keratin, covering and protecting the end of the foot of hoofed mammals or UNGULATES; ▷ p. 110 (box).

hoofed mammal, ▷ UNGULATE.

hookworm, any of various blood-sucking parasitic ROUNDWORMS, an extremely significant cause of disease in humans; ▷ p. 55.

hoopoe (*Upupa epops*), the only species in the family Upupidae (order Upupiformes). A distinctive ground-dweller of Africa and the Palaearctic, the hoopoe is pink, black and white, and has a large swept back head crest and a long bill, which it uses to probe the ground for insect larvae; ▷ p. *91*. See also WOOD HOOPOE.

hop (*Humulus lupulus*), a climbing perennial herb in the family Cannabidaceae. It is cultivated for the female inflorescences, which are used in brewing beer as they contain bitter and resinous substances that give a characteristic flavour.

hopper, common name for any jumping insect, especially juvenile LOCUSTS. See also PLANTHOPPER, FROGHOPPER, LEAFHOPPER, TREEHOPPER.

horehound, 1. white horehound (*Marrubium vulgare*), a Eurasian and North African perennial herb in the family Lamiaceae (formerly Labiatae). **2.** black horehound (*Ballota nigra*), a related Eurasian perennial herb. **3.** water horehound (*Lycopus europeus*), a Eurasian herb in the family Lamiaceae.

horizon, ▷ PROFILE.

hormone, a chemical secretion produced by an ENDOCRINE cell or gland and carried by the bloodstream to affect the activity of cells elsewhere in the body; ▷ p. 183. For plant hormones ▷ pp. 32–3.

horn, a tough fibrous material consisting chiefly of KERATIN. It forms the hard outgrowths on the heads of some animals, e.g. cattle (▷ p. 114).

hornbeam, any of around 35 species of shrubs and trees in the north temperate and Central American genus *Carpinus* in the family Corylaceae. The hornbeam of the Old World is *C. betulus* and its strong timber is used for turning and tool handles.

hornbill, any of 56 species of the order Bucerotiformes, slender-bodied birds with long eyelashes, long tails and massive downwardly curved bills, used to crush insects and fruit. Most live in trees in the Old World, but a few are ground-dwellers and grow to a large size; the southern ground hornbill (*Bucorvus leadbeateri*) is 1 m (39 in) tall. The nesting behaviour of most species is interesting because the female is sealed in the nest with the chicks and is dependent upon the male 'posting' food through a narrow entrance slit; ▷ pp. 91, 95, **161**, *201*.

hornero, any of seven species of OVENBIRD, thrush-like birds that build oven-shaped nests of twigs consolidated with clay; ▷ p. 95.

hornet (*Vespa crabo***),** a large European social WASP with brown and yellow markings, which usually nests in hollow trees.

horntail or **woodwasp,** any large, conspicuous SAWFLY of the family Siricidae, of worldwide distribution. They have a powerful ovipositor, which is used to bore into timber. An egg is laid in each hole, and the larvae feed and grow within the tree.

hornwort, any member of the bryophyte class Anthocerotae; ▷ p. 37.

horse, the common name given to two PERISSODACTYL mammals of the family Equidae: the domesticated horse (*Equus caballus*), believed to be descended from the now-extinct tarpan (*E. c. gomelini*); and Przewalski's horse (*E. przewalskii*), the only true wild horse that still survives. In a broader sense, the name 'horse' may be used to refer to any other member of the family, i.e. asses and zebras; ▷ p. *9*, 111, 154 (box).

horse chestnut (*Aesculus hippocastanum***),** a large spreading tree with spectacular flower spikes native to the Balkans but introduced elsewhere as an ornamental. See also BUCKEYE.

horsefly, any robust, medium to large DIPTERAN insect of the family Tabanidae, worldwide in distribution with over 2000 species known. They are strong fliers, particularly active on warm sunny days. Females are blood-feeders, especially on cattle and horses, and humans can also be attacked. They are the carriers of several tropical diseases. They are sometimes referred to as 'clegs' or 'gadflies'.

horseradish (*Armoracia rusticana***),** a temperate perennial herb in the cabbage family, Brassicaceae (formerly Cruciferae), cultivated for its strongly flavoured root, which contains mustard oil.

horseshoe bat, any of over 50 microchiropteran BATS of the family Rhinolophidae, with a widespread distribution throughout the Old World, especially in the tropics. These insectivorous bats derive their name from the distinctively shaped nose-leaf used in ECHOLOCATION.

horseshoe crab or **king crab,** any marine ARTHROPOD of the class Merostomata. They are not in fact crabs at all, but related to the extinct TRILOBITES; ▷ p. 69 (box).

horsetail, any member of the plant division Sphenophyta; ▷ p. 37.

host, the exploited species in a parasitic or commensal association; ▷ p. 168.

Hosta, an oriental genus of perennial herbs in the lily family, Liliaceae. They are also known as plantain lilies. A number of species and cultivars are grown for their ornamental foliage.

housefly (*Musca domestica***),** a DIPTERAN insect of the family Muscidae, of worldwide distribution and commonly associated with man. The larvae ('maggots') develop in excrement or manure. Adults frequently feed on food prepared for human consumption and are important vectors of microorganisms causing a variety of intestinal diseases; ▷ p. 59 (box).

houseleek, any of around 40 species of succulent herbs in the Eurasian and North African genus *Sempervivum* in the family Crassulaceae, but especially *S. tectorum*, grown in rock gardens.

hoverfly, any medium-sized DIPTERAN insect of the family Syrphidae. The adults are usually brightly coloured (mimicking the coloration of distasteful species such as bees and wasps; ▷ pp. 153, 167, box) and feed on flowers. They are expert fliers, able to hover and fly backwards. The larval life styles are very variable. Some are important predators of aphids and COCCIDS.

howler monkey, any large New World monkey of the genus *Alouatta* (family Cebidae) with a loud, booming call; ▷ pp. **138**, 154.

huckleberry, any of a number of species of shrubs in the New World genus *Gaylussacia* in the heather family, Ericaceae. Many bear edible fruits similar to the BLUEBERRY.

human (*Homo sapiens sapiens***),** a HOMINID ape of widespread distribution, the only (almost) hairless primate; ▷ pp. 14, 15, 17, **18**, 141 (box). For the relations of humans with other animals ▷ pp. 52 (sponges), 55 (worms), 64 (insects), 75, 76 (fishes), 78 (amphibians), 89 (reptiles), 97 (birds), 109 (elephants), 110–11 (horses, rhinos), 114 (bovids), 119 (bats), 122 (carnivores), 132 (rabbits), 134 (rodents), 145 (whales). For the impact of humans on natural ecosystems and the biosphere ▷ pp. 194–233. See also MAN.

humidity, the concentration of water vapour in the atmosphere. It is often expressed as relative humidity, which describes the degree to which air is saturated under given conditions of temperature and pressure.

hummingbird, any of 319 species of tiny birds, often with iridescent colours, belonging to the family Trochilidae (order Trochiliformes); they are found throughout the Americas, predominantly in tropical regions. Hummingbirds include the smallest birds: the bee hummingbird (*Mellisuga helenae*) is only 6.5 cm (2½ in) long and weighs 2.5 gm (¹⁄₁₀ oz). The short wings can beat up to 70 times per second and are modified to create lift on the upstroke as well as the downstroke. This enables them to hover or even to fly backwards. Their feet are tiny, and hummingbirds can perch but cannot walk. Most are nectar-feeders and have long bills and protrusible tongues; ▷ pp. 43, *91*, 92, 95, *96*, 171.

humpback whale (*Megaptera novaeangliae***),** a large baleen whale (CETACEAN), similar to other RORQUALS, but with an arched back and longer flippers; ▷ pp. 142, *144*, **145**, 154.

humus, a complex organic material resulting from the decomposition of plant and animal detritus in soil. A major component is humic acid. It is good at retaining water and mineral ions.

hunting dog, African hunting dog or **wild dog** (*Lycaon pictus*), a powerful, carnivorous and highly social CANID of African savannah/open woodland; ▷ pp. 127 (box), 165.

hutia, any of 13 stocky, short-legged, rat-like caviomorph RODENTS of the family Capromyidae, found only in the West Indies.

Huxley, Thomas Henry (1825–95), English biologist, one of Darwin's greatest contemporary supporters (▷ box, p. 13). He also introduced the term 'agnostic'.

hyacinth (*Hyacinthus orientalis***),** a Mediterranean herb growing from a bulb with sweet scented flowers in the lily family, Liliaceae. Many large-flowered cultivars have been derived from the wild species. The BLUEBELL is also called 'wild hyacinth', although it belongs to a different genus. See also GRAPE HYACINTH.

hyaena, variant spelling of HYENA.

hybrid, an individual produced by crossing genetically different parents. There may be several big differences between the parents, or just a single gene difference.

hydra, any small freshwater HYDROZOAN of the genus *Hydra*, common in ponds and streams. The POLYP phase is dominant throughout the life cycle; ▷ pp. **53**, *178*, *183*.

hydrangea, any of around 25 species of shrubs and woody climbers in the oriental and New World genus *Hydrangea* in the Hydrangeaceae family. A number are grown as ornamentals.

hydrocarbon, any compound containing only hydrogen and carbon, especially fossil fuels (oil, natural gas, COAL); ▷ p. 196.

hydrogen, the lightest of all the elements (symbol H), and the most abundant in the universe. In its molecular form (H_2) it is an inflammable gas, found in small quantities in the atmosphere (▷ p. 6). It is a constituent of water (H_2O) and of most organic compounds, and is thus essential to life (▷ p. 6).

hydroid, name given to many tiny solitary or colonial HYDROZOAN coelenterates. They are structurally simple, with a two-layered wall, blind gut, and a ring of tentacles armed with stinging cells and used for food capture; ▷ p. 53.

hydrological cycle, the cycle by which water circulates in the biosphere. Precipitation (rain, snow, etc.) from clouds falls to the ground, where some of it is stored as groundwater (water held in soil and pervious rocks), some in lakes, and some runs off into rivers. Some evaporation occurs at this stage, but most water evaporates from the oceans, where it is carried by rivers. As evaporated water (water vapour) accumulates in the atmosphere it forms clouds, so completing the cycle. However, there is also a biotic component to the cycle, by which plants absorb water through their roots, and also lose some (in the form of water vapour) through their leaves by the process of transpiration (▷ p. 31). Water also cycles through animals, but in much smaller quantities than through plants. Human activity often has a radical effect on natural hydrological cycles, e.g. by building dams, draining wetlands, etc. (▷ p. 217).

hydroseral succession, the process by which ponds or lakes eventually fill in; ▷ p. *219*.

hydrostatic skeleton, an internal, fluid-filled skeleton found in invertebrates such as molluscs and many worms; ▷ p. 176.

hydrothermal vent, a vent found deep in the oceans from which sulphurous water heated by volcanic activity is released; ▷ pp. 7, 221.

hydrotropism, the response in growth direction of a root or stem to the presence of water; roots are normally positively hydrotropic, i.e. they grow towards water. See also TROPISM.

hydrozoan, any COELENTERATE of the class Hydrozoa, a very diverse group comprising some 2700 species, including the solitary freshwater hydras and the colonial siphonophores, such as the Portuguese man-of-war; ▷ p. 53

hyena, any mammal of the family Hyaenidae (order Carnivora), which contains three 'true' hyenas – the spotted (*Crocuta crocuta*), the striped (*Hyaena*

hyaena) and the brown (*H. brunnea*) – and the unusual aardwolf (*Proteles cristatus*), which feeds mainly on termites; ➪ pp. 122, **125** (box), 165, 203.

hymenopteran, any insect of the order Hymenoptera (➪ p. *61*), a large group that is divided into two suborders. The larger suborder, Apocrita, is itself subdivided into the 'stinging' forms (Aculeata), which include the ANTS, WASPS and BEES; and the 'parasitic' forms (Parasitica), most of which are small or tiny insects (often referred to as 'wasps') that live parasitically in or on other insects (➪ ICHNEUMON WASP). All the insects of this group are distinguished by a 'waist' or constriction between the abdomen and the thorax. The members of the smaller suborder, Symphyta – the SAWFLIES – do not have a marked constriction of this kind.

hyperparasite, an organism that is parasitic in or on another parasite.

hypha (pl. hyphae), a tubular filament, aggregates of which make up the vegetative body (mycelium) of most fungi and some algae. Hyphae are either septate (divided into compartments) or aseptate (not so divided); ➪ p. 35.

hypnosis or **tonic immobility,** (in animals) a paralysis-like state of general unresponsiveness induced by the proximity of a predator and presumed to have a defensive function similar to that of death-feigning; ➪ p. 167.

hypodermis, a distinct layer of cells beneath the EPIDERMIS of roots and leaves. It often has protective and storage functions.

hypothalamus, part of the vertebrate brain, controlling the PITUITARY GLAND and important as an integrating centre for many autonomic (involuntary) functions; ➪ p. 183.

hyrax or **dassie,** any mammal of the family Procaviidae (order Hyracoidea), comprising 11 species of small, rodent-like herbivore of Africa and the Middle East. Some (rock hyraxes) live in arid, rocky terrain, others (tree hyraxes) are arboreal; ➪ pp. *101, 108*, **109**, 205.

hyssop, any of a number of species in the labiate genera Hyssopus, Origanum and Agastache, particularly *H. officinalis, O. syriacum* and anise hyssop (*A. foeniculum*).

Iberis, a Mediterranean genus of around 30 species of annual or perennial herbs or dwarf shrubs in the cabbage family, Brassicaceae (formerly Crucifereae), e.g. candytuft (*I. amara* and *I. umbellata*).

ibex (*Capra ibex*), a nimble, brown wild GOAT with massive, curved horns. It inhabits steep, alpine areas of Eurasia and northeast Africa; ➪ pp. 115, 210.

ibis, any of 25 species of the family Threskiornithidae (order Ciconiiformes), wading birds occurring in most temperate parts of the world. Characteristically they have long downwardly curved bills. The glossy ibis (*Plegadis falcinellus*) is a worldwide species. They are closely related to the SPOONBILLS; ➪ pp. *91*, 217.

ice age, ➪ GLACIAL PERIOD.

ice bug, another name for GRYLLOBLATTID.

ice-fish, any fish belonging to one of two families, both of which inhabit freezing Antarctic waters. Members of the family Nototheniidae (related to the SALMONIDS) make up 68% of the Antarctic fish fauna, while those of the family Channichthyidae (PERCIFORM fishes) lack haemoglobin in the blood; ➪ p. 73.

ice plant (*Mesembrianthemum crystallinum*), a succulent herb with leaves covered with glistening papillae. The name has been extended to other *Mesembrianthemum* spp. and species of SEDUM.

ichneumon wasp or **fly,** any parasitic WASP of the family Ichneumonidae, of worldwide distribution. These insects generally have long ovipositors, with which they lay eggs in or on the bodies of larvae or pupae of other insects; the developing larvae then feed on the body fluids of the host.

ichthy-, prefix indicating fish.

Ichthyornis, ➪ p. 90.

ichthyosaur, any extinct marine REPTILE of the order Ichthyosauria; ➪ pp. *17*, 82, **84**.

-idae, a standard ending indicating subclass status in plants (➪ p. 45) and family status in the animal kingdom (➪ p. 18).

-iformes, a standard ending indicating order status in some groups of animals; for example, Passeriformes, the order of perching birds (passerines).

iguana, any of several mainly large IGUANID lizards. The common iguana of America is a herbivorous tree-dweller, distinguished by a crest of spines on its back.

iguanid, any of around 600 species of LIZARD of the mainly New World family Iguanidae, including IGUANAS, BASILISKS and the CHUCKWALLA. Most species are agile tree- or ground-dwellers, feeding on insects and other small invertebrates, but the larger species are herbivorous and the marine iguana of the Galápagos Islands dives for seaweed.

iguanodon, a gigantic herbivorous bipedal ORNITHOPOD dinosaur of the genus *Iguanodon*; ➪ p. 83.

ileum, the lower part of the small intestine; ➪ p. 179.

imago, an insect in its final mature (winged) state; ➪ p. 63.

immigration, a movement of animals into an area, usually in response to more or less unpredictable factors, such as food availability, or as a consequence of EMIGRATION from elsewhere; ➪ pp. 172–3.

immunoglobulin, ➪ ANTIBODY.

impala (*Aepyceros melampus*), a common, gregarious African ANTELOPE, notable for its enormous leaps when alarmed; ➪ p. 115.

Impatiens, a mainly tropical Asian and African genus of over 500 species of annual and perennial herbs in the family Balsaminaceae. Himalayan balsam (*I. glandulifera*) is cultivated as an ornamental and busy Lizzie (*I. wallerana*) is a popular house plant.

imperfect stage or **state,** the phase in a fungal life cycle in which reproduction is asexual. The fungi imperfecti (➪ p. 35) have no known perfect (sexual) stage.

implantation, in placental mammals, the attachment of the embryo (blastocyst) to the wall of the uterus. In some species (e.g. roe deer, seals), implantation may be delayed for a prolonged period after fertilization; ➪ pp. 107, 143, 185, *186*.

imprinting, a process during particular developmental periods by which animals learn characteristics of their surroundings (e.g. their mother, siblings or habitat). Imprinting affects immediate attachment behaviour and later mate choice; ➪ p. 95.

impulse, a tiny electrical pulse acting as a signal between nerve cells; ➪ p. 182 (box).

-inae, a standard ending indicating subtribe status in plants, and subfamily status in animals (➪ p. 18).

inbreeding, breeding between close relatives, involving the fusion of genetically related gametes (sex cells). Inbreeding reduces genetic variability, and increases the chances that harmful or lethal genes will be expressed in offspring.

incisor, a tooth adapted for cutting, especially one of the cutting fore teeth of mammals (➪ box, p. 98).

incus, one of the three auditory ossicles in the mammalian middle ear (the 'anvil'); ➪ pp. 98 (box), *191*.

indehiscent, (of plant structures) not spontaneously opening to release their contents. Compare DEHISCENT.

indigenous, (of a species) native to the area. Compare ACCIDENTAL, ENDEMIC, EXOTIC.

indigo, any of a number of species in the leguminous genus Indigofera, e.g. *I. tinctoria*. The dye, indigo, is extracted from the leaves.

Indo-Pacific, describing the region of the Indian and Pacific Oceans lying off Southeast Asia.

indri or **babakoto** (*Indri indri*), a woolly LEMUR (family Indriidae), the largest of the lemurs and the largest living primate (except man) on Madagascar; ➪ p. 137.

inflorescence, a collection of flowers on the same stalk (called a peduncle). The way the flowers are arranged in an inflorescence varies considerably, from the 'lamb's tail' of the male catkin (➪ p. *42*) to the globe shape of onion and garlic flowers. Varieties of inflorescence include RACEME, SPADIX and UMBEL.

infrared radiation, electromagnetic radiation at a lower frequency than visible light, perceptible as heat. It is detectable by various animals; ➪ pp. 55, 88, 189 (box). For its role in the greenhouse effect ➪ p. 224.

infructescence, a collection of fruits arising from an INFLORESCENCE. A pineapple is an example.

ingestion, intake of food; ➪ p. 178.

inheritance, ➪ GENETICS.

-ini, a standard ending indicating tribe status in animals; for example, Cyprinini, minnows and carps.

ink cap, any of around 100 species of the cosmopolitan *AGARIC* genus, Coprinus. In most species the gills become a liquid suspension of black spores, which has been used as ink.

innate, or **instinctive behaviour,** genetically determined behaviour that is not learned or modified by environmental influences in contrast to behaviour acquired by LEARNING; ➪ pp. 148, **150**.

inner cell mass, a concentration of cells in an animal blastula, part of which gives rise to the embryo proper; ➪ pp. 186–7.

inner ear, a complex, composite organ in vertebrates by which sound is perceived, balance maintained, etc.; ➪ pp. 190–1.

insect, any ARTHROPOD of the class Insecta, an enormously successful group comprising several million species (only a portion of which have been described). Conventionally the class is divided into wingless insects (APTERYGOTES, 4 orders) and winged insects (PTERYGOTES, 25 orders), but this is not universally accepted (➪ HEXAPOD); ➪ pp. 42–3, **58–65**, 230–1.

insecticide, ➪ PESTICIDE.

insectivore, 1. any animal that feeds on insects. **2.** any mammal of the order Insectivora, a diverse and successful group that contains around 345 species and includes hedgehogs, shrews, moles, solenodons, tenrecs, and various related species. The order is represented on every continent except Australasia and Antarctica; ➪ pp. **120–1**.

instar, any of the successive growth stages of a developing (juvenile) insect, between any two moults; ➪ p. 62.

instinctive, ➪ INNATE.

instrumental conditioning, ➪ CONDITIONING and pp. 150–1.

insulin, in vertebrates, a HORMONE secreted by the PANCREAS that regulates blood-sugar levels; ➪ pp. 25, **183**.

intelligence, attempts to measure the relative intelligence of different species of animal have generally relied on testing their ability to solve various kinds of problem. Some of these involve an animal learning to find a hidden food item, the detection of which depends on the ability to discern some kind of general rule; in others animals are set the task of obtaining food that can only be reached by the use of sticks or by piling up boxes. In spite of the limitations of such tests, tentative comparisons have been possible – many birds, for instance, appear to have comparable ability to many mammals, while many primates (especially higher primates) show the greatest aptitude; ⇨ pp. 98 (box), 141 (box), **150–1**.

intention movement, the initial phase of a behaviour pattern, which may serve to inform others that an animal is about to perform the full sequence of behaviour, as when a bird performs a pre-flight intention movement prior to taking off (the failure to perform such a movement being interpreted by other birds as an ALARM RESPONSE). Often such movements are left incomplete and have become ritualized as COMMUNICATION signals; ⇨ p. 155 (box).

interglacial, a warmer period between glacial periods or ice ages, of which there have been many over the last million years; ⇨ pp. **5, 10**.

interneurone, a nerve cell in the central nervous system responsible for integrating and processing incoming signals; ⇨ p. 182.

internode, a length of stem between NODES; ⇨ p. 32.

interstitial fluid, ⇨ LYMPHATIC SYSTEM and p. 181.

intertidal zone, the zone between the low- and high-tide marks; ⇨ pp. 220–1.

intromittent organ, any male copulatory organ, such as a penis, that can be inserted into the female's genital tract; ⇨ p. 185.

inversion, the freeing of a piece of chromosome by breakage and its reintegration into the chromosome the other way round.

invertebrate, any animal lacking a backbone, i.e. all animals except the vertebrates (fishes, amphibians, reptiles, birds and mammals); ⇨ p. 51 (box).

Iridaceae, a monocotyledonous plant family, including irises, crocuses and gladioli; ⇨ p. 45.

iris, 1. any of around 300 species of perennial herbs with rhizomes, fleshy roots or bulbs in the north temperate genus *Iris* in the family Iridaceae. They have two ranked narrow leaves and a characteristic flower form popular for ornament. See also FLAG and ORRIS ROOT. **2.** an opaque muscular diaphragm in the vertebrate eye that surrounds the pupil and alters its size; ⇨ p. 189.

iron, a metallic element (symbol Fe) essential to chlorophyll synthesis in plants (⇨ p. 31) and in the HAEMOGLOBIN of animal blood.

ironbark, ⇨ EUCALYPTUS.

iron pan, a band of red material in the soil profiles found in boreal forests; ⇨ p. 209.

ironwood, a name used for many different species of tree and shrub which have in common a hard, tough timber which is so dense that it may be heavier than water. Giant ironwood (*Choriocarpia subargentia*) is an example.

island ecosystems, ⇨ p. 214.

island model, ⇨ p. 233.

isolating mechanism, any characteristic of a species that serves to distinguish it from other species, so preventing it from mating with them; ⇨ p. 17.

isopod, any CRUSTACEAN of the class Isopoda, which includes the woodlice and various aquatic relatives; ⇨ p. 67.

ivy, any of around 15 species of mostly evergreen climbers or shrubs in the mainly north temperate genus *Hedera*. They are cultivated for their glossy and often variegated foliage. Common ivy (*H. helix*) is hardy and tolerant of shade and has many cultivars.

jaçana, any of eight water birds with long legs and very long toes and claws, belonging to the family Jacanidae (order Ciconiiformes), occurring worldwide in tropics and subtropics. They habitually walk on floating vegetation, and are sometimes known as 'lily-trotters'; ⇨ pp. 158, 159 (box).

jacamar, any of 18 species of the family Galbulidae (order Galbuliformes). They are small, long-tailed birds with long sword-like bills, living in the Americas. Jacamars are the New World equivalents of the Old World BEE-EATERS; ⇨ p. *91*.

jacaranda, any of around 50 species of mainly blue flowered trees and shrubs in the tropical American genus *Jacaranda*.

jackal, any of four small fox-like CANIDS of the genus *Canis*, opportunistic scavengers and predators mainly found in dry, open areas in Africa, southern Asia and southeastern Europe; ⇨ pp. 126, **127**, 203.

jackdaw (*Corvus monedula*), a medium-sized social CROW of the Palaearctic, famous for its habit of storing objects and food in caches; ⇨ p. 213.

jack rabbit, name given to various North American HARES; ⇨ p. 205.

Jacobsen's organ, ⇨ VOMERONASAL ORGAN.

jaeger, ⇨ SKUA.

jaguar (*Panthera onca*), a large, stocky, yellowish-brown CAT with dark rosette markings. It inhabits tropical forest, swamps and savannah in South and Central America and southwest USA; ⇨ p. 201.

japonica, 1. ⇨ QUINCE. **2.** *Camellia japonica*, a popular ornamental CAMELLIA with around 2000 named cultivars.

jasmine, any of around 450 species of scrambling shrubs and climbers in the tropical to temperate genus *Jasminum* in the olive family, Oleaceae. A number of species are popular ornamentals, e.g. winter jasmine (*J. nudiflorum*). See also GARDENIA.

jawless fish or **agnathan,** any FISH of a primitive group characterized by the absence of (true) jaws, sometimes assigned to the class Agnatha but probably in fact of more diverse origins. All the earliest fossil fishes were jawless (⇨ p. 70), but the group is represented today only by the HAGFISHES and LAMPREYS.

jay, any of 40 relatively small and colourful CROWS, living mostly in the Americas but also in the Palaearctic. Some, such as the American blue jay (*Cyanocitta cristatus*) or the European jay (*Garrulus glandarius*), are familiar urban birds; ⇨ p. 158.

jejunum, the part of the small intestine between the duodenum and the ileum.

jellyfish, any marine COELENTERATE of the class Scyphozoa, in which the free-swimming MEDUSA phase is dominant. Some of the smaller jellyfishes are classified as hydrozoans; ⇨ pp. **53**, 220.

jerboa, any small, mouse-like myomorph RODENT of the family Dipodidae, with long hind limbs and a long tail. They live in burrows in the deserts of North Africa, the Middle East and Central Asia.

joint, a point or junction between two body parts at which movement is possible, as in insects and arthropods (⇨ p. 58) and in vertebrates (⇨ box, p. 177).

jonquil, ⇨ NARCISSUS.

Joshua tree, ⇨ YUCCA.

Judas tree (*Cercis siliquastrum*), a leguminous tree with inflorescences often growing directly from the branches. Judas is said to have hanged himself from one.

Juglandales, a dicotyledonous plant order, including walnut and hickory; ⇨ p. *40*.

jumping, as a means of locomotion, ⇨ pp. 104, 177.

jumping mouse, any of some 14 mouse-like myomorph RODENTS of the family Zapodidae, with long hind limbs and tails, and colourful fur. They are found in North America and Eurasia.

jungle, a vernacular word usually applied to tropical rainforest. Its image of a tangled mass of vegetation really applies only to forest-edge and river-edge communities, or regenerating areas.

jungle fowl, any of four GAMEBIRDS of the family Phasianidae, living in Asian forests. The red jungle fowl (*Gallus gallus*) is the ancestor of the domestic chicken; ⇨ p. 97 (box).

juniper, any of around 60 species of conifer in the northern hemisphere genus *Juniperus*. They have two sorts of leaf: needle-like and spreading in whorls of three (juvenile form), and scale-like, pressed closely to the stem and opposite (mature form). The timber is resistant to insect attack owing to the presence of oils. It is commercially known as cedar wood; ⇨ p. 209 (box).

Jurassic period, a geological period lasting from 205 to 135 million years ago; ⇨ p. 4.

jute, a fibre (from which hessian cloth is made) obtained principally from white jute (*Corchorus capsularis*) and upland jute (*C. olitorius*). India and Bangladesh are the main producers.

kagu (*Rhynochetes jubatus*), a small, superficially heron-like bird with grey body, red bill and legs, of the family Rhynochetidae (order Gruiformes). This threatened species is confined to the dense forests of New Caledonia.

kakapo or **owl parrot** (*Strigops habroptilus*), a rare flightless parrot, feeding on tussock grass, mosses and berries on New Zealand's South Island; ⇨ pp. 97 (box), 214.

kale, 1. a cultivar of BRASSICA *napus*, the leaves of which are eaten as a vegetable. Curly kale has curly leaves. **2.** a cultivar of BRASSICA *oleracea* var. *ancephala* used for animal feed.

kangaroo, any Australasian heribivorous MARSUPIAL mammal of the genus *Macropus* and related genera in the family Macropodidae, which also includes various kangaroo-like animals, such as the smaller WALLABIES, PADEMELONS and QUOKKA. Kangaroos typically have powerful hind limbs adapted for jumping, but the tree kangaroos are adept climbers; ⇨ pp. *101*, 103, **104**, 177, 205. See also RAT KANGAROO.

kangaroo rat, name variously applied to New World sciuromorph RODENTS related to the POCKET MICE, and to similar leaping myomorph rodents (*Notomys*) found in arid central Australia.

kapok, ⇨ SILK COTTON TREE.

katydid, another name for BUSH CRICKET.

kauri or **kauri pine,** any of around 21 species of conifer in the principally Australasian genus *Agathis*, family Araucariaceae. They have large leathery broad leaves and are a useful source of timber; ⇨ p. 39.

keck, ⇨ COW PARSLEY and HOGWEED.

keel, ⇨ CARINA.

kelp, any species of large seaweed, particularly brown algae such as the OARWEEDS and giant kelps (*Macrocystis* spp.); ⇨ pp. **34**, 221.

kelp fly, any small DIPTERAN insect of the family Coelopidae; the larvae of these seashore insects breed in rotting seaweed. See also SHORE FLY.

keratin, a fibrous sulphur-based protein, a structural component of nails, claws, hooves, hair, feathers, the outer skin layer (epidermis), etc.; ⇨ pp. 92, 93, 110 (box).

kernel, the part of a seed lying within the seed coat (the TESTA).

kestrel, any of 13 FALCONS of the genus *Falco*,

occurring worldwide. The northern-hemisphere kestrel (*F. tinnunculus*) is often seen hovering above roads.

kiang, a variety of Asiatic wild ASS.

kidney, in vertebrates, one of a pair of organs responsible for excretion and regulation of the water–salt balance of body fluids; ▷ p. 179 (box).

kidney bean, any cultivar of the legume, *Phaseolus vulgaris* (family Fabaceae), the seeds of which are harvested for human consumption while immature. See also FRENCH BEAN, HARICOT BEAN and RUNNER BEAN.

killer whale (*Orcinus orca*), a robust black-and-white toothed whale (CETACEAN) with a tall dorsal fin, a ferocious predator found worldwide, especially in cooler coastal waters; ▷ pp. **145**, 165.

killifish or **toothcarp,** any small, carp-like fish of the family Cyprinodontidae, comprising perhaps 268 species (although precise membership is disputed); they are mainly American brackish/freshwater fishes. Like the related RIVULINES, they are frequently brightly coloured and popular aquarium fishes. Many practise internal fertilization.

king, ▷ REPRODUCTIVE.

kingbird, any of 10 species of TYRANT FLYCATCHER of the genus *Tyrannus*, living throughout the Americas. Although small, these birds can be very agressive towards larger species.

king crab, an alternative name for HORSESHOE CRAB.

kingcup or **marsh marigold** (*Caltha palustris*), a north temperate and arctic perennial herb in the buttercup family, Ranunculaceae. It is a marsh plant with yellow flowers.

kingdom, the highest classificatory group. Most authorities today recognize five kingdoms: Animalia (animals; ▷ pp. 50–145), Plantae (plants; ▷ pp. 30–47), Fungi (fungi; ▷ p. 35), Protista (protists, i.e. all single-celled eukaryote organisms; ▷ pp. 26–7), and Monera (bacteria and all other single-celled prokaryote organisms; ▷ p. 18.

kingfisher, any of 94 compact-bodied birds, usually with short tails, large heads and long, heavy, pointed bills, belonging to the families Alcedinidae, Dacelonidae and Cerylidae (order Coraciiformes). Kingfishers occur worldwide, most species living in Southeast Asia. The third and fourth toes are joined to one another, and many species are brightly coloured, often with iridescent hues; ▷ pp. **91**, 97 (box). See also KOOKABURRA.

kinkajou, potto or **honey bear, 1.** (*Potos flavus*), a fruit-eating, arboreal PROCYONID mammal of Central and South American tropical forest, with a long prehensile tail; ▷ p. 130. **2.** ▷ POTTO.

kin selection, evolutionary process by which a gene is selected that causes its carrier to behave in a way that reduces its own chances of reproducing while increasing those of a related animal; ▷ pp. 17, **149** (box), 153. See also ALTRUISM.

kite, any of 21 BIRDS OF PREY in the family Accipitridae (order Ciconiiformes). True kites such as *Milvus milvus*, the red kite of the western Palaearctic, have broad forked tails; ▷ pp. 213, 217.

kittiwake, either of two GULLS of the genus *Rissa*, a small gregarious species of the northern hemisphere. The common name is in imitation of the constant calling of these birds.

kiwi, any of three species of the family Apterygidae (order Struthioniformes), small flightless birds that live in New Zealand as nocturnal forest-dwellers. Although they have poor eyesight, their sense of smell is good, and they use their long bills to probe for insect larvae and worms; ▷ pp. 97, 214.

kiwi fruit or **Chinese gooseberry** (*Actinidia chinensis*), a climbing shrub native to East Asia

and the Himalaya, cultivated in New Zealand for its edible fruits.

klipspringer (*Oreotragus oreotragus*), a small, yellowish-grey, browsing ANTELOPE with very short horns, inhabiting rocky places in Africa south of the Sahara.

knapweed, any of a number of species of biennial or perennial herbs in the mainly European and Mediterranean genus *Centaurea* in the daisy family, Asteraceae (formerly Compositae), e.g. lesser knapweed (*C. nigra*). See also CORNFLOWER.

knifefish, name given to several unrelated groups of fishes that have long anal fins, supposedly resembling a knife blade, including the FEATHERBACKS and members of the family Gymnotidae (order Gymnotiformes). The latter have no dorsal fins and can produce weak electrical charges for navigation. The electric eel (*Electrophorus electricus*) – not a true eel but a member of the CARP order – is sometimes called the electric knifefish; it also lacks a dorsal fin and is capable of producing electrical discharges in excess of 500 volts, sufficient to stun prey fishes (▷ p. 77).

knot, either of two small WADERS of the family Scolopacidae, with a medium-length bill and dumpy body, occurring throughout much of the world. Huge flocks aggregate from time to time to fly in a coordinated fashion, producing spectacular aerobatic displays.

knotweed, ▷ POLYGONUM.

koala (*Phascolarctos cinereus*), a MARSUPIAL mammal of eastern Australia, highly specialized as a herbivorous tree-dweller; ▷ pp. *104*, *105*, 162.

kob (*Kobus kob*), an elegant, reddish-brown grazing ANTELOPE with long, curved horns, inhabiting floodplains and moist areas near water in Africa; ▷ p. 157.

Koch, Robert (1843–1910), German bacteriologist, who established the links between specific bacteria and specific diseases.

Kodiak bear, a subspecies of BROWN BEAR; ▷ pp. **128**, *129*.

koel, either of two black Australian and Oriental CUCKOOS, with particularly long tails.

kohlrabi, a cultivar of *BRASSICA oleracea* with a swollen green or purple stem that is eaten as a vegetable.

kola, ▷ COLA.

Komodo dragon, a MONITOR lizard, the largest of living lizards (▷ pp. 85, 89, 214).

kookaburra or **laughing jackass** (*Dacelo novaeguineae*), one of the largest KINGFISHERS, growing to a length of 45 cm (18 in). It is drably coloured to blend in with the Australian scrubland, and has a distinctive, loud raucous laugh.

kouprey (*Bos sauveli*), a grey or dark-brown species of wild CATTLE, inhabiting woodland in Indonesia.

kowari (*Dasyuroides byrnei*), a small, burrowing, desert-dwelling, carnivorous DASYURID marsupial, whose long tail ends in a black brush tip. It is now restricted to southwestern Queensland.

krait, any brightly coloured ELAPID snake of the genus *Bungarus* of southern Asia. Although venomous, kraits tend to flee from danger if possible.

Krebs cycle, ▷ TCA CYCLE.

krill, any small shrimp-like CRUSTACEAN of the order Euphausiacea, the principal food of baleen whales; ▷ pp. ▷ pp. **67**, 220.

k-strategist, any species adapted to CLIMAX communities; ▷ p. 43.

kudu, either of two large, brown or grey ANTELOPES of the genus *Tragelaphus*, with white stripes and long, spiral horns. They inhabit woodland and thicket in Africa.

kumquat, any of several species of shrubs and trees in the East Asian genus *Fortunella*. Their citrus fruits are like small oranges.

Labiatae, ▷ LAMIACEAE.

laburnum, either of two species of trees and shrubs in the central and southeastern European leguminous genus *Laburnum* . They are cultivated for their attractive hanging yellow inflorescences, but their seeds and leaves are highly toxic.

labyrinth, another name for the inner ear; ▷ pp. 190–1.

lacertid, any of over 150 species of LIZARD of the family Lacertidae, widely distributed in a broad range of habitats throughout most of Europe, Asia and Africa. Generally slender and long-tailed, most lacertids are ground-dwelling and feed on small prey, but there is considerable variation in size and appearance. The group includes the common European wall lizard (*Lacerta muralis*) and the sand lizard (*L. agilis*; ▷ p. 85).

lacewing, any NEUROPTERAN insect of the families Hemerobiidae and Chrysopidae, respectively the brown and green lacewings, characterized by delicate, lace-like wings (▷ p. 61). The predatory larvae are important natural enemies of aphids and a number of other insect pests.

lac insect (*Laccifer lacca*), an Indian SCALE INSECT formerly cultivated for its thick resinous scales, which formed the basis of the varnish shellac; ▷ p. 64.

lacquer, an exudate from the lacquer or varnish tree (*Rhus verniciflua*) native to East Asia and cultivated in Japan. It is used in oriental wood decoration.

lactation, the secretion of milk from the mammary glands of mammals; ▷ pp. 98 (box), 100.

lactose, a disaccharide SUGAR found in milk.

ladies' fingers, ▷ OKRA.

ladybird or **ladybug,** any BEETLE of the family Coccinellidae, familiar rounded beetles, often brightly coloured in red, yellow, white and black. Both adults and larvae are carnivorous, especially on APHIDS and COCCIDS; ▷ pp. 58, 63 (box).

lady fern (*Athyrium filix-femina*), a large fern found in damp soils in temperate regions.

lady's mantle, any of around 250 species of perennial herbs in the temperate and mountain genus *Alchemilla* in the rose family, Rosaceae. Several are grown in gardens; ▷ p. 211.

lady's smock, ▷ CARDAMINE.

lagomorph, any mammal of the order Lagomorpha, which comprises just two families – the RABBITS and HARES (Leporidae, about 44 species) and the PIKAS (Ochotonidae, 14 species); ▷ p. 132.

lagoon, an area of water sheltered from the open sea by a coral reef; ▷ p. 215.

lake ecosystems, ▷ p. 218.

Lamarck, Jean Baptiste (1744–1829), French naturalist, who proposed a theory of evolution by the inheritance of acquired characteristics; ▷ p. 12.

lamb's ears (*Stachys byzantina*), a herb related to BETONY with woolly grey leaves and grown as an ornamental.

lamellibranch, another name for BIVALVE.

lamellicorn beetle, ▷ SCARABAEOID BEETLE.

Lamiaceae, a dicotyledonous plant family (formerly Labiatae), including mints, thyme, basil, etc.; ▷ pp. 45, 46.

lamina, the blade of a leaf.

lammergeier (*Gyptaetus barbatus*), a long-tailed VULTURE frequenting high mountains of North Africa and the Orient. It drops bones from a height to break them and extract marrow.

lamprey, any JAWLESS FISH of family Petromyzonidae, a small group (20–30 species) of mainly freshwater, eel-like fishes. They feed parasitically on the tissue and body fluids of living fishes; ▷ pp. 70, 75.

lamp shell or **brachiopod,** any marine invertebrate of the phylum Brachiopoda. Lamp shells resemble bivalve molluscs (clams, oysters, etc.) but are distinguishable by having unequal valves and a hole in the larger valve for the stucture by which they attach themselves to surfaces; they are filter-feeders, sieving particles by means of a lophophore (crown of tentacles). This ancient group flourished in the Mesozoic era, but now have a restricted distribution, most being found on underwater cliffs; ▷ p. 5.

lampyrid beetle, ▷ FIREFLY.

lancelet, any CHORDATE of the subphylum Cephalo-chordata, characterized by a notochord and dorsal nerve cord, well-developed segmental muscles, a tail and dorsal fin for swimming, but no head, eyes or other fins. They mostly live buried in gravel or sand, feeding by drawing water through the gill slits in the body wall and extracting fine particles; ▷ p. 51 (box).

land crab, any DECAPOD crustacean of the family Gecarcinidae, the largest terrestrial crustaceans; ▷ p. 67.

language, a system of COMMUNICATION, conventionally defined as symbolic and able to denote abstract ideas, events distant in time and place, etc. Much debate has centred on whether animal communication systems qualify as language; ▷ pp. 141 (box), 153 (box), **155**.

langur, any Old World colobine monkey of the genus *Presbytis* and related genera. Like the related African COLOBUS monkeys, the Asian langurs are of slender build and highly arboreal, and are specialized as herbivores. The PROBOSCIS MONKEY is closely related; ▷ p. 139.

lantern fish, any of about 105 deep-sea fishes of the family Myctophidae (order Myctophiformes). They are extremely abundant worldwide, and although deep-sea species, they are familiar due to their migrations to the surface at night. They are then visible from ships because of the rows of light organs on the belly and flanks; ▷ p. 74.

lappet moth, any LEPIDOPTERAN insect (moth) of the family Lasiocampidae. They are generally medium to large and stout-bodied; adults do not have mouthparts. The group includes the eggars.

lapwing or **peewit,** any of 20 gregarious PLOVERS, with long legs, delicate bills and head crests, and unusually broad, rounded wings. They occur in the Old World and South America.

larch, any of around 10 species of conifer in the north temperate genus *Larix*, family Pinaceae. Various North American species (e.g. *L. laricina*) are known as tamaracks. They are deciduous trees with two kinds of twig: long shoots with spirally arranged leaves, and short side shoots with a tuft of leaves. Several species are grown as ornamentals or for timber; ▷ pp. 39, 208–9.

larder, ▷ HOARDING.

lark, any of 91 PASSERINE birds of the family Alaudidae, medium-sized, ground-dwelling and nesting birds, which have a very long hind claw and brown-streaked plumage. They feed on seeds and insects in open habitats worldwide, but mostly in Africa; ▷ p. 97 (box). See also MAGPIE LARK.

larkspur, any of around 40 species of the Mediterranean and Central Asian genus *Consolida*, which is closely related to DELPHINIUM.

larva (pl. -vae), an animal in an immature but active state, often markedly different in structure and life style from the adult. Larvae typically undergo a dramatic METAMORPHOSIS to become adult (e.g. tadpoles to frogs).

larynx, the modified upper part of the trachea (windpipe) containing the vocal cords.

lateral-line system, a sensory system in fishes and various amphibians involved in the detection of pressure, vibration, etc.; ▷ pp. 73 (box), 80, *190*, **191**.

lateritic soil, a red-coloured soil type found in tropical forests; ▷ p. 200.

latex, a fluid, often milky in appearance, produced by special cells (laticifers) in certain plants, especially the RUBBER tree; ▷ p. 45.

laughing jackass, ▷ KOOKABURRA.

Laurasia, an ancient supercontinent; ▷ pp. 8–9.

laurel, any of a number of small trees and shrubs chiefly in the family Lauraceae, e.g. BAY LAUREL. The unrelated cherry laurel (*Prunus laurocerasus*) is a popular evergreen garden shrub.

Laurentia, an ancient continent; ▷ pp. 8–9.

Lavatera, a mainly Mediterranean genus of annual to perennial herbs and shrubs in the MALLOW family, Malvaceae, e.g. tree mallow (*L. arborea*).

lavender, any of around 28 species of perennial herbs and shrubs in the warm temperate genus *Lavandula* in the family Lamiaceae (formerly Labiatae), but principally *L. angustifolia*, used in perfumery. Sea lavender (*Limonium vulgare*) is an unrelated perennial herb of salt marshes.

lawyer's wig or **shaggy ink cap** (*Coprinus comatus*), an edible species of INK CAP FUNGUS with a tall white shaggy cap.

leaching, the washing of elements, minerals, nutrients and other materials through the soil; ▷ pp. **196**, 200, 207, 209.

leaf, a lateral organ on a plant stem, in higher plants often consisting of a stalk (petiole) and blade (lamina). Leaves play a crucial role in transpiration, respiration and photosynthesis (▷ pp. 31–3). For leaf growth and development ▷ p. 32.

leaf beetle, any BEETLE of the family Chrysomelidae, a large group containing over 20 000 species distributed worldwide. Both adults and larvae are plant-feeders, and a number of species are very destructive to crops. See also COLORADO BEETLE.

leafcutter ant, any tropical ANT of the genus *Atta*. They cut sections from leaves and transport them back to their nest, where they are used as 'bedding material' for the cultivation of a particular fungus; the fungus then serves as the ants' food supply; ▷ p. 65.

leafhopper, any HOMOPTERAN insect (bug) of the family Cicadellidae (or Jassidae). Resembling slim FROGHOPPERS, they are extremely abundant, with over 8500 species worldwide.

leaf insect, any EXOPTERYGOTE insect of the family Phyllidae (order Phasmida). Occurring in tropical Southeast Asia, leaf insects have rather stubby, depressed bodies with flattened outgrowths on the limbs, so that they resemble leaves, usually those on which they feed; ▷ p. 166.

leaf miner, any small DIPTERAN insect of the family Agromyzidae, whose larvae bore into leaves and feed on the tissues. The name is also given to various other insects that mine leaves, e.g. in the LEPIDOPTERAN families Gracillariidae and Cosmopterygidae.

leaf-nosed bat, 1. (Old World leaf-nosed bats), any of about 40 microchiropteran BATS of the family Hipposideridae, of widespread distribution throughout the Old World tropics. These insectivorous bats have a highly elaborate nose-leaf similar to that of a horseshoe bat.

2. (New World leaf-nosed or spear-nosed bats), any of about 140 microchiropteran BATS of the family Phyllostomatidae, widely distributed in the New World from southern USA to northern Argentina. These bats feed both on insects and on the nectar and pollen of flowers, and some prey on vertebrates. In contrast to their Old World namesakes, the New World leaf-nosed bats have a

simple nose-flap shaped like a spearhead. Although generally small, the American false vampire is the largest New World bat, with a wingspan that may exceed 1 m (39 in); ▷ p. *118*.

learning, (in animals) the acquisition of patterns of behaviour other than by genetic means, or the (beneficial) modification of behaviour as a result of experience. Such learning may result from a variety of processes, including HABITUATION, CONDITIONING and CULTURAL TRANSMISSION; ▷ pp. 57, 95, 103, 136, 141 (box), **150–1**, 152 (box), 162 (box), 167 (box).

leatherback (*Dermochelys coriacea*), the largest species of marine turtle (CHELONIAN), growing to a length of over 1.5 m (5 ft) and weighing up to around 600 kg (1323 lb). Its shell is buried in leathery skin and it feeds mainly on jellyfish.

leatherjacket, ▷ CRANE FLY, TRIGGERFISH.

lechwe (*Kobus leche*), a chestnut to black grazing ANTELOPE with long horns, inhabiting floodplains and swamps in central and southern Africa.

leech, any predatory or parasitic ANNELID of the class Hirudinea; ▷ pp. 55, 176, 219.

leek (*Allium ampeloprasum*), a relative of the ONION with closely packed concentric leaves that are used as a vegetable and for flavouring.

Leewenhoek, Anton van (1632–1723), Dutch shopkeeper who ground lenses to construct microscopes that could magnify up to 400 times. With these he became the first person to describe red blood cells, sperm, protozoans, bacteria, and the microscopic structure of nerves and plant tissues.

legionary ant, ▷ ARMY ANT.

legume, any member of the family Fabaceae (formerly Leguminosae, which also included what are now the Mimosaceae and Caesalpinaceae), including peas, beans, clover, gorse, etc., nearly all of which are symbiotic with nitrogen-fixing bacteria; ▷ p. 32.

lek, (mainly in various bird species) a traditional mating ground in which small territories are defended; ▷ pp. *94*, **156–7**.

lemming, any of several small myomorph RODENTS of the genus *Lemmus* (family Muridae) of northern Eurasia and America, such as the Norway lemming (*L. lemmus*). They are similar in appearance and closely related to VOLES; ▷ pp. **135**, *172* (box), 211.

lemon (*Citrus limon*), a commercially very important acid CITRUS FRUIT cultivated in areas with a Mediterranean climate. Lemon oil is extracted from the rind.

lemur, any of 21 species of lower PRIMATE in 5 families, mainly restricted to forested areas of Madagascar, including the true lemurs, dwarf and mouse lemurs, woolly lemurs (sifakas and the indri), sportive lemurs, and the aye-aye. The flying lemurs or colugos (▷ p. 137) are not related; ▷ pp. 136, **137**.

lens, the part of the (vertebrate) eye responsible for focusing incoming light on the retina. In the compound eyes of insects (etc.) there are many separate lenses; ▷ pp. 189–90.

lentil, any of five species of climbing herbs in the Mediterranean and southwestern Asian genus *Lens* in the pea family, Fabaceae (formerly Leguminosae). *L. culinaris* is cultivated for its seeds (lentils).

leopard (*Panthera pardus*), a solitary and versatile big CAT noted for its fine spotted coat, still extensively (if patchily) distributed over much of Africa and Asia. The black form is known as a panther; ▷ pp. **125**, *148*.

leopard cat (*Felis bengalensis*), a small, nocturnal, yellowish, black-spotted CAT found in forest and scrub in parts of Southeast Asia and Japan.

lepidopteran, any INSECT of the order Lepidoptera,

the BUTTERFLIES and MOTHS, comprising about 150 000 known species occurring worldwide (▷ p. *61*). They are frequently highly colourful, for purposes of camouflage or mimicry, or as warning coloration. The adults are covered in small scales, which reduce turbulence as they slowly flap their large wings and provide the individual spots of colour that contribute to the total wing pattern. With the exception of one small family of moths, which have mandibles (biting mouthparts) and eat pollen, adults feed on liquids sucked up through a coiled proboscis. Larvae (caterpillars), again with few exceptions, have mandibles and feed on plants.

lesser panda, ▷ RED PANDA.

lettuce, any of a number of species of herb in the widely distributed genus *Lactuca* in the family Asteraceae (formerly Compositae). The cultivated *L. sativa,* which is used in salads, does not occur in the wild.

leucocyte, a white BLOOD cell.

leucoplast, a type of organelle; ▷ p. 20 (box).

liana, ▷ CLIMBER.

liberty cap (*Psilocybe semilanceata*), a hallucogenic basidiomycete fungus; ▷ p. 35.

lichee, ▷ LITCHI.

lichen, any composite organism formed by a symbiotic relationship of a FUNGUS and an ALGA or a CYANOBACTERIUM; ▷ pp. 30, **34** (box), 204, 208, 211, 221.

licorice, an alternative name for ▷ LIQUORICE.

life cycle, the complete series of changes that occur in an animal or plant from its origin to the origin of its offspring. See also GROWTH, REPRODUCTION.

life history, the series of changes that occur in an animal or plant from its origin to its death.

lignin, an important strengthening component of cell walls in the woody tissues of plants (▷ XYLEM). It is a polymer of sugars, phenols, amino acids and alcohols.

lignum vitae, either of two species of tropical and subtropical American evergreen trees, *Guaiacum officinale* and *G. sanctum.* They yield the hardest commercial timber and the medicinal resin guaiacum.

lilac, any of around 30 species of shrubs and small trees in the Asian and European genus *Syringa* in the olive family, Oleaceae. Many cultivars are grown for their fragrant attractive flowers.

Liliaceae, a monocotyledonous plant family, including lilies, tulips and bluebells; ▷ p. 45.

Liliidae, a monocotyledonous plant subclass; ▷ p. 45.

Liliopsida, a plant class – the MONOCOTYLEDONS; ▷ p. 45.

lily, any of around 90 species of bulbous herbs in the temperate genus *Lilium* in the family Liliaceae. Many species of other genera are also referred to as lilies because of their lily-like flowers, e.g. day lily.

lily of the valley (*Convallaria majalis*), a north temperate perennial herb with creeping rhizomes belonging to the lily family, Liliaceae. It is grown in gardens for its fragrant flowers.

lily-trotter, ▷ JACANA.

lima bean or **butter bean** (*Phaseolus lunatus*), a tropical legume grown as a vegetable. Either the whole pod or, more usually, the seeds being eaten.

lime, 1. (also called linden) any of around 30 species of trees in the north temperate genus *Tilia.* The leaves are usually heart-shaped and the yellow flowers are an important nectar source. Basswood is the American *T. americana.* **2.** the lemon-like CITRUS FRUIT of the unrelated *Citrus aurantifolia,* native to India and Malaysia, but cultivated throughout the tropics. **3.** ▷ CALCIUM CARBONATE.

limestone, a sedimentary rock. Most types are composed almost entirely of CALCIUM CARBONATE from the shells of unicellular organisms that originally accumulated at the bottom of seas and that have subsequently been compacted and uplifted. Limestone soils are BASIC; ▷ p. 197.

limnetic zone, the upper, open part of a body of fresh water, often restricted to the zone away from the lake bottom or shore; ▷ p. 218.

limpet, name given to various marine GASTROPOD molluscs with conical shells, belonging to different groups. They are found at the water's edge, where they typically live by rasping small algae off rocks with their strong radula (feeding organ); ▷ pp. *56,* 219, 221.

limpkin (*Aramus guarauna*), a superficially heron-like bird, found wading in the freshwater marshes of the southeastern USA and the Neotropics. It belongs to the family Heliornithidae (order Gruiformes) and is closely related to the FINFOOTS. The bill is long and curves slightly downwards.

linden, ▷ LIME.

ling, 1. (*Molva molva*), a deep-water relative of the COD, found in the northeast Atlantic, valuable as a food fish. It has a very elongated body and grows to a length of 1.8 m (6 ft). **2.** an alternative name for HEATHER.

linkage, 1. sex linkage, in which certain genes are linked to a sex chromosome, mostly the X, but sometimes the Y (▷ p. 23). **2.** genetic linkage, in which there is a greater likelihood of groups of genes being passed on together the closer they are on the chromosome; the further apart they are, the more likely they are to be separated by crossing-over (▷ p. 23).

Linnaeus, Carl (1707–78), Swedish naturalist, who first developed the binomial system of biological classification; ▷ p. 18.

linnet (*Acanthis cannabina*), a grey/pink FINCH, living in scrublands of the western Palaearctic. It is rather susceptible to the use of pesticides.

linsang, any of several small, spotted semi-arboreal cat-like VIVERRIDS with long tails and retractile claws. They are found in West Africa, and eastern and Southeast Asia; ▷ p. 123.

linseed, ▷ FLAX.

lion (*Panthera leo*), a large, social big CAT, now largely confined to the open savannah of sub-Saharan Africa, varying in colour from tawny to reddish-brown; ▷ pp. *124–5,* **125,** 159, 165, 203. See also PUMA.

lipid, any of a group of simple molecules that are essential components of cell membranes and of FATS; ▷ pp. **6** (box), 20.

liquorice, an extract of the rhizomes of the perennial herb *Glycyrrhiza glabra,* native from the Mediterranean to Central Asia. It is used in confectionery and medicine.

litchi, lichee or **lychee** any of around 12 species of trees and shrubs in the genus *Litchi.* The evergreen *L. chinensis* is native to China and has a fruit with a juicy edible ARIL.

lithosphere, ▷ CRUST, MANTLE and p. *8.*

litter, the undecomposed organic matter, largely but not exclusively of plant origin, that falls to the soil surface on the death of an organism or some part of it. It may accumulate in a layer on the soil before incorporation into lower layers.

little brown bat, ▷ VESPERTILIONID BAT.

littoral, relating to the shore. The littoral fringe can thus mean the coastline of an island or the shoreline of a lake. In marine situations it may be periodically emersed as the tide ebbs.

livebearer, any of 150 brackish/freshwater fishes of the family Poeciliidae, from eastern USA to northeastern Argentina. They practise internal fertilization and bear live young. The GUPPY and other livebearers are popular aquarium fishes.

liver, a large well-vascularized organ of vertebrates. Amongst other metabolic functions, it secretes bile, stores minerals, vitamins and glycogen, manufactures and destroys proteins (producing UREA), and removes many toxins from blood; ▷ p. 179.

liverwort, any member of the bryophyte class Hepaticae; ▷ p. 37.

lizard, any REPTILE of the suborder Sauria, containing some 3000 species and constituting (with snakes and amphisbaenians) the most successful group of living reptiles. The major families include AGAMIDS, CHAMELEONS, GECKOS, IGUANIDS, LACERTIDS, MONITORS, SKINKS and TEIIDS; ▷ pp. **82–9.**

llama (*Lama glama* or *peruana*), a domesticated, ruminating South American ARTIODACTYL mammal (family Camelidae), descended from the wild GUANACO; ▷ p. 115 (box).

loach, any small, cylindrical, freshwater fish of the family Cobitidae, belonging to the same order as the CARP. There are at least 175 species distributed in Eurasia and Africa. They are often distinctively marked and are popular aquarium fishes; ▷ p. 73.

loam, a soil type in which the texture consists of an even mixture of sand, silt and clay. It is texturally well balanced with good drainage properties, but also has the ability to retain water.

lobelia, any of around 250 species of herbs and shrubs in the widely distributed genus *Lobelia* in the family Campanulaceae. A number are popular garden plants; ▷ p. *211.*

lobster, any large DECAPOD crustacean of the genus *Homarus,* such as *H. americanus,* the heaviest living crustacean. All lobsters have the first pair of limbs modified into large pincers; ▷ pp. *67,* 221.

locomotion, or movement in animals, ▷ pp. **176–7.**

locust, 1. any short-horned GRASSHOPPER that in certain conditions is able to undergo a dramatic change in behaviour, morphology and physiology. The change from grasshopper to locust is PHEROMONE-induced when the animals are crowded together, such crowding usually occurring as a result of specific weather conditions. Locusts move together in huge swarms and cause enormous damage to crops in Africa; ▷ pp. 173, 203. **2.** a name given to a number of different plant species, e.g. CAROB (*Ceratonia siliqua*), honey locust (*Gleditsia triacanthos*) and black locust (*Robinia pseudacacia*). See also ROBINIA.

loganberry (*Rubus* x *loganbaccus*), a natural hybrid between the European RASPBERRY (*R. idaeus*) and the American blackberry (*R. vitifolius*). See also BRAMBLE.

logrunner, any of about 19 PASSERINE birds of the family Orthonychidae, Southeast Asian and Australian thrush-like species that live on or very near the ground. They use powerful feet to scrape leaf litter to search for insects.

London pride (*Saxifraga umbrosa* and *S. urbium*), two perennial SAXIFRAGES with basal rosettes of fleshy leaves and pink or white flowers popular in gardens.

long-horned or **longicorn beetle,** any medium to large BEETLE of the family Cerambycidae, a large group containing over 20 000 species distributed worldwide. The name of the family comes from the characteristically long antennae. The virtually legless larvae are wood-boring.

long-legged bat, ▷ FUNNEL-EARED BAT.

long-tailed tit, any of eight PASSERINE birds of the family Aegithalidae, tiny species of the northern hemisphere, which tend to move around in large flocks. They build complex dome-shaped nests with side entrances. See also TIT.

loofa, sponge gourd or **luffa** (*Luffa aegyptiaca*), a tropical climbing herb in the cucumber family,

Cucurbitaceae. The fibrous skeleton of the fruit is used for bath sponges.

loon, ⇨ DIVER.

looper, ⇨ GEOMETRID.

loosestrife, 1. any of several species of herb in the genus *Lysimachia* in the primrose family, Primulaceae, e.g. yellow loosestrife (*L. vulgaris*); ⇨ p. 42. **2.** purple loosestrife (*Lythrum salicaria*), a perennial herb in the family Lythraceae.

lophophore, a tentacled feeding structure found in a number of marine animals, including BRYOZOANS, and LAMP SHELLS.

lords-and-ladies, cuckoopint; ⇨ ARUM.

Lorenz, Konrad (1903–88), Austrian zoologist, one of the founders of modern ethology (⇨ p. 148). For his studies on courtship and imprinting in ducks and geese (⇨ p. 95) he shared the Nobel Prize in 1973 with von FRISCH and TINBERGEN.

lorikeet, a general term applied to some 20 Australasian species of smaller LORIES.

loris, any of several slow-moving, nocturnal lower PRIMATES of the family Lorisidae, found in forested habitats in southern and Southeast Asia; ⇨ pp. 136, 137. See also POTTO and ANGWANTIBO.

lory, any of 55 PARROTS with a brush-tipped tongue used to feed on nectar. They live in the southern Orient, Australasia and the Pacific.

lotus, 1. sacred lotus (*Nelumbo nucifer*) and American lotus (*N. lutea*), aquatic plants with large leaves and large fragrant flowers. **2.** ziziphus lotus (*Ziziphus mauritania*), thought to have produced the lotus fruits of ancient times. **3.** ⇨ TREFOIL.

louse, a general term applied to many insects, especially those that suck blood (e.g. body louse, head louse), but the term has no scientific precision. See also SUCKING LOUSE, BITING LOUSE.

lovage (*Levisticum officinale*), a perennial herb of southern Europe cultivated for its aromatic fruits and its leaves, belonging to the carrot family, Apiaceae (formerly Umbelliferae).

lovebird, any of nine small African PARROTS of the genus *Agapornis*. They have a relatively large head and short tail, and are gregarious.

love-in-a-mist (*Nigella damascina*), a southern European annual herb in the buttercup family, Ranunculaceae. It has blue flowers surrounded by finely dissected leaves.

love-lies-bleeding, ⇨ ARAMANTH.

lucerne, ⇨ ALFALFA.

lugworm, any POLYCHAETE of the genus *Arenicola*, marine worms with tufted gills that live in deep burrows on sandy shores.

lumpfish or **lumpsucker,** any of 26 cool-water northern-hemisphere fishes of the subfamily Cyclopterinae (family Cyclopteridae), in the same order as the SCORPION FISHES. They are deep-bodied, with rows of isolated spines in the skin, and have a powerful suction disc on the underside. Their unfertilized eggs are eaten as a substitute for caviar.

lunar cycle, the time it takes for the moon to make a complete revolution around the earth, important as a cue for the cyclical activity patterns of various marine organisms; ⇨ p. 175.

lung, a respiratory structure used for GAS EXCHANGE in air; ⇨ pp. 73, 79, 93, *180*, **181**.

lungfish or **dipnoan,** any freshwater BONY FISH of the orders Ceratodontiformes (Australian lungfish, 1 species) and Lepidosireniformes (South American and African lungfishes, 5 species). The lungfishes are now generally believed to belong to the immediate ancestral stock from which land vertebrates arose; ⇨ pp. 70–1, **73**, 78.

lungwort, any of around 20 species of perennial herbs in the European and west Asian genus *Pulmonaria* in the family Bordginaceae.

lupin, any of around 200 species of annual and perennial herbs and shrubs in the American and Mediterranean genus *Lupinus* in the pea family, Fabaceae (formerly Leguminosae). Some are grown as ornamentals and some for fodder or green manure.

lure, a device used by an AMBUSH predator to entice prey within range; ⇨ pp. 75, 164, 221 (box).

lycaenid butterfly, any LEPIDOPTERAN insect of the family Lycaenidae, mainly found in the Palaearctic region. The larvae of a number of species are carnivorous, feeding on ants, aphids or coccids.

lychee, ⇨ LITCHI.

lycophyte, any member of the division Licophyta, comprising the quillworts and club mosses; ⇨ pp. 30, 37.

lycopod, another name for CLUB MOSS (⇨ p. 37), especially of the genus *Lycopodium*.

Lyell, Charles (1797–1875), Scottish geologist, who in his *Principles of Geology* (1830–3) proposed the theory of UNIFORMITARIANISM; ⇨ p. 12.

lymphatic system, in vertebrates, a network of small vessels by which the interstitial fluid that surrounds cells and through which GAS EXCHANGE takes place is collected and returned to the blood; ⇨ p. **181**.

lynx (*Felis lynx*), a pale brown CAT with dark spots, prominent black ear tufts and a short tail. It inhabits coniferous forest and scrub in Eurasia and northern North America; ⇨ pp. *125*, 172 (box), 205, 209.

lyrebird, either of two PASSERINE birds of the family Menuridae, closely related to the SCRUB WRENS. They are superficially pheasant-like in appearance, living in dense forest of southeast Australia. The outer feathers of the male's long tail are in the shape of a Greek lyre. The superb lyrebird (*Menura novaehollandiae*) feeds by scraping at the forest floor, searching for insects and frogs. It has a remarkable ability to mimic the calls of other birds.

Lysenko, Trofim Denisovich (1898–1976), Russian plant biologist who attempted to apply Lamarck's theory of the inheritance of acquired characteristics (⇨ p. 12) to crop breeding. His genetic theories found favour with Stalin, echoing as they did the Soviet doctrine of the improvability of humanity by changing the environment. For years Lysenkoism was scientific orthodoxy in the USSR.

lysosome, a type of organelle in eukaryote cells; ⇨ pp. **20** (box), *21*.

macadamia nut, the expensive and much prized nut from the Australian trees *Macadamia integrifolia* and *M. tetraphylla* in the family Proteaceae.

macaque, any Old World cercopithecine monkey of the genus *Macaca*, widely distributed in southern Asia and into northern Africa, often in rocky or woody areas. The macaques include the rhesus monkey (*M. mulatta*) and the barbary ape (*M. sylvanus*); ⇨ p. 139, 151.

macaw, any of 17 large, brightly coloured South American PARROTS, with very deep bills and tails longer than the body. They tend to perch upright and eat fruit; ⇨ pp. *93*, *97*.

mace, ⇨ NUTMEG.

mackerel, any of six marine PERCIFORM fishes of the family Scombridae (which also includes the TUNAS), occurring in all tropical and temperate seas. Reaching lengths of 50 cm (20 in) or so, they are powerful swimmers, with spindle-shaped bodies and a series of finlets on the tail stem; their backs are marked with iridescent bars, spots and lines of blue, green and gold. They are valuable food fishes, with oily flesh. The Spanish mackerel is related and also important commercially.

macrobiota, soil organisms exceeding about 4 cm (1½ in) in size.

macroparasite, a relatively large parasitic organism (typically visible to the naked eye), such as a plant or an insect; ⇨ p. 168.

madder (*Rubia tinctorum*), a Mediterranean evergreen perennial herb from the roots of which a red dye is extracted. Indian madder (*R. cordifolia*) is similarly used.

maggot, a soft-bodied legless LARVA (grub) of a DIPTERAN fly, e.g. a housefly.

magnetism (sensitivity to), ⇨ ELECTROMAGNETIC SENSE.

magnolia, any of around 85 species of evergreen or deciduous trees and shrubs in the Asian and New World genus *Magnolia*. Because of their large showy but unspecialized flowers many are popular ornamentals, e.g. large-leaved cucumber tree (*M. macrophylla*) and sweet bay (*M. virginiana*). Some of the earliest flowering plants are thought to have resembled magnolias (⇨ p. 40), and *Magnolia* is the type genus for all the flowering plants, the division Magnoliophyta; ⇨ p. 45 (box).

Magnoliidae, a dicotyledonous plant subclass; ⇨ pp. *40*, 45.

Magnoliopsida, a plant class – the DICOTYLEDONS; ⇨ p. 45.

magpie, any of about 12 long-tailed, sometimes brightly coloured CROWS found throughout the northern hemisphere; ⇨ p. 11. The familar black and white western Palaearctic magpie (*Pica pica*) has become increasingly urbanized. The Australian magpie is not closely related.

magpie lark (*Grallina cyanoleuca*), a thrush-sized, glossy black and white PASSERINE bird with a long tail, living near water in Australia. Mates continually call to each other, and the nest is a mud bowl perched in a tree.

mahogany, a strong red brown timber obtained from a number of different genera, but in the strict sense from species of the tropical New World genus *Swietenia* and the African genus *Khaya*.

maidenhair fern, any of around 200 species of fern in the widely distributed genus *Adiantum*. These delicate ferns have leaves with glossy black stalks and fan-shaped segments.

maidenhair tree, ⇨ GINKGO.

maize or **corn** (*Zea mays*), an important cereal crop originating in America, where it was cultivated by the native Americans. It is used as an animal feed and in the manufacture of starch and the distillation of whisky. SWEET CORN is a popular human food.

mako, any fast-swimming predatory marine SHARK of the genus *Isurus* (family Isuridae), such as *I. glaucus*, of Indo-Pacific and Australian seas.

malaria, ⇨ p. 168.

male fern, *Dryopteris felix-mas*, a common, north temperate fern of woodlands and hedgerows.

mallard (*Anas platyrhynchos*), a large DUCK of the Holarctic. The male has a green neck, plump grey body, and curled tail feathers; ⇨ p. 97 (box).

mallee fowl (*Leipoa ocellata*), a large mound-nesting MEGAPODE with cryptic colouring; ⇨ p. 95.

malleus, one of the three auditory ossicles in the mammalian middle ear (the 'hammer'); ⇨ pp. 98 (box), *191*.

mallow, any of a number of species of plant in the mallow family, Malvaceae, but particularly in the genus *Malva*, e.g. common mallow (*M. sylvestris*). See also LAVATERA and MARSH MALLOW.

Malpighian tubules, the excretory organs of insects and some other arthropods; ⇨ p. 59.

Malthus, Thomas (1766–1834), English clergyman. His *Essay on the Principle of Population* – in which he argued that human populations inevitably outstrip the food supply until reduced by factors such as famine, war and disease – was read by Darwin in 1838, and influenced his ideas on natural selection (⇨ p. 13).

maltose, a disaccharide SUGAR found in high concentrations in germinating seeds. It is used in brewing.

Malus, a genus of the family Rosaceae, including APPLES.

mamba, any tropical African ELAPID snake of the genus *Dendroaspis*. Notoriously venomous and often aggressive, mambas spend much of their time in trees; they may be predominantly green or black.

Mammalia, the class compromising the mammals.

mammal-like reptile, any extinct SYNAPSID reptile of a group that dominated the earth before the age of the dinosaurs and is thought to represent the ancestral stock from which the mammals arose. The group includes the early PELYCOSAURS, which were replaced by the more advanced dicynodonts and cynodonts; ⊳ pp. 82, **98.**

mammals, ⊳ pp. 98–101 (evolution), 102–45 (major groups), 148–73 (behaviour), 174–91 (physiology).

mammary gland, a breast (or similar compound modified skin glands) in female mammals that secretes MILK (⊳ box, p. 98).

mammoth, any member of an extinct group of trunked mammals (proboscideans) closely related to the living elephants; ⊳ pp. *100, 108.*

man, any APE of the genus *Homo* (family Hominidae, which includes the gorilla and chimpanzees), comprising modern man (*H. sapiens sapiens*) and various extinct species such as *H. habilis* and *H. erectus* (⊳ pp. *9, 18*). See also AUSTRALOPITHECINE and HUMAN.

manatee, any of three species of herbivorous marine mammal of the family Trichechidae (order Sirenia, which also includes the DUGONG). Manatees are found in tropical and subtropical coastal waters on either side of the Atlantic; ⊳ p. 142.

mandarin (*Citrus reticulata*), any of a number of cultivars of this species of CITRUS FRUIT such as clementine, satsuma and tangerine. The small orange-like fruits are characteristically easy to peel.

mandible, 1. the lower jaw of vertebrates, separate from the skull. **2.** the external mouthparts of certain arthropods (e.g. insects) used for holding or biting food (⊳ p. 59).

mandrake, any of around 6 species of herbs in the Mediterranean genus *Mandragon* in the potato family, Solanaceae, but particularly *M. officinarum.* The roots are said to resemble the human form.

mandrill (*Mandrillus sphinx*), a baboon-like Old World cercopithecine monkey, a large West African forest-living species with brightly coloured red and blue muzzle and hindquarters; ⊳ p. 139. See also DRILL.

maned wolf (*Chrysocyon brachyurus*), a large, long-legged, fox-like CANID whose reddish coat has long, dark hair above the shoulders. It inhabits grasslands and scrub forest in central South America; ⊳ p. 9.

mangabey, any of four species of Old World cercopithecine monkey of the genus *Cercocebus*, long-limbed, agile monkeys with long tails found in the canopy or floor of tropical rainforests in central and eastern Africa; ⊳ p. 139.

mangelwurzel or **mangold,** ⊳ BEET.

mango (*Mangifera indica*), an evergreen tree cultivated throughout the tropics particularly for its fruits, which may be eaten raw or cooked or used in chutneys.

mangold, ⊳ BEET.

mangosteen (*Garcinia mangostana*), a Southeast Asian tree in the family Clusiaceae. It is reputed to bear one of the best tropical fruits.

mangrove, any of various woody plants growing in tropical maritime swamps. The term includes species of a number of genera, e.g. *Ceriops* and *Rhizophora*. The bark of some species is used for tanning leather; ⊳ pp. 215, **216–17.**

manioc, ⊳ CASSAVA.

manna, an edible, usually sweet, plant exudate that hardens on drying. A modern source is manna ASH (*Fraxinus ornus*). The 'manna' referred to in Exodus may have been tamarisk or lichen.

mannakin, ⊳ WAXBILL.

man-of-war bird, ⊳ FRIGATE BIRD.

manta, any large RAY of the family Mobulidae, especially *Manta birostris*, the largest of all rays, which may reach a width of 6.5 m (21 ft) between the tips of the pectoral fins. It is a plankton-feeder, cruising near the surface and filtering out food organisms as water passes over its gills. The members of this family are sometimes also known as 'devil rays'.

mantid or **mantis** (praying mantid or mantis), any EXOPTERYGOTE insect of the suborder Mantodea (order Dictyoptera), comprising some 2000 species known mainly from the tropics (⊳ p. *61*). The head is mobile with large well-spaced eyes, and the prothorax is long. The spiny fore legs hinge back to form an efficient device for grabbing prey, which ranges from other insects to small birds, lizards and frogs. They are often winged, and may be camouflaged to prevent detection by potential prey; ⊳ p. 164.

mantis, ⊳ MANTID.

mantispid, any NEUROPTERAN insect of the small family Mantispidae. The adults bear a strong resemblance to small MANTIDS, with strong fore legs for capturing prey (⊳ p. 59). The juveniles are parasitic on the eggs or young of spiders or other insects.

mantle, the layer of the earth below the CRUST, extending to a depth of 2900 km (1800 mi). The upper, solid part of the mantle, together with the crust, forms the lithosphere (to a depth of 250 km/155 miles), and the lower, molten part forms the asthenosphere; ⊳ p. *8.*

maple, any of around 200 species of deciduous and evergreen trees and shrubs in the north temperate and tropical genus *Acer*, family *Aceraceae*. The leaves may be entire, but are more often lobed and many have attractive autumn colouring making them useful ornamentals, e.g. the Japanese maple (*A. palmatum*). Maple syrup is obtained from the sugar maple (*A. saccharum*). See also SYCAMORE.

maquis, degraded vegetation dominated by a heathland assemblage of dwarf shrubs, found in areas of mediterrranean climate. It is usually maintained by fire or grazing, and is rich in bulbous plant species; ⊳ p. 205.

mara or **Patagonian cavy** (*Dolichotis patagonum*), a large, long-legged, blunt-nosed caviomorph RODENT in the same family as the CAVIES. It is herbivorous, inhabiting arid, grassy areas in South America.

marble gall, a smooth marble-shaped gall found on OAK. It is caused by the cynipid wasp, *Andricus kollari*, which lays its eggs in buds. Popularly and mistakenly referred to as OAK APPLE.

mare's tail (*Hippuris vulgaris*), a perennial, usually aquatic, herb of temperate and cold regions. The name arises from the appearance of submerged shoots.

margay (*Felis wiedi*), a yellow-brown CAT with black markings. A good climber, it inhabits forest and scrub in South and Central America.

marigold (*Calendula officinalis*), a cultivated ornamental annual herb in the daisy family, Asteraceae (formerly Compositae). There are many cultivars, e.g. hen-and-chickens. Some other plants are called marigolds, e.g. KINGCUP and African marigold (*Tagetes erecta*).

marijuana, ⊳ CANNABIS.

marine ecosystems, ⊳ pp. 220–1.

marjoram, any of a number of species of aromatic perennial herbs in the genus ORIGANUM, e.g. sweet marjoram (*O. marjorana*) and oregano or marjoram (*O. vulgare*) used for culinary purposes.

marlin, either of two species of BILLFISH.

marmoset, any of several small tree-dwelling New World monkeys of the families Callitrichidae (which also includes the closely related tamarins) and Callimiconidae (Goeldi's marmoset). They are social and active by day; ⊳ pp. 136, **138,** 158.

marmot, any large, squirrel-like, burrowing, colonial RODENT of the genus *Marmota* in the SQUIRREL family, inhabiting pastures and alpine meadows in North America and Eurasia; ⊳ pp. *133,* 211. Marmots hibernate in winter. See also WOODCHUCK.

marram grass (*Ammophila arenaria*), a European and Mediterranean grass found on maritime sands, which it stabilizes with its spreading rhizomes. The closely related beachgrass (*A. breviligulata*) is the American equivalent.

marrow (*Cucurbita pepo*), a trailing annual herb in the cucumber family, Cucurbitaceae. It is cultivated in temperate regions for its more or less cylindrical fruit, which is used as a vegetable. Varieties include COURGETTE and PUMPKIN.

marsh mallow (*Althaea officinalis*), a European and Mediterranean perennial herb in the mallow family, Malvaceae. The roots yield a mucilage used in confectionery.

marsh, 1. (in the UK) a wetland environment in which mineral soils are waterlogged for much of the year, but are not generally submerged. Peat formation is very low. **2.** (in the USA) sites dominated by reed and sedge with a high water table (termed 'swamp' in Europe).

marsupial, any mammal of the order Marsupialia, which comprises some 266 species in 18 families and includes opossums, kangaroos, wombats, bandicoots, the koala, and many others; the majority of species are found in Australasia, but the group is also well represented in South America (chiefly by the opossums). Marsupials are distinguished from the PLACENTAL MAMMALS primarily on the basis of differences in the reproductive system (⊳ pp. 100, **103**). Evolving largely in isolation from the placentals, the marsupials represent a whole spectrum of animal types, adapted to a wide range of habitat and diet; ⊳ pp. 17 (box), 100, *101,* **103–5.**

marsupial cat (*Satanellus albopunctatus*), a large, cat-like carnivorous DASYURID marsupial, with brown fur spotted with white. It is restricted to New Guinea. For the extinct marsupial cat (*Thylacasmilus*) ⊳ p. 99.

marsupial frog, name given to various frogs (ANURANS) that brood their young in pockets in the skin. The eggs are either pushed into the pockets by adults as part of their mating behaviour (as in certain South American species), or newly hatched tadpoles squirm into the pockets (as in the Australian marsupial or hip-pocket frog); ⊳ p. 81.

marsupial mole (*Notoryctes typhlops*), the sole member of the MARSUPIAL family Notoryctidae, and the only Australian mammal specialized for a subterranean life. It has a tough, horny nose-shield and large fore claws for digging; ⊳ p. 104.

marsupial mouse, any of several very small, mouse-like, insectivorous DASYURID marsupials. They are restricted to New Guinea.

marsupium, the abdominal pouch of a marsupial mammal in which the very immature young develop after birth; ⊳ p. 103.

marten, any of eight small, stocky, brown MUSTELIDS of the genus *Martes*, with short legs and long, bushy tails, found in forests in North America and Eurasia. They resemble polecats and are excellent climbers; ⊳ p. 131.

martin, any SWALLOW of the genus *Delichon* (etc.), such as the Eurasian house martin (*D. urbica*); ⊳ p. 95.

mastic, a resin used as a varnish or as a chewing gum obtained from a number of different plants, e.g. *Pistacia lentiscus* and American mastic or pepper tree (*Schinus molle*).

Mastigomycota, one of the two divisions of true fungi; ⇨ p. 35.

mastodon, any member of an extinct group of trunked mammals (proboscideans) related to the living elephants; ⇨ pp. 108–9.

matamata (*Chelus fimbriata*), bizarre-looking South American freshwater turtle (CHELONIAN), distinguished by its rough shell and flat, triangular head. Lurking underwater, it feeds by opening its mouth suddenly and sucking in both water and prey.

maté or **mate,** the leaves of the HOLLY *Ilex paraguayensis* used to make a tea-like drink in South America.

mating, the act of copulation that leads to FERTILIZATION. It may be preceded by COURTSHIP behaviour and occurs within various mating systems (⇨ pp. 157–8). For references ⇨ REPRODUCTION.

may, ⇨ HAWTHORN.

mayfly, any EXOPTERYGOTE insect of the order Ephemeroptera. Almost the entire life cycle is spent as an aquatic nymph, which may be vegetarian or carnivorous. The non-feeding adult, which has no functional mouthparts or gut, lives only a few hours, during which mating and egg-laying take place. The adults are copied by fly fishermen to bait their hooks; ⇨ pp. *61*, 62, 219.

meadow grass, an English name for species of the genus *POA*.

meadowlark, any of seven medium-sized terrestrial AMERICAN BLACKBIRDS, living in the American grasslands and feeding on ground-dwelling insects.

meadow mouse, ⇨ VOLE.

mealworm beetle, any of various medium-sized, dark brown to black TENEBRIONID BEETLES of the genus *Tenebrio*. They are occasionally pests of stored foods, and are also found in birds' nests. The larvae – mealworms – are used as food for birds and as fishing bait.

mealy bug, any COCCID insect (bug) with a white waxy coating covering the females. Many are important pests of plant crops.

measuring worm, ⇨ GEOMETRID.

mechanoreception, sensitivity to mechanical energy, underlying the senses of hearing, balance, etc.; ⇨ pp. 190–1.

mediterranean, characterized by mediterranean climate and vegetation; ⇨ pp. 199 (map), **205**.

medlar (*Mespilus germanica*), a small tree belonging to the rose family, Rosaceae, from southeastern Europe and Asia. Preserves may be made from the apple-like fruits.

medulla oblongata, in vertebrates, part of the brain associated with involuntary actions such as breathing and heart beat; ⇨ p. 183.

medusa, one of the two life forms of a COELENTERATE, essentially a free-swimming inverted POLYP. A jellyfish is a typical medusoid coelenterate; ⇨ p. 53.

meerkat or **suricate** (*Suricata suricatta*), a small, greyish, social, mongoose-like VIVERRID with dark markings and a long tail. It is found in dry, open areas in southern Africa; ⇨ p. 123.

megachiropteran, a member of the smaller of the two suborders of bats, Megachiroptera (fruit bats); ⇨ pp. 118–19.

megapode, mound-builder or **scrubfowl,** any of 19 birds of the family Megapodiidae (order Craciformes); they incubate their eggs in mounds of sand; ⇨ p. 95. See also BRUSH TURKEY, MALLEE FOWL.

megaspore, in heterosporous spore-bearing plants and in seed-bearing plants, the larger type of spore containing the female egg; ⇨ p. 38. Compare MICROSPORE.

meiosis, the process by which sex cells are produced from body cells; ⇨ p. 22.

melanin, a dark brown/black pigment, usually found in skin and hair in special cells called melanocytes; ⇨ p. 96 (box).

melatonin, ⇨ PINEAL GLAND.

melilot or **sweet clover,** any of around 20 species of herbs in the genus *Melilotus* in the family Fabaceae. They produce plentiful nectar and several species are used as fodder crops.

melon, 1. sweet melon (*Cucumis melo*), an annual vine in the same genus as the CUCUMBER. The fruits have a juicy, often scented flesh and the many cultivars, include the cantaloupe (var. *cantalupensis*). **2.** water melon (*Citrullus lanatus*), an annual vine in the cucumber family, Cucurbitaceae. It is cultivated in tropical and subtropical areas for its juicy fruit.

membrane, 1. a thin, pliable sheet of tissue that surrounds or connects organs. **2.** the material that surrounds cells and provides compartments within them; ⇨ pp. 20–1.

memory, the retention of a learned response or a perception over time; ⇨ pp. 57, **150** (box).

Mendel, Gregor (1822–84), Austrian monk, the founder of modern genetics. He devised his laws of inheritance in 1865, publishing them the next year, but his work remained unnoticed until rediscovered in 1900 by de VRIES; ⇨ pp. 13, 22–3.

menharden (*Brevoortia tyrannus*), a relative of the HERRING, found in huge shoals on the Atlantic coasts of the northern USA. It has very oily flesh, and is processed for fishmeal oil and fertilizer.

menstrual cycle, ⇨ p. 185.

merganser or **sawbill,** either of two small DUCKS with thin, saw-toothed bills. The red breasted merganser (*Mergus serrator*) lives on northern rivers and lakes; ⇨ p. 96.

meristem, a group of cells in plants that divide by mitosis (⇨ p. 22) and thereby contribute to plant growth and organ formation. Meristematic activity may be generalized (as in the developing embryo) or localized (as in the apexes of stems, roots or the edges of developing leaves). Primary meristems are those that have always been meristematic; secondary meristems (e.g. the CAMBIUM, producing the wood in trees) develop from differentiated cells; ⇨ p. 32.

merlin or **pigeon hawk** (*Falco columbarius*), a small FALCON occurring in the northern hemisphere, with dark plumage and reddish underparts.

mermaid's purse, a tough protective capsule containing fertilized eggs produced by various cartilaginous fishes; ⇨ p. 74.

mescal or **peyote** (*Lophophora williamsii*), a spineless cactus native to southern North America. It contains mescaline, an alkaloid, and has been used for many thousands of years for its hallucinogenic properties.

mesoderm, the middle cell layer in a developing animal embryo, giving rise to the circulatory system, muscles, etc.; ⇨ p. 186.

mesosome, an area of infolded plasma membrane in a prokaryote cell; ⇨ p. 24 (box).

mesozoan, any member of the small phylum Mesozoa, whose representatives are simple, minute, flatworm-like creatures that live parasitic lives in other soft-bodied animals. They are so simple in structure that their relationship to other animals is obscure, yet their life history is complex.

Mesozoic era, a geological era lasting from 250 to 65 million years ago; ⇨ p. 4.

mesquite (*Prosopis glandulosa*), a leguminous tree native to southern North America. It is a source of cattle food, timber, fuel and gum.

metabolic pathway, any sequence of chemical reactions that are catalysed by enzymes and that result in the breakdown (catabolism) or build-up (anabolism) of compounds.

metabolism, collective term for all the chemical processes that occur in living things, involving e.g. growth, energy production, respiration, etc.

metamorphosis, a radical structural transformation in an animal, as in the change of a tadpole to a frog (⇨ p. 80), or of an insect larva to an adult (⇨ pp. 60, 62–3). In insects, two kinds of metamorphosis are recognized. ENDOPTERYGOTES such as beetles, butterflies and bees pass through larval and pupal stages before reaching the adult state (complete metamorphosis), while EXOPTERYGOTES such as grasshoppers and dragonflies have juvenile stages (nymphs) that are essentially similar to adults (incomplete metamorphosis); ⇨ p. **187**.

metatherians, the MARSUPIALS; ⇨ pp. 100–1.

metazoan, any animal belonging to the Metazoa, a group consisting of all multicellular animals except for sponges (known as parazoans; ⇨ p. 52). The single-celled animal-like protists are grouped together as the protozoans (⇨ p. 26).

meteorite, a rock-like object, originating as a meteroid orbiting the sun, that has fallen to earth. Meteorites have been suggested as sources of amino acids – and hence of the beginning of life on earth – and as a possible explanation for the extinction of the dinosaurs; ⇨ pp. 7, 84 (box).

methane, a colourless, odourless inflammable gas (CH_4) whose presence in the early atmosphere probably contributed to the beginning of life (⇨ p. 7). Methane is expelled as a waste product by ruminating animals (⇨ box, p. 112), and is also a greenhouse gas (⇨ p. 225).

Michaelmas daisy, ⇨ ASTER.

microbe, any microscopic organism; sometimes restricted to bacteria and viruses.

microbiology, the study of the biology of microorganisms.

microbody, a type of organelle; ⇨ p. 20 (box).

microchiropteran, a member of the larger of the two suborders of bats, Microchiroptera; ⇨ pp. 118–19.

microfilaments, fine contractile strands present in the cytoplasm of cells, along which organelles such as mitochondria are moved (⇨ pp. 20–1).

microorganism, any microscopic organism.

microparasite, a microscopic parasitic organism; ⇨ p. 168.

micropyle, the structure in the ovules of seed-bearing plants through which pollen (or the pollen tube) reaches the nucellus; ⇨ p. 38.

microsaur, any member of a group of extinct amphibians that may be the ancestors of the living CAECILIANS; ⇨ p. 78.

microspore, in heterosporous spore-bearing plants and in seed-bearing plants, the smaller type of spore containing the male gamete (sex cell); ⇨ p. 38. In seed-bearing plants, the microspores become the pollen grains.

midge, common name given to insects of several DIPTERAN families (eg. Dixiidae, Chironimidae). Their larvae are frequently aquatic. Adults are similar to mosquitoes but smaller. They are often common near water and may be a pest through sheer weight of numbers. Species known as midges are often referred to (more or less indiscriminately) as 'gnats'.

midshipman, any TOADFISH of the genus *Porichthys*. They are unusual in being shallow-water fishes that possess light organs, usually a feature of deeper-water fishes. They are known for their choruses of growls and whistles, which are clearly audible from the shore.

midwife toad, any of three species of ANURAN amphibian of the family Discoglossidae (which also includes the fire-bellied toads), especially *Alytes obstetricans*. Distributed in western Europe and northern Africa, these toads have a unique method of parental care, the male carrying a string of eggs entwined around his back legs; ⇨ pp. 79, 81.

mignonette, any of a number of annual to perennial herbs in the European and Mediterranean genus *Reseda* in the family Resedaceae. Common mignonette (*R. odorata*) is grown for its scent.

migration, a movement from one habitat to another, usually cyclical and triggered by seasonal or other factors that recur at more or less fixed intervals; ⇨ pp. 170–1, 172. Other references: crustaceans 67; fishes 76; amphibians 80; reptiles 86; birds 96, 97 (box); antelopes 115; bats 118; whales 145.

mildew, 1. a fungus growing on the surface of plants and causing disease. **2.** a fungus or bacterium staining and degrading cloth and paint, etc.

milk, a nutritious fluid containing protein and fat secreted by mammals for the suckling of their young; ⇨ pp. 98 (box), 100, 185.

milkweed, 1. any of a number of perennial herbs and shrubby plants with a milky latex in the New World and African genus *Asclepias* in the family Asclepiadaceae. **2.** (*Danaus plexippus*), a NYMPHALID butterfly, widely distributed outside Africa and Asia. It is well known for its long migrations in North America (⇨ p. 170).

millet, any of a number of different cereal grasses having in common a short growing season and an ability to produce grain in hot dry conditions on poor soils. They are cultivated principally in Asia and parts of Africa for human consumption and as a fodder crop.

millipede, any of some 8000 herbivorous ARTHROPODS of the class Diplopoda, characterized by a body comprising numerous segments, each bearing a pair of legs; ⇨ pp. 68 (box), 177, 201.

milt, the soft ROE of male fishes.

mimicry, a form of deception, usually defensive, that involves adopting the appearance of something else. Unpalatable species such as bees and wasps tend to mimic one another (Müllerian mimicry), while some harmless species such as hoverflies look like unpalatable ones (Batesian mimicry), in both cases to reduce the impact of predation. Some animals mimic the appearance of inanimate or inedible objects, again as a form of defence; ⇨ pp. 153, **166–7**.

mimosa, any of around 400 species of herbs, lianas, shrubs and trees in the mainly New World genus *Mimosa* in the leguminous family, Mimosaceae, notably the SENSITIVE PLANT. See also ACACIA.

mimulus or **monkey flower,** any of around 150 species of herbs and shrubs in the South African, Asian and American genus *Mimulus* in the family Scrophulariaceae, e.g. musk (*M. moschatus*).

mink, either of two MUSTELIDS of the genus *Mustela*, resembling polecats, with a glossy brown fur, elongated bodies and short legs. They are found near water in North America (*M. vison*) and Europe (*M. lutreola*); ⇨ p. 131.

minke whale or **lesser rorqual** (*Balaenoptera acutorostrats*), a baleen whale, the smallest of the RORQUALS; ⇨ p. 145.

minnow, common name for small fishes of the CARP family, especially the European species *Phoxinus phoxinus*, a very abundant fish of rivers and streams. The name is sometimes loosely applied to small stream fishes from other groups.

mint, any of around 25 species of aromatic mainly perennial herbs in the Old World temperate genus *Mentha* in the family Lamiaceae (formerly Labiatae), including the widely cultivated peppermint (*M. aquatica* x *spicata*) and spearmint (*M. spicata*).

Miocene epoch, a geological epoch lasting from 23 to 5·3 million years ago; ⇨ p. 4.

Mississippian period, the earlier of the two parts into which the Carboniferous period is divided in the USA; ⇨ p. 4.

mistletoe, any of a number of semiparasitic shrubs in the family Loranthaceae. *Viscum album* is the mistletoe native to Britain and *Phoradendron serotinum* is the commercial mistletoe in North America.

mite, any tiny parasitic ARACHNID of the order Acari, including gall mites (family Eriophydae), chicken mites (Dermanyssidae), itch mites (Sarcoptidae), harvest mites (Trombidiidae), SPIDER MITES, and water mites (Hydrachnidae); ⇨ pp. 68–9.

mitochondrion (pl. -dria), a type of organelle found in eukaryote cells; ⇨ pp. 20 (box), *21*.

mitosis, the mechanism by which normal body cells divide; ⇨ p. 22. Compare MEIOSIS.

mixed forest, forest consisting of a mixture of broad-leaved deciduous and needle-leaved coniferous trees. It is typical of the Great Lakes area of North America; ⇨ pp. 207, 208.

moa, any recently extinct flightless birds of New Zealand, belonging to the family Diornithidae (order Struthioniformes), known by Maoris and perhaps by early European settlers. One species (*Diornis maximus*) stood over 3 m (10 ft) tall; ⇨ pp. 97 (box), 214.

mobbing, harassment of a predator (e.g. an owl) by a group of potential prey (e.g. small birds). Such behaviour often involves mock attacks and loud vocalizations; ⇨ pp. 148 (box), 167.

mockingbird, any of 34 PASSERINE birds of the tribe Mimini, in the same family as the STARLINGS. They are brown/grey, thrush-like birds with long tails, living in scrubland of North and South America. All species are strongly territorial, and many, such as the northern mockingbird (*Mimus polyglottus*), are superb vocal mimics; ⇨ p. 15. See also CATBIRD.

mola or **ocean sunfish,** any of three large oceanic fishes of the family Molidae, belonging to the same order as the PUFFERS. The largest, *Mola mola*, grows up to 4 m (13 ft) in length; its body is disc-shaped, with very tall dorsal and anal fins projecting from top and bottom. The other species are more elongated but of the same general form. Large specimens are rare, and their biology is little known.

molar, one of the large flat cheek teeth in mammals, adapted for grinding and chewing; ⇨ p. 98 (box).

molasses, ⇨ SUGAR CANE.

mold, ⇨ MOULD.

mole, the common name given to all but 2 of the 29 species in the mammalian family Talpidae (order Insectivora), the others being the DESMANS. Moles are highly adapted as burrowers, feeding mainly on earthworms. They are widely distributed in Europe, Asia and North America; ⇨ pp. 120, **121**. See also GOLDEN MOLE, MARSUPIAL MOLE.

mole cricket, any ORTHOPTERAN insect of the family Gryllotalpidae. Mole crickets are highly modified for subterranean life, and have powerful shovel-like fore limbs (⇨ p. 59). Sound produced by rubbing the wings together is amplified by a pair of horn-shaped tunnels (⇨ p. 154).

molecular biology, the study of biological molecules, especially DNA, RNA and proteins.

molecular clock, a chronology of evolutionary developments based on the degree of change in amino-acid sequences; ⇨ p. 15 (box).

mole rat, any burrow-living caviomorph RODENT of the family Bathyergidae of sub-Saharan Africa. Most are solitary or live in small groups, but the naked mole rat (*Heterocephalus glaber*) is highly social (⇨ box, p. 158); ⇨ pp. **135**, 155, 206. See also BLIND MOLE RAT and ZOKOR.

mole salamander, any SALAMANDER of the family Ambystomatidae, comprising 31 species from North and Central America. They spend much of their time underground, using natural crevices and burrows. There are several neotenic forms, the best known being the AXOLOTL.

mollusc, any invertebrate of the phylum Mollusca, comprising over 50 000 species in seven classes. The group includes CHITONS, gastropods, bivalves (including clams, mussels, oysters, etc.), TUSK SHELLS and cephalopods (squid, cuttlefish, etc.); ⇨ pp. 56–7, 215, 218.

molting, variant spelling of MOULTING.

monarch flycatcher, any of 98 PASSERINE birds of the tribe Monarchini, in the CROW family (Corvidae). They are small, mostly forest-dwelling birds with small head crests and long tails, which in most species are flicked constantly. They live in Africa, the Orient and Australia.

Monera, one of the five kingdoms of living things, including the archaebacteria, bacteria and mycoplasms; ⇨ pp. 24–5.

moneywort, ⇨ CREEPING JENNY.

mongoose, any small VIVERRID carnivore of the genus *Herpestes* and related genera, opportunistic predators found in a range of habitats in Africa and Asia; ⇨ p. **123**. See also SALANO.

monito del monte or **colocolo** (*Dromiciops australis*), a small, arboreal, insectivorous MARSUPIAL found only in cool forests in Chile, where it hibernates in winter. It is the sole member of the family Microbiotheriidae.

monitor or **varanid,** any of about 30 species of LIZARD of the family Varanidae, found in tropical and subtropical regions of the Old World. All are predatory carnivores, with long necks and rapidly darting snake-like tongues. Monitors are generally large, and the Komodo dragon (*Varanus komodoensis*) is the largest of all lizards (⇨ pp. 85, **89**).

monkey, common name given to all higher PRIMATES except the tarsiers, apes and man. Two major groups are recognized: the New World monkeys (marmosets, tamarins and the cebid monkeys, such as capuchins and howlers) and the Old World monkeys (macaques, baboons, langurs, colobus, etc.); ⇨ pp. 136, **138–9**, 201.

monkey flower, ⇨ MIMULUS.

monkey nut, ⇨ PEANUT.

monkey puzzle, ⇨ ARAUCARIA.

monkfish, alternative name for either shallow-water ANGLER FISHES or ANGEL SHARKS.

monkshood, ⇨ ACONITUM.

monocotyledon (often abbreviated to **monocot**), any member of the flowering-plant class Liliopsida. The main distinguishing feature is a single seed leaf (cotyledon); ⇨ p. 45. Compare DICOTYLEDON.

monoculture, a stand of vegetation, usually cultivated, consisting of a single species – either an annual crop, such as corn, or trees, such as spruce; ⇨ pp. 212, 230.

monoecious, (of a plant species) having male and female organs on the same plant, but in UNISEXUAL flowers, e.g. hazel (⇨ p. *42*).

monogamy, a mating system in which one male mates with a single female; the most common system in birds; ⇨ p. 158.

monophyletic, (of a classificatory group) including an ancestral species and all its known descendants; ⇨ p. 19.

monotreme, any mammal of the order Monotremata, the 'egg-laying mammals'. This exclusively Australasian group contains just 3 living species, the platypus and 2 species of echidna; ⇨ pp. 100, *101*, **102**.

monsoon forest, mixed deciduous and evergreen subtropical forest found in areas where the rainfall is confined to certain seasons of the year, dictated by monsoon conditions, as in northern India (▷ map, p. 199).

Monstera, a tropical New World genus of around 30 species of lianas in the arum family, Araceae. The Swiss cheese plant (*M. deliciosa*) is a popular house plant.

montane regions, regions of mountain areas that lie below the tree line. These may be forested or may be kept clear of trees by human activity, such as the grazing of domestic animals.

mooneye, either of two North American freshwater fishes of the family Hiodontidae, in the same order as the BONY TONGUES.

moon-phase spawning, a breeding cycle that is linked to the LUNAR CYCLE, occurring in some marine animals; ▷ p. 175.

moonrat or **gymnure,** the common name given to 5 of the 17 species in the mammalian family Erinaceidae (order Insectivora), the other 12 being HEDGEHOGS. Moonrats resemble spineless hedgehogs and are found in China and Southeast Asia, generally in forested habitats; ▷ p. 120.

moorhen (*Gallinula chloropus*), a water bird with a dark body, and red and yellow bill with a bright red patch at the base, belonging to the family Rallidae (order Gruiformes).

moorland, dwarf-shrub vegetation, usually dominated by heather, developed at high altitude and under relatively high-precipitation conditions after human deforestation (▷ p. 213). Peat usually develops on top of the soil. Moorland merges with lowland heath.

moose or **elk** (*Alces alces*), a stocky DEER with a broad muzzle, shoulder hump, and palmate antlers in males. The largest member of the deer family, it inhabits forest in North America (moose) and Eurasia (elk); ▷ pp. *196,* 209.

moray, any EEL of the family Muraenidae (110 species); these eels lack both pelvic and pectoral fins, and have sharp teeth in the jaws; ▷ p. 75 (box).

Morgan, Thomas Hunt (1866–1945), American biologist, regarded as the father of modern genetics. Working on the fruit fly *Drosophila,* he established that particular sets of characteristics are often inherited together (▷ LINKAGE), and also discovered crossing-over (▷ p. 23).

morning glory (*Ipomoea purpurea*), a twining vine native to tropical America, in the family Convolvulaceae. It is grown as an ornamental. The flowers wither by midday. See also SWEET POTATO.

morphology, the study of the form and structure of organisms; the basis of much comparative anatomy (▷ p. 14).

morula, in embryology, a solid ball of cells resulting from the initial cleavages of a fertilized animal egg (zygote); ▷ p. 186.

mosasaur, any extinct marine REPTILE of the genus *Mosasaurus,* related to the MONITOR lizards, which flourished in the Cretaceous period; ▷ pp. 82, **84.**

mosquito, any medium to small DIPTERAN insect of the family Culicidae. With a few exceptions, the females are blood-feeders while males feed at flowers. They are more common in the tropics, where they are important as carriers of disease, but they have a worldwide distribution. Some species in tundra regions become extremely numerous in the brief arctic summers. Over 1600 species are known. The larvae are aquatic, feeding on particulate material; ▷ pp. *59* (box), *64,* 211, 230.

moss, any member of the bryophyte class Musci; ▷ pp. **36–37,** 207, 208, 211, 216. See also CLUB MOSS.

mossy forest, any forests in high-precipitation environments, often at high altitude or in oceanic

climates, where bryophytes (mosses, liverworts, etc.) play an important role as EPIPHYTES living on the branches of the trees.

moth, any LEPIDOPTERAN that is not a BUTTERFLY (i.e. does not belong to the superfamilies Papilionoidea or Hesperioidea). They are often nocturnal species, resting with the wings in the same plane along the length of the abdomen, but the distinction between butterfly and moth has little taxonomical significance; ▷ pp. *59* (box), 64, 166.

mother-in-law's tongue, ▷ *DIEFFENBACHIA.*

mother-of-thousands, 1. *Saxifraga stolonifera,* a SAXIFRAGE popular as a house plant. **2.** ivy-leaved TOADFLAX or pennywort (*Cymbalaria muralis*), a creeping perennial common on old walls. **3.** mind-your-own-business (*Soleirolia soleirolii*), a creeping carpet-forming plant grown in greenhouses.

motile, able to move (usually applied to micro-organisms and sperm that have this ability).

motivation, the internal state of an animal (e.g. hunger) that affects its response or responsiveness to external stimuli (e.g. food); ▷ p. 149.

motor nerve, ▷ EFFERENT NERVE and p. 182.

mouflon (*Ovis musimon*), a small, dark-brown wild SHEEP, with large, curved horns. It inhabits rugged, mountainous areas in Corsica, Sardinia, Cyprus, Asia Minor and Iran.

mould or **mold,** the woolly or furry growth of a fungus (▷ p. 35) such as ASPERGILLUS. See also SLIME MOULDS.

moulting, in arthropods, reptiles, birds and mammals, the process of shedding the body covering (cuticle, scales, feathers, fur). In insects and other arthropods (▷ pp. 58, 62, 66) moulting is an integral part of growth, and is initiated by the hormone ECDYSONE.

mound-builder, ▷ MEGAPODE.

mountain ash, 1. ▷ SORBUS. **2.** *EUCALYPTUS regnans.*

mountain beaver or **sewellel** (*Aplodontia rufa*), a terrestrial, burrowing sciuromorph RODENT resembling a small beaver, inhabiting coniferous forests in eastern North America. It is the sole member of the family Aplodontidae.

mountain lion, another name for PUMA.

mouse, common name given to numerous myomorph RODENTS of the large family Muridae, the difference between a mouse and a rat (as such) being largely a matter of size. Unqualified, the name 'mouse' often refers to various species of the genus *Mus,* especially the house mouse (*M. musculus*), which are pests throughout the world (▷ box, p. 134). Many other more or less mouse-like rodents (not necessarily closely related) are also known as mice; ▷ pp. 133, 134 (box), **135,** *195,* 213. See also DORMOUSE, FIELDMOUSE, HARVEST MOUSE, JUMPING MOUSE, VOLE.

mouse deer, another name for CHEVROTAIN.

mouse-tailed bat, any of three microchiropteran BATS of the family Rhinopomatidae. These small, insectivorous bats derive their name from their long, thin, furless tails; they are found in arid lands from the Middle East through to Southeast Asia.

moustached bat, ▷ GHOST-FACED BAT.

mouthbrooder, ▷ CICHLID.

mouth-brooding, (in fishes, e.g. cichlids) the use of the mouth to transport and shelter the eggs and young, particularly in times of danger; ▷ p. 74.

mouthparts, in arthropods and other invertebrates, the paired feeding structures surrounding the mouth; ▷ pp. *59,* 178.

movement or **locomotion,** ▷ pp. 27 (protists), 176–7 (animals).

mucin, any of various nitrogen-based proteins that form viscous, sticky solutions when dissolved in water, as in saliva, mucus, etc.; ▷ p. 160.

mucus, a sticky protective fluid secreted by the mucous membranes that lines the digestive tract; ▷ p. 179.

mud dauber, any WASP of the genus *Sceliphron* (family Sphecidae), of tropical and subtropical distribution. They make nests of 10–50 cells out of mud or clay, which they knead into shape; the cells are provisioned with spiders.

mud flat, a coastal, usually estuarine, ecosystem in which the muddy substrate is derived from organic matter carried downstream from terrestrial ecosystems. Mudflats are rich in invertebrate life and therefore in their predators, such as wading birds (▷ p. 96).

mud puppy, any SALAMANDER of the genus *Necturus* (family Proteidae), especially *N. maculosus,* occurring in eastern USA. The other 5 species of the same genus are often referred to as 'waterdogs'. All are neotenic, i.e. sexually mature in the larval state (▷ p. 81). See also OLM.

mudskipper, any of various fish of the GOBY genus *Periophthalmus* capable of movement over mud flats; ▷ p. 217.

mugwort, ▷ *ARTEMISIA.*

mulberry, any of around 12 species of trees and shrubs in the mainly tropical genus *Morus.* The juicy part of the edible blackberry-like fruit is derived from the perianth parts of the flower. In China the white mulberry (*M. alba*) was grown to feed silkworms and for its timber. The related paper mulberry (*Broussonetia papyrifolia*) provides fibre used in paper making in Japan.

mulgara (*Dasyurus cristicauda*), a small, burrowing, desert-dwelling, carnivorous DASYURID marsupial widespread throughout inland Australia. Its short, fat tail has a black crest.

mullein, any of around 360 species of mostly biennial herbs in the mainly Mediterranean genus *Verbascum* in the family Scrophulariaceae. Some are garden ornamentals. See also AARON'S ROD.

Müllerian mimicry, ▷ MIMICRY.

mullet, any of about 280 PERCIFORM fishes of the family Mugilidae, predominantly marine and brackish-water fishes from all temperate and tropical seas. They have cylindrical bodies and blunt snouts, and feed from the sea bed. See also GOATFISH.

multi-male group, a form of social organization in which a number of males and females live together, some or all of which may breed; ▷ p. 159.

mung bean (*Phaseolus aureus*), a tropical and subtropical legume grown for its edible pods and seeds. Chinese bean sprouts are frequently germinated mung beans.

muntjac, any of five small, brown, forest-dwelling DEER with small, simple antlers, belonging to the genus *Muntiacus.* Native to India, and eastern and Southeast Asia, the muntjac has been introduced elsewhere.

muriqui, another name for the woolly SPIDER MONKEY.

murre, ▷ GUILLEMOT.

Musci, the bryophyte class comprising the mosses; ▷ p. 37.

muscle, body tissue, usually composed of long, thin cells or fibres, that is capable of contraction and thus movement, when anchored to some fixed point; ▷ pp. **176,** *177.*

museum beetle, ▷ DERMESTID BEETLE.

mushroom, any species of AGARIC or BOLETE fungus. Some people would confine the term to the common mushroom and its close relatives but a court case in the 1950s decided that soup containing *Boletus edulis* could be described as mushroom soup.

musk, ▷ MIMULUS.

musk deer, any of three small, deer-like, ruminating

ARTIODACTYL mammals of the family Moschidae, typically found in forests in central and eastern Asia; ⇨ p. 116.

musk ox (*Ovibos moschatus*), a large, massively built, ox-like ARTIODACTYL (related to sheep and goats, not cattle), with long, dark, shaggy fur, a short tail and curved horns. It lives in herds in the arctic tundra of North America and Greenland; ⇨ p. 115.

muskrat (*Ondatra zibethicus*), a large, amphibious, rat-like myomorph RODENT of the family Muridae, indigenous to North America but introduced to Europe. It has a blunt muzzle, long, naked tail and partially webbed hind feet.

mussel, any BIVALVE mollusc of the family Mytilidae. Principally intertidal, mussels are found worldwide in temperate latitudes, providing an important source of food; several species are cultivated. They attach themselves to rocks, other shells, ropes, etc. by means of strong threads (byssi); ⇨ p. 56.

mustard, black mustard (*BRASSICA nigra*), brown mustard (*B. juncea*) or the related white mustard (*Sinapis alba*), the seeds of all of which can be ground to make a condiment. The mustard in mustard and cress is either white mustard or RAPE (*B. napus*).

mustelid, any mammal of the family Mustelidae (order Carnivora), the largest group of CARNIVORES, comprising 67 species and including weasels, badgers, skunks, otters and others. Generally small to medium-sized solitary mammals, mustelids have adapted to a wide range of terrestrial and aquatic habitats, and are only absent from Australasia, Madagascar and Antarctica; ⇨ pp. 122, **131**.

musth, in mature male elephants, a condition of heightened sexual excitement, often accompanied by increased aggression; ⇨ p. 109.

mutation, a change occurring in DNA as it replicates itself, a key mechanism in evolution; ⇨ pp. 13, 15 (box), **16**, **22**.

mutualism, another name for SYMBIOSIS.

mycelium, the collection of hyphae (filaments) constituting the vegetative body of a fungus; ⇨ p. 35.

Myceteae or **Fungi,** the kingdom comprising the fungi; ⇨ p. **35**, and FUNGUS.

mycoplasm, any of a group of minute prokaryote organisms; ⇨ p. 24.

mycorrhizal, (of a fungus) living in a symbiotic association with the roots of a higher plant. The fungus obtains carbohydrates from the plant, and supplies it with minerals such as phosphates; ⇨ p. 35.

mynah, any of 10 Oriental and Australian PASSERINE birds in the family Sturnidae, similar to the closely related STARLINGS. They are common and urbanized in southern Asia. Some are used as vocal cagebirds.

myomorph, any RODENT of the suborder Myomorpha – the 'mouse-like' rodents; ⇨ p. 133.

myriapod, name given to several groups of ARTHROPODS characterized by a long segmented body and many legs, the most important of which are the centipedes and millipedes; ⇨ p. 68 (box).

myrrh, 1. a constituent of incense, perfumes and medicines, obtained from species of the Arabian and northeast African genus *Commiphora*, e.g. *C. molmol*. 2. sweet cicely (*Myrrhus odorata*), a European herbaceous perennial with aromatic leaves.

Myrtaceae, a dicotyledonous plant family, including eucalyptuses; ⇨ p. 45.

myrtle (*Myrtus communis*), an aromatic evergreen shrub, native to the Mediterranean region and western Asia, but introduced elsewhere for use in scents and medicines. The unrelated bog myrtle (*Myrica gale*) was used to flavour beer.

myxomycete, any member of the protistan group Myxomycetes; ⇨ p. 26 (box).

NAD, nicotinamide adenine dinucleotide, which in its reduced form, NADH$^+$, provides cells with a utilizable form of energy; ⇨ p. 21.

naiad, ⇨ NYMPH.

narcissus, any of around 60 species of bulbous herbs in the European, west Asian and North African genus *Narcissus* in the family Amaryllidaceae. Those with large trumpets (coronas) are referred to as daffodils, e.g. wild daffodil (*N. pseudonarcissus*). Many cultivars and species, e.g. jonquil (*N. jonquilla*) are grown as ornamentals.

narwhal (*Monodon monoceros*), a medium-sized (up to 6 m/20 ft) toothed whale (CETACEAN) of Arctic waters, distinguished (in the male) by a long spiral tusk; ⇨ p. 145.

nasturtium, 1. *Nasturtium*, a genus of around 6 species of usually perennial herbs; ⇨ CRESS. 2. any of a number of annual or perennial herbs in the unrelated genus *Tropaeolum*, particularly *T. majus*, popular in gardens.

national park, an area of a country set aside by the government for conservation (and sometimes recreational) purposes. There are many different approaches in different countries; ⇨ p. 232.

native, ⇨ INDIGENOUS.

native companion, ⇨ BROLGA.

natterjack toad (*Bufo calamita*), a robust European toad (ANURAN) with a loud booming call; it has short hind legs, and scampers rather than hops.

natural selection, the theory that evolution occurs through the survival and breeding of individuals better adapted to their environments than other members of the same species; ⇨ pp. 13, 16–17.

nature reserve, an area set aside for wild species, to which (in theory) human access is severely restricted; ⇨ p. 232.

nature versus nurture, long-standing debate among students of behaviour (animal and human) over the relative importance of genetic endowment and interaction with the environment in determining an individual's behaviour, intelligence, etc.; ⇨ p. 150.

nauplius, the primary larval stage of a crustacean; ⇨ p. 66.

nautiloid, any CEPHALOPOD mollusc of the group Nautiloidea, including the NAUTILUS and numerous extinct species (⇨ p. 5).

nautilus, any of several CEPHALOPOD molluscs of the genus *Nautilus*, especially *N. pompilius* (the pearly nautilus), a spectacular remnant of a once-important group of shelled cephalopods (nautiloids). Its protective chambered shell provides buoyancy, and by changing the amount of gas in the chambers, it can rise or sink in the water; it is predatory, catching crustaceans with its numerous tentacles. See also ARGONAUT.

navigation, an advanced form of ORIENTATION.

Nearctic, a faunal region; ⇨ p. 50 (map).

nectar, a sugary fluid produced by flowers to attract pollinators; ⇨ p. 42.

nectar guides, spots or lines on the petals of a flower to guide pollinators to the source of nectar; ⇨ p. 42.

nectarine (*Prunus persica* var. *nucipersica*), a very long-established variety of peach with a smooth skin and quite distinct flavour.

needle, any thin, elongated, pointed structure, such as a spine or the leaf of a conifer; ⇨ p. 39.

needlefish or **garfish,** any of 32 mainly marine fishes of the family Belonidae. They are long and slender and both of their jaws are elongated.

nekton, the collective term for all aquatic animals that swim freely in the middle depths, as opposed to

bottom-dwelling (BENTHIC) animals, and PLANKTON, which drift with the currents; ⇨ pp. 218, 220.

nektonic, 1. (of an aquatic animal) swimming freely in the middle depths; such animals are collectively known as NEKTON. **2.** (of an aquatic zone) the middle zone, between the surface and the bottom.

nematocyst, a specialized type of stinging cell found in coelenterates; ⇨ pp. 53, 56.

nematode, a ROUNDWORM; ⇨ p. 55.

nemertean, a RIBBON WORM.

Nemesia, a genus of around 65 species of herbs and shrubs native mainly to South Africa. Garden cultivars are derived from *N. strumosa*.

neo-Darwinism, a term sometimes applied to the modern genetic approach to evolution; ⇨ pp. 16–17.

neoteny, the ability to breed while still in the larval form, or the persistence of larval features in the adult form; ⇨ pp. 79, 81. See also PAEDOGENESIS.

Neotropical, 1. a faunal region; ⇨ p. 50 (map). **2.** a floral region; ⇨ p. 30 (map).

nephridium (pl. -dia), in many invertebrates, a simple tubular excretory organ; ⇨ p. 179 (box).

nephron, the tiny functional unit of the vertebrate kidney; ⇨ p. 179 (box).

neritic zone, the coastal zone of a marine ecosystem; ⇨ pp. 220, 221.

neritid, ⇨ PERIWINKLE.

nerve cell or **neurone,** the functional unit of a nervous system; ⇨ pp. 182–3.

nerve cord, in many invertebrates, a bundle of nerve fibres, usually one of a pair, running the length of the body and providing the main neural pathway for communication between the head and the rest of the body. In vertebrates, the nerve cord is the SPINAL CORD; ⇨ pp. 59, 182, *183*.

nerve net, a simple, uncentralized nervous system found in coelenterates such as hydras; ⇨ pp. **182**, *183*.

nervous system, a more or less complex network of interconnecting nerve cells that provides a rapid means of communication between cells and hence a system by which the activity and functioning of the body as a whole can be coordinated; ⇨ pp. 50–1, 55, 57, 59, **182–3**.

nest, any area or structure in which eggs or young are produced and protected; ⇨ pp. 95, 160–1, 185.

nestling, a young bird before it leaves the nest.

net, any of a group of protists belonging to the Myxomycetes; ⇨ p. 26 (box).

nettle, any of around 50 species of annual or perennial herbs, usually with stinging hairs, in the temperate genus *Urtica* in the family Urticaceae, e.g. stinging nettle (*U. dioica*). See also DEAD-NETTLE.

neural groove/tube, a trough in an animal embryo that closes over to form a tube, the basis of the central nervous system; ⇨ pp. 186–7.

neuromast, in the lateral-line system, ⇨ p. *190*.

neuromodulator, ⇨ p. 182 (box).

neurone, a nerve cell; ⇨ pp. 182–3.

neuropteran, any ENDOPTERYGOTE insect of the order Neuroptera, which includes DOBSONFLIES, ALDER FLIES, SNAKE FLIES, LACEWINGS, ANTLIONS and MANTISPIDS; ⇨ p. *61*.

neurosecretory cell, a hormone-secreting nerve cell; ⇨ p. 183.

neurotransmitter, ⇨ p. 182 (box).

neuston, small animals that live at or just below the surface film of water in lakes or ponds.

newt, any of 22 species of highly aquatic SALA-MANDER, included in the 'true' salamander family (Salamandridae). They are found in Europe, western Asia, North America, China and Japan. Newts are notable for their highly ritualized courtship behaviour; ⇨ pp. 80–1.

niche, ecological, the role that an organism has within an ecosystem; ⇨ p. **195**.

Nicotiana, a genus of around 70 species of herbs and a few shrubs in the potato family, Solanaceae. Some species are the source of TOBACCO and others are grown as ornamentals.

nicotine, ⇨ TOBACCO.

nictitating membrane, the 'third eyelid', a protective membrane that can be drawn across the eye of some mammals and most amphibians, reptiles and birds; ⇨ p. 79.

nidicolous, (of a hatchling bird) blind, defenceless, and nest-bound for several weeks; ⇨ p. 95.

nidifugous, (of a hatchling bird) well developed at hatching and soon able to leave the nest; ⇨ p. 95.

nighthawks, any of eight birds, closely related to the American NIGHTJARS. They have long wings and fly like bats around the tops of trees at dusk, catching insects.

nightingale (*Erithacus megarhynchos*), a small, shy, olive-green THRUSH, usually found in Palaearctic woodland and famed for its rich song, often sung well into the summer night.

nightjar, any of 68 cryptically coloured birds of the family Caprimulgidae, in the same order as the OWLS; they occur worldwide. The head is flattened and the mouth wide and surrounded by sensory hairs. They tend to feed at dusk or at night by catching insects on the wing; ⇨ pp. 90, *91*, 97, *201*. See also OWLET-NIGHTJAR, POORWILL.

night monkey or **douroucouli,** any large-eyed, thickly furred New World monkey of the genus *Aotus* (family Cebidae), the only genus of nocturnal monkeys in this family; ⇨ p. 138.

nightshade, 1. any of a number of species of the genus SOLANUM, family Solanaceae, e.g. black nightshade (*S. nigrum*). **2.** ⇨ DEADLY NIGHT-SHADE. **3.** enchanter's nightshade (*Circaea lutetiana*), a north temperate Old World perennial herb.

nilgai or **blue bull** (*Boselaphus tragocamelus*), a large, grey ANTELOPE with a short mane, and horns in the male only. It inhabits woodlands in India.

ningaui, either of two mouse-like carnivorous DASYURID marsupials of the genus *Ningaui*, with large ears and long tails. They are restricted to central and northwestern Australia; ⇨ p. 105.

nitrate, any NO$_3^-$ compound, absorbed by plants from the soil; an important stage in the nitrogen cycle; ⇨ pp. 32 (box), **197**. Nitrates are also synthesized artificially to make fertilizers; ⇨ p. 226.

nitrification, the oxidation of ammonia by soil bacteria to form nitrates; ⇨ p. 197.

nitrogen, an element (symbol N) that in its molecular form (N$_2$) comprises 78% of the atmosphere. It is an essential component of nucleic acids and proteins (⇨ p. 6), and is converted into these by plants (⇨ pp. 31–2), from which animals obtain their proteins as part of the nitrogen cycle (⇨ p. 197). For the polluting effects of the oxides of nitrogen ⇨ p. 225.

nitrogen cycle, the biogeochemical cycle involving nitrogen; ⇨ p. 197.

nitrogen-fixing bacteria, a group of bacteria, such as *Rhizobium*, capable of fixing nitrogen; ⇨ pp. 25, 32 (box), 169.

noctuid, any nocturnal LEPIDOPTERAN insect (moth)

of the very large family Noctuidae, including UNDERWINGS and CUTWORMS.

nocturnal, associated with the night, as opposed to the daylight hours. Compare DIURNAL.

noddy, any of four species of sooty-brown, tern-like birds, with rounded tails and slow flight (family Laridae, order Ciconiiformes). They are mainly restricted to the southern hemisphere. The nodding movements of the head during mating account for their name.

node, the region of a plant stem at which a leaf is attached; ⇨ p. 32.

non-renewable resource, any raw material required by mankind that is finite in quantity and cannot be replaced once exhausted. Examples include fossil fuels and peats.

noradrenaline, ⇨ p. 182 (box).

nose-leaf, a more or less complex skin structure on the face of many bats using ECHOLOCATION; ⇨ p. *118*.

Nothofagus, ⇨ BEECH (2).

nothosaur, any extinct marine REPTILE of the suborder Nothosauria, which flourished in the Triassic period, to be replaced in the Jurassic by plesiosaurs; ⇨ p. 84.

nucellus, the structure within the ovules of seed-bearing plants that encloses the megaspore, within which is the female ovum; ⇨ pp. 38, 41.

nuclear envelope, the double membrane surrounding the nucleus of a cell (⇨ p. 21).

nucleic acids, DNA and RNA; ⇨ p. 6 (box).

nucleocapsid, the nucleic acid and protein layer of a virus; ⇨ p. 24.

nucleoid, an area within prokaryote cells in which the DNA resides; ⇨ pp. **20**, *24*.

nucleoplasm, the PROTOPLASM in the nucleus of eukaryote cells (⇨ p. 20).

nucleus, the structure within eukaryote cells within which the chromosomes are contained; ⇨ pp. 20, *21*, 27.

nudibranch, ⇨ SEA SLUG.

numbat or **banded anteater** (*Myrmecobius fasciatus*), a pouchless MARSUPIAL, the sole member of the family Myrmecobiidae. It possesses strongly clawed fore feet and a long tongue, and feeds on termites. It is now restricted to Western Australia.

nuptial pad, a thickening on the ball of each thumb of male frogs, used for grasping the female during mating.

nut, strictly a dry, INDEHISCENT (i.e. not opening), one-seeded fruit with a woody PERICARP (e.g. hazel). Almonds and peanuts, for example, are thus not true nuts; ⇨ p. 43.

nuthatch, any of 24 PASSERINE birds in the family Sittidae, stocky, arboreal, insectivorous birds, usually with blue-grey backs and straight bills, patchily distributed throughout the northern hemisphere and Australasia. They have the unique ability to run up and down trees with equal ease.

nutmeg (*Myristica fragrans*), a Moluccan tree the seed and aril of which are dried to provide the spices nutmeg and mace respectively.

nutrient, 1. any mineral substance necessary for a plant's functioning that is absorbed through its roots; ⇨ pp. **31–2**, **196**. For the availability of nutrients in different ecosystems, ⇨ pp. 200–11. **2.** any organic nutritious substance present in food; ⇨ pp. 178–9.

nutrient cycling, the movement of nutrients within an ecosystem; ⇨ pp. 196–7.

nyala (*Tragelaphus angasi*), a grey or brown ANTELOPE with white stripes, a conspicuous mane and short horns. It inhabits dense cover near water in southeast Africa.

nymph, any of the juvenile stages of an EXOPTERY-GOTE insect, such as a dragonfly or a grasshopper. Nymphs (especially aquatic ones) are sometimes referred to as 'naiads'; ⇨ p. 60.

Nymphaeaceae, a dicotyledonous plant family, including water lilies; ⇨ p. 45.

nymphalid butterfly, any LEPIDOPTERAN insect of the family Nymphalidae, one of the largest and most colourful of butterfly families, comprising about 5000 species worldwide. The adults have small non-functional fore legs. The group includes FRITILLARIES, TORTOISESHELLS and ADMIRALS.

oak, 1. any of around 450 species of deciduous or evergreen trees in the principally northern hemisphere genus *Quercus* in the family Fagaceae. The leaves are usually cut or lobed and the nut or acorn is held in a cup. The timber is very strong and durable and used in building and furniture making; ⇨ p. 205, 223. See also MARBLE GALL and OAK APPLE. **2.** Australian oak (*EUCALYPTUS regnans*), an evergreen tree reaching 100 m (330 ft) or more in height. **3.** African oak (*Lophira lanceolata*), a tropical West African tree important for its timber.

oak apple, a knobbly gall found on OAK. It is caused by the cynipid wasp, *Biorhiza pallida*, which lays its eggs at the base of the buds; ⇨ MARBLE GALL.

oarfish, either of two species of the family Regalecidae, related to the RIBBONFISHES, occurring in all oceans but seldom encountered. The flat, ribbon-shaped body reaches 7 m (23 ft) in length, and is bright silver; a tall scarlet dorsal fin runs the length of the body, and the first few rays of this fin are elongated to form a crest on the head.

oarweed, any species of *Laminaria*, a genus of large brown algae (seaweeds) found at and below low water on coasts in northern temperate and cold regions; ⇨ p. 221. See also KELP.

oat, any of several cereals of the genus *Avena* grown principally in Europe, parts of Central Asia and North America. The most important species are the common oat (*A. sativa*), which is well adapted to cool moist conditions, and red or Algerian oat (*A. byzantina*), which can tolerate hot dry conditions. The grain is used for human consumption and both grain and straw for animal feed.

ocean ecosystems, ⇨ pp. **220–1**, 233 (conservation).

ocean-floor spreading, the process by which new oceanic crust is created; ⇨ p. 8.

Oceanian, a biogeographical region covering the islands of the Pacific.

ocellus (pl. -li), a simple type of EYE found in insects and many other invertebrates; ⇨ p. 189. See also SIGHT.

ocelot (*Felis pardalis*), a yellowish CAT with black markings. It inhabits forest and steppe in South and Central America; ⇨ p. 9.

octocoral, any colonial ANTHOZOAN of the subclass Octocorallia, including the horny corals (sea feathers and SEA FANS), SEA PENS, whip corals and pipe corals. They differ from sea anemones and stony corals in having polyps bearing eight feather-like tentacles; ⇨ p. 53.

octodont, another name for DEGU.

octopus, any marine CEPHALOPOD mollusc of the order Octopoda, mainly small bottom-living predators; ⇨ pp. 56–7.

odd-toed ungulate, ⇨ PERISSODACTYL.

odonate, any EXOPTERYGOTE insect of the order Odonata, comprising DRAGONFLIES and DAMSELFLIES; ⇨ p. *61*.

odour specialist, a specialized type of olfactory cell; ⇨ p. 188.

oesophagus, the muscular tube leading from the back of the mouth to the stomach, along which food passes by PERISTALSIS; ⇨ p. *178*.

oestrogen, a group of sex hormones (e.g oestradiol) that stimulate the development of secondary sexual characteristics in female vertebrates.

oestrous cycle, in many female mammals, the hormonally controlled cycle of sexual receptivity, associated with the production of ova (▷ OVUM); ▷ pp. 135, **185.**

-oidea, a standard ending indicating superfamily status in animals (▷ p. 18).

oil, 1. a liquid FAT. **2.** a liquid HYDROCARBON formed hundreds of millions of years ago from the remains of unicellular organisms; ▷ pp. 196, 227 (as a pollutant).

oil beetle, any BEETLE of the family Meloidae, a remarkable, widely distributed group of over 2000 species. The larvae live in the nests of bees and wasps feeding on larvae, or in the egg masses of grasshoppers. The BLISTER BEETLE belongs to this family.

oilbird (*Steatornis caripensis*), a nocturnal gregarious bird similar to a NIGHTJAR, the only member of the family Steatornithidae. Cave-dwelling oilbirds hunt and navigate by means of ECHOLOCATION; ▷ p. 97.

oil palm, either of two species of palm in the genus *Elaeis*. The fruits are a source of palm oil.

okapi (*Okapia johnstoni*), a solitary ruminating ARTIODACTYL mammal of tropical forest in central Africa, the closest relative of the giraffe; ▷ pp. 50, **117.**

okra or **ladies' fingers** (*Abelmoschus esculentus*), a tropical and subtropical annual herb in the mallow family, Malvaceae, cultivated for its pods, which are used for food and medicinal purposes.

Old Caledonian Forest, a type of forest dominated by Scots pine; ▷ p. 209 (box).

old man's beard, 1. *Clematis vitalba*, a perennial woody climber of hedges in calcareous districts. It has beard-like fruits. See also CLEMATIS. **2.** any lichen in the genus *Usnea*; ▷ p. 34.

oleander or **rosebay** (*Nerium oleander*), an ornamental but highly toxic shrub, native from the Mediterranean region to Japan.

oleaster (*Elaeagnus angustifolia*), an ornamental shrub native to southeastern Europe and western Asia.

olfaction (adj. olfactory), the sense of SMELL.

olfactory bulb or **lobe,** a part of the brain associated with smell; ▷ p. *183.*

Oligocene epoch, a geological epoch lasting from 34 to 23 million years ago; ▷ p. 4.

oligochaete, any ANNELID of the class Oligochaeta, which includes the EARTHWORMS and their freshwater relatives; ▷ p. 55.

oligotrophic, (of a lake) having nutrient-poor water; ▷ p. 218.

olingo (*Bassaricyon gabbi*), an omnivorous, arboreal PROCYONID mammal of Central and South American tropical forest; ▷ p. 130.

olive (*Olea europea*), a tree much cultivated in the Mediterranean region for its fruits, from which olive oil is pressed. The olives are inedible when picked and must first be treated, for example with salt, to achieve a lactic fermentation.

olm (*Proteus anguinus*), a European salamander placed in the same family (Proteidae) as the North American MUD PUPPY. It is found in caves along the Adriatic seaboard of northeastern Italy and Croatia.

omasum or **psalterium,** the third chamber of a ruminant's stomach; ▷ p. *112.*

ommatidium (pl. -dia), a tiny optical unit, many of which constitute the compound eyes of insects (etc.); ▷ p. *190.*

omnivore, any animal that has unspecialized feeding habits, e.g. feeding on both plants and other animals. Human beings, for example, are omnivores; ▷ pp. 162, **195.**

onager, a variety of Asiatic wild ASS.

onion (*Allium cepa*), a biennial herb of the Alliaceae family, cultivated for its pungent bulbs, which are used as a vegetable.

ontogeny, the developmental sequence of an individual organism from zygote (fertilized egg) to maturity; ▷ pp. 14, 186–7.

onychophoran, a VELVET WORM.

oogenesis, the process of formation and maturation of female gametes (ova or eggs) within the ovaries.

operculum, in bony fishes, the hard but flexible outer flap covering the gill chamber.

opium, 1. the dried juice of the fruit capsule of the opium POPPY, containing codeine and morphine. It is an addictive narcotic drug. **2.** (in ecology) a parasite community.

opossum, any MARSUPIAL mammal of the family Didelphidae, a successful group that comprises about 70 species and is widely distributed throughout much of South and Central America and into North America; ▷ pp. 9, *99, 103,* **104.** See also POSSUM, RAT OPOSSUM, YAPOK.

opportunist, a species that is able to expand its population rapidly to take advantage of unpredictable occurrences that prove suitable for its survival (▷ pp. 172–3). Often these are small, produce many young (or seeds), and have a short generation time. Many are pests (▷ p. 230).

opposable, (of a thumb or other digit) capable of being placed opposite and against one or more of the remaining digits, enabling precise manipulation of objects (as in primates; ▷ p. 136).

-opsida, a standard ending indicating class status in plants (▷ p. 45).

optic lobe, a part of the brain associated with vision; ▷ pp. 59, *183.*

optic nerve, a group of nerve fibres by which sensory nerve impulses are conveyed from the eye to the brain; ▷ pp. *189, 190.*

optimality theory, the supposition that natural selection will favour behaviour that gives maximum gain (e.g. in reproductive success) at minimum cost (e.g. in time or energy expended); ▷ p. 148.

orange, 1. sweet orange (*Citrus sinensis*), the most important CITRUS FRUIT. Many cultivars are grown in tropical and subtropical parts of the world, some selected for juicing and others for ease of peeling to eat fresh. **2.** bitter or Seville orange (*Citrus aurantium*), an orange with a rough skin and very acid juice. Marmalade is made from the fruit and neroli oil is extracted from the flowers. Oil of BERGAMOT is extracted from a subspecies.

orange-tip (*Anthocharis cardamines*), a European PIERID butterfly, with orange tips to the fore wings.

orang-utan (*Pongo pygmaeus*), the only non-hominid great APE and hence the most distantly related to man, a large primate restricted to forested habitats in Sumatra and Borneo; ▷ pp. 136, **140–1.**

orb weaver, a general term for spiders that produce complex orb webs; ▷ pp. 69, 155.

orchid, any of around 18 000 species of plant in the widely distributed family Orchidaceae; pp. *42, 43, 44,* **46** (box). See also HELLEBORE, VANILLA.

order, a classificatory group between class and family. For example, the order Primates in the class Mammalia contains all the primates. The names of all plant orders end in -ales, and those of several animal orders in -iformes; ▷ p. 18.

Ordovician period, a geological period lasting from 510 to 438 million years ago; ▷ p. 4.

oregano, ▷ MARJORAM.

organelle, any of a number of miniature organs with specialized functions within eukaryote cells; ▷ pp. 7, **20,** 26–7.

organic, associated with life. Organic compounds are chemicals containing carbon; ▷ p. 6.

organogenesis, the process whereby organs form in an animal embryo; ▷ p. 186.

oribi (*Ourebia ourebi*), a small, reddish-fawn, grazing ANTELOPE with white underparts, and short tail and horns. It inhabits grasslands in Africa.

Oriental, a faunal region; ▷ p. 50 (map).

orientation, the ability to direct one's course to a distant goal by referring to external cues such as landmarks, scent, the position of the sun, etc. The most sophisticated form of orientation – navigation – involves the ability to reach a distant destination irrespective of its direction and without reference to landmarks; ▷ pp. 118 (box), **171.**

Origanum, a north temperate Old World genus of around 15 species of aromatic perennial herbs and shrubs in the family Lamiaceae (formerly Labiatae). Several are used as kitchen herbs (▷ MARJORAM) or as ornamentals.

oriole, 1. any of about 30 PASSERINE birds in the tribe Oriolini, in the same family as the CROWS. They are Old World and Australasian birds of medium size, with powerful pointed bills for eating fruit and insects. They live in topmost dense foliage and have rich babbling songs. **2.** any of about 20 Neotropical AMERICAN BLACKBIRDS, which are tree-dwelling and build pendulous woven nests.

ornithischian, any herbivorous DINOSAUR of the order Ornithischia, one of the two major groups of dinosaurs. The order includes a range of large and small herbivores, including the ORNITHOPODS, such as HADROSAURS, triceratops and iguanadon; ▷ pp. 83–4.

ornithopod, any herbivorous ORNITHISCHIAN dinosaur of the suborder Ornithopoda. Initially small and bipedal, the ornithopods later gave rise to large herbivores such as hadrosaurs, iguanodon and triceratops; ▷ pp. 83–4.

orogeny, a period of mountain-building.

orthopteran, any EXOPTERYGOTE insect of the order Orthoptera, which includes GRASSHOPPERS, LOCUSTS and CRICKETS. This large order comprises about 18 000 species worldwide (best represented in the tropics), and contains carnivorous and vegetarian members. Orthopterans usually have enlarged hind legs for jumping, and some can fly. The fore wings are narrow and hardened, protecting the large, folding, triangular hind wings. Sound production (▷ STRIDULATION) is characteristic of the order; ▷ p. 61.

oryx, any of three elegant, pale-coloured, long-horned ANTELOPES of the genus *Oryx*, inhabiting semi-desert areas of Africa and the Arabian peninsula (where the Arabian oryx, *O. leucoryx*, has been reintroduced); ▷ pp. 115, 204.

osier, ▷ WILLOW.

osmoregulation, regulation of the water–salt balance of body fluids; ▷ pp. 73, 87–8, 93, 179 (box).

osmosis, the movement or DIFFUSION of water (or other solvent) from a less concentrated solution to a more concentrated solution; ▷ pp. 27 (protists), 73 (fishes).

osprey (*Pandion haliateus*), a large cosmopolitan BIRD OF PREY in the family Accipitridae (order Ciconiiformes). It dives for fish, which are held in powerful, roughened talons; ▷ p. 96.

ossicle, a small bone, especially (in mammals) any one of the three auditory ossicles – malleus, incus and stapes – by which vibrations are passed from the tympanic membrane (eardrum) to the inner ear; ▷ pp. 98 (box), *191.*

osteichthyan, another name for BONY FISH.

osteocyte, ▷ BONE.

osteostracan, ⊳ p. 70.

ostracod, any small bean-shaped CRUSTACEAN of the class Ostracoda; they have bivalved shells, and are common in freshwater and marine habitats.

ostrich (*Struthio camelus*), the world's largest bird (family Struthionidae, order Struthioniformes); males stand 2.75 m (9 ft) tall. They have long, bare necks and long, powerful running legs with bare thighs and two-toed feet. They live in the savannah of Africa (extinct in the Middle East); ⊳ pp. *91*, 96, 97 (box).

otolith organ, one of a pair of organs in the vertebrate inner ear by which gravity and movement are detected; ⊳ pp. 190–1.

otter, any of several aquatic MUSTELIDS, typically with streamlined bodies and webbed feet, such as the freshwater Eurasian otter (*Lutra lutra*). The sea otter (*Enhydris lutra*) is the largest mustelid; ⊳ pp. 131, *151*.

otter civet (*Cynogale bennetti*), a brown, otter-like, semi-aquatic VIVERRID with partially webbed feet. It inhabits swampy areas and stream-sides in Southeast Asia; ⊳ p. 123.

otter shrew, the common name given to 3 of the 34 species in the mammalian family Tenrecidae (order Insectivora), the others being the TENRECS. They are essentially aquatic tenrecs, occurring in western and central Africa; ⊳ pp. 120, *121*.

oval window, a small membrane in the vertebrate ear; ⊳ p. 191.

ovary, 1. (in animals) a female organ (gonad), usually paired, in which egg cells or ova (⊳ OVUM) and various hormones (e.g. OESTROGEN) are produced; ⊳ p. 184. **2.** (in flowering plants) a protective structure surrounding the OVULE, within which is the ovum. The ovary is part of the CARPEL, and there are usually several carpels in each flower; ⊳ p. 41.

ovenbird, common name given to a large and diverse group of 230 species of small, usually brown PASSERINE of forest, scrub and grassland of central and South America. They belong to the family Furnariidae, and are closely related to the WOODCREEPERS. Some, such as species of the genus *Upucerthia*, are secretive ground-dwellers (earthcreepers), while another group of 50 species are called spinetails because of their long, pointed tails. The group name comes from the habit shown by many species of building dome-shaped nests (⊳ p. 95). See also HORNERO.

overgrazing, the degradation of an ecosystem by excessive pressure from grazing animals; ⊳ pp. 203, 204, 205, 206.

oviduct, the muscular and ciliated tube through which ova (eggs) pass from an ovary to the uterus, cloaca or exterior. See also FALLOPIAN TUBE.

oviparous, laying (fertilized) eggs (e.g. birds, turtles, crocodiles); ⊳ p. 185.

ovipositor, the egg-laying structure of female insects, at the tip of abdomen. In some insects (bees, ants, etc.) it is modified into a sting; ⊳ p. 59.

ovoviviparous, producing eggs that hatch within the mother's body (e.g. many insects, fishes, snakes and lizards).

ovulation, (in animals) the release of one of more ova (eggs) from the ovary.

ovule, the structure in the flowering-plant ovary (⊳ p. 41) or on scales in gymnosperm (e.g. conifer) cones (⊳ p. 38) in which the ovum – the female egg – forms.

ovum (pl. **ova**) or **egg cell,** the female gamete in sexually reproducing organisms. For references ⊳ REPRODUCTION.

Owen, Richard (1804–94), British comparative anatomist and palaeontologist. By the 1850s Owen had demonstrated that fossil animals do not show a simple linear ascent towards the more advanced members of a group, but rather a radiation outwards of many different lines of specialization – a process known as adaptive radiation (⊳ p. 17). Owen also originated the concept of anatomical homologies, which have provided important evidence for evolution (⊳ p. 14). He is also remembered for his naming of the dinosaurs (from the Greek meaning 'terrible lizards'; ⊳ p. 83), and for his analytical work on the first bird, *Archaeopteryx* (⊳ box, p. 90).

owl, any of about 175 BIRDS OF PREY belonging to the order Strigiformes and occurring worldwide. Owls have large heads, rounded tails, and broad wings, the feathers of which are modified for silent flight. The face is flattened so that the large eyes point forwards to give good binocular vision. The powerful, hooked bill is used together with talons to capture and hold prey. The 17 species of barn owl (family Tytonidae) have bare heart-shaped faces and asymmetrical ears to give stereophonic hearing and so facilitate the location of prey; the tail is shorter than the legs. The 161 species of true or typical owls (family Strigidae) have tails that are longer than the legs, and many have ear tufts, ⊳ pp. 90, *91*, 93, 97, 211.

owlet-nightjar, any of eight small, solitary nocturnal inhabitants of the forests of Australia belonging to the family Aegothelidae, in the same order as the OWLS. They are similar to NIGHTJARS but have smaller, partially hidden bills.

Oxalis, a temperate and tropical genus of around 800 species of annual to perennial herbs in the family Oxalidaceae. Procumbent yellow sorrel (*O. corniculata*) is a common garden weed. Wood sorrel (*O. acetosella*) has white flowers and is common in Eurasian woodlands.

oxlip, ⊳ PRIMULA.

oxpecker or **tick bird,** either of two slender, brown PASSERINE birds in the family Sturnidae, similar to the closely related STARLINGS. They live in association with herds of large African mammals; ⊳ p. 96.

oxygen, a highly reactive element (symbol O), and the most abundant in the earth's crust. In its molecular form (O_2) it is a gas, which composes 21% of the atmosphere. In this form it is essential to the respiration of plants (⊳ pp. 32–3) and animals (⊳ pp. 180–1), and in the energy-producing processes (cell respiration; ⊳ p. 21) of virtually all cells apart from certain bacteria (⊳ p. 25). It is also crucial to all life as a constituent of water (H_2O) and many organic compounds (⊳ p. 6).

oyster, name given to a number of sedentary BIVALVE molluscs, especially those of the family Oestridae. Oysters live firmly attached to hard surfaces such as rocks, mangrove roots or shells, and have been widely used for food and cultivated for that purpose for many centuries. They are also the source of most commercially important pearls; ⊳ p. 56.

oystercatcher, any of 11 WADERS, white birds with brightly coloured bills, legs and eyes, occurring worldwide. The bill has a chisel-like end, used to pry open shellfish; ⊳ p. 96.

oyster mushroom (*Pleurotus ostreatus*), an edible basidiomycete fungus; ⊳ p. 35.

oyster plant, ⊳ SALSIFY.

ozone, a colourless, irritant gas (symbol O_3) with a distinctive smell. It is formed when an electrical spark is passed through oxygen, and also by strong sunlight acting on the fumes from cars. It forms a layer high in the atmosphere that cuts out most of the harmful ultraviolet radiation from the sun; ⊳ p. 225.

paca, either of two large, stocky, herbivorous RODENTS in the family Dasyproctidae. They have very short tails and brown fur spotted with white, and are found in Central and South America.

pacarana (*Dinomys branickii*), a stocky, forest-dwelling, paca-like caviomorph RODENT with small ears, long claws, and brown fur spotted with white (family Dinomyidae). It is found in mountainous areas in eastern South America.

paccinian corpuscle, ⊳ p. 190.

pachycephalosaur, any of a group of thick-skulled herbivorous ORNITHOPOD dinosaurs; ⊳ p. 84.

paddlefish, any primitive RAY-FINNED FISH of the family Polyodontidae, similar to the related STURGEONS but with smooth scaleless bodies and long, paddle-shaped snouts. They feed on plankton in fresh waters in North America and China; ⊳ pp. 70–1.

paddleworm, any POLYCHAETE of the genus *Phyllodoce*, active worms with paddle-shaped swimming appendages, often found under stones on beaches.

paddy (*Oryza sativa*), ⊳ RICE.

pademelon or **scrub wallaby,** any of four wallaby-like MARSUPIALS of the genus *Thylogale* in the KANGAROO family. They are found in Australia, Tasmania and New Guinea.

paedogenesis, reproduction by larval or juvenile forms, or by animals resembling juveniles stages of related species (e.g. the axolotl; ⊳ pp. 78, 81).

painted lady (*Vanessa cardui*), a widespread NYMPHALID butterfly (subfamily Nymphalinae); larvae feed on thistles and nettles.

painted snipe, either of two species of very colourful wader with long toes and a medium-length tail, belonging to the family Rostratulidae (order Ciconiiformes) and occurring in the Neotropics, Old World and Australia. They are secretive birds in which, unusually, the female is brighter. See also SNIPE.

Palaearctic, a faunal region; ⊳ p. 50 (map).

Palaeocene epoch, a geological epoch that lasted from 65 to 53 million years ago; ⊳ p. 4.

palaeontology, the study of extinct life, especially of the evidence provided by fossils (⊳ pp. 4, 12).

Palaeotropical, a floral region; ⊳ p. 30 (map).

Palaeozoic era, a geological era lasting from 570 to 250 million years ago; ⊳ p. 4.

palate, the horizontal partition in the roof of the mouth separating the nasal and oral cavities.

Pallas's cat (*Felis manul*), an orange-grey CAT with black and white head markings, found in mountainous and rocky steppe and woodland across much of central Asia.

palm, any member of the monocotyledonous family Arecaceae (formerly Palmae). Most are trees, but some (e.g. rattans) are climbers; ⊳ p. 45.

Palmae, ⊳ ARECACEAE.

palmate, (of a leaf or antler) lobed like the outspread fingers of a hand (as in the leaf of a horse chestnut).

palm civet, any small civet-like VIVERRID carnivore of the genus *Nandinia* and related genera, typically arboreal and omnivorous (often fruit-eating), found in forested areas of Africa and southern Asia; ⊳ pp. 123, *201*.

palp, a segmented sense organ of touch and taste, attached in pairs to the mouth parts of insects (⊳ p. 59) and crustaceans.

pampas, a type of temperate grassland found in South America; ⊳ pp. 199 (map), **206**.

pampas cat (*Felis colocolo*), a grey-brown CAT with brown spots, found in grassland and forest in South America.

pampas grass (*Cortaderia selloana*), a large ornamental grass native to South America. Related species are to be found in New Zealand.

Panama land bridge, a land bridge that has intermittently connected North and South America; ▷ pp. **9**, 103, 106.

pancreas, a large compound gland near the stomach in vertebrates. It secretes digestive enzymes and bicarbonate into the intestine to help digestion and secretes the hormones insulin and glucagon into the blood to regulate the levels of blood sugar.

panda or **giant panda** (*Ailuropoda melanoleuca*), a large, black-and-white member of the BEAR family, inhabiting bamboo forests in central China; ▷ pp. **129**. See also RED PANDA.

Pangaea, an ancient supercontinent; ▷ pp. 8–9.

pangolin or **scaly anteater,** any mammal of the family Manidae (order Pholidota), comprising seven species distributed in Africa and southern Asia. The ecological equivalents of the New World ANTEATERS, pangolins are specialist ant- and termite-feeders. Their body armour of overlapping horny scales is unique among mammals; ▷ pp. *101*, *107* (box).

pansy, 1. wild pansy or heartsease (*VIOLA tricolor*), an annual or perennial herb found on cultivated and waste ground and short grassland. **2.** garden pansy (*VIOLA* x *wittrockiana*), a hybrid with large showy flowers and with many varieties. It is thought to be derived from *V. tricolor*, *V. lutea* and *V. altaica*.

panther, a black form of LEOPARD; ▷ p. 125.

papaw, papaya or **pawpaw** (*Carica papaya*) a small tropical tree cultivated for its large edible fruits, weighing up to 9 kg (20 lbs). Papain, extracted from unripe fruit, is used as a meat tenderizer.

paper wasp, any of several social WASPS of the family Vespidae, of worldwide distribution. They make nests from chewed wood fragments and saliva to form a strong light 'paper'; ▷ p. 161.

papilla (pl. -ae), **1.** (in animals) a small, blunt projection at the base of a tooth, hair or feather. **2.** (in plants) a similar protuberance, a type of hair.

paprika, ▷ CAPSICUM.

papyrus (*Cyperus papyrus*), a grass-like herb that grows by rivers and lakes in Africa. Papyrus paper was made by pressing the split stems while wet; ▷ p. 217 (box).

parabronchus, in birds, a minute respiratory tube; ▷ 181.

parakeet, any of about 30 small, slender, pointed-tailed PARROTS, mostly living in Africa and South America. Their plumage is smooth, unlike that of other parrots; ▷ p. 213.

parallel evolution, the process by which unrelated organisms evolve in isolation to fill a similar range of ecological niches; ▷ p. 17 (box).

paramammal, a variant name for a MAMMAL-LIKE REPTILE.

Paramecium, a common genus of freshwater ciliate protozoans; ▷ pp. 26–7.

Paraná pine, ▷ ARAUCARIA.

parapodium (pl. -dia), **1.** a jointless lateral appendage occurring in pairs on each segment of polychaete worms; ▷ p. 55. **2.** a swimming organ in some molluscs, formed as a lateral expansion of the foot.

parapsid, any extinct REPTILE of a group defined by a single, upper opening in the temporal region of the skull (▷ p. 82), including ICHTHYOSAURS and PLESIOSAURS. The group is now generally thought to be artificial, comprising modified DIAPSID reptiles.

parasite, ▷ PARASITISM.

parasitism, a close association between different species in which one species (the parasite) benefits from living in or on another species (the host), which is itself harmed by the association; ▷ pp. 14 (box), 162, **168–9**, 178. Other references: fungi 35; plants 47 (box); worms 55; insects 60 (box), 63, 64;

ticks and mites 68–9; fishes 74, 75, 76. See also BLOOD-FEEDING and BROOD PARASITE.

parasitoid, the larva of a parasitic insect that kills a host (also a larval insect) by destroying its soft tissues before it can metamorphose.

parazoan, ▷ METAZOAN.

pardalote, any of four Australasian PASSERINE birds in the family Pardalotidae, brightly coloured birds that nest in underground tunnels and feed almost exclusively on the exudate of leaf-eating insects.

parental care, protective behaviour by one or both parents that serves to increase the chances of the offspring surviving; ▷ pp. 157–8, **185**.
Other references: arachnids 68–9; fishes 74; amphibians 81; reptiles 86; birds 95; mammals 98 (box), 100, 103, 136.

parrot, any of approximately 350 often brightly coloured birds of the order Psittaciformes, with large heads and long tail feathers. Parrots and their relatives have powerful bills for eating fruits and seeds, which they also use to manoeuvre themselves through branches. The upper jaw is hooked over the front of the lower, and is movable on the skull. At the base of the bill there is a cere (a soft, waxy swelling) surrounding the nostrils. The feet have two toes pointing forwards and two backwards, and are suitable for climbing, perching and manipulating food. Most live in the tropics and subtropics of the southern hemisphere, a few occur in India and Central America; ▷ pp. *91*, *93*, 96, *97* (box), 201.

parrotfish, any of 68 marine PERCIFORM fishes of the family Scaridae, found in tropical and warm-temperate seas. They are similar in many respects to the closely related WRASSES, but their jaw teeth are coalesced into blades, giving the mouth the appearance of a beak, which they use to scrape algae from coral (▷ p. 75). Many are among the most brilliantly coloured of all animals.

parsley (*Petroselinum crispum*), a temperate biennial herb from the carrot family, Apiaceae (formerly Umbilliferae), grown for flavouring. The aromatic leaves are rich in vitamin C.

parsnip (*Pastinaca sativa*), a biennial herb of the carrot family, Apiaceae (formerly Umbilliferae), grown in temperate regions for its sweet, edible root.

parson's nose, ▷ PYGOSTYLE.

parthenogenesis, a form of ASEXUAL REPRODUCTION in which unfertilized egg cells (▷ OVUM) develop directly into new individuals; ▷ pp. 27 (protists), 63 (aphids), **184**.

partridge, any of 60 gregarious, plump, short-tailed GAMEBIRDS of the family Phasianidae, living in open country in Africa, the Palaearctic and the Orient; ▷ pp. *91*, *97* (box).

pasqueflower (*Pulsatilla vulgaris*), a European and western Asian perennial herb in the buttercup family, Ranunculaceae. This species, *P. alpina* and *P. vernalis* are grown in gardens.

passerine, the common name for members of the order Passeriformes, the 'perching birds'. Passerines include over 5500 species, nearly 60% of all birds. Passerine feet have three forward-pointing toes, all of the same length, and one that is directed backwards. The leg and foot ligaments are arranged so that as the bird crouches, its grip on a branch is tightened. The order is divided into two suborders. In birds of the suborder Tyranni, sometimes called the 'suboscines', such as TYRANT FLY-CATCHERS, ANTBIRDS and OVENBIRDS, the voicebox or SYRINX is not well developed. The second suborder, Passeri, includes all other passerines (sometimes called the 'oscines'); these birds – the 'songbirds' – have a complex syrinx, and most produce melodious song.

passionflower, any of around 400 species of vines in

the tropical genus *Passiflora*. The name derives from missionaries' interpretation of the flower as symbolic of the Crucifixion. A number are cultivated as ornamentals, e.g. red passion flower (*P. coccinea*), or for their fruit (passion fruit), e.g. giant granadilla (*P. quadrangularis*); ▷ p. *42*.

Pasteur, Henri (1822–95), French scientist, and founder of the science of microbiology. He proposed that disease was caused by microorganisms, developed inoculation techniques, and invented the pasteurization of milk.

patagium, the wing membrane or fold of skin connecting the fore and hind limbs in bats (▷ p. 118) and gliding animals (▷ pp. *134*; 137, box).

patas (*Erythrocebus patas*), an Old World cercopithecine monkey (▷ p. 139), a large, ground-dwelling, omnivorous species found from Senegal to Sudan and Tanzania; ▷ p. 158.

pathogen, any disease-causing agent.

pauropod, any ARTHROPOD of the class Pauropoda, tiny soft-bodied millipede-like animals living in forest litter and soil.

Pavlov, Ivan (1849–1936), Russian physiologist and Nobel prizewinner (1904) for his work on digestive physiology. He is better known today for his studies of classical conditioning in dogs (▷ p. 150).

pawpaw, ▷ PAPAW.

pea, any of a number of leguminous plants in the family Fabaceae with generally round seeds. The garden pea (*Pisum sativum*) has been cultivated since ancient times for its edible seeds. In the sugar pea (*P. sativum* var. *saccharatum*) the pods are also eaten. The pods of the asparagus pea (*Psophocarpus tetragonolobus*) taste of ASPARAGUS. The field pea (*P. sativum* var. *arvense*) is grown as a fodder crop. See also CHICK PEA and SWEET PEA.

peach (*Prunus persica*), a tree of the rose family, Rosaceae, native to China but now widely cultivated in warm temperate regions of the world for its fruit. A number of cultivars have been developed; ▷ pp. 43, 46.

peacock, 1. ▷ PEAFOWL. **2.** (*Inacis io*) a NYMPHALID butterfly (subfamily Nymphalinae), adults of which have large 'eye' marks on the wings.

peafowl or **peacock,** any of three large, pheasant-like GAMEBIRDS of the family Phasianidae. The males of *Pavo cristatus* are the familiar peacocks, originally from India and Sri Lanka; ▷ pp. 17, 94, 96, 97 (box), 159 (box).

peanut, groundnut, goober or **monkey nut** (*Arachis hypogea*), an annual tropical and subtropical herb in the pea family, Fabaceae, cultivated for its fruits, which bury themselves underground. The seeds are rich in oil and protein. They are not true NUTS.

peanut worm, any marine invertebrate of the phylum Sipunculida, rather drab worms found in sand, mud, decaying plants and boring into coral. They have a simple sac-like body with tentacles for deposit-feeding.

pear, any of around 30 species of shrubs and trees in the north temperate Old World genus *Pyrus* of the rose family, Rosaceae. Commercial fruit varieties have been selected from a number of species, e.g. the European pear (*P. communis*).

pearl, a hard accumulation of shell material (mainly calcium carbonate), secreted in layers around a particle of grit, sand, etc. by oysters and mussels; ▷ p. 56.

pearlfish, any of about 24 small fishes of the family Carapidae, occurring in tropical and temperate seas. They have relatively large heads, very short bodies, and long, thin tails tapering to a point. Many live as parasites within the bodies of marine invertebrates, including pearl oysters and sea cucumbers, feeding on the internal organs.

peat, undecomposed organic matter with low mineral content formed largely through low microbial activity in waterlogged sites. Peat builds up as stratified layers and can attain depths of 10 m (33 ft) or more; ▷ pp. 208–9, 210–11, **216–17**, *219*.

pecan (*Carya illinoiensis*), a New World tree in the same genus as HICKORY, cultivated for its edible nuts, which have a high fat content.

peccary, any of three non-ruminating, pig-like ARTI-ODACTYL mammals of the family Tayassuidae, occupying essentially the same kinds of ecological niche in Central and South America as PIGS do in the Old World; ▷ pp. 112, **113**, 233.

pecking order, a linear social HIERARCHY, as in gregarious birds such as chickens; ▷ p. 157 (box).

pectoral, relating to or situated on the chest, breast or thorax.

pectoral girdle, the bony or cartilaginous arch comprising the shoulder blade and collar bone that supports the fore limbs of vertebrates.

pedicel, 1. the stalk of a single flower. Compare PEDUNCLE. **2.** a narrow basal attachment (stalk), either of an animal (e.g. sea anemone) or of an organ (e.g. a crab's eye).

pedipalp, one of a pair of appendages in an arachnid immediately in front of the mouth, often modified for sensory or other functions.

pedology, the study of SOILS.

peduncle, 1. a narrow stalk-like structure by which part or the whole of an organism's body is attached, e.g. the thorax–abdomen connection in insects. **2.** the stalk of an INFLORESCENCE. Compare PEDICEL.

peewit, ▷ LAPWING.

pelagic environment, that part of a marine ecosystem comprising the water itself, as opposed to the bottom (the BENTHIC ENVIRONMENT); ▷ p. 220.

Pelagornis, ▷ p. 90.

pelargonium, any of around 250 species of herbs or subshrubs in the warm temperate to tropical genus *Pelargonium* in the geranium family, Geraniaceae. Many are grown as ornamentals under the name GERANIUM.

pelecypod, another name for BIVALVE.

pelican, any of seven large, heavy water birds of the family Pelecanidae (order Ciconiiformes), occurring worldwide. They have long necks and flattened, hooked bills underslung by an extensible throat pouch. They feed on fish along coasts and inland lakes, and are strong soaring fliers; ▷ pp. 90, *91*.

pellicle, a hard protective outer layer found in some protists, such as euglenoids and *Paramecium*; ▷ p. 27.

pelvis, the basin-shaped skeletal structure of vertebrates to which the hind limbs are attached and which is formed by the pelvic girdle and adjoining bones of the spine.

pelycosaur, any member of an early group of MAMMAL-LIKE REPTILES that flourished some 300 million years ago. Many had sail-like dorsal fins, believed to be involved in regulation of body temperature; ▷ p. 5.

penduline tit, any of 12 PASSERINE birds of the subfamily Remizinae (family Paridae); large TITS with rather square tails, living in the Holarctic but mostly found in Africa. The name is derived from their habit of building hanging purse-like nests at the tips of branches.

penguin, any of 17 species of the family Spheniscidae (order Ciconiiformes); they are distinctively shaped social birds, distributed throughout the cooler waters of the southern oceans. The short, stubby wings are highly efficient underwater flippers, used to chase fish, squid and crustaceans. The emperor (*Apterodytes forsteri*) of Antarctica, standing 1 m (39 in) tall, is the largest, and like the king penguin, it lays a single egg incubated on top

of the feet. The smallest, the fairy penguin (*Eudyptula minor*) of Australia and New Zealand, is only 35 cm (14 in) tall and, like many penguins, lives in burrows; ▷ pp. *17*, 90, *91*, 92, *95*, 97.

penicillin, an antibiotic effective against many bacteria, obtained from ascomycete fungi of the genus *Penicillium*.

penis, ▷ INTROMITTENT ORGAN.

Pennsylvanian period, the later of the two parts into which the Carboniferous period is divided in the USA; ▷ p. 4.

pentadactyl, possessing five digits; ▷ p. 14.

peony, any of around 33 species of perennial herbs and shrubs in the mainly north temperate genus *Paeonia*, family Paeoniaceae. Some are popular ornamentals.

pepper, 1. any of around 1500 species of tropical shrubs, woody climbers and small trees in the genus *Piper* in the family Piperaceae. Black pepper is produced by grinding the whole dried fruit (peppercorn) of the pepper plant (*P. nigrum*). For white pepper the flesh is removed from the fruit before drying. The leaves of betel pepper (*P. betle*) are chewed with the BETEL NUT. **2.** ▷ CAPSICUM. **3.** Jamaica pepper; ▷ ALLSPICE.

peppered moth (*Biston betularia*), a GEOMETRID moth; ▷ p. 16 (box).

peppermint, ▷ MINT.

peptide, any of a group of compounds consisting of two or more amino acids linked in a certain way. See also POLYPEPTIDE.

perception, ▷ SENSES and pp. 188–91.

perch, common name applied somewhat loosely to PERCIFORM fishes of the family Percidae (about 150 species). The family comprises freshwater fishes from the northern hemisphere, although some have been introduced in Australia, New Zealand and South Africa. Like many other perciforms, they have two dorsal fins, the front one spiny, the one nearer the tail soft-rayed; most are small, but a few grow to around 1 m (39 in) and are commercially important. In Europe the name 'perch' usually refers to *Perca fluviatilis*, a lake and river fish whose yellow-green body is marked with vertical black bars.

perching bird, ▷ PASSERINE.

perciform, any fish of the order Perciformes, the 'PERCH-like fishes', by far the largest order of fishes (or indeed of any other vertebrates), comprising between 6000 and 7000 species in around 150 families. They are abundant in fresh waters (roughly 15% of species) and in the sea (85%), particularly in tropical and temperate coastal waters. The perciforms include several important food fishes (e.g. TUNA, MACKEREL), sport fishes (e.g. SWORDFISH, MARLIN), and aquarium fishes. The largest families are the SEA BASSES (about 370 species), the CICHLIDS (700 species), the combtooth BLENNIES (300 species), the WRASSES (500 species), and the GOBIES (about 800 species).

perennation, the ability of certain plants to survive through the winter by means of perennating organs, usually underground food-storage organs such as tap roots (e.g. dandelion), bulbs (daffodil), corms (crocus), and tubers (potato); ▷ p. 44.

pereopod, the segmented thoracic walking leg of a crustacean.

perfect state or **stage, 1.** a flower with both male and female organs. **2.** the phase in a fungal life cycle in which reproduction is sexual. Compare IMPERFECT STAGE.

perianth, the collective term for the petals and sepals (▷ p. 41), i.e. the leaf-like organs of a flower surrounding the sex organs.

pericarp, the part of a fruit developing from the wall of the ovary. Pericarps can vary greatly in size,

texture and composition, e.g. fleshy (peach), woody (nuts), succulent (grape), papery (poppy capsule).

period, the unit of geological time into which eras are divided; ▷ p. 4.

perissodactyl, any hoofed mammal (UNGULATE) of the order Perissodactyla (the odd-toed ungulates), comprising 16 species in 3 families: horses, asses and zebras (family Equidae, 7 species); rhinoceroses (Rhinocerotidae, 5 species); and tapirs (Tapiridae, 4 species). All perissodactyls are strictly herbivorous, and most are adapted to open grassland habitats; ▷ pp. **110–11**.

peristalsis, the succession of rhythmical contractions in the wall of the digestive tract by which food and waste products are moved along. In simple circulatory systems, peristalsis in the walls of blood vessels is responsible for blood flow; ▷ pp. 178, 181.

periwinkle, 1. (also called winkle) any small GASTROPOD of the genus *Littorina*. They are found in huge numbers in the intertidal zone, where they live by grazing tiny algae from rocky surfaces; ▷ p. 221. They can survive for long periods out of water, and several species are evolving towards life on land. Periwinkles are replaced by a similar group, the neritids, in tropical waters. **2.** any of around five species of evergreen creeping shrubs or perennial herbs in the genus *Vinca*.

permafrost, the permanently frozen lower soil layers found at high latitudes; ▷ pp. 210–11.

Permian period, a geological period lasting from 300 to 250 million years ago; ▷ p. 4.

peroxisome, a type of organelle; ▷ p. 20 (box).

persimmon, the edible fruit of species of the genus *Diospyros*, particularly *D. kaki*, grown in subtropical regions. The American persimmon (*D. verginiana*) also yields North American EBONY.

pest, any species that competes with humans for resources, or that otherwise has a negative impact on humans; ▷ pp. 213, **230–1**.

pesticide, any natural or synthetic substance that controls pests. Insecticides are effective against insects, herbicides (weedkillers) against plants, fungicides against fungi; ▷ pp. 213, 227, **230–1**.

petal, any member of the inner whorl of the PERIANTH of a flower, surrounding the sex organs; ▷ p. 41. They are often brightly coloured to attract pollinators; ▷ p. 42.

petiole, a slender stalk, usually of a leaf.

petrel, any of 50 species of oceanic sea bird (family Procellariidae, order Ciconiiformes), with slender wings, webbed feet and short tails. Representatives occur worldwide, breeding on oceanic islands and feeding mostly on fish and squid. Most are small but the southern giant petrel (*Macronectes giganteus*) has a wing span of 2.5 m (over 8 ft) and a very heavily built body; ▷ p. *91*. See also SHEARWATER, PRION and FULMAR.

petunia, any of around 35 species of annual or perennial herbs in the South American genus *Petunia* in the potato family, Solanaceae.

peyote, ▷ MESCAL.

Phaeophyta, a plant division comprising the brown algae; ▷ p. 34.

phage, ▷ BACTERIOPHAGE.

phagocytosis, the process by which amoebas and other protists engulf food; ▷ p. 26.

phalanger or **cuscus,** any of 10 arboreal MARSUPIALS of the genus *Phalanger* (family Phalangeridae), with pointed fore claws, opposable big toes and prehensile tails for climbing. Mostly nocturnal leaf-eaters, phalangers inhabit forests in Australia and New Guinea. See also HONEY POSSUM.

phalarope, any of three small, colourful WADERS, found worldwide. They swim using lobed toes.

pharynx, the cavity forming the upper part of the

alimentary canal, lying behind the mouth and in front of the larynx and oesophagus. In fishes gill slits develop in this area.

phasic response, in sensory perception, ⇨ p. 188.

phasitonic response, in sensory perception, ⇨ p. 188.

phasmid, any EXOPTERYGOTE insect of the order Phasmida, which includes STICK INSECTS and LEAF INSECTS; ⇨ p. *61.*

pheasant, any of 50 large ground-dwelling GAME-BIRDS belonging to the family Phasianidae. All except one African species are native to the forests of Asia. The foot bears a spur. The male and female are very differently coloured, the male usually spectacularly so, with a long tail; ⇨ pp. *91,* 94, 97 (box), 155 (box). See also PEAFOWL.

phenetic classification, a system of classification based solely on the observed or measurable characteristics of organisms (i.e. their phenotypes) as distinct from the genes they possess; ⇨ p. 18.

phenotype, the observable features of an individual organism resulting from the alleles that are expressed (⇨ p. 23). Comparing phenotypes is the basis of phenetic classification (⇨ p. 18). Compare GENOTYPE.

pheromone, a chemical signal released by one animal that has a specific behavioural or physiological effect on another member of the same species; ⇨ pp. **153–4,** 183, 188–9. Other references: insects 62, 63, 156 (box); crustaceans 66; amphibians 80.

Philadelphus, a north temperate and subtropical genus of around 70 species of shrubs. Some species are grown in gardens for their fragrant flowers, e.g. mock orange (*P. coronarius*), misleadingly also called syringa (LILAC).

philodendron, any of around 500 species of usually epiphytic lianas in the tropical American genus *Philodendron* in the arum family, Araceae. Some are grown as foliage plants.

phloem, the part of the vascular tissue of plants responsible for conducting substances manufactured by the plant, such as sugars and amino acids. The conducting cells are the sieve elements, and other 'general purpose' cells are also present. Primary phloem is formed in the region of the apex of roots and shoots; secondary phloem develops from the secondary CAMBIUM, and in some cases forms part of the bark; ⇨ p. 32.

phlox, any of around 65 species of annual and perennial herbs and shrubs in the mainly North American genus *Phlox* in the family Polemoniaceae. A number are grown as ornamentals.

phospholipids, a group of lipids important in cell membranes; ⇨ p. 20 (box).

phosphorus, a non-metallic element (symbol P), a constituent of many important compounds in living organisms, such as co-enzymes, lipids, nucleic acids (RNA and DNA), and those involved in energy metabolism.

photo-, prefix indicating light.

photoperiodism, the ability of plants to detect seasonal changes in daylight hours; ⇨ p. 33.

photophore, a light-emitting organ, especially any of the luminous spots on various marine (mostly deep-water) fishes; ⇨ p. 221 (box). See also BIO-LUMINESCENCE.

photopigment, any molecule that is sensitive to light. The CHLOROPHYLL in green plants absorbs the light that drives photosynthesis (⇨ box, p. 33). Other photopigments allow light energy to be encoded as nerve signals, such as rhodopsin in the retina of vertebrates; ⇨ pp. 96 (box), **189.**

photoreception, sensitivity to light, the basis of the sense of SIGHT; ⇨ pp. 189–90.

photorespiration, a type of apparent respiration in

plants in which oxygen combines with an intermediate of the photosynthetic carbon cycle to give a product that is further oxidized.

photosynthesis, the process by which plants and various bacteria convert simple inorganic molecules into the complex compounds necessary for life; ⇨ pp. 25, **33** (box), 194.

phototrophic bacteria, a group of AUTOTROPHIC bacteria capable of photosynthesis; ⇨ p. 25.

phototropism, the ability of plants to alter growth direction in response to light; ⇨ p. 33.

phylloxeran, any HOMOPTERAN insect (bug) of the family Phylloxeridae. These aphid-like insects are found on deciduous trees, where they often form galls. The grape phylloxera (*Viteus vitifolii*) has caused considerable losses in the wine industry.

phylogeny, the assumed evolutionary history of a species or group, showing its ancestral relations; ⇨ pp. 14, **18** (phylogenetic classification).

phylum, any of several high-ranking classificatory groups into which the animal kingdom is divided. For example, the chordates (all vertebrates plus various obscure animals such as sea squirts) are grouped in the phylum Chordata, while the vertebrates themselves are grouped into the subphylum Vertebrata (⇨ pp. 18, **51**). The equivalent of phylum in the plant kingdom is DIVISION.

physiology, the study of the processes and functions of living organisms; ⇨ pp. 20–1 (cells), 24–7 (single-celled organisms), 31–3 (plants), 174–91 (animals).

-phyta, a standard ending indicating division status in plants (⇨ pp. 30, 45).

phytochrome, a light-absorbing molecule (pigment) in plants; ⇨ p. 33.

phytogeographic region, ⇨ FLORAL REGION.

phytoplankton, ⇨ PLANKTON.

pichiciego, ⇨ ARMADILLO.

pierid butterfly, any LEPIDOPTERAN insect of the family Pieridae, common worldwide. The larvae are sometimes pests of agricultural crops.

pig, any of eight stockily built, non-ruminating ARTIODACTYL mammals of the family Suidae, widespread in the Old World, principally in woodland and forest habitats. Pigs are distinguished by their flexible, muscular snouts, and males by their long, curved tusks. The domestic pig is descended from the WILD BOAR; ⇨ pp. **112–13,** *201,* 207.

pigeon, any of approximately 140 of the larger species belonging to the order Columbiformes, which also includes DOVES. They are found worldwide. Both pigeons and doves feed on seeds and fruits, and have dense, soft plumage. Parent birds produce 'milk' within a specialized crop to feed their chicks; ⇨ pp. *91,* 97 (box), 213, 230.

pigeon hawk, ⇨ MERLIN.

pika, any of 14 species of the mammalian family Ochotonidae (order Lagomorpha). Small relations of the rabbits and hares, pikas are herbivorous burrow- or steppe-dwellers in Asia and North America; ⇨ p. 132.

pike (*Esox lucius*), the northern pike, a predatory freshwater fish of northern circumpolar distribution, related to the SALMONIDS. A large species (up to 1.5 m/5 ft) with dorsal and anal fins positioned close to the tail, it hunts by darting from cover at passing fishes. It is a valued food fish, especially in western Europe; ⇨ p. 75.

pilchard, common name given to several members of the HERRING family, especially *Sardina pilchardus*, an abundant European species. Similar to small herring in appearance, they shoal in the surface waters, and are very important as a food fish. Their flesh is oily and they do not keep well, and although they are eaten fresh locally, they are more often marketed in cans as 'sardines'.

pillbug, another name for WOODLOUSE (especially US).

pilot whale, any of several black toothed whales (CETACEANS) of the genus *Globicephala*, found in all oceans outside the polar regions; ⇨ p. 145.

pimento, ⇨ ALLSPICE, CAPSICUM.

pimiento, ⇨ CAPSICUM.

pimpernel, any of several species of annual or perennial herb in the genera *Lysimachia* and *Anagalis* in the primrose family, Primulaceae, e.g. yellow pimpernel (*L. nemorum*) and scarlet pimpernel (*A. arvensis*).

pine, any of around 96 species of conifer in the northern hemisphere genus *Pinus*, family Pinaceae, such as Scots pine (*Pinus sylvestris*). In the tropics they are to be found mainly at high altitude. They are characterized by having needle-like leaves in clusters usually of two, three or five from short shoots. Several species are important sources of timber (⇨ DEAL). Resin in the timber helps to preserve it and is a source of terpentine and rosin; ⇨ pp. **39,** 205, 208–9. Some other coniferous trees are also referred to as pines; e.g. Chile pine and Paraná pine (⇨ ARAUCARIA) and KAURI PINE. See also SCREW PINE, which is not a conifer.

pineal gland or **organ,** a small GLAND in the brain that secretes the hormone melatonin into the blood, believed to be important in regulating biological rhythms; ⇨ pp. 171, 174, 175.

pineapple (*Ananas comosus*), a tropical American plant of the family Bromeliaceae; ⇨ pp. 43, 46.

pink, any of a number of species in the genera DIANTHUS (principally) in the Caryophyllaceae family, and SILENE, e.g. Cheddar pink (*D. gratianopolitanus*) and California Indian pink (*S. californica*). See also CARNATION, CAMPION and CATCHFLY.

pin mould, any of around 50 species of the widely distributed fungal genus *Mucor*. They give the appearance of a pin cushion with their spores contained in the head of the 'pin'.

pinnate, feather-like; applied to leaves (such as those of acacias) with leaflets in two rows on the stalk.

pinniped, any carnivorous marine mammal of the order Pinnipedia, which comprises 33 species in three families – two families of SEALS and the WALRUS; ⇨ pp. 142, **143.**

pinnule, a subdivision of a leaflet.

Pinus, ⇨ PINE.

pinworm, ⇨ THREADWORM.

pipefish, any of about 200 marine fishes of the subfamily Syngnathinae (family Syngnathidae, which includes the very similar SEA HORSES), from the Atlantic, Indian and Pacific Oceans. Pipefishes have elongated thread-like bodies encased in bony rings, and tubular snouts; they move by undulating their fins; ⇨ p. 77.

pipistrelle, ⇨ VESPERTILIONID BAT.

pipit, any of about 50 PASSERINE birds of the subfamily Motacillinae, in the SPARROW family, closely related to WAGTAILS. They are small, brown-streaked insectivorous birds with long tails flanked with white feathers, occurring worldwide, mostly in Africa. Male and female plumage is alike. Like larks, pipits sing while flying.

piranha, name given to various stocky, deep-bodied CHARACIN fishes of the American tropics, especially those of the genus *Serrasalmus*. These carnivorous freshwater fishes move in large shoals and feed principally on fish, but they will eat larger animals that fall into the water, particularly if they are injured; ⇨ pp. 75–6.

pisci-, prefix indicating fish.

pistachio (*Pistacia vera*), a small tree native to the Near East and west Asia. The fruit is a DRUPE, the stone of which is the pistachio nut, for which the tree is cultivated; ➪ p. 205.

pistil, the female part of the flower, usually comprising two or more fused carpels (➪ pp. 40–1).

pitcher plant, any carnivorous plant belonging to the family Nepenthaceae, Sarraceniaceae and Cephalotaceae; ➪ p. 47 (box).

pith, structural material in the centre of some stems; ➪ p. *32.*

pituitary gland, a small ENDOCRINE gland, attached to the base of the brain, consisting of two lobes (anterior and posterior). It secretes a number of important hormones that affect other endocrine glands and control growth, metabolism, reproductive function, stress responses and water balance; ➪ pp. 171, 175, **183.**

pit viper, any of around 120 species of highly venomous SNAKE of the family Crotalidae, largely confined to the New World. They are closely related to the true VIPERS and have a similar venom-injection system (➪ p. 87), but are distinguished from them by special heat-sensitive organs located in pits on the snout (➪ pp. 88, 165, 191). The group includes RATTLESNAKES, the COTTONMOUTH, the SIDEWINDER and the FER-DE-LANCE.

placenta, in placental mammals, an organ that forms in the uterus during pregnancy, linking the embryo to the maternal blood supply and providing it with oxygen, nutrients and a means of disposing of wastes; ➪ pp. 100, 185, **186–7.**

placental mammals, the dominant group of living mammals (infraclass Eutheria), accounting for all species except the MARSUPIALS and the MONOTREMES; ➪ pp. 100, *101,* 106–45.

placoderm, ➪ p. 70.

placodont, any extinct marine REPTILE of the suborder Placodontia ('plated tooth') of the Triassic period. The placodonts had broad heads and flat crushing teeth for smashing shellfish.

plaice (*Pleuronectes platessa*), a common European FLATFISH (family Pleuronectidae), distinguishable by bright orange spots on a brown background on the side bearing the eyes. It reaches a length of up to 90 cm (35 in), and is a very important food fish. In US and Canadian usage, the name 'plaice' is given to various other species in the same family, especially *Hippoglossoides platessoides.* See also FLOUNDER.

plains wanderer (*Pedionomus torquatus*), a long-legged, quail-like bird belonging to its own family within the order Ciconiiformes. It lives in short grassland of Australasia.

planarian, any turbellarian flatworm of the subclass Tricladida, most of which are aquatic and have a three-branched gut; ➪ p. 55.

plane, any of around 10 species of trees in the genus *Platanus,* family Platanaceae, native to southeast Europe, southwest Asia and North America (where they are known as SYCAMORES). They are characterized by their flaking bark. The hybrid London plane (*P.* x *hybrida*) is thought to have arisen about 1700 and is now a common street tree in both Europe and North America. Buttonwood or American sycamore (*P. occidentalis*) is grown as an ornamental and for its timber in the USA.

planigale, any of five small, mouse-like DASYURID marsupials of the genus *Planigale,* which feed on insects and other invertebrates. They are widespread in Australia, with one species in New Guinea.

plankton, the collective term for the minute aquatic organisms (in either fresh or salt water) that drift with the currents. Phytoplankton is the 'plant' constituent of plankton (in fact mostly unicellular algae), and zooplankton the animal constituent; ➪ pp. 66–7, 170, 195, 197, 215, **218, 220.** Compare NEKTON.

plant, any member of the kingdom Plantae; ➪ pp. 30, 31–47 (biology), 194–221 (role in ecosystems).

Plantae, the plant kingdom; ➪ pp. 30, 31–47.

plantain, 1. any of around 260 species of annual and perennial herbs and shrubs in the temperate genus *Plantago* in the family Plantaginaceae. Several species, e.g. common plantain (*P. major*), are frequent weeds in lawns. **2.** *Musa* x *paradisiaca,* a BANANA cultivar which is starchy rather than sweet and cooked for eating while still green. **3.** See also WATER PLANTAIN.

plantain-eater, ➪ TURACO.

plantain lily, ➪ HOSTA.

plant association, an assemblage of plant species that consistently appears together. It can be defined by certain characteristic species and provides a unit of vegetation that can be used for description and mapping.

plant bug, general term for plant-feeding HEMIPTERAN insects.

plant hopper, general term for LEAFHOPPERS, TREEHOPPERS and FROGHOPPERS.

plant louse, another name for APHID.

plasma, 1. the fluid portion of BLOOD; ➪ p. 180. **2.** the PROTOPLASM within the PLASMAMEMBRANE.

plasmamembrane, another word for cell membrane; ➪ p. 21.

plasmids, small, circular pieces of DNA/RNA found in many bacteria in addition to the main chromosome; ➪ p. 25.

plasmodium (pl. -dia), an amoeboid mass formed by some of the Myxomycetes; ➪ p. 26 (box).

plastid, a type of organelle found in most plant cells; ➪ p. 20 (box).

plastron, the lower part of a turtle's shell, beneath the belly; ➪ p. 85.

plate tectonics, the theory that explains continental drift; ➪ p. 8.

platyhelminth, a FLATWORM; ➪ p. 55.

platypus (*Ornithorhynchus anatinus*), a small, semi-aquatic monotreme mammal of eastern Australia and Tasmania that forages underwater for freshwater invertebrates; ➪ pp. 98 (box), 100, *101,* **102,** 191 (box).

play, an activity most commonly seen in young mammals and birds involving repetition, exaggeration and reordering of fragments of behaviour such as fighting, locomotion and object manipulation, apparently for practice purposes.

Pleistocene epoch, a geological epoch that lasted from 1.6 to 0.01 million years ago; ➪ p. 4.

pleopod, a swimming leg on the abdomen of crustaceans.

plesiosaur, any extinct marine REPTILE of the suborder Plesiosauroidea, which flourished in the Jurassic and Cretaceous periods; ➪ pp. 82, **84.**

pleurodire, any member of the smaller of the two suborders of CHELONIAN reptiles (turtles, etc.), which withdraw the head into the shell by making a sideways bend in the neck. The other suborder is the CRYPTODIRES.

Pliocene epoch, a geological epoch that lasted from 5.3 to 1.6 million years ago; ➪ p. 4.

pliosaur, a short-necked PLESIOSAUR; ➪ p. **84.**

plover, any of 67 species of WADER, occurring worldwide. Plovers are usually small with stout bodies, short bills and long pointed wings. Most are camouflaged against the open ground in which they live. Many are gregarious; ➪ p. 167. See also CROCODILE BIRD and LAPWING.

plum, any of several species of the genus *Prunus* of the rose family, Rosaceae. Many commercial fruit varieties have been derived from the European plum (*P. domestica domestica*) and in North America from *P. americana.* Cultivars of the Japanese plum (*P. salicina*) are grown in frost-free areas such as South Africa; ➪ pp. 43, 46. See also DAMSON and GREENGAGE.

plume moth, LEPIDOPTERAN insect of the family Pterophoridae. They are easily identified by the divided wings, which resemble plumes rather than a single solid structure.

Poa, a genus of grasses, family Poaceae, with around 300 species, commonly known as meadow grasses in Britain and as bluegrasses in North America. Some species are important for lawns and pastures.

Poaceae, a monocotyledonous plant family (formerly Graminae), including grasses, bamboo and reeds; ➪ p. 45.

pochard (*Athya ferina*), a diving DUCK with a long bill and high-domed head, living in the temperate regions of the Palaearctic. The male has a chestnut head and black chest.

pocket gopher, any of some 34 burrowing RODENTS of the family Geomyidae, inhabiting North and Central America. They possess large fore claws and incisor teeth for digging, and fur-lined cheek pouches.

pocket mouse, any small, mouse-like sciuromorph RODENT of the genus *Perognathus* and related genera (family Heteromyidae). They have long, hairy tails and cheek pouches, and are mainly found in arid habitats in North and Central America. The related kangaroo rats (*Dipodomys,* etc.) are found in similar habitats and bound along, propelled by their long hind limbs.

pod, popular name for the fruit (usually elongated and DEHISCENT) of legumes (peas, beans, etc.) and for the similar fruit of other plants such as wallflowers.

podocarp, any member of the coniferous family Podocarpaceae; ➪ p. 39.

podzol, a type of soil profile developed under certain acid vegetation types, such as heather or pine and having a distinct pattern of horizons (➪ PROFILE), the upper being HUMUS, then a bleached layer, followed by a horizon of deposition in which first humus, then iron and aluminium are deposited to form a hard 'pan'; ➪ p. 209.

pogonophoran, any marine invertebrate of the phylum Pogonophora, worm-like animals with long three-part bodies, tentacles but no gut. Most are small and inhabit tubes in deep-water mud, but a recently discovered group of giant forms is associated with areas of deep-sea volcanic activity (➪ p. 7). They feed by taking dissolved organic substances through the body wall or on the waste products of symbiotic bacteria in the body.

poikilothermic or **ectothermic** ('cold-blooded'), relating to an animal whose body temperature varies with and is largely dependent on the environmental temperature. All animals except birds and mammals are poikilothermic or 'poikilotherms'; ➪ pp. 73 (fishes), 79 (amphibians), 87 (reptiles).

poinsettia, ➪ EUPHORBIA.

poison-dart frog, any ANURAN amphibian of the family Dendrobatidae, most of which have skin poisons and bright warning coloration (➪ p. 167). Their name derives from the use of toxic secretions from various species by South American Indians for tipping their blow-pipe darts.

poison ivy or **poison oak,** either of two American species of lianas, shrubs or trees, (*Rhus radicans* and *R. toxicodendron*), related to SUMAC.

pokeweed, any of around 25 species of herbs, shrubs and trees in the tropical and warm temperate genus

Phytolacca. A dye from the berries of Virginian pokeweed (*P. americana*) is used to colour ink and wine.

polar bear (*Thalarctos maritimus*), a very large, mainly omnivorous, white to yellowish-white BEAR, inhabiting coastal regions of Alaska; ⊳ pp. **128–9**.

polar desert or **cold desert**, the areas at very high latitudes where precipitation is extremely low; ⊳ p. 210.

polecat, any of three small, stocky, brownish MUST-ELIDS with elongated bodies and short legs, found in forest or steppe in Eurasia; ⊳ p. *131*. See also FERRET, ZORILLA.

pollack or **pollock** (*Pollachius pollachius*), a relative of the COD found in North Atlantic waters, a food fish.

pollen, the spore-like structures (MICROSPORES) containing the male sex cells in seed-bearing plants. In the gymnosperms (conifers, etc.) the pollen grains are formed in male cones (⊳ p. 38), and in the flowering plants in the anthers (⊳ pp. 40–1). They are taken to the female organs by wind or animal pollinators (⊳ p. 42).

pollen basket, a trough-like structure on the hind leg of a bee, specialized for carrying pollen.

pollen sac, the structure that produces male gametes (sex cells) in seed-bearing plants (the anther in flowering plants); ⊳ pp. 38, 41.

pollen tube, a filamentous structure extending from the germinated pollen grain. It grows to the ovule, serving to bring the male sex cells into the proximity of the ovum (female sex cell); ⊳ pp. 38 (conifers and allies), 40–1 (flowering plants).

pollination, the process by which pollen is brought from the male organs to the female organs of seed-bearing plants; wind and animals (especially insects) are the principal agents of pollination; ⊳ pp. 38, 40–1, **42**, 65, 96, 119.

pollution, the introduction of harmful substances into the environment through human activity; ⊳ pp. 224–7.

polyandry, an unusual mating system, occurring in a few birds, in which a female mates with more than one male; ⊳ p. 158.

polyanthus (*Primula* x *polyantha*), any of a number of cultivars derived from hybridization involving common primrose, cowslip and oxslip (⊳ PRIMULA).

polychaete (or **bristleworm**), any ANNELID worm of the class Polychaeta, which includes PADDLE-WORMS, LUGWORMS, RAGWORMS and FAN WORMS; ⊳ p. 55.

polyembryony, ⊳ p. 63.

polygamy, a mating system in which an individual mates with several members of the opposite sex – usually a single male with several females (polygyny) but occasionally a single female with several males (polyandry); ⊳ pp. 157–8.

polygonum, any of around 260 species of annual or perennial herbs in the widely distributed genus *Polygonum* and the closely related genera *Fallopia* and *Reynoutria* in the family Polygonaceae. Some are grown as ornamentals, e.g. Russian vine (*F. baldschmanica*). Some species are referred to as knotweeds, e.g. Japanese knotweed (*R. japonica*), which is a pernicious weed.

polygyny, a mating system in which a male mates with more than one female, usually leaving her to care for the young alone; the most common system in mammals; ⊳ pp. 157–8.

polymer, any natural or synthetic compound (e.g. STARCH, PROTEIN) made up of a large number of repeated units.

polymorphism, a phenomenom in some animal species in which several structurally and functionally distinct types exist; ⊳ p. 53.

polyp, one of the two life forms of a COELENTERATE, usually consisting of a hollow cylindrical body with tentacles growing around the mouth. A sea anemone is a typical polypoid coelenterate; ⊳ p. 53. See also MEDUSA.

polypeptide, any POLYMER (compound with repeated units) made up of amino acids, e.g. proteins (⊳ box, p. 6).

polyploid, (of a cell) having three or more copies of one, several (aneuploidy) or all chromosomes (euploidy).

polypody, any of various climbing ferns of the genus *Polypodium*, family Polypodiaceae (⊳ box, p. 37).

polysaccharide, any of a class of large carbohydrate molecules, including CELLULOSE, GLYCOGEN and STARCH.

polyzoan, ⊳ BRYOZOAN.

pomegranate (*Punica granatum*), a small tree cultivated in the tropics and subtropics for its edible fruits, which have tough skins surrounding a mass of seeds, each surrounded by juicy pulp.

pond, a small body of fresh water; ⊳ p. 218.

pond-skater or **water strider**, any HETEROPTERAN insect of the family Gerridae, of worldwide distribution. These common insects live on the water surface, feeding on dead insects. Members of one genus, *Halobates*, live on the surface of tropical oceans; ⊳ p. 155.

pondweed, any of around 100 species of aquatic herbs in the widely distributed genus *Potamogeton* with floating and/or submerged leaves; ⊳ p. *219*. The unrelated Canadian pondweed (*Elodea canadensis*) is entirely submerged.

poorwill, any of four American NIGHTJARS. They are well camouflaged and rarely seen, but they are known by their call.

poplar, any of around 35 species of trees in the north temperate genus *Populus*, in the willow family, Salicaceae. The light timber of these fast-growing trees is used for wood pulp, matches and packing cases. Many, such as the ASPENS, cottonwood (*P. deltoides*) and Lombardy poplar (*P. nigra 'Italica'*), are grown as ornamentals.

poppy, any species of annual to perennial herbs in the genera *Papaver* and *Mecanopsis* and any of a number of other plants in the same family Papaveraceae, such as the Californian poppy (*Eschscholzia californica*). Poppy MALLOW (*Callirhoe papaver*) and water poppy (*Hydrocleys nymphoides*) are unrelated. Opium is obtained from the opium poppy (*P. somniferum somniferum*). *P. somniferum hortense* is grown as an ornamental; ⊳ pp. 42, 45.

population, **1.** a group of interbreeding members of a species in a particular location (⊳ pp. 16–17, 194). Numbers may vary according to various complex circumstances; ⊳ pp. 172–3. **2.** the total numbers of a species.

porbeagle, any fast-swimming predatory marine SHARK of the genus *Lamna* (family Isuridae), such as *L. nasus*, of northern waters.

porcupine, any large caviomorph RODENT of the families Erithizontidae (New World porcupines) and Hystricidae (Old World porcupines). All species are covered in stiff protective spines and most are nocturnal. The New World species are largely tree-dwelling; ⊳ pp. 9, **135**, 167, *201*.

porcupine fish, any of 15 marine fishes of the family Diodontidae; they are related to the PUFFERS, and like them are able to inflate their bodies in defence. However, they have long spines, which are either fixed, or lie flat against the body and become erect when the fish is inflated, making them virtually impossible for a predator to seize (⊳ pp. 76–7). Species with fixed spines are often called 'burrfishes'.

porgy or **sea bream**, any of 100 PERCIFORM fishes of the family Sparidae, marine fishes from tropical and temperate seas worldwide. They are numerous and support many local and multinational fisheries.

porpoise, any small toothed whale (CETACEAN) of the family Phocoenidae (32 species), inhabiting coastal waters, especially in the northern hemisphere; ⊳ pp. 142, 144, **145**, 165.

Portuguese man-of-war, a surface-dwelling medusa-like SIPHONOPHORE, a colonial coelenterate with a float at the top and a long, trailing body covered with polyps bearing powerful stinging cells for protection and food capture; ⊳ p. 53.

possum, the common name given to a large number of small, arboreal, herbivorous or omnivorous, nocturnal MARSUPIALS from Australia, Tasmania and New Guinea. These include RINGTAIL POSSUMS, PYGMY POSSUMS, the HONEY POSSUM, and several PHALANGERS and GLIDERS. See also OPOSSUM.

posterior, at the rear of a body or structure.

potassium, a metallic element (symbol K) essential to various organic processes; ⊳ pp. 31 (box), 179.

potato (*SOLANUM tuberosum*), a herb belonging to the family Solanaceae, originating in Central and South America but now widely grown as a staple crop in temperate and subtropical countries. The tubers are not only rich in starch but also contain protein and vitamin C; ⊳ p. 46.

Potentilla, a northern hemisphere genus of around 500 annual to perennial herbs and small shrubs in the rose family, Rosaceae. Silverweed (*P. anserina*) has silvery leaves. A number of species are called cinquefoils; ⊳ p. 211.

potato blight fungus, ⊳ p. 35.

potoroo, another name for RAT KANGAROO.

potter wasp, any solitary WASP of the family Eumenidae, of temperate and tropical distribution. They make small vase-shaped nests out of mud (⊳ p. 161).

potto or **kinkajou** (*Perodicticus potto*), a slow-moving, nocturnal lower PRIMATE (family Lorisidae) inhabiting forested habitats in Africa. The related ANGWANTIBO is sometimes referred to as the golden potto; ⊳ p. 137. See also KINKAJOU (1).

prairie, a type of temperate grassland found in North America; ⊳ pp. 199 (map), **206**.

prairie chicken, either of two grouse-like GAME-BIRDS of the North American grasslands, belonging to the family Phasianidae; ⊳ p. 94.

prairie dog, any highly social sciuromorph RODENT of the genus *Cynomys* (family Sciuridae), such as the black-tailed prairie dog (*C. ludovicianus*), that lives in extensive burrow systems in the North American grasslands (prairies); ⊳ pp. 134, **161**, 206.

prairie wolf, another name for the COYOTE.

pratincole, any of nine species of Old World and Australasian plover-like birds belonging to the family Glareolidae (order Ciconiiformes), closely related to the COURSERS. They have long tails and short bills.

prawn, name given to numerous DECAPOD crustaceans, especially those of the genus *Palaemon* (etc.), but the distinction between a prawn and a SHRIMP is inexact. Prawns typically have flattened bodies and two pairs of pincers; ⊳ p. 67.

praying mantid or **mantis** ⊳ MANTID.

Precambrian era, the earliest geological era, lasting from the formation of the earth 4600 million years ago to 570 million years ago; ⊳ p. 4.

predation, behaviour that involves one species (the predator) killing and eating a member of another species (the prey) as a means of obtaining food (FORAGING); ⊳ pp. 162–3, **164–5**, 166–7.

predator, ⊳ PREDATION.

preen gland or **uropygial gland**, (in birds) a GLAND

at the base of the tail feathers, secreting an oily fluid that is spread with the beak over the feathers to condition and waterproof them; ▷ p. 93.

preening, (in birds) grooming behaviour involving the use of the bill to clean and arrange the feathers and often to distribute oil from a posterior PREEN GLAND; ▷ p. 93.

pregnancy, ▷ GESTATION.

prehensile, (of a structure such as a limb or tail) able to grasp. One of the distinctions between Old World and New World monkeys is that many of the latter have prehensile tails (▷ pp. 138–9).

premolar, in many mammals, any of the eight two-cusped teeth (two pairs in each jaw) lying between the incisors and the first molars.

prey, ▷ PREDATION.

prickly pear, any of a number of species of CACTUS in the genus *Opuntia*. Some species, such as Indian fig (*O. ficus-indica*), are cultivated for their edible fruit.

primary, in birds, any one of the outer flight feathers; ▷ pp. 90 (box), *92*.

primary productivity, the gain in weight by autotrophs (principally vegetation) in an ecosystem over a given period of time, so giving a measure of the input of energy into that ecosystem. It is often given in kilograms per square metre per annum; ▷ pp. **194**, *195*. For the primary productivity of different terrestrial ecosystems, ▷ pp. 200–10. See also BIOMASS.

primary reproductive, ▷ REPRODUCTIVE.

primate, any mammal of the order Primates, which comprises some 180 species and includes the Old World lemurs and lorises (lower primates) and monkeys, apes and man (higher primates). Primates are typically versatile, intelligent and social animals; most species are tree-dwelling and herbivorous, but there is considerable variation in both habitat and diet; ▷ pp. **136–41**, 154 (box).

primitive streak, ▷ p. *187*.

primrose, ▷ PRIMULA.

primula, any of around 500 species of perennial herbs in the mainly north temperate genus *Primula* in the family Primulaceae. Common primrose (*P. vulgaris*) has flowers on individual stems, but COWSLIP (*P. veris*) and oxslip (*P. elatior*), both also known as paigle, have umbels of flowers at the apex of a leafless stalk. Many species and varieties are cultivated, including POLYANTHUS.

prion, any of six species of small southern-hemisphere PETRELS. The Antarctic prion (*Pachyptila desolata*) scoops up fishes by hydroplaning, using its breast as a keel and its feet to drive it along.

privet, any of around 50 species of shrubs and trees in the genus *Ligustrum* in the olive family, Oleaceae. They are native to Europe, through Asia and Malaysia to Australia. Common privet (*L. vulgare*) is used for hedges.

problem-solving, ▷ INTELLIGENCE.

proboscidean, any trunked mammal of the order Proboscidea, a formerly successful group that is today represented by just two species of elephant; ▷ pp. *108–9*.

proboscis monkey (*Nasalis larvatus*), a large Old World colobine monkey of Borneo, closely related to the LANGURS, remarkable for its long, bulbous nose; ▷ pp. *139*, 217.

procambium, the cells of a primary meristem giving rise to the primary vascular tissue of plants (▷ p. 32).

procyonid, any CARNIVORE of the family Procyonidae, comprising 16 species and including raccoons, the coati, the olingo, the kinkajou, the ringtail, and the red panda. They are generally small, omnivorous and ground-dwelling, and (except for the red panda) found only in the New World; ▷ pp. 122, **130**.

producer, ▷ AUTOTROPH.

productivity, ▷ PRIMARY PRODUCTIVITY.

profile, a section through a soil observed by digging a soil pit. Its description entails the recording of the distinct layers (horizons) that may be visible.

profundal zone, the lower zone in a freshwater ecosystem; ▷ p. *218*.

prokaryote, any unicellular organism whose cells are prokaryotic, i.e. in which there is no nucleus. Bacteria are the principal group of prokaryotes; ▷ pp. 7, **20**, 24–5. Compare EUKARYOTE.

prolactin, in vertebrates, a HORMONE secreted by the PITUITARY GLAND, associated in mammals with milk production; ▷ p. 183.

prominent, any LEPIDOPTERAN insect (moth) of the family Notodontidae, characterized by tufts of scales on the hind edge of the fore wings, which stand up when the moth is at rest. The group includes the puss moth (*Cerura vinula*), whose larvae are bright green with a red head.

pronghorn or **American antelope** (*Antilocapra americana*), an antelope-like ruminating ARTIODACTYL mammal of North America, the ecological equivalent of the Old World antelopes; ▷ pp. **117**, *149*, 206.

prosimian, a lower primate (lorises, lemurs, etc.); ▷ p. 136.

protea, any of around 100 species of evergreen shrubs and trees in the mainly South African genus *Protea*. They have colourful flower heads with decorative bracts; ▷ p. 205.

protein, any of a large number of POLYMERS made up of POLYPEPTIDE chains of AMINO ACIDS; ▷ pp. **6** (box), 20, 31, *179*.

prothorax, the first segment of an insect's THORAX, bearing the first pair of walking legs.

protist, any unicellular eukaryote organism. Protists are grouped into the kingdom Protista; ▷ pp. 7, **26–7**.

protochordate, ▷ CHORDATE and p. 51 (box).

protoplasm, the jelly-like material of a cell within the membrane. In eukaryote cells it is differentiated into the nucleoplasm (in the nucleus) and the CYTOPLASM; ▷ p. 20.

protostome, ▷ p. 51 (box).

prototherians, the MONOTREMES; ▷ pp. 100–1.

protozoan or **protozoon** (pl. -zoa), any animal-like unicellular organism. Some authorities restrict the term to eukaryotic animal-like protists, while others group these with animal-like prokaryotes in the phylum Protozoa; ▷ pp. 7, **26–7**. Compare METAZOAN.

proturan or **telsontail,** any APTERYGOTE insect of the order Protura, a worldwide group of very small, obscure insects living in soil, leaf litter and similar habitats, where they extract fluid from fungal hyphae by means of piercing mouthparts. They have no eyes or antennae, but the first pair of legs are elongated and held in front of the body as feelers. Unlike any other insect, they grow by adding extra abdominal segments at each moult until mature; ▷ pp. 60–1.

prune, a dried PLUM of a variety with a high sugar content. The plums are usually dried after falling from the tree.

Prunus, a genus of the family Rosaceae, including PLUMS, PEACHES, APRICOTS and CHERRIES.

psalterium, ▷ OMASUM.

pseudocarp, a fruit-like structure incorporating tissues other than the ovary (from which the true fruits develop). A strawberry 'fruit' is an example of a pseudocarp, being predominantly RECEPTACLE tissue.

pseudopodium (pl. -dia), literally 'false foot', a temporary extension caused by the protoplasm of a unicellular organism or phagocytic cell that serves to move the cell or take in food; ▷ p. 27.

pseudoscorpion, any tiny scorpion-like predatory ARACHNID of the order Pseudoscorpiones. Found abundantly in leaf litter, they are distinguishable by their size and by the absence of the long abdomen and sting.

Pseudotsuga, ▷ DOUGLAS FIR.

psilophyte, any member of the division Psilophyta, comprising the whisk ferns; ▷ pp. 30, **37**.

psocopteran, ▷ BOOKLOUSE.

ptarmigan, any of three GROUSE of the genus *Lagopus*, especially *L. mutus*, a bird of the Arctic and northern-temperate zone, with plumage mottled brown-black in summer, and white in winter; ▷ p. 67 (box).

pteridophyte, any member of the plant division Pteridophyta, the ferns; ▷ pp. 30, **36–7**.

pterodactyl, ▷ PTEROSAUR.

pterosaur, any of a group of flying DIAPSID reptiles that emerged towards the end of the Triassic period, 205 million years ago, and became extinct at the end of the Cretaceous. The group includes the pterodactyls; ▷ pp. 82, 84, *176*.

pterygote, any INSECT of the group Pterygota, the 'winged insects', accounting for the vast majority of living insect species. The group is subdivided into the EXOPTERYGOTES (16 orders), such as grasshoppers, dragonflies and aphids, which undergo incomplete METAMORPHOSIS; and the ENDOPTERYGOTES (9 orders), such as beetles, butterflies and bees, which undergo complete metamorphosis. Although belonging to fewer orders, the endopterygotes are far more numerous in terms of species; ▷ pp. 60–1, 62–3.

pudu, either of two DEER of the genus *Pudu*, with simple, unbranched antlers. The smallest members of the deer family, they inhabit dense forest in the Andean region of South America.

puffball, a basidiomycete fungus, e.g. *Lycoperdon perlatum*, in which the spores are contained in a papery thin 'ball' and are released in clouds through an opening by a puffing action when the fungus is hit by a rain drop or passing animal; ▷ p. *35*.

puffbird, any of 33 small, large-headed, drab-coloured birds living in the middle tree levels of South American rainforests (family Bucconidae, order Galbuliformes). These insectivorous birds are so called because of their fluffy plumage; ▷ p. *91*.

puffer, any of about 120 species in the family Tetraodontidae (order Tetraodontiformes), mainly marine or brackish-water fishes found in the tropical and subtropical parts of the Atlantic, Indian and Pacific Oceans. They can inflate their body cavity with air or water so as to become virtually spherical, and in many species this causes small spines in the skin to project radially outwards. There are four large teeth in the jaws (two upper, two lower), which are united to form a sharp beak. The flesh of many species is highly toxic, causing paralysis or death when eaten, but in Japan (where they are known as 'fugu') they are considered a delicacy and eaten after special preparation; ▷ pp. 73 (box), 76, **77**.

puffin, any of three small AUKS, with large heads and very deep, brightly coloured bills. They are found in the Holarctic; ▷ p. 96.

pug, name given to various GEOMETRID moths.

pulse, 1. (in animals) the regular expansion and contraction of an artery synchronized with each heart beat. **2.** a common name for the fruit of LEGUMES (e.g. peas, beans).

puma, cougar or **mountain lion** (*Felis concolor*), a widely distributed New World CAT, with a plain, tawny to greyish-brown coat and a long tail; ▷ pp. **125**, 205.

pumpkin, any of a number of cultivars of species of the genus *Cucurbita* such as *C. pepo* (MARROW) and *C. moschata*. The fruits are generally rounded rather than cylindrical and are used in pies and for animal feed.

punctuated equilibrium, a theory first propounded in 1971 by Stephen Jay Gould and Niles Eldredge concerning the pace of evolution. They proposed that natural selection acts on species as a whole rather than on individuals, thus leading to long periods of equilibrium (stasis) punctuated by a series of jumps (saltations) during which change is rapid. The fossil evidence available at the time supported this interpretation, but since then further work on the fossil record has lent support to the currently accepted theory of punctuated gradualism, which states that evolutionary change is gradual, but at certain times the rate of change speeds up rapidly; however, this change is still brought about by the normal process (as described by Darwin) of selection on individuals (▷ pp. 13, 16–17).

punctuated gradualism, ▷ PUNCTUATED EQUILIBRIUM.

pupa, the intermediate, usually inactive, form of an insect between the larval and imago (adult) stages; ▷ pp. 60, *63*, 65.

pupil, the central, usually circular aperture in the iris of the vertebrate eye, the size of which dictates the amount of light entering the eye; ▷ p. 189.

puss moth, ▷ PROMINENT.

pygmy chimpanzee, ▷ CHIMPANZEE.

pygmy possum, any of seven small, arboreal, forest-dwelling MARSUPIALS of the family Burramyidae, found in Australia, Tasmania and New Guinea. They have prehensile tails and opposable big toes. See also FEATHERTAIL POSSUM.

pygostyle ('parson's nose'), in birds, the remnant of the tail, from which the tail feathers emanate; ▷ pp. 90 (box), *92*.

Pyracantha, a mainly Asian and European genus of around six species of ornamental thorny evergreen shrubs in the rose family, Rosaceae.

pyralid moth, any LEPIDOPTERAN insect of the widely distributed family Pyralidae. Their larvae feed on dried materials, and several species are pests (e.g. *Ephestia* species feed on flour, *Plodia* species on dried fruits and nuts).

pyrethrum, ▷ CHRYSANTHEMUM.

python, any non-venomous constricting snake (CONSTRICTOR) of the subfamily Pythoninae (family Boidae), largely confined to the Old World tropics. The reticulated python (*Python reticulatus*) is reputedly the longest of all snakes, one specimen measuring 10 m (33 ft); ▷ pp. **88**, **89** and BOA.

quadrupedal, moving on four limbs.

quagga, a recently extinct species of ZEBRA; ▷ p. 111.

quail, any of 40 small, rounded GAMEBIRDS with short legs, belonging to the families Phasianidae and Odontophoridae (the New World quails). They occur worldwide but most live in Africa and Asia.

Quaternary period, the most recent geological period, lasting from 1.6 million years ago to the present; ▷ p. 4.

queen, ▷ REPRODUCTIVE.

quelea, any of three small WEAVERS. They live in vast colonies in the African bushland and are one of the most serious grain pests; ▷ pp. 97 (box), *230*.

Quercus, ▷ OAK.

quetzel (*Pharomachrus mocinno*), a beautiful green, white and red TROGON, with very long tail streamers. It was considered sacred by the Aztecs.

quill, the shaft of a feather (▷ p. 93), or the hollow spine of a porcupine (▷ p. 135).

quillwort, any species of the lycophyte family Isoetaceae, the leaves resembling quills; ▷ p. 37.

quince, 1. common quince (*Cydonia oblonga*), a tree in the rose family, Rosaceae, with pear-shaped fruits used in preserves and to flavour apples. **2.** japonica, any of 3 species of spring-flowering shrubs in the genus *Chaenomeles* in the rose family, Rosaceae. The fruits are used in preserves.

quinine, an alkaloid extracted from the bark of trees and shrubs of the South American genus *Cinchona*. It has various medicinal uses, especially in the treatment of malaria, and is also used to flavour tonic water.

quokka (*Setonix brachyurus*), a kangaroo-like MARSUPIAL in the same family as the KANGAROO. It is relatively small with shorter hind legs and tail, and smaller ears than the true kangaroos, and is restricted to Western Australia.

quoll or **tiger cat,** the name given to the 'native cats' of Australia and Tasmania. Quolls are amongst the largest living DASYURID marsupial carnivores; ▷ p. 105.

rabbit, common name given to those species in the family Leporidae (which also includes HARES) that live socially in burrows and give birth to undeveloped young. The distinction is not strictly applied, however – the American jack rabbits are in fact hares, and some so-called 'hares' are really rabbits; ▷ pp. 17, **132**, *183*, *195*. See also COTTONTAIL.

rabbit-eared bandicoot, greater bilby or **dalgyte** (*Macrotis lagotis*), a rare desert-dwelling, burrowing Australian MARSUPIAL of the family Thylacomyidae, with exceptionally long ears.

rabbitfish, any of 25 marine PERCIFORM fishes of the family Siganidae, from the Indo-Pacific and eastern Mediterranean. They have blunt snouts and enlarged upper lips, vaguely evocative of those of a rabbit; their sharp fin spines are venomous. The name 'rabbitfish' is also sometimes applied to the CHIMAERAS.

raccoon, any New World PROCYONID mammal of the genus *Procyon*, such as the common raccoon (*P. lotor*); ▷ pp. *130*, *213*.

raccoon dog (*Nyctereutes procyonoides*), a stocky, short-legged CANID inhabiting woodland in far eastern Asia. It is raccoon-like in appearance and omnivorous.

race, ▷ SUBSPECIES.

raceme, an INFLORESCENCE in which flowers are borne on an elongated axis, the younger uppermost, the older below, as in the lupin.

racerunner (*Cnemidophorus lemniscatus*), an extremely fast-moving TEIID lizard, slenderly built with a long tail.

radiocarbon dating, the use of radioactive carbon-14 to date material; ▷ p. 4.

radiolarian, any of a group of marine protozoa; ▷ p. 160.

radish (*Raphanus sativus*), a root crop related to the true BRASSICAS grown throughout the world. The pungent roots are used in salads.

radula, a horny tongue-like feeding organ found in molluscs; ▷ pp. 56–7.

raffia, a soft fibre obtained from the young leaflets of palms in the genus *Raphia*. It may be dyed and used as a weaving material.

Rafflesia, a genus of tropical plants, including the stinking-corpse lily; ▷ pp. *30*, *42*, *47* (box).

ragged robin or **cuckooflower** (*Lychnis flos-cuculi*), a perennial herb of the pink family, Caryophyllaceae, that grows in damp places. It has divided pink petals and is sometimes cultivated. See also CATCHFLY.

ragweed, ▷ AMBROSIA.

ragworm, any POLYCHAETE of the genus *Nereis*, marine worms with flattened bodies and fleshy limbs that live in burrows on sandy shores.

ragwort, any of a number of species of annual to perennial herbs in the genus *SENECIO*, such as ragwort (*S. jacobaea*), which is poisonous to grazing animals, and Oxford ragwort (*S. squalidus*).

rail, any of over 50 slender, dull-coloured birds with long bills, small bodies and long legs and toes, belonging to the family Rallidae (order Gruiformes), found worldwide in marshy habitats. They are often found on islands, where many have become flightless and some are now close to extinction; ▷ pp. *91*, 97 (box). See also CRAKE.

rainforest, forest where precipitation is high, i.e. more than 1500 mm (60 in) per year; ▷ pp. 199 (map), **200–1** (tropical), **207** (temperate).

raised bog, a peat-forming ecosystem whose mass of peat accumulation is so considerable that its surface is raised above the influence of ground water and is dependent entirely on precipitation for water supply; ▷ pp. 216, *219*.

raisin, a dried grape.

ramsons or **wild garlic** (*Allium ursinum*), a north temperate Old World bulbous herb in the family Alliaceae. The crushed leaves smell of GARLIC.

range, the geographical area, or the range of environmental conditions, within which a species of plant or animal normally occurs. The term is generally used with the geographical significance, more or less interchangeably with 'distribution'.

rank, 1. a position within a social HIERARCHY. **2.** a position within a taxonomical hierarchy (▷ TAXON).

Ranunculaceae, a dicotyledonous plant family, including buttercups, clematis, anemones, etc.; ▷ p. 45.

Ranunculus, a widely distributed genus of around 400 species of annual and perennial herbs in the buttercup family, Ranunculaceae. Examples are the meadow buttercup (*R. acris*), lesser CELANDINE and common water crowfoot (*R. aquatilis*).

rape, either of two BRASSICAS (*B. napus* and *B. campestris*), grown as an oilseed crop to produce oil for human consumption and rapeseed meal for animal feed.

raptor, another name for BIRD OF PREY.

raspberry, any of several of the species of the genus *Rubus* in the rose family, Rosaceae, particularly *R. idaeus*. The raspberry produces biennial prickly canes, which fruit in their second year.

rat, common name given to numerous myomorph RODENTS of the large family Muridae (the 'mouse' family), the difference between a rat and a mouse (as such) being largely a matter of size. Unqualified, the name 'rat' usually refers to species of the genus *Rattus*, especially the brown or Norway rat (*R. norvegicus*) and the black, ship or roof rat (*R. rattus*), both originally native to Asia but now pests on a worldwide basis (▷ box, p. 134). Many caviomorph rodents, more or less rat-like in appearance, are also known as rats; ▷ pp. 133, 134 (box), **135**, 150–1, 153, 154 (box), *162* (box), 213. See also CANE RAT, CHINCHILLA RAT, MOLE RAT, MUSKRAT, ROCK RAT, SPINY RAT, WATER RAT.

ratel or **honey badger** (*Mellivora capensis*), a fierce, stocky MUSTELID, resembling a small badger, with white, grey and black markings. It is found in Africa, Arabia, central and southern Asia. Its alternative name comes from its fondness for honey; ▷ p. 169.

ratite, the collective name given to 10 species of flightless, running birds without a well-developed keelbone or flight feathers; ▷ pp. 92, 97 (box). See also MOA, OSTRICH, RHEA, EMU, CASSOWARY and KIWI.

rat kangaroo or **potoroo,** any of several small kangaroo-like MARSUPIALS of the family Potoroidae which are widespread in Australia. They have generalized herbivorous diets and nocturnal habits.

rat opossum or **shrew opossum**, any of seven small, insectivorous, South American MARSUPIALS of the family Caenolestidae. They have elongated snouts, small eyes, rat-like tails and no pouch.

rat-tail, ⇨ GRENADIER.

rattan, any species of around 10 genera of climbing palms that are important for their flexible stems, which are used for CANE baskets and wickerwork. Malacca cane is the stem of some species of *Calamus*; ⇨ p. 45.

rattlesnake, name given to several species of New World PIT VIPER, so called because of the warning sound they produce when threatened; ⇨ pp. **89**, 167.

raven, any of six carrion-feeding CROWS of the New and Old Worlds. The Holarctic raven (*Corvus corax*) grows to 65 cm (26 in).

ray, 1. any of the supporting structures (spines) of a fish's FINS. **2.** any of about 350 marine CARTILAGINOUS FISHES of the subclass Selachii or Elasmobranchii, which also includes the SHARKS. Rays are vertically flattened, with the sides of the body expanded into large pectoral 'wings', which are undulated to provide propulsion. They are mainly bottom-dwelling. Their teeth are united into crushing plates, and many feed on crustaceans. Some lay eggs in horny capsules, but most give birth to live young; ⇨ pp. 73, 74, *75*. See also MANTA, SKATE, STINGRAY and TORPEDO.

ray-finned fish or **actinopterygian**, any BONY FISH of the subclass Actinopterygii, the dominant group of living fishes. Apart from a few primitive orders (⇨ STURGEON, PADDLEFISH, GAR, BOWFIN, BICHIR), all modern ray-finned fishes belong to the subdivision (or infraclass) Teleostei, the TELEOSTS; ⇨ pp. 70–1.

Ray, John (1628–1705), English naturalist, who developed an early system of biological classification; ⇨ p. 18.

razorbill (*Alca torda*), a medium-sized black and white AUK, with a deep but narrow bill (hence its common name); ⇨ p. 97 (box).

Recent epoch, another name for the HOLOCENE EPOCH.

receptacle, the base of a flower to which the sepals, petals, stamens and carpels are attached; ⇨ pp. 40, *41*.

recessive, (of an allele) only expressed when paired with another recessive allele rather than with a DOMINANT allele; ⇨ p. 23.

recombination, the rearrangement of genes that occurs during meiosis; ⇨ p. 23.

rectum, the final, short segment of the intestine (between the colon and the anus in vertebrates) where faeces are stored prior to defecation; ⇨ pp. *178*, 179.

red algae, the Rhodophyta, a division of seaweeds; ⇨ pp. 34, 221.

redcurrant (*Ribes rubrum*), a small shrub like the related BLACKCURRANT, but with red sharp-flavoured berries.

red dog, ⇨ DHOLE.

red fox (*Vulpes vulpes*), a CANID widely distributed in Eurasia and North America. It is usually reddish-brown with a long, bushy tail and black stockings; ⇨ p. 127.

red hot poker, any of around 65 species of perennial herbs in the African and Madagascan genus *Kniphofia* in the lily family, Liliaceae.

red mullet, ⇨ GOATFISH.

red panda, lesser panda or **cat bear** (*Ailurus fulgens*), a largely herbivorous, semi-arboreal PROCYONID, the only procyonid to occur naturally outside the New World; ⇨ p. **130** (box).

red pepper, ⇨ CAPSICUM.

redpoll (*Acanthis flammea*), a Holarctic species of FINCH, preferring conifer woodlands as well as stands of alders and birches.

redshank, either of two plump-bodied WADERS of the family Scolopacidae, with bright red legs and a red base to the bill. They inhabit marshes in the Palaearctic; ⇨ p. 96.

red spider (mite), ⇨ SPIDER MITE.

redstart, any of 13 small slim THRUSHES of the genus *Phoenicurus*, with constantly flickering red tails; they live in the Palaearctic and the Orient. The unrelated American redstart is a WOOD WARBLER.

red wolf, ⇨ WOLF.

redwood, 1. coast redwood (*Sequoia sempervirens*), a coniferous tree native to the 'fog belt' on the Californian coast. It may grow to 110 m (361 ft) or more; ⇨ pp. 30, 228. **2.** Sierra or giant redwood (*Sequoiadendron giganteum*), a coniferous tree native to the western slopes of the Sierra Nevada, California. It may grow to 100 m (328 ft) or more; ⇨ p. 30, 228. **3.** dawn redwood (*Metasequoia glyptostroboides*), a coniferous tree native to China.

reed, any of a number of grasses of wet places such as reed-grass (*Phalaris arundinacea*) and species of the widely distributed genus *Phragmites*. Norfolk Reed (*Phragmites communis*) is much prized for thatching, as a well-maintained roof of this reed may last 80 years; ⇨ p. *219*.

reedbed, a swamp dominated by one or a few species of emergent aquatic plant, most typically by the reed, but also sometimes by bulrush or some sedges. It forms an important stage in the succession from open water; ⇨ p. 219.

reedbuck, any of three grey or brown ANTELOPES of the genus *Redunca*, inhabiting savannah, reedbeds and montane grassland in Africa.

reedfish (*Calamoichthys calabaricus*), a primitive African freshwater RAY-FINNED FISH, closely related to the BICHIRS but with a longer, more eel-like body and no pelvic fins; ⇨ pp. 70–1.

reedmace (*Typha latifolia*), a grass-like marsh plant belonging to the family Typhaceae, with a distinctive compact cylindrical flower head. Although not strictly a rush it is popularly and confusingly referred to as BULRUSH (2).

reef, ⇨ CORAL REEF.

refection or **coprophagy** the practice of eating the faeces after their first passage through the gut in order to extract extra nutrition; ⇨ pp. 132, 133.

reforestation, the re-establishment of forest vegetation in an area once bearing a tree cover but subsequently cleared by human activity or natural disaster.

regeneration, in plants and animals, the process of replacing lost or damaged parts (organs, tissues or cells); ⇨ pp. 54, 55, 89, 184, 187.

reindeer, the name commonly used for the CARIBOU in the Eurasian part of its range.

reindeer moss, any of several species of arctic lichen, such as *Cladonia alpestris* and *C. rangiferina*, on which reindeer and caribou feed.

reinforcement, the perpetuation or strengthening of a behavioural response by reward or punishment; ⇨ p. 150.

releasing stimulus or **releaser** ⇨ SIGN STIMULUS.

relict, a species, or group of species, now occupying only a small remnant of a former widespread distribution, often as a result of climatic changes, such as glaciation and then deglaciation (⇨ pp. 10–11).

remex (pl. remiges), one of the large primary or secondary feathers on a bird's wing (⇨ p. *92*).

remora or **sharksucker**, any of eight marine PERCIFORM fishes of the family Echeneidae, widely distributed in the Atlantic, Indian and Pacific Oceans. Remoras have elongated bodies and flat heads, on the top of which there is a large suction disc with which they attach themselves to sharks, larger bony fishes, sea turtles and marine mammals; ⇨ p. 76.

renewable resource, raw material required by mankind that is constantly being regenerated by natural processes and that, with careful management, could be sustained, e.g. timber production, fisheries.

replication, the doubling of the DNA in a cell prior to the cell dividing; ⇨ p. 22.

reproduction, the process by which an organism produces one or more offspring essentially similar to itself. In both plants and animals, reproduction may be ASEXUAL or SEXUAL. For the latter, a number of specialized organs are required, which together make up the individual's REPRODUCTIVE SYSTEM. Text references: **133**, **184–7**; genetics 22; viruses 24; bacteria 25; protists 27; vegetative reproduction 33; algae 34; fungi 35; spore-bearing plants 36; conifers, etc. 38; flowering plants 40–1; sponges 52; coelenterates 53; echinoderms 54; worms 55; molluscs 56–7; insects 62–3; crustaceans 66; arachnids 68–9; fishes 74–5; amphibians 80–1; reptiles 86; birds 94–5; mammals 100; monotremes 102; marsupials 103; armadillos 107; elephants 108; pigs 113; deer 116; insectivores 120–1; rabbits 132; rodents 135; primates 136; seals, etc. 143; whales 144.

reproductive, in social insects, a fertile male or female. In social bees, wasps and ants, there is only a female reproductive (queen), while in termites there is a king and a queen. The founding king and queen are known as the primary reproductives. For references ⇨ CASTE.

reproductive system, the group of associated organs (genitals or genitalia in animals, SPORANGIA in plants) by means of which an individual is capable of REPRODUCTION.

reptile, any member of the vertebrate class Reptilia, whose living representatives are classified in four orders: Crocodilia (crocodiles, alligators and the gavial), CHELONIA (turtles, tortoises and terrapins), Squamata (SNAKES, LIZARDS and AMPHISBAENIANS) and Rhynchocephalia (the TUATARA). It also includes numerous extinct groups, including dinosaurs, pterosaurs and ichthyosaurs; ⇨ pp. 82–9.

residence time, the length of time an atom of an element spends (on average) in a particular form, e.g. the time spent by an atom of carbon in the atmosphere during the course of the carbon cycle (⇨ pp. 196–7).

resin, a water-soluble exudate (often consisting of phenols and terpenes) of woody plants, especially conifers, in which it is produced in resin ducts. Conifer resins are used as tar pitch in boat building, as turpentines for paint solvents, and as pinenes for scents in disinfectants and foods; ⇨ pp. 39, 209.

respiration, 1. in animals and plants, the process whereby oxygen is taken in from the environment (air or water) and carbon dioxide (the waste product) given out. In all but small, simple animals this is carried out through a specialized respiratory system; ⇨ pp. 32 (plants), 180–1 (animals). Other references: protists 27; insects 59, *180*; fishes 72–3; amphibians 78, 79; birds 93; marine mammals 142, 143 (box), 144. **2.** cell respiration, the complex chemical processes by which glucose is converted into a utilizable form of energy within cells; ⇨ p. 21.

respiratory pigment, a special molecule in BLOOD that has a high oxygen-carrying capacity and which combines with oxygen at the respiratory surfaces and releases it to tissues within the body; ⇨ p. 180.

retaliation, fighting back, often an animal's last line of defence against attack by a predator; ⇨ p. 167.

reticulum, the second chamber of a ruminant's stomach; ⇨ p. *112*.

retina, a light-sensitive membrane at the back of the vertebrate eye; ⇨ pp. **189–90**.

retrix (pl. retrices), one of the large feathers forming the tail of a bird (⇨ p. *92*).

retrovirus, any of a group of RNA viruses; ⇨ p. 24.

rhabdom, in compound eyes, ⇨ p. *190*.

rhea, either of two flightless birds of the family Rheidae (order Struthioniformes), sometimes called the South American ostriches, from which they differ by their feathered thighs and three-toed feet. The greater rhea (*Rhea americanus*) grows to 1.5 m (6 ft) and lives in grassland; ⇨ pp. *91, 97* (box).

rhebok (*Pelea capreolus*), an elegant, long-necked, grey-brown, soft-furred ANTELOPE with short horns. It inhabits grassy mountains and plateaux in southern Africa.

rhesus monkey, a southern Asian MACAQUE, often used in medical research; ⇨ p. 139, 154 (box).

rhinoceros, any massively built, horned PERISSO-DACTYL mammal (family Rhinocerotidae) of sub-Saharan Africa (2 species) and Southeast Asia (3 species); ⇨ pp. 16, **110–11**, 203. For the extinct woolly rhino ⇨ p. *100*.

rhinoceros beetle, any SCARABAEOID BEETLE of the subfamily Dynastinae (family Scarabaeidae). These tropical, medium-sized to large beetles have one or more long, curved 'horns' on their head. The larvae of some species are pests of rice, sugar cane and coconut crops.

Rhizobium, a genus of bacteria involved in nitrogen fixation in leguminous plants; ⇨ p. 32 (box).

rhizoid, any of a number of fine hair-like structures acting as roots in various spore-bearing plants and fungi; ⇨ pp. 36–7.

rhizome, an underground horizontal stem; ⇨ p. 44.

rhododendron, any of around 500–600 species of the genus *Rhododendron* in the heath family, Ericaceae, originating mainly in East Asia. They are shrubs, rarely trees, although *R. arboreum* can attain 30 m (98 ft) in height. Many species and hybrids are cultivated, some under the name AZALEA. *R. ponticum* has become naturalized in Britain, smothering the natural vegetation in some places; ⇨ p. 211.

Rhodophyta, a plant division comprising the red algae; ⇨ pp. 34.

rhodopsin, a red PHOTOPIGMENT occurring in the rods of the vertebrate retina; ⇨ pp. 96 (box), 189.

rhubarb (*Rheum rhubarbarum*), a perennial temperate herb in the family Polygonaceae cultivated for its edible leaf stalks. The leaves are poisonous because of their high calcium oxalate content.

ribbonfish, any of eight species of the family Trachipteridae (order Lampridiformes), silver mid-water fishes found in all oceans. The body is elongated and laterally compressed, and the dorsal fin runs the length of the body; the tail fin is turned upwards like a fan.

ribbon worm, any invertebrate of the phylum Nemertea, solid, elongated, slow-moving worms rather similar to flatworms in structure. Mostly marine, all ribbon worms are predatory, and some grow to astonishing lengths, reaching 10 m (33 ft).

ribonucleic acid, ⇨ RNA.

ribosome, any of many small particles, composed of protein and r-RNA, found in both prokaryote and eukaryote cells. They are sites of protein synthesis; ⇨ pp. **6** (box), 20 (box), *21*, **22**, *24*.

rice, any of around 20 species of the cereal grass genus *Oryza*. The many cultivars attributed to *O. sativa* are also referred to as paddy. Rice forms the basic food for about 60% of the world's population and most is grown and eaten in Asia and the Far East, where there is evidence for its cultivation for over 5000 years. Rice is a semi-aquatic plant frequently grown in standing water, but there are varieties suitable for drier conditions. There are two types of grain: the sugary 'pudding' rice, which tends to stick together when cooked, and a more starchy rice, which is not sticky.

richness, the number of species found in an ecosystem; ⇨ p. 195.

ridley, either of two small marine turtles (CHELONIANS) of the genus *Lepidochelys*. One species is found in the Atlantic, the other in the Pacific.

right whale, any large baleen whale (CETACEAN) of the family Balaenidae (3 species), such as the bowhead or Greenland right (*Balaena mysticetus*); ⇨ pp. *142*, 144, **145**.

ringtail (*Bassariscus astutus*), a brown, cat-like PROCYONID mammal with a long, black-ringed tail. It inhabits rocky places in Mexico and southwestern USA. It is also known as the ring-tailed cat, cacomistle and civet cat; ⇨ p. *130*.

ring-tailed cat, another name for RINGTAIL.

ringtail possum, any of 16 arboreal forest-dwelling MARSUPIALS of the family Pseudocheiridae, including the common ringtail possum (*Pseudocheirus peregrinus*), a nocturnal leaf-eater with grasping hands, opposable big toes and a long prehensile tail.

ripening, the termination of development of certain organs or parts of a plant, especially the final changes in a fruit; ⇨ pp. 41, 43.

ritualization, an evolutionary process by which many DISPLAYS and other forms of COMMUNICATION behaviour have developed from other, non-communicatory behaviour. Ritualized signals are generally stereotyped, often exaggerated forms of the original behaviour patterns; ⇨ p. 155 (box).

river dolphin, any aquatic CETACEAN mammal of the family Platanistidae (5 species), found in various South American and southern Asian rivers; ⇨ pp. *142*, **144–5**, 165.

river ecosystems, ⇨ pp. 218–19.

rivuline, any of 210 small freshwater fishes of the family Aplocheilidae, related to the FLYING FISHES. These brightly coloured, hardy fishes are found in Africa, southern Asia and the Americas, and are very popular in aquaria.

RNA, ribonucleic acid. RNA occurs in different forms – m-RNA (messenger RNA), t-RNA (transfer RNA) and r-RNA (ribosomal RNA) – all of which are involved in the processes by which DNA leads to the construction of proteins; ⇨ pp. **6** (box), 15 (box), 19, **22**, 24 (viruses). See also TRANSLATION.

roach (*Rutilus rutilus*), a widespread European freshwater fish of the CARP family. It is popular with anglers and eaten in many parts of Europe.

roadrunner, either of two long-tailed, crested cuckoo-like birds, belonging to the family Neomorphidae (order Cuculiformes). They live in deserts of the southwestern USA, where they are famous for their running speed and agility at catching lizards and snakes.

robber fly, any medium to large bristly DIPTERAN insect of the family Asilidae; over 4000 species of this worldwide family have been described. The adults have strong legs and a stiff proboscis to catch and feed on their insect prey. The larvae are soil- or litter-dwellers.

robin, name given to many unrelated birds with red breasts. The familar western Palaearctic robin is a member of the genus *Erithacus* and is one of the smaller THRUSHES (⇨ p. 159). The much larger American robin (*Turdus migratorius*) is a typical thrush.

rocket, any of a number of plants of the family Brassicaceae (formerly Crucifereae), e.g. garden rocket (*Eruca vesicaria sativa*) and dame's rocket (*Hesperis matronalis*). A yellow dye is obtained from the unrelated dyer's rocket (*Reseda luteola*).

rock rat, 1. (*Aconaemys fuscus*), a stocky, burrowing, rat-like RODENT in the same family as the DEGUS, found in the Andes. **2.** (*Petromys typicus*), African rock rat or dassie rat, a squirrel-like rodent with silky fur and a long tail, inhabiting rocky hills in northeastern and southern Africa.

rockrose, any species of herb or shrub in the genera *Cistus*, *Tuberaria* and *Helianthemum* in the family Cistaceae. A number are grown in gardens; ⇨ p. 205.

rod, any of numerous rod-shaped cells in the retina of many vertebrates, sensitive to dim light but not to colour; ⇨ pp. 97, 189, **190**. Compare CONE.

rodent, any mammal of the order Rodentia, the largest group of mammals with nearly 1700 species. Highly versatile and opportunistic in their feeding, rodents are generally small to medium-sized mammals and are found in every terrestrial habitat on every continent except Antarctica. The group is divided into three suborders: the sciuromorphs or 'squirrel-like rodents' (squirrels, marmots, beavers, etc.), the caviomorphs or 'cavy-like rodents' (cavies, porcupines, coypu, etc.), and the myomorphs or 'mouse-like rodents' (rats, mice, voles, etc.); ⇨ pp. **133–5**. See also RAT, MOUSE.

roe, the eggs of a female fish (hard roe) or the sperm of a male fish (soft roe).

roller, any of 12 species of conspicuous, brightly coloured, raucous bird belonging to the family Coraciidae (order Coraciiformes), with broad, hooked bills used to capture insects. Rollers occur in the Old World and get their name from their aerobatic courtship displays; ⇨ pp. 90, *91*.

rook (*Corvus frugilegus*), a large CROW of the Palaearctic with a pale grey bill and face, which tends to breed early in large communal rookeries.

rookery, a communal breeding ground or colony of various birds and mammals, including penguins (⇨ p. 97), seals and walruses (⇨ p. *159*).

root, any of a number of (usually underground) leafless outgrowths of a plant, used for anchorage and the absorption of water and mineral nutrients; ⇨ p. 31. In some groups of spore-bearing plants (⇨ p. 37) the functions of roots are carried out by RHIZOIDS or RHIZOMES (the latter also being found in some flowering plants). There are many variations on the basic root form, e.g. the TUBERS of dahlias, TAPROOTS of carrots and dandelions, buttress roots in some tropical trees, aerial roots in EPIPHYTES.

root crop, any plant cultivated for its roots, which are used either for human or animal consumption, such as CARROT, PARSNIP, SWEDE and TURNIP.

rorqual or **finback,** any large to very large baleen whale (CETACEAN) of the family Balaenopteridae (6 species), including the humpback, minke, fin, sei and blue whale; ⇨ pp. 142, *144*, **145**.

Rosaceae, a dicotyledonous plant family, including roses, strawberry, apple, cherry, hawthorn, etc.; ⇨ p. 45.

rose, any of around 250 species of usually prickly shrubs in the north temperate and subtropical genus *Rosa* in the family Rosaceae. Some wild species such as sweet brier are cultivated, but most of the thousands of ornamental cultivars have arisen by selection and hybridization. These can be classified as climbers, old shrub and new shrub roses; ⇨ p. 43.

rosebay, ⇨ OLEANDER.

rosebay willowherb, ⇨ WILLOWHERB.

rosella, any of eight medium-sized, long-tailed Australian PARROTS of the genus *Platycercus*, usually red or orange. They are often seen flying in pairs.

rosemary (*Rosmarinus officinalis*), a Mediterranean aromatic evergreen shrub in the family

Lamiaceae (formerly Labiatae). It is grown as an ornamental and for use in flavouring and potpourris.

roseroot (*Rhodiola rosea* = SEDUM *rosea*), a north temperate perennial herb grown as an ornamental.

rosette, a radial arrangement of leaves on a shortened, compact stem (e.g. in dandelions).

rosewood, originally the rose-scented and highly prized timber from the slender leguminous tree, Brazilian rosewood (*Dalbergia nigra*), but extended to include a number of other similar timbers.

Rosidae, a dicotyledonous plant subclass; ⊳ pp. *40*, 45.

rostrum, a beak, or a body part shaped like a beak, especially of an insect (⊳ p. *162*) or arachnid.

rotifer, any microscopic ASCHELMINTH of the class Rotifera. They are common in ponds and ditches, and although the same size as protozoans, they have advanced structures and organ systems, including a complex ciliate feeding apparatus.

round dance, ⊳ DANCE LANGUAGE and p. 153.

round window, in the vertebrate ear, ⊳ p. *191*.

roundworm, any invertebrate animal of the class Nematoda, including EELWORMS, THREADWORMS and HOOKWORMS; only 15 000 have been described but this is probably a fraction of those that actually exist. Some are extremely important human parasites; ⊳ pp. **55**, 168.

rove beetle, any BEETLE of the family Staphylinidae, comprising over 27 000 described species. They have characteristically short wing coverings (elytra), and many are small and dark-coloured. They are often predatory.

rowan, ⊳ SORBUS.

r-strategist, any species that is particularly successful at colonizing new or disturbed habitats. They are often regarded as pests or weeds by humans; ⊳ pp. 43, **230**.

rubber, a long-chain hydrocarbon obtained from the LATEX of a number of plants. Para rubber (*Hevea brasiliensis*), a tropical tree in the family Euphorbiaceae, is the principal source and is grown commercially in many countries. The latex is tapped by making oblique incisions in the bark. Minor sources include ceara rubber (*Manihot glaziovii*), also in the family Euphorbiaceae, India rubber (*Ficus elastica*) and Panama rubber (*Castilla elastica*), both in the family Moraceae. See also GUTTA-PERCHA.

rudd (*Scardinius erythrophthalmus*), a European freshwater fish of the CARP family, a colourful species popular in aquaria.

rue (*Ruta graveolens*), a Mediterranean aromatic shrub in the family Rutaceae. It is used in flavouring and perfumery. Species of the unrelated genus, *Thalictrum*, are also called rue.

ruff (*Philomachus pugnax*), a long-necked WADER of the family Scolopacidae, inhabiting lowland marshes of the Palaearctic. In spring the male develops an elaborate brown ruff and ear tufts; ⊳ p. 94.

rumen, the large, first compartment of the stomach of a ruminant mammal; ⊳ p. 112 (box).

ruminant, ⊳ RUMINATION.

rumination, or 'chewing the cud', a complex digestive process characteristic of most artiodactyl mammals (cattle, sheep, antelope, etc.). Rumination is carried out in a specialized multi-chambered stomach, from which partially digested food ('cud') is regurgitated for further processing and then reswallowed; ⊳ pp. 112 (box), 113–17, 169.

runner bean, 1. a usually climbing cultivar of the legume, *Phaseolus vulgaris*, the pods of which are eaten before maturity. See also FRENCH BEAN, HARICOT BEAN and KIDNEY BEAN. **2.** scarlet runner bean (*Phaseolus coccineus* = *P. multiflorus*), a climbing legume, the pods of which are eaten before maturity.

rush, principally any member of the family Juncaceae, such as the soft rush (*Juncus effusus*), and some members of the family Cyperaceae, e.g. BULRUSH; but also including the flowering rush (*Butomus umbellatus*) in the family Butomaceae. The stems of some species are used for making baskets, mats and chair seats. The pith of BOG RUSHES (*Juncus spp.*) was used for making rush lights.

rust, a plant disease in which leaves appear rusty, caused by species of basidiomycete fungus (⊳ p. 35) in the order Uredinales, e.g. *Puccinia graminis* on cereals.

rye (*Secale cereale*), a cereal grown in Europe principally for its grain, which is used in the production of black bread, crispbread, starch production and animal feed, and in North America for the production of rye whisky.

rye grass, any of around 12 species of the grass genus *Lolium* to be found in Europe, Asia and North Africa. The perennial rye grass (*L. perenne*) is an important fodder crop in temperate regions.

sable (*Martes zibellina*), a small, stocky, brown MUSTELID with a long body and short legs and tail, found in forests in northern Asia; ⊳ p. 131.

saccharide, any SUGAR or other carbohydrate (e.g. POLYSACCHARIDES).

sac-winged bat, ⊳ SHEATH-TAILED BAT.

saddle fungus, any of around 25 species of the north temperate ascomycete (⊳ p. 35) genus *Helvella*, having a hollow stem and a saddle like cap.

saffron (*Crocus sativus*), a species of CROCUS cultivated for its stigmas, from which saffron is extracted for use as a dye and in cooking.

sage, any of several species of herb in the genus *Salvia* in the family Lamiaceae (formerly Labiatae). *S. officinalis* is cultivated as a flavouring.

sagebrush, ⊳ ARTEMISIA.

sago palm, any of several species of palm in the genus *Metroxylon*, e.g. spineless sago palm (*M. sagus*). An almost pure starch product is obtained from the pith of the trunks; ⊳ p. 45.

saguaro (*Carnegia gigantea*), a CACTUS larger than all others, growing to 20 m (66 ft) in height and 12 tonnes (tons) in weight; ⊳ p. *204*.

saiga (*Saiga tatarica*), a goat-like ANTELOPE with pale brown, woolly fur and a trunk-like nose. It occurs on arid, grassy steppes in Central Asia; ⊳ p. *206*.

sailfish, either of two species of BILLFISH.

St John's wort, any of around 400 species of herbs and shrubs in the temperate and tropical genus *Hypericum* in the family Hypercaceae, e.g. perforate St John's wort (*H. perforatum*).

saithe (*Pollachius virens*), a relative of the COD, alternatively known as the coalfish or coalie, found throughout the North Atlantic. It is a valuable commercial fish, though judged inferior to the cod as a table fish.

saki, any New World monkey of the genera *Chiropotes* and *Pithecia* (family Cebidae), medium-sized shaggy monkeys with long, bushy tails; ⊳ p. 138.

salamander or **urodele,** any AMPHIBIAN of the order Urodela (or Caudata), comprising some 350 species of tailed amphibian in 9 families. Salamanders are typically lizard-like in general appearance, but they do not have a scaly skin. They are found in Europe, Asia (north of the Himalaya) and North America, with one group extending into the tropics of South America. The largest family (Plethodontidae) comprises the lungless salamanders (⊳ p. 79). The term 'salamander' is sometimes used in a restricted sense to refer to members of the family Salamandridae ('true' salamanders and NEWTS); ⊳ pp. **78–81**. See also CONGO EEL, GIANT SALAMANDER, MOLE SALAMANDER, MUD PUPPY and SIREN.

salano or **Malagasy brown-tailed mongoose** (*Salanoia concolor*), a brown, mongoose-like VIVERRID with dark or pale spots, inhabiting forest in Madagascar.

Salicaceae, a dicotyledonous plant family, including willows, poplars and aspen; ⊳ p. 45.

salinity, the concentration of salt in a volume of water. Usually the salt is mainly sodium chloride (familiar as table salt), but other salts, such as calcium sulphate, may also be present. Salinity is measured in parts per thousand.

saliva, a fluid secretion from the salivary glands, containing some digestive enzymes and acting as a lubricant within the digestive tract; ⊳ p. 178.

sallow, ⊳ WILLOW.

salmon (*Salmo salar*), the Atlantic salmon, a SALMONID of the northern Atlantic, one of the world's best-known food and sport fishes. It makes strenuous efforts to penetrate rivers to reach its spawning grounds, often involving spectacular leaps over obstacles; ⊳ pp. 73, *74*, 76 (box), 170.

Salmonella, a genus of bacteria including *S. typhimurium*, a cause of food poisoning; ⊳ p. 24.

salmonid, any fish of the 'salmon family', Salmonidae (order Salmoniformes), a large and diverse group that includes not only SALMON but TROUT, CHARR, WHITEFISH and GRAYLING. Salmonids are distributed widely in fresh waters of the northern hemisphere and have been introduced elsewhere. Many species live in the sea and migrate to fresh water to spawn. They have a small, fatty, rayless fin placed well back behind the main dorsal fin. The group includes some of the most valuable of all commercial and sport fishes, which are the object of both intensive fishing and large-scale conservation programmes. To meet the huge demand for salmonids as luxury food items, several species are now farmed; ⊳ pp. 71.

salp, ⊳ TUNICATE.

salsify or **oyster plant** (*Tragopogon porrifolius*), a hardy biennial plant native to the Mediterranean region. Salsify is cultivated for its edible roots and shoots.

salt, a basic taste quality; ⊳ p. 188.

saltation, ⊳ PUNCTUATED EQUILIBRIUM.

salt marsh, a maritime wetland ecosystem found on sheltered coastal sites, often in estuaries or in areas sheltered by shingle ridges. They are established on a muddy substrate and experience the full range of tidal flow, the vegetation depending on the elevation of the mud surface.

samphire (*Crithmum maritimum*), a perennial herb in the Apiaceae family (formerly Umbelliferae) found on maritime coasts of Europe and the Mediterranean. The fleshy leaves are used in pickle or salads. Species of the unrelated genus *Salicornia* are known as marsh samphire.

sandalwood, a fragrant wood obtained principally from white sandalwood (*Santalum album*), a tree native to India, Southeast Asia and Australia.

sand flea, a SAND HOPPER.

sand fly, any small DIPTERAN insect of the genera *Phlebotomus* or *Lutzomyia* in the family Psychodidae. They are blood-feeders and transmit some forms of the disease leishmaniasis; ⊳ p. 64.

sandgrouse, any of 16 camouflaged, grouse-like birds of the family Pteroclidae (order Ciconiiformes), adapted for life in deserts and semi-arid areas of Africa and Eurasia. They are strong fliers and have feathered legs, which the male soaks with water to carry back to the young; ⊳ pp. 91, 93.

sand hopper, any small shrimp-like CRUSTACEAN of the class Amphipoda; ⊳ pp. 67, 170.

sand lizard (*Lacerta agilis*), a fast-moving LACERTID lizard; ▷ p. 85.

sandpiper, any of 12 lightly built, camouflaged WADERS, occurring worldwide. The tip of the medium-length bill is extremely sensitive and used to probe the sand and mud for food; ▷ pp. 158, 171.

sap, the fluids in the vascular system (xylem and phloem) of plants; ▷ p. 31–2. The term is sometimes also applied to the fluids in parts of cells.

sapling, a young tree or shrub.

saprophyte, any species of plant that derives its nutrition by feeding on the dead remains of other organisms.

saprotroph, any organism – plant, animal, fungus (▷ p. 35) or microbe – that obtains its energy by feeding on the dead remains of other organisms.

sapsucker, any of four WOODPECKERS of the genus *Sphyrapicus*, living in the Americas. They make holes in trees, returning from time to time to suck the sap.

sapwood, the outer, youngest region of wood in trees, involved in the conduction of SAP. Compare HEARTWOOD.

sardine, ▷ PILCHARD.

sarsaparilla, the dried rhizomes of some tropical American species of climbing plants in the genus *Smilax*. Sarsaparilla is used in flavouring and medicines.

sassaby or **topi** (*Damaliscus lunatus*), a reddish-brown, grazing ANTELOPE with long, curved horns. It inhabits savannah in Africa.

sassafras, any of three species of aromatic trees in the genus *Sassafras*. Oil of sassafras is obtained from the North American *S. officinale*.

satinwood, the timber of a number of unrelated tropical trees, e.g. West Indian satinwood (*Xanthoxylem americanum*), which has golden yellow wood used in cabinet making.

satsuma, ▷ MANDARIN.

saturniid, any LEPIDOPTERAN insect (moth) of the family Saturniidae, mainly found in the tropics. Several species of the genus *Antheraea* (e.g. the Japanese oak silkworm, *A. yamamai*) are used for silk production.

saurischian, any DINOSAUR of the order Saurischia, one of the two major groups of dinosaurs. The order includes the carnivorous bipedal THEROPODS and the gigantic herbivorous SAUROPODS; ▷ pp. 83–4.

sauropod, any giant herbivorous quadrupedal SAUR-ISCHIAN dinosaur of the suborder Sauropoda, including brontosaurs, diplodocus and brachiosaurs; ▷ p. 83.

saury, any of four species of the family Scombere-socidae. They are closely related to the NEEDLE FISHES, and like them have elongated bodies and beak-like jaws.

savannah, tropical grassland, especially but not exclusively that in Africa; ▷ pp. 199 (map), **202–3**.

savannah baboon, any savannah-living BABOON, such as *Papio cynocephalus*, the CHACMA, and the olive baboon (*P. anubis*); ▷ p. 139.

sawbill, ▷ MERGANSER.

sawfly, any HYMENOPTERAN insect of the suborder Symphyta. Their name comes from the saw-teeth along the ovipositor, which is used to cut into plant tissues. They have herbivorous, caterpillar-like larvae. See also HORNTAIL.

Saxifragales, a dicotyledonous plant order, including saxifrages; ▷ p. 40.

saxifrage, any of around 370 species of usually perennial herbs in the north temperate, arctic and Andean genus *Saxifraga* in the family Saxifra-gaceae. Most are mountain plants and many grow well in rock gardens; ▷ pp. 10, 211. See also LONDON PRIDE and MOTHER-OF-THOUSANDS.

scabious, any of around 100 species of annual to perennial herbs in the mainly Mediterranean genus *Scabiosa* in the family Dipsacaceae, which also includes field scabious (*Knautia arvensis*) and devil's bit scabious (*Succisa pratensis*).

scale, 1. any of numerous small protective plates (made of various materials) covering the bodies of animals such as fishes and reptiles; ▷ pp. 72 (fishes), 107 (pangolins). **2.** a cone scale, carrying the ovule or pollen sac in conifers; ▷ p. 38.

scale insect, any COCCID insect (bug) with a hardened resinous scale covering the female (▷ box, p. 63). Although usually pests, some have been cultivated. The food colouring cochineal comes from the scale insect *Dactylopius*. See also LAC INSECT.

scallop, any marine BIVALVE mollusc of the family Pectinidae; they have fan-shaped shells with radiating ribs.

scaly anteater, another name for PANGOLIN.

scaly-tailed squirrel, any of seven arboreal, squirrel-like RODENTS of the family Anomaluridae, inhabiting tropical forests in West and central Africa. They have a furry membrane between the fore and hind limbs, which is used for gliding.

scapula, the shoulder blade, a large flat triangular bone on the upper back forming half of the PEC-TORAL GIRDLE.

scarab, any SCARABAEOID BEETLE of the subfamily Scarabaeinae (or sometimes applied more broadly to any member of the family Scarabaeidae). The true scarabs feed on dung, some species making balls of excrement, which they roll away to bury in secluded spots. The sacred scarab (*Scarabaeus sacer*) was regarded as divine by the ancient Egyptians.

scarabaeoid beetle, (or lamellicorn beetle), any BEETLE of the superfamily Scarabaeoidea. Adults typically have antennae composed of flattened segments called lamellae, which can be fanned open to expose a large sensory surface area. The so-called 'scarabaeiform' larvae are C-shaped, with three pairs of short legs, and are relatively inactive; they live within their food of rotting wood, dung or decaying vegetation. See also CHAFER, DUNG BEETLE, GOLIATH BEETLE, HERCULES BEETLE, RHINOCEROS BEETLE, SCARAB, STAG BEETLE.

scavenger, any animal that lives off the dead remains of plants and other animals; ▷ pp. 194, *195*.

scent, any chemical from a specialized scent gland, usually deposited onto objects in the environment in order to mark trails and territorial boundaries, and for other COMMUNICATION purposes; ▷ pp. 110–17 (hoofed mammals), 122–31 (carnivores), **153–4**, 163 (box), 188–9. See also SMELL and PHERO-MONE.

Schistosoma, a genus of FLATWORM blood fluke causing the disease schistosomiasis; ▷ pp. 168–9.

schooling or **shoaling,** behaviour in fishes that results in the formation and coordination of groups through mutual social attraction, often for defence against predators; ▷ pp. 77, **156** (box).

Scilla, ▷ SQUILL.

sciuromorph, any RODENT of the suborder Sciuro-morpha – the 'squirrel-like' rodents; ▷ p. 133.

sclera, in the mammalian eye, ▷ p. 189.

sclerite, a thickened area or plate of the CUTICLE of an insect.

sclerophyllous, (of a plant) having hard, leathery leaves. Such plants are usually found in mediterranean or semi-arid conditions.

scorpion, any of about 800 predatory ARACHNIDS of the order Xiphosura, characterized by large pincers, a squat eight-legged body, and a long abdomen with a prominent pointed sting; ▷ p. 68.

scorpion fish, any of about 300 tropical and temperate marine fishes of the family Scorpaenidae (order Scorpaeniformes). They are mostly bottom-living and armed with many sharp spines. Many are important commercially.

scorpionfly, any INSECT of the small order Mecoptera, containing about 400 species. They are easily identified by the downward-pointing elongate head, and the large male genital organ, which is held scorpion-like above the abdomen; ▷ p. 61.

scorpion grass, ▷ FORGET-ME-NOT.

scoter, any of three stocky northern diving DUCKS. They have swellings at the base of the bill. The male common scoter (*Melanitta nigra*) is the only all-black duck.

screamer, any of three large, heavily built turkey-like birds of the family Anhimidae (order Anseriformes) which live in Neotropical wetlands. The horned screamer (*Anhima cornuta*) has a horny head crest and partially webbed toes.

screw pine, the name given to several species of trees and shrubs in the tropical Old World genus *Pandanus* because their fruits resemble cones.

scrotum/scrotal sac, ▷ TESTIS.

scrub, a type of vegetation dominated by small shrubs and sometimes stunted trees, found in semi-desert areas (▷ pp. 199, map; **204**). Scrub may also invade other areas as a result of environmental degradation, such as overgrazing; ▷ pp. 203, 205. See also ARCTIC SCRUB.

scrub bird, either of two PASSERINE birds of the family Menuridae, closely related to the LYREBIRDS. They are brown, stocky birds with erect tails, living in dense Australian scrubland, where they run fast on powerful legs and feet. They are good mimics; ▷ p. 95.

scrubfowl, ▷ MEGAPODE.

scrubwren, any of 26 Australasian PASSERINE birds in the subfamily Acanthizinae (family Pardaloti-dae), small, dull-plumaged, terrestrial seed-eaters.

scurvy grass, any of around 24 species of annual to perennial, often maritime, herbs in the northern hemisphere genus *Cochleria* in the cabbage family, Brassicaceae (formerly Cruciferae). *C. officinalis* was used against scurvy.

sea, ▷ pp. 220–1.

sea anemone, any ANTHOZOAN of the order Actiniaria (etc.), which are solitary and live as polyps throughout their adult life. They typically have an oval ring of tentacles around the mouth; ▷ pp. 1, **53**, 215.

sea bass, any of about 370 PERCIFORM fishes of the family Serranidae. They share the same range as BASSES, except that there are few freshwater species; they are large, heavy-bodied fishes, with stout spines in the fins.

sea bream, ▷ PORGY.

sea cat, ▷ WOLFFISH.

sea cow, another name for SIRENIAN.

sea cucumber, any worm- or sausage-like ECHINODERM of the class Holothuroidea, comprising over 1100 known species; ▷ p. 54.

sea dollar, a short-spined, sand-dwelling SEA URCHIN; ▷ p. 54.

sea fan, any of about 500 OCTOCORALS of the genus *Gorgonia* (etc.). They are sedentary filter-feeders with a fan-like core skeleton of horny flexible material; ▷ pp. 53, 215.

sea gooseberry, ▷ COMB JELLY.

sea horse, any of about 30 marine fishes of the subfamily Hippocampinae (family Syngnathidae). They are closely related to the PIPEFISHES, and similar to them in range and basic structure (though the body is deeper). They swim with the body held vertically and the head angled downwards. As with the pipefishes, the males care for the fertilized eggs, carrying them in a belly pouch; ▷ pp. 72, 77.

seal, any carnivorous marine mammal of the families Phocidae (earless or 'true' seals; 18 species) and Otariidae (eared seals, including fur seals and sea lions; 14 species). All seals breed on land, so although highly adapted for life in water, they retain some ability to move on land; ▷ pp. *14,* 142, *143, 185.*

sea lettuce, any of around 30 species of GREEN ALGAE in the genus *Ulva. U. lactuca* is a delicate leaf-like seaweed of the upper shore.

sea lily or **feather star,** any ECHINODERM of the class Crinoidea, comprising about 80 living species and many thousands of extinct ones; ▷ p. 54.

sea lion, any of various large eared SEALS, such as the Californian sea lion (*Zalophus californianus*). familiar as a zoo and circus animal; ▷ p. **143,** 158.

sea mat, ▷ BRYOZOAN.

sea pen, any of about 300 OCTOCORALS of the genus *Pennatula* (etc.). They live anchored in muddy bottoms, extending a feather-like structure into water currents to intercept food particles; ▷ pp. 53.

sea potato, a short-spined, sand-dwelling SEA URCHIN; ▷ p. 54.

sea robin or **gurnard,** any marine fish of the family Trigilidae (86 species), related to the SCORPION FISHES and distributed worldwide. They have broad heads encased in bony plates, and the gill covers are often armed with sharp spines. The pectoral fins are usually greatly expanded, with the first few rays developed into finger-like feelers, which are used to test the sea bottom. The family is noted for sound production, mostly grunts and squeaks.

sea scorpion, 1. any extinct ARTHROPOD of the order Eurypterida. Flourishing during the Silurian period, they grew up to 3 m (10 ft) in length and somewhat resembled scorpions and horseshoe crabs, and may have been ancestral to the arachnids; ▷ p. 5. **2.** any marine fish of the family Cottidae, related to the SCORPION FISHES, with a tapering, plated body and spines.

sea slug or **nudibranch,** any predatory marine GASTROPOD of the order Nudibranchia. They lack an external shell and are frequently brightly coloured; ▷ pp. 56, *167.*

sea snake, any of around 50 species of SNAKE of the family Hydrophiidae. Closely related to the ELAPID snakes, all sea snakes are highly venomous, with fixed fangs at the front of the mouth (▷ p. 87). They occur in the tropical Indian and Pacific Oceans and swim by beating the flattened tail from side to side. Most species give birth to live young; ▷ p. 164.

sea spiders, a group of slow-moving marine ARACHNIDS with eight to twelve long legs and a small body. They look like the terrestrial harvestmen, but their true affinities are uncertain.

sea squirt, any TUNICATE of the class Ascidiacea; ▷ p. 51 (box).

sea star, an alternative name for STARFISH.

sea urchin, any of about 700 ECHINODERMS of the class Echinoidea; ▷ pp. 54, 215, 221.

sea walnut, ▷ COMB JELLY.

seaweed, any multicellular alga, including KELPS and WRACKS; ▷ pp. 34, 221.

sea whips, a group of OCTOCORALS with a core skeleton of long whip-like horny material, which is extended into water currents to intercept food particles; ▷ pp. 53, 215.

sebaceous, secreting SEBUM (a similar oily material).

sebum, (in mammals) a fatty, waxy secretion from the sebaceous glands that lubricates and water-proofs the skin and hair.

secondary, in birds, any one of the inner flight feathers; ▷ pp. 90 (box), *92.*

secretary bird (*Sagittarius serpentarius*), a long-legged, eagle-like BIRD OF PREY, the only representative of the family Sagittariidae (order Ciconiiformes). It chases prey, often snakes, and stamps on them; ▷ p. 203.

sedge, 1. any of around 2000 species of the widely distributed genus *Carex* in the family Cyperaceae. They are grass-like perennial herbs with stems often triangular in section; ▷ pp. 216, *219.* **2.** any of certain species of the related genera, *Cladium,* e.g. great sedge (*C. mariscus,* and *Rhynchospora,* e.g. white beak-sedge (*R. alba*).

sedimentary rocks, a large group of rocks, including limestone, sandstone and shales, formed at the bottom of ancient seas by slow deposition of small particles. They often contain fossils; ▷ p. 4.

Sedum, a mainly north temperate genus of succulent herbs and small shrubs in the family Crassulaceae. A number are grown as ornamentals. See also STONECROP and ROSEROOT.

seed, the unit of dispersal containing the embryo of the new sporophyte plant in SEED-BEARING PLANTS. It develops from the fertilized OVULE, and usually contains food reserves. The kernel of the seed is protected by the seed coat (TESTA), usually making the seed resistant to harsh conditions; ▷ pp. 38–41.

seed-bearing plants, plants that produce seeds rather than spores as the units of dispersal. Sometimes referred to as the spermatophytes, the seed-bearing plants comprise the gymnosperms (conifers and their allies; ▷ pp. 38–9) and the angiosperms (flowering plants; ▷ pp. 40–1). Compare SPORE-BEARING PLANTS.

seed fern, any species of fossil fern-like plants of the Carboniferous period bearing gymnospermous seeds on their fronds.

segmented worm, ▷ ANNELID.

sei whale (*Balaenoptera borealis*), a baleen whale, one of the RORQUALS; ▷ p. 145.

selection, ▷ ARTIFICIAL SELECTION, KIN SELECTION, NATURAL SELECTION, SEXUAL SELECTION.

semicircular canal, any one of three fluid-filled looped tubes in the vertebrate inner ear by which angular acceleration is detected (and hence balance maintained); ▷ pp. 190–1.

semidesert, an ecosystem in which vegetation is open and scattered, often shrubby in form (▷ SCRUB). Conditions are too arid, however, for farming to be successful; ▷ p. 204.

Sempervivum, ▷ HOUSELEEK.

Senecio, a widely distributed genus of around 1500 species of herbs, shrubs and trees in the daisy family, Asteraceae (formerly Compositae), including GROUNDSEL and RAGWORT. A number of closely related genera such as *Pericallis* were once included (▷ CINERARIA).

senna, any of around 500 species of leguminous trees, shrubs and herbs in the tropical and warm temperate genus *Cassia.* The pods and leaves of some species are used as a laxative.

senses, the various faculties by which animals gain information about the external environment and the internal state of their body. The senses are dependent on the sensitivity of specialized sensory cells, which are often grouped together in sense organs.

Text references: 188–91; insects 59; fishes 73 (box); amphibians 80; reptiles 88; birds 97. See also BALANCE, ECHOLOCATION, ELECTROMAGNETIC SENSE, HEARING, SIGHT, SMELL, TASTE, THERMORECEPTION, TOUCH.

sensillum (pl. -la), a kind of sensory organ found in insects and other invertebrates; ▷ pp. 188, 191.

sensitive plant (*Mimosa pudica*), a shrub native to tropical America. The leaflets rapidly droop if touched; ▷ MIMOSA.

sensory, relating to the SENSES.

sepal, any member of the outermost whorl of a flower; ▷ p. 41.

septate, ▷ HYPHA.

sequoia, ▷ REDWOOD (1).

sere, a stage in an ecological SUCCESSION. By definition therefore seres are unstable and of limited duration.

serow, either of two GOAT-like artiodactyls of the genus *Capricornis,* with grey to black fur, a short mane and short horns. They inhabit montane forest and bush in eastern and Southeast Asia; ▷ p. 115.

serpent star, an alternative name for BRITTLESTAR; ▷ p. 54.

serval (*Felis serval*), a slim, long-legged, orange-brown CAT with black spots, found in savannah in Africa.

service tree, 1. wild service tree (*SORBUS torminalis*). Its edible berries were once sold in England as 'chequers berries'. **2.** true service tree (*S. domestica*).

sesame (*Sesamum indicum*), a tropical herb in the family Pedaliaceae cultivated for its seeds which are used in confectionary and as a source of oil.

sessile, 1. (of an animal such as a barnacle) fixed in one position. **2.** (of a leaf or flower), having no stalk, i.e. growing from the stem.

seta, ▷ CHAETA.

sewellel, another name for MOUNTAIN BEAVER.

sex, the quality of being male or female, involving the ability to produce one of two types of gamete (often called sperm and eggs respectively). Sex may be determined by special chromosomes (▷ p. 23) or by environmental factors such as temperature or food availability (▷ p. 74), or it may depend on whether or not the eggs are fertilized (▷ p. 63). See also REPRODUCTION.

sex cell, ▷ GAMETE.

sexton beetle, alternative name for CARRION BEETLE.

sexual dimorphism, a phenomenon in many animal species in which one of the sexes (usually the male) is larger and more striking in appearance than the other, e.g. birds of paradise, elephant seals; ▷ p. **159** (box). Other references: insects 63 (box); fishes 74; birds 94; cattle, etc. 114; deer 116, *152;* seals, etc. 143.

sexual reproduction, a means of REPRODUCTION that involves the fusion of male and female sex cells (gametes), usually derived from two sexually differentiated individuals (▷ SEX) and involving some form of REPRODUCTIVE SYSTEM.

sexual selection, the evolutionary process whereby characteristics are favoured for their sexual attractiveness, rather than their value in increasing the survival prospects of the individual concerned; ▷ pp. 17, **159** (box).

Seychelles coconut, ▷ COCO DE MER.

shad, any of 31 temperate-water fishes of the subfamily Alosinae, in the HERRING family. They are like large herrings in appearance (up to 60 cm/24 in); some migrate into rivers to spawn, while others are entirely freshwater.

shag (*Phalacrocorax aristotelis*), a northern-hemisphere CORMORANT. Adults are glossy bluish-green and have a distinctive head crest.

shaggy ink cap, ▷ LAWYER'S WIG.

shallot (*Allium ascalonicum*), a perennial herb allied to and used in a similar way to the ONION.

shamrock, a plant with a trefoil leaf used as the national emblem of Ireland. This is principally black medick (*Medicago lupulina*), but white CLOVER may be substituted.

shark, any of about 250 CARTILAGINOUS FISHES of the subclass Selachii or Elasmobranchii, which also includes the RAYS. Most are active marine predators, with long, streamlined bodies, usually two dorsal fins, and five to seven gill slits on either side of the head; their mouths are typically armed with rows of sharp teeth; ▷ pp. 73, 74, **75**, 76. See also ANGEL SHARK, BASKING SHARK, COW SHARK, DOGFISH, HAMMERHEAD, MAKO, PORBEAGLE, THRESHER and WHALE SHARK.

sharksucker, ▷ REMORA.

shearwater, any of 20 species of oceanic sea bird in the family Procellariidae (order Ciconiiformes). They fly low above the water on stiffly held wings, tipping from side to side and appearing to 'shear' the water; ▷ pp. 91, 170.

sheatfish, any of 70 or so CATFISHES of the family Siluridae. *Siluris glanis*, also known as the wels, is a large species (up to 5 m/16 ft) occurring in European fresh waters.

sheath, an enclosing structure, such as the leaf sheath surrounding the stem in grasses.

sheathbill, either of two white, stout, gull-like WADERS, with a heavy bill covered with a warty sheath, belonging to the family Chionididae (order Ciconiiformes). They are found scavenging around seal and penguin colonies in Antarctica; ▷ p. 96.

sheath-tailed bat or **sac-winged bat,** any of around 40 microchiropteran BATS of the family Emballonuridae, ranging throughout the tropical and subtropical parts of the world. The group is characterized by a loose membrane or sheath spanning the hind legs, through which the small tail projects. Many species also have glandular sacs on the wings, the scents from which may assist in the search for mates.

sheep, common name given to various BOVID mammals of the genus *Ovis* and related genera in the subfamily Caprinae (which also includes GOATS), such as the American bighorn (*O. canadensis*) and the mouflon (*O. musimon*); ▷ pp. 114, **115**, *157*, 212.

sheep ked, any DIPTERAN insect of the family Hippoboscidae, strange, wingless species that live as external parasites on sheep. The young are laid as larvae.

shelduck, either of two species of large, goose-like DUCKS, living near coasts. The common shelduck (*Tadorna tadorna*) of the Palaearctic has a striking chestnut and white body and greenish-black head.

shell, any protective, usually hard outer covering; ▷ pp. 27 (protists), 56–7 (molluscs), 66 (crustaceans), 85 (turtles), 86 (reptilian egg), 95 (bird's egg).

shepherd's purse (*Capsella bursa-pastoris*), a widely distributed weed in the cabbage family, Brassicaceae (formerly Cruciferae), with fruits resembling the purse carried by shepherds.

shield bug or stink bug, any HETEROPTERAN insect (bug) of the family Pentatomidae, of worldwide distribution and mainly plant-feeding. They are easily recognized by their shield-like shape. Some species are capable of producing a foul-smelling liquid, hence their alternative name.

shiitake (*Lentinus edodes*), an edible species of AGARIC fungus which is much cultivated in Japan.

shinbone, ▷ TIBIA (1).

shiner, any of various small, brightly coloured North American freshwater fishes of the CARP family.

ship worm, any BIVALVE mollusc of the genus *Teredo*, a specialized group that bore into wood by scraping at it with the shell valves. They were once a major danger to wooden sailing vessels.

shoaling, ▷ SCHOOLING.

shoebill (*Balaeniceps rex*), a large stork-like bird, with a massive shoe-shaped bill and a small head tuft, living in central African swamps. It belongs to the same family as the PELICANS.

shorebird, ▷ WADER.

shore fly, any small DIPTERAN insect of the family Ephydridae, inhabiting the shores of seas and lakes. Some species are predatory. See also KELP FLY.

short-tailed bat (*Mystacina tuberculata*), a microchiropteran BAT (the only species in its family) known only from New Zealand. It has wings that fold away into pockets, allowing the bat to move agilely on the ground on all fours. It feeds on a variety of food, including ground-dwelling insects, fruit, pollen and flowers.

shrew, any mammal of the family Soricidae (order Insectivora), the most numerous and diverse of all insectivore groups, with nearly 270 species of worldwide distribution outside Australasia and Antarctica. Shrews are generally small, active predators feeding on invertebrate prey; ▷ pp. 120–1. See also ELEPHANT SHREW, OTTER SHREW and TREE SHREW.

shrew opossum, another name for RAT OPOSSUM.

shrike, any of 30 PASSERINE birds of the family Laniidae, medium to large-sized predatory birds with long rounded tails, large heads and hooked and notched bills, living in Africa and the northern hemisphere. Many, such as the great grey shrike (*Lanius excubitor*), swoop down on insects, small birds and mammals, and often impale food on thorns and store food in 'larders' (▷ p. *163*). See also CUCKOO SHRIKE.

shrimp, common name given to numerous small DECAPOD crustaceans, especially those of the genus *Crangon* (etc.), although the distinction between shrimps and PRAWNS is imprecise. Shrimps typically have flattened transparent bodies and a fan-like tail. The name is also given to any small edible crustacean with a shrimp-like appearance; ▷ pp. **67**, 219. See also FAIRY SHRIMP.

shrub or **bush,** a small or medium-sized woody plant with numerous main stems. Compare TREE.

siamang, the largest of the GIBBONS; ▷ p. 140.

sidewinder (*Crotalus cerastes*), a desert-dwelling North American PIT VIPER, so named because of its highly distinctive mode of locomotion; ▷ p. 176 (box).

sifaka, either of two species of the genus *Propithecus* (family Indriidae), medium-sized LEMURS active by day in the forests of Madagascar; ▷ p. 137.

sight, one of the basic SENSES, a complex form of photoreception (sensitivity to light) involving the formation of an image of the light source by means of groups of associated photosensory cells (EYES). Text references: **189–90**; cephalopods 57; insects 59; spiders 69; fishes 73; amphibians 80; reptiles 88; birds 97; cats 124; dogs 126; bears 128; primates 136; whales 144.

sign stimulus, the part of a stimulus that elicits a certain behavioural response in an animal, e.g. the red belly of a male three-spined stickleback elicits aggression in another male. When a sign stimulus comes from another animal and is involved in some form of social interaction, it is sometimes called a 'releasing stimulus' or 'releaser'; ▷ p. 149.

Silene, a genus of around 300 species of annual to perennial herbs or, rarely, dwarf shrubs in the family Caryophyllaceae. They are distributed in Europe and the non-tropical regions of Asia, Africa and the New World. See also CAMPION, CATCHFLY and PINK.

silk, a fine protein-based fibre secreted by insects and spiders, for construction of cocoons, prey capture, etc.; ▷ pp. 65, 69, **160**, *164*.

silk cotton tree (*Ceiba pentandra*), a large tropical tree cultivated for its cotton-like seed hairs known as kapok and used as a stuffing material for quilts and insulation.

silkworm moth (*Bombyx mori*), a LEPIDOPTERAN insect of the family Bombycidae, one of several species that are farmed for their silk. It is now totally domesticated and not found in the wild; ▷ pp. 64, **65**, 160. See also SATURNIID.

Silurian period, a geological period lasting from 438 to 410 million years ago; ▷ p. 4.

silverfish (*Lepisma saccharina*), a silvery THYSANURAN insect commonly found at night in domestic kitchens and among books, where they cause damage to paper and bindings. They are fast-moving when disturbed; ▷ p. 61.

silverside, any of 160 small, slender fishes of the family Atherinidae (order Atheriniformes), including grunion (▷ p. *175*). They shoal in temperate and tropical seas, and support commercial fisheries.

silversword, ▷ TARWEED.

silverweed, ▷ *PONTENTILLA*.

simian, a higher primate (monkeys, apes, man); ▷ p. 136.

sinus, a cavity or channel, e.g. the air cavities in the cranial bones, or the dilated channels for venous blood in the cranium.

siphon, any of the various tubular organs, especially in molluscs, which are used to inhale or expel water; ▷ pp. 56–7.

siphonophore, any large, free-swimming colonial HYDROZOAN of the order Siphonophora, in which there is extreme division of function between the constituent parts (polyps and medusae). Siphonophores typically have a float at the top and a long trailing stem that bears feeding, protective and reproductive polyps. Most species are tropical, but the PORTUGUESE MAN-OF-WAR is found in northern European waters and may be abundant in summer; ▷ p. 53.

siren, any SALAMANDER of the family Sirenidae, comprising three species from southeastern USA and Mexico. Permanently aquatic and eel-like, sirens have external gills and tiny fore limbs but no hind limbs. They are highly distinctive and are sometimes placed in an order of their own; ▷ p. 78.

sirenian or **sea cow,** any herbivorous marine mammal of the order Sirenia, which comprises just 4 living species in 2 families – 3 species of MANATEE and the DUGONG; ▷ p. 142.

sisal, a fibre suitable for ropes and carpets obtained from the leaves principally of sisal AGAVE (*Agave sisalana*) and also from some other species of agave.

siskin, any of 13 fine-billed FINCHES. The western Palaearctic siskin (*Carduelis spinus*) is dependent upon the seeds of pines and spruces.

sit-and-wait predator, an AMBUSH predator.

sitatunga (*Tragelaphus spekei*), a semi-aquatic ANTELOPE with brown, shaggy fur and white markings, and short horns. It inhabits marshes and swamps in central and southern Africa; ▷ p. 115.

skate, any large RAY of the family Rajidae, especially *Raja batis*, a large species from the Atlantic, which may reach 2.5 m (over 8 ft) between the tips of its pectoral fins; ▷ p. 75.

skeleton, a hard structure that supports the body, protects the soft inner organs, and provides anchorage for muscles. Skeletons may be internal, as in vertebrates, or external, as in insects and other arthropods (EXOSKELETON); ▷ pp. 72, 92, *142*, **176**. See also CYTOSKELETON.

skimmer, any of three superficially tern-like birds of the family Laridae (order Ciconiiformes), living in the seas around India, Africa and the Americas. The lower half of the bill is extended and blade-like, and used to plough just beneath the water surface to catch fish; ▷ p. 93.

skin, the external covering of an animal, important for protecting the inner bodily tissues from damage by environmental agents. It also restricts loss of water and other substances from the body; in homoiothermic (warm-blooded) animals, it regulates heat loss. It has an important sensory function, with many sense organs to detect changes in the external environment. See also DERMIS and EPIDERMIS.

skin beetle, ▷ DERMESTID BEETLE.

skin breathing, respiration by means of simple DIFFUSION across the body surface, normal in very small animals and common in amphibians; ▷ pp. 79, **180**.

skink, any LIZARD of the family Scincidae, the largest family of lizards with around 800 species. Occurring nearly worldwide, skinks are mainly insectivorous ground-dwellers or burrowers, with long, round, smooth-scaled bodies and tapering tails; most have short legs and a few have lost their limbs altogether.

Skinner, Burrhus Frederic (1904–), American psychologist; he pioneered the study of learning through operant CONDITIONING, for which he devised a special automated chamber, the 'Skinner box'.

skipper, any LEPIDOPTERAN insect of the widespread family Hesperiidae, small, stout-bodied butterflies with an erratic flight.

skua or **jaeger,** any of seven species of fast-flying, brown, gull-like sea birds of the family Laridae (order Ciconiiformes), migrating throughout the world's oceans. They are powerful and aggressive, and will take the eggs and young of other species; ▷ pp. 96, *164*.

skull, (in vertebrates) the part of the skeleton enclosing the brain; ▷ pp. *92, 110, 122, 133*.

skunk, any of several New World MUSTELIDS, typically with black and white markings and a bushy tail, well known for their ability to eject a foul-smelling fluid at an attacker; ▷ pp. **131**, 167.

slash and burn, a form of shifting cultivation practised in many forested regions, especially in the tropics, involving the opening of a clearing by cutting and burning, the growing of crops for a few seasons while soil fertility lasts, then abandonment.

slater, ▷ WOODLOUSE.

sleep, ▷ 174 (box).

sleeper, any of about 150 PERCIFORM fishes of the family Eleotrididae, marine, brackish, and freshwater species from tropical and warm-temperate regions. They closely resemble GOBIES, but their pelvic fins are not united into a sucking disc; ▷ p. 73.

sleeping sickness, ▷ p. 168.

slime mould, any of a group of protists belonging to the Myxomycetes; ▷ p. 26 (box).

slit-faced bat, any of about 10 microchiropteran BATS of the family Nycteridae, found chiefly in the tropical and subtropical parts of Africa and the Middle East, where they feed on a variety of invertebrate prey including scorpions. The group is characterized by a slit running from the nose to a point between the eyes, which is formed from a fleshy NOSE-LEAF thought to be used in ECHOLOCATION.

sloe, ▷ BLACKTHORN.

sloth, any edentate mammal of the families Megalonychidae (2 species of two-toed sloth) and Bradypodidae (3 species of three-toed sloth). Adapted for life in the trees of the tropical rainforest of South and Central America, all sloths are slow-moving and strictly herbivorous; ▷ pp. 106, **107**.

sloth bear (*Melursus ursinus*), a moderately large, black, shaggy BEAR of southern India and Sri Lanka, a specialized feeder on ants and termites; ▷ p. 129.

slow worm (*Anguis fragilis*), a limbless, snake-like lizard of Europe, southwest Asia and northern Africa; it feeds on worms and other small invertebrates.

slug, any terrestrial GASTROPOD mollusc in which the shell is absent or greatly reduced. In contrast to SEA SLUGS, land slugs are herbivorous; ▷ pp. **56**, 176.

smell or **olfaction,** one of the basic SENSES, dependent on sensitivity to remote/distant chemical stimuli; ▷ p. 188.
Other references: leeches 55; fishes 73; amphibians 80; reptiles 88; birds 97; carnivores 122; dogs 126; bears 128; primates 136.

smelt, any fish of the family Osmeridae, northern-hemisphere relatives of the SALMONIDS; the well-known European species *Osmerus eperlanus* smells strongly of cucumber. The name 'smelt' is also given to various unrelated groups of fishes. It is the Australian name for small marine fishes of the family Mugiloididae, while New Zealand smelts are fresh- and brackish-water species of the family Retropinnidae. There are also deep-sea smelts (Bathylagidae) and the herring smelts (▷ ARGENTINE).

smoky bat or **thumbless bat,** either of two microchiropteran BATS of the family Furipteridae, found in the American tropics. Both species are insectivorous, smoky-coloured, and lack the clawed thumb of other bats.

smut fungus, a species of basidiomycete fungus (▷ p. 35) of the order Ustilaginales. Many are important diseases of cereal crops, reducing the flower spike to a sooty mass.

snail, any of numerous GASTROPOD molluscs, most of which belong to the subclass Pulmonata (freshwater and terrestrial snails); they typically have a lung in place of gills, and a spirally coiled shell. *Helix aspersa* is the common garden snail; ▷ pp. 56, 176, 218, 219. See also WHELK, CONCH.

snake, any REPTILE of the suborder Serpentes (or Ophidia), containing over 2000 species and constituting (with lizards and amphisbaenians) the most successful group of living reptiles. The major families include COLUBRIDS, CONSTRICTORS, ELAPIDS, PIT VIPERS, SEA SNAKES and VIPERS; ▷ pp. 14–15, **82–9**, **176**, 191 (box).

snakebird, ▷ DARTER.

snake fly, any NEUROPTERAN insect of the family Raphidiidae. Adults are easily recognized by a greatly elongated 'neck' (thorax). Juvenile stages are terrestrial, living on tree surfaces and feeding on small insects.

snapdragon (*Antirrhinum majus*), a Mediterranean perennial herb in the family Lamiaceae (formerly Labiatae) with many garden cultivars. The strongly two-lipped flowers give rise to the common name, which may also be applied to other members of the genus.

snapper, any of about 185 PERCIFORM fishes of the family Lutjanidae, found throughout tropical seas except the eastern Pacific. They are fully scaled fishes, with spiny fins and large teeth at the front of the mouth; many are commercially valuable. The name 'snapper' is also used for various other groups of fishes, especially marketable species.

snapping turtle, any of several large, predatory North American freshwater turtles (CHELONIANS) of the family Chelydridae. The alligator snapping turtle lurks in murky water and snaps at prey attracted by a fleshy lure in its mouth.

snipe, any of 15 WADERS of the genus *Gallinago*, occurring worldwide in marshes and along sea shores. The common snipe (*G. gallinago*) has a very long bill used to probe deep for worms. See also PAINTED SNIPE.

snook, any of 35 PERCIFORM fishes of the family Centropomidae. They are principally marine or brackish-water fishes found in the Atlantic, Indian and Pacific Oceans, but there are several fresh-water species, particularly in Africa. They are robust predatory fishes, with powerful spines in the fins.

snowdrop, any of around 12 species of bulbous herbs in the European to Middle Eastern genus *Galanthus* in the family Amaryllidaceae. *G. nivalis* has many garden cultivars.

snow leopard (*Panthera uncia*), a large, greyish CAT with dark markings. It inhabits mountainous steppe and forest in the region of the Himalaya.

soapwort, any of around 20 species in the mainly Mediterranean genus *Saponaria* in the family Caryophyllaceae. The name derives from its use as soap, the leaves containing saponins.

Social Darwinism, a movement in later 19th-century sociology that attempted to apply Darwinian ideas of natural selection to society. It became associated with the eugenics movement, which argued the superiority of the white race and advocated the maintenance of racial purity – aims that became absorbed by the Nazis with monstrous results.

social organization, ▷ pp. 158–9.

sociobiology, the study of social behaviour, particularly its causation, development, and evolution through natural selection of the benefits to individuals.

sodium, a metallic element (symbol Na) essential to nerve function in animals; ▷ p. *196*, 209.

soft water, water with a low concentration of inorganic salt ions; ▷ p. 218.

softwood, wood from conifer tress (especially spruces, firs), generally of poorer quality than the HARDWOODS of broadleaved trees; ▷ p. 45.

soil, the substrate of most ecosystems, consisting of fragmented mineral material, air, water, organic matter and living organisms (including plant roots), mixed in variable proportions. It is the source of most of a plant's water and mineral uptake. For the soils of different ecosystems ▷ pp. 200–11. See also CHERNOZEM, HUMUS, LOAM, PEAT, PODZOL.

Solanaceae, a dicotyledonous plant family, including potatoes, tomatoes, aubergines (egg plant), peppers (CAPSICUM), tobacco and some nightshades; ▷ pp. 45, 46.

Solanum, a genus of around 1500 species of mainly tropical herbs or shrubs in the family, Solanaceae. Some species, such as AUBERGINE and POTATO are grown for food, and some for medical or ornamental purposes.

soldier, in ants and termites, a special CASTE of sterile members specialized for defence of the colony.

soldier beetle, any BEETLE of the family Cantharidae (over 3500 species). Adults, some of which are brightly coloured, are common on flowers and vegetation.

sole, any FLATFISH of the family Soleidae, especially *Solea solea*, which is very common in Europe. It is a migratory species, aggregating in great numbers in shallow water in the summer; it is perhaps the most commercially valuable of all the flatfishes.

solenodon, either of two mammalian species of the family Solenodontidae (order Insectivora). Relatively large, shrew-like animals, solenodons were formerly among the dominant carnivores in the West Indies, but the two surviving species are now highly endangered and restricted to Cuba and Hispaniola; ▷ pp. 120, **121**.

solitaire, 1. any of nine small THRUSHES with dark backs and white chests; they are forest birds living mainly in the Caribbean and South America. **2.** ▷ DODO.

somatic or **body cell,** any plant or animal cell except for the GERM CELLS.

somite, ▷ p. *187*.

song, a species-specific pattern of sounds, often serving to attract mates, used by a variety of animals, including birds (▷ BIRDSONG), whales (▷ p. 145) and crickets; ▷ p. 154.

sooty mould, a microscopic fungus that grows on honeydew on leaves of trees, causing the leaves to blacken and possibly contributing to the dark colour of honeydew or 'forest' honey.

Sorbus, a north temperate genus of the rose family, Rosaceae, with around 100 species of trees and shrubs. Some are grown as ornamentals, e.g. dogberry or American mountain ash (*S. americana*), rowan or European mountain ash (*S. aucuparia*) and the true SERVICE TREE (*S. domestica*).

soredium (pl. soredia), the structure by which lichens achieve asexual reproduction; ▷ p. 34 (box).

sorghum, any of around 60 species of annual and perennial tropical and subtropical grasses in the genus *Sorghum.* Many cultivars of *S. bicolor* are grown for grain or fodder.

sorrel, 1. any of a number of species of herbs in the genus *Rumex.* See also DOCK. Sheep's sorrell (*R. acetosella*) is a serious weed. **2.** ▷ OXALIS.

sorus (pl. sori), a collection of sporangia (spore-producing organs), occurring in large numbers on the underside of fern leaves; ▷ p. *36*. Similar structures are found on certain fungi.

sound (perception of), ▷ HEARING and p. 191.

sour, a basic taste quality; ▷ p. 188.

soursop (*Annona muricata*), a tropical American small evergreen tree in the family Annonaceae. It is cultivated for its acid fruits used in drinks and ice cream. See also CUSTARD APPLE.

South African, a floral region; ▷ p. 30 (map).

soybean or **soya bean** (*Glycine max*), a leguminous plant of the family Fabaceae, native to Asia. The seeds are cultivated for their oil, and a protein-rich food produced from the flour is used as a meat substitute. They are also a source of soy sauce and beansprouts.

spadefoot toad, any ANURAN amphibian of the family Pelobatidae, occurring in Europe, northern Africa and North America; these toads have a spade-like outgrowth on the foot used in burrowing (▷ p. 79). The name is sometimes used to describe various other burrowing species.

spadix, an INFLORESCENCE in which small, inconspicuous flowers are borne on an elongated, swollen axis, as in *Arum maculatum* (cuckoopint or lords-and-ladies).

Spanish fly, ▷ BLISTER BEETLE.

sparrow, any of 33 small, cryptically coloured FINCHES of the subfamily Passerinae (family Passeridae, the 'sparrow' family). They are native to the Old World, and some have become cohabitants with man. The house sparrow (*Passer domesticus*) probably came from the Middle East about 7000 years ago with the spread of agriculture. Since then it has spread worldwide; ▷ p. 97 (box). See also ACCENTOR, BUNTING.

sparrowhawk, any of 25 small birds of prey (▷ ACCIPITER), usually less than 40 cm (15 in) long, found in woodlands and forests worldwide.

spathe, a relatively large BRACT enclosing a SPADIX (e.g. in arum lily).

spawn, (to deposit) large numbers of eggs or young in water, e.g. in fishes, frogs, molluscs and crustaceans.

spearmint, ▷ MINT.

spear-nosed bat, ▷ LEAF-NOSED BAT (2).

specialist, an animal that is highly adapted (specialized) for a particular way of life, or one that feeds on a narrow range of foods; ▷ pp. 162–3.

speciation, the process by which new species arise; ▷ pp. 8–9, 10–11, **16–17**.

species, the smallest classificatory group (although sometimes divided into subspecies or races). The members of a species are able to breed among themselves; ▷ p. 18 and SPECIATION.

spectacled bear (*Tremarctos ornatus*), a small BEAR with black-brown fur and pale markings around the eyes. It inhabits forests in South America.

speedwell, ▷ VERONICA.

Spencer, Herbert (1820–1903), English philosopher. An important early supporter of evolutionary ideas, he coined the phrase 'the survival of the fittest' (▷ p. 13), and attempted to apply Darwinian ideas to social theory (▷ SOCIAL DARWINISM).

sperm, ▷ SPERM CELL.

spermaceti, a fine oil obtained from the head of the sperm whale; ▷ p. 145.

spermatheca, (in female insects, etc.) a sac in which sperm is stored prior to fertilization; ▷ pp. **62**, 63.

spermatium (pl. -tia), a non-motile male gamete (sex cell) produced by certain fungi and by red algae.

spermatophore, a package or capsule of SPERM picked up from the ground by the female or transferred to her directly, a feature of various animal groups practising internal FERTILIZATION, including many crustaceans, insects and salamanders; ▷ pp. 62, 68, 80, **184**.

spermatophytes, a term sometimes applied to the SEED-BEARING PLANTS.

spermatozoid, a motile male gamete (sex cell) produced by most brown seaweeds; ▷ p. 34.

spermatozoon (pl. -zoa), ▷ SPERM CELL.

sperm cell, a male gamete (sex cell). In animals they are produced in a TESTIS, and in plants in various structures (ANTHERIDIA in spore-bearing plants, POLLEN SACS in seed-bearing plants). The sperms of some algae (SPERMATOZOIDS), all spore-bearing plants (▷ p. 36) and most animals (spermatozoa) are motile, having a dense round or elongated head carrying the genetic material and a long tail-like flagellum with which they swim to fertilize an OVUM. The sperm of some fungi and red algae (SPERMATIA) and of seed-bearing plants (contained within POLLEN grains) are non-motile, and rely on water, wind or animals to take them to the ovum. For references ▷ REPRODUCTION.

sperm whale, any marine CETACEAN mammal of the family Physeteridae (3 species), found in all oceans and including the largest of the toothed whales, the true sperm whale or cachalot (*Physeter catodon*); ▷ pp. 142, 143 (box), 144, **145**.

sphagnum moss, ▷ BOG MOSS.

sphenophyte, any member of the division Sphenophyta, comprising the horsetails; ▷ pp. 30, **37**.

sphinx moth, ▷ HAWK MOTH.

spider, any predatory ARACHNID of the order Araneae, characterized by eight legs, a large abdomen, and a combined head and thorax (cephalothorax). Of the 30 000 described species, most are terrestrial; ▷ pp. **69**, 160, 164.

spider beetle, any BEETLE of the family Ptinidae; they are spider-like in appearance, and mostly associated with stored food.

spider crab, any DECAPOD crustacean of the family Maiidae, with a small triangular body and extremely long legs; ▷ p. 67.

spider-hunting wasp, any solitary WASP of the family Pompilidae, of worldwide distribution, which collect spiders to feed their offpring. The biggest (*Pepsis formosa*) stores large tarantulas in a burrow for its larval food.

spider mite, any herbivorous MITE of the family Tetranychidae, such as the red spider mite, which is a pest in orchards.

spider monkey, any agile, long-limbed New World monkey of the genus *Ateles* (family Cebidae). The related woolly spider monkey or muriqui (*Brachyteles arachnoides*) is the largest and probably the rarest of the New World monkeys; ▷ p. *138*.

spider plant (*Chlorophytum comosum*), a herb in the lily family, Liliaceae, with often white striped grass-like leaves. It is a popular house plant.

spiderwort, ▷ *TRADESCANTIA*.

spike, a RACEME inflorescence bearing stalkless (sessile) flowers, as in plantain.

spinach, any of several species of plant in the family Chenopodiaceae grown for their edible leaves, e.g. spinach (*Spinacia oleracea*) and spinach BEET or Swiss CHARD (*Beta vulgaris*).

spinal column, ▷ VERTEBRA.

spinal cord, (in vertebrates) a cord of nervous tissue that extends from the brain along the spinal column in the spinal canal (▷ box, p. 51). It consists of many nerve cells and nerve tracts, and carries impulses to and from the brain (▷ p. 183). It is constructed of a core of grey matter and outer layer of white matter.

spine, any hard pointed structure, such as the modified hair of a hedgehog (▷ p. 120) or the hardened fin ray of a fish. See also SPINAL CORD.

spinetail, ▷ OVENBIRD.

spinneret, any of several organs through which spiders exude silk; ▷ p. 69.

spiny anteater, another name for ECHIDNA.

spiny eel, 1. any of 10 species of marine, mostly deep-sea fish of the families Notacanthidae and Lipogenyidae (order Notacanthiformes). They typically have a long row of short, sharp spines along the back, and no tail fin; they are bottom-dwelling, browsing with a head-down posture.

2. any PERCIFORM fish of the family Mastacembelidae (60 species), occurring in tropical fresh/estuarine waters of Africa, Syria, southern Asia, and China. They are scaled, and have a row of sharp spines on the back preceding the soft dorsal fin, but the pelvic fins are absent.

spiny-headed worm, any worm-like invertebrate of the phylum Acanthocephala, the adults of which are gut parasites of fishes and other vertebrates (including man). They are unusual in having no gut at the young or adult stage.

spiny lobster, any DECAPOD crustacean of the genus *Palinurus*; they are lobster-like in appearance but have very large antennae and simple legs instead of claws; ▷ p. 67.

spiny rat, any of some 55 rat-like caviomorph RODENTS of the family Echimyidae with spiny or bristly coats. They are widespread in South America.

spiracle, 1. any of a series of small, paired openings in the cuticle of an insect by which air enters and leaves the TRACHEAE (2); ▷ pp. 59, 180–1. **2.** a rudimentary GILL SLIT visible as an opening behind the eye of sharks, rays and some primitive bony fishes.

spiraea, any of around 100 species of shrub in the north temperate genus *Spiraea* in the rose family, Rosaceae. Many, such as bridal wreath (*S. prunifolia*), are grown as ornamentals.

spirochaete, any corkscrew-shaped bacterium; ▷ p. 24.

Spirogyra, a genus of filamentous GREEN ALGAE with a spiral chloroplast. It is common in freshwater lakes and ponds.

spittlebug, ▷ FROGHOPPER.

sponge, any primitive aquatic animal of the phylum Porifera, which comprises some 5000 species, the great majority of which are marine. Sponges are the simplest of multicellular animals, with little coordination between individual cells; ▷ pp. 5, **52,** 160, 215.

spoonbill, any of six species of wading bird in the family Threskiornithidae (order Ciconiiformes), which also includes the IBISES. The bill is straight but ends in a spoon-like expansion, which is dragged from side to side in the water and snapped shut on prey; ▷ pp. **91,** 97 (box).

sporangium (pl. -gia), an organ in which spores are produced by meiosis (▷ p. 22). Sporangia are borne on leaves called SPOROPHYLLS; ▷ pp. 34 (algae), 35 (fungi), 36–7 (spore-bearing plants). Seed-bearing plants also have sporangia in the form of the pollen sac and the embryo sac (▷ pp. 38, 41).

spore, a reproductive unit of one or a few cells, usually microscopic, produced by all plants (including algae), fungi, bacteria and protozoans. They can be sexual or non-sexual in origin. In hostile environments bacteria produce spores that can remain dormant and resist extreme conditions for long periods (▷ p. 25). The spores of protists perform a similar function (▷ p. 27), although in both groups this is only one form of reproduction. In plants they are produced asexually by the SPOROPHYTE generation in the ALTERNATION OF GENERATIONS, and develop into the GAMETOPHYTE generation. In algae and most spore-bearing plants the spores are undifferentiated and these plants are called homosporous (▷ p. 36), but in heterosporous spore-bearing plants (▷ p. 36) and in the seed-bearing plants (▷ p. 38), two spore types of different sexes are produced.

spore-bearing plants, a group of primitive land plants that produce spores rather than seeds as the unit of dispersal. The group includes mosses, liverworts, club mosses, horsetails and ferns; ▷ pp. 36–7. Compare SEED-BEARING PLANTS.

sporophyll, a leaf bearing a SPORANGIUM (spore-producing organ). In plants such as ferns the sporophylls are the ordinary leaves, whereas in flowering plants the STAMENS and CARPELS are the radically modified sporophylls. In gymnosperms (e.g. conifers) they are the scales of the cones.

sporophyte, the stage in a plant life cycle that produces SPORES by asexual means. It alternates with a sexual GAMETOPHYTE stage (▷ ALTERNATION OF GENERATIONS). In most plants (except for some seaweeds and the bryophytes; ▷ pp. 34, 36) it is the sporophyte form that is the obvious plant; ▷ pp. 34 (algae), 36–7 (spore-bearing plants), 41 (flowering plants).

springbok or **springbuck** (*Antidorcas marsupialis*), a small, reddish-fawn, herd-dwelling ANTELOPE with a dark lateral band, white underparts and short horns. It inhabits dry plains in southern Africa.

spring hare or **springhaas** (*Pedetes capensis*), a hopping, burrowing, hare-like sciuromorph RODENT, with very short fore limbs, long hind limbs and a long, furry tail. Inhabiting arid habitats in southern Africa, it is the sole member of the family Pedetidae.

springtail, another name for COLLEMBOLAN.

sprout, a young shoot and/or root of a newly germinated seed or bud.

spruce, any of around 50 species of conifer in the north temperate genus *Picea,* family Pinaceae. The narrow leaves are attached to peg-like projections that remain on the twig after the leaves fall. Norway spruce (*P. abies*) is commonly used as a Christmas tree and is an important source of timber; ▷ pp. 39, 208.

spurge, ▷ *EUPHORBIA.*

squash, 1. summer squash, any cultivar of *Cucurbita pepo* (e.g. COURGETTE) the fruits of which are eaten while immature. **2.** winter squash, any of a number of cultivars of species of Cucurbita (e.g. *C. pepo* and *C. moschata*) the fruits of which are harvested when mature and stored for winter consumption. See also MARROW.

squawfish, either of two large North American species of the genus *Ptychocheilus* in the CARP family.

squid, any marine CEPHALOPOD mollusc of the order Teuthoidea, fast-moving predatory animals found in most seas; ▷ pp. **56–7,** 215, 220.

squill, any of around 80 species of bulbous herbs in the Old World genus *Scilla* in the lily family, Liliaceae.

squirrel, any of some 270 sciuromorph RODENTS of the family Sciuridae. Many squirrels are arboreal seed-eaters, but others are ground-living and burrowing (GROUND SQUIRRELS, CHIPMUNKS, WOODCHUCK, PRAIRIE DOGS). The African SCALY-TAILED SQUIRRELS are not closely related; ▷ pp. **133–4,** *201, 207, 209.*

squirrel monkey, either of two species of New World monkey of the genus *Saimiri* (family Cebidae), small, lively monkeys with short, brightly coloured coats; ▷ p. 138.

stag beetle, any SCARABAEOID BEETLE of the family Lucanidae. The males have large antler-like mandibles, but – unlike the shorter, more powerful ones of the female – they are largely ornamental; ▷ p. 63.

stag's-horn fern, any of various epiphytic ferns of the genus *Platycerium*; ▷ p. 37 (box).

stag's-horn moss (*Lycopodium clavatum*), a creeping species of club moss (▷ p. 37).

stalking, the use of stealth by predators to approach within striking range of prey without being detected; ▷ pp. 164–5.

stamen, the male reproductive organ of a flower, consisting of a pollen-producing anther on a stalk (the filament); ▷ p. 41.

stand, a term used of an undefined area of vegetation selected for descriptive study.

stapes, one of the three auditory ossicles in the mammalian middle ear (the 'stirrup'); ▷ pp. 98 (box) 191.

starch, a POLYSACCHARIDE carbohydrate into which glucose is converted for storage in plants; ▷ pp. 21, **32,** 33 (box), 179.

starfish or **sea star,** any of over 1800 ECHINODERMS of the class Asteroidea; ▷ pp. 5, **54,** 215.

stargazer, any of 25 tropical marine PERCIFORM fishes of the family Uranoscopidae, from the Atlantic, Indian and Pacific Oceans. Small, stocky-bodied fishes with eyes on the top of the head, they lie buried in the sand and possess venomous spines on the gill covers. They also have electric organs behind the eyes that can generate about 50 volts.

starling, any of about 100 PASSERINE birds in the family Sturnidae, which also includes the MOCKINGBIRDS. They are medium-sized birds, most with glossy plumage and long pointed bills, often used for digging for insect larvae; they are native to the Old World, and have been introduced to North America and Australia. Many starlings are noisy and gregarious; ▷ pp. 96, 97 (box), 213, 230.

statocyst, a relatively simple organ of balance found in many invertebrates; ▷ p. 190.

statolith, a granule of calcareous (or other) matter found in a statocyst organ; ▷ p. 190.

steelhead, ▷ TROUT.

stegosaur, any heavily armoured quadrupedal ORNITHISCHIAN dinosaurs of the genus *Stegosaurus*; ▷ p. 83.

stem, the part of a plant bearing leaves, buds and flowers; ▷ pp. 31–3.

stem cell, an undifferentiated cell set aside to replace or increase already specialized cells; ▷ pp. 186–7.

steppe, a type of temperate grassland found in eastern Europe and central Asia; ▷ pp. 199 (map), **206.**

sternum, the breastbone in vertebrates (▷ p. 92), or a functionally similar part of an invertebrate, such as the plate covering the underside of an insect.

stick insect, any tropical EXOPTERYGOTE insect of the family Phasmatidae (order Phasmida). They are long, thin and cylindrical, and closely resemble the twigs and plants on which they live; even their eggs resemble plant seeds. Some species are winged; ▷ pp. 61, 63 (box).

stickleback, any member of the family Gasterosteidae (order Gasterosteiformes), including marine, brackish and freshwater forms, all in the northern hemisphere. They are small fishes with rows of sharp dorsal spines. Nest-building, courtship and nest-guarding behaviour have been extensively studied, especially in the three-spined stickleback (*Gasterosteus aculeatus*); ▷ pp. 74, 77, 149, 155 (box), 161.

sticky willie, ▷ CLEAVERS.

stigma, the specialized surface for pollen reception at the tip of the carpel of a flower; ▷ p. 41.

stilt, any of nine species of long-legged WADERS, with long needle-like bills used to probe for insects and small crustaceans. They occur worldwide, and are closely related to AVOCETS.

stimulus, ▷ SIGN STIMULUS.

sting, a sharp organ in invertebrates such as bees and jellyfishes which is adapted to wound a predator or prey by piercing and injecting poison; ▷ pp. 53, *55,* 56, 59, 68, 77.

stingray, any RAY of the family Dasyatidae, which bear a long, venomous spine on the tail. Although usually harmless, stingrays will drive their spine into a leg when stepped on, causing excruciating pain. There is a small but significant number of reports of apparently aggressive attacks on divers; ▷ p. 75 (box).

stink bug, ▷ SHIELD BUG.

stinkhorn, any species (*e.g. Phallus impudicus*) of basidiomycete fungus (▷ p. 35) in the widely distributed family Phallaceae. The fruit body is shaped like a phallus, hence the scientific name, and the spores are dispersed by flies that come to feast on the foetid jelly at the apex.

stint, any of four small (less than 15 cm / 6 in) round-bodied WADERS of the family Scolopacidae, with short necks and a relatively short bill. They are found near water in the Nearctic, Palaearctic and the Orient.

stipe, the stalk of large kelps (brown seaweeds; ▷ p. 34) or of the caps of mushrooms and toadstools.

stipule, an outgrowth from a leaf base; it can be leaf-like, spiny, or a tendril.

stitchwort, any of a number of species of herb in the genus *Stellaria* in the family Caryophyllaceae, e.g. greater stitchwort (*S. holostea*). See also CHICKWEED.

stoat (*Mustela erminea*), a weasel-like MUSTELID with an elongated body, short legs and red-brown fur turning white in winter, widespread in North America and Eurasia. It is one of the smallest members of the order Carnivora; ▷ p. 131.

stock, any of several species of herbs in the cabbage family, Brassicaceae (formerly Cruciferae) cultivated in gardens, namely *Matthiola incana* (the biennial form of which is called Brompton stock), night-scented stock (*M. longipetala*) and Virginia stock (*Malcolmia maritima*).

stoma, ⊳ STOMATA.

stomach, a sac-like secretory part of the digestive tract in which the early stages of digestion occurs; ⊳ pp. *178*, **179**.

stomata (sing. stoma), the microscopic pores in the surface of leaves; ⊳ pp. 31 (box), 33 (box).

stonecrop, any of a number of species of SEDUM. Biting stonecrop or wall pepper (*S. acre*), common on walls, was used in herbal medicines; ⊳ p. 211.

stone curlew or **thick-knees,** any of nine species of slender scrubland birds belonging to the family Burhinidae (order Ciconiiformes). They are typically nocturnal and occur worldwide. Their alternative name comes from their distinctively swollen tarsal joints; ⊳ p. 213.

stonefish, any of around 20 tropical Indo-Pacific marine fishes of the family Synanceiidae, related to the SCORPION FISHES. Stonefishes have extremely venomous dorsal fin spines, and lie on the bottom in shallow water, superbly camouflaged by their colour and warty outgrowths on the body; ⊳ p. 77.

stonefly, any EXOPTERYGOTE insect of the order Plecoptera, of worldwide distribution. Adults are often drab, although some Australian forms are colourful; they are poor fliers and never found far from water. Nymphs are aquatic, usually in running water, and have small, tuft-like tracheal gills for respiration and a pair of long tail filaments. They take several years to develop and may be vegetarian or carnivorous; ⊳ pp. *61*, 219.

stoneroller, name given to several North American fishes of the CARP family; they are found in the major rivers and in the Great Lakes.

stork, any of 17 species of the family Ciconiidae (order Ciconiiformes) closely related to the New World VULTURES. They are large birds found in most temperate and tropical parts of the world, and are usually black and white with brightly coloured legs and long bills, held outstretched when flying. The white stork (*Ciconia ciconia*) of Eurasia is strongly migratory, spending the winter in Africa; ⊳ pp. *91*, 203.

storm petrel, any of 20 small oceanic sea birds (family Procellariidae, order Ciconiiformes), with small wings and fluttering flight. Some, such as Wilson's storm petrel (*Oceanites oceanicus*), migrate from the Antarctic to sub-Arctic waters; ⊳ p. 96.

stratigraphy, the study of the strata (layers) of sedimentary rocks, important in dating the fossils contained within them (⊳ p. 4).

strawberry, any of around 15 species of perennial temperate and subtropical herbs in the genus *Fragaria* in the rose family, Rosaceae. Cultivars of *F.* x *ananassa* are grown for their fruit; ⊳ p. 43.

strawberry tree, ⊳ ARBUTUS.

stream, ⊳ pp. 218–19.

streptococcus (pl. -ci), any member of the bacterial genus *Streptococcus*. They are globular eubacteria growing in chains and responsible for a variety of diseases such as scarlet fever and rheumatic fever; ⊳ p. 24.

stridulation, the production of shrill creaking noises by rubbing together special body structures, as in crickets and grasshoppers; ⊳ pp. 59, **62**.

string bean, ⊳ FRENCH BEAN.

stromatolite, a fossilized colony of cyanobacteria; ⊳ p. *6*.

sturgeon, any primitive RAY-FINNED FISH of the family Acipenseridae, a mainly marine group of the northern hemisphere, although some species spawn in fresh water. They have plate-like scales on the flanks, and lay numerous eggs, sold as caviar; ⊳ pp. 70–1, 72, *76*.

style, the part of the carpel between the ovary and the stigma; ⊳ p. 41.

stylopid or **stylops,** any INSECT of the small and highly unusual order Strepsiptera. Females are bag-like, and (together with the larvae) are internal parasites of other insects, especially HOMOPTERANS, DIPTERANS, bees and wasps (⊳ p. 63). Males have short, strap-like fore wings, but can fly by means of their fan-shaped hind wings.

subduction, the process by which oceanic crust is destroyed; ⊳ p. 8.

sublittoral, the lowest zone on a shore profile, consisting of those parts below the mean low-water mark down to the lowest level at which plants can grow.

submission, form of behaviour shown by subordinate animals to dominant ones in social encounters or within social HIERARCHIES, often involving APPEASEMENT or some other kind of display; ⊳ pp. 153, 154 (box).

subordination, a state of inferiority or low status (in relation to a higher-ranking or dominant individual) within a social grouping or HIERARCHY.

subspecies, a subdivision of a species, forming a group whose members resemble one another and differ from other members of the same species in one or more characteristics. Nevertheless, because reproductive isolation is incomplete, groups of subspecies nearly always merge into each other and can interbreed; ⊳ pp. 17 (gull subspecies), 18 (table). The term 'race' is sometimes used for subspecies in zoology and botany, but in human beings the so-called races have very little genetic distinction and are in no way as differentiated as subspecies. The two subspecies of *Homo sapiens* are the extinct *H. s. neanderthalensis* (Neanderthal man) and *H. s. sapiens* (modern man); ⊳ p. 18.

substrate, 1. the non-living material (e.g. soil, rock, mud, shingle) on which an animal or plant lives and grows. **2.** the substance on which an enzyme acts.

succession, a process of change in which communities of organisms gradually replace one another until eventually a stable (CLIMAX) condition is achieved. The process is directional and, to a certain extent, predictable. For an example ⊳ p. 219.

succulent, any plant adapted to arid environments by storing water in thick, fleshy leaves. The adaptation is found in various groups, e.g. stonecrops, aloes, and agaves.

sucker, 1. a shoot produced from an underground root; a means of vegetative reproduction (⊳ p. 33) common in roses and in certain trees. **2.** any freshwater fish of the family Catostomidae, in the same order as the CARP. There are some 60 species occurring in China, Siberia and North America. They have thick lips and no teeth in the jaws.

sucker-footed bat or **golden bat** (*Myzopoda aurita*), a microchiropteran BAT (the only species in its family), known only from Madagascar. Like the New World DISC-WINGED BATS, this bat has a suction disc on each limb, but they are immobile; it roosts inside hollow stems or curled leaves.

sucking louse, any EXOPTERYGOTE insect of the order Anoplura (about 300 species). Wingless blood-feeding parasites of mammals, these insects have fine piercing mouthparts and well-developed claws for grasping hairs. The human head louse (*Pediculus humanus capitis*) commonly affects schoolchildren, while the human body louse (*P. h. corporis*) is only a problem when clothes are not regularly laundered. The crab louse (*Pthirus pubis*) affects pubic and underarm regions, and is sexually transmitted; ⊳ pp. *61*, 64.

suckling, (in mammals) feeding (or being fed) on milk; ⊳ pp. 98 (box), 100.

sucrose, a type of SUGAR – familiar as table sugar – into which glucose is converted in plants for transport; ⊳ pp. **32**, 33 (box), 179.

sugar, any of a group of carbohydrates of low molecular weight, including mono-, di- and some oligosaccharides. Sugars include GLUCOSE, FRUCTOSE, LACTOSE, MALTOSE and SUCROSE. Sugars are a source of cellular energy, and also act as structural molecules (e.g. for plant cell walls) and as signalling molecules; ⊳ pp. 6 (box), 21, 31–3, **179**.

sugar beet, ⊳ BEET.

sugar cane, any of a number of giant perennial species and cultivars of grass within the genus *Saccharum* cultivated mainly in tropical areas for the production of sugar, which is extracted from the stem or CANE – hence 'cane sugar' as distinct from BEET sugar. Other products include molasses and cane syrup; ⊳ p. 45.

sugar palm (*Arenga pinnata*), a palm cultivated in the tropics for its sugary sap, harvested for the production of palm sugar.

sulphur, 1. a non-metallic element (symbol S), essential in proteins; ⊳ p. 31. **2.** any PIERID butterfly of the genus *Gonepteryx*.

sulphur bacteria, a group of bacteria that use sulphur as a source of energy; ⊳ pp. 7, 25.

sulphur dioxide, a gas (SO_2) released from the combustion of fossil fuels. It contributes to acid rain; ⊳ p. 225.

sultana, a dried seedless grape.

sumac, any of a number of lianas, shrubs and trees in the temperate and subtropical genus *Rhus*. Several are grown as ornamentals; ⊳ POISON IVY.

sun bear (*Helarctos malayanus*), or 'honey bear', an omnivorous, climbing BEAR with black-brown fur. The smallest of the bears, it inhabits forests in Southeast Asia; ⊳ p. *201*.

sunbird, any of 115 PASSERINE birds of the tribe Nectariniini (family Nectariniidae). They are very colourful, active species, living mostly in Africa but extending throughout the Old World tropics to Australia. They have downwardly curved bills, tubular tongues, and many are nectar-eaters; ⊳ p. 157.

sunbittern (*Eurypya helias*), a bird with a long orange bill and legs and a barred body, belonging to its own family within the order Gruiformes. It lives in dense Neotropical forests, where it dances in sunny clearings, spreading its multicoloured wings.

sundew, any of several carnivorous plants of the genus *Drosera*, family Droseraceae, found in bogs; ⊳ pp. 47 (box), 219.

sunfish, any of about 30 PERCIFORM fishes of the family Centrarchidae, North American freshwater fishes, with deep bodies and spiny fins. Most are nest-builders, the male guarding the fertilized eggs. They are kept as aquarium fishes and have been widely introduced beyond their natural range. See also MOLA.

sunflower, any of around 110 species of tall annual and perennial herbs in the North American genus *Helianthus* in the daisy family, Asteraceae (formerly Compositae). Cultivars of *H. annuus* are grown for their seeds, which are rich in oil. See also ARTICHOKE.

sunstar, a many-armed STARFISH; ⊳ p. 54.

surgeon fish or **tang,** any of 77 tropical marine PERCIFORM fishes of the family Acanthuridae. They are colourful, deep-bodied fishes that graze on algae. Their name derives from the pair of scalpel-shaped spines on either side of the tail stem, which can be erected in defence; ⊳ p. 77.

suricate, another name for MEERKAT; ⊳ p. 123.

suspension-feeding, ⊳ FILTER-FEEDING.

swallow, any of 87 PASSERINE birds of the subfamily Hirundinidae, small, gregarious, aerial-feeding insect-eaters with a small bill surrounded by bristles, occurring worldwide. Swallows have long pointed wings and forked tails, often with streamers; their legs and feet are very small. They usually make cup-shaped nests of mud pellets and straw; ⊳ pp. 17 (box), 94, 95, 211.

swallowtail, any LEPIDOPTERAN insect of the family Papilionidae, comprising 600 known species, mainly from the tropics. The hind wings often have a tail-like extension.

swamp, ⊳ MARSH (2).

swamp cypress, any of various conifers of the genus *Taxodium*, especially *T. distichum*; ⊳ p. 39.

swamp monkey (*Allenopithecus nigroviridis*), a stockily built Old World cercopithecine monkey (⊳ p. 139), sparsely distributed in swamp forests of the Congo and Zaïre.

swan, any of seven species of heavily built water birds of the family Anatidae, which also includes DUCKS and GEESE. Usually white with a long slender neck, the mute swan (*Cygnus olor*) is a widespread species in the Palaearctic. The male has a prominent swelling at the base of the bill; ⊳ pp. *91, 93*.

sweat gland, (in mammals) a tubular gland in the DERMIS of the skin secreting sweat through pores on the skin surface. Evaporation of sweat cools the skin and the blood in surface vessels.

swede, a cultivar of *BRASSICA napus* with a swollen root and stem used either as a vegetable or as winter feed for animals.

sweet, a basic taste quality; ⊳ p. 188.

sweet bay, 1. ⊳ BAY LAUREL. **2.** ⊳ MAGNOLIA.

sweet brier or **eglantine** (*Rosa rubiginosa*), a prickly shrub with glandular leaves that smell of apple when crushed. See also ROSE.

sweet cicely, ⊳ MYRRH (2).

sweet clover, ⊳ MELILOT.

sweet corn, a variety of MAIZE in which conversion of sugar to starch is reduced, so giving rise to grain that is sweeter and more tender.

sweet pea (*Lathyrus odoratus*), an ornamental climbing legume with scented flowers. There are cultivars showing a wide variety of flower colours.

sweet pepper, ⊳ CAPSICUM.

sweet potato (*Ipomoea batatus*), a perennial herbaceous vine in the family Convolvulaceae. It is cultivated in tropical, subtropical and warm temperate regions for its edible tubers. See also MORNING GLORY.

sweetsop, ⊳ CUSTARD APPLE.

sweet william, ⊳ *DIANTHUS*.

swift, any of about 100 small, usually brown birds with long, scythe-like wings. They spend much of their time flying fast, catching insects with their small wide mouths. They belong to the family Apodidae (order Apodiformes) – a reference to their small feet. They occur worldwide except in high latitudes; ⊳ pp. 17 (box), 90, *91*, 95, 96. See also SWIFTLET, TREE SWIFT.

swiftlet, any of 28 cave-dwelling SWIFTS, occurring in the Orient and Australasia. Their nests are made of plant material consolidated with copious saliva and are the basis of bird's-nest soup; ⊳ p. 95.

swim bladder or air bladder, a gas-filled sac in the body cavity of most bony fishes, usually acting as a buoyancy aid; ⊳ pp. *72*, 73 (box).

swimmeret, any of a series of small unspecialized appendages under the abdomen of many crustaceans, used for swimming or carrying eggs.

Swiss cheese plant, ⊳ *MONSTERA*.

swordfish (*Xiphias gladius*), a large PERCIFORM fish,

the sole member of the family Xiphiidae, occurring in tropical and temperate seas worldwide. It is a powerful swimmer, and reaches lengths of up to 4.9 m (16 ft). The long blade-like snout, flattened above and below, is thought to be used to knock out prey fishes. The adults are scaleless and have no teeth or pelvic fins.

sycamore, 1. *Acer pseudoplatanus*, a large MAPLE native to continental Europe, but introduced to Britain about 500 years ago and now naturalized. **2.** North American name for PLANE. **3.** biblical name for FIG, possibly the sycamore fig (*Ficus sycamorus*).

symbiont, ⊳ SYMBIOSIS.

symbiosis or **mutualism,** a close association between two different species (symbionts) from which both derive benefit; ⊳ pp. 167, **168–9**. Other references: lichens 34 (box); fungi/plants 35, 39, 43; pollination 42-3; algae/coral 53, 215; insects 65; crustaceans 67; fishes 76. See also COMMENSALISM.

synapse, a specialized junction between nerve cells; ⊳ p. 182 (box).

synapsid, any REPTILE of a group defined by a single, lower opening in the temporal region of the skull (⊳ pp. 82, *84*), comprising the MAMMAL-LIKE REPTILES.

syrinx, the singing organ of a bird, situated at the base of the trachea (windpipe); ⊳ p. 93.

tabasco, ⊳ CAPSICUM.

tadpole, the aquatic larva of an amphibian; ⊳ p. 81.

tahr, any of three dark-brown wild GOATS of the genus *Hemitragus*, with shaggy manes and short horns. They inhabit rugged mountains in Oman, the Himalaya and southern India; ⊳ p. 115.

taiga, ⊳ BOREAL FOREST.

tail flash, a form of ALARM RESPONSE, as when pronghorn antelope expose their white rump-patch as a signal for others to flee; ⊳ pp. 149 (box), 153, 167.

tailorbird, any of 10 species of WARBLER living in India and Southeast Asia. They make nests by stitching two leaves together and building the nest inside.

takahe (*Notornis mantelli*), a large, rare flightless RAIL. The takahe has a deep red bill, a green back and a blue head and breast. It eats tussock grass in the high valleys of New Zealand's South Island.

takin (*Budorcas taxicolor*), a stocky, light- to dark-brown, GOAT-like artiodactyl with short, robust horns. It inhabits high altitude thickets in eastern Central Asia.

talapoin, either of two species of Old World cercopithecine monkey of the genus *Miopithecus* (⊳ p. 139), small monkeys that live in lowland forest in west/central Africa and are known as crop-raiders.

tamandua, another name for the lesser ANTEATERS; ⊳ p. 107.

tamarack, ⊳ LARCH.

tamarin, any of several small tree-dwelling New World monkeys of the family Callitrichidae (which also includes the closely related marmosets). They are social and active by day; ⊳ pp. 136, **138**, *229* (box).

tamarind (*Tamarindus indica*), a tropical leguminous tree. Round the seeds is an acid pulp, also called tamarind, used in Indian cooking.

tamarisk, any of around 54 species of maritime trees or shrubs in the mainly Mediterranean genus *Tamarix* in the family Tamaricaceae. See also MANNA.

tanager, any of around 250 PASSERINE birds of the tribe Thraupini, related to the BUNTINGS. They are brightly coloured fruit- and insect-eaters, living in open woodlands of the New World.

tangerine, ⊳ MANDARIN.

tannin, any of various phenolic compounds found in

plants, sometimes in special tannin cells. They are used in the preparation of leather and medically as astringents, and are present in tea; ⊳ pp. 47, 209.

tapetum, a pigmentary layer of the retina in certain species, e.g. cats (⊳ p. 124), that reflects light back through the retina to enhance visual acuity in low-light conditions.

tapeworm, any of some 3000 ribbon-like FLATWORMS of the class Cestoda, which live as parasites in vertebrates; ⊳ pp. 55, 178.

tapioca, ⊳ CASSAVA.

tapir, any sturdily built, forest-dwelling PERISSODACTYL mammal (family Tapiridae) of Central and South America (3 species) and Southeast Asia (1 species), with a distinctive short, flexible trunk; ⊳ pp. 110, **111**, *201*.

taproot, a large single root growing vertically downwards, and from which smaller lateral roots extend; ⊳ p. 44.

tarantula, any large, hairy SPIDER of the family Theraphosidae, mainly restricted to the New World tropics, which feeds on large insects and small vertebrates. The group includes the so-called 'bird-eating spiders' (⊳ p. *69*). Although tarantulas have a strong bite, they are rarely dangerous to humans. The name is also given to the unrelated *Lycosa tarentula* of southern Europe.

tardigrade, a WATER BEAR.

tarpan, an extinct wild HORSE, believed to be the ancestor of the domesticated horse; ⊳ p. 111.

tarpon, either of two species of the genus *Elops* in the family Albulidae, in the same order as the TENPOUNDERS. They are large and attractively coloured, with bright silver flanks and dark green-blue on the back, and are popular as game fishes.

tarragon, ⊳ ARTEMISIA.

tarsier, any of three small higher PRIMATES of the family Tarsiidae, nocturnal tree-dwellers in the dense forests of Borneo, Sumatra and the Philippines; ⊳ pp. 136, 138 (box).

tarsus, 1. the small bones that support the back part of the foot of a vertebrate, including the ankle and heel. **2.** the part of an arthropod's limb furthest from the body (⊳ p. *59*).

tarweed, any of various plants of the genera *Madia* (from western South America) and *Argyroxiphium* (the Hawaiian tarweeds, including the silverswords; ⊳ p. *214*).

Tasmanian devil (*Sarcophilus harrisi*), a squat, powerfully built DASYURID marsupial, now confined to Tasmania. Although persecuted as a sheep-killer, it is largely a scavenger of carrion; ⊳ p. 105.

Tasmanian wolf or **tiger,** another name for THYLACINE; ⊳ p. 104.

taste or **gustation,** one of the basic SENSES, dependent on sensitivity to directly contacted chemical stimuli; ⊳ pp. 59, 73 (box), **188**.

taste bud, in vertebrates, an associated group of gustatory cells, usually found on the tongue; ⊳ p. 188.

Taxodiaceae, the redwood family (⊳ pp. 30, 39).

taxon (pl. taxa), any classificatory group, such as kingdom, phylum, division, class, family, genus and species (⊳ p. 18).

taxonomy, the science of biological classification; ⊳ p. 18.

tayra (*Eira barbata*), a dark brown MUSTELID resembling a long-legged marten. It inhabits forests in Central and South America.

TCA cycle or **Krebs cycle,** a metabolic pathway involving the transformation of glucose into a utilizable form of energy; ⊳ p. 21.

teak, the timber from the teak tree (*Tectona grandis*), a native of Southeast Asia. It has a high resin content and is therefore resistant to decay.

teal, any of nine small dabbling DUCKS, with short necks and a fast erratic flight. The Palaearctic common teal (*Anas crecca*) has a colourful chestnut and green head.

teaplant (*Camellia sinensis*), ⊳ pp. 46–7.

teasel, any of around 15 species of biennial herbs in the Old World genus *Dipsacus*. Fuller's teasel (*D. sativus*) was used for teasing cloth.

tectorial membrane, in the vertebrate ear, ⊳ p. 191.

tegu, any large American TEIID lizard of the genus *Tupinambis*. Female tegus lay their eggs in a hole torn in the wall of an arboreal termites' nest; the termites thereupon repair the damage, thus sealing the eggs safely within.

teiid, any of over 200 species of LIZARD of the exclusively New World family Teiidae, mainly of tropical and subtropical distribution. Many teiids are ground-dwelling and feed on small animals, but there is considerable variation within the group, which includes TEGUS and RACERUNNERS.

teleost, any RAY-FINNED FISH belonging to the subdivision (or infraclass) Teleostei, comprising over 20 000 species, or about 95% of all living fish species. Teleosts typically have SWIM BLADDERS and tails in which the upper and lower halves are equal, but the group is enormously diverse and not easily characterized; ⊳ pp. 70–1, 72.

telsontail, common name for a PROTURAN insect.

temnospondyl, any member of a group of extinct amphibians that are thought to be the ancestors of frogs and toads, and possibly also of salamanders; ⊳ p. 78.

temperate, characterized by moderate temperatures. The temperate zones are those between the Arctic Circle and the tropic of Cancer, and between the Antarctic Circle and the tropic of Capricorn (i.e. between latitudes 66° 33′ and 23½° N and S), although there are considerable variations in climate and vegetation within these zones, even at similar latitudes; ⊳ pp. 198–9 (maps). For temperate ecosystems ⊳ pp. **206–7**.

temperature (sensitivity to), ⊳ THERMORECEPTION.

tench (*Tinca tinca*), a freshwater fish of the CARP family, originally European but introduced to North America and elsewhere. It is eaten in many parts of Europe, and folklore attributes healing properties to the copious mucus produced from its skin.

tendril, an elongated organ originating from a leaf, stem or bud, which attaches itself to a support in response to physical contact (an ability called haptotropism). Tendrils are found in many climbing plants, e.g. peas; ⊳ p. 45.

tenebrionid beetle, any BEETLE of the very large family Tenebrionidae, containing some 15 000 species. Among the more important members of this group are the FLOUR BEETLES and the MEALWORMS.

tenpounder, any large, tropical or subtropical (mainly) marine fish of the family Elopidae (order Elopiformes). They are attractively coloured (silver-blue above with brilliant silver flanks) and popular as game fishes, but the flesh is not good to eat.

tenrec, the common name given to all but 3 of the 34 species in the mammalian family Tenrecidae (order Insectivora), the others usually being referred to as OTTER SHREWS. Tenrecs were amongst the first mammals to colonize Madagascar, where they are largely restricted today, and they are diverse in form and habit, having evolved to fill a wide range of insectivorous niches; ⊳ pp. 120, **121**.

tentacle, (in many marine invertebrates) a slender, elongated flexible organ, usually on the head or around the mouth, used for feeling, grasping or catching prey; ⊳ pp. 53, 54, *55*, 56–7, 121.

termite or **white ant,** any EXOPTERYGOTE insect of the order Isoptera, comprising about 2000 known species, mainly from the tropics. They are social, living in colonies with a well-defined caste system; ⊳ pp. *61*, 63, 64–5, 153, **156** (box), **160** (box), 169, 202, 203.

tern, any of about 40 species of predominantly grey sea birds (family Laridae, order Ciconiiformes), worldwide in distribution. They are smaller and more slender than true gulls and have smaller feet. The bill is usually sharply pointed and the tail forked; ⊳ p. 96, **171** (box).

terrapin, name given to a number of freshwater turtles (CHELONIANS) of the family Emydidae, of Old and New World distribution. They have flattened, webbed limbs and mainly feed on small aquatic animals.

territoriality, ⊳ TERRITORY.

territory, an area defended by an animal or group of animals, often as a mating area (⊳ LEK) or as a source of vital resources such as food and nesting material. Territorial behaviour or territoriality – establishing, marking and defending a territory – occurs in many species, and may vary within a species depending on the availability of resources; ⊳ pp. 153, **156–7**, 163. Other references: insects 62; amphibians 80; reptiles 86; birds 94, 152 (box); shrews 121; tigers 125; dogs 126–7; primates 137–41; seals, etc. 143.

Tertiary period, a geological period lasting from 65 to 1.6 million years ago; ⊳ p. 4.

test, (in various invertebrates) a hard outer covering.

testa, the seed coat. It is derived from the peripheral cell layer of the ovule (the integument), and is commonly hard and protective when dry; ⊳ pp. 38, 41.

testis, in male animals an organ (gonad), usually paired, in which SPERM and often hormones (e.g. testosterone) are produced. In most animals the testes are situated internally, but in mammals they are generally contained in a scrotal sac; ⊳ p. 184.

testosterone, ⊳ TESTIS.

Tethys Sea, an ancient ocean that during the Mesozoic era lay to the south of Europe and much of Asia, and to the north of Africa and India; ⊳ p. 9 (map).

tetra, name given to various CHARACIN fishes, for example those of the genus *Alestes*. Small and brightly coloured, they are very popular aquarium fishes; ⊳ p. 74.

tetrapod, any (ancestrally) four-limbed vertebrate, i.e. the amphibians, reptiles, birds and mammals. In many tetrapods, some or all of the limbs have been lost or modified (e.g. the fore limbs of birds into wings).

thallus (adj. thalloid), a plant body showing no differentiation into stem, roots and leaves, and without a vascular system. All algae and fungi are thalloid (⊳ pp. 34–5), as are some liverworts (⊳ p. 37) and the gametophytes of ferns (⊳ p. 36).

thanatosis, the technical name for DEATH-FEIGNING; ⊳ p. 167.

thecodont, any member of an extinct group of DIAPSID reptiles that had teeth set in sockets. The group gave rise to the dinosaurs, pterosaurs, crocodiles and birds; ⊳ pp. **83**, *84*.

thelodont, ⊳ p. 70.

therians, the MARSUPIALS and PLACENTAL MAMMALS; ⊳ pp. 100–1

thermoreception, sensitivity to temperature or heat. A related sense (sensitivity to infrared radiation) is found in some snakes; ⊳ pp. 88, 189, 191 (box).

theropod, any carnivorous bipedal SAURISCHIAN dinosaur of the suborder Theropoda, including tyrannosaurs and deinonychosaurs (clawed dinosaurs); ⊳ pp. 83–4.

thistle, any species of prickly annual to perennial herbs in the genera *Carduus* (around 120 Old World species) and *Cirsium* (around 150 northern hemisphere species) in the daisy family, Asteraceae (formerly Compositae). A number of other thistle-like plants are also referred to as thistles, e.g. carline thistle (*Carlina vulgaris*), and sow thistle (*Sonchus* spp.).

thorax, the part of an insect's body lying between the head and the abdomen, to which the legs and wings are attached (⊳ p. 59). In crustaceans and arachnids the thorax is often fused with the head to form a CEPHALOTHORAX. The part of a vertebrate's body bounded by the ribs (roughly the 'chest') may also be termed the thorax.

thorn, a sharply pointed structure modified from a shoot or root; in gorse almost all the shoot system is thorny.

thornbill, 1. any of five species of South American HUMMINGBIRD belonging to the genus *Chalcostigma*, with a short, sharp bill. **2.** any of 13 Australasian species of tiny, cryptically coloured 'finches', with a warbling song. They belong to the genus *Acanthiza* (family Pardalotidae, order Passeriformes) and build dome-shaped nests.

thorn forest, a tropical vegetation type dominated by thorn-bearing shrubs and trees (such as acacias). Often modified by fire and grazing.

thorny devil (*Moloch horridus*), an Australian desert-dwelling AGAMID lizard, with an extraordinary body covering of protective spines.

threadworm, name given to various thread-like ROUNDWORMS, including some that are parasitic in the human gut ('pinworm').

threat display, a form of DISPLAY occurring between members of the same species engaged in an AGGRESSIVE encounter (⊳ p. 157, box), or a defensive ploy that aims to intimidate a predator into withdrawing (⊳ p. 167).

thresher, any of five predatory marine SHARKS of the family Alopiidae. They may reach a length of nearly 6 m (20 ft), but roughly half of this is accounted for by the extremely elongated upper lobe of the tail. They use the tail to round up the schooling fishes on which they feed, and possibly to stun larger fishes.

threshold level, the lowest level of stimulation to which a sensory cell responds; ⊳ p. 188.

thrift, any of around 80 species of perennial herbs in the maritime, arctic and alpine genus *Armeria* in the family Plumbaginaceae.

thrips, any EXOPTERYGOTE insect of the worldwide order Thysanoptera. They are mostly minute in size and feed on the contents of plant cells. Being so small, they 'row' themselves through the air using fringed, feather-like wings; ⊳ p. *61*.

thrush, any of about 180 PASSERINE birds of the subfamily Turdinae, in the 'thrush' family Turdidae; they are found worldwide. Thrushes are good songsters (e.g. NIGHTINGALE). Larger, 'typical' thrushes are about 23 cm (10 in) long, usually brown/grey with a pale mottled breast. The smaller thrushes are about 15 cm (6 in) long and usually more brightly coloured; ⊳ p. 97 (box). See also ROBIN, BLUEBIRD, CHAT.

thumb, the first digit of the hand (especially in primates), often opposable to the other fingers; ⊳ p. 136.

thumbless bat, ⊳ SMOKY BAT.

thylacine, Tasmanian wolf or **Tasmanian tiger** (*Thylacinus cynocephalus*), the largest of the modern MARSUPIAL carnivores, almost certainly hunted to extinction in the present century; ⊳ p. 105 (box).

thyme, any of around 300 species of small aromatic shrubs in the temperate Eurasian genus *Thymus* in the family Lamiaceae (formerly Labiatae). Common thyme (*T. vulgaris*) is used in flavouring.

thyroid gland, in vertebrates, an endocrine GLAND that secretes HORMONES (principally thyroxine) involved in body growth and metabolism; ⇨ p. 183.

thyroxine, ⇨ THYROID GLAND.

thysanuran or **bristletail** any APTERYGOTE insect of the order Thysanura, containing over 550 species of worldwide distribution. Thysanurans are mostly small, and found in soil, leaf litter, rotting wood, and ant and termite nests, where they feed on decaying plant material. Their name refers to the presence of three fringed tail filaments. Each abdominal segment bears a pair of short appendages, which are probably sensory; ⇨ pp. 60–1.

tibia, 1. (or shinbone) the inner and usually the larger of the two bones of the vertebrate hind limb, between the knee and ankle. **2.** the fourth joint of an insect's leg (⇨ p. 59).

tick, any tiny parasitic ARACHNID of the families Ixodidae (hard ticks) and Argasidae (soft ticks) in the order Acari. Many thousands of species have been described and many others await discovery; ⇨ pp. 68–9.

tick-bird, ⇨ OXPECKER.

tidal cycle, the rhythmical ebb and flow of the tides, important as a cue for the cyclical activity patterns of various marine organisms; ⇨ p. 175.

tiger (*Panthera tigris*), the largest of the CATS, a solitary, nocturnal predator, now largely confined to reserves in southern Asia; ⇨ pp. 124–5, *201*.

tiger beetle, any GROUND BEETLE of the subfamily Cicindelinae. Active predators with well-developed eyes and large mandibles, they are fast-moving and often fly. The larvae are also predatory, lying in wait for prey in tunnels dug vertically into the soil.

tiger cat, 1. (*Felis tigrana*), a small CAT with a dark striped coat, found in Central and South America. **2.** another name for QUOLL.

tiger moth, any LEPIDOPTERAN insect (moth) of the genus *Arctia* in the family Arctiidae (over 3500 species). They are often brightly coloured, in black, orange and yellow. They (and various other arctiids) produce the densely haired larvae known as 'woolly bears'.

timber wolf, another name for the grey WOLF; ⇨ pp. 126–7.

tinamou, any of 47 species in the order Tinamiformes, living in forest and barren land of Mexico, Central and South America. Although closely related to the flightless RATITES, they do occasionally fly. Most, such as the great tinamou (*Tinamous major*), are camouflaged ground-dwellers that feed on plants; ⇨ pp. 91, 97 (box).

Tinbergen, Niko, (1907–89), Dutch-born ethologist (moved to Oxford in 1949), who performed meticulous investigations into the behaviour of many species, including gulls and sticklebacks, for which he shared the Nobel Prize in 1973 with LORENZ and von FRISCH; ⇨ pp. 148–9.

tissue, any group of cells, often similar in form, that together perform specific functions for the organism, e.g. muscle tissue.

tit, a name given to birds of two different families, typically small round-bodied songbirds with rounded wings. Tits tend to be arboreal and gregarious and feed mostly on insects and buds; ⇨ LONG-TAILED TIT, PENDULINE TIT, TITMOUSE.

titi, any New World monkey of the genus *Callicebus* (family Cebidae), long-furred species with long, non-prehensile tails; ⇨ p. 138, 155.

titlark, ⇨ PIPIT.

titmouse or **chickadee,** any of 53 PASSERINE birds of the subfamily Parinae (family Paridae), living mostly in mixed forests of Africa and throughout the northern hemisphere. They nest in trees, and some, such as the blue tit (*Parus caeruleus*), have become urbanized; ⇨ p. 150. See also TIT.

toad, the common name given to many of the ANURAN amphibians, generally to warty-skinned species, but the distinction between toad and FROG is often arbitrary. In a restricted sense, the name 'toad' may be used for the 'true' toads of the family Bufonidae, such as the common European toad *Bufo bufo* (⇨ p. 80) and the NATTERJACK; ⇨ pp. 78–81, 167, 231 (box).

toadfish, any of 64 bottom-dwelling, mainly marine fishes of the family Batrachoididae (order Batrachoidiformes). They have short, robust bodies, wide mouths, and rather upward-facing eyes. Some can produce sounds (e.g. MIDSHIPMEN), and many have light-producing organs.

toadflax, any of around 150 species of herbs in the mainly Mediterranean genus *Linaria* in the family Scrophulariaceae. Ivy-leaved toadflax or MOTHER-OF-THOUSANDS is closely related.

toadstool, a non-technical term for the visible fruiting body of various basidiomycete fungi; sometimes restricted to poisonous species; ⇨ p. 35.

tobacco, any of several species of herbs in the genus NICOTIANA. Cultivars of *N. tabaccum* are grown commercially in subtropical and maritime regions. The leaves, which contain the alkaloid nicotine, are harvested and dried as tobacco.

toddy cat, a name sometimes given to a PALM CIVET; ⇨ p. 123.

tomato (*Lycopersicon esculentum*), a perennial herb in the potato family, Solanaceae. It is cultivated for its edible fruits, which are used as a vegetable; ⇨ pp. 43, 46.

tongue, (in most vertebrates) a fleshy, muscular movable organ in the mouth used especially in tasting and swallowing food, and in speech in humans.

tonic immobility, ⇨ HYPNOSIS.

tonic response, in sensory perception, ⇨ p. 188.

tool, (in animal behaviour) any external object used as a functional extension of the body to achieve a short-term goal; ⇨ pp. 131, 141 (box), **151** (box).

tooth, a hard bone-like structure in the mouth, used for tearing or mastication of food. Teeth are often adapted as weapons.

toothed whale, ⇨ CETACEAN.

tooth shell, ⇨ TUSK SHELL.

topi, another name for SASSABY.

torpedo or **electric ray,** any of 38 RAYS of the family Torpedinidae. They have electricity-generating organs in the head, which can give a powerful shock (up to 220 volts); ⇨ p. 77.

tortoise, any of several slow-moving terrestrial CHELONIAN reptiles (⇨ pp. **82–9**), especially of the family Testudinidae. They are typically herbivorous, with a high-domed shell and clawed feet.

tortoiseshell, any NYMPHALID butterfly of the genus *Nymphalis* (subfamily Nymphalinae), widespread in temperate areas.

totipotency, the condition of an undifferentiated cell that can form any cell type; ⇨ pp. 21, 187.

toucan, any of 41 medium to large-sized birds of Neotropical rainforests, in the family Ramphastidae, related to the BARBETS. They have brightly coloured bills, which in some species can be as long as the rest of the body; they are used to eat fruit and insects. The eye is surrounded by a brightly coloured patch of bare skin; ⇨ p. 201. See also TOUCANET.

touch, one of the basic SENSES; ⇨ pp. 59, 80, **190**.

touraco, ⇨ TURACO.

trachea, 1. (or windpipe) in air-breathing vertebrates, the tube by which air is conveyed from the throat to the lungs; ⇨ p. 180. **2.** in insects, any of the numerous tiny tubes by which air is conveyed from the SPIRACLES, via even smaller tubes (tracheoles), to the tissues; ⇨ pp. 59, **180–1**.

tracheid, the conducting cells in XYLEM.

tracheole, ⇨ TRACHEA (2) and p. 180.

tracheophytes, the vascular plants, comprising all higher land plants; ⇨ p. 36.

Tradescantia, a New World genus of around 40 species of perennial herbs in the family Commelinaceae. Spiderwort (*T. virginiana*) is grown in gardens and wandering Jews (*T. fluminensis* and *T. albiflora*) are popular house plants.

tradition, ⇨ CULTURAL TRANSMISSION.

trait, another term for CHARACTERISTIC.

transcription, the process by which an RNA copy is made of a gene (a section of DNA); ⇨ p. 22.

translation, the process by which the DNA code, carried by m-RNA, is converted into a peptide chain on the ribosomes, using amino acids brought to the ribosomes by t-RNA (⇨ box, p. 6).

translocation, the transport of synthesized materials around a plant; ⇨ p. 32.

transpiration, the evaporation of water from plants through the stomata of the leaves; ⇨ p. 31.

tree, a large, woody plant having one or a few main stems; distinguished from a shrub or bush by its greater size and by having fewer stems; ⇨ pp. 39, **45**, 200–1, 207, 208–9.

treecreeper, any small, brown, arboreal, insectivorous bird belonging to one of two different groups. The subfamily Certhiinae (family Certhiidae) contains seven species of northern-hemisphere treecreepers, which have stiff tail feathers used for support. The family Climacteridae contains seven species of Australasian treecreeper, in which the tail is not used for support.

tree fern, a fern with a tree-like habit. Examples may be found in a number of families, e.g. Cyathaceae and Dicksoniaceae. The stems (which are woodless) have a tough outer cortex and support is also assisted by the remains of leaf bases; ⇨ p. 37 (box).

tree frog, a name loosely applied to any arboreal ANURAN, not all of which live in trees, some living in bushes, tall grasses and waterside plants; tree frogs often have sticky pads on the tips of the fingers and toes for clinging onto leaves. Thus FLYING FROGS and GLASS FROGS may be termed tree frogs, although the name is sometimes used specifically to refer to the tree frogs of the diverse and extensive family Hylidae; ⇨ p. 201.

treehopper, any HOMOPTERAN insect (bug) of the mainly tropical family Membracidae. These plant-feeders have enlarged thoracic shields, which extend backwards over the abdomen and may resemble plant thorns, ants, etc. They often live in a symbiotic association with ants.

tree line, the point on an environmental gradient at which trees cease to be able to grow, owing to an increase in either latitude or altitude (⇨ pp. 208–11).

tree shrew, any mammal of the family Tupaiidae (order Scandentia), comprising 18 species of small, shrew- or squirrel-like animal that occur in the tropical forests of Southeast Asia, where they feed on insects or fruit; ⇨ pp. 101, 136 (box).

tree swift, any of four Australasian and Oriental birds superficially resembling SWIFTS but with long erectile crests. They belong to the family Hemiprocnidae (order Apodiformes). Unlike true swifts, they spend much of the time perching in trees.

trefoil, any of around 100 species of annual or perennial herbs in the temperate genus *Lotus* in the pea family, Fabaceae (formerly Leguminosae). Some other leguminous plants with trifoliate leaves are also referred to as trefoils.

trematode, a FLUKE; ⊳ p. 55.

trial-and-error learning, ⊳ CONDITIONING and pp. 150–1.

Triassic period, a geological period lasting from 250 to 205 million years ago; ⊳ p. 4.

tribe, a term sometimes used for a classificatory group between genus and subfamily (⊳ p. 18).

triceratops, any herbivorous quadrupedal ORNITHOPOD dinosaur of the genus *Triceratops;* ⊳ pp. *83,* **84.**

triggerfish, any marine fish of the family Balistidae, related to the PUFFERS, occurring in the Atlantic, Indian and Pacific Oceans. Their name derives from the mechanism that locks the large, robust first dorsal spine in an erect position; they are deep-bodied fishes, with tough, closely fitting scales. Some triggerfishes, particularly smaller, strikingly marked species, are known as 'leatherjackets' or 'filefishes'; ⊳ p. 77.

trilobites, an extinct group of primitive marine ARTHROPODS, familiar as fossils from the Palaeozoic era, which were at their most abundant some 500 million years ago. They had oval bodies divided into three parts by longitudinal furrows and flanked with typical arthropod limbs, and they had compound eyes. A wide diversity in body shape and appendage structure indicates a wide range of life styles; ⊳ p. 5.

triplet, ⊳ CODON.

triton, any of several tropical GASTROPODS of the family Cymatiidae; these very large snails are predatory in habit.

trogon, any of 39 secretive insectivorous and frugivorous forest birds, belonging to the order Trogoniformes. Occurring in all tropical regions, trogons usually have iridescent colours on the head, breast and back, and in many the long tail is barred; ⊳ pp. *91,* 93. See also QUETZEL.

Trollius, ⊳ GLOBEFLOWER.

trophectoderm, ⊳ p. *186.*

trophic level, a feeding level in a food chain; ⊳ pp. 194–5.

tropical forest, ⊳ pp. 200–1.

tropicbird, any of three species of the family Phaethontidae (order Ciconiformes), large white oceanic birds with distinctive tail streamers. Their webbed feet are placed far back on the body, and they walk awkwardly on land.

tropics, the zone between the tropic of Cancer and the tropic of Capricorn, spanning the equator, i.e. between 23½° N and S. Although generally characterized by high temperatures, there are considerable variations in vegetation and climate; ⊳ pp. 198–9 (maps). For tropical ecosystems ⊳ pp. 200–3.

tropism, the ability of a plant stem or root to change the direction of its growth in response to a unidirectional stimulus. It does this by growing unequally on either side. Response to light is called phototropism (⊳ p. 33), to gravity gravitropism (⊳ p. 33), to touch haptotropism (⊳ TENDRIL), and to water hydrotropism.

trout, common name given to many fishes from various groups, especially to several SALMONIDS, such as the European trout (*Salmo trutta*) and the rainbow trout (*S. gairdneri*) of North America and the eastern Pacific (both of which have been widely introduced elsewhere). These two species are river fishes but have migratory forms that move between fresh and salt water, known as the sea trout and steelhead respectively; ⊳ p. 76 (box).

trout-perch, either of two small freshwater fishes of the family Percopsidae (order Percopsiformes), of

limited distribution in North American fresh waters. They have a small adipose (fatty) fin in addition to the spine-bearing fins.

truffle, a subterranean fruit body of some species of ascomycete or basidiomycete fungi (⊳ p. 35). Some highly prized edible species are sought with the aid of trained pigs or dogs.

trumpeter, any of three species of plump, noisy, crane-like birds of the family Psophiidae (order Gruiformes). They have short bills and inhabit lowland rainforests of the Neotropics. The white-winged trumpeter (*Psophia leucoptera*) stands about 50 cm (20 in) tall.

trumpet flower (*Bignonia capreolata*), an evergreen, woody climber in the family Bignoniaceae. It is native to southeastern USA, and is grown as an ornamental.

trunkfish, ⊳ BOXFISH.

trypanosome, a parasitic protozoan of the genus *Trypanosoma;* ⊳ p. 26.

tsetse fly, any medium-sized DIPTERAN insect of the family Glossinidae. Species of the African genus *Glossina* transmit several forms of sleeping sickness (trypanosomiasis). They lay one larva at a time, which is fully grown and immediately pupates; ⊳ pp. 62, 64, 168, 203.

Tsuga, ⊳ HEMLOCK SPRUCE.

tuatara (*Sphenodon punctatus*), the only species in the order Rhynchocephalia, this lizard-like reptile of New Zealand is believed to be the relict of a group that flourished some 220–150 million years ago; ⊳ pp. 82, **85.**

tube foot, any of the small flexible tubular structures in echinoderms (e.g. starfishes), used in locomotion, grasping, respiration, etc.; ⊳ p. 54.

tuber, a much-swollen underground stem (e.g. potato) or root (e.g. dahlia); ⊳ pp. 44, 206.

tuco-tuco, any of several burrowing, cavy-like RODENTS of the genus *Ctenomys* (family Ctenomyidae) from South America. They have blunt heads, long tails, and large incisors and claws for digging; ⊳ p. 161.

tufted hair-grass or **tussock grass** (*Deschampsia caespitosa*), a coarse grass of badly drained soils in Europe and mountains of North Africa.

tulip, any of around 100 species of bulbous herbs in the Eurasian and North African genus *Tulipa* in the lily family, Liliaceae. Tulips are popular garden plants with many cultivars of complex origin.

tulip tree, either of two species of the genus *Liriodendron* related to MAGNOLIA. Its tulip-like flowers make it an attractive ornamental.

tumbleweed, any plant that may become detached from the substrate, be blown about as a ball and regrow in a new position, e.g. the CLUB MOSS, *Selaginella lepidophylla.* Some, such as various AMARANTHS of the western USA, become detached on withering.

tuna, any of 13 marine PERCIFORM fishes of the family Scombridae (which also includes the MACKERELS), occurring in tropical and warm-temperate seas worldwide. They are large, powerful swimmers (⊳ p. 73), and very important as food fish. The blue-fin tuna (*Thunnus thynnus*) reaches 4 m (13 ft) in length, and is among the biggest of all bony fishes.

tundra, a vegetation type characteristic of high latitudes and altitudes; ⊳ pp. 199 (map), **210–11.**

tunicate, any CHORDATE of the subphylum Urochordata, a major group containing several classes of important marine animals, including sea squirts, salps (⊳ p. 220) and larvaceans. They are of interest because of their close links with VERTEBRATES and their importance in planktonic communities. All adult tunicates are filter-feeders; ⊳ p. 51 (box).

tunny, alternative name for TUNA.

turaco, touraco or **plantain-eater,** any of 17 plump-bodied African birds of the family Musophagidae (order Musophagiformes), with long tails and erectile head crests. The outer toe projects sideways and can be moved forwards or backwards; ⊳ pp. *91,* 96.

turbellarian, any free-living FLATWORM of the class Turbellaria, including the PLANARIANS; ⊳ p. 55.

turbot, name given to several FLATFISHES, especially *Scophthalmus maximus* (family Bothidae), a European species that may reach up to 1 m (39 in) in length; it is a well-camouflaged species found on gravel bottoms in very shallow water, and excellent as a food fish. See also FLOUNDER.

turkey, either of two typical GAMEBIRDS of the family Phasianidae, large, bare-headed, ground-dwelling birds. *Meleagris gallopavo,* the wild turkey of North America, is the species from which the domesticated varieties were derived; ⊳ p. 97 (box).

turmeric, the powdered rhizome of *Curcuma longa* (= *C. domestica*) in the ginger family, Zingiberaceae. Turmeric is used to flavour curries and as a dye.

turnip, a cultivar of *BRASSICA campestris rapifera* with a swollen root and stem used as a vegetable. The leaves may be used in salads or cooked.

turnover, the rate of replacement of one generation (for example, of animals or of leaves) by the next.

turnstone, either of two species of small, short-legged WADER of the family Scolopacidae, found nearly worldwide. Their name is derived from their habit of turning stones to find food.

turpentine, an essential oil obtained from the RESIN of certain pines. It is used as a paint solvent.

turtle, any of several aquatic CHELONIAN reptiles, especially of the marine family Chelonidae. In US and scientific usage, the term may refer to any chelonian reptile, including tortoises and terrapins; ⊳ pp. **82–9,** *170.*

tusk, a long, greatly enlarged tooth (e.g. in elephants and walruses) that projects when the mouth is closed and is used in digging for food, as a weapon, in display, etc.

tusk shell or **tooth shell,** any small MOLLUSC of the class Scaphopoda. Their popular name derives from the shape and colour of the open-ended shell. They live buried head down in sand, feeding by means of sticky tentacles.

tussock grass ⊳ TUFTED HAIR-GRASS.

twintail, common name for a DIPLURAN insect.

twitch grass, ⊳ COUCH GRASS.

tympanic membrane, or **eardrum,** in terrestrial vertebrates, a thin membrane that is made to vibrate by sound waves, the vibrations then being transmitted to the inner ear; ⊳ p. *191.*

tyrannosaur, any gigantic scavenging bipedal THEROPOD dinosaur of the genus *Tyrannosaurus,* especially *T. rex;* ⊳ pp. *83,* **84.**

tyrant flycatcher, any of 340 PASSERINE birds of the subfamily Tyranninae (family Tyrannidae), small insectivorous birds with broad, hook-tipped bills, living in forests and woodland of the Americas. They are the New World equivalent of the unrelated Old World FLYCATCHERS. They are generally drab in colour, but the males of many species have a bright erectile crest displayed when they are excited.

uakari, any New World monkey of the genus *Cacajao* (family Cebidae), the only short-tailed cebid monkeys; ⊳ p. 138.

ugli, a CITRUS FRUIT resulting from a cross between a GRAPEFRUIT and a MANDARIN, but larger and rougher skinned than the latter.

ultraviolet light, electromagnetic radiation at a

higher frequency than visible light, although detectable by some animals such as bees (⇨ p. 42). The ultraviolet light from the sun is harmful, causing e.g. mutations (⇨ p. 7); most of it is blocked by the ozone layer (⇨ p. 225).

umbel, a flat-topped INFLORESCENCE in which the flower 'stalks' are all attached to the same point on the axis (like umbrella ribs), as in cow parsley.

umbelliferous, belonging to the plant family APIACEAE (formerly Umbelliferae).

umbilical cord, in placental mammal, a long, tube-like structure linking the fetus with the PLACENTA; ⇨ p. 187.

umbrella bird, any of three species of COTINGA of the genus *Cephalopterus*. They have feathered wattles. The species are so called because the males are able to raise and lower a tuft of head feathers.

understorey, the lower stratum of plant cover in stratified vegetation where there is an upper leaf CANOPY; ⇨ p. 201.

underwing, any NOCTUID moth of the genus *Noctua*, abundant species with brightly coloured hind wings in yellow or white. These make the moths very obvious in flight, but they appear to vanish on landing; ⇨ p. 167.

ungulate, any mammal that has hooves. The group of mammals so defined is divided into two sub-groups (orders): the PERISSODACTYLS or 'odd-toed ungulates' (horses, tapirs, rhinoceroses) and the ARTIODACTYLS or 'even-toed ungulates' (pigs, sheep, cattle, deer, antelope, etc.); ⇨ p. 110 (box).

uniformitarianism, a geological theory that states that all geological change is the result of gradual processes. It is opposed to CATASTROPHISM; ⇨ p. 12.

unisexual, (of part or all of an organism) possessing either male or female reproductive organs, not both. Some flowers are unisexual, but most are HERMAPHRODITE. See also DIOECIOUS, MONOECIOUS.

unit membrane, ⇨ p. 20.

Uralian Sea, an ancient sea separating Europe and Asia; ⇨ p. 9.

urbanization, the process by which human populations become concentrated in towns and cities, which spread into the surrounding countryside as a consequence. Many wild animals and plants have adapted to urban living; ⇨ p. 213.

urea, one of the water-soluble end products of protein metabolism. Made by the LIVER, it is a major nitrogenous excretory product and (in mammals) is excreted in the urine; ⇨ p. 73, 88, 93.

urine, a fluid excreted by the kidneys, containing UREA and other waste products collected from body fluids; ⇨ p. 179 (box).

urochordate, a TUNICATE.

urodele, any AMPHIBIAN of the order Urodela, comprising the SALAMANDERS.

uropygial gland, ⇨ PREEN GLAND.

Urticaceae, a dicotyledonous plant family, including nettles; ⇨ p. 45.

uterus (or womb), in female mammals, a hollow muscular cavity in the pelvic region that houses the developing fetus; ⇨ pp. *185, 186.*

vacuole, a membrane-bound compartment in some animal and protist and all plant cells. Many types of plant cell have a large central vacuole responsible for controlling cell turgor (rigidity); ⇨ pp. **20** (box), *21, 27, 32.*

vagina, the elastic, muscular tube in female mammals extending from the cervix to the vulva, that receives the penis during intercourse and serves as the birth canal.

valerian, any of around 200 species of perennial herbs in the Eurasian, South African and New World genus *Valeriana* in the family Valerianaceae. Red valerian – sometimes known as Jupiter's beard – is the related *Centranthus ruber*.

valve, a structure (especially in the heart and veins) that closes temporarily to obstruct the passage of fluid in one direction only.

vampire bat, any of three microchiropteran BATS of the family Desmodontidae, specialized blood-feeding species occurring in the tropical and subtropical parts of the New World; ⇨ pp. 119, 178.

vanilla, flavouring extracted from the seed pods of various members of the climbing orchid genus *Vanilla*; ⇨ p. 46 (box).

varanid lizard, ⇨ MONITOR.

variegation, patches or spots of different colours on plants, generally on petals or leaves (as in ivy).

variety, (in plants) a subdivision of a species. Names of naturally occurring varieties are usually Latin, e.g. *Phaseolus vulgaris* var. *humilis* (bush bean), whereas the names of cultivated 'varieties' (CULTIVARS) are vernacular, e.g. *Phaseolus vulgaris*, Canadian Wonder.

vascular plants or **tracheophytes,** all plants with a VASCULAR SYSTEM, characteristic of all higher land plants; ⇨ p. 36.

vascular system, (in plants) the system of tissues comprising the XYLEM (for conducting dissolved mineral sales and water) and the PHLOEM (for transporting chemicals synthesized within the plant), together with various other specialized cells (e.g. for strengthening); ⇨ pp. 31–2, 36.

vector, an organism that carries and transmits a disease-causing organism.

vegetable, any part of a plant that is used in salads or savoury dishes; ⇨ p. 46.

vegetation, the plant constituent of an ecosystem; flora.

vegetation zone, any of several divisions of the earth on the basis of vegetation type; ⇨ p. 199 (map).

vegetative body, the non-sexual parts of a plant, e.g. stem, roots, leaves.

vegetative reproduction, a form of asexual reproduction in plants, by which new (genetically identical) plants bud from specialized stems, leaves and occasionally roots of the parent plant; ⇨ pp. 33, 44.

vein, 1. a blood vessel by which deoxygenated blood is carried to the heart; ⇨ 181. **2.** a vascular vessel in a leaf.

veld, a type of temperate grassland found in South Africa; ⇨ p. 206.

veliger, the free-swimming larval stage of many gastropods; ⇨ p. 56.

velvet ant, any solitary WASP of the family Mutillidae, occurring worldwide. Both sexes are covered with black and reddish short fine hairs. The females are wingless (accounting for the misleading name of the group), and are parasites of bees and wasps.

velvet worm, any ARTHROPOD of the class Onchychopora, small, worm-like animals with short jointless limbs; ⇨ p. 58 (box).

venation, the arrangement or system of veins in an insect's wing or in a leaf.

ventilation, in respiration, the means by which oxygen-carrying air or water is brought into contact with the respiratory surfaces; ⇨ pp. 180–1.

ventral, relating to or situated on the front or stomach.

ventricle, a chamber in a heart that receives blood from an atrium and pumps it out through the arteries; ⇨ *181.*

Venus flytrap (*Dionaea muscipula*), an insectiverous plant of the family Droseraceae native to southeast North America; ⇨ p. 47 (box).

verbena, any of around 250 species of annual to perennial herbs and shrubs in the New World genus *Verbena* in the family Verbenaceae. Several species are grown as ornamentals. Vervain is the name of some wild species, e.g. *V. officinalis.*

Veronica, a mainly temperate genus of around 300 species of annual or perennial herbs and low shrubs in the family Scrophulariaceae. Many are called speedwells and some are grown as ornamentals. *V. beccabunga* – brooklime – is used in salads.

vertebra (pl. -brae), any of the independent bony or cartilaginous segments of the spinal column, through which the SPINAL CORD passes.

vertebrate, any CHORDATE of the subphylum Vertebrata – i.e. fishes, amphibians, reptiles, birds and mammals, all of which have a backbone (box, ⇨ p. 51).

vervain, ⇨ VERBENA.

vervet, ⇨ GUENON.

vesicle, 1. a small membranous, spherical sac, often involved in secretory processes. **2.** a pocket of embryonic tissue that is the beginning of an organ, e.g. the primary cavity of the vertebrate brain.

vespertilionid bat, or 'evening bat', any of about 275 microchiropteran BATS of the family Vespertilionidae, of worldwide distribution outside the Antarctic; in northern latitudes many species hibernate for five or six months. These extremely numerous bats are almost entirely insectivorous, consuming vast quantities of midges, mosquitoes and other pests. Included in the family are the little brown bat (*Myotis lucifugus*), a common species in North America, and the common pipistrelle (*Pipistrellus pipistrellus*), probably the commonest European bat. See also BULLDOG BAT.

vestigial structures, ⇨ pp. 14–15.

vetch, any of around 140 species of annual or perennial herbs in the north temperate and South American genus *Vicia* in the pea family, Fabaceae (formerly Leguminoseae). See also BROAD BEAN.

vibrio, any comma-shaped bacteria; ⇨ p. 24.

vibrissa (pl. -issae), any of the stiff sensory hairs ('whiskers') on the face of many mammals (⇨ p. 133), or the bristle-like feathers around the beak of various birds (⇨ p. 97).

viburnum, any of around 200 species of shrubs and small trees in the north temperate genus *Viburnum* in the family Caprifoliaceae. Many such as the GUELDER ROSE are popular ornamentals.

vicariance, the existence of closely related groups of organisms in different geographical areas that have been separated by the formation of a natural barrier such as a body of water (⇨ pp. 8–9, 10–11, 15, 16–17).

vicuna (*Vicugna vicugna*), a ruminating South American ARTIODACTYL mammal (family Camelidae), related to the Old World camels; ⇨ pp. 9, 115 (box).

villus (pl. villi), in vertebrates, any of numerous tiny, finger-like projections in the small intestine that provide a large surface area for ABSORPTION; ⇨ p. 179.

vine, any climbing or trailing plant with slender stems. It is often used to refer strictly to the GRAPEVINE, but is extended to include other plants with similar habit, e.g. HOP and RUNNER BEAN.

Viola, a mainly temperate genus of around 500 species of herbs and, rarely, small shrubs in the family Violaceae. The flowers have a spur in which nectar accumulates. A number of species are cultivated, e.g. sweet violet (*V. odorata*), from which an oil used in perfumes is extracted. See also PANSY.

violet, ⇨ VIOLA.

viper, any SNAKE of the family Viperidae, containing some 40 species with a widespread distribution throughout the Old World. All vipers are venomous, with a sophisticated venom-injection system involving fangs set at the front of the jaw that rotate into striking position ($\Rightarrow$ p. *87*). The adder or common viper (*Vipera berus*) is the only poisonous snake in northern Europe, while the large puff adder (*Bitis arietans*) is one of the commonest snakes throughout Africa. See also PIT VIPER.

viperfish, any of six deep-sea fishes of the family Chauliodontidae, of worldwide distribution. They have luminous organs, and long fangs in the jaws; $\Rightarrow$ p. *221*.

vireo, any of about 50 PASSERINE birds of the family Vireonidae, small, usually dull-plumaged finch-like birds of the Americas. The bill is notched and hook-tipped. They tend to be very vocal. Greenlets are a group of 17 small Neotropical vireos.

Virginia creeper (*Parthenocissus quinquefolia*), a climbing vine in the family Vitaceae. It is native to North America and is grown for its autumn colour.

virion, the inert, extracellular phase of a virus, consisting of the nucleic acid enclosed in a protein coat ($\Rightarrow$ p. 24).

viroid, disease-causing particles even smaller than viruses; $\Rightarrow$ p. 24.

virus, any of a group of minute organisms that lack a cell structure; $\Rightarrow$ p. 24.

viscacha or **plains viscacha** (*Lagostomus maximus*), a large, long-tailed, soft-furred caviomorph RODENT in the CHINCHILLA family. Viscachas are burrowing rodents living in colonies in South America; $\Rightarrow$ p. 161.

vision, $\Rightarrow$ SIGHT.

Vitaceae, a dicotyledonous plant family, including grape vine and virginia creeper; $\Rightarrow$ p. 45.

vitamin, any chemical substance that is essential in small quantities for normal metabolic functioning; $\Rightarrow$ p. 179.

vitreous humour, in the mammalian eye, $\Rightarrow$ p. *189*.

viverrid, any mammal of the family Viverridae (order Carnivora), comprising some 66 species and including civets, genets, mongooses, linsangs and the fossa. Viverrids are typically small, lithe and long-bodied, with long, bushy tails. Most species are opportunistic and omnivorous, inhabiting forested habitats in the Old World; $\Rightarrow$ pp. 122, **123**.

viviparous, producing live young (as in most mammals); $\Rightarrow$ p. 185.

vole or **meadow mouse,** any of numerous small, burrowing, herbivorous myomorph RODENTS of the family Muridae, especially those of the genus *Microtus*. They have blunt muzzles, small ears and short tails, and are found in grassy habitats in North America and Eurasia; $\Rightarrow$ p. 135. See also WATER RAT.

vomeronasal organ or **Jacobsen's organ,** an olfactory organ found (in addition to the nose) in various mammals, snakes, etc.; $\Rightarrow$ p. 189.

Vries, Hugo de (1848–1935), Dutch botanist, who in 1900 rediscovered MENDEL's work on genetics, having worked out the laws of inheritance independently.

vulture, 1. any of seven large scavenging birds of the subfamily Cathartine (the New World vultures), in the same family as the STORKS. They have broad wings, fan-like at the tips, on which they soar while looking for carrion; they have no feathers on the head or neck; $\Rightarrow$ pp. *91*, 92, 96, 97. See also CONDOR. **2.** any of 11 scavenging birds of the Old World in the family Accipitridae, which also includes HAWKS and EAGLES. They are similar to the New World species, but usually have feathered necks; $\Rightarrow$ pp. 92, 96, 97, 203, 233. See also GRIFFON, LAMMERGEIER.

wader or **shore bird,** a general term applied to birds that live in close association with water. The name is given to many birds in the order Ciconiiformes,

especially those of the families Charadriidae ($\Rightarrow$ AVOCET, LAPWING, OYSTERCATCHER, PLOVER, STILT) and Scolopacidae ($\Rightarrow$ CURLEW, GODWIT, PHALAROPE, SANDPIPER, SNIPE, WOODCOCK); $\Rightarrow$ pp. *91*, 96, 216.

waggle dance, $\Rightarrow$ DANCE LANGUAGE and p. 153.

wagtail, any of 12 or so PASSERINE birds of the subfamily Motacillinae, in the SPARROW family, very closely related to PIPITS. They are small, mostly Old World insectivorous birds, with long tails bearing outer white feathers, often found near water. The sexes are differently coloured, with the male brighter.

wahoo, $\Rightarrow$ BURNING BUSH (3).

walking stick, another name for STICK INSECT.

wallaby, any herbivorous MARSUPIAL mammal of the genus *Petrogale* and related genera in the family Macropodidae, which also includes KANGAROOS and PADEMELONS. Wallabies are similar to kangaroos but smaller; $\Rightarrow$ pp. 103, **104**.

Wallace, Alfred Russel (1823–1913), British naturalist who struck upon the idea of evolution by natural selection independently of Darwin ($\Rightarrow$ p. 13). He also made important early contributions to zoogeography, proposing the evolutionary separation of the faunas of Asia and Australasia (Wallace's line).

wallflower, any of around 10 species of perennial herbs and subshrubs in the genus *Cheiranthus* in the cabbage family, Brassicaceae (formerly Cruciferae). The most popular garden wallflower is *C. cheiri* from the Mediterranean region.

wall lizard (*Lacerta muralis*), a common European LACERTID lizard.

walnut, any of around 15 species of tree in the genus *Juglans*, the timber of which is much prized. *J. regia* is cultivated commercially for its nuts in countries such as the USA.

walrus (*Odobenus rosmarus*), a large carnivorous mammal found in Arctic coastal waters, the sole species in the family Odobenidae (order Pinnipedia, which also includes the SEALS); $\Rightarrow$ pp. 142, **143**, *159*.

wandering Jew, $\Rightarrow$ *TRADESCANTIA*.

wapiti (*Cervus canadensis*), a large DEER with branched antlers in the males, closely resembling the red deer. It inhabits grassland and forest edge in eastern Central Asia, and in North America, where it is also known as an elk.

warble fly, any of various DIPTERAN insects of the family Oestridae, which also includes the BOTFLIES. The larvae are internal parasites of mammals; the eggs are laid on the host skin, and the larvae burrow to the gut before returning to the skin, where they form a swelling known as a 'warble'.

warbler, any of about 360 PASSERINE birds of the families Cisticolidae (the African warblers) and Sylviidae. They are primarily small, slim, elegant birds, typically with dull plumage, often with pale underparts, living in woodlands and reedbeds of the Old World, with a few species in Australia. Most are insectivorous, sometimes eating berries, and many are fine songsters, even on the wing. Many are strongly migratory, with some, such as the willow warbler (*Phylloscopus trochilus*), undertaking round trips of 25 000 km (15 500 mi); $\Rightarrow$ pp. 152 (box), 171, 205, 209, 211, 217, 233. See also WOOD WARBLER.

warm-blooded, $\Rightarrow$ HOMOIOTHERMIC.

warm receptor, a specialized type of temperature-sensitive cell; $\Rightarrow$ p. 189.

warning coloration, bright and conspicuous coloration in an unpalatable animal (e.g. a venomous snake or a stinging bee) that serves to discourage potential predators; $\Rightarrow$ pp. 56, 65, *66*, 77, *81*, 89, 153, **167** (box).

wart hog (*Phacochoerus aethiopicus*), a wild PIG of

sub-Saharan savannah, with heavy tusks and wart-like outgrowths on its face; $\Rightarrow$ pp. 112–13.

wasp, any HYMENOPTERAN insect of the superfamilies Vespoidea (the 'true' wasps) and Sphecoidea (the DIGGER WASPS), as well as numerous parasitic forms (in the Parasitica). The life style of wasps is very variable. Parasitic wasps parasitize other insects at any stage of the (parasite's) life cycle, while some species lay their eggs in plants and cause the production of galls ($\Rightarrow$ GALL WASP). Other wasps are predatory and provision their solitary nests with paralyzed prey on which their offspring feed. Social wasps build nests and have a worker caste of sterile females. Nests are constructed from chewed wood, and the larvae are tended in a paper comb; $\Rightarrow$ p. 63. See also DIGGER WASP, FIG WASP, GALL WASP, HORNTAIL WASP, MUD DAUBER, PAPER WASP, POTTER WASP, SPIDER-HUNTING WASP.

water, a simple molecule (H_2O) essential to life; $\Rightarrow$ pp. 6, 31, 33 (box). See also HYDROLOGICAL CYCLE. For wetland and aquatic ecosystems $\Rightarrow$ pp. 216–21.

water arum (*Calla palustris*), a north temperate and subarctic perennial herb of ponds and swamps.

water bear, any tiny aquatic ARTHROPOD of the class Tardigrada, bizarre creatures with eight legs ending in claws or discs.

water boatman, any HETEROPTERAN insect of the family Corixiidae. These boat-shaped insects swim using long oarlike hind legs. They cling to underwater vegetation and feed on particulate material such as diatoms and algae. They breathe from an air bubble carried under the wing covers.

waterbuck (*Kobus ellipsiprymnus*), a large, stocky, dark grey or reddish-brown, shaggy ANTELOPE, which inhabits savannah and woodland near water in Africa; $\Rightarrow$ p. 115.

water bug, any HETEROPTERAN insect of the family Belastomatidae. This tropical family contains the largest of all bugs ($\Rightarrow$ HEMIPTERANS), and some of the biggest insects (*Lethocorus grandis* is over 11 cm/4⅓ in long). They readily fly and are attracted to lights at night. The fore legs are modified for grasping prey, which includes fish, tadpoles, frogs and insects; $\Rightarrow$ p. 59.

water chestnut, any of a number of species of floating plants in the genus *Trapa* grown for their edible starchy seeds, e.g. the Eurasian and African *T. natans*. The unrelated Chinese water chestnut (*Eleocharis dulcis*) has edible tubers.

water civet or **aquatic genet** (*Osbornictis piscivora*), a very rare, cat-like VIVERRID with reddish-brown fur and a long, bushy tail. Found near water in Zaïre, it feeds on fish and other small vertebrates; $\Rightarrow$ p. 123.

watercress (*Nasturtium officinale*), a perennial European herb in the cabbage family, Brassicaceae (formerly Cruciferae). It is cultivated for use in salads.

waterdog, $\Rightarrow$ MUD PUPPY.

water dragon, name given to a number of aquatic or semi-aquatic AGAMID lizards.

water fleas, a group of tiny freshwater CRUSTACEANS, closely related to the FAIRY SHRIMPS; $\Rightarrow$ p. 67.

waterfowl, ducks, geese and swans collectively.

water lily, any of around 75 species of perennial herbs in the widely distributed family Nymphaceae, which has three genera. The amazon water lily (*Victoria amazonica*) has floating leaves up to 2 m (7 ft) in diameter; $\Rightarrow$ p. *219*.

water moccasin, another name for COTTONMOUTH.

water opossum, another name for YAPOK; $\Rightarrow$ p. 104.

water plantain, any of several species of the genus *Alisma*, family Alismataceae. They are marsh or swamp plants of north temperate regions and

Australia, and are characterized by broad pointed leaves.

water rat, any of several semi-aquatic myomorph RODENTS of the subfamily Hydromyinae (family Muridae), inhabiting rivers and swamps in Australia, New Guinea and the Philippines. The name is sometimes also used of other species, such as the North American MUSKRAT and the Eurasian water vole (*Arvicola terrestris*).

water scorpion, any HETEROPTERAN insect of the worldwide family Nepidae, resembling small WATER BUGS. A pair of grooved terminal filaments are used together as a tube to penetrate the water film, so allowing the insects to breathe air.

water strider, ▷ POND-SKATER.

Watson, James (1928–), American geneticist, who with CRICK established the structure of the DNA molecule (▷ p. 22). He went on to become the director of the project to sequence the entire human genome (▷ p. 23).

wattle, 1. a fleshy protuberance under the throat of some birds, lizards, etc. **2.** ▷ ACACIA.

wattlebird, either of two PASSERINE birds of the family Callaeatidae, ground-dwelling, medium-sized, crow-like birds of New Zealand forests, with prominent wattles at the base of the bill. Male and female have differently shaped bills; this dimorphism was extreme in the huia, which became extinct in 1907.

wave, name given to various GEOMETRID moths.

wax, any of various compounds (esters) of fatty acids and long-chain alcohols. Waxes are secreted by honeybees to construct combs (▷ p. 160) and by many plants to form a waterproof, protective covering on leaves, fruits and stems.

waxbill, any of about 140 tiny, colourful FINCHES belonging to the subfamily Estrildinae, in the sparrow family (Passeridae), living in grasslands throughout the Old World and Australia. They build globe-shaped nests, usually timed to coincide with maximum availability of grass seeds. Many Australian representatives are called mannakins.

waxwing, any of three PASSERINE birds in the family Bombycillidae, arboreal fruit-eaters with crested heads, slightly hooked bills and square tails, living in the northern hemisphere. Their plumage is soft and silky, and the tips of some of the wing feathers have red terminal tufts resembling wax droplets.

weasel, the common name given to many small MUSTELID carnivores of the genus *Mustela* and related genera, such as the European or least weasel (*M. nivalis*), the smallest of all the carnivores; ▷ p. **131**.

weathering, the gradual release of elements from rocks; ▷ p. 196.

weaver, any of 117 medium-sized FINCHES of the subfamily Ploceinae, in the sparrow family (Passeridae), some of which show complex social behaviour and many of which are colonial. Most species live in Africa, with a few in Asia and Australia. Weavers are so called because their nests are made from grass woven together like a basket. The nests are often built in inaccessible places (e.g. over-hanging water) and have complex entrance tunnels; ▷ pp. 94, 95, *161*. See also QUELEA.

web, a network of fine silken threads built by a spider for the purpose of prey capture; ▷ pp. 69, 160, *164*.

webspinner, any EXOPTERYGOTE insect of the small order Embioptera. These tropical insects have silk-producing glands on their enlarged fore legs, which they swing to and fro in front of them to 'knit' protective silken tunnels on the bark of trees. Living communally within their tunnels, they emerge at night to feed on plant material; ▷ pp. *61*.

weed, any species of plant that grows in a location

where it is not required by human beings; ▷ pp. 206, **230–1**.

weedkiller, ▷ PESTICIDE.

weeverfish, any of four marine PERCIFORM fishes of the family Trachinidae, from the eastern Atlantic, the Black Sea, and off Chile. Small bottom-dwelling fishes with eyes on the upper surface of the head, they lie half-buried in the sand. Their dorsal fin spines and those on the gill covers are highly venomous, and can present a hazard to bathers in some areas; ▷ p. 77.

weevil, any BEETLE of the immensely large family Curculionidae, containing over 60 000 species. Adults have mandibulate (biting) mouthparts on the end of a 'beak' (rostrum) of variable length and thickness, and the antennae are often elbowed. They feed on grains and other hard vegetable matter, and many are very damaging plant pests (▷ pp. 64, *162*). The larvae are grub-like and legless. Weevils are sometimes referred to as 'snout beetles'.

Wegener, Alfred (1880–1930), German meteorologist who first came up with serious evidence to support the idea of continental drift; ▷ p. 8.

Weissman, August (1834–1914), German biologist. In 1890 he proposed that, as germ cells (that give rise to sperm and eggs) are separate from the cells of the rest of the body, acquired characteristics cannot be inherited. In fact, in plants this distinction is not as clear as it is in animals (▷ p. 21), and it has been shown that chemical and physical factors can have an effect on the chromosomes. Today most geneticists accept the 'central dogma', enunciated by Francis Crick and based on Weissman's ideas. This states that information can only flow in one direction, from DNA to protein (via RNA) and not the other way round (▷ p. 22). However, it has been suggested that even this may not hold true in all circumstances.

wels, ▷ SHEATFISH.

Welwitschia, a gnetophyte genus, which now includes only one species, *W. mirabilis;* ▷ p. 39.

western hemlock, ▷ HEMLOCK SPRUCE.

western red cedar, ▷ ARBOR VITAE.

wetland, any ecosystem where water accumulates and yet emergent vegetation can maintain itself; ▷ pp. **216–17**, 229.

wet rot, timber decay in buildings due to one of a number of fungi (e.g. species of Coniophora) that can become established and maintain growth only if the timber is kept continuously wet. See also DRY ROT.

whale, common name given to most of the larger CETACEANS and sometimes to any cetacean mammal; ▷ pp. 14, **144–5**, 154, 220, 233 (box).

whalebone, ▷ BALEEN.

whale shark (*Rhincodon typus*), an extremely large marine SHARK, mainly inhabiting surface waters in warm seas. This slow-moving, inoffensive creature is the largest of all fishes, said to reach sizes over 18 m (59 ft); ▷ p. 76.

wheat, any one of several species of cereal grass in the temperate genus *Triticum.* Flour produced from hard (high-gluten) wheats is used for bread making and from soft (low-gluten) wheats for biscuits and pastry.

wheatear, any of 16 medium-sized colourful, migratory THRUSHES of the genus *Oenanthe*, living in the northern hemisphere.

wheel animalcule, a ROTIFER.

whelk, any GASTROPOD of the family Buccinidae, shallow-water marine snails that are either scavenging or predatory, using a long siphon to seek food by chemosensory means (smell or taste); the shell has a notch at the tip for the siphon.

whin, ▷ GORSE.

whip scorpion, any ARACHNID of the order Uropygi, resembling true scorpions except that the tail is thin without a sting and the first pair of legs are slender and whip-like. Whip scorpions are harmless unless provoked, when they may squirt acid at the attacker.

whip snake, any harmless COLUBRID snake of the genus *Coluber* of Old and New World distribution. They are so named because they resist vigorously when attacked.

whirligig beetle, any BEETLE of the family Gyrinidae, comprising about 700 species and occurring worldwide. Adults live on the water surface, often circling rapidly and swimming in swarms. Their eyes are divided into two parts, one part functional in air, the other underwater.

whisk fern, any member of the division Psilophyta of spore-bearing plants; ▷ p. 37.

whistler, any of about 50 PASSERINE birds of the tribe Pachycephalini, in the CROW family (Corvidae). They are stout, strong-billed species, which dig and probe for insects in Australasia and the Pacific. They are good songsters, and with a few spectacular exceptions dull-coloured.

white, any PIERID butterfly of the genus *Pieris*, such as the Eurasian cabbage white (*P. brassicae*), the larvae of which feed on cabbages, etc.

white ant, ▷ TERMITE.

whitebeam, either Swedish whitebeam (SORBUS *intermedia*) or common whitebeam (*S. aria*).

white-eye, any of over 90 PASSERINE birds of the family Zosteropidae, small, usually green arboreal birds with brush-tipped tongues and a ring of white feathers around the eyes. The live in the Old World tropics, Australasia and the Pacific.

whitefish, any SALMONID of the genus *Coregonus.* They are principally lake-dwelling fishes, spawning in shallow water over gravel bottoms.

whitefly, any HOMOPTERAN insect (bug) of the family Aleyrodidae. These very small insects have a dusting of white wax over the wings and body surface, and are a major problem for greenhouse-grown crops.

white fox, ▷ ARCTIC FOX.

white matter, ▷ SPINAL CORD.

white whale, another name for BELUGA (1).

whiting (*Merlangius merlangus*), an important food fish related to the COD, common in shallow European offshore waters. The young often hide among the tentacles of jellyfishes.

whorl, a set of leaves, petals, sepals, etc., arising from and encircling the stem or floral axis; ▷ p. 40.

whortleberry, ▷ BILBERRY.

whydah or **widow bird,** any of 15 large African FINCHES of the subfamily Estrildinae, closely related to the WAXBILLS. Males have long tail feathers in the breeding season to attract females, which then lay their eggs in the nests of other finch species; ▷ p. 95.

widow bird, ▷ WHYDAH.

wigeon, any of four compact DUCKS, occurring worldwide. Their feathers have fine bar-like markings. The American wigeon (*Anas americanus*) lives on ponds and marshes, and is a colourful green and black bird.

wild boar (*Sus scrofa*), a wild PIG of Eurasia and northern Africa, the ancestor of the domesticated pig; ▷ pp. *112*, **113**, 153, 205.

wild cat (*Felis sylvestris*), a small, stocky, nocturnal CAT with brown fur and black stripes. It inhabits forest and grassland from western Europe to India and Africa; ▷ pp. 125, 205.

wild dog, another name for DHOLE, HUNTING DOG and DINGO.

wildebeest or **gnu,** either of two stocky ANTELOPES

of the genus *Connochaetes*, both inhabiting the African savannah: the common blue wildebeest or brindled gnu (*C. taurinus*) and the rarer white-tailed gnu (*C. gnou*); ⇨ pp. 115, *171*, 203.

wilderness, an area of the earth's surface in which human influence is minimal. It is unlikely that any area can be regarded as totally devoid of such influence.

willow, any of around 500 species of trees and shrubs in the mainly north temperate and arctic genus *Salix*, family Salicaceae. They are to be found mainly in damp situations. Cricket-bat willow (*S. alba* var. *calva*) provides timber for cricket bats. The flexible suckers from a number of species (e.g. *S. viminalis*), referred to as osiers, are used for wickerwork. Sallows are also a type of willow; ⇨ pp. 42, *208*, 211, *219*.

willowherb, any of a number of species of herbs in the genera *Epilobium* and *Chamaenerion* in the family Onagraceae. Rosebay willowherb or fireweed (*C. angustifolium*) rapidly colonizes waste ground; ⇨ p. *43*.

wilting, the consequence of excessive water loss by leaves, which become limp or flaccid; ⇨ pp. 31–2.

windflower, ⇨ ANEMONE.

windpipe, another name for TRACHEA (1).

wing, a flight structure in birds (⇨ p. 92), insects (⇨ p. 59), bats (⇨ p. 118), and pterosaurs (⇨ pp. 84, *176*). See also FLIGHT.

winkle, ⇨ PERIWINKLE.

wintergreen, any of a number of species of plant in the unrelated genera *Gaultheria* and *Pyrola*. Spicy wintergreen (*G. procumbens*) was the source of wintergreen oil, now obtained from the bark of American black BIRCH (*Betula lenta*).

wisent, another name for the European BISON.

wishbone, ⇨ FURCULA.

wisteria, any of around nine species of woody climbers in the East Asian and North American leguminous genus *Wisteria*. Several species are grown as ornamentals.

witch hazel, any of around five species of winter-flowering shrubs and small trees in the East Asian and North American genus *Hamamelis*.

wolf, either of two dog-like carnivores (CANIDS) of the genus *Canis* – the grey or timber wolf (*C. lupus*) and the red wolf (*C. rufus*); ⇨ pp. 122, **126–7**, *154*, 165, 205, 207, 209. See also COYOTE, MANED WOLF, THYLACINE.

wolfbane, ⇨ ACONITUM.

wolffish, any of six marine PERCIFORM fishes of the family Anarhichadidae, inhabiting deep waters in the North Pacific and North Atlantic. They have elongated bodies, up to 2 m (6½ ft) in some species; the head is rounded, and the jaws are armed with both pointed and blunt crushing teeth, rather crookedly arranged. They are also known as 'sea cats' or 'catfish'.

wolverine or **glutton** (*Gulo gulo*), a fierce, stocky, brown MUSTELID with powerful jaws, resembling a small bear. It inhabits tundra and taiga in North America and Eurasia, scavenging or killing vertebrate prey; ⇨ p. 209.

womb, ⇨ UTERUS.

wombat, any of three short-legged, stocky, burrowing MARSUPIALS of the family Vombatidae, found in mainland Australia and Tasmania. Wombats are nocturnal and herbivorous, feeding mostly on grasses.

wood, the hard fibrous tissues of the stems and roots of woody plants (shrubs, bushes, many climbers, and trees). Wood is largely composed of secondary XYLEM (vascular tissue) produced by the CAMBIUM; ⇨ p. 45. See also HEARTWOOD, SAPWOOD, SOFTWOOD, HARDWOOD.

woodchuck or **groundhog** (*Marmota monax*), a ground-dwelling, burrowing RODENT in the SQUIRREL family; they are found in North America. The woodchuck usually hibernates in winter.

woodcock, any of six species of plump-bodied, short-tailed WADER in the genus *Scolopax* in the family Scolopacidae. They are found mostly in woodlands of the Old World, with one American species.

woodcreeper, any of 49 PASSERINE birds of the subfamily Dendrocolaptinae, closely related to the OVENBIRDS. They are small, woodpecker-like birds of Central and South America, probing tree bark for insects with their powerful bills. The tail is long and used as a brace, and the ends of the tail feathers have curved tips for grip.

wood hoopoe, any of five birds of the family Phoeniculidae (order Upupiformes), typically living in trees in the African savannah. They are slender, with long stiff tails used as a brace while they probe the bark for insects with their long and slightly curved bills. See also HOOPOE.

woodland, a small area of temperate forest, often under human management.

woodlouse, any CRUSTACEAN of the class Isapoda, the principal group of terrestrial crustaceans. They have flattened bodies, simple antennae and seven pairs of similar legs; they live as scavengers. The closely related species found on rocky shores are usually called 'slaters'; ⇨ pp. 66, **67**.

wood mouse, another name for FIELDMOUSE.

woodpecker, any of over 200 species of the family Picidae (order Piciformes), occurring in the Americas and the Old World. They are slender birds with very flexible necks, strong, chisel-tipped bills, and very long protrusible tongues. Most live in trees, using their long claws to grip the bark and long stiff tails to brace the body while they probe for insects or excavate a nest; ⇨ pp. *91*, 93, *95*. See also SAPSUCKER.

woodswallow, any of 10 PASSERINE birds of the tribe Artamini, in the same family as the CROWS and closely related to the CURRAWONGS. They are aggressive grey and white birds of open country in Australia and the Orient, often seen perching together.

wood warbler or **American warbler**, any of 115 PASSERINE birds of the tribe Parulini, related to the BUNTINGS. They are small, active, predominantly insect-eating birds of the Americas, in some respects analogous to the Old World warblers. American warblers are usually brightly coloured.

woodwasp, ⇨ HORNTAIL.

woodworm or **furniture beetle** (*Anobium punctatum*), a BEETLE (family Anobiidae) that causes considerable damage, especially to softwoods used in furniture and building construction; ⇨ p. 64.

woolly bear, ⇨ TIGER MOTH and DERMESTID BEETLE.

woolly monkey, either of two species of New World monkey of the genus *Lagothrix* (family Cebidae), thickly furred monkeys that live in the forest canopy in the Amazon basin; ⇨ p. 138.

worker, a sterile CASTE in social insects that supports and maintains the colony. Workers are female in bees, wasps and ants, but of either sex in termites.

worm, common name given to numerous (typically) slender, long-bodied invertebrates of very diverse origins; ⇨ pp. 55, 215, 221.

worm lizard, another name for AMPHISBAENIAN.

wormwood, ⇨ ARTEMISIA.

wrack, any of various brown algae (SEAWEEDS) of the genus *Fucus*, e.g. BLADDERWRACK; ⇨ pp. 34, 221.

wrasse, any of about 500 marine PERCIFORM fishes of the family Labridae, found in tropical and temperate seas worldwide. They are often brightly coloured, with thick lips and strong teeth in the jaws as well as crushing teeth in the throat. They exhibit elaborate courtship and nesting behaviour, and under certain conditions they may change sex (⇨ p. 74). They are known for sleeping at night lodged in crevices or buried in the sand. Some species are CLEANER FISHES.

wren, any of 75 PASSERINE birds of the subfamily Troglodytinae (family Certhiidae), small brown, stockily built birds with thin, slightly curved bills used for capturing insects, short rounded wings, and short uptilted tails. Most live in dense vegetation of Central and South America. The Holarctic wren (*Troglodytes troglodytes*), like most, is strongly territorial, defending its territory by song. The four species of New Zealand wren (family Acanthisittidae) are unrelated; they are superficially similar, with short tails. See also SCRUB-WREN.

wrybill (*Anarhynchus frontalis*), an unusual PLOVER of New Zealand in which the bill is twisted to the right and used for turning over stones.

wryneck, either of two dull-coloured birds of the genus *Jynx*, closely related to the WOODPECKERS and occurring in Africa and Eurasia. They use natural cavities as nest spaces rather than excavating.

xeromorphic (of a plant), drought-resistant.

xerophyte, any plant modified in its structure or its physiology in such a way that it is able to tolerate drought.

xylem, the part of the vascular tissue of plants responsible for conducting water and dissolved mineral salts. Characteristically, the cells are elongated, with relatively thick walls strengthened by LIGNIN. The conducting cells are the vessels and tracheids. The xylem formed just below the apex is the primary xylem; that formed from the lateral secondary MERISTEM is secondary xylem, making up the principal component of wood in trees, shrubs, etc.; ⇨ pp. 31–2.

yak (*Bos mutus*), a massively built dark-brown species of wild CATTLE with dense, shaggy fur. Inhabiting the high plateaux and mountains of Central Asia, it is now rare in the wild but common in domestication; ⇨ p. 114.

yam, any of around 600 species of herbaceous twining climbers in the tropical and subtropical genus *Dioscorea*. A number are cultivated for their edible tubers, e.g. Asiatic yam (*D. alata*). Some are a source of diosgenin, used in the production of contraceptive drugs.

yapok or **water opossum** (*Chironectes minimus*), a semi-aquatic New World marsupial of the OPOSSUM family; ⇨ p. 104.

yarrow, ⇨ ACHILLEA.

yeast, a unicellular fungus that reproduces by budding, e.g. baker's or brewer's yeast (*Saccharomyces cerevisiae*), special strains of which are also used in wine making; ⇨ p. 35.

yellow, any PIERID butterfly of the genus *Colias*, the males of which have yellow wings.

yellow flag, ⇨ FLAG.

yellow-green algae, species of ALGAE belonging to the division Xanthophyta. There are both unicellular and filamentous forms.

yew, any of around 20 species of coniferous trees and shrubs in the principally north temperate genus *Taxus*. They have linear leaves arranged in two ranks. The seed is enclosed in a fleshy ARIL. Although the aril is edible, the seed and all other parts of the plant are highly toxic.

yolk, the store of nutrients in an EGG (mainly protein and fat) that supports the development of an animal embryo. In the eggs of birds, reptiles, some fishes, etc., the yolk is contained in a yolk sac, a structure that is also present in mammals, but contains little or no yolk; ⇨ pp. *86*, **186**.

yolk sac, ⊳ YOLK.

Young, Thomas (1773–1829), English physicist; ⊳ p. 189 (box).

Yucca, a North and Central American genus of around 40 species of xerophytic trees and shrubs in the family Agavaceae, e.g. the Joshua tree (*Y. brevifolia*), which may grow 15 m (50 ft) high.

zebra, any of three distinctively striped African PERISSODACTYL mammals of the family Equidae – the common, plains, or Burchell's zebra (*Equus burchelli*), Grevy's zebra (*E. grevyi*), and the mountain zebra (*E. zebra*). A fourth species, the quagga (*E. quagga*) of southern Africa, was hunted to extinction in the 1880s; ⊳ pp. 110, **111**, 154 (box), 158, 203.

zeitgeber, any external cue, such as the change from light to darkness, that allows the synchonization of internal bodily rhythms or BIOLOGICAL CLOCKS; ⊳ p. 174.

Zingiberidae, a monocotyledonous plant subclass; ⊳ p. 45.

zokor or **mole rat,** any of six vole-like, burrowing RODENTS of the genus *Myospalax* (family Muridae). Possessing large fore claws and incisor teeth, and short tails, they inhabit the steppes of eastern Asia.

zonation, a banded pattern of plant and animal communities caused by variation in a physical factor of the environment, such as tidal rise and fall, or simply a gradient, such as moisture availability around the edge of a lake, or changes in altitude.

zoogeographical region, ⊳ FAUNAL REGION.

zoogeography, the study of the patterns of animal distribution around the world.

zoology, the study of animals, often divided up into specialized fields such as ichthyology (fishes), ornithology (birds), entomology (insects), etc.

zooplankton, ⊳ PLANKTON.

zorapteran, any EXOPTERYGOTE insect of the order Zoraptera, comprising just 22 species of tropical distribution. These small insects (up to 3 mm/^{1}–10 in) may be winged or wingless, and live in moist decaying wood. They share some features with ORTHOPTERANS, others with the BOOKLOUSE; ⊳ p. 61.

zorilla or **African polecat** (*Ictonyx striatus*), a skunk-like MUSTELID with black and white fur and a long bushy tail. It inhabits semi-arid areas in Africa; ⊳ p. 131.

zucchini, ⊳ COURGETTE.

zygomycete, any member of the fungal class Zygomycetes; ⊳ p. 35.

Zygomycotina, a subdivision of the fungal division Amastigomycota; ⊳ p. 35.

zygospore, the type of spore produced sexually by zygomycete fungi; ⊳ p. 35.

zygote, a DIPLOID cell resulting from the fusion of a SPERM and an OVUM; ⊳ pp. 184, 186.

PICTURE ACKNOWLEDGEMENTS

The Publishers would like to thank the following for permission to reproduce pictures and for their assistance in providing pictures for the Encyclopedia:

Jacana, pp. 16, 26, 43 (top right), 46, 53 (top), 54 (bottom right), 55 (top), 65 (bottom), 69, 75, 86, 88 (bottom right), 94 (right), 95, 96, 104, 107 (bottom), 112, 113, 114, 115, 118 (top and bottom), 119, 120, 121 (top), 123, 125, 126 (bottom), 128–9, 130 (top), 131, 140, 142, 144–5, 146–7, 151 (bottom), 152 (top), 154, 156 (bottom), 157, 169 (top), 175 (bottom), 211, 213 (bottom), 216, 232.

Oxford Scientific Films, pp. 30 (Kjell B. Sanoved), 134 (Animals Animals), 138 (top, D. Wechsler; bottom, Michael Fogden), 158 (D. Curl), 184 (Mantis Wildlife Films), 205 (M. Chillmaid), 206 (M. Colbeck), 226 (top, J. Dermid).

Planet Earth Pictures, pp. 1, 7, 34 (bottom), 35 (top), 39 (bottom), 44, 45 (top and bottom), 47 (right), 50–1, 52, 57 (left), 62, 63, 66 (bottom), 68, 72, 76, 88 (bottom left), 102, 126 (top), 132 (top and bottom), 133, 135, 148 (bottom), 149, 151 (top), 152 (bottom), 156 (top), 163 (top), 164 (bottom), 165, 167 (bottom), 169 (bottom), 171, 192–3, 200, 204, 212, 213 (top), 224, 226 (bottom), 227, 229 (left and right), 233.

Bridgeman Art Library, p. 12.

Bruce Coleman Limited, pp. 6, 15, 35 (centre), 37 (top), 47 (left), 64, 65 (top), 67, 74, 77, 80, 81, 88 (top), 94 (left), 99, 103, 105, 108, 109, 116, 121 (bottom), 124/125, 128, 130 (bottom), 136, 137, 139, 141, 148 (top), 150, 155, 159, 161, 162, 163 (bottom), 164 (top), 166, 168, 174, 175 (top), 188, 202–3, 207, 208–9, 210, 214, 217, 219, 228, 231 (top and bottom).

J.W. Cowan / M.E. Collinson, pp. 34 (top), 35 (bottom), 37 (bottom), 39 (top), 42, 43 (left and bottom).

Images Colour Library, pp. 222–3.

Incafo Archivo Fotografico, S.L. / F. Cardela, p. 215.

Peale Museum, Philadelphia, p. 12.

Science Photo Library, pp. 2–3, 24, 25, 27 (top and bottom), 31, 32, 225.

C.M. Young / R.H. Emson, pp. 51, 53 (bottom), 54 (top left and right, bottom left), 55 (bottom), 56 (top and bottom), 57 (right), 167 (top).

Zefa Picture Library (UK) Ltd., pp. i (title), 28–9, 48–9, 124.

Special thanks to: Françoise Mestre, Jennifer Jeffrey